Mathematical Applications

for the Management, Life, and Social Sciences

Volume II

Eleventh Edition

Ronald J. Harshbarger | James J. Reynolds

CENGAGE
Learning·

Australia • Brazil • Japan • Korea • Mexico • Singapore • Spain • United Kingdom • United States

CENGAGE
Learning·

**Mathematical Applications for the
Management, Life, and Social Sciences**

Mathematical Applications for the Management, Life, and Social Sciences,
Eleventh Edition
Hershbarger / Reynolds

For product information and technology assistance, contact us at
Cengage Learning Customer & Sales Support, 1-800-354-9706

For permission to use material from this text or product,
submit all requests online at **cengage.com/permissions**
Further permissions questions can be emailed to
permissionrequest@cengage.com

This book contains select works from existing Cengage Learning resources and
was produced by Cengage Learning Custom Solutions for collegiate use. As such,
those adopting and/or contributing to this work are responsible for editorial
content accuracy, continuity and completeness.

Compilation © 2015 Cengage Learning

ISBN: 978-1-305-75176-7

WCN: 01-100-101

Cengage Learning
20 Channel Center Street
Boston, MA 02210
USA

Cengage Learning is a leading provider of customized learning solutions with
office locations around the globe, including Singapore, the United Kingdom,
Australia, Mexico, Brazil, and Japan. Locate your local office at:
www.international.cengage.com/region.

Cengage Learning products are represented in Canada by Nelson Education, Ltd.

For your lifelong learning solutions, visit **www.cengage.com/custom.**

Visit our corporate website at **www.cengage.com.**

9   Derivatives 533

9.1   Limits 535
9.2   Continuous Functions; Limits at Infinity 549
9.3   Rates of Change and Derivatives 559
9.4   Derivative Formulas 573
9.5   The Product Rule and the Quotient Rule 583
9.6   The Chain Rule and the Power Rule 590
9.7   Using Derivative Formulas 597
9.8   Higher-Order Derivatives 602
9.9   Applications: Marginals and Derivatives 608
        Key Terms and Formulas 617
        Review Exercises 618
        Chapter 9 Test 622
        Extended Applications & Group Projects 624
            I. Marginal Return to Sales • II. Energy from Crude Oil (Modeling) • III. Tangent Lines and Optimization in Business and Economics

10   Applications of Derivatives 628

10.1   Relative Maxima and Minima: Curve Sketching 630
10.2   Concavity; Points of Inflection 643
10.3   Optimization in Business and Economics 655
10.4   Applications of Maxima and Minima 667
10.5   Rational Functions: More Curve Sketching 675
        Key Terms and Formulas 684
        Review Exercises 685
        Chapter 10 Test 688
        Extended Applications & Group Projects 690
            I. Production Management • II. Room Pricing in the Off-Season (Modeling)

11   Derivatives Continued 692

11.1   Derivatives of Logarithmic Functions 694
11.2   Derivatives of Exponential Functions 702
11.3   Implicit Differentiation 709
11.4   Related Rates 717
11.5   Applications in Business and Economics 723
        Key Terms and Formulas 731
        Review Exercises 732
        Chapter 11 Test 734
        Extended Applications & Group Projects 735
            I. Inflation • II. Renewable Electric Power (Modeling)

12   Indefinite Integrals 737

12.1   Indefinite Integrals 739
12.2   The Power Rule 746
12.3   Integrals Involving Exponential and Logarithmic Functions 755
12.4   Applications of the Indefinite Integral in Business and Economics 764
12.5   Differential Equations 771
        Key Terms and Formulas 780
        Review Exercises 781
        Chapter 12 Test 783
        Extended Applications & Group Projects 784
            I. Employee Production Rate • II. Supply and Demand

13   Definite Integrals: Techniques of Integration 786

13.1   Area Under a Curve 788
13.2   The Definite Integral: The Fundamental Theorem of Calculus 797
13.3   Area Between Two Curves 807
13.4   Applications of Definite Integrals in Business and Economics 816
13.5   Using Tables of Integrals 825
13.6   Integration by Parts 829
13.7   Improper Integrals and Their Applications 834
13.8   Numerical Integration Methods: The Trapezoidal Rule and Simpson's Rule 841
        Key Terms and Formulas 849
        Review Exercises 850
        Chapter 13 Test 853
        Extended Applications & Group Projects 855
            I. Purchasing Electrical Power (Modeling) • II. Retirement Planning

14   Functions of Two or More Variables 858

14.1   Functions of Two or More Variables 860
14.2   Partial Differentiation 866
14.3   Applications of Functions of Two Variables in Business and Economics 875
14.4   Maxima and Minima 882
14.5   Maxima and Minima of Functions Subject to Constraints: Lagrange Multipliers 892
        Key Terms and Formulas 899
        Review Exercises 899
        Chapter 14 Test 902
        Extended Applications & Group Projects 903
            I. Advertising • II. Competitive Pricing

Appendix A   Graphing Calculator Guide AP-1

Appendix B   Excel Guide AP-27

        Part 1 Excel 2003 AP-27
        Part 2 Excel 2007 and 2010 AP-45

Appendix C   Areas Under the Standard Normal Curve AP-65

Answers A-1
Index I-1

© bibiphoto/Shutterstock.com

Derivatives

If a firm receives $90,000 in revenue during a 30-day month, its average revenue per day is $90,000/30 = $3000. This does not necessarily mean the actual revenue was $3000 on any one day, just that the average was $3000 per day. Similarly, if a person drove 50 miles in one hour, the average velocity was 50 miles per hour, but the driver could still have received a speeding ticket for traveling 70 miles per hour.

The smaller the time interval, the nearer the average velocity will be to the instantaneous velocity (the speedometer reading). Similarly, changes in revenue over a smaller number of units can give information about the instantaneous rate of change of revenue. The mathematical bridge from average rates of change to instantaneous rates of change is the limit.

This chapter is concerned with *limits* and *rates of change*. We will see that the *derivative* of a function can be used to determine instantaneous rates of change.

The topics and some representative applications discussed in this chapter include the following.

SECTIONS		APPLICATIONS
9.1	Limits	Cost-benefit models
9.2	Continuous Functions; Limits at Infinity	Federal income taxes
9.3	Rates of Change and Derivatives	Marginal revenue, velocity, elderly in the workforce
9.4	Derivative Formulas	Personal income, world tourism, revenue
9.5	The Product Rule and the Quotient Rule	Sensitivity to a drug, marginal revenue
9.6	The Chain Rule and the Power Rule	Demand, allometric relationships
9.7	Using Derivative Formulas	Revenue
9.8	Higher-Order Derivatives	Cellular subscribers, acceleration
9.9	Applications: Marginals and Derivatives	Marginals for cost, revenue, and profit; competitive markets

Chapter   **Warm-Up**

Prerequisite Problem Type	For Section	Answer	Section for Review
If $f(x) = \dfrac{x^2 - x - 6}{x + 2}$, then find: (a) $f(-2.1)$ (b) $f(-2)$	9.1–9.9	(a) -5.1 (b) undefined	1.2 Function notation
Let $f(x) = \begin{cases} x^2 + 1 & \text{if } x \le 1 \\ x + 2 & \text{if } x > 1 \end{cases}$, find: (a) $f(0.99)$ (b) $f(1.01)$	9.1	(a) 1.9801 (b) 3.01	2.4 Piecewise defined functions
Factor: (a) $x^2 - x - 6$ (b) $x^2 - 4$ (c) $x^2 + 3x + 2$	9.1 9.7	(a) $(x + 2)(x - 3)$ (b) $(x - 2)(x + 2)$ (c) $(x + 1)(x + 2)$	0.6 Factoring
Write as a power: (a) $\sqrt{t}$ (b) $\dfrac{1}{x}$ (c) $\dfrac{1}{\sqrt[3]{x^2 + 1}}$	9.4–9.8	(a) $t^{1/2}$ (b) x^{-1} (c) $(x^2 + 1)^{-1/3}$	0.3, 0.4 Exponents and radicals
Simplify: (a) $\dfrac{4(x + h)^2 - 4x^2}{h}$, if $h \ne 0$ (b) $(2x^3 + 3x + 1)(2x) + (x^2 + 4)(6x^2 + 3)$ (c) $\dfrac{x(3x^2) - x^3(1)}{x^2}$, if $x \ne 0$	9.3 9.5 9.7	(a) $8x + 4h$ (b) $10x^4 + 33x^2 + 2x + 12$ (c) $2x$	0.5 Simplifying algebraic expressions
Simplify: (a) $\dfrac{x^2 - x - 6}{x + 2}$ if $x \ne -2$ (b) $\dfrac{x^2 - 4}{x - 2}$ if $x \ne 2$	9.1 9.7	(a) $x - 3$ (b) $x + 2$	0.7 Simplifying fractions
If $f(x) = 3x^2 + 2x$, find $\dfrac{f(x + h) - f(x)}{h}$, if $h \ne 0$	9.3	$6x + 3h + 2$	1.2 Function notation
Find the slope of the line passing through $(1, 2)$ and $(2, 4)$.	9.3	2	1.3 Slopes
Write the equation of the line passing through $(1, 5)$ with slope 8.	9.3 9.4 9.6	$y = 8x - 3$	1.3 Point-slope equation of a line

OBJECTIVES

9.1

- To use graphs and numerical tables to find limits of functions, when they exist
- To find limits of polynomial functions
- To find limits of rational functions
- To find limits of piecewise defined functions

Limits

| APPLICATION PREVIEW |

Although everyone recognizes the value of eliminating any and all particulate pollution from smokestack emissions of factories, company owners are concerned about the cost of removing this pollution. Suppose that USA Steel has shown that the cost C of removing p percent of the particulate pollution from the emissions at one of its plants is

$$C = C(p) = \frac{7300p}{100 - p}$$

To investigate the cost of removing as much of the pollution as possible, we can evaluate the limit as p (the percent) approaches 100 from values less than 100. (See Example 6.) Using a limit is important in this case, because this function is undefined at $p = 100$ (it is impossible to remove 100% of the pollution).

In various applications we have seen the importance of the slope of a line as a rate of change. In particular, the slope of a linear total cost, total revenue, or profit function for a product tells us the marginals or rates of change of these functions. When these functions are not linear, how do we define marginals (and slope)?

We can get an idea about how to extend the concept of slope (and rate of change) to functions that are not linear. Observe that for many curves, if we take a very close (or "zoom-in") view near a point, the curve appears straight. See Figure 9.1. We can think of the slope of the "straight" line as the slope of the curve. The mathematical process used to obtain this "zoom-in" view is the process of taking limits. The concept of limit is essential to the study of calculus.

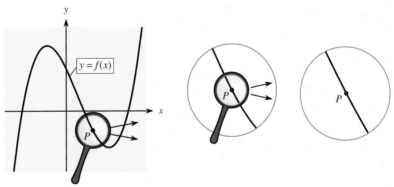

Figure 9.1   When we zoom in near point P, the curve appears straight.

Concept of a Limit

We have used the notation $f(c)$ to indicate the value of a function $f(x)$ at $x = c$. If we need to discuss a value that $f(x)$ approaches as x approaches c, we use the idea of a *limit*. For example, if

$$f(x) = \frac{x^2 - x - 6}{x + 2}$$

then we know that $x = -2$ is not in the domain of $f(x)$, so $f(-2)$ does not exist even though $f(x)$ exists for every value of $x \neq -2$. Figure 9.2 on the next page shows the graph of $y = f(x)$ with an open circle where $x = -2$. The open circle indicates that $f(-2)$ does not exist but shows that points near $x = -2$ have functional values that lie on the line on either side of the open circle. Even though $f(-2)$ is not defined, the figure shows that as x approaches -2 from either side of -2, the graph approaches the open circle at $(-2, -5)$ and the values of $f(x)$ approach -5. Thus -5 is the limit of $f(x)$ as x approaches -2, and we write

$$\lim_{x \to -2} f(x) = -5, \quad \text{or} \quad f(x) \to -5 \quad \text{as} \quad x \to -2$$

| TABLE 9.1 |

Left of −2	
x	$f(x) = \dfrac{x^2 - x - 6}{x + 2}$
−3.000	−6.000
−2.500	−5.500
−2.100	−5.100
−2.010	−5.010
−2.001	−5.001

Right of −2	
x	$f(x) = \dfrac{x^2 - x - 6}{x + 2}$
−1.000	−4.000
−1.500	−4.500
−1.900	−4.900
−1.990	−4.990
−1.999	−4.999

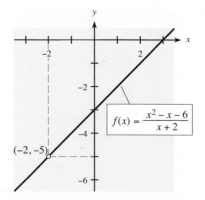

$$f(x) = \frac{x^2 - x - 6}{x + 2}$$

$(-2, -5)$

Figure 9.2

This conclusion is fairly obvious from the graph, but it is not so obvious from the equation for $f(x)$.

We can use the values near $x = -2$ in Table 9.1 to help verify that $f(x) \to -5$ as $x \to -2$. Note that to the left of −2, the values of $f(x)$ get very close to −5 as x gets very close to −2, and to the right of −2, the values of $f(x)$ get very close to −5 as x gets very close to −2. Hence, Table 9.1 and Figure 9.2 indicate that the value of $f(x)$ approaches −5 as x approaches −2 from both sides of $x = -2$.

From our discussion of the graph in Figure 9.2 and Table 9.1, we see that as x approaches −2 from either side of −2, the limit of the function is the value L that the function approaches. This limit L is not necessarily the value of the function at $x = -2$. This leads to our intuitive definition of **limit.**

Limit

Let $f(x)$ be a function defined on an open interval containing c, except perhaps at c. Then

$$\lim_{x \to c} f(x) = L$$

is read "the **limit** of $f(x)$ as x approaches c equals L." The number L exists if we can make values of $f(x)$ as close to L as we desire by choosing values of x sufficiently close to c. When the values of $f(x)$ do not approach a single finite value L as x approaches c, we say the limit does not exist.

As the definition states, a limit as $x \to c$ can exist only if the function approaches a single finite value as x approaches c from both the left and right of c.

EXAMPLE 1   Limits

Figure 9.3 shows three functions for which the limit exists as x approaches 2. Use this figure to find the following.

(a) $\lim\limits_{x \to 2} f(x)$ and $f(2)$ (if it exists)

(b) $\lim\limits_{x \to 2} g(x)$ and $g(2)$ (if it exists)

(c) $\lim\limits_{x \to 2} h(x)$ and $h(2)$ (if it exists)

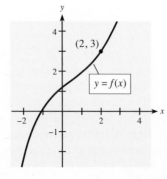

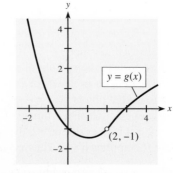

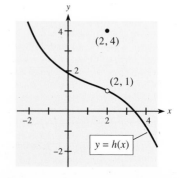

Figure 9.3   (a)   (b)   (c)

Solution

(a) From the graph in Figure 9.3(a), we see that as x approaches 2 from both the left and the right, the graph approaches the point $(2, 3)$. Thus $f(x)$ approaches the single value 3. That is,

$$\lim_{x \to 2} f(x) = 3$$

The value of $f(2)$ is the y-coordinate of the point on the graph at $x = 2$. Thus $f(2) = 3$.

(b) Figure 9.3(b) shows that as x approaches 2 from both the left and the right, the graph approaches the open circle at $(2, -1)$. Thus

$$\lim_{x \to 2} g(x) = -1$$

The figure also shows that at $x = 2$ there is no point on the graph. Thus $g(2)$ is undefined.

(c) Figure 9.3(c) shows that

$$\lim_{x \to 2} h(x) = 1$$

The figure also shows that at $x = 2$ there is a point on the graph at $(2, 4)$. Thus $h(2) = 4$, and we see that $\lim_{x \to 2} h(x) \neq h(2)$. ∎

As Example 1 shows, the limit of the function as x approaches c may or may not be the same as the value of the function at $x = c$.

In Example 1 we saw that the limit as x approaches 2 meant the limit as x approaches 2 from both the left and the right. We can also consider limits only from the left or only from the right; these are called **one-sided limits.**

One-Sided Limits

Limit from the Right: $\lim\limits_{x \to c^+} f(x) = L$

means the values of $f(x)$ approach the value L as $x \to c$ but $x > c$.

Limit from the Left: $\lim\limits_{x \to c^-} f(x) = M$

means the values of $f(x)$ approach the value M as $x \to c$ but $x < c$.

Note that when one or both one-sided limits fail to exist, then the limit does not exist. Also, when the one-sided limits differ, such as if $L \neq M$ above, then the values of $f(x)$ do not approach a *single* value as x approaches c, and $\lim_{x \to c} f(x)$ does not exist.

EXAMPLE 2   One-Sided Limits

Using the functions graphed in Figure 9.4, determine why the limit as $x \to 2$ does not exist for

(a) $f(x)$.

(b) $g(x)$.

(c) $h(x)$.

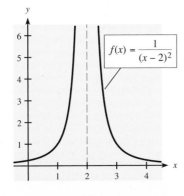
(a)

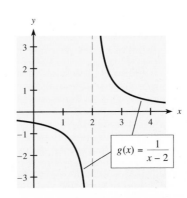
(b)

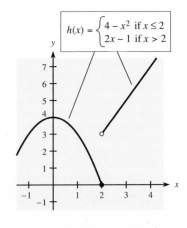
(c)

Figure 9.4

Solution

(a) As $x \to 2$ from the left side and the right side of $x = 2$, $f(x)$ increases without bound, which we denote by saying that $f(x)$ approaches ∞ as $x \to 2$. In this case, $\lim_{x \to 2} f(x)$ does not exist [denoted by $\lim_{x \to 2} f(x)$ DNE] because $f(x)$ does not approach a finite value as $x \to 2$. In this case, we write

$$f(x) \to \infty \text{ as } x \to 2$$

The graph has a vertical asymptote at $x = 2$.

(b) As $x \to 2$ from the left, $g(x)$ approaches $-\infty$, and as $x \to 2$ from the right, $g(x)$ approaches ∞, so $g(x)$ does not approach a finite value as $x \to 2$. Therefore, the limit does not exist. The graph of $y = g(x)$ has a vertical asymptote at $x = 2$.
In this case we summarize by writing

$$\lim_{x \to 2^-} g(x) \text{ DNE} \qquad \text{or} \qquad g(x) \to -\infty \text{ as } x \to 2^-$$

$$\lim_{x \to 2^+} g(x) \text{ DNE} \qquad \text{or} \qquad g(x) \to \infty \text{ as } x \to 2^+$$

and

$$\lim_{x \to 2} g(x) \text{ DNE}$$

(c) As $x \to 2$ from the left, the graph approaches the point at $(2, 0)$, so $\lim_{x \to 2^-} h(x) = 0$. As $x \to 2$ from the right, the graph approaches the open circle at $(2, 3)$, so $\lim_{x \to 2^+} h(x) = 3$. Because these one-sided limits differ, $\lim_{x \to 2} h(x)$ does not exist. ◼

Examples 1 and 2 illustrate the following two important facts regarding limits.

The Limit

1. The limit $\lim_{x \to c} f(x) = L$ only if the following two conditions are satisfied:

(a) The limit L is a finite value (real number).
(b) The limit as x approaches c from the left equals the limit as x approaches c from the right. That is, we must have

$$\lim_{x \to c^-} f(x) = \lim_{x \to c^+} f(x)$$

Figure 9.4 and Example 2 illustrate cases where $\lim_{x \to c} f(x)$ does not exist.

2. The limit $\lim_{x \to c} f(x)$ and the function value $f(c)$ are independent. When $\lim_{x \to c} f(x)$ exists, $f(c)$ may be (i) the same as the limit, (ii) undefined, or (iii) defined but different from the limit (see Figure 9.3 and Example 1).

✓ **CHECKPOINT**

1. Can $\lim_{x \to c} f(x)$ exist if $f(c)$ is undefined?
2. Does $\lim_{x \to c} f(x)$ exist if $f(c) = 0$?
3. Does $f(c) = 1$ if $\lim_{x \to c} f(x) = 1$?
4. If $\lim_{x \to c} f(x) = 0$, does $\lim_{x \to c} f(x)$ exist?
5. Let $\lim_{x \to c} f(x) = 4$.
    (a) Does $\lim_{x \to c^+} f(x) = 4$?   (b) Find $\lim_{x \to c^-} f(x)$.

Properties of Limits; Algebraic Evaluation

We have seen that the value of the limit of a function as $x \to c$ will not always be the same as the value of the function at $x = c$. However, there are many functions for which the limit and the functional value agree [see Figure 9.3(a)], and for these functions we can easily evaluate limits, as the following properties indicate.

Properties of Limits

If k is a constant, $\lim\limits_{x \to c} f(x) = L$, and $\lim\limits_{x \to c} g(x) = M$, then the following are true.

I. $\lim\limits_{x \to c} k = k$

II. $\lim\limits_{x \to c} x = c$

III. $\lim\limits_{x \to c} [f(x) \pm g(x)] = L \pm M$

IV. $\lim\limits_{x \to c} [f(x) \cdot g(x)] = LM$

V. $\lim\limits_{x \to c} \dfrac{f(x)}{g(x)} = \dfrac{L}{M}$ if $M \neq 0$

VI. $\lim\limits_{x \to c} \sqrt[n]{f(x)} = \sqrt[n]{\lim\limits_{x \to c} f(x)} = \sqrt[n]{L}$,
     provided that $L > 0$ when n is even.

If f is a polynomial function, then Properties I–IV imply that $\lim\limits_{x \to c} f(x)$ can be found by evaluating $f(c)$. Moreover, if h is a rational function whose denominator is not zero at $x = c$, then Property V implies that $\lim\limits_{x \to c} h(x)$ can be found by evaluating $h(c)$. The following summarizes these observations and recalls the definitions of polynomial and rational functions.

Function	Definition	Limit
Polynomial function	The function $f(x) = a_n x^n + a_{n-1} x^{n-1} + \cdots + a_1 x + a_0,$ where $a_n \neq 0$ and n is a positive integer, is called a **polynomial function** of degree n.	$\lim\limits_{x \to c} f(x) = f(c)$ for all values c (by Properties I–IV)
Rational function	The function $h(x) = \dfrac{f(x)}{g(x)}$ where both $f(x)$ and $g(x)$ are polynomial functions, is called a **rational function.**	$\lim\limits_{x \to c} h(x) = \lim\limits_{x \to c} \dfrac{f(x)}{g(x)} = \dfrac{f(c)}{g(c)}$ when $g(c) \neq 0$ (by Property V)

EXAMPLE 3   Limits

Find the following limits, if they exist.

(a) $\lim\limits_{x \to -1} (x^3 - 2x)$ (b) $\lim\limits_{x \to 4} \dfrac{x^2 - 4x}{x - 2}$

Solution

(a) Note that $f(x) = x^3 - 2x$ is a polynomial, so

$$\lim\limits_{x \to -1} f(x) = f(-1) = (-1)^3 - 2(-1) = 1$$

Figure 9.5(a) on the next page shows the graph of $f(x) = x^3 - 2x$.

(b) Note that this limit has the form

$$\lim\limits_{x \to c} \dfrac{f(x)}{g(x)}$$

where $f(x)$ and $g(x)$ are polynomials and $g(c) \neq 0$. Therefore, we have

$$\lim\limits_{x \to 4} \dfrac{x^2 - 4x}{x - 2} = \dfrac{4^2 - 4(4)}{4 - 2} = \dfrac{0}{2} = 0$$

Figure 9.5(b) shows the graph of $g(x) = \dfrac{x^2 - 4x}{x - 2}$.

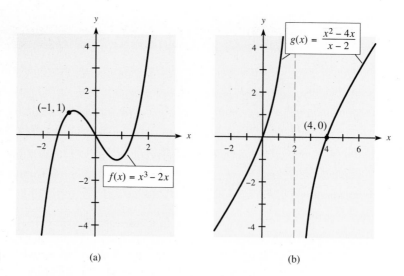

Figure 9.5   (a)   (b)

We have seen that we can use Property V to find the limit of a rational function $f(x)/g(x)$ as long as the denominator is *not* zero. If the limit of the denominator of $f(x)/g(x)$ *is* zero, then there are two possible cases.

Rational Functions: Evaluating Limits of the Form $\lim\limits_{x \to c} \dfrac{f(x)}{g(x)}$ where $\lim\limits_{x \to c} g(x) = 0$

Type I. If $\lim\limits_{x \to c} f(x) = 0$ and $\lim\limits_{x \to c} g(x) = 0$, then $\lim\limits_{x \to c} \dfrac{f(x)}{g(x)}$ has the

0/0 indeterminate form at $x = c$. We can factor $x - c$ from $f(x)$ and $g(x)$, reduce the fraction, and then find the limit of the resulting expression, if it exists.

Type II. If $\lim\limits_{x \to c} f(x) \neq 0$ and $\lim\limits_{x \to c} g(x) = 0$, then $\lim\limits_{x \to c} \dfrac{f(x)}{g(x)}$ does not exist. In this case, the values of $f(x)/g(x)$ become unbounded as x approaches c; the line $x = c$ is a vertical asymptote.

EXAMPLE 4 **0/0 Indeterminate Form**

Evaluate the following limits, if they exist.

(a) $\lim\limits_{x \to 2} \dfrac{x^2 - 4}{x - 2}$ (b) $\lim\limits_{x \to 1} \dfrac{x^2 - 3x + 2}{x^2 - 1}$

Solution

(a) This limit has the 0/0 indeterminate form at $x = 2$ because both the numerator and denominator equal zero when $x = 2$. Thus we can factor $x - 2$ from both the numerator and the denominator and reduce the fraction. (We can divide by $x - 2$ because $x - 2 \neq 0$ while $x \to 2$.)

$$\lim_{x \to 2} \frac{x^2 - 4}{x - 2} = \lim_{x \to 2} \frac{(x - 2)(x + 2)}{x - 2} = \lim_{x \to 2} (x + 2) = 4$$

Figure 9.6(a) shows the graph of $f(x) = (x^2 - 4)/(x - 2)$. Note the open circle at $(2, 4)$.

(b) By substituting 1 for x in $(x^2 - 3x + 2)/(x^2 - 1)$, we see that the expression has the 0/0 indeterminate form at $x = 1$, so $x - 1$ is a factor of both the numerator and the denominator. (We can then reduce the fraction because $x - 1 \neq 0$ while $x \to 1$.)

$$\lim_{x \to 1} \frac{x^2 - 3x + 2}{x^2 - 1} = \lim_{x \to 1} \frac{(x - 1)(x - 2)}{(x - 1)(x + 1)} = \lim_{x \to 1} \frac{x - 2}{x + 1}$$
$$= \frac{1 - 2}{1 + 1} = \frac{-1}{2} \qquad \text{(by Property V)}$$

Figure 9.6(b) shows the graph of $g(x) = (x^2 - 3x + 2)/(x^2 - 1)$. Note the open circle at $(1, -\frac{1}{2})$.

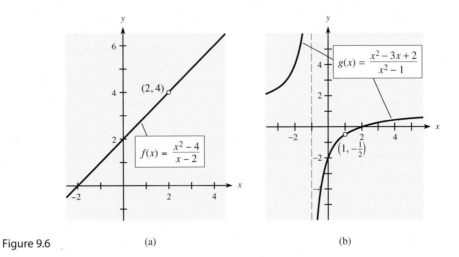

Figure 9.6
(a)
(b)

Note that although both problems in Example 4 had the 0/0 indeterminate form, they had different answers.

EXAMPLE 5   Limit with $a/0$ Form

Find $\lim\limits_{x \to 1} \dfrac{x^2 + 3x + 2}{x - 1}$, if it exists.

Solution

Substituting 1 for x in the function results in 6/0, so this limit has the form $a/0$, with $a \neq 0$, and is like the Type II form discussed previously. Hence the limit does not exist. Because the numerator is not zero when $x = 1$, we know that $x - 1$ is *not* a factor of the numerator, and we cannot divide numerator and denominator as we did in Example 4. Table 9.2 confirms that this limit does not exist, because the values of the expression become unbounded as x approaches 1.

■ **| TABLE 9.2 |** ■

Left of $x = 1$		Right of $x = 1$	
x	$\dfrac{x^2 + 3x + 2}{x - 1}$	x	$\dfrac{x^2 + 3x + 2}{x - 1}$
0.5	−7.5	1.5	17.5
0.7	−15.3	1.2	35.2
0.9	−55.1	1.1	65.1
0.99	−595.01	1.01	605.01
0.999	−5995.001	1.001	6005.001
0.9999	−59,995.0001	1.0001	60,005.0001
$\lim\limits_{x \to 1^-} \dfrac{x^2 + 3x + 2}{x - 1}$ DNE		$\lim\limits_{x \to 1^+} \dfrac{x^2 + 3x + 2}{x - 1}$ DNE	
$(f(x) \to -\infty$ as $x \to 1^-)$		$(f(x) \to \infty$ as $x \to 1^+)$	

The left-hand and right-hand limits do not exist. Thus $\lim\limits_{x \to 1} \dfrac{x^2 + 3x + 2}{x - 1}$ does not exist. ■

In Example 5, even though the left-hand and right-hand limits do not exist (see Table 9.2), knowledge that the functional values are unbounded (that is, that they become infinite) is helpful in graphing. The graph is shown in Figure 9.7 on the next page. We see that $x = 1$ is a vertical asymptote.

Figure 9.7

$f(x)$ unbounded ($f(x) \to +\infty$) as $x \to 1^+$

$$y = \frac{x^2 + 3x + 2}{x - 1}$$

$f(x)$ unbounded ($f(x) \to -\infty$) as $x \to 1^-$

EXAMPLE 6   Cost-Benefit | APPLICATION PREVIEW |

USA Steel has shown that the cost C of removing p percent of the particulate pollution from the smokestack emissions at one of its plants is

$$C = C(p) = \frac{7300p}{100 - p}$$

Investigate the cost of removing as much of the pollution as possible.

Solution

First note that the costs of removing 90% and 99% of the pollution are found as follows:

Removing 90%: $C(90) = \dfrac{7300(90)}{100 - 90} = \dfrac{657{,}000}{10} = 65{,}700$ (dollars)

Removing 99%: $C(99) = \dfrac{7300(99)}{100 - 99} = \dfrac{722{,}700}{1} = 722{,}700$ (dollars)

The cost of removing 100% of the pollution is undefined because the denominator of the function is 0 when $p = 100$. To see what the cost approaches as p approaches 100 from values smaller than 100, we evaluate $\lim\limits_{x \to 100^-} \dfrac{7300p}{100 - p}$. This limit has the Type II form for rational functions. Thus $\dfrac{7300\,p}{100 - p} \to \infty$ as $x \to 100^-$, which means that as the amount of pollution that is removed approaches 100%, the cost increases without bound. (That is, it is cost prohibitive to remove 100% of the pollution.)

© iStockphoto.com/steba2

✓ CHECKPOINT

6. Evaluate the following limits (if they exist).

   (a) $\lim\limits_{x \to -3} \dfrac{2x^2 + 5x - 3}{x^2 - 9}$ (b) $\lim\limits_{x \to 5} \dfrac{x^2 - 3x - 3}{x^2 - 8x + 1}$ (c) $\lim\limits_{x \to -3/4} \dfrac{4x}{4x + 3}$

In Problems 7–10, assume that f, g, and h are polynomials.

7. Does $\lim\limits_{x \to c} f(x) = f(c)$?

8. Does $\lim\limits_{x \to c} \dfrac{g(x)}{h(x)} = \dfrac{g(c)}{h(c)}$?

9. If $g(c) = 0$ and $h(c) = 0$, can we be certain that

   (a) $\lim\limits_{x \to c} \dfrac{g(x)}{h(x)} = 0$?   (b) $\lim\limits_{x \to c} \dfrac{g(x)}{h(x)}$ exists?

10. If $g(c) \neq 0$ and $h(c) = 0$, what can be said about $\lim\limits_{x \to c} \dfrac{g(x)}{h(x)}$ and $\lim\limits_{x \to c} \dfrac{h(x)}{g(x)}$?

Limits of Piecewise Defined Functions

As we noted in Section 2.4, "Special Functions and Their Graphs," many applications are modeled by piecewise defined functions. To see how we evaluate a limit involving a piecewise defined function, consider the following example.

EXAMPLE 7 **Limits of a Piecewise Defined Function**

Find $\lim_{x \to 1^-} f(x)$, $\lim_{x \to 1^+} f(x)$, and $\lim_{x \to 1} f(x)$, if they exist, for

$$f(x) = \begin{cases} x^2 + 1 & \text{if } x \le 1 \\ x + 2 & \text{if } x > 1 \end{cases}$$

Solution

Because $f(x)$ is defined by $x^2 + 1$ when $x < 1$,

$$\lim_{x \to 1^-} f(x) = \lim_{x \to 1^-} (x^2 + 1) = 2$$

Because $f(x)$ is defined by $x + 2$ when $x > 1$,

$$\lim_{x \to 1^+} f(x) = \lim_{x \to 1^+} (x + 2) = 3$$

And because

$$2 = \lim_{x \to 1^-} f(x) \ne \lim_{x \to 1^+} f(x) = 3$$

$\lim_{x \to 1} f(x)$ does not exist.

Table 9.3 and Figure 9.8 show these results numerically and graphically.

| TABLE 9.3 |

Left of 1	
x	**$f(x) = x^2 + 1$**
0.9	1.81
0.99	1.98
0.999	1.998
0.9999	1.9998

Right of 1	
x	**$f(x) = x + 2$**
1.01	3.01
1.001	3.001
1.0001	3.0001
1.00001	3.00001

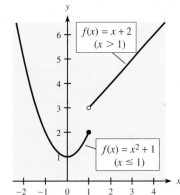

Figure 9.8

Calculator Note

We have used graphical, numerical, and algebraic methods to understand and evaluate limits. Graphing calculators can be especially effective when exploring limits graphically or numerically. See the following example and Appendix A, Section 9.1, for details.

EXAMPLE 8 **Limits: Graphically, Numerically, and Algebraically**

Consider the following limits.

(a) $\lim_{x \to 5} \dfrac{x^2 + 2x - 35}{x^2 - 6x + 5}$ (b) $\lim_{x \to -1} \dfrac{2x}{x + 1}$

Investigate each limit by using the following methods.

(i) Graphically: Graph the function with a graphing calculator and trace near the limiting x-value.
(ii) Numerically: Use the table feature of a graphing calculator to evaluate the function very close to the limiting x-value.
(iii) Algebraically: Use properties of limits and algebraic techniques.

Solution

(a) $\lim\limits_{x \to 5} \dfrac{x^2 + 2x - 35}{x^2 - 6x + 5}$

    (i) Figures 9.9(a) and 9.9(b) show the graph of $y = (x^2 + 2x - 35)/(x^2 - 6x + 5)$. Tracing near $x = 5$ will show y-values getting close to 3.

    (ii) Figure 9.9(c) shows a table for $y_1 = (x^2 + 2x - 35)/(x^2 - 6x + 5)$ with x-values approaching 5 from both sides (note that the function is undefined at $x = 5$). Again, the y-values approach 3 as x approaches 5 from both sides.

Both (i) and (ii) strongly suggest $\lim\limits_{x \to 5} \dfrac{x^2 + 2x - 35}{x^2 - 6x + 5} = 3$.

    (iii) Algebraic evaluation of this limit confirms what the graph and the table suggest.

$$\lim\limits_{x \to 5} \frac{x^2 + 2x - 35}{x^2 - 6x + 5} = \lim\limits_{x \to 5} \frac{(x + 7)(x - 5)}{(x - 1)(x - 5)} = \lim\limits_{x \to 5} \frac{x + 7}{x - 1} = \frac{12}{4} = 3$$

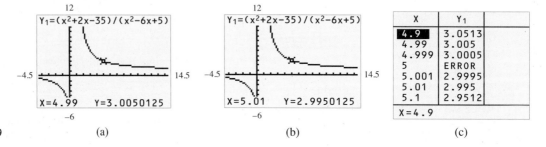

Figure 9.9   (a)   (b)   (c)

(b) $\lim\limits_{x \to -1} \dfrac{2x}{x + 1}$

    (i) Figure 9.10(a) shows the graph of $y = 2x/(x + 1)$; it indicates a break in the graph near $x = -1$. Evaluation confirms that the break occurs at $x = -1$ and also suggests that the function becomes unbounded near $x = -1$. In addition, we can see that as x approaches -1 from opposite sides, the function is headed in different directions. All this suggests that the limit does not exist.

    (ii) Figure 9.10(b) shows a graphing calculator table of values for $y_1 = 2x/(x + 1)$ and with x-values approaching $x = -1$ from both sides. The table reinforces our preliminary conclusion from the graph that the limit does not exist, because the function values are not approaching a single value and are becoming unbounded near $x = -1$.

    (iii) Algebraically we see that this limit has the form $-2/0$. Thus $\lim\limits_{x \to -1} \dfrac{2x}{x + 1}$ DNE.

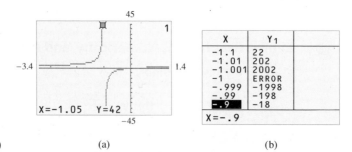

Figure 9.10   (a)   (b)

1. Yes. See Figure 9.2 and Table 9.1.
2. Not necessarily. See Figure 9.4(c).
3. Not necessarily. See Figure 9.3(c).
4. Not necessarily. See Figure 9.4(c).
5. (a) Yes   (b) 4
6. (a) $\lim\limits_{x\to-3} \dfrac{2x^2 + 5x - 3}{x^2 - 9} = \dfrac{7}{6}$

    (b) $\lim\limits_{x\to 5} \dfrac{x^2 - 3x - 3}{x^2 - 8x + 1} = -\dfrac{1}{2}$

    (c) $\lim\limits_{x\to-3/4} \dfrac{4x}{4x + 3}$ does not exist.

7. Yes, Properties I–IV yield this result.
8. Not necessarily. If $h(c) \neq 0$, then this is true. Otherwise, it is not true.
9. For both (a) and (b), $g(x)/h(x)$ has the $0/0$ indeterminate form at $x = c$. In this case we can make no general conclusion about the limit.
10. $\lim\limits_{x\to c} \dfrac{g(x)}{h(x)}$ does not exist and $\lim\limits_{x\to c} \dfrac{h(x)}{g(x)} = 0$

| EXERCISES | 9.1

In Problems 1–6, a graph of $y = f(x)$ is shown and a c-value is given. For each problem, use the graph to find the following, whenever they exist.

(a) $\lim\limits_{x\to c} f(x)$ and   (b) $f(c)$

In Problems 7–10, use the graph of $y = f(x)$ and the given c-value to find the following, whenever they exist.

(a) $\lim\limits_{x\to c^-} f(x)$ (b) $\lim\limits_{x\to c^+} f(x)$

(c) $\lim\limits_{x\to c} f(x)$ (d) $f(c)$

1. $c = 4$

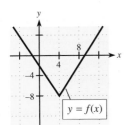

2. $c = 6$

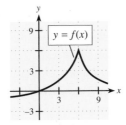

7. $c = -10$

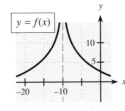

8. $c = 2$

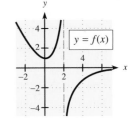

3. $c = 20$

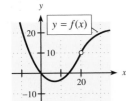

4. $c = -10$

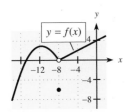

9. $c = -4\frac{1}{2}$

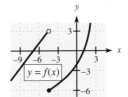

10. $c = 2$

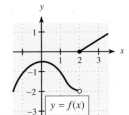

5. $c = -8$

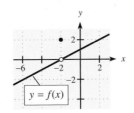

6. $c = -2$

In Problems 11–14, complete each table and predict the limit, if it exists.

11. $f(x) = \dfrac{2 - x - x^2}{x - 1}$

$\lim\limits_{x \to 1} f(x) = ?$

x	f(x)
0.9	
0.99	
0.999	
1.001	
1.01	
1.1	

12. $f(x) = \dfrac{2x + 1}{\frac{1}{4} - x^2}$

$\lim\limits_{x \to -0.5} f(x) = ?$

x	f(x)
−0.51	
−0.501	
−0.5001	
−0.4999	
−0.499	
−0.49	

13. $f(x) = \begin{cases} 5x - 1 & \text{for } x < 1 \\ 8 - 2x - x^2 & \text{for } x \geq 1 \end{cases}$

$\lim\limits_{x \to 1} f(x) = ?$

x	f(x)
0.9	
0.99	
0.999	
1.001	
1.01	
1.1	

14. $f(x) = \begin{cases} 4 - x^2 & \text{for } x \leq -2 \\ x^2 + 2x & \text{for } x > -2 \end{cases}$

$\lim\limits_{x \to -2} f(x) = ?$

x	f(x)
−2.1	
−2.01	
−2.001	
−1.999	
−1.99	

In Problems 15–38, use properties of limits and algebraic methods to find the limits, if they exist.

15. $\lim\limits_{x \to -35} (34 + x)$

16. $\lim\limits_{x \to 80} (82 - x)$

17. $\lim\limits_{x \to -1} (4x^3 - 2x^2 + 2)$

18. $\lim\limits_{x \to 3} (2x^3 - 12x^2 + 5x + 3)$

19. $\lim\limits_{x \to -1/2} \dfrac{4x - 2}{4x^2 + 1}$

20. $\lim\limits_{x \to -1/3} \dfrac{1 - 3x}{9x^2 + 1}$

21. $\lim\limits_{x \to 3} \dfrac{x^2 - 9}{x - 3}$

22. $\lim\limits_{x \to -4} \dfrac{x^2 - 16}{x + 4}$

23. $\lim\limits_{x \to 7} \dfrac{x^2 - 8x + 7}{x^2 - 6x - 7}$

24. $\lim\limits_{x \to -5} \dfrac{x^2 + 8x + 15}{x^2 + 5x}$

25. $\lim\limits_{x \to 10} \dfrac{3x^2 - 30x}{x^2 - 100}$

26. $\lim\limits_{x \to -6} \dfrac{2x^2 - 72}{3x^2 + 18x}$

27. $\lim\limits_{x \to -2} \dfrac{x^2 + 4x + 4}{x^2 + 3x + 2}$

28. $\lim\limits_{x \to 10} \dfrac{x^2 - 8x - 20}{x^2 - 11x + 10}$

29. $\lim\limits_{x \to 3} f(x)$, where $f(x) = \begin{cases} 10 - 2x & \text{if } x < 3 \\ x^2 - x & \text{if } x \geq 3 \end{cases}$

30. $\lim\limits_{x \to 5} f(x)$, where $f(x) = \begin{cases} 7x - 10 & \text{if } x < 5 \\ 25 & \text{if } x \geq 5 \end{cases}$

31. $\lim\limits_{x \to -1} f(x)$, where $f(x) = \begin{cases} x^2 + \dfrac{4}{x} & \text{if } x \leq -1 \\ 3x^3 - x - 1 & \text{if } x > -1 \end{cases}$

32. $\lim\limits_{x \to 2} f(x)$, where $f(x) = \begin{cases} \dfrac{x^3 - 4}{x - 3} & \text{if } x \leq 2 \\ \dfrac{3 - x^2}{x} & \text{if } x > 2 \end{cases}$

33. $\lim\limits_{x \to 2} \dfrac{x^2 + 6x + 9}{x - 2}$

34. $\lim\limits_{x \to 5} \dfrac{x^2 - 6x + 8}{x - 5}$

35. $\lim\limits_{x \to -1} \dfrac{x^2 + 5x + 6}{x + 1}$

36. $\lim\limits_{x \to 3} \dfrac{x^2 + 2x - 3}{x - 3}$

37. $\lim\limits_{h \to 0} \dfrac{(x + h)^3 - x^3}{h}$

38. $\lim\limits_{h \to 0} \dfrac{2(x + h)^2 - 2x^2}{h}$

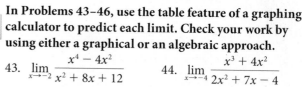

In Problems 39–42, graph each function with a graphing calculator and use it to predict the limit. Check your work either by using the table feature of the calculator or by finding the limit algebraically.

39. $\lim\limits_{x \to 10} \dfrac{x^2 - 19x + 90}{3x^2 - 30x}$

40. $\lim\limits_{x \to -3} \dfrac{x^4 + 3x^3}{2x^4 - 18x^2}$

41. $\lim\limits_{x \to -1} \dfrac{x^3 - x}{x^2 + 2x + 1}$

42. $\lim\limits_{x \to 5} \dfrac{x^2 - 7x + 10}{x^2 - 10x + 25}$

In Problems 43–46, use the table feature of a graphing calculator to predict each limit. Check your work by using either a graphical or an algebraic approach.

43. $\lim\limits_{x \to -2} \dfrac{x^4 - 4x^2}{x^2 + 8x + 12}$

44. $\lim\limits_{x \to -4} \dfrac{x^3 + 4x^2}{2x^2 + 7x - 4}$

45. $\lim\limits_{x \to 4} f(x)$, where $f(x) = \begin{cases} 12 - \dfrac{3}{4}x & \text{if } x \leq 4 \\ x^2 - 7 & \text{if } x > 4 \end{cases}$

46. $\lim\limits_{x \to 7} f(x)$, where $f(x) = \begin{cases} 2 + x - x^2 & \text{if } x \leq 7 \\ 13 - 9x & \text{if } x > 7 \end{cases}$

47. Use values 0.1, 0.01, 0.001, 0.0001, and 0.00001 to approximate

$$\lim\limits_{a \to 0} (1 + a)^{1/a}$$

to three decimal places. This limit equals the special number e that is discussed in Section 5.1, "Exponential Functions," and Section 6.2, "Compound Interest; Geometric Sequences."

48. (a) If $\lim\limits_{x \to 2^+} f(x) = 5$, $\lim\limits_{x \to 2^-} f(x) = 5$, and $f(2) = 0$, find $\lim\limits_{x \to 2} f(x)$, if it exists. Explain your conclusions.

(b) If $\lim\limits_{x \to 0^+} f(x) = 3$, $\lim\limits_{x \to 0^-} f(x) = 0$, and $f(0) = 0$, find $\lim\limits_{x \to 0} f(x)$, if it exists. Explain your conclusions.

49. If $\lim\limits_{x \to 3} f(x) = 4$ and $\lim\limits_{x \to 3} g(x) = -2$, find

(a) $\lim\limits_{x \to 3} [f(x) + g(x)]$

(b) $\lim\limits_{x \to 3} [f(x) - g(x)]$

(c) $\lim\limits_{x \to 3} [f(x) \cdot g(x)]$

(d) $\lim\limits_{x \to 3} \dfrac{g(x)}{f(x)}$

50. If $\lim_{x \to -2} f(x) = 6$ and $\lim_{x \to -2} g(x) = 3$, find

(a) $\lim_{x \to -2} [5f(x) - 4g(x)]$ (b) $\lim_{x \to -2} [g(x)]^2$

(c) $\lim_{x \to -2} [4 - xf(x)]$ (d) $\lim_{x \to -2} \left[\dfrac{f(x)}{g(x)} \right]$

51. If $\lim_{x \to 2} [f(x) + g(x)] = 5$ and $\lim_{x \to 2} g(x) = 11$, find

(a) $\lim_{x \to 2} f(x)$

(b) $\lim_{x \to 2} \{[f(x)]^2 - [g(x)]^2\}$

(c) $\lim_{x \to 2} \dfrac{3g(x)}{f(x) - g(x)}$

52. If $\lim_{x \to 5} [f(x) - g(x)] = 8$ and $\lim_{x \to 5} g(x) = 2$, find

(a) $\lim_{x \to 5} f(x)$ (b) $\lim_{x \to 5} \{[g(x)]^2 - f(x)\}$

(c) $\lim_{x \to 5} \left[\dfrac{2xg(x)}{4 - f(x)} \right]$

APPLICATIONS

53. **Revenue** The total revenue for a product is given by

$$R(x) = 1600x - x^2$$

where x is the number of units sold. What is
$\lim_{x \to 100} R(x)$?

54. **Profit** If the profit function for a product is given by

$$P(x) = 92x - x^2 - 1760$$

find $\lim_{x \to 40} P(x)$.

55. **Sales and training** The average monthly sales volume (in thousands of dollars) for a firm depends on the number of hours x of training of its sales staff, according to

$$S(x) = \dfrac{4}{x} + 30 + \dfrac{x}{4}, \quad 4 \le x \le 100$$

(a) Find $\lim_{x \to 4^+} S(x)$.   (b) Find $\lim_{x \to 100^-} S(x)$.

56. **Sales and training** During the first 4 months of employment, the monthly sales S (in thousands of dollars) for a new salesperson depend on the number of hours x of training, as follows:

$$S = S(x) = \dfrac{9}{x} + 10 + \dfrac{x}{4}, \quad x \ge 4$$

(a) Find $\lim_{x \to 4^+} S(x)$.   (b) Find $\lim_{x \to 10} S(x)$.

57. **Advertising and sales** Suppose that the daily sales S (in dollars) t days after the end of an advertising campaign are

$$S = S(t) = 400 + \dfrac{2400}{t + 1}$$

(a) Find $S(0)$.   (b) Find $\lim_{t \to 7} S(t)$.

(c) Find $\lim_{t \to 14} S(t)$.

58. **Advertising and sales** Sales y (in thousands of dollars) are related to advertising expenses x (in thousands of dollars) according to

$$y = y(x) = \dfrac{200x}{x + 10}, \quad x \ge 0$$

(a) Find $\lim_{x \to 10} y(x)$.   (b) Find $\lim_{x \to 0} y(x)$.

59. **Productivity** During an 8-hour shift, the rate of change of productivity (in units per hour) of infant activity centers assembled after t hours on the job is

$$r(t) = \dfrac{128(t^2 + 6t)}{(t^2 + 6t + 18)^2}, \quad 0 \le t \le 8$$

(a) Find $\lim_{t \to 4} r(t)$.   (b) Find $\lim_{t \to 8} r(t)$.

(c) Is the rate of productivity higher near the lunch break (at $t = 4$) or near quitting time (at $t = 8$)?

60. **Revenue** If the revenue for a product is $R(x) = 100x - 0.1x^2$, and the average revenue per unit is

$$\overline{R}(x) = \dfrac{R(x)}{x}, \quad x > 0$$

find (a) $\lim_{x \to 100} \dfrac{R(x)}{x}$ and   (b) $\lim_{x \to 0^+} \dfrac{R(x)}{x}$

61. **Cost-benefit** Suppose that the cost C of obtaining water that contains p percent impurities is given by

$$C(p) = \dfrac{120{,}000}{p} - 1200$$

(a) Find $\lim_{p \to 100^-} C(p)$, if it exists. Interpret this result.

(b) Find $\lim_{p \to 0^+} C(p)$, if it exists.

(c) Is complete purity possible? Explain.

62. **Cost-benefit** Suppose that the cost C of removing p percent of the particulate pollution from the smokestacks of an industrial plant is given by

$$C(p) = \dfrac{730{,}000}{100 - p} - 7300$$

(a) Find $\lim_{p \to 80} C(p)$

(b) Find $\lim_{p \to 100^-} C(p)$ if it exists.

(c) Can 100% of the particulate pollution be removed? Explain.

63. **Federal income tax** The following table shows part of the April 15, 2015, tax rate schedule for single filers. Use this schedule and create a table of values that could be used to find the following limits, if they exist. Let x represent the amount of taxable income, and let $T(x)$ represent the tax due.

(a) $\lim_{x \to 36{,}900^-} T(x)$ (b) $\lim_{x \to 36{,}900^+} T(x)$

(c) $\lim_{x \to 36{,}900} T(x)$

Single Filers

If taxable income is over—	But not over—	The tax is:
$0	$9075	10% of the amount over $0
$9075	$36,900	$907.50 plus 15% of the amount over 9075
$36,900	$89,350	$5081.25 plus 25% of the amount over 36,900
$89,350	$186,350	$18,193.75 plus 28% of the amount over 89,350

Source: Internal Revenue Service

64. *Parking costs* The Ace Parking Garage charges $5.00 for parking for 2 hours or less and $1.50 for each extra hour or part of an hour after the 2-hour minimum. The parking charges for the first 5 hours could be written as a function of the time as follows:

$$f(t) = \begin{cases} \$5.00 & \text{if } 0 < t \leq 2 \\ \$6.50 & \text{if } 2 < t \leq 3 \\ \$8.00 & \text{if } 3 < t \leq 4 \\ \$9.50 & \text{if } 4 < t \leq 5 \end{cases}$$

 (a) Find $\lim_{t \to 1} f(t)$, if it exists.

 (b) Find $\lim_{t \to 2} f(t)$, if it exists.

65. *Municipal water rates* The Corner Water Corp. of Shippenville, Pennsylvania, has the following rates per 1000 gallons of water used.

Usage (x)	Cost per 1000 Gallons ($C(x)$)
First 10,000 gallons	$15.96
Next 110,000 gallons	13.56
Over 120,000 gallons	Additional 11.04

If Corner Water has a monthly service fee of $12.76, write a function $C = C(x)$ that models the charges (where x is thousands of gallons) and find $\lim_{x \to 10} C(x)$ (that is, as usage approaches 10,000 gallons).

66. *Airport parking* Long-term parking at Savannah Airport is free for the first half-hour and costs $1.00 for each hour or part of an hour thereafter. If $C = C(t)$ is the charge for t hours in Savannah's long-term parking, create a table of values for parking costs close to $t = 1/2$ and $t = 2$ and use them to find the following limits, if they exist.

 (a) $\lim_{t \to 0.5^-} C(t)$ (b) $\lim_{t \to 0.5^+} C(t)$

 (c) $\lim_{t \to 0.5} C(t)$ (d) $\lim_{t \to 2} C(t)$

Dow Jones Industrial Average The graph in the figure shows the Dow Jones Industrial Average (DJIA) on a particularly tumultuous day in August 2011. Use the graph for Problems 67 and 68, with t as the time of day and $D(t)$ as the DJIA at time t.

67. Estimate $\lim_{t \to 9.30\text{AM}^+} D(t)$, if it exists. Explain what this limit corresponds to.

68. Estimate $\lim_{t \to 4.00\text{PM}^-} D(t)$, if it exists. Explain what this limit corresponds to.

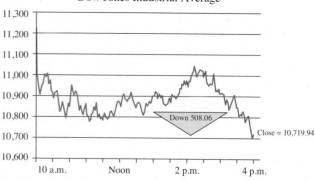

Dow Jones Industrial Average

Source: Google Finance, August 10, 2011

Obesity **Obesity (BMI ≥ 30) is a serious problem in the United States and expected to get worse. Being overweight increases the risk of diabetes, heart disease, and many other ailments, but the severely obese (BMI ≥ 40) are most at risk and the most expensive to treat. The following table shows the percent of obese Americans who are severely obese for selected years from 1990 and projected to 2030.**

Percent Obese Who Are Severely Obese

Year	1990	2000	2010	2015	2020	2025	2030
Percent	6.30	9.95	15.9	18.6	21.1	23.8	26.3

Source: American Journal of Preventive Medicine 42 (June 2012): 563–570. ajpmonline.org

The percent of obese American adults who are severely obese can be modeled by the function

$$S(x) = \frac{0.264x^2 + 10.7x - 66.9}{-0.00850x^2 + 1.25x + 0.854}$$

where x is the number of years after 1980. Use this function in Problems 69 and 70.

69. (a) Find $\lim_{x \to 60} S(x)$, if it exists.

 (b) What does this limit predict?

 (c) Does this prediction seem plausible? Explain.

70. (a) Find $\lim_{x \to 40} S(x)$, if it exists.

 (b) What does this limit estimate?

 (c) Is the function accurate as $x \to 40$? Explain.

OBJECTIVES
9.2

- To determine whether a function is continuous or discontinuous at a point
- To determine where a function is discontinuous
- To find limits at infinity and horizontal asymptotes

Continuous Functions; Limits at Infinity

▮ | APPLICATION PREVIEW |

Suppose that a friend of yours and her husband have a taxable income of $148,850 and she tells you that she doesn't want to make any more money because that would put them in a higher tax bracket. She makes this statement because the tax rate schedule for married taxpayers filing a joint return (part of which is shown in the table) appears to have a jump in taxes for taxable income at $148,850.

Married Filing Jointly or Qualifying Widow(er)

If taxable income is over	But not over	The tax is:
$0	$18,150	10% of the amount over $0
$18,150	$73,800	$1,815.00 plus 15% of the amount over $18,150
$73,800	$148,850	$10,162.50 plus 25% of the amount over $73,800
$148,850	$226,850	$28,925.00 plus 28% of the amount over $148,850
$226,850	$405,100	$50,765.00 plus 33% of the amount over $226,850

Source: Internal Revenue Service

To see whether the couple's taxes would jump to some higher level, we will write the function that gives income tax for married taxpayers with taxable incomes up to $405,100 as a function of taxable income and show that the function is continuous (see Example 3). That is, we will see that the tax paid does not jump at $148,850 even though the tax on income above $148,850 is collected at a higher rate. In this section, we will show how to determine whether a function is continuous, and we will investigate some different types of discontinuous functions.

Continuous Functions

We have found that $f(c)$ is the same as the limit as $x \rightarrow c$ for any polynomial function $f(x)$ and any real number c. Any function for which this special property holds is called a **continuous function.** The graphs of such functions can be drawn without lifting the pencil from the paper, and graphs of others may have holes, vertical asymptotes, or jumps that make it impossible to draw them without lifting the pencil. In general, we define continuity of a function at the value $x = c$ as follows.

Continuity at a Point

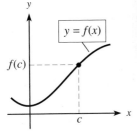

The function f is **continuous at $x = c$** if *all* of the following conditions are satisfied.

1. $f(c)$ exists   2. $\lim\limits_{x \to c} f(x)$ exists   3. $\lim\limits_{x \to c} f(x) = f(c)$

The figure at the left illustrates these three conditions.
If one or more of the conditions above do not hold, we say the function is **discontinuous at $x = c$.**

If a function is discontinuous at one or more points, it is called a **discontinuous function.** Figure 9.11 on the next page shows graphs of some functions that are discontinuous at $x = 2$.

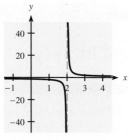

(a) $f(x) = \dfrac{1}{x-2}$

$\lim\limits_{x\to 2} f(x)$ and $f(2)$ do not exist.

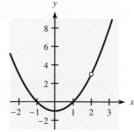

(b) $f(x) = \dfrac{x^3 - 2x^2 - x + 2}{x - 2}$

$f(2)$ does not exist.

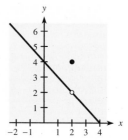

(c) $f(x) = \begin{cases} 4 - x & \text{if } x \neq 2 \\ 4 & \text{if } x = 2 \end{cases}$

$\lim\limits_{x\to 2} f(x) = 2 \neq 4 = f(2)$

Figure 9.11

We have seen that $\lim\limits_{x\to c} f(x) = f(c)$ for every real number c, if $f(x)$ is a polynomial, and $\lim\limits_{x\to c} h(x) = h(c)$ if $h(x) = \dfrac{f(x)}{g(x)}$ is a rational function and $g(c) \neq 0$. So, we have the following.

Polynomial and Rational Functions

Every polynomial function is continuous for all real numbers.
Every rational function is continuous at all values of x except those that make the denominator 0.

EXAMPLE 1 Discontinuous Functions

For what values of x, if any, are the following functions continuous?

(a) $h(x) = \dfrac{3x + 2}{4x - 6}$ (b) $f(x) = \dfrac{x^2 - x - 2}{x^2 - 4}$

Solution

(a) This is a rational function, so it is continuous for all values of x except for those that make the denominator, $4x - 6$, equal to 0. Because $4x - 6 = 0$ at $x = 3/2$, $h(x)$ is continuous for all real numbers except $x = 3/2$. Figure 9.12(a) shows a vertical asymptote at $x = 3/2$.

(b) This is a rational function, so it is continuous everywhere except where the denominator is 0. To find the zeros of the denominator, we factor $x^2 - 4$.

$$f(x) = \frac{x^2 - x - 2}{x^2 - 4} = \frac{x^2 - x - 2}{(x - 2)(x + 2)}$$

Because the denominator is 0 for $x = 2$ and for $x = -2$, $f(2)$ and $f(-2)$ do not exist (recall that division by 0 is undefined). Thus the function is continuous except at $x = 2$ and $x = -2$. The graph of this function (see Figure 9.12(b)) shows a hole at $x = 2$ and a vertical asymptote at $x = -2$.

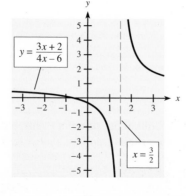

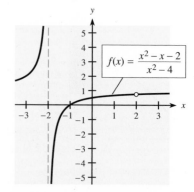

Figure 9.12   (a)   (b)

✓ CHECKPOINT

1. Find any x-values where the following functions are discontinuous.

   (a) $f(x) = x^3 - 3x + 1$ (b) $g(x) = \dfrac{x^3 - 1}{(x-1)(x+2)}$

If the pieces of a piecewise defined function are polynomials, the only values of x where the function might be discontinuous are those at which the definition of the function changes.

EXAMPLE 2   Piecewise Defined Functions

Determine the values of x, if any, for which the following functions are discontinuous.

(a) $g(x) = \begin{cases} (x+2)^3 + 1 & \text{if } x \leq -1 \\ 3 & \text{if } x > -1 \end{cases}$ (b) $f(x) = \begin{cases} 4 - x^2 & \text{if } x < 2 \\ x - 2 & \text{if } x \geq 2 \end{cases}$

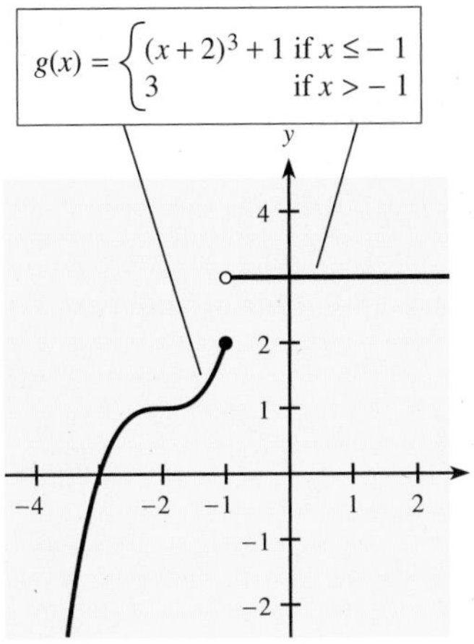

Solution

(a) $g(x)$ is a piecewise defined function in which each part is a polynomial. Thus, to see whether a discontinuity exists, we need only check the value of x for which the definition of the function changes—that is, at $x = -1$. Note that $x = -1$ satisfies $x \leq -1$, so $g(-1) = (-1+2)^3 + 1 = 2$. Because $g(x)$ is defined differently for $x < -1$ and $x > -1$, we use left- and right-hand limits.

For $x \to -1^-$, we know that $x < -1$, so $g(x) = (x+2)^3 + 1$:

$$\lim_{x \to -1^-} g(x) = \lim_{x \to -1^-} [(x+2)^3 + 1] = (-1+2)^3 + 1 = 2$$

Similarly, for $x \to -1^+$, we know that $x > -1$, so $g(x) = 3$:

$$\lim_{x \to -1^+} g(x) = \lim_{x \to -1^+} 3 = 3$$

Because the left- and right-hand limits differ, $\lim_{x \to -1} g(x)$ does not exist, so $g(x)$ is discontinuous at $x = -1$. This result is confirmed by examining the graph of g, shown in Figure 9.13.

(b) As with $g(x)$, $f(x)$ is continuous everywhere except perhaps at $x = 2$, where the definition of $f(x)$ changes. Because $x = 2$ satisfies $x \geq 2$, $f(2) = 2 - 2 = 0$. The left- and right-hand limits are

Left: $\lim_{x \to 2^-} f(x) = \lim_{x \to 2^-} (4 - x^2) = 4 - 2^2 = 0$

Right: $\lim_{x \to 2^+} f(x) = \lim_{x \to 2^+} (x - 2) = 2 - 2 = 0$

Because the right- and left-hand limits are equal, we conclude that $\lim_{x \to 2} f(x) = 0$. The limit is equal to the functional value, or:

$$\lim_{x \to 2} f(x) = f(2)$$

so we conclude that f is continuous at $x = 2$ and thus f is continuous for all values of x. This result is confirmed by the graph of f, shown in Figure 9.14.

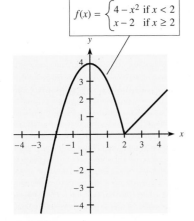

$g(x) = \begin{cases} (x+2)^3 + 1 & \text{if } x \leq -1 \\ 3 & \text{if } x > -1 \end{cases}$

Figure 9.13

$f(x) = \begin{cases} 4 - x^2 & \text{if } x < 2 \\ x - 2 & \text{if } x \geq 2 \end{cases}$

Figure 9.14

EXAMPLE 3   Taxes | APPLICATION PREVIEW |

A partial tax rate schedule for married taxpayers filing a joint return (shown in the table) appears to have a jump in taxes for taxable income at $148,850.

Married Filing Jointly or Qualifying Widow(er)

If taxable income is over—	But not over—	The tax is:
$0	$18,150	10% of the amount over $0
$18,150	$73,800	$1,815.00 plus 15% of the amount over $18,150
$73,800	$148,850	$10,162.50 plus 25% of the amount over $73,800
$148,850	$226,850	$28,925.00 plus 28% of the amount over $148,850
$226,850	$405,100	$50,765.00 plus 33% of the amount over $226,850

Source: Internal Revenue Service

(a) Use the table and write the function that gives income tax for married taxpayers in this income range as a function of taxable income, x.

(b) Is the function in part (a) continuous at $x = 148,850$?

(c) A married friend of yours and her husband have a taxable income of $148,850, and she tells you that she doesn't want to make any more money because doing so would put her in a higher tax bracket. What would you tell her to do if she is offered a raise?

Solution

(a) The function that gives the tax due for married taxpayers with $0 \leq x \leq 405,100$ is

$$T(x) = \begin{cases} 0.10x & \text{if } 0 \leq x \leq 18,150 \\ 1815.00 + 0.15(x - 18,150) & \text{if } 18,150 < x \leq 73,800 \\ 10,162.50 + 0.25(x - 73,800) & \text{if } 73,800 < x \leq 148,850 \\ 28,925.00 + 0.28(x - 148,850) & \text{if } 148,850 < x \leq 226,850 \\ 50,765.00 + 0.33(x - 226,850) & \text{if } 226,850 < x \leq 405,100 \end{cases}$$

(b) If this function is continuous at $x = 148,850$, there is no jump in taxes at $148,850. We examine the three conditions for continuity at $x = 148,850$:

(i) $T(148,850) = 28,925$ so $T(148,850)$ exists.

(ii) Because the function is piecewise defined near $x = 148,850$, we evaluate $\lim_{x \to 148,850} T(x)$ by evaluating one-sided limits:

From the left, we evaluate $\lim_{x \to 148,850^-} T(x)$:

$$\lim_{x \to 148,850^-} [10,162.50 + 0.25(x - 73,800)] = 28,925$$

From the right, we evaluate $\lim_{x \to 148,850^+} T(x)$:

$$\lim_{x \to 148,850^+} [28,925.00 + 0.28(x - 148,850)] = 28,925$$

Because these one-sided limits agree, the limit exists and is

$$\lim_{x \to 148,850} T(x) = 28,925.$$

(iii) Because $\lim_{x \to 148,850} T(x) = T(148,850) = 28,925$, the function is continuous at $x = 148,850$.

(c) If your friend earned more than $148,850, she and her husband would pay taxes at a higher rate *only* on the amount of money *above* $148,850. Thus she should take any raise that is offered. ∎

✓ CHECKPOINT

2. If $f(x)$ and $g(x)$ are polynomials, $h(x) = \begin{cases} f(x) & \text{if } x \leq a \\ g(x) & \text{if } x > a \end{cases}$ is continuous everywhere except perhaps at _____.

Limits at Infinity

We have seen that the graph of $y = 1/x$ has a vertical asymptote at $x = 0$ (shown in Figure 9.15(a)). By graphing $y = 1/x$ and evaluating the function for very large x-values, we can see that $y = 1/x$ never becomes negative for positive x-values regardless of how large the x-value is. Although no value of x makes $1/x$ equal to 0, it is easy to see that $1/x$ approaches 0 as x gets very large. This is denoted by

$$\lim_{x \to \infty} \frac{1}{x} = 0$$

and means that the line $y = 0$ (the x-axis) is a horizontal asymptote for $y = 1/x$. We also see that $y = 1/x$ approaches 0 as x decreases without bound, and we denote this by

$$\lim_{x \to -\infty} \frac{1}{x} = 0$$

These limits for $f(x) = 1/x$ can also be established with numerical tables.

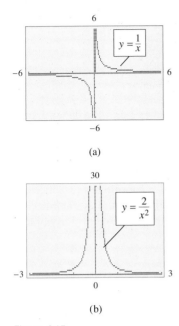

(a)

(b)

Figure 9.15

x	$f(x) = 1/x$	x	$f(x) = 1/x$
100	0.01	-100	-0.01
100,000	0.00001	$-100,000$	-0.00001
100,000,000	0.00000001	$-100,000,000$	-0.00000001
$\downarrow$	$\downarrow$	$\downarrow$	$\downarrow$
∞	0	$-\infty$	0

$$\lim_{x \to \infty} \frac{1}{x} = 0 \qquad\qquad \lim_{x \to -\infty} \frac{1}{x} = 0$$

We can use the graph of $y = 2/x^2$ in Figure 9.15(b) to see that the x-axis ($y = 0$) is a horizontal asymptote and that

$$\lim_{x \to \infty} \frac{2}{x^2} = 0 \qquad \text{and} \qquad \lim_{x \to -\infty} \frac{2}{x^2} = 0$$

By using graphs and/or tables of values, we can generalize the results for the functions shown in Figure 9.15 and conclude the following.

Limits at Infinity

If c is any constant, then

1. $\displaystyle\lim_{x \to \infty} c = c$ and $\displaystyle\lim_{x \to -\infty} c = c.$

2. $\displaystyle\lim_{x \to \infty} \frac{c}{x^p} = 0$, where $p > 0$.

3. $\displaystyle\lim_{x \to -\infty} \frac{c}{x^n} = 0$, where $n > 0$ is any integer.

In order to use these properties for finding the limits of rational functions as x approaches ∞ or $-\infty$, we first divide each term of the numerator and denominator by the highest power of x present and then determine the limit of the resulting expression.

EXAMPLE   4   Limits at Infinity

Find each of the following limits, if they exist.

(a) $\displaystyle\lim_{x \to \infty} \frac{2x - 1}{x + 2}$ (b) $\displaystyle\lim_{x \to -\infty} \frac{x^2 + 3}{1 - x}$

Solution

(a) The highest power of x present is x^1, so we divide each term in the numerator and denominator by x and then use the properties for limits at infinity.

$$\lim_{x \to \infty} \frac{2x - 1}{x + 2} = \lim_{x \to \infty} \frac{\dfrac{2x}{x} - \dfrac{1}{x}}{\dfrac{x}{x} + \dfrac{2}{x}} = \lim_{x \to \infty} \frac{2 - \dfrac{1}{x}}{1 + \dfrac{2}{x}}$$

$$= \frac{2 - 0}{1 + 0} = 2 \quad \text{(by Properties 1 and 2)}$$

Figure 9.16(a) on the next page shows the graph of this function with the y-coordinates of the graph approaching 2 as x approaches ∞ and as x approaches $-\infty$. That is, $y = 2$ is a horizontal asymptote. Note also that there is a discontinuity (vertical asymptote) where $x = -2$.

(b) We divide each term in the numerator and denominator by x^2 and then use the properties.

$$\lim_{x \to -\infty} \frac{x^2 + 3}{1 - x} = \lim_{x \to -\infty} \frac{\dfrac{x^2}{x^2} + \dfrac{3}{x^2}}{\dfrac{1}{x^2} - \dfrac{x}{x^2}} = \lim_{x \to -\infty} \frac{1 + \dfrac{3}{x^2}}{\dfrac{1}{x^2} - \dfrac{1}{x}}$$

This limit does not exist because the numerator approaches 1 and the denominator approaches 0 through positive values. Thus

$$\frac{x^2 + 3}{1 - x} \to \infty \text{ as } x \to -\infty$$

The graph of this function, shown in Figure 9.16(b), has y-coordinates that increase without bound as x approaches $-\infty$ and that decrease without bound as x approaches ∞. (There is no horizontal asymptote.) Note also that there is a vertical asymptote at $x = 1$.

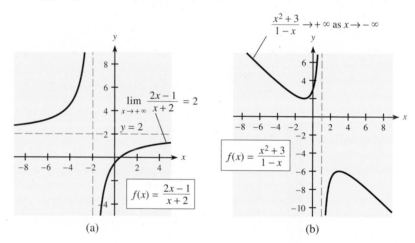

Figure 9.16   (a)   (b)

In our work with limits at infinity, we have mentioned horizontal asymptotes several times. The connection between these concepts follows.

Limits at Infinity and Horizontal Asymptotes	If $\lim\limits_{x \to \infty} f(x) = b$ or $\lim\limits_{x \to -\infty} f(x) = b$, where b is a constant, then the line $y = b$ is a horizontal asymptote for the graph of $y = f(x)$. Otherwise, $y = f(x)$ has no horizontal asymptotes.

✓ **CHECKPOINT**

3. (a) Evaluate $\lim\limits_{x \to \infty} \dfrac{x^2 - 4}{2x^2 - 7}$.

  (b) What does part (a) say about horizontal asymptotes for
  $f(x) = (x^2 - 4)/(2x^2 - 7)$?

Calculator Note We can use the graphing and table features of a graphing calculator to help locate and investigate discontinuities and limits at infinity (horizontal asymptotes). A graphing calculator can be used to focus our attention on a possible discontinuity and to support or suggest appropriate algebraic calculations. See the following example.

EXAMPLE 5 **Limits with Technology**

Use a graphing utility to investigate the continuity of the following functions.

(a) $f(x) = \dfrac{x^2 + 1}{x + 1}$ (b) $g(x) = \dfrac{x^2 - 2x - 3}{x^2 - 1}$

(c) $h(x) = \dfrac{|x + 1|}{x + 1}$ (d) $k(x) = \begin{cases} \dfrac{-x^2}{2} - 2x & \text{if } x \le -1 \\ \dfrac{x}{2} + 2 & \text{if } x > -1 \end{cases}$

Solution

(a) Figure 9.17(a) shows that $f(x)$ has a discontinuity (vertical asymptote) near $x = -1$. Because $f(-1)$ DNE, we know that $f(x)$ is not continuous at $x = -1$.

(b) Figure 9.17(b) shows that $g(x)$ is discontinuous (vertical asymptote) near $x = 1$, and this looks like the only discontinuity. However, the denominator of $g(x)$ is zero at $x = 1$ and $x = -1$, so $g(x)$ must have discontinuities at both of these x-values. Evaluating or using the table feature confirms that $x = -1$ is a discontinuity (a hole, or missing point). The figure also shows a horizontal asymptote; evaluation of $\lim_{x \to \infty} g(x)$ confirms this is the line $y = 1$.

(c) Figure 9.17(c) shows a discontinuity (jump) at $x = -1$. We also see that $h(-1)$ DNE, which confirms the observations from the graph.

(d) The graph in Figure 9.17(d) appears to be continuous. The only "suspicious" x-value is $x = -1$, where the formula for $k(x)$ changes. Evaluating $k(-1)$ and examining a table near $x = -1$ indicates that $k(x)$ is continuous there. Algebraic evaluations of the two one-sided limits confirm this.

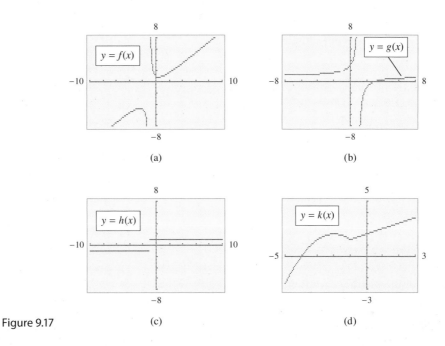

Figure 9.17

(a)   (b)

(c)   (d)

Continous Functions and Limits at Infinity

- The following information is useful in discussing continuity of functions.

A. A polynomial function is continuous everywhere.

B. A rational function is a function of the form $\dfrac{f(x)}{g(x)}$, where $f(x)$ and $g(x)$ are polynomials.

1. If $g(x) \neq 0$ at any value of x, the function is continuous everywhere.

2. If $g(c) = 0$, the function is discontinuous at $x = c$.

(a) If $g(c) = 0$ and $f(c) \neq 0$, then there is a vertical asymptote at $x = c$.

(b) If $g(c) = 0$ and $\lim_{x \to c} \dfrac{f(x)}{g(x)} = L$, then the graph has a missing point at (c, L).

C. A piecewise defined function may have a discontinuity at any x-value where the function changes its formula. One-sided limits must be used to see whether the limit exists.

- The following steps are useful when we are evaluating limits at infinity for a rational function $f(x) = p(x)/q(x)$.

1. Divide both $p(x)$ and $q(x)$ by the highest power of x found in either polynomial.

2. Use the properties of limits at infinity to complete the evaluation.

1. (a) This is a polynomial function, so it is continuous at all values of x (discontinuous at none).

   (b) This is a rational function. It is discontinuous at $x = 1$ and $x = -2$ because these values make its denominator 0.

2. $x = a$

3. (a) $\lim\limits_{x \to \infty} \dfrac{x^2 - 4}{2x^2 - 7} = \dfrac{1}{2}$

   (b) The line $y = 1/2$ is a horizontal asymptote.

| EXERCISES | 9.2

In Problems 1 and 2, refer to the figure. For each given x-value, use the figure to determine whether the function is continuous or discontinuous at that x-value. If the function is discontinuous, state which of the three conditions that define continuity is not satisfied.

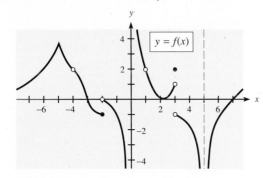

1. (a) $x = -5$ (b) $x = 1$ (c) $x = 3$ (d) $x = 0$
2. (a) $x = 2$ (b) $x = -4$ (c) $x = -2$ (d) $x = 5$

In Problems 3–8, determine whether each function is continuous or discontinuous at the given x-value. Examine the three conditions in the definition of continuity.

3. $f(x) = \dfrac{x^2 - 4}{x - 2}$, $x = -2$

4. $y = \dfrac{x^2 - 9}{x + 3}$, $x = 3$

5. $y = \dfrac{x^2 - x - 12}{x^2 + 3x}$, $x = -3$

6. $f(x) = \dfrac{x^2 - 6x + 8}{x^2 + x - 20}$, $x = 4$

7. $f(x) = \begin{cases} x - 3 & \text{if } x \le 2 \\ 4x - 7 & \text{if } x > 2 \end{cases}$, $x = 2$

8. $f(x) = \begin{cases} x^2 + 1 & \text{if } x \le 1 \\ 2x^2 - 1 & \text{if } x > 1 \end{cases}$, $x = 1$

In Problems 9–16, determine whether the given function is continuous. If it is not, identify where it is discontinuous and which condition fails to hold. You can verify your conclusions by graphing each function with a graphing utility.

9. $f(x) = 4x^2 - 1$

10. $y = 5x^2 - 2x$

11. $g(x) = \dfrac{4x^2 + 3x + 2}{x + 2}$

12. $y = \dfrac{4x^2 + 4x + 1}{x + 1/2}$

13. $y = \dfrac{x}{x^2 + 1}$

14. $y = \dfrac{2x - 1}{x^2 + 3}$

15. $f(x) = \begin{cases} 3 & \text{if } x \le 1 \\ x^2 + 2 & \text{if } x > 1 \end{cases}$

16. $f(x) = \begin{cases} x^3 + 10 & \text{if } x \le -2 \\ 2 & \text{if } x > -2 \end{cases}$

 In Problems 17–20, use the trace and table features of a graphing calculator to investigate whether each of the following functions has any discontinuities.

17. $y = \dfrac{x^2 - 5x - 6}{x + 1}$

18. $y = \dfrac{x^2 - 5x + 4}{x - 4}$

19. $f(x) = \begin{cases} x - 4 & \text{if } x \le 3 \\ x^2 - 8 & \text{if } x > 3 \end{cases}$

20. $f(x) = \begin{cases} x^2 + 4 & \text{if } x \ne 4 \\ 8 & \text{if } x = 4 \end{cases}$

Each of Problems 21–24 contains a function and its graph. For each problem, answer parts (a) and (b).

(a) Use the graph to determine, as well as you can,
   (i) vertical asymptotes. (ii) $\lim\limits_{x \to \infty} f(x)$.
   (iii) $\lim\limits_{x \to -\infty} f(x)$.   (iv) horizontal asymptotes.

(b) Check your conclusions in (a) by using the functions to determine items (i)–(iv) analytically.

21. $f(x) = \dfrac{8}{x + 2}$

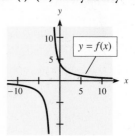

22. $f(x) = \dfrac{x - 3}{x - 2}$

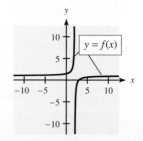

23. $f(x) = \dfrac{2(x+1)^3(x+5)}{(x-3)^2(x+2)^2}$

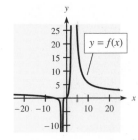

24. $f(x) = \dfrac{4x^2}{x^2 - 4x + 4}$

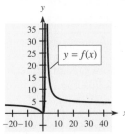

In Problems 25–32, complete (a) and (b).
(a) Use analytic methods to evaluate each limit.
(b) What does the result from part (a) tell you about horizontal asymptotes?
 You can verify your conclusions by graphing the functions with a graphing calculator.

25. $\displaystyle\lim_{x \to \infty} \frac{3}{x+1}$

26. $\displaystyle\lim_{x \to -\infty} \frac{4}{x^2 - 2x}$

27. $\displaystyle\lim_{x \to \infty} \frac{x^3 - 1}{x^3 + 4}$

28. $\displaystyle\lim_{x \to -\infty} \frac{3x^2 + 2}{x^2 - 4}$

29. $\displaystyle\lim_{x \to -\infty} \frac{5x^3 - 4x}{3x^3 - 2}$

30. $\displaystyle\lim_{x \to \infty} \frac{4x^2 + 5x}{x^2 - 4x}$

31. $\displaystyle\lim_{x \to \infty} \frac{3x^2 + 5x}{6x + 1}$

32. $\displaystyle\lim_{x \to -\infty} \frac{5x^3 - 8}{4x^2 + 5x}$

In Problems 33 and 34, use a graphing calculator to complete (a) and (b).
(a) Graph each function using a window with $0 \le x \le 300$ and $-2 \le y \le 2$. What does the graph indicate about $\displaystyle\lim_{x \to \infty} f(x)$?
(b) Use the table feature with x-values larger than 10,000 to investigate $\displaystyle\lim_{x \to \infty} f(x)$. Does the table support your conclusions in part (a)?

33. $f(x) = \dfrac{x^2 - 4}{3 + 2x^2}$

34. $f(x) = \dfrac{5x^3 - 7x}{1 - 3x^3}$

In Problems 35 and 36, complete (a)–(c). Use analytic methods to find (a) any points of discontinuity and (b) limits as $x \to \infty$ and $x \to -\infty$. (c) Then explain why, for these functions, a graphing calculator is better as a support tool for the analytic methods than as the primary tool for investigation.

35. $f(x) = \dfrac{1000x - 1000}{x + 1000}$

36. $f(x) = \dfrac{3000x}{4350 - 2x}$

For Problems 37 and 38, let
$$f(x) = \frac{a_n x^n + a_{n-1}x^{n-1} + \cdots + a_1 x + a_0}{b_m x^m + b_{m-1}x^{m-1} + \cdots + b_1 x + b_0}$$
be a rational function.

37. If $m = n$, show that $\displaystyle\lim_{x \to \infty} f(x) = \frac{a_n}{b_n}$, and hence that $y = \dfrac{a_n}{b_n}$ is a horizontal asymptote.

38. (a) If $m > n$, show that $\displaystyle\lim_{x \to \infty} f(x) = 0$ and hence that $y = 0$ is a horizontal asymptote.
 (b) If $m < n$, find $\displaystyle\lim_{x \to \infty} f(x)$. What does this say about horizontal asymptotes?

APPLICATIONS

39. **Sales volume** Suppose that the weekly sales volume (in thousands of units) for a product is given by
$$y = \frac{32}{(p+8)^{2/5}}$$
where p is the price in dollars per unit. Is this function continuous
(a) for all values of p? (b) at $p = 24$?
(c) for all $p \ge 0$?
(d) What is the domain for this application?

40. **Worker productivity** Suppose that the average number of minutes M that it takes a new employee to assemble one unit of a product is given by
$$M = \frac{40 + 30t}{2t + 1}$$
where t is the number of days on the job. Is this function continuous
(a) for all values of t? (b) at $t = 14$?
(c) for all $t \ge 0$?
(d) What is the domain for this application?

41. **Demand** Suppose that the demand for a product is defined by the equation
$$p = \frac{200,000}{(q+1)^2}$$
where p is the price and q is the quantity demanded.
(a) Is this function discontinuous at any value of q? What value?
(b) Because q represents quantity, we know that $q \ge 0$. Is this function continuous for $q \ge 0$?

42. **Advertising and sales** The sales volume y (in thousands of dollars) is related to advertising expenditures x (in thousands of dollars) according to
$$y = \frac{200x}{x + 10}$$
(a) Is this function discontinuous at any points?
(b) Advertising expenditures x must be nonnegative. Is this function continuous for these values of x?

43. **Annuities** If an annuity makes an infinite series of equal payments at the end of the interest periods, it is called a **perpetuity**. If a lump sum investment of A_n is needed to result in n periodic payments of R when the interest rate per period is i, then
$$A_n = R \left[\frac{1 - (1+i)^{-n}}{i} \right]$$

(a) Evaluate $\lim_{n \to \infty} A_n$ to find a formula for the lump sum payment for a perpetuity.

(b) Find the lump sum investment needed to make payments of $100 per month in perpetuity if interest is 12%, compounded monthly.

44. *Response to adrenalin* Experimental evidence suggests that the response y of the body to the concentration x of injected adrenalin is given by

$$y = \frac{x}{a + bx}$$

where a and b are experimental constants.

(a) Is this function continuous for all x?

(b) On the basis of your conclusion in part (a) and the fact that in the context of the application $x \geq 0$ and $y \geq 0$, must a and b be both positive, be both negative, or have opposite signs?

45. *Cost-benefit* Suppose that the cost C of removing p percent of the impurities from the waste water in a manufacturing process is given by

$$C(p) = \frac{9800p}{101 - p}$$

Is this function continuous for all those p-values for which the problem makes sense?

46. *Pollution* Suppose that the cost C of removing p percent of the particulate pollution from the exhaust gases at an industrial site is given by

$$C(p) = \frac{8100p}{100 - p}$$

Describe any discontinuities for $C(p)$. Explain what each discontinuity means.

47. *Pollution* The percent p of particulate pollution that can be removed from the smokestacks of an industrial plant by spending C dollars is given by

$$p = \frac{100C}{7300 + C}$$

Find the percent of the pollution that could be removed if spending C were allowed to increase without bound. Can 100% of the pollution be removed? Explain.

48. *Cost-benefit* The percent p of impurities that can be removed from the waste water of a manufacturing process at a cost of C dollars is given by

$$p = \frac{100C}{8100 + C}$$

Find the percent of the impurities that could be removed if cost were no object (that is, if cost were allowed to increase without bound). Can 100% of the impurities be removed? Explain.

49. *Federal income tax* The tax owed by a married couple filing jointly and their tax rates can be found in the following tax rate schedule.

Married Filing Jointly or Qualifying Widow(er)

If taxable income is over—	But not over—	The tax rate is:
$0	$18,150	10%
$18,150	$73,800	15%
$73,800	$148,850	25%
$148,850	$226,850	28%
$226,850	$405,100	33%
$405,100	$457,600	35%
$457,600	no limit	39.6%

Source: Internal Revenue Service

From this schedule, the tax rate $R(x)$ is a function of taxable income x, as follows.

$$R(x) = \begin{cases} 0.10 & \text{if} & 0 \leq x \leq 18{,}150 \\ 0.15 & \text{if} & 18{,}150 < x \leq 73{,}800 \\ 0.25 & \text{if} & 73{,}800 < x \leq 148{,}850 \\ 0.28 & \text{if} & 148{,}850 < x \leq 226{,}850 \\ 0.33 & \text{if} & 226{,}850 < x \leq 405{,}100 \\ 0.35 & \text{if} & 405{,}100 < x \leq 457{,}600 \\ 0.396 & \text{if} & x > 457{,}600 \end{cases}$$

Identify any discontinuities in $R(x)$.

50. *Calories and temperature* Suppose that the number of calories of heat required to raise 1 gram of water (or ice) from $-40°C$ to $x°C$ is given by

$$f(x) = \begin{cases} \frac{1}{2}x + 20 & \text{if } -40 \leq x < 0 \\ x + 100 & \text{if } 0 \leq x \end{cases}$$

(a) What can be said about the continuity of the function $f(x)$?

(b) What happens to water at $0°C$ that accounts for the behavior of the function at $0°C$?

51. *Electrical usage costs* The monthly charge in dollars for x kilowatt-hours (kWh) of electricity used by a residential consumer of Excelsior Electric Membership Corporation from November through June is given by the function

$$C(x) = \begin{cases} 20 + 0.188x & \text{if} & 0 \leq x \leq 100 \\ 38.80 + 0.15(x - 100) & \text{if} & 100 < x \leq 500 \\ 98.80 + 0.10(x - 500) & \text{if} & x > 500 \end{cases}$$

(a) What is the monthly charge if 1100 kWh of electricity is consumed in a month?

(b) Find $\lim_{x \to 100} C(x)$ and $\lim_{x \to 500} C(x)$, if the limits exist.

(c) Is C continuous at $x = 100$ and at $x = 500$?

52. *Postage costs* First-class postage for a standard letter is 49 cents for the first ounce or part of an ounce that a letter weighs plus an additional 21 cents for each additional ounce or part of an ounce. Use the table or graph of the postage function, $f(x)$, to determine the following.

(a) $\lim_{x \to 2.5} f(x)$

(b) $f(2.5)$

(c) Is $f(x)$ continuous at 2.5?
(d) $\lim\limits_{x \to 4} f(x)$
(e) $f(4)$
(f) Is $f(x)$ continuous at 4?

Weight x	Postage $f(x)$
$0 < x \le 1$	$0.49
$1 < x \le 2$	0.70
$2 < x \le 3$	0.91
$3 < x \le 4$	1.12
$4 < x \le 5$	1.33

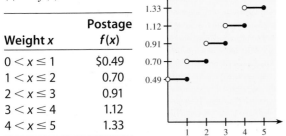
Postage Function

53. Modeling *U.S. workforce* Since 1950 the U.S. workforce has seen the arrival of the Baby Boomers and the changing role of women, among other influences. The data in the table show the millions of men and women in the U.S. workforce for selected years from 1950 and projected to 2050. Complete the following to explore the dynamics of the changing roles of men and women in the workforce.

(a) With x representing the number of years past 1950, use the data in the table to find a linear model for the number of men in the workforce, $m(x)$, and a model for the number of women in the workforce, $w(x)$. Report each model with three significant digits.

(b) Use the results from part (a) to find the function $r(x)$ that gives the ratio of men to women in the U.S. workforce.

(c) Find $\lim\limits_{x \to 0} r(x)$ and $\lim\limits_{x \to 100} r(x)$, and interpret their meanings.

(d) Find $\lim\limits_{x \to \infty} r(x)$ and interpret its meaning.

Year	Workforce (in millions)	
	Men	Women
1950	43.8	18.4
1960	46.4	23.2
1970	51.2	31.5
1980	61.5	45.5
1990	69.0	56.8
2000	75.2	65.7
2010	82.2	75.5
2015	84.2	78.6
2020	85.4	79.3
2030	88.5	81.6
2040	94.0	86.5
2050	100.3	91.5

Source: U.S. Bureau of Labor Statistics

54. *U.S. smart phones* Using data from 2010 and projected to 2016, the fraction of U.S. cell phone users who are smart phone users can be modeled by the function

$$f(t) = \frac{21.35t + 69.59}{4.590t + 233.1}$$

where t is equal to the number of years after 2010 (*Source:* emarketer.com).

(a) Is the function continuous for years from 2010 onward?

(b) Evaluate $\lim\limits_{t \to \infty} f(t)$ if it exists to find a horizontal asymptote.

(c) When can we be sure this model is no longer valid for this application?

(d) Does the $\lim\limits_{t \to \infty} f(t)$ give the long-term projection of the fraction of smart phone users?

OBJECTIVES

9.3

- To define and find average rates of change
- To define the derivative as a rate of change
- To use the definition of derivative to find derivatives of functions
- To use derivatives to find slopes of tangents to curves

Rates of Change and Derivatives

■ | APPLICATION PREVIEW |■

In Chapter 1, "Linear Equations and Functions," we studied linear revenue functions and defined the marginal revenue for a product as the rate of change of the revenue function. For linear revenue functions, this rate is also the slope of the line that is the graph of the revenue function. In this section, we will define marginal revenue as the rate of change of the revenue function, even when the revenue function is not linear.

Thus, if an oil company's revenue (in thousands of dollars) is given by

$$R = 100x - x^2, \quad x \ge 0$$

where x is the number of thousands of barrels of oil sold per day, we can find and interpret the marginal revenue when 20,000 barrels are sold (see Example 4).

We will discuss the relationship between the *instantaneous rate of change* of a function at a given point and the slope of the line tangent to the graph of the function at that point. We will see how the derivative of a revenue function can be used to find both the slope of its tangent line and its marginal revenue.

Average Rates of Change

For linear functions, we have seen that the slope of the line measures the average rate of change of the function and can be found from any two points on the line. However, for a function that is not linear, the slope between different pairs of points no longer always gives the same number, but it can be interpreted as an **average rate of change.**

Average Rate of Change

The **average rate of change** of a function $y = f(x)$ from $x = a$ to $x = b$ is defined by

$$\text{Average rate of change} = \frac{f(b) - f(a)}{b - a}$$

The figure shows that this average rate is the same as the slope of the segment (or secant line) joining the points $(a, f(a))$ and $(b, f(b))$.

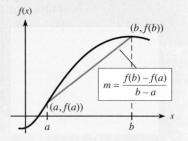

EXAMPLE 1 Total Cost

Suppose a company's total cost in dollars to produce x units of its product is given by

$$C(x) = 0.01x^2 + 25x + 1500$$

Find the average rate of change of total cost for the second 100 units produced (from $x = 100$ to $x = 200$).

Solution

The average rate of change of total cost from $x = 100$ to $x = 200$ units is

$$\frac{C(200) - C(100)}{200 - 100} = \frac{[0.01(200)^2 + 25(200) + 1500] - [0.01(100)^2 + 25(100) + 1500]}{100}$$

$$= \frac{6900 - 4100}{100} = \frac{2800}{100} = 28 \text{ dollars per unit}$$

EXAMPLE 2 Elderly in the Workforce

Figure 9.18 shows the percents of elderly men and of elderly women in the workforce in selected years from 1970 and projected to 2040. Find and interpret the average rate of change of the percent of (a) elderly men in the workforce and (b) elderly women in the workforce from 1970 to 2040.

Elderly in the Workforce, 1970–2040

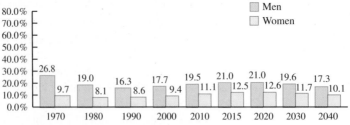

Figure 9.18 *Source:* Bureau of the Census, U.S. Department of Commerce

Solution

(a) From 1970 to 2040, the annual average rate of change in the percent of elderly men in the workforce is

$$\frac{\text{Change in men's percent}}{\text{Change in years}} = \frac{17.3 - 26.8}{2040 - 1970} = \frac{-9.5}{70} \approx -0.136 \text{ percentage points per year}$$

This means that from 1970 to 2040, *on average*, the percent of elderly men in the workforce dropped by 0.136 percentage points per year.

(b) Similarly, the average rate of change for women is

$$\frac{\text{Change in women's percent}}{\text{Change in years}} = \frac{10.1 - 9.7}{2040 - 1970} = \frac{0.4}{70} \approx 0.0057 \text{ percentage points per year}$$

In like manner, this means that from 1970 to 2040, *on average*, the percent of elderly women in the workforce increased by 0.0057 percentage points each year. ◼

Instantaneous Rates of Change: Velocity

Another common rate of change is velocity. For instance, if we travel 200 miles in our car over a 4-hour period, we know that we averaged 50 mph. However, during that trip there may have been times when we were traveling on an Interstate at faster than 50 mph and times when we were stopped at a traffic light. Thus, for the trip we have not only an average velocity but also instantaneous velocities (or instantaneous speeds as displayed on the speedometer). Let's see how average velocity can lead us to instantaneous velocity.

Suppose a ball is thrown straight upward at 64 feet per second from a spot 96 feet above ground level. The equation that describes the height y of the ball after x seconds is

$$y = f(x) = 96 + 64x - 16x^2$$

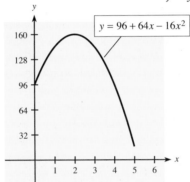

$y = 96 + 64x - 16x^2$

Figure 9.19

Figure 9.19 shows the graph of this function for $0 \le x \le 5$. The average velocity of the ball over a given time interval is the change in the height divided by the length of time that has passed. Table 9.4 shows some average velocities over time intervals beginning at $x = 1$.

| TABLE 9.4 |

AVERAGE VELOCITIES

Time (seconds)			Height (feet)			Average Velocity (ft/sec)
Beginning	Ending	Change (Δx)	Beginning	Ending	Change (Δy)	($\Delta y / \Delta x$)
1	2	1	144	160	16	16/1 = 16
1	1.5	0.5	144	156	12	12/0.5 = 24
1	1.1	0.1	144	147.04	3.04	3.04/0.1 = 30.4
1	1.01	0.01	144	144.3184	0.3184	0.3184/0.01 = 31.84

In Table 9.4, the smaller the time interval, the more closely the average velocity approximates the instantaneous velocity at $x = 1$. Thus the instantaneous velocity at $x = 1$ is closer to 31.84 feet per second than to 30.4 feet per second.

If we represent the change in time by h, then the average velocity from $x = 1$ to $x = 1 + h$ approaches the instantaneous velocity at $x = 1$ as h approaches 0. (Note that h can be positive or negative.) This is illustrated in the following example.

EXAMPLE   3   Velocity

Suppose a ball is thrown straight upward so that its height $f(x)$ (in feet) is given by the equation

$$f(x) = 96 + 64x - 16x^2$$

where x is time (in seconds).
(a) Find the average velocity from $x = 1$ to $x = 1 + h$.
(b) Find the instantaneous velocity at $x = 1$.

Solution

(a) Let h represent the change in x (time) from 1 to $1 + h$. Then the corresponding change in $f(x)$ (height) is

$$f(1 + h) - f(1) = [96 + 64(1 + h) - 16(1 + h)^2] - [96 + 64 - 16]$$
$$= 96 + 64 + 64h - 16(1 + 2h + h^2) - 144$$
$$= 16 + 64h - 16 - 32h - 16h^2 = 32h - 16h^2$$

The average velocity V_{av} is the change in height divided by the change in time.

$$V_{av} = \frac{f(1 + h) - f(1)}{1 + h - 1} = \frac{32h - 16h^2}{h} = 32 - 16h$$

(b) The instantaneous velocity V is the limit of the average velocity as h approaches 0.

$$V = \lim_{h \to 0} V_{av} = \lim_{h \to 0} (32 - 16h) = 32 \text{ feet per second} \qquad \blacksquare$$

Note that average velocity is found over a time interval. Instantaneous velocity is usually called **velocity,** and it can be found at any time x, as follows.

Velocity	Suppose that an object moving in a straight line has its position y at time x given by $y = f(x)$. Then the **velocity** function for the object at time x is $$V = \lim_{h \to 0} \frac{f(x + h) - f(x)}{h}$$ provided that this limit exists.

The instantaneous rate of change of any function (commonly called *rate of change*) can be found in the same way we find velocity. The function that gives this instantaneous rate of change of a function f is called the **derivative** of f.

Derivative	If f is a function given by $y = f(x)$, then the **derivative** of $f(x)$ at any value x, denoted $f'(x)$, is a new function defined by $$f'(x) = \lim_{h \to 0} \frac{f(x + h) - f(x)}{h}$$ if this limit exists. If $f'(c)$ exists, we say that f is **differentiable** at c.

The following procedure illustrates how to find the derivative of a function $y = f(x)$ at any value x.

Derivative Using the Definition

Procedure	**Example**
To find the derivative of $y = f(x)$ at any value x:	Find the derivative of $f(x) = 4x^2$.
1. Let h represent the change in x from x to $x + h$.	1. The change in x from x to $x + h$ is h.
2. The corresponding change in $y = f(x)$ is $$f(x + h) - f(x)$$	2. The change in $f(x)$ is $$\begin{aligned} f(x + h) - f(x) &= 4(x + h)^2 - 4x^2 \\ &= 4(x^2 + 2xh + h^2) - 4x^2 \\ &= 4x^2 + 8xh + 4h^2 - 4x^2 \\ &= 8xh + 4h^2 \end{aligned}$$

(Continued)

3. Form the **difference quotient** $\dfrac{f(x+h)-f(x)}{h}$ and simplify.

3. $\dfrac{f(x+h)-f(x)}{h} = \dfrac{8xh+4h^2}{h}$

$= 8x + 4h$

4. Find $\lim\limits_{h\to 0}\dfrac{f(x+h)-f(x)}{h}$ to determine $f'(x)$, the derivative of $f(x)$.

4. $f'(x) = \lim\limits_{h\to 0}\dfrac{f(x+h)-f(x)}{h}$

$f'(x) = \lim\limits_{h\to 0}(8x+4h) = 8x$

Note that in the example above, we could have found the derivative of the function $f(x) = 4x^2$ at a particular value of x, say $x = 3$, by evaluating the derivative formula at that value:

$$f'(x) = 8x \quad \text{so} \quad f'(3) = 8(3) = 24$$

In addition to $f'(x)$, the derivative of $y = f(x)$ may be denoted by

$$\frac{dy}{dx}, \quad y', \quad \frac{d}{dx}f(x), \quad D_x y, \quad \text{or} \quad D_x f(x)$$

We can, of course, use variables other than x and y to represent functions and their derivatives. For example, we can represent the derivative of the function $p = 2q^2 - 1$ by dp/dq.

✓ CHECKPOINT

1. Find the average rate of change of $f(x) = 30 - x - x^2$ over $[1, 4]$.

2. For the function $y = f(x) = x^2 - x + 1$, find

   (a) $f(x+h) - f(x)$.

   (b) $\dfrac{f(x+h)-f(x)}{h}$.

   (c) $f'(x) = \lim\limits_{h\to 0}\dfrac{f(x+h)-f(x)}{h}$.

   (d) $f'(2)$.

For linear functions, we defined the **marginal revenue** for a product as the rate of change of the total revenue function for the product. If the total revenue function for a product is not linear, we define the marginal revenue for the product as the instantaneous rate of change, or the derivative, of the revenue function.

Marginal Revenue

Suppose that the total revenue function for a product is given by $R = R(x)$, where x is the number of units sold. Then the **marginal revenue** at x units is

$$\overline{MR} = R'(x) = \lim_{h\to 0}\frac{R(x+h)-R(x)}{h}$$

provided that the limit exists.

Note that the marginal revenue (derivative of the revenue function) can be found by using the steps in the Procedure/Example box beginning on the preceding page. These steps can also be combined, as they are in Example 4.

EXAMPLE 4   Revenue | APPLICATION PREVIEW |

Suppose that an oil company's revenue (in thousands of dollars) is given by the equation

$$R = R(x) = 100x - x^2, \quad x \geq 0$$

where x is the number of thousands of barrels of oil sold each day.

(a) Find the function that gives the marginal revenue at any value of x.
(b) Find the marginal revenue when 20,000 barrels are sold (that is, at $x = 20$).

Solution

(a) The marginal revenue function is the derivative of $R(x)$.

$$R'(x) = \lim_{h \to 0} \frac{R(x + h) - R(x)}{h} = \lim_{h \to 0} \frac{[100(x + h) - (x + h)^2] - (100x - x^2)}{h}$$

$$= \lim_{h \to 0} \frac{100x + 100h - (x^2 + 2xh + h^2) - 100x + x^2}{h}$$

$$= \lim_{h \to 0} \frac{100h - 2xh - h^2}{h} = \lim_{h \to 0} (100 - 2x - h) = 100 - 2x$$

Thus, the marginal revenue function is $\overline{MR} = R'(x) = 100 - 2x$.

(b) The function found in part (a) gives the marginal revenue at *any* value of x. To find the marginal revenue when 20 units are sold, we evaluate $R'(20)$.

$$R'(20) = 100 - 2(20) = 60$$

Hence the marginal revenue at $x = 20$ is 60 thousand dollars per thousand barrels of oil. Because the marginal revenue is used to approximate the revenue from the sale of one additional unit, we interpret $R'(20) = 60$ to mean that the expected revenue from the sale of the next thousand barrels (after 20,000) will be approximately $60,000. [*Note:* The actual revenue from this sale is $R(21) - R(20) = 1659 - 1600 = 59$ thousand dollars.] ∎

Tangent to a Curve   Just as average rates of change are connected with slopes, so are instantaneous rates (derivatives). In fact, the slope of the graph of a function at any point is the same as the derivative at that point. To show this, we define the slope of a curve at a point on the curve as the slope of the line tangent to the curve at the point.

In geometry, a **tangent** to a circle is defined as a line that has one point in common with the circle. [See Figure 9.20(a).] This definition does not apply to all curves, as Figure 9.20(b) shows. Many lines can be drawn through the point A that touch the curve only at A. One of the lines, line l, looks like it is tangent to the curve, in the same sense as a line is tangent to a circle.

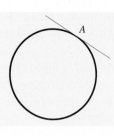

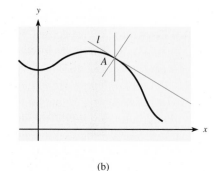

Figure 9.20   (a)   (b)

In Figure 9.21, the line l represents the tangent line to the curve at point A and shows that secant lines (s_1, s_2, etc.) through A approach line l as the second points (Q_1, Q_2, etc.) approach A. (For points and secant lines to the left of A, there would be a similar figure and conclusion.) This means that as we choose points on the curve closer and closer to A (from both sides of A), the limiting position of the secant lines through A is the **tangent line** to the curve at A.

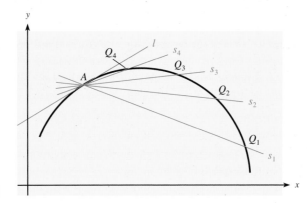

Figure 9.21

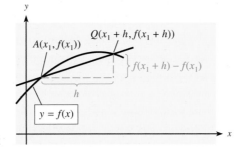

Figure 9.22

From Figure 9.22, we see that the slope of an arbitrary secant line through $A(x_1, f(x_1))$ and $Q(x_1 + h, f(x_1 + h))$ is given by

$$m_{AQ} = \frac{f(x_1 + h) - f(x_1)}{h}$$

Thus, as Q approaches A, the slope of the secant line AQ approaches the **slope of the tangent** line at A, and we have the following.

Slope of the Tangent

The **slope of the tangent** to the graph of $y = f(x)$ at point $A(x_1, f(x_1))$ is

$$m = \lim_{h \to 0} \frac{f(x_1 + h) - f(x_1)}{h}$$

if this limit exists. That is, $m = f'(x_1)$, the derivative at $x = x_1$.

EXAMPLE 5   Slope of the Tangent

Find the slope of $y = f(x) = x^2$ at the point $A(2, 4)$.

Solution
The formula for the slope of the tangent to $y = f(x)$ at $(2, 4)$ is

$$m = f'(2) = \lim_{h \to 0} \frac{f(2 + h) - f(2)}{h}$$

Thus for $f(x) = x^2$, we have

$$m = f'(2) = \lim_{h \to 0} \frac{(2 + h)^2 - 2^2}{h}$$

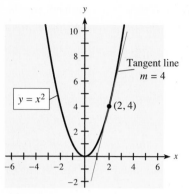

Figure 9.23

Taking the limit immediately would result in both the numerator and the denominator approaching 0. To avoid this, we simplify the fraction before taking the limit.

$$m = \lim_{h \to 0} \frac{4 + 4h + h^2 - 4}{h} = \lim_{h \to 0} \frac{4h + h^2}{h} = \lim_{h \to 0} (4 + h) = 4$$

Thus the slope of the tangent to $y = x^2$ at (2, 4) is 4 (see Figure 9.23). ∎

The statement "the slope of the tangent to the curve at (2, 4) is 4" is frequently simplified to the statement "the slope of the curve at (2, 4) is 4." Knowledge that the slope is a positive number on an interval tells us that the function is increasing on that interval, which means that a point moving along the graph of the function rises as it moves to the right on that interval. If the derivative (and thus the slope) is negative on an interval, the curve is decreasing on the interval; that is, a point moving along the graph falls as it moves to the right on that interval.

EXAMPLE 6　Tangent Line

Given $y = f(x) = 3x^2 + 2x + 11$, find

(a)　the derivative of $f(x)$ at any point $(x, f(x))$.
(b)　the slope of the tangent to the curve at (1, 16).
(c)　the equation of the line tangent to $y = 3x^2 + 2x + 11$ at (1, 16).

Solution

(a)　The derivative of $f(x)$ at any value x is denoted by $f'(x)$ and is

$$y' = f'(x) = \lim_{h \to 0} \frac{f(x + h) - f(x)}{h}$$

$$= \lim_{h \to 0} \frac{[3(x + h)^2 + 2(x + h) + 11] - (3x^2 + 2x + 11)}{h}$$

$$= \lim_{h \to 0} \frac{3(x^2 + 2xh + h^2) + 2x + 2h + 11 - 3x^2 - 2x - 11}{h}$$

$$= \lim_{h \to 0} \frac{6xh + 3h^2 + 2h}{h} = \lim_{h \to 0} (6x + 3h + 2) = 6x + 2$$

(b)　The derivative is $f'(x) = 6x + 2$, so the slope of the tangent to the curve at (1, 16) is $f'(1) = 6(1) + 2 = 8$.
(c)　The equation of the tangent line uses the given point (1, 16) and the slope $m = 8$. Using $y - y_1 = m(x - x_1)$ gives $y - 16 = 8(x - 1)$, or $y = 8x + 8$. ∎

The discussion in this section indicates that the derivative of a function has several interpretations.

Interpretations of the Derivative	For a given function, each of the following means "find the **derivative**."
	1.　Find the **velocity** of an object moving in a straight line.
	2.　Find the **instantaneous rate of change** of a function.
	3.　Find the **marginal revenue** for a given revenue function.
	4.　Find the **slope of the tangent** to the graph of a function.

That is, all the terms printed in boldface are mathematically the same, and the answers to questions about any one of them give information about the others.

Note in Figure 9.23 that near the point of tangency at (2, 4), the tangent line and the function look coincident. In fact, if we graphed both with a graphing calculator and repeatedly zoomed in near the point (2, 4), the two graphs would eventually appear as one. Try this for yourself. Thus the derivative of $f(x)$ at the point where $x = a$ can be approximated by finding the slope between $(a, f(a))$ and a second point that is nearby.

In addition, we know that the slope of the tangent to $f(x)$ at $x = a$ is defined by

$$f'(a) = \lim_{h \to 0} \frac{f(a + h) - f(a)}{h}$$

Hence we could also estimate $f'(a)$—that is, the slope of the tangent at $x = a$—by evaluating

$$\frac{f(a + h) - f(a)}{h} \quad \text{when } h \approx 0 \text{ and } h \neq 0$$

EXAMPLE 7   Approximating the Slope of the Tangent Line

(a) Let $f(x) = 2x^3 - 6x^2 + 2x - 5$. Use $\dfrac{f(a + h) - f(a)}{h}$ and two values of h to make estimates of the slope of the tangent to $f(x)$ at $x = 1$ on opposite sides of $x = 1$.

(b) Use the following table of values of x and $g(x)$ to estimate $g'(3)$.

x	1	1.9	2.7	2.9	2.999	3	3.002	3.1	4	5
$g(x)$	1.6	4.3	11.4	10.8	10.513	10.5	10.474	10.18	6	−5

Solution
A graphing calculator can facilitate the following calculations.

(a) We can use $h = 0.0001$ and $h = -0.0001$ as follows:

With $h = 0.0001$: $f'(1) \approx \dfrac{f(1 + 0.0001) - f(1)}{0.0001} = \dfrac{f(1.0001) - f(1)}{0.0001} \approx -4$

With $h = -0.0001$: $f'(1) \approx \dfrac{f(1 + (-0.0001)) - f(1)}{-0.0001} = \dfrac{f(0.9999) - f(1)}{-0.0001} \approx -4$

(b) We use the given table and measure the slope between $(3, 10.5)$ and another point that is nearby (the closer, the better). Using $(2.999, 10.513)$, we obtain

$$g'(3) \approx \frac{y_2 - y_1}{x_2 - x_1} = \frac{10.5 - 10.513}{3 - 2.999} = \frac{-0.013}{0.001} = -13$$

Calculator Note  Most graphing calculators have a feature called the **numerical derivative** that can approximate the derivative of a function at a point. To find the numerical derivative of $f(x)$ at $x = c$ with a graphing calculator, choose the MATH menu and select 8:nDeriv. Next enter the function, x, and the value, c, so that the display shows nDeriv($f(x), x, c$), and press ENTER. The numerical derivative will appear. The numerical derivative of $f(x) = 2x^3 - 6x^2 + 2x - 5$ with respect to x at $x = 1$ found in part (a) of Example 7 can also be found as follows on many graphing calculators:

$$\text{nDeriv}(2x^3 - 6x^2 + 2x - 5, x, 1) \approx -3.999998 \approx -4$$

Notice both methods give the same approximation.

Differentiability and Continuity   So far we have talked about how the derivative is defined, what it represents, and how to find it. However, there are functions for which derivatives do not exist at every value of x. Figure 9.24 on the next page shows some common cases where $f'(c)$ does not exist but where $f'(x)$ exists for all other values of x. These cases occur where there is a discontinuity, a corner, or a vertical tangent line.

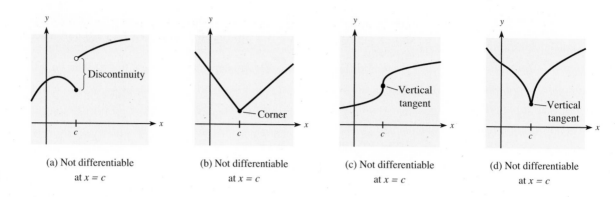

Figure 9.24

(a) Not differentiable
at $x = c$

(b) Not differentiable
at $x = c$

(c) Not differentiable
at $x = c$

(d) Not differentiable
at $x = c$

From Figure 9.24 we see that a function may be continuous at $x = c$ even though $f'(c)$ does not exist. Thus continuity does not imply differentiability at a point. However, differentiability does imply continuity.

Differentiability Implies Continuity	If a function f is differentiable at $x = c$, then f is continuous at $x = c$.

✓ **CHECKPOINT**

3. Which of the following are given by $f'(c)$?
   (a) The slope of the tangent when $x = c$
   (b) The y-coordinate of the point where $x = c$
   (c) The instantaneous rate of change of $f(x)$ at $x = c$
   (d) The marginal revenue at $x = c$, if $f(x)$ is the revenue function
4. Must a graph that has no discontinuity, corner, or cusp at $x = c$ be differentiable at $x = c$?

Calculator Note

We can use a graphing calculator to explore the relationship between secant lines and tangent lines. For example, if the point (a, b) lies on the graph of $y = x^2$, then the equation of the secant line to $y = x^2$ from $(1, 1)$ to (a, b) has the equation

$$y - 1 = \frac{b - 1}{a - 1}(x - 1), \quad \text{or} \quad y = \frac{b - 1}{a - 1}(x - 1) + 1$$

Figure 9.25 illustrates the secant lines for three different choices for the point (a, b).

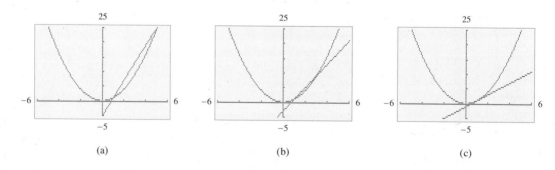

Figure 9.25   (a)   (b)   (c)

We see that as the point (a, b) moves closer to $(1, 1)$, the secant line looks more like the tangent line to $y = x^2$ at $(1, 1)$. Furthermore, (a, b) approaches $(1, 1)$ as $a \to 1$, and the slope of the secant approaches the following limit.

$$\lim_{a \to 1} \frac{b - 1}{a - 1} = \lim_{a \to 1} \frac{a^2 - 1}{a - 1} = \lim_{a \to 1}(a + 1) = 2$$

This limit, 2, is the slope of the tangent line at $(1, 1)$. That is, the derivative of $y = x^2$ at $(1, 1)$ is 2. [Note that a graphing calculator's calculation of the numerical derivative of $f(x) = x^2$ with respect to x at $x = 1$ gives $f'(1) = 2$.]

✓ CHECKPOINT ANSWERS

1. -6
2. (a) $f(x + h) - f(x) = 2xh + h^2 - h$

   (b) $\dfrac{f(x + h) - f(x)}{h} = 2x + h - 1$

   (c) $f'(x) = 2x - 1$

   (d) $f'(2) = 3$
3. Parts (a), (c), and (d) are given by $f'(c)$. The y-coordinate where $x = c$ is given by $f(c)$.
4. No. See Figure 9.24(c).

| EXERCISES | 9.3

In Problems 1–4, for each given function find the average rate of change over each specified interval.

1. $f(x) = x^2 + x - 12$ over (a) $[0, 5]$ and (b) $[-3, 10]$
2. $f(x) = 6 - x - x^2$ over (a) $[-1, 2]$ and (b) $[1, 10]$
3. For $f(x)$ given by the table, over (a) $[2, 5]$ and (b) $[3.8, 4]$

x	0	2	2.5	3	3.8	4	5
$f(x)$	14	20	22	19	17	16	30

4. For $f(x)$ given in the table, over (a) $[3, 3.5]$ and (b) $[2, 6]$

x	1	2	3	3.5	3.7	6
$f(x)$	40	25	18	15	18	38

5. Given $f(x) = 2x - x^2$, find the average rate of change of $f(x)$ over each of the following pairs of intervals.
   (a) $[2.9, 3]$ and $[2.99, 3]$ (b) $[3, 3.1]$ and $[3, 3.01]$
   (c) What do the calculations in parts (a) and (b) suggest the instantaneous rate of change of $f(x)$ at $x = 3$ might be?
6. Given $f(x) = x^2 + 3x + 7$, find the average rate of change of $f(x)$ over each of the following pairs of intervals.
   (a) $[1.9, 2]$ and $[1.99, 2]$
   (b) $[2, 2.1]$ and $[2, 2.01]$
   (c) What do the calculations in parts (a) and (b) suggest the instantaneous rate of change of $f(x)$ at $x = 2$ might be?
7. In the Procedure/Example box in this section, we were given $f(x) = 4x^2$ and found $f'(x) = 8x$. Find
   (a) the instantaneous rate of change of $f(x)$ at $x = 4$.
   (b) the slope of the tangent to the graph of $y = f(x)$ at $x = 4$.
   (c) the point on the graph of $y = f(x)$ at $x = 4$.

8. In Example 6 in this section, we were given $f(x) = 3x^2 + 2x + 11$ and found $f'(x) = 6x + 2$. Find
   (a) the instantaneous rate of change of $f(x)$ at $x = 6$.
   (b) the slope of the tangent to the graph of $y = f(x)$ at $x = 6$.
   (c) the point on the graph of $y = f(x)$ at $x = 6$.
9. Let $f(x) = 3x^2 - 2x$.
   (a) Use the definition of derivative and the Procedure/Example box in this section to verify that $f'(x) = 6x - 2$.
   (b) Find the instantaneous rate of change of $f(x)$ at $x = -1$.
   (c) Find the slope of the tangent to the graph of $y = f(x)$ at $x = -1$.
   (d) Find the point on the graph of $y = f(x)$ at $x = -1$.
10. Let $f(x) = 9 - \dfrac{1}{2}x^2$.
   (a) Use the definition of derivative and the Procedure/Example box in this section to verify that $f'(x) = -x$.
   (b) Find the instantaneous rate of change of $f(x)$ at $x = 2$.
   (c) Find the slope of the tangent to the graph of $y = f(x)$ at $x = 2$.
   (d) Find the point on the graph of $y = f(x)$ at $x = 2$.

In Problems 11–14, the tangent line to the graph of $f(x)$ at $x = 1$ is shown. On the tangent line, P is the point of tangency and A is another point on the line.
(a) **Find the coordinates of the points P and A.**
(b) **Use the coordinates of P and A to find the slope of the tangent line.**
(c) **Find $f'(1)$.**
(d) **Find the instantaneous rate of change of $f(x)$ at P.**

11.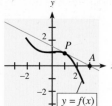

12.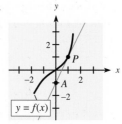

13.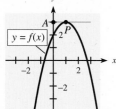

14.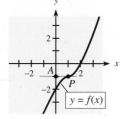

For each function in Problems 15–18, find
(a) the derivative, by using the definition.
(b) the instantaneous rate of change of the function at any value and at the given value.
(c) the slope of the tangent at the given value.
15. $f(x) = 5x^2 + 6x - 11$; $x = -2$
16. $f(x) = 16x^2 - 4x + 2$; $x = 1$
17. $p(q) = 2q^2 + q + 5$; $q = 10$
18. $p(q) = 2q^2 - 4q + 5$; $q = 2$

For each function in Problems 19–22, approximate $f'(a)$ in the following ways.
(a) Use the numerical derivative feature of a graphing calculator.
(b) Use $\dfrac{f(a+h) - f(a)}{h}$ with $h = 0.0001$.
(c) Graph the function on a graphing calculator. Then zoom in near the point until the graph appears straight, pick two points, and find the slope of the line you see.
19. $f'(2)$ for $f(x) = 3x^4 - 7x - 5$
20. $f'(-1)$ for $f(x) = 2x^3 - 11x + 9$
21. $f'(4)$ for $f(x) = (2x - 1)^3$
22. $f'(3)$ for $f(x) = \dfrac{3x + 1}{2x - 5}$

In Problems 23 and 24, use the given tables to approximate $f'(a)$ as accurately as you can.

23.

x	12.0	12.99	13	13.1	a = 13
f(x)	1.41	17.42	17.11	22.84	

24.

x	−7.4	−7.50	−7.51	−7	a = −7.5
f(x)	22.12	22.351	22.38	24.12	

In the figures given in Problems 25 and 26, at each point A and B draw an approximate tangent line and then use it to complete parts (a) and (b).
(a) Is $f'(x)$ greater at point A or at point B? Explain.
(b) Estimate $f'(x)$ at point B.

25.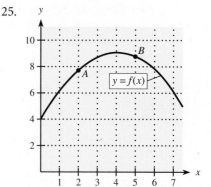

26.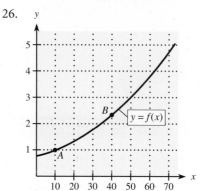

In Problems 27 and 28, a point (a, b) on the graph of $y = f(x)$ is given, and the equation of the line tangent to the graph of $f(x)$ at (a, b) is given. In each case, find $f'(a)$ and $f(a)$.
27. $(4, -11)$; $7x - 3y = 61$
28. $(-1, 6)$; $x + 10y = 59$
29. If the instantaneous rate of change of $f(x)$ at $(2, -4)$ is 5, write the equation of the line tangent to the graph of $f(x)$ at $x = 2$.
30. If the instantaneous rate of change of $g(x)$ at $(-1, -2)$ is $1/2$, write the equation of the line tangent to the graph of $g(x)$ at $x = -1$.

Because the derivative of a function represents both the slope of the tangent to the curve and the instantaneous rate of change of the function, it is possible to use information about one to gain information about the other. In Problems 31 and 32, use the graph of the function $y = f(x)$ given in Figure 9.26.

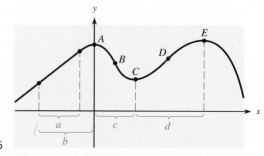

Figure 9.26

31. (a) Over what interval(s) (a) through (d) is the rate of change of $f(x)$ positive?
    (b) Over what interval(s) (a) through (d) is the rate of change of $f(x)$ negative?

(c) At what point(s) A through E is the rate of change of $f(x)$ equal to zero?

32. (a) At what point(s) A through E does the rate of change of $f(x)$ change from positive to negative?
    (b) At what point(s) A through E does the rate of change of $f(x)$ change from negative to positive?

33. Given the graph of $y = f(x)$ in Figure 9.27, determine for which x-values A, B, C, D, or E the function is
    (a) continuous.   (b) differentiable.

34. Given the graph of $y = f(x)$ in Figure 9.27, determine for which x-values F, G, H, I, or J the function is
    (a) continuous.   (b) differentiable.

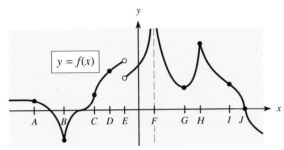

Figure 9.27

In Problems 35–38, (a) find the slope of the tangent to the graph of $f(x)$ at any point, (b) find the slope of the tangent at the given point, (c) write the equation of the line tangent to the graph of $f(x)$ at the given point, and (d) graph both $f(x)$ and its tangent line (use a graphing utility).

35. $f(x) = x^2 + x;\ (2, 6)$
36. $f(x) = x^2 + 3x;\ (-1, -2)$
37. $f(x) = x^3 + 3;\ (1, 4)$
38. $f(x) = 5x^3 + 2;\ (-1, -3)$

APPLICATIONS

39. *Total cost* Suppose total cost in dollars from the production of x printers is given by

$$C(x) = 0.0001x^3 + 0.005x^2 + 28x + 3000$$

Find the average rate of change of total cost when production changes
(a) from 100 to 300 printers.
(b) from 300 to 600 printers.
(c) Interpret the results from parts (a) and (b).

40. *Average velocity* If an object is thrown upward at 64 feet per second from a height of 20 feet, its height S after t seconds is given by

$$S(x) = 20 + 64t - 16t^2$$

What is the average velocity in the
(a) first 2 seconds after it is thrown?
(b) next 2 seconds?

41. *Demand* If the demand for a product is given by

$$D(p) = \frac{1000}{\sqrt{p}} - 1$$

what is the average rate of change of demand when p increases from
(a) 1 to 25?
(b) 25 to 100?

42. *Revenue* If the total revenue function for a blender is

$$R(x) = 36x - 0.01x^2$$

where x is the number of units sold, what is the average rate of change in revenue $R(x)$ as x increases from 10 to 20 units?

43. *Total cost* Suppose the figure shows the total cost graph for a company. Arrange the average rates of change of total cost from A to B, B to C, and A to C from smallest to greatest, and explain your choice.

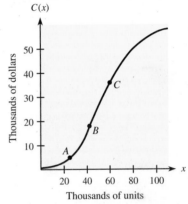

44. *Foreign-born population* The figure shows the percent of the U.S. population that was foreign-born for selected years from 1910 and projected to 2020.
    (a) Use the figure to find the average rate of change in the percent of the U.S. population that was foreign-born from 1960 to 2020. Interpret your result.
    (b) From the figure, determine for which two consecutive data points the average rate of change in the percent of foreign-born was (i) closest to zero and (ii) furthest from zero.

Foreign-Born Population Percent: 1910–2020

Source: U.S. Bureau of the Census

45. *Marginal revenue* The revenue function for a sound system is

$$R(x) = 300x - x^2 \text{ dollars}$$

where x denotes the number of units sold.
    (a) What is the function that gives marginal revenue?
    (b) What is the marginal revenue if 50 units are sold, and what does it mean?
    (c) What is the marginal revenue if 200 units are sold, and what does it mean?

(d) What is the marginal revenue if 150 units are sold?

(e) As the number of units sold passes through 150, what happens to revenue?

46. *Marginal revenue* Suppose the total revenue function for a blender is

$$R(x) = 36x - 0.01x^2 \text{ dollars}$$

where x is the number of units sold.

(a) What function gives the marginal revenue?

(b) What is the marginal revenue when 600 units are sold, and what does it mean?

(c) What is the marginal revenue when 2000 units are sold, and what does it mean?

(d) What is the marginal revenue when 1800 units are sold, and what does it mean?

47. *Labor force and output* The monthly output at the Olek Carpet Mill is

$$Q(x) = 15,000 + 2x^2 \text{ units},   (40 \le x \le 60)$$

where x is the number of workers employed at the mill. If there are currently 50 workers, find the instantaneous rate of change of monthly output with respect to the number of workers. That is, find $Q'(50)$.

48. *Consumer expenditure* Suppose that the demand for x units of a product is

$$x = 10,000 - 100p$$

where p dollars is the price per unit. Then the consumer expenditure for the product is

$$E(p) = px = p(10,000 - 100p) = 10,000p - 100p^2$$

What is the instantaneous rate of change of consumer expenditure with respect to price at

(a) any price p?   (b) $p = 5$?   (c) $p = 20$?

In Problems 49–52, find derivatives with the numerical derivative feature of a graphing calculator.

49. *Profit* Suppose that the profit function for the monthly sales of a car by a dealership is

$$P(x) = 500x - x^2 - 100$$

where x is the number of cars sold. What is the instantaneous rate of change of profit when

(a) 200 cars are sold? Explain its meaning.

(b) 300 cars are sold? Explain its meaning.

50. *Profit* If the total revenue function and the total cost function for a toy are

$$R(x) = 2x \text{ and } C(x) = 100 + 0.2x^2 + x$$

what is the instantaneous rate of change of profit if 10 units are produced and sold? Explain its meaning.

51. *Heat index* The highest recorded temperature in the state of Alaska was 100°F and occurred on June 27, 1915, at Fort Yukon. The *heat index* is the apparent temperature of the air at a given temperature and humidity level. If x denotes the relative humidity (in percent), then the heat index (in degrees Fahrenheit) for an air temperature of 100°F can be approximated by the function

$$f(x) = 0.009x^2 + 0.139x + 91.875$$

(a) At what rate is the heat index changing when the humidity is 50%?

(b) Write a sentence that explains the meaning of your answer in part (a).

52. *Receptivity* In learning theory, receptivity is defined as the ability of students to understand a complex concept. Receptivity is highest when the topic is introduced and tends to decrease as time passes in a lecture. Suppose that the receptivity of a group of students in a mathematics class is given by

$$g(t) = -0.2t^2 + 3.1t + 32$$

where t is minutes after the lecture begins.

(a) At what rate is receptivity changing 10 minutes after the lecture begins?

(b) Write a sentence that explains the meaning of your answer in part (a).

53. *Marginal revenue* Suppose the graph shows a manufacturer's total revenue, in thousands of dollars, from the sale of x cellular telephones to dealers.

(a) Is the marginal revenue greater at 300 cell phones or at 700? Explain.

(b) Use part (a) to decide whether the sale of the 301st cell phone or the 701st brings in more revenue. Explain.

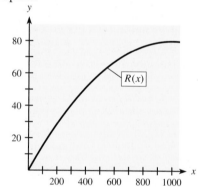

54. *Social Security beneficiaries* The graph shows a model for the number of millions of Social Security beneficiaries projected to 2030. The model was developed with data from the Social Security Trustees Report.

(a) Was the instantaneous rate of change of the number of beneficiaries with respect to the year greater in 1960 or in 1980? Justify your answer.

(b) Is the instantaneous rate of change of the number of beneficiaries projected to be greater in 2000 or in 2030? Justify your answer.

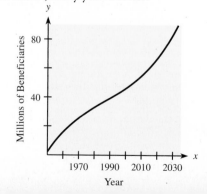

OBJECTIVES

9.4

- To find derivatives of powers of x
- To find derivatives of constant functions
- To find derivatives of functions involving constant coefficients
- To find derivatives of sums and differences of functions

Derivative Formulas

▮| APPLICATION PREVIEW |▮

For more than 50 years, U.S. total personal income has experienced steady growth. With Bureau of Economic Analysis, U.S. Department of Commerce data for selected years from 1960 and projected to 2018, U.S. total personal income I, in billions of current dollars, can be modeled by

$$I = I(t) = 6.29t^2 - 51.7t + 601$$

where t is the number of years past 1960. We can find the rate of growth of total U.S. personal income in 2018 by using the derivative $I'(t)$ of the total personal income function. (See Example 8.)

As we discussed in the previous section, the derivative of a function can be used to find the rate of change of the function. In this section we will develop formulas that will make it easier to find certain derivatives.

Derivative of $f(x) = x^n$ We can use the definition of derivative to show the following:

$$\text{If } f(x) = x^2, \text{ then } f'(x) = 2x.$$
$$\text{If } f(x) = x^3, \text{ then } f'(x) = 3x^2.$$
$$\text{If } f(x) = x^4, \text{ then } f'(x) = 4x^3.$$
$$\text{If } f(x) = x^5, \text{ then } f'(x) = 5x^4.$$

Do you recognize a pattern here? We can use the definition of derivative and the binomial formula to verify the pattern and find a formula for the derivative of $f(x) = x^n$. If n is a positive integer, then

$$f'(x) = \lim_{h \to 0} \frac{f(x+h) - f(x)}{h} = \lim_{h \to 0} \frac{(x+h)^n - x^n}{h}$$

Because we are assuming that n is a positive integer, we can use the binomial formula, developed in Section 8.3, to expand $(x + h)^n$. This formula can be stated as follows:

$$(x + h)^n = x^n + nx^{n-1}h + \frac{n(n-1)}{1 \cdot 2} x^{n-2} h^2 + \cdots + h^n$$

Thus the derivative formula yields

$$f'(x) = \lim_{h \to 0} \frac{\left[x^n + nx^{n-1}h + \dfrac{n(n-1)}{1 \cdot 2} x^{n-2}h^2 + \cdots + h^n \right] - x^n}{h}$$

$$= \lim_{h \to 0} \left[nx^{n-1} + \frac{n(n-1)}{1 \cdot 2} x^{n-2}h + \cdots + h^{n-1} \right]$$

Note that each term after nx^{n-1} contains h as a factor, so all terms except nx^{n-1} will approach 0 as $h \to 0$. Thus

$$f'(x) = nx^{n-1}$$

Even though we proved this derivative rule only for the case when n is a positive integer, the rule applies for any real number n.

| **Powers of x Rule** | If $f(x) = x^n$, where n is a real number, then $f'(x) = nx^{n-1}$. |

EXAMPLE 1   **Powers of x Rule**

Find the derivatives of the following functions.

(a) $y = x^{14}$ (b) $f(x) = x^{-2}$ (c) $y = x$ (d) $g(x) = x^{1/3}$

Solution

(a) If $y = x^{14}$, then $dy/dx = 14x^{14-1} = 14x^{13}$.

(b) The Powers of x Rule applies for all real values. Thus for $f(x) = x^{-2}$, we have

$$f'(x) = -2x^{-2-1} = -2x^{-3} = \frac{-2}{x^3}$$

(c) If $y = x$, then $dy/dx = 1x^{1-1} = x^0 = 1$. (Note that $y = x$ is a line with slope 1.)

(d) The Powers of x Rule applies to $y = x^{1/3}$.

$$g'(x) = \frac{1}{3}x^{1/3-1} = \frac{1}{3}x^{-2/3} = \frac{1}{3x^{2/3}}$$

In Example 1 we took the derivative with respect to x of *both sides* of each equation. We denote the operation "take the derivative with respect to x" by $\dfrac{d}{dx}$. Thus for $y = x^{14}$, in part (a), we can use this notation on both sides of the equation.

$$\frac{d}{dx}(y) = \frac{d}{dx}(x^{14}) \quad \text{gives} \quad \frac{dy}{dx} = 14x^{13}$$

Similarly, we can use this notation to indicate the derivative of an expression.

$$\frac{d}{dx}(x^{-2}) = -2x^{-3}$$

The differentiation rules are stated and proved for the independent variable x, but they also apply to other independent variables. The following examples illustrate differentiation involving variables other than x and y.

EXAMPLE 2   **Derivatives**

Find the derivatives of the following functions.

(a) $u(s) = s^8$ (b) $p = q^{2/3}$ (c) $C(t) = \sqrt{t}$ (d) $s = \dfrac{1}{\sqrt{t}}$

Solution

(a) If $u(s) = s^8$, then $u'(s) = 8s^{8-1} = 8s^7$.

(b) If $p = q^{2/3}$, then

$$\frac{dp}{dq} = \frac{2}{3}q^{2/3-1} = \frac{2}{3}q^{-1/3} = \frac{2}{3q^{1/3}}$$

(c) Writing $\sqrt{t}$ in its equivalent form, $t^{1/2}$, using the derivative formula, and writing the derivative in radical form give $C'(t) = \dfrac{1}{2}t^{1/2-1} = \dfrac{1}{2}t^{-1/2} = \dfrac{1}{2} \cdot \dfrac{1}{t^{1/2}} = \dfrac{1}{2\sqrt{t}}$

(d) Writing $1/\sqrt{t}$ as $\dfrac{1}{t^{1/2}} = t^{-1/2}$, taking the derivative, and writing the derivative in a form similar to that of the original function give

$$\frac{ds}{dt} = -\frac{1}{2}t^{-1/2-1} = -\frac{1}{2}t^{-3/2} = -\frac{1}{2} \cdot \frac{1}{t^{3/2}} = -\frac{1}{2\sqrt{t^3}}$$

EXAMPLE 3 Tangent Line

Write the equation of the tangent line to the graph of $y = x^3$ at $x = 1$.

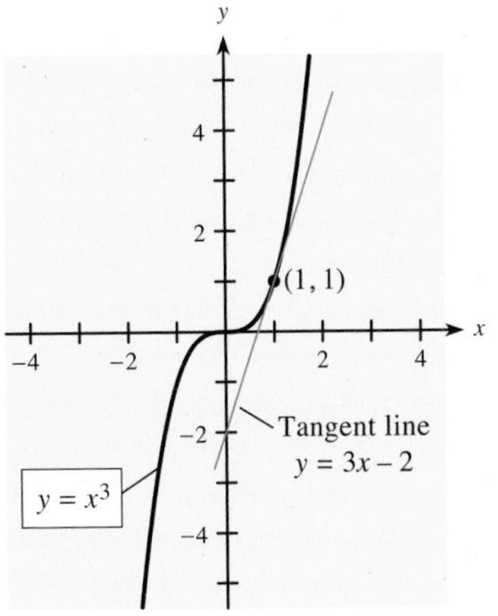

Solution

Writing the equation of the tangent line to $y = x^3$ at $x = 1$ involves three steps.

1. Evaluate the function to find the point of tangency.
   At $x = 1$: $y = (1)^3 = 1$, so the point is $(1, 1)$

2. Evaluate the derivative to find the slope of the tangent.
   At any point: $m_{\text{tan}} = y' = 3x^2$
   At $x = 1$: $m_{\text{tan}} = y'|_{x=1} = 3(1^2) = 3$

3. Use $y - y_1 = m(x - x_1)$ with the point $(1, 1)$ and slope $m = 3$.

$$y - 1 = 3(x - 1) \Rightarrow y = 3x - 3 + 1 \Rightarrow y = 3x - 2$$

Figure 9.28 shows the graph of $y = x^3$ and the tangent line at $x = 1$.

Figure 9.28

Derivative of a Constant A function of the form $y = f(x) = c$, where c is a constant, is called a **constant function.** We can show that the derivative of a constant function is 0, as follows.

$$f'(x) = \lim_{h \to 0} \frac{f(x + h) - f(x)}{h} = \lim_{h \to 0} \frac{c - c}{h} = \lim_{h \to 0} 0 = 0$$

We can state this rule formally.

Constant Function Rule   If $f(x) = c$, where c is a constant, then $f'(x) = 0$.

EXAMPLE 4 Derivative of a Constant

Find the derivative of the function defined by $y = 4$.

Solution

Because 4 is a constant, $\dfrac{dy}{dx} = 0$.

Recall that the function defined by $y = 4$ has a horizontal line as its graph. Thus the slope of the line (and the derivative of the function) is 0.

Derivative of $y = c \cdot u(x)$ We now can take derivatives of constant functions and powers of x. But we do not yet have a rule for taking derivatives of functions of the form $f(x) = 4x^5$ or $g(t) = \frac{1}{2} t^2$. The following rule provides a method for handling functions of this type.

Coefficient Rule   If $f(x) = c \cdot u(x)$, where c is a constant and $u(x)$ is a differentiable function of x, then $f'(x) = c \cdot u'(x)$.

The Coefficient Rule says that the derivative of a constant times a function is the constant times the derivative of the function.

Using Properties of Limits II and IV (from Section 9.1, "Limits"), we can show

$$\lim_{h \to 0} c \cdot g(h) = c \cdot \lim_{h \to 0} g(h)$$

We can use this result to verify the Coefficient Rule. If $f(x) = c \cdot u(x)$, then

$$f'(x) = \lim_{h \to 0} \frac{f(x + h) - f(x)}{h} = \lim_{h \to 0} \frac{c \cdot u(x + h) - c \cdot u(x)}{h}$$

$$= \lim_{h \to 0} c \cdot \left[\frac{u(x + h) - u(x)}{h} \right] = c \cdot \lim_{h \to 0} \frac{u(x + h) - u(x)}{h}$$

so $f'(x) = c \cdot u'(x)$.

EXAMPLE 5   Coefficient Rule for Derivatives

Find the derivatives of the following functions.

(a) $f(x) = 4x^5$ (b) $g(t) = \dfrac{1}{2}t^2$ (c) $p = \dfrac{5}{\sqrt{q}}$

Solution

(a) $f'(x) = 4(5x^4) = 20x^4$

(b) $g'(t) = \dfrac{1}{2}(2t) = t$

(c) $p = \dfrac{5}{\sqrt{q}} = 5q^{-1/2}$,   so $\dfrac{dp}{dq} = 5\left(-\dfrac{1}{2}q^{-3/2} \right) = -\dfrac{5}{2\sqrt{q^3}}$ ◼

EXAMPLE 6   World Tourism

World tourism has grown into one of the world's major industries. Since 1990 the receipts from world tourism y, in billions of dollars, can be modeled by the function

$$y = 15.9x^{1.18}$$

where x is the number of years past 1980 (*Source:* World Tourism Organization).

(a) Name the function that models the rate of change of the receipts from world tourism.
(b) Find the function from part (a).
(c) Find the rate of change in world tourism in 2020.

Solution

(a) The rate of change of world tourism receipts is modeled by the derivative.

(b) $\dfrac{dy}{dx} = 15.9(1.18x^{1.18-1}) = 18.762x^{0.18}$

(c) $\dfrac{dy}{dx}\bigg|_{x=40} = 18.762(40^{0.18}) \approx 36.45$

Thus, the model estimates that world tourism receipts will change by about $36.45 billion per year in 2020.   ◼

Derivatives of Sums and Differences   In Example 6 of Section 9.3, "Rates of Change and Derivatives," we found the derivative of $f(x) = 3x^2 + 2x + 11$ to be $f'(x) = 6x + 2$. This result, along with the results of several of the derivatives calculated in the exercises for that section, suggest that we can find the derivative of a function by finding the derivatives of its terms and combining them. The following rules state this formally.

Sum Rule   If $f(x) = u(x) + v(x)$, where u and v are differentiable functions of x, then
$f'(x) = u'(x) + v'(x)$.

We can prove this rule as follows. If $f(x) = u(x) + v(x)$, then

$$f'(x) = \lim_{h \to 0} \frac{f(x+h) - f(x)}{h} = \lim_{h \to 0} \frac{[u(x+h) + v(x+h)] - [u(x) + v(x)]}{h}$$

$$= \lim_{h \to 0} \left[\frac{u(x+h) - u(x)}{h} + \frac{v(x+h) - v(x)}{h} \right]$$

$$= \lim_{h \to 0} \frac{u(x+h) - u(x)}{h} + \lim_{h \to 0} \frac{v(x+h) - v(x)}{h}$$

$$= u'(x) + v'(x)$$

A similar rule applies to the difference of functions.

Difference Rule

If $f(x) = u(x) - v(x)$, where u and v are differentiable functions of x, then
$f'(x) = u'(x) - v'(x)$.

EXAMPLE 7   Sum and Difference Rules

Find the derivatives of the following functions.

(a) $y = 3x + 5$

(b) $f(x) = 4x^3 - 2x^2 + 5x - 3$

(c) $p = \frac{1}{3}q^3 + 2q^2 - 3$

(d) $u(x) = 5x^4 + x^{1/3}$

(e) $y = 4x^3 + \sqrt{x}$

(f) $s = 5t^6 - \dfrac{1}{t^2}$

Solution

(a) $y' = 3 \cdot 1 + 0 = 3$

(b) The rules regarding the derivatives of sums and differences of two functions also apply if more than two functions are involved. We may think of the functions that are added and subtracted as terms of the function f. Then it would be correct to say that we may take the derivative of a function term by term. Thus,

$$f'(x) = 4(3x^2) - 2(2x) + 5(1) - 0 = 12x^2 - 4x + 5$$

(c) $\dfrac{dp}{dq} = \frac{1}{3}(3q^2) + 2(2q) - 0 = q^2 + 4q$

(d) $u'(x) = 5(4x^3) + \frac{1}{3}x^{-2/3} = 20x^3 + \dfrac{1}{3x^{2/3}}$

(e) We may write the function as $y = 4x^3 + x^{1/2}$, so

$$y' = 4(3x^2) + \frac{1}{2}x^{-1/2} = 12x^2 + \frac{1}{2x^{1/2}} = 12x^2 + \frac{1}{2\sqrt{x}}$$

(f) We may write $s = 5t^6 - 1/t^2$ as $s = 5t^6 - t^{-2}$, so

$$\frac{ds}{dt} = 5(6t^5) - (-2t^{-3}) = 30t^5 + 2t^{-3} = 30t^5 + \frac{2}{t^3}$$

Each derivative in Example 7 has been *simplified*. This means that the final form of the derivative contains no negative exponents and the use of radicals or fractional exponents matches the original problem.

Also, in part (a) of Example 7, we saw that the derivative of $y = 3x + 5$ is 3. Because the slope of a line is the same at all points on the line, it is reasonable that the derivative of a linear equation is a constant. In particular, the slope of the graph of the equation $y = mx + b$ is m at all points on its graph because the derivative of $y = mx + b$ is $y' = f'(x) = m$.

EXAMPLE 8   Personal Income | APPLICATION PREVIEW |

Suppose that t is the number of years past 1960 and that

$$I = I(t) = 6.29t^2 - 51.7t + 601$$

models the U.S. total personal income in billions of current dollars. For 2018, find the model's prediction for

(a) the U.S. total personal income.
(b) the rate of change of U.S. total personal income.

Solution

(a) For 2018, $t = 58$ and

$$I(58) = 6.29(58^2) - 51.7(58) + 601 \approx 18{,}762 \text{ billion (current dollars)}$$

(b) The rate of change of U.S. total personal income is given by

$$I'(t) = 12.58t - 51.7$$

The predicted rate for 2018 is

$$I'(58) = \$677.94 \text{ billion (current dollars) per year}$$

This predicts that U.S. total personal income will change by about \$677.94 billion from 2018 to 2019.

✓ CHECKPOINT

1. True or false: The derivative of a constant times a function is equal to the constant times the derivative of the function.
2. True or false: The derivative of the sum of two functions is equal to the sum of the derivatives of the two functions.
3. True or false: The derivative of the difference of two functions is equal to the difference of the derivatives of the two functions.
4. Does the Coefficient Rule apply to $f(x) = x^n/c$, where c is a constant? Explain.
5. Find the derivative of each of the following functions.

   (a) $f(x) = x^{10} - 10x + 5$ (b) $s = \dfrac{1}{t^5} - 10^7 + 1$

6. Find the slope of the line tangent to $f(x) = x^3 - 4x^2 + 1$ at $x = -1$.

EXAMPLE 9   Horizontal Tangents

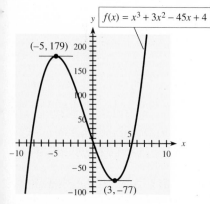

Figure 9.29

Find all points on the graph of $f(x) = x^3 + 3x^2 - 45x + 4$ where the tangent line is horizontal.

Solution

A horizontal line has slope equal to 0. Thus, to find the desired points, we solve $f'(x) = 0$.

$$f'(x) = 3x^2 + 6x - 45$$

We solve $3x^2 + 6x - 45 = 0$ as follows:

$$3x^2 + 6x - 45 = 0 \Rightarrow 3(x^2 + 2x - 15) = 0 \Rightarrow 3(x + 5)(x - 3) = 0$$

Solving $3(x + 5)(x - 3) = 0$ gives $x = -5$ and $x = 3$. The y-coordinates for these x-values come from $f(x)$. The desired points are $(-5, f(-5)) = (-5, 179)$ and $(3, f(3)) = (3, -77)$. Figure 9.29 shows the graph of $y = f(x)$ with these points and the tangent lines at them indicated.

Marginal Revenue   The marginal revenue $R'(x)$ is used to estimate the change in revenue caused by the sale of one additional unit.

EXAMPLE 10   **Revenue**

Suppose that a manufacturer of a product knows that because of the demand for this product, his revenue is given by

$$R(x) = 1500x - 0.02x^2, \qquad 0 \le x \le 1000$$

where x is the number of units sold and $R(x)$ is in dollars.

(a) Find the marginal revenue at $x = 500$.
(b) Find the change in revenue caused by the increase in sales from 500 to 501 units.

Solution

(a) The marginal revenue for any value of x is

$$R'(x) = 1500 - 0.04x$$

The marginal revenue at $x = 500$ is

$$R'(500) = 1500 - 20 = 1480 \,(\text{dollars per unit})$$

We can interpret this to mean that the approximate revenue from the sale of the 501st unit will be \$1480.

(b) The revenue at $x = 500$ is $R(500) = 745{,}000$, and the revenue at $x = 501$ is $R(501) = 746{,}479.98$, so the change in revenue is

$$R(501) - R(500) = 746{,}479.98 - 745{,}000 = 1479.98 \,(\text{dollars})$$

Notice that the marginal revenue at $x = 500$ is a good estimate of the revenue from the 501st unit.

Calculator Note

We have mentioned that graphing calculators have a numerical derivative feature that can be used to estimate the derivative of a function at a specific value of x. This feature can also be used to check the derivative of a function that has been computed with a formula. See Appendix A, Section 9.4, for details. We graph both the derivative calculated with a formula and the numerical derivative. If the two graphs lie on top of one another, the computed derivative agrees with the numerical derivative. Figure 9.30 illustrates this idea for the derivative of $f(x) = \frac{1}{3}x^3 - 2x^2 + 4$. Figure 9.30(a) shows $f'(x) = x^2 - 4x$ as y_1 and the calculator's numerical derivative of $f(x)$ as y_2. Figure 9.30(b) shows the graphs of both y_1 and y_2 (the graphs are coincident).

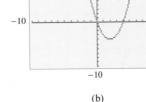

Figure 9.30   (a)   (b)

   EXAMPLE 11   **Comparing $f(x)$ and $f'(x)$**

(a) Graph $f(x) = x^3 - 3x + 3$ and its derivative $f'(x)$ on the same set of axes so that all values of x that make $f'(x) = 0$ are in the x-range.
(b) Investigate the graph of $y = f(x)$ near values of x where $f'(x) = 0$. Does the graph of $y = f(x)$ appear to turn at values where $f'(x) = 0$?
(c) Compare the interval of x values where $f'(x) < 0$ with the interval where the graph of $y = f(x)$ is decreasing from left to right.
(d) What is the relationship between the intervals where $f'(x) > 0$ and where the graph of $y = f(x)$ is increasing from left to right?

Solution

(a) The graphs of $f(x) = x^3 - 3x + 3$ and $f'(x) = 3x^2 - 3$ are shown in Figure 9.31.

(b) The values where $f'(x) = 0$ are the x-intercepts, $x = -1$ and $x = 1$. The graph of $y = x^3 - 3x + 3$ appears to turn at both these values.

(c) $f'(x) < 0$ where the graph of $y = f'(x)$ is below the x-axis, for $-1 < x < 1$. The graph of $y = f(x)$ appears to be decreasing on this interval.

(d) They appear to be the same intervals.

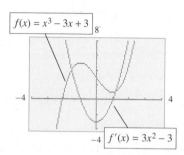

Figure 9.31

---

1. True, by the Coefficient Rule.
2. True, by the Sum Rule.
3. True, by the Difference Rule.
4. Yes, $f(x) = x^n/c = (1/c)x^n$, so the coefficient is $(1/c)$.
5. (a) $f'(x) = 10x^9 - 10$

   (b) $\dfrac{ds}{dt} = \dfrac{-5}{t^6}$

6. The slope of the tangent at $x = -1$ is $f'(-1) = 11$.

| EXERCISES | 9.4

Find the derivatives of the functions in Problems 1–14.

1. $y = 4$
2. $f(s) = 6$
3. $f(t) = t$
4. $s = t^2$
5. $y = 6 - 8x + 2x^2$
6. $y = 12 + 2x - 7x^3$
7. $f(x) = 3x^4 - x^6$
8. $f(x) = 3x^4 - x^9$
9. $y = 10x^5 - 3x^3 + 5x - 11$
10. $y = 3x^5 - 5x^3 - 8x + 8$
11. $w = z^7 - 3z^6 + 13$
12. $u = 2t^{10} - 5t^5 - 9$
13. $g(x) = 2x^{12} - 5x^6 + 9x^4 + x - 5$
14. $h(x) = 12x^{20} + 8x^{10} - 2x^7 + 17x - 9$

In Problems 15–18, at the indicated points, find
(a) the slope of the tangent to the curve, and
(b) the instantaneous rate of change of the function.

15. $y = 7x^2 + 2x + 1, \quad x = 2$
16. $C(x) = 3x^2 - 5, \quad (3, 22)$
17. $P(x) = x^3 - 6x, \quad (2, -4)$
18. $R(x) = 16x + x^2, \quad x = 1$

In Problems 19–26, find the derivative of each function.

19. $y = x^{-5} + x^{-8} - 3$
20. $y = x^{-1} - x^{-2} + 13$
21. $z = 3t^{11/3} - 2t^{7/4} - t^{1/2} + 8$
22. $w = 5u^{8/5} - 3u^{5/6} + u^{1/3} + 5$

23. $f(x) = 5x^{-4/5} + 2x^{-4/3}$
24. $f(x) = 6x^{-8/3} - x^{-2/3}$

25. $g(x) = \dfrac{3}{x^5} + \dfrac{2}{x^4} + 6\sqrt[3]{x}$

26. $h(x) = \dfrac{7}{x^7} - \dfrac{3}{x^3} + 8\sqrt{x}$

In Problems 27–30, write the equation of the tangent line to each curve at the indicated point. As a check, graph both the function and the tangent line.

27. $y = x^3 - 5x^2 + 7 \quad$ at $x = 1$
28. $y = x^4 - 4x^3 - 2 \quad$ at $x = 2$

29. $f(x) = 4x^2 - \dfrac{1}{x} \quad$ at $x = -\dfrac{1}{2}$

30. $f(x) = \dfrac{x^3}{3} - \dfrac{3}{x^3} \quad$ at $x = -1$

In Problems 31–34, find the coordinates of points where the graph of $f(x)$ has horizontal tangents. As a check, graph $f(x)$ and see whether the points you found look as though they have horizontal tangents.

31. $f(x) = -x^3 + 9x^2 - 15x + 6$

32. $f(x) = \dfrac{1}{3}x^3 - 3x^2 - 16x + 8$

33. $f(x) = x^4 - 4x^3 + 9$

34. $f(x) = 3x^5 - 5x^3 + 2$

In Problems 35 and 36, find each derivative at the given x-value (a) with the appropriate rule and (b) with the numerical derivative feature of a graphing calculator.

35. $y = 5 - 2\sqrt{x}$ at $x = 4$

36. $y = 1 + 3x^{2/3}$ at $x = -8$

In Problems 37–40, complete the following.

(a) **Calculate the derivative of each function with the appropriate formula.**

(b) **Check your result from part (a) by graphing your calculated derivative and the numerical derivative of the given function with respect to x evaluated at x.**

37. $f(x) = 2x^3 + 5x - \pi^4 + 8$

38. $f(x) = 3x^2 - 8x + 2^5 - 20$

39. $h(x) = \dfrac{10}{x^3} - \dfrac{10}{\sqrt[5]{x^2}} + x^2 + 1$

40. $g(x) = \dfrac{5}{x^{10}} + \dfrac{4}{\sqrt[4]{x^3}} + x^5 - 4$

The tangent line to a curve at a point closely approximates the curve near the point. In fact, for x-values close enough to the point of tangency, the function and its tangent line are virtually indistinguishable. Problems 41 and 42 explore this relationship. Use each given function and the indicated point to complete the following.

(a) **Write the equation of the tangent line to the curve at the indicated point.**

(b) **Use a graphing calculator to graph both the function and its tangent line. Be sure your graph shows the point of tangency.**

(c) **Repeatedly zoom in on the point of tangency. Do the function and the tangent line eventually become indistinguishable?**

41. $f(x) = 3x^2 + 2x$ at $x = 1$

42. $f(x) = 4x - x^2$ at $x = 5$

For each function in Problems 43–46, do the following.

(a) **Find $f'(x)$**

(b) **Graph both $f(x)$ and $f'(x)$ with a graphing utility.**

(c) **Use the graph of $f'(x)$ to identify x-values where $f'(x) = 0$, $f'(x) > 0$, and $f'(x) < 0$.**

(d) **Use the graph of $f(x)$ to identify x-values where $f(x)$ has a maximum or minimum point, where the graph of $f(x)$ is rising, and where the graph of $f(x)$ is falling.**

43. $f(x) = 8 - 2x - x^2$

44. $f(x) = x^2 + 4x - 12$

45. $f(x) = x^3 - 12x - 5$

46. $f(x) = 7 - 3x^2 - \dfrac{x^3}{3}$

APPLICATIONS

47. *Revenue* Suppose that a wholesaler expects that his monthly revenue, in dollars, for an electronic game will be

$$R(x) = 100x - 0.1x^2, \quad 0 \le x \le 800$$

where x is the number of units sold. Find his marginal revenue and interpret it when the quantity sold is

(a) $x = 300$.   (b) $x = 600$.

48. *Revenue* The total revenue, in dollars, for a commodity is described by the function

$$R = 300x - 0.02x^2$$

(a) What is the marginal revenue when 40 units are sold?

(b) Interpret your answer to part (a).

49. *Metabolic rate* According to Kleiber's law the metabolic rate q of the vast majority of animals is related to the animal's mass M according to

$$q = kM^{3/4}$$

where k is a constant. This means that a cat, with mass about 100 times that of a mouse, has a metabolism about $100^{3/4} \approx 32$ times greater than that of a mouse. Find the function that describes the rate of change of the metabolic rate with respect to mass.

50. *Capital investment and output* The monthly output of a certain product is

$$Q(x) = 800x^{5/2}$$

where x is the capital investment in millions of dollars. Find dQ/dx, which can be used to estimate the effect on the output if an additional capital investment of $1 million is made.

51. *Demand* The demand for q units of a product depends on the price p (in dollars) according to

$$q = \frac{1000}{\sqrt{p}} - 1, \quad \text{for } p > 0$$

Find and explain the meaning of the instantaneous rate of change of demand with respect to price when the price is

(a) $25.   (b) $100.

52. *Demand* Suppose that the demand for a product depends on the price p according to

$$D(p) = \frac{50{,}000}{p^2} - \frac{1}{2}, \quad p > 0$$

where p is in dollars. Find and explain the meaning of the instantaneous rate of change of demand with respect to price when

(a) $p = 50$.   (b) $p = 100$.

53. *Cost and average cost* Suppose that the total cost function, in dollars, for the production of x units of a product is given by

$$C(x) = 4000 + 55x + 0.1x^2$$

Then the average cost of producing x items is

$$\overline{C(x)} = \frac{\text{total cost}}{x} = \frac{4000}{x} + 55 + 0.1x$$

(a) Find the instantaneous rate of change of average cost with respect to the number of units produced, at any level of production.
(b) Find the level of production at which this rate of change equals zero.
(c) At the value found in part (b), find the instantaneous rate of change of cost and find the average cost. What do you notice?

54. *Cost and average cost* Suppose that the total cost function, in dollars, for a certain commodity is given by

$$C(x) = 40{,}500 + 190x + 0.2x^2$$

where x is the number of units produced.
(a) Find the instantaneous rate of change of the average cost

$$\overline{C} = \frac{40{,}500}{x} + 190 + 0.2x$$

for any level of production.
(b) Find the level of production where this rate of change equals zero.
(c) At the value found in part (b), find the instantaneous rate of change of cost and find the average cost. What do you notice?

55. *Cost-benefit* Suppose that for a certain city the cost C, in dollars, of obtaining drinking water that contains p percent impurities (by volume) is given by

$$C = \frac{120{,}000}{p} - 1200$$

(a) Find the rate of change of cost with respect to p when impurities account for 10% (by volume).
(b) Write a sentence that explains the meaning of your answer in part (a).

56. *Cost-benefit* Suppose that the cost C, in dollars, of processing the exhaust gases at an industrial site to ensure that only p percent of the particulate pollution escapes is given by

$$C(p) = \frac{8100(100 - p)}{p}$$

(a) Find the rate of change of cost C with respect to the percent of particulate pollution that escapes when $p = 2$ (percent).
(b) Write a sentence interpreting your answer to part (a).

57. *Wind chill* One form of the formula that meteorologists use to calculate wind chill temperature (WC) is

$$WC = 35.74 + 0.6215t - 35.75s^{0.16} + 0.4275t\, s^{0.16}$$

where s is the wind speed in mph and t is the actual air temperature in degrees Fahrenheit. Suppose temperature is constant at 15°.
(a) Express wind chill WC as a function of wind speed s.
(b) Find the rate of change of wind chill with respect to wind speed when the wind speed is 25 mph.
(c) Interpret your answer to part (b).

58. *Allometric relationships—crabs* For fiddler crabs, data gathered by Thompson[*] show that the allometric relationship between the weight C of the claw and the weight W of the body is given by

$$C = 0.11W^{1.54}$$

Find the function that gives the rate of change of claw weight with respect to body weight.

Recall that for all modeling problems, use the unrounded model for any calculations unless instructed otherwise.

59. Modeling *Consumer price index* The table below gives the U.S. consumer price index (CPI) for selected years from 2012 and projected to 2050. With the reference year as 2012, a 2020 CPI = 120.56 means goods and services that cost \$100.00 in 2012 are expected to cost \$120.56 in 2020.
(a) Find the quadratic function that is the best fit for the data, with x as the number of years past 2010 and y as the CPI in dollars. Report the model as $y = f(x)$ with three significant digit coefficients.
(b) Use the data to find the average rate of change of the CPI from 2012 to 2020.
(c) Find the derivative of the reported model found in part (a).
(d) Find the instantaneous rate of change of the CPI for the year 2020.
(e) Use the rate of change from part (d) to predict the CPI for 2022.

Year	CPI	Year	CPI
2012	100.00	2030	158.90
2014	104.00	2035	182.43
2016	108.58	2040	209.44
2018	114.09	2045	240.45
2020	120.56	2050	276.05
2025	138.41		

Source: Social Security Administration

60. Modeling *E-commerce* The following table gives the online sales, in billions of dollars, from 2000 and projected to 2017.
(a) Model these data with a power function $E(t)$, where t is the number of years past 1990. Report the model with three significant digit coefficients.

[*]d'Arcy Thompson, *On Growth and Form* (Cambridge, England: Cambridge University Press, 1961).

(b) Use the reported model from part (a) to find the function that models the rate of change of online sales.

(c) Use the result from part (b) to find and interpret the rate of change of online sales in 2020.

Year	Sales ($billion)	Year	Sales ($billion)
2000	28	2009	134
2001	34	2010	171
2002	44	2011	202
2003	56	2012	226
2004	70	2013	259
2005	87	2014	297
2006	106	2015	339
2007	126	2016	386
2008	132	2017	434

Source: U.S. Census Bureau, Forrester Research, Inc.

61. **Modeling *U.S. population*** The table gives the U.S. population to the nearest million for selected years from 1950 and projected to 2050.

Year	Total	Year	Total
1950	160	2010	315
1960	190	2015	328
1970	215	2020	343
1980	235	2030	370
1990	260	2040	391
2000	288	2050	409

Source: Social Security Administration

(a) Find a cubic function $P(t)$ that models these data, where P is the U.S. population in millions and t is the number of years past 1950. Report the model with three significant digit coefficients.

(b) Use the part (a) result to find the function that models the instantaneous rate of change of the U.S. population.

(c) Find and interpret the instantaneous rates of change in 2000 and 2025.

62. **Modeling *Gross domestic product*** The table shows U.S. gross domestic product (GDP) in billions of dollars for selected years from 2000 to 2070 (actual and projected).

Year	GDP	Year	GDP
2000	9143	2040	79,680
2005	12,145	2045	103,444
2010	16,174	2050	133,925
2015	21,270	2055	173,175
2020	27,683	2060	224,044
2025	35,919	2065	290,042
2030	46,765	2070	375,219
2035	61,100		

Source: Social Security Administration Trustees Report

(a) Model these data with a cubic function $g = g(t)$, where g is in billions of dollars and t represents the number of years past 2000. Report the model with three significant digit coefficients.

(b) Use the reported model to find the predicted instantaneous rate of change of the GDP in 2025.

(c) Interpret your answer to part (b).

- To use the Product Rule to find the derivatives of certain functions
- To use the Quotient Rule to find the derivatives of certain functions

9.5

The Product Rule and the Quotient Rule

| APPLICATION PREVIEW |

When medicine is administered, reaction (measured in change of blood pressure or temperature) can be modeled by

$$R = m^2\left(\frac{c}{2} - \frac{m}{3}\right)$$

where c is a positive constant and m is the amount of medicine absorbed into the blood.* The rate of change of R with respect to m is the sensitivity of the body to medicine. To find an expression for sensitivity as a function of m, we calculate dR/dm. We can find this derivative with the Product Rule for derivatives. See Example 5.

Source: R. M. Thrall et al., *Some Mathematical Models in Biology*, U.S. Department of Commerce, 1967.

Product Rule   We have simple formulas for finding the derivatives of the sums and differences of functions. But we are not so lucky with products. The derivative of a product is *not* the product of the derivatives. To see this, we consider the function $f(x) = x \cdot x$. Because this function is $f(x) = x^2$, its derivative is $f'(x) = 2x$. But the product of the derivatives of x and x would give $1 \cdot 1 = 1 \neq 2x$. Thus we need a different formula to find the derivative of a product. This formula is given by the **Product Rule.**

Product Rule

If $f(x) = u(x) \cdot v(x)$, where u and v are differentiable functions of x, then

$$f'(x) = u(x) \cdot v'(x) + v(x) \cdot u'(x)$$

Thus the derivative of a product of two functions is the first function times the derivative of the second plus the second function times the derivative of the first.

We can prove the Product Rule as follows. If $f(x) = u(x) \cdot v(x)$, then

$$\lim_{h \to 0} \frac{f(x+h) - f(x)}{h} = \lim_{h \to 0} \frac{u(x+h) \cdot v(x+h) - u(x) \cdot v(x)}{h}$$

Subtracting and adding $u(x+h) \cdot v(x)$ in the numerator gives

$$f'(x) = \lim_{h \to 0} \frac{u(x+h) \cdot v(x+h) - u(x+h) \cdot v(x) + u(x+h) \cdot v(x) - u(x) \cdot v(x)}{h}$$

$$= \lim_{h \to 0} \left\{ u(x+h) \left[\frac{v(x+h) - v(x)}{h} \right] + v(x) \left[\frac{u(x+h) - u(x)}{h} \right] \right\}$$

Properties III and IV of limits (from Section 9.1, "Limits") give

$$f'(x) = \lim_{h \to 0} u(x+h) \cdot \lim_{h \to 0} \frac{v(x+h) - v(x)}{h} + \lim_{h \to 0} v(x) \cdot \lim_{h \to 0} \frac{u(x+h) - u(x)}{h}$$

Because u is differentiable and hence continuous, it follows that $\lim_{h \to 0} u(x+h) = u(x)$, so we have the formula we seek:

$$f'(x) = u(x) \cdot v'(x) + v(x) \cdot u'(x)$$

EXAMPLE 1   Product Rule

(a) Find dy/dx if $y = (2x^3 + 3x + 1)(x^2 + 4)$.
(b) At the point where $x = 1$, find the slope of the tangent to the graph of

$$y = f(x) = (4x^3 + 5x^2 - 6x + 5)(x^3 - 4x^2 + 1)$$

Solution

(a) Using the Product Rule with $u(x) = 2x^3 + 3x + 1$ and $v(x) = x^2 + 4$, we have

$$\frac{dy}{dx} = (2x^3 + 3x + 1)(2x) + (x^2 + 4)(6x^2 + 3)$$

$$= 4x^4 + 6x^2 + 2x + 6x^4 + 3x^2 + 24x^2 + 12$$
$$= 10x^4 + 33x^2 + 2x + 12$$

(b) $f'(x) = (4x^3 + 5x^2 - 6x + 5)(3x^2 - 8x) + (x^3 - 4x^2 + 1)(12x^2 + 10x - 6)$
   If we substitute $x = 1$ into $f'(x)$, we find that the slope of the curve at $x = 1$ is
   $f'(1) = 8(-5) + (-2)(16) = -72$.

Quotient Rule   For a quotient of two functions, we might be tempted to take the derivative of the numerator divided by the derivative of the denominator; but this is incorrect. With the example $f(x) = x^3/x$ (which equals x^2 if $x \neq 0$), this approach would give $3x^2/1 = 3x^2$ as the derivative, rather than $2x$. Thus, finding the derivative of a function that is the quotient of two functions requires the **Quotient Rule.**

Quotient Rule

If $f(x) = \dfrac{u(x)}{v(x)}$, where u and v are differentiable functions of x, with $v(x) \neq 0$, then

$$f'(x) = \frac{v(x) \cdot u'(x) - u(x) \cdot v'(x)}{[v(x)]^2}$$

The preceding formula says that the derivative of a quotient is the denominator times the derivative of the numerator minus the numerator times the derivative of the denominator, all divided by the square of the denominator.

To see that this rule is reasonable, again consider the function $f(x) = x^3/x$, $x \neq 0$. Using the Quotient Rule, with $u(x) = x^3$ and $v(x) = x$, we get

$$f'(x) = \frac{x(3x^2) - x^3(1)}{x^2} = \frac{3x^3 - x^3}{x^2} = \frac{2x^3}{x^2} = 2x$$

We see that $f'(x) = 2x$ is the correct derivative. The proof of the Quotient Rule is left for the student in the exercises in this section.

EXAMPLE 2   **Quotient Rule**

(a) If $f(x) = \dfrac{x^2 - 4x}{x + 5}$, find $f'(x)$.

(b) If $f(x) = \dfrac{x^3 - 3x^2 + 2}{x^2 - 4}$, find the instantaneous rate of change of $f(x)$ at $x = 3$.

Solution
(a) Using the Quotient Rule with $u(x) = x^2 - 4x$ and $v(x) = x + 5$, we get

$$f'(x) = \frac{(x + 5)(2x - 4) - (x^2 - 4x)(1)}{(x + 5)^2}$$

$$= \frac{2x^2 + 6x - 20 - x^2 + 4x}{(x + 5)^2} = \frac{x^2 + 10x - 20}{(x + 5)^2}$$

(b) We evaluate $f'(x)$ at $x = 3$ to find the desired rate of change. Using the Quotient Rule with $u(x) = x^3 - 3x^2 + 2$ and $v(x) = x^2 - 4$, we get

$$f'(x) = \frac{(x^2 - 4)(3x^2 - 6x) - (x^3 - 3x^2 + 2)(2x)}{(x^2 - 4)^2}$$

$$= \frac{(3x^4 - 6x^3 - 12x^2 + 24x) - (2x^4 - 6x^3 + 4x)}{(x^2 - 4)^2} = \frac{x^4 - 12x^2 + 20x}{(x^2 - 4)^2}$$

Thus, the instantaneous rate of change at $x = 3$ is $f'(3) = 33/25 = 1.32$

EXAMPLE 3   **Quotient Rule**

Use the Quotient Rule to find the derivative of $y = 1/x^3$.

Solution
Letting $u(x) = 1$ and $v(x) = x^3$, we get

$$y' = \frac{x^3(0) - 1(3x^2)}{(x^3)^2} = -\frac{3x^2}{x^6} = -\frac{3}{x^4}$$

Note that we could have found the derivative more easily by rewriting y.

$$y = 1/x^3 = x^{-3} \quad \text{gives} \quad y' = -3x^{-4} = -\frac{3}{x^4}$$

Recall that we proved the Powers of x Rule for positive integer powers and assumed that it was true for all real number powers. In Problem 38 of the exercises in this section, you will be asked to use the Quotient Rule to show that the Powers of x Rule applies to negative integers.

It is not necessary to use the Quotient Rule when the denominator of the function in question contains only a constant. For example, the function $y = (x^3 - 3x)/3$ can be written $y = \frac{1}{3}(x^3 - 3x)$, so the derivative is $y' = \frac{1}{3}(3x^2 - 3) = x^2 - 1$.

✓ CHECKPOINT

1. True or false: The derivative of the product of two functions is equal to the product of the derivatives of the two functions.
2. True or false: The derivative of the quotient of two functions is equal to the quotient of the derivatives of the two functions.
3. Find $f'(x)$ for each of the following.
    (a) $f(x) = (x^{12} + 8x^5 - 7)(10x^7 - 4x + 19)$ Do not simplify.
    (b) $f(x) = \dfrac{2x^4 + 3}{3x^4 + 2}$ Simplify.
4. If $y = \frac{4}{3}(x^2 + 3x - 4)$, does finding y' require the Product Rule? Explain.
5. If $y = f(x)/c$, where c is a constant, does finding y' require the Quotient Rule? Explain.

EXAMPLE 4   Marginal Revenue

Suppose that the revenue function for a flash drive is given by

$$R(x) = 10x + \frac{100x}{3x + 5}$$

where x is the number of flash drives sold and R is in dollars.

(a) Find the marginal revenue function.
(b) Find the marginal revenue when $x = 15$.

Solution

(a) We must use the Quotient Rule to find the marginal revenue (the derivative).

$$\overline{MR} = R'(x) = 10 + \frac{(3x + 5)(100) - 100x(3)}{(3x + 5)^2}$$

$$= 10 + \frac{300x + 500 - 300x}{(3x + 5)^2} = 10 + \frac{500}{(3x + 5)^2}$$

(b) The marginal revenue when $x = 15$ is $R'(15)$.

$$R'(15) = 10 + \frac{500}{[(3)(15) + 5]^2} = 10 + \frac{500}{(50)^2} = 10.20 \text{ (dollars per unit)}$$

Recall that $R'(15)$ estimates the revenue from the sale of the 16th flash drive.

16th $10.20

$R'(15) = 10.20

EXAMPLE 5   **Sensitivity to a Drug** | APPLICATION PREVIEW |

When medicine is administered, reaction (measured in change of blood pressure or temperature) can be modeled by

$$R = m^2 \left(\frac{c}{2} - \frac{m}{3} \right)$$

where c is a positive constant and m is the amount of medicine absorbed into the blood. The rate of change of R with respect to m is the sensitivity of the body to medicine. Find an expression for sensitivity s as a function of m.

Solution

The sensitivity is the rate of change of R with respect to m, or the derivative. Thus

$$s = \frac{dR}{dm} = m^2 \left(0 - \frac{1}{3} \right) + \left(\frac{c}{2} - \frac{1}{3}m \right)(2m)$$

$$= -\frac{1}{3}m^2 + mc - \frac{2}{3}m^2 = mc - m^2$$

✓ CHECKPOINT
ANSWERS

1. False. The derivative of a product is $\dfrac{d}{dx}\,(fg) = f \cdot \dfrac{dg}{dx} + g \cdot \dfrac{df}{dx}$

2. False. The derivative of a quotient is $\dfrac{d}{dx}\left(\dfrac{f}{g} \right) = \dfrac{g \cdot f' - f \cdot g'}{g^2}$

3. (a) $f'(x) = (x^{12} + 8x^5 - 7)(70x^6 - 4) + (10x^7 - 4x + 19)(12x^{11} + 40x^4)$

    (b) $f'(x) = \dfrac{-20x^3}{(3x^4 + 2)^2}$

4. No; y' can be found with the Coefficient Rule: $y' = \dfrac{4}{3}\,(2x + 3)$

5. No; y' can be found with the Coefficient Rule: $y' = \left(\dfrac{1}{c} \right) f'(x)$

| **EXERCISES** | **9.5**

In Problems 1–4, find the derivative and simplify.

1. $y = (5x + 3)(x^2 - 2x)$
2. $s = (t^4 + 1)(t^3 - 1)$
3. $f(x) = (x^{12} + 3x^4 + 4)(2x^3 - 1)$
4. $y = (3x^7 + 4)(8x^6 - 6x^4 - 9)$

In Problems 5–8, find the derivative, but do not simplify your answer.

5. $y = (7x^6 - 5x^4 + 2x^2 - 1)(4x^9 + 3x^7 - 5x^2 + 3x)$
6. $y = (9x^9 - 7x^7 - 6x)(3x^5 - 4x^4 + 3x^3 - 8)$
7. $y = (x^2 + x + 1)(\sqrt[3]{x} - 2\sqrt{x} + 5)$
8. $y = (\sqrt[5]{x} - 2\sqrt[4]{x} + 1)(x^3 - 5x - 7)$

In Problems 9 and 10, at each indicated point find

(a) the slope of the tangent line, and

(b) the instantaneous rate of change of the function.

9. $y = (x^2 + 1)(x^3 - 4x)$ at $(-2, 0)$
10. $y = (x^3 - 3)(x^2 - 4x + 1)$ at $(2, -15)$

In Problems 11–20, find the indicated derivatives and simplify.

11. $\dfrac{dp}{dq}$ for $p = \dfrac{q^2 + 3}{2q - 1}$

12. $C'(x)$ for $C(x) = \dfrac{2x^3}{3x^4 + 2}$

13. $\dfrac{dy}{dx}$ for $y = \dfrac{1 - 2x^2}{x^4 - 2x^2 + 5}$

14. $\dfrac{ds}{dt}$ for $s = \dfrac{t^3 - 4}{t^3 - 2t^2 - t - 5}$

15. $\dfrac{dz}{dx}$ for $z = x^2 + \dfrac{x^2}{1 - x - 2x^2}$

16. $\dfrac{dy}{dx}$ for $y = 200x - \dfrac{100x}{3x + 1}$

17. $\dfrac{dp}{dq}$ for $p = \dfrac{3\sqrt[3]{q}}{1 - q}$

18. $\dfrac{dy}{dx}$ for $y = \dfrac{2\sqrt{x} - 1}{1 - 4\sqrt{x^3}}$

19. y' for $y = \dfrac{x(x^2 + 4)}{x - 2}$

20. $f'(x)$ for $f(x) = \dfrac{(x + 1)(x - 2)}{x^2 + 1}$

In Problems 21 and 22, at the indicated point for each function, find
(a) **the slope of the tangent line, and**
(b) **the instantaneous rate of change of the function.**

21. $y = \dfrac{x^2 + 1}{x + 3}$　at $(2, 1)$

22. $y = \dfrac{x^2 - 4x}{x^2 + 2x}$　at $\left(2, -\dfrac{1}{2}\right)$

In Problems 23–26, write the equation of the tangent line to the graph of the function at the indicated point. Check the reasonableness of your answer by graphing both the function and the tangent line.

23. $y = (9x^2 - 6x + 1)(1 + 2x)$　at $x = 1$
24. $y = (4x^2 + 4x + 1)(7 - 2x)$　at $x = 0$
25. $y = \dfrac{3x^4 - 2x - 1}{4 - x^2}$　at $x = 1$
26. $y = \dfrac{x^2 - 4x}{2x - x^3}$　at $x = 2$

In Problems 27–30, use the numerical derivative feature of a graphing calculator to find the derivative of each function at the given x-value.

27. $y = \left(4\sqrt{x} + \dfrac{3}{x}\right)\left(3\sqrt[3]{x} - \dfrac{5}{x^2} - 25\right)$　at $x = 1$

28. $y = (3\sqrt[4]{x^5} + \sqrt[5]{x^4} - 1)\left(\dfrac{2}{x^3} - \dfrac{1}{\sqrt{x}}\right)$　at $x = 1$

29. $f(x) = \dfrac{4x - 4}{3x^{2/3}}$　at $x = 1$

30. $f(x) = \dfrac{3\sqrt[3]{x} + 1}{x + 2}$　at $x = -1$

In Problems 31–34, complete the following.
(a) **Find the derivative of each function, and check your work by graphing both your calculated derivative and the numerical derivative of the function.**
(b) **Use your graph of the derivative to find points where the original function has horizontal tangent lines.**
(c) **Use a graphing calculator to graph the function and indicate the points found in part (b) on the graph.**

31. $f(x) = (x^2 + 4x + 4)(x - 7)$
32. $f(x) = (x^2 - 14x + 49)(2x + 1)$
33. $y = \dfrac{x^2}{x - 2}$
34. $y = \dfrac{x^2 - 7}{4 - x}$

In Problems 35 and 36,
(a) **find $f'(x)$.**
(b) **graph both $f(x)$ and $f'(x)$ with a graphing utility.**
(c) **identify the x-values where $f'(x) = 0$, $f'(x) > 0$, and $f'(x) < 0$.**
(d) **identify x-values where $f(x)$ has a maximum point or a minimum point, where $f(x)$ is increasing, and where $f(x)$ is decreasing.**

35. $f(x) = \dfrac{10x^2}{x^2 + 1}$　　36. $f(x) = \dfrac{8 - x^2}{x^2 + 4}$

37. Prove the Quotient Rule for differentiation. (*Hint:* Add $[-u(x) \cdot v(x) + u(x) \cdot v(x)]$ to the expanded numerator and use steps similar to those used to prove the Product Rule.)

38. Use the Quotient Rule to show that the Powers of x Rule applies to negative integer powers. That is, show that $(d/dx)x^n = nx^{n-1}$ when $n = -k$, $k > 0$, by finding the derivative of $f(x) = 1/(x^k)$.

APPLICATIONS

39. *Cost-benefit*　If the cost C (in dollars) of removing p percent of the particulate pollution from the exhaust gases at an industrial site is given by

$$C(p) = \dfrac{8100p}{100 - p}$$

find the rate of change of C with respect to p.

40. *Cost-benefit*　If the cost C (in dollars) of removing p percent of the impurities from the waste water in a manufacturing process is given by

$$C(p) = \dfrac{9800p}{101 - p}$$

find the rate of change of C with respect to p.

41. *Revenue*　Suppose the revenue (in dollars) from the sale of x units of a product is given by

$$R(x) = \dfrac{60x^2 + 74x}{2x + 2}$$

Find the marginal revenue when 49 units are sold. Interpret your result.

42. *Revenue*　The revenue (in dollars) from the sale of x units of a product is given by

$$R(x) = \dfrac{3000}{2x + 2} + 80x - 1500$$

Find the marginal revenue when 149 units are sold. Interpret your result.

43. *Revenue*　A travel agency will plan a group tour for groups of size 25 or larger. If the group contains exactly 25 people, the cost is $300 per person. If each person's cost is reduced by $10 for each additional person above the 25, then the revenue is given by

$$R(x) = (25 + x)(300 - 10x)$$

where x is the number of additional people above 25. Find the marginal revenue if the group contains 30 people. Interpret your result.

44. **Revenue** McRobert's Electronics sells 200 TVs per month at a price of $400 per unit. Market research indicates that the store can sell one additional TV for each $1 it reduces the price. In this case the total revenue is

$$R(x) = (200 + x)(400 - x)$$

where x is the number of additional TVs beyond the 200. If the store sells a total of 250 TVs, find the marginal revenue. Interpret your result.

45. **Response to a drug** The reaction R to an injection of a drug is related to the dosage x (in milligrams) according to

$$R(x) = x^2\left(500 - \frac{x}{3}\right)$$

where 1000 mg is the maximum dosage. If the rate of reaction with respect to the dosage defines the sensitivity to the drug, find the sensitivity.

46. **Nerve response** The number of action potentials produced by a nerve, t seconds after a stimulus, is given by

$$N(t) = 25t + \frac{4}{t^2 + 2} - 2$$

Find the rate at which the action potentials are produced by the nerve.

47. **Test reliability** If a test having reliability r is lengthened by a factor n, the reliability of the new test is given by

$$R = \frac{nr}{1 + (n - 1)r}, \quad 0 < r \leq 1$$

Find the rate at which R changes with respect to n.

48. **Advertising and sales** The sales of a product s (in thousands of dollars) are related to advertising expenses (in thousands of dollars) by

$$s = \frac{200x}{x + 10}$$

Find and interpret the meaning of the rate of change of sales with respect to advertising expenses when
(a) $x = 10$.   (b) $x = 20$.

49. **Candidate recognition** Suppose that the proportion P of voters who recognize a candidate's name t months after the start of the campaign is given by

$$P(t) = \frac{13t}{t^2 + 100} + 0.18$$

(a) Find the rate of change of P when $t = 6$, and explain its meaning.
(b) Find the rate of change of P when $t = 12$, and explain its meaning.
(c) One month prior to the election, is it better for $P'(t)$ to be positive or negative? Explain.

50. **Endangered species population** It is determined that a wildlife refuge can support a group of up to 120 of a certain endangered species. If 75 are introduced onto the refuge and their population after t years is given by

$$p(t) = 75\left(1 + \frac{4t}{t^2 + 16}\right)$$

find the rate of population growth after t years. Find the rate after each of the first 7 years.

51. **Wind chill** In January 2014 the so-called "Polar Vortex" of dense, frigid air plunged deep into the United States and resulted in record cold temperatures and dangerous wind chills. If s is the wind speed in miles per hour and $s \geq 5$, then the wind chill (in degrees Fahrenheit) for an air temperature of 0°F can be approximated by the function

$$f(s) = \frac{289.173 - 58.5731s}{s + 1}$$

(a) At what rate is the wind chill changing when the wind speed is 20 mph?
(b) Explain the meaning of your answer to part (a).

52. **Response to injected adrenalin** Experimental evidence has shown that the response y of a muscle is related to the concentration of injected adrenaline x according to the equation

$$y = \frac{x}{a + bx}$$

where a and b are constants. Find the rate of change of response with respect to the concentration.

53. **Social Security beneficiaries** The table gives the number of millions of Social Security beneficiaries (actual and projected) for selected years from 1950 through 2030.

Year	Number of Beneficiaries (millions)	Year	Number of Beneficiaries (millions)
1950	2.9	2000	44.8
1960	14.3	2010	53.3
1970	25.2	2020	68.8
1980	35.1	2030	82.7
1990	39.5		

Source: Social Security Trustees Report

With $B(t)$ representing the number of beneficiaries (in millions) t years past 1950, these data can be modeled by the function

$$B(t) = (0.01t + 3)(0.0238t^2 - 9.79t + 3100) - 9290$$

(a) Find the function that gives the instantaneous rate of change of the number of beneficiaries.
(b) Find and interpret the instantaneous rate of change in 2020.
(c) Use the data to determine which of the average rates of change (from 2010 to 2020, from 2020 to 2030, or from 2010 to 2030) best approximates the instantaneous rate from part (b).

54. *Emissions* The table shows data for sulfur dioxide emissions from electricity generation (in millions of short tons) for selected years from 2000 and projected to 2035. These data can be modeled by the function

$$E(x) = (0.001x - 0.062)(-0.18x^2 + 8.2x - 200)$$

where x is the number of years past 2000.
(a) Find the function that models the rate of change of these emissions.
(b) Find and interpret $E'(20)$.

Year	Short Tons (in millions)
2000	11.4
2005	10.2
2008	7.6
2015	4.7
2020	4.2
2025	3.8
2030	3.7
2035	3.8

Source: U.S. Department of Energy

55. *Females in the workforce* For selected years from 1950 and with projections to 2050, the table shows the percent of total U.S. workers who were female.

Year	% Female	Year	% Female
1950	29.6	2010	47.9
1960	33.4	2015	48.3
1970	38.1	2020	48.1
1980	42.5	2030	48.0
1990	45.2	2040	47.9
2000	46.6	2050	47.7

Source: U.S. Bureau of the Census

Assume these data can be modeled with the function

$$p(t) = \frac{78.6t + 2090}{1.38t + 64.1}$$

where $p(t)$ is the percent of the U.S. workforce that is female and t is the number of years past 1950.
(a) Find the function that models the instantaneous rate of change of the percent of U.S. workers who were female.
(b) Use the function from part (a) to find the instantaneous rates of change in 2005 and in 2020.
(c) Interpret each of the rates of change in part (b).

56. *Alzheimer's* As the Baby Boom generation ages and the proportion of the U.S population over 65 increases, the number of Americans with Alzheimer's disease and other dementia is projected to grow each year. With data from the Social Security Administration and the Alzheimer's Association for selected years from 2000 and projected to 2050, the percent of the U.S. population with Alzheimer's can be modeled with the function

$$A(t) = \frac{0.34t^2 + 1.3t + 450}{-0.012t^2 + 3.3t + 260}$$

where t is the number of years past 1990.
(a) Find the function that gives the instantaneous rate of change of $A(t)$.
(b) Use your answer in part (a) to find and interpret the instantaneous rate of change of the percent of the U.S. population with Alzheimer's in 2025.

OBJECTIVES

9.6

• To use the Chain Rule to differentiate functions
• To use the Power Rule to differentiate functions

The Chain Rule and the Power Rule

■| APPLICATION PREVIEW |■

The demand x for a product is given by

$$x = \frac{98}{\sqrt{2p + 1}} - 1$$

where p is the price per unit. To find how fast demand is changing when price is $24, we take the derivative of x with respect to p. If we write this function with a power rather than a radical, it has the form

$$x = 98(2p + 1)^{-1/2} - 1$$

The formulas learned so far cannot be used to find this derivative. We use a new formula, the Power Rule, to find this derivative. (See Example 6.) In this section we will discuss the Chain Rule and the Power Rule, which is one of the results of the Chain Rule, and we will use these formulas to solve applied problems.

Composite Functions

Recall from Section 1.2, "Functions," that if f and g are functions, then the composite functions g of f (denoted $g \circ f$) and f of g (denoted $f \circ g$) are defined as follows:

$$(g \circ f)(x) = g(f(x)) \quad \text{and} \quad (f \circ g)(x) = f(g(x))$$

EXAMPLE 1   **Composite Function**

If $f(x) = 3x^2$ and $g(x) = 2x - 1$, find $F(x) = f(g(x))$.

Solution

Substituting $g(x) = 2x - 1$ for x in $f(x)$ gives

$$f(g(x)) = f(2x - 1) = 3(2x - 1)^2 \quad \text{or} \quad F(x) = 3(2x - 1)^2$$

Chain Rule

We could find the derivative of the function $F(x) = 3(2x - 1)^2$ by expanding the expression $3(2x - 1)^2$. Then

$$F(x) = 3(4x^2 - 4x + 1) = 12x^2 - 12x + 3$$

so $F'(x) = 24x - 12$. But we can also use a very powerful rule, called the **Chain Rule,** to find derivatives of composite functions. If we write the composite function $y = f(g(x))$ in the form $y = f(u)$, where $u = g(x)$, we state the Chain Rule as follows.

Chain Rule

If f and g are differentiable functions with $y = f(u)$, and $u = g(x)$, then y is a differentiable function of x, and

$$\frac{dy}{dx} = f'(u) \cdot g'(x)$$

or, equivalently,

$$\frac{dy}{dx} = \frac{dy}{du} \cdot \frac{du}{dx}$$

Note that dy/du represents the derivative of $y = f(u)$ *with respect to u* and du/dx represents the derivative of $u = g(x)$ *with respect to x*. For example, if $y = 3(2x - 1)^2$, then the outside function, f, is the squaring function, and the inside function, g, is $2x - 1$, so we may write $y = f(u) = 3u^2$, where $u = g(x) = 2x - 1$. Then the derivative is

$$\frac{dy}{dx} = \frac{dy}{du} \cdot \frac{du}{dx} = 6u \cdot 2 = 12u$$

To write this derivative in terms of x, we substitute $2x - 1$ for u. Thus

$$\frac{dy}{dx} = 12(2x - 1) = 24x - 12$$

Note that we get the same result by using the Chain Rule as we did by expanding $f(x) = 3(2x - 1)^2$. The Chain Rule is important because it is not always possible to rewrite the function as a polynomial. Consider the following example.

EXAMPLE 2   **Chain Rule**

If $y = \sqrt{x^2 - 1}$, find $\dfrac{dy}{dx}$.

Solution

If we write this function as $y = f(u) = \sqrt{u} = u^{1/2}$, with $u = x^2 - 1$, we can find the derivative.

$$\frac{dy}{dx} = \frac{dy}{du} \cdot \frac{du}{dx} = \frac{1}{2} \cdot u^{-1/2} \cdot 2x = u^{-1/2} \cdot x = \frac{1}{\sqrt{u}} \cdot x = \frac{x}{\sqrt{u}}$$

To write this derivative in terms of x alone, we substitute $x^2 - 1$ for u. Then

$$\frac{dy}{dx} = \frac{x}{\sqrt{x^2 - 1}}$$

Note that we could not find the derivative of a function like that of Example 2 by the methods learned previously.

EXAMPLE 3   Allometric Relationships

The relationship between the length L (in meters) and weight W (in kilograms) of a species of fish in the Pacific Ocean is given by $W = 10.375L^3$. The rate of growth in length is given by $\frac{dL}{dt} = 0.36 - 0.18L$, where t is measured in years.

(a) Determine a formula for the rate of growth in weight $\frac{dW}{dt}$ in terms of L.

(b) If a fish weighs 30 kilograms, approximate its rate of growth in weight using the formula found in part (a).

Solution

(a) The rate of change uses the Chain Rule, as follows:

$$\frac{dW}{dt} = \frac{dW}{dL} \cdot \frac{dL}{dt} = 31.125L^2\,(0.36 - 0.18L) = 11.205L^2 - 5.6025L^3$$

(b) From $W = 10.375L^3$ and $W = 30$ kg, we can find L by solving

$$30 = 10.375L^3$$

$$\frac{30}{10.375} = L^3 \quad \text{so} \quad L = \sqrt[3]{\frac{30}{10.375}} \approx 1.4247 \text{ m}$$

Hence, the rate of growth in weight is

$$\frac{dW}{dt} = 11.205(1.4247)^2 - 5.6025(1.4247)^3 \approx 6.542 \text{ kilograms/year}$$

Power Rule   The Chain Rule is very useful and will be extremely important with functions that we will study later. A special case of the Chain Rule, called the **Power Rule,** is useful for the algebraic functions we have studied so far, composite functions where the outside function is a power.

Power Rule

If $y = u^n$, where u is a differentiable function of x, then

$$\frac{dy}{dx} = nu^{n-1} \cdot \frac{du}{dx}$$

EXAMPLE 4   Power Rule

(a) If $y = (x^2 - 4x)^6$, find $\dfrac{dy}{dx}$.   (b) If $p = \dfrac{4}{3q^2 + 1}$, find $\dfrac{dp}{dq}$.

Solution

(a) This has the form $y = u^n = u^6$, with $u = x^2 - 4x$. Thus, by the Power Rule,

$$\frac{dy}{dx} = nu^{n-1} \cdot \frac{du}{dx} = 6u^5\,(2x - 4)$$

Substituting $x^2 - 4x$ for u gives

$$\frac{dy}{dx} = 6(x^2 - 4x)^5 (2x - 4) = (12x - 24)(x^2 - 4x)^5$$

(b) We can use the Power Rule to find dp/dq if we write the equation in the form

$$p = 4(3q^2 + 1)^{-1}$$

Then

$$\frac{dp}{dq} = 4[-1(3q^2 + 1)^{-2} (6q)] = \frac{-24q}{(3q^2 + 1)^2}$$

The derivative of the function in Example 4(b) can also be found by using the Quotient Rule, but the Power Rule provides a more efficient method.

EXAMPLE 5   Power Rule with Radicals

Find the derivatives of (a) $y = 3\sqrt[3]{x^2 - 3x + 1}$ and (b) $g(x) = \dfrac{1}{\sqrt{(x^2 + 1)^3}}$.

Solution
(a) Because $y = 3(x^2 - 3x + 1)^{1/3}$, we can make use of the Power Rule with $u = x^2 - 3x + 1$.

$$y' = 3\left(nu^{n-1} \frac{du}{dx} \right) = 3\left[\frac{1}{3} u^{-2/3} (2x - 3) \right]$$

$$= (x^2 - 3x + 1)^{-2/3} (2x - 3) = \frac{2x - 3}{(x^2 - 3x + 1)^{2/3}}$$

(b) Writing $g(x)$ as a power gives $g(x) = (x^2 + 1)^{-3/2}$. Then

$$g'(x) = -\frac{3}{2} (x^2 + 1)^{-5/2}(2x) = -3x \cdot \frac{1}{(x^2 + 1)^{5/2}} = \frac{-3x}{\sqrt{(x^2 + 1)^5}}$$

✓ **CHECKPOINT**

1. (a) If $f(x) = (3x^4 + 1)^{10}$, does $f'(x) = 10(3x^4 + 1)^9$?
    (b) If $f(x) = (2x + 1)^5$, does $f'(x) = 10(2x + 1)^4$?
    (c) If $f(x) = \dfrac{[u(x)]^n}{c}$, where c is a constant, does $f'(x) = \dfrac{n[u(x)]^{n-1} \cdot u'(x)}{c}$?

2. (a) If $f(x) = \dfrac{12}{2x^2 - 1}$, find $f'(x)$ by using the Power Rule (not the Quotient Rule).

    (b) If $f(x) = \dfrac{\sqrt{x^3 - 1}}{3}$, find $f'(x)$ by using the Power Rule (not the Quotient Rule).

EXAMPLE 6   Demand | APPLICATION PREVIEW |

The demand for x hundred units of a product is given by

$$x = 98(2p + 1)^{-1/2} - 1$$

where p is the price per unit in dollars. Find the rate of change of the demand with respect to price when $p = 24$.

Solution
The rate of change of demand with respect to price is

$$\frac{dx}{dp} = 98\left[-\frac{1}{2} (2p + 1)^{-3/2}(2) \right] = -98(2p + 1)^{-3/2}$$

When $p = 24$, the rate of change is

$$\left.\frac{dx}{dp}\right|_{p=24} = -98(48 + 1)^{-3/2} = -98 \cdot \frac{1}{49^{3/2}} = -\frac{2}{7}$$

This means that when the price is $24, demand is changing at the rate of $-2/7$ hundred units per dollar, or if the price changes by $1, demand will change by about $-200/7$ units.

✓ CHECKPOINT ANSWERS	

1. (a) No, $f'(x) = 10(3x^4 + 1)^9 (12x^3)$.   (b) Yes   (c) Yes

2. (a) $f'(x) = \dfrac{-48x}{(2x^2 - 1)^2}$ (b) $f'(x) = \dfrac{x^2}{2\sqrt{x^3 - 1}}$

| EXERCISES | 9.6

In Problems 1–4, find $\dfrac{dy}{du}, \dfrac{du}{dx},$ and $\dfrac{dy}{dx}.$

1. $y = u^3$ and $u = x^2 + 1$
2. $y = u^4$ and $u = x^2 + 4x$
3. $y = u^4$ and $u = 4x^2 - x + 8$
4. $y = u^{10}$ and $u = x^2 + 5x$

Differentiate the functions in Problems 5–22.

5. $f(x) = (3x^5 - 2)^{20}$
6. $g(x) = (3 - 2x)^{10}$
7. $h(x) = \frac{3}{4}(x^5 - 2x^3 + 5)^8$
8. $k(x) = \frac{5}{7}(2x^3 - x + 6)^{14}$
9. $s(t) = 5t - 3(2t^4 + 7)^3$
10. $p(q) = 4(3q^2 - 1)^4 - 13q$
11. $g(x) = (x^4 - 5x)^{-2}$ 12. $p = (q^3 + 1)^{-5}$
13. $f(s) = \dfrac{3}{(2s^5 + 1)^4}$ 14. $g(t) = \dfrac{1}{4t^3 + 1}$
15. $g(x) = \dfrac{1}{(2x^3 + 3x + 5)^{3/4}}$
16. $y = \dfrac{1}{(3x^3 + 4x + 1)^{3/2}}$
17. $y = \sqrt{3x^2 + 4x + 9}$ 18. $y = \sqrt{x^2 + 3x}$
19. $y = \dfrac{11(x^3 - 7)^6}{9}$ 20. $y = \dfrac{5\sqrt{1 - x^3}}{6}$
21. $z = \dfrac{(3w + 1)^5 - 3w}{7}$
22. $y = \dfrac{\sqrt{2x - 1} - \sqrt{x}}{2}$

At the indicated point, for each function in Problems 23–26, find
(a) the slope of the tangent line, and
(b) the instantaneous rate of change of the function.
You may use the numerical derivative feature on a graphing calculator to check your work.
23. $y = (x^3 + 2x)^4$ at $x = 2$
24. $y = \sqrt{5x^2 + 2x}$ at $x = 1$
25. $y = \sqrt{x^3 + 1}$ at $(2, 3)$
26. $y = (4x^3 - 5x + 1)^3$ at $(1, 0)$

In Problems 27–30, write the equation of the line tangent to the graph of each function at the indicated point. As a check, graph both the function and the tangent line you found to see whether it looks correct.
27. $y = (x^2 - 3x + 3)^3$ at $(2, 1)$
28. $y = (x^2 + 1)^3$ at $(2, 125)$
29. $y = \sqrt{3x^2 - 2}$ at $x = 3$
30. $y = \dfrac{1}{(x^3 - x)^3}$ at $x = 2$

 In Problems 31 and 32, complete the following for each function.
(a) Find $f'(x)$.
(b) Check your result in part (a) by graphing both it and the numerical derivative of the function.
(c) Find x-values for which the slope of the tangent is 0.
(d) Find points (x, y) where the slope of the tangent is 0.
(e) Use a graphing utility to graph the function and locate the points found in part (d).
31. $f(x) = (x^2 - 4)^3 + 12$
32. $f(x) = 10 - (x^2 - 2x - 8)^2$

 In Problems 33 and 34, do the following for each function $f(x)$.
(a) Find $f'(x)$.
(b) Graph both $f(x)$ and $f'(x)$ with a graphing utility.
(c) Determine x-values where $f'(x) = 0, f'(x) > 0,$ $f'(x) < 0.$
(d) Determine x-values for which $f(x)$ has a maximum or minimum point, where the graph is increasing, and where it is decreasing.
33. $f(x) = 12 - 3(1 - x^2)^{4/3}$
34. $f(x) = 3 + \dfrac{1}{16}(x^2 - 4x)^4$

In Problems 35 and 36, find the derivative of each function.
35. (a) $y = \dfrac{2x^3}{3}$ (b) $y = \dfrac{2}{3x^3}$
    (c) $y = \dfrac{(2x)^3}{3}$ (d) $y = \dfrac{2}{(3x)^3}$

36. (a) $y = \dfrac{3}{(5x)^5}$ (b) $y = \dfrac{3x^5}{5}$

    (c) $y = \dfrac{3}{5x^5}$ (d) $y = \dfrac{(3x)^5}{5}$

APPLICATIONS

37. **Ballistics** Ballistics experts are able to identify the weapon that fired a certain bullet by studying the markings on the bullet. Tests are conducted by firing into a bale of paper. If the distance s, in inches, that the bullet travels into the paper is given by

$$s = 27 - (3 - 10t)^3$$

for $0 \le t \le 0.3$ second, find the velocity of the bullet one-tenth of a second after it hits the paper.

38. **Population of microorganisms** Suppose that the population of a certain microorganism at time t (in minutes) is given by

$$P = 1000 - 1000(t + 10)^{-1}$$

Find the rate of change of population.

39. **Revenue** The revenue from the sale of a product is

$$R = 1500x + 3000(2x + 3)^{-1} - 1000 \text{ dollars}$$

where x is the number of units sold. Find the marginal revenue when 100 units are sold. Interpret your result.

40. **Revenue** The revenue from the sale of x units of a product is

$$R = 15(3x + 1)^{-1} + 50x - 15 \text{ dollars}$$

Find the marginal revenue when 40 units are sold. Interpret your result.

41. **Pricing and sales** Suppose that the weekly sales volume y (in thousands of units sold) depends on the price per unit (in dollars) of the product according to

$$y = 32(3p + 1)^{-2/5}, \quad p > 0$$

(a) What is the rate of change in sales volume when the price is $21?
(b) Interpret your answer to part (a).

42. **Pricing and sales** A chain of auto service stations has found that its monthly sales volume S (in thousands of dollars) is related to the price p (in dollars) of an oil change according to

$$S = \dfrac{98}{\sqrt{p + 5}}, \quad p > 10$$

(a) What is the rate of change of sales volume when the price is $44?
(b) Interpret your answer to part (a).

43. **Demand** Suppose that the demand for q units of a product priced at p per unit is described by

$$p = \dfrac{200,000}{(q + 1)^2}$$

(a) What is the rate of change of price with respect to the quantity demanded when $q = 49$?
(b) Interpret your answer to part (a).

Stimulus-response **The relation between the magnitude of a sensation y and the magnitude of the stimulus x is given by**

$$y = k(x - x_0)^n$$

where k is a constant, x_0 is the threshold of effective stimulus, and n depends on the type of stimulus. Find the rate of change of sensation with respect to the amount of stimulus for each of Problems 44–46.

44. For the stimulus of visual brightness $y = k(x - x_0)^{1/3}$
45. For the stimulus of warmth $y = k(x - x_0)^{8/5}$
46. For the stimulus of electrical stimulation $y = k(x - x_0)^{7/2}$
47. **Demand** If the demand for q units of a product priced at p per unit is described by the equation

$$p = \dfrac{100}{\sqrt{2q + 1}}$$

find the rate of change of p with respect to q.

48. **Advertising and sales** The daily sales S (in thousands of dollars) attributed to an advertising campaign are given by

$$S = 1 + \dfrac{3}{t + 3} - \dfrac{18}{(t + 3)^2}$$

where t is the number of weeks the campaign runs. What is the rate of change of sales at
(a) $t = 8$?   (b) $t = 10$?
(c) Should the campaign be continued after the 10th week? Explain.

49. **Body-heat loss** The description of body-heat loss due to convection involves a coefficient of convection, K_c, which depends on wind velocity according to the following equation.

$$K_c = 4\sqrt{4v + 1}$$

Find the rate of change of the coefficient with respect to the wind velocity.

50. **Data entry speed** The data entry speed (in entries per minute) of a data clerk trainee is

$$S = 10\sqrt{0.8x + 4}, \quad 0 \le x \le 100$$

where x is the number of hours of training he has had. What is the rate at which his speed is changing and what does this rate mean when he has had
(a) 15 hours of training?
(b) 40 hours of training?

51. **Investments** If an IRA is a variable-rate investment for 20 years at rate r percent per year, compounded monthly, then the future value S that accumulates from an initial investment of $1000 is

$$S = 1000\left[1 + \dfrac{0.01r}{12}\right]^{240}$$

What is the rate of change of S with respect to r and what does it tell us if the interest rate is (a) 6%? (b) 12%?

52. *Concentration of body substances* The concentration C of a substance in the body depends on the quantity of the substance Q and the volume V through which it is distributed. For a static substance, this is given by

$$C = \frac{Q}{V}$$

For a situation like that in the kidneys, where the fluids are moving, the concentration is the ratio of the rate of change of quantity with respect to time and the rate of change of volume with respect to time.
   (a) Formulate the equation for concentration of a moving substance.
   (b) Show that this is equal to the rate of change of quantity with respect to volume.

53. *National health expenditures* The table shows the total national expenditures for health (both actual and projected, in billions of dollars) for the years from 2001 to 2018. (These data include expenditures for medical research and medical facilities construction.)

Year	Amount	Year	Amount
2001	$1469	2010	$2624
2002	1602	2011	2770
2003	1735	2012	2931
2004	1855	2013	3111
2005	1981	2014	3313
2006	2113	2015	3541
2007	2241	2016	3790
2008	2379	2017	4062
2009	2509	2018	4353

Source: U.S. Centers for Medicare and Medicaid Services

Assume these data can be modeled with the function

$$A(t) = 445(0.1t + 1)^3 - 2120(0.1t + 1)^2$$
$$+ 4570(0.1t + 1) - 1600$$

where $A(t)$ is in billions of dollars and t is the number of years past 2000.
   (a) Use this model to determine and interpret the instantaneous rates of change of the total national health expenditures in 2008 and 2015.
   (b) Use the data to find an average rate of change that approximates the 2015 instantaneous rate.

54. *Energy use* Energy use per dollar of GDP indexed to 1980 means that energy use for any year is viewed as a percent of the use per dollar of GDP in 1980. The following data show the energy use per dollar of GDP, as a percent, for selected years from 1985 and projected to 2035.

Energy Use per Dollar of GDP

Year	Percent	Year	Percent
1985	83	2015	51
1990	79	2020	45
1995	75	2025	41
2000	67	2030	37
2005	60	2035	34
2010	56		

Source: U.S. Department of Energy

These data can be modeled with the function

$$E(t) = 0.0039(0.4t + 2)^3 - 0.13(0.4t + 2)^2$$
$$- 1.4(0.4t + 2) + 91$$

where $E(t)$ is the energy use per dollar of GDP (indexed to 1980) and t is the number of years past 1980.
   (a) Use this model to find and interpret the instantaneous rates of change of energy use per dollar of GDP in 2000 and 2025.
   (b) Use the data in the table to find an average rate of change that approximates the 2025 instantaneous rate.

55. *Gross domestic product* The table shows U.S. gross domestic product (GDP) in billions of dollars for selected years from 2000 to 2070 (actual and projected).

Year	GDP	Year	GDP
2000	9143	2040	79,680
2005	12,145	2045	103,444
2010	16,174	2050	133,925
2015	21,270	2055	173,175
2020	27,683	2060	224,044
2025	35,919	2065	290,042
2030	46,765	2070	375,219
2035	61,100		

Source: Social Security Administration Trustees Report

Assume the GDP can be modeled with the function

$$G(t) = 212.9(0.2t + 5)^3 - 5016(0.2t + 5)^2$$
$$+ 8810.4t + 104,072$$

where $G(t)$ is in billions of dollars and t is the number of years past 2000.
   (a) Use the model to find and interpret the instantaneous rates of change of the GDP in 2005 and 2015.
   (b) Use the data in the table to find the average rate of change of the GDP from 2005 to 2015.
   (c) How well does your answer from part (b) approximate the instantaneous rate of change of GDP in 2010?

56. *Diabetes* The figure shows the percent of the U.S. population with diabetes (diagnosed and undiagnosed) for selected years from 2010 and projections to 2050. Assume this percent can be modeled by

$$y = 6.97(0.5x + 5)^{0.495}$$

where y is the percent and x is the number of years past 2010.

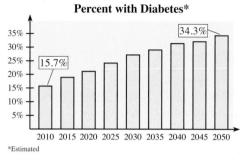

Percent with Diabetes*

35%

34.3%

30%

25%   15.7%

20%

15%

10%

5%

2010 2015 2020 2025 2030 2035 2040 2045 2050

*Estimated

Source: Centers for Disease Control and Prevention

(a) Use the figure to find the average rate of change from 2010 to 2050.

(b) Use the model to find the instantaneous rate of change of the U.S. population with diabetes in 2020. Is the average rate found in part (a) a good approximation of the 2020 instantaneous rate?

(c) Use the model to find and interpret the instantaneous rate of change of the percent of the U.S. population with diabetes in 2050.

OBJECTIVE

9.7

- To use derivative formulas separately and in combination with each other

Using Derivative Formulas

▌| APPLICATION PREVIEW |▌

Suppose the weekly revenue function for a product is given by

$$R(x) = \frac{36{,}000{,}000x}{(2x + 500)^2}$$

where $R(x)$ is the dollars of revenue from the sale of x units. We can find marginal revenue by finding the derivative of the revenue function. This revenue function contains both a quotient and a power, so its derivative is found by using both the Quotient Rule and the Power Rule, but in which order are these rules applied? (See Example 4.) In this section we consider functions whose derivatives require more than one derivative formula and discuss how we decide the order in which to apply the formulas.

We have used the Power Rule to find the derivatives of functions like

$$y = (x^3 - 3x^2 + x + 1)^5$$

but we have not found the derivatives of functions like

$$y = [(x^2 + 1)(x^3 + x + 1)]^5$$

This function is different because the function u (which is raised to the fifth power) is the product of two functions, $(x^2 + 1)$ and $(x^3 + x + 1)$. The equation is of the form $y = u^5$, where $u = (x^2 + 1)(x^3 + x + 1)$. This means that the Product Rule should be used to find du/dx. Then

$$\frac{dy}{dx} = 5u^4 \cdot \frac{du}{dx}$$

$$= 5[(x^2 + 1)(x^3 + x + 1)]^4 \left[(x^2 + 1)(3x^2 + 1) + (x^3 + x + 1)(2x)\right]$$

$$= 5[(x^2 + 1)(x^3 + x + 1)]^4 (5x^4 + 6x^2 + 2x + 1)$$

$$= (25x^4 + 30x^2 + 10x + 5)[(x^2 + 1)(x^3 + x + 1)]^4$$

A different type of problem involving the Power Rule and the Product Rule is finding the derivative of $y = (x^2 + 1)^5 (x^3 + x + 1)$. We may think of y as the *product* of two functions, one of which is a power. Thus the fundamental formula we should use is the Product

Rule. The two function8s are $u(x) = (x^2 + 1)^5$ and $v(x) = x^3 + x + 1$. The Product Rule gives

$$\frac{dy}{dx} = u(x) \cdot v'(x) + v(x) \cdot u'(x)$$

$$= (x^2 + 1)^5 (3x^2 + 1) + (x^3 + x + 1)[5(x^2 + 1)^4(2x)]$$

Note that the Power Rule was used to find $u'(x)$, since $u(x) = (x^2 + 1)^5$.
We can simplify dy/dx by factoring $(x^2 + 1)^4$ from both terms:

$$\frac{dy}{dx} = (x^2 + 1)^4[(x^2 + 1)(3x^2 + 1) + (x^3 + x + 1) \cdot 5 \cdot 2x]$$

$$= (x^2 + 1)^4(13x^4 + 14x^2 + 10x + 1)$$

EXAMPLE 1   Power of a Quotient

If $y = \left(\dfrac{x^2}{x - 1}\right)^5$, find y'.

Solution
We again have an equation of the form $y = u^n$, but this time u is a quotient. Thus we will need the Quotient Rule to find du/dx.

$$y' = nu^{n-1} \cdot \frac{du}{dx} = 5u^4 \frac{(x - 1) \cdot 2x - x^2 \cdot 1}{(x - 1)^2}$$

Substituting for u and simplifying give

$$y' = 5\left(\frac{x^2}{x - 1}\right)^4 \cdot \frac{2x^2 - 2x - x^2}{(x - 1)^2} = \frac{5x^8(x^2 - 2x)}{(x - 1)^6} = \frac{5x^{10} - 10x^9}{(x - 1)^6}$$

EXAMPLE 2   Quotient of Two Powers

Find $f'(x)$ if $f(x) = \dfrac{(x - 1)^2}{(x^4 + 3)^3}$.

Solution
This function is the quotient of two functions, $(x - 1)^2$ and $(x^4 + 3)^3$, so we must use the Quotient Rule to find the derivative of $f(x)$, but taking the derivatives of $(x - 1)^2$ and $(x^4 + 3)^3$ will require the Power Rule.

$$f'(x) = \frac{[v(x) \cdot u'(x) - u(x) \cdot v'(x)]}{[v(x)]^2}$$

$$= \frac{(x^4 + 3)^3[2(x - 1)(1)] - (x - 1)^2[3(x^4 + 3)^2 \, 4x^3]}{[(x^4 + 3)^3]^2}$$

$$= \frac{2(x^4 + 3)^3(x - 1) - 12x^3(x - 1)^2(x^4 + 3)^2}{(x^4 + 3)^6}$$

We see that 2, $(x^4 + 3)^2$, and $(x - 1)$ are all factors in both terms of the numerator, so we can factor them from both terms and reduce the fraction.

$$f'(x) = \frac{2(x^4 + 3)^2(x - 1)[(x^4 + 3) - 6x^3(x - 1)]}{(x^4 + 3)^6}$$

$$= \frac{2(x - 1)(-5x^4 + 6x^3 + 3)}{(x^4 + 3)^4}$$

EXAMPLE 3   **Product with a Power**

Find $f'(x)$ if $f(x) = (x^2 - 1)\sqrt{3 - x^2}$.

Solution

The function is the product of two functions, $x^2 - 1$ and $\sqrt{3 - x^2}$. Therefore, we will use the Product Rule to find the derivative of $f(x)$, but the derivative of $\sqrt{3 - x^2} = (3 - x^2)^{1/2}$ will require the Power Rule.

$$f'(x) = u(x) \cdot v'(x) + v(x) \cdot u'(x)$$

$$= (x^2 - 1)\left[\frac{1}{2}(3 - x^2)^{-1/2}(-2x)\right] + (3 - x^2)^{1/2}(2x)$$

$$= (x^2 - 1)[-x(3 - x^2)^{-1/2}] + (3 - x^2)^{1/2}(2x)$$

$$= \frac{-x^3 + x}{(3 - x^2)^{1/2}} + 2x(3 - x^2)^{1/2}$$

We can combine these terms over the common denominator $(3 - x^2)^{1/2}$ as follows:

$$f'(x) = \frac{-x^3 + x}{(3 - x^2)^{1/2}} + \frac{2x(3 - x^2)^1}{(3 - x^2)^{1/2}} = \frac{-x^3 + x + 6x - 2x^3}{(3 - x^2)^{1/2}} = \frac{-3x^3 + 7x}{(3 - x^2)^{1/2}}$$

We should note that in Example 3 we could have written $f'(x)$ in the form

$$f'(x) = (-x^3 + x)(3 - x^2)^{-1/2} + 2x(3 - x^2)^{1/2}$$

Now the factor $(3 - x^2)$, to different powers, is contained in both terms of the expression. Thus we can factor $(3 - x^2)^{-1/2}$ from both terms. (We choose the $-1/2$ power because it is the smaller of the two powers.) Dividing $(3 - x^2)^{-1/2}$ into the first term gives $(-x^3 + x)$, and dividing it into the second term gives $2x(3 - x^2)^1$. (Why?) Thus we have

$$f'(x) = (3 - x^2)^{-1/2}[(-x^3 + x) + 2x(3 - x^2)] = \frac{-3x^3 + 7x}{(3 - x^2)^{1/2}}$$

which agrees with our previous answer.

✓ CHECKPOINT
1. If a function has the form $y = [u(x)]^n \cdot v(x)$, where n is a constant, we begin to find the derivative by using the _____ Rule and then use the _____ Rule to find the derivative of $[u(x)]^n$.
2. If a function has the form $y = [u(x)/v(x)]^n$, where n is a constant, we begin to find the derivative by using the _____ Rule and then use the _____ Rule.
3. Find the derivative of each of the following and simplify.

   (a) $f(x) = 3x^4(2x^4 + 7)^5$ (b) $g(x) = \dfrac{(4x + 3)^7}{2x - 9}$

EXAMPLE 4   **Revenue | APPLICATION PREVIEW |**

Suppose that the weekly revenue function for a product is given by

$$R(x) = \frac{36{,}000{,}000x}{(2x + 500)^2}$$

where $R(x)$ is the dollars of revenue from the sale of x units.

(a) Find the marginal revenue function.
(b) Find the marginal revenue when 50 units are sold.

Solution

(a) $\overline{MR} = R'(x)$

$$= \frac{(2x + 500)^2(36{,}000{,}000) - 36{,}000{,}000x[2(2x + 500)^1(2)]}{(2x + 500)^4}$$

$$= \frac{36{,}000{,}000(2x + 500)(2x + 500 - 4x)}{(2x + 500)^4} = \frac{36{,}000{,}000(500 - 2x)}{(2x + 500)^3}$$

(b) $\overline{MR}\,(50) = R'(50) = \dfrac{36{,}000{,}000(500 - 100)}{(100 + 500)^3} = \dfrac{36{,}000{,}000(400)}{(600)^3} = \dfrac{200}{3} \approx 66.67$

The marginal revenue is $66.67 when 50 units are sold. That is, the predicted revenue from the sale of the 51st unit is approximately $66.67.

It may be helpful to review the formulas needed to find the derivatives of various types of functions. Table 9.5 presents examples of different types of functions and the formulas needed to find their derivatives.

▌▌| TABLE 9.5 |▌▌

SUMMARY OF DERIVATIVE FORMULAS

Examples	Formulas
$f(x) = 14$	If $f(x) = c$, then $f'(x) = 0$.
$y = x^4$	If $f(x) = x^n$, then $f'(x) = nx^{n-1}$.
$g(x) = 5x^3$	If $g(x) = cf(x)$, then $g'(x) = cf'(x)$.
$y = 3x^2 + 4x$	If $f(x) = u(x) + v(x)$, then $f'(x) = u'(x) + v'(x)$.
$y = (x^2 - 2)(x + 4)$	If $f(x) = u(x) \cdot v(x)$, then $f'(x) = u(x) \cdot v'(x) + v(x) \cdot u'(x)$.
$f(x) = \dfrac{x^3}{x^2 + 1}$	If $f(x) = \dfrac{u(x)}{v(x)}$ then $f'(x) = \dfrac{v(x) \cdot u'(x) - u(x) \cdot v'(x)}{[v(x)]^2}$.
$y = (x^3 - 4x)^{10}$	If $y = u^n$ and $u = g(x)$, then $\dfrac{dy}{dx} = nu^{n-1} \cdot \dfrac{du}{dx}$.
$y = \left(\dfrac{x - 1}{x^2 + 3}\right)^3$	Power Rule, then Quotient Rule to find $\dfrac{du}{dx}$, where $u = \dfrac{x - 1}{x^2 + 3}$.
$y = (x + 1)\sqrt{x^3 + 1}$	Product Rule, then Power Rule to find $v'(x)$, where $v(x) = \sqrt{x^3 + 1}$.
$y = \dfrac{(x^2 - 3)^4}{x + 1}$	Quotient Rule, then Power Rule to find the derivative of the numerator.

✓ **CHECKPOINT ANSWERS**

1. Product, Power
2. Power, Quotient
3. (a) $f'(x) = 12x^3(12x^4 + 7)(2x^4 + 7)^4$

    (b) $g'(x) = \dfrac{2(24x - 129)(4x + 3)^6}{(2x - 9)^2}$

| EXERCISES | 9.7

Find the derivatives of the functions in Problems 1–32. Simplify and express the answer using positive exponents only.

1. $f(x) = \pi^4$

2. $f(x) = \dfrac{1}{4}$

3. $g(x) = \dfrac{4}{x^4}$

4. $y = \dfrac{x^4}{4}$

5. $g(x) = 5x^3 + \dfrac{4}{x}$

6. $y = 3x^2 + 4\sqrt{x}$

7. $y = (x^2 - 2)(x + 4)$

8. $y = (x^3 - 5x^2 + 1)(x^3 - 3)$

9. $f(u) = \dfrac{u^3 + 1}{u^2}$

10. $C(w) = \dfrac{1 + w^2 - w^4}{1 + w^4}$

11. $y = \dfrac{(x^3 - 4x)^{10}}{10}$

12. $y = \dfrac{5}{2}\,(3x^4 - 6x^2 + 2)^5$

13. $y = \dfrac{5}{3}\,x^3(4x^5 - 5)^3$

14. $y = 3x^4(2x^5 + 1)^7$

15. $y = (x - 1)^2(x^2 + 1)$

16. $f(x) = (5x^3 + 1)(x^4 + 5x)^2$

17. $y = \dfrac{(x^2 - 4)^3}{x^2 + 1}$

18. $y = \dfrac{(x^2 - 3)^4}{x}$

19. $p = [(q + 1)(q^3 - 3)]^3$

20. $s = [(4 - t^2)(t^2 + 5t)]^4$

21. $R(x) = [x^2(x^2 + 3x)]^4$

22. $c(x) = [x^3(x^2 + 1)]^{-3}$

23. $y = \left(\dfrac{2x - 1}{x^2 + x}\right)^4$

24. $y = \left(\dfrac{5 - x^2}{x^4}\right)^3$

25. $g(x) = (8x^4 + 3)^2(x^3 - 4x)^3$

26. $y = (3x^3 - 4x)^3(4x^2 - 8)^2$

27. $f(x) = \dfrac{\sqrt[3]{x^2 + 5}}{4 - x^2}$

28. $g(x) = \dfrac{\sqrt[3]{2x - 1}}{2x + 1}$

29. $y = x^2\sqrt[3]{4x - 3}$

30. $y = 3x\sqrt[3]{4x^4 + 3}$

31. $c(x) = 2x\sqrt{x^3 + 1}$

32. $R(x) = x\sqrt[3]{3x^3 + 2}$

In Problems 33 and 34, find the derivative of each function.

33. (a) $F_1(x) = \dfrac{3(x^4 + 1)^5}{5}$

(b) $F_2(x) = \dfrac{3}{5(x^4 + 1)^5}$

(c) $F_3(x) = \dfrac{(3x^4 + 1)^5}{5}$

(d) $F_4(x) = \dfrac{3}{(5x^4 + 1)^5}$

34. (a) $G_1(x) = \dfrac{2(x^3 - 5)^3}{3}$

(b) $G_2(x) = \dfrac{(2x^3 - 5)^3}{3}$

(c) $G_3(x) = \dfrac{2}{3(x^3 - 5)^3}$

(d) $G_4(x) = \dfrac{2}{(3x^3 - 5)^3}$

APPLICATIONS

35. *Physical output* The total physical output P of workers is a function of the number of workers, x. The function $P = f(x)$ is called the physical productivity function. Suppose that the physical productivity of x construction workers is given by

$$P = 10(3x + 1)^3 - 10$$

Find the marginal physical productivity, dP/dx.

36. *Revenue* Suppose that the revenue function for a certain product is given by

$$R(x) = 15(2x + 1)^{-1} + 30x - 15$$

where x is in thousands of units and R is in thousands of dollars.
(a) Find the marginal revenue when 2000 units are sold.
(b) How is revenue changing when 2000 units are sold?

37. *Revenue* Suppose that the revenue in dollars from the sale of x campers is given by

$$R(x) = 60{,}000x + 40{,}000(10 + x)^{-1} - 4000$$

(a) Find the marginal revenue when 10 units are sold.
(b) How is revenue changing when 10 units are sold?

38. *Production* Suppose that the production of x items of a new line of products is given by

$$x = 200[(t + 10) - 400(t + 40)^{-1}]$$

where t is the number of weeks the line has been in production. Find the rate of production, dx/dt.

39. *National consumption* If the national consumption function is given by

$$C(y) = 2(y + 1)^{1/2} + 0.4y + 4$$

find the marginal propensity to consume, dC/dy.

40. *Demand* Suppose that the demand function for q units of an appliance priced at $\$p$ per unit is given by

$$p = \dfrac{400(q + 1)}{(q + 2)^2}$$

Find the rate of change of price with respect to the number of appliances.

41. *Volume* When squares of side x inches are cut from the corners of a 12-inch-square piece of cardboard, an open-top box can be formed by folding up the sides. The volume of this box is given by

$$V = x(12 - 2x)^2$$

Find the rate of change of volume with respect to the size of the squares.

42. *Advertising and sales* Suppose that sales (in thousands of dollars) are directly related to an advertising campaign according to

$$S = 1 + \dfrac{3t - 9}{(t + 3)^2}$$

where t is the number of weeks of the campaign.
(a) Find the rate of change of sales after 3 weeks.
(b) Interpret the result in part (a).

43. *Advertising and sales* An inferior product with an extensive advertising campaign does well when it is released, but sales decline as people discontinue use of the product. If the sales S (in thousands of dollars) after t weeks are given by

$$S(t) = \dfrac{200t}{(t + 1)^2}, \quad t \geq 0$$

what is the rate of change of sales when $t = 9$? Interpret your result.

44. *Advertising and sales* An excellent film with a very small advertising budget must depend largely on word-of-mouth advertising. If attendance at the film after t weeks is given by

$$A = \frac{100t}{(t + 10)^2}$$

what is the rate of change in attendance and what does it mean when (a) $t = 10$? (b) $t = 20$?

45. *Per capita expenditures for U.S. health care* The dollars spent per person per year for health care (projected to 2018) are shown in the table. These data can be modeled by

$$y = \frac{4.38(x - 10)^2 + 78(x - 10) + 1430}{0.0029x + 0.25}$$

where x is the number of years past 1990 and y is the per capita expenditures for health care.

(a) Find the instantaneous rate of change of per capita health care expenditures in 2005 and 2015.

(b) Interpret the rate of change for 2015 found in part (a).

(c) Use the data to find the average rate of change of per capita health care expenditures from 2004 to 2006. How well does this approximate the instantaneous rate of change in 2005?

Year	$ per Person	Year	$ per Person
2000	4789	2010	8465
2002	5563	2012	9275
2004	6331	2014	10,289
2006	7091	2016	11,520
2008	7826	2018	12,994

Source: U.S. Medicare and Medicaid Services

OBJECTIVE   9.8

- To find second derivatives and higher-order derivatives of certain functions

Higher-Order Derivatives

| APPLICATION PREVIEW |

Since cell phones were introduced, their popularity has increased enormously. Figure 9.32(a) shows a graph of the billions of worldwide cellular subscribers (actual and projected) as a function of the number of years past 1990 (*Source:* International Telecommunications Union and Key Global Telecom Indicators). Note that the number of subscribers is always increasing and that the rate of change of that number (as seen from tangent lines to the graph) is always positive. However, the tangent lines shown in Figure 9.32(b) indicate that the rate of change of the number of subscribers is greater at B than at either A or C.

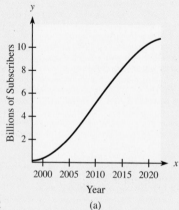

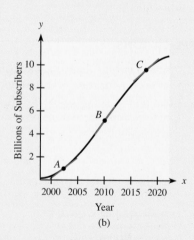

Figure 9.32   (a)   (b)

Furthermore, the rate of change of the number of subscribers (the slopes of tangents) increases from A to B and then decreases from B to C. To learn how the rate of change of the number of subscribers is changing, we are interested in finding the derivative of the rate of change of the number of subscribers—that is, the derivative of the derivative of the number of subscribers. (See Example 4.) This is called the second derivative. In this section we will discuss second and higher-order derivatives.

Second Derivatives Because the derivative of a function is itself a function, we can take a derivative of the derivative. The derivative of a first derivative is called a **second derivative.** We can find the second derivative of a function f by differentiating it twice. If f' represents the first derivative of a function, then f'' represents the second derivative of that function.

EXAMPLE 1 Second Derivative

(a) Find the second derivative of $y = x^4 - 3x^2 + x^{-2}$.

(b) If $f(x) = 3x^3 - 4x^2 + 5$, find $f''(x)$.

Solution

(a) The first derivative is $y' = 4x^3 - 6x - 2x^{-3}$.

   The second derivative, which we may denote by y'', is

$$y'' = 12x^2 - 6 + 6x^{-4}$$

(b) The first derivative is $f'(x) = 9x^2 - 8x$.

   The second derivative is $f''(x) = 18x - 8$.

It is also common to use $\dfrac{d^2y}{dx^2}$ and $\dfrac{d^2f(x)}{dx^2}$ to denote the second derivative of a function.

EXAMPLE 2 Second Derivative

If $y = \sqrt{2x - 1}$, find d^2y/dx^2.

Solution

The first derivative is

$$\frac{dy}{dx} = \frac{1}{2}(2x - 1)^{-1/2}(2) = (2x - 1)^{-1/2}$$

The second derivative is

$$\frac{d^2y}{dx^2} = -\frac{1}{2}(2x - 1)^{-3/2}(2) = -(2x - 1)^{-3/2}$$

$$= \frac{-1}{(2x - 1)^{3/2}} = \frac{-1}{\sqrt{(2x - 1)^3}}$$

Higher-Order Derivatives We can also find third, fourth, fifth, and higher derivatives, continuing indefinitely. The third, fourth, and fifth derivatives of a function f are denoted by f''', $f^{(4)}$, and $f^{(5)}$, respectively. Other notations for the third and fourth derivatives include

$$y''' = \frac{d^3y}{dx^3} = \frac{d^3f(x)}{dx^3}, \qquad y^{(4)} = \frac{d^4y}{dx^4} = \frac{d^4f(x)}{dx^4}$$

EXAMPLE 3 Higher-Order Derivatives

Find the first four derivatives of $f(x) = 4x^3 + 5x^2 + 3$.

Solution

$$f'(x) = 12x^2 + 10x, \qquad f''(x) = 24x + 10, \qquad f'''(x) = 24, \qquad f^{(4)}(x) = 0$$

Just as the first derivative, $f'(x)$, can be used to determine the rate of change of a function $f(x)$, the second derivative, $f''(x)$, can be used to determine the rate of change of $f'(x)$.

EXAMPLE   4   Worldwide Cellular Subscriberships | APPLICATION PREVIEW |

By using International Telecommunications Union and Key Global Telecom Indicators data from 1990 and projected to 2020, the billions of world cell phone subscriberships can be modeled by the function

$$C(t) = -0.000728t^3 + 0.0414t^2 - 0.296t + 0.340$$

where t is the number of years past 1990.

(a) Find the instantaneous rate of change of world cell phone subscriberships.
(b) Find the instantaneous rate of change of the function found in part (a).
(c) For what t-value does the second derivative of $C(t)$ equal zero?
(d) Find the instantaneous rate of change of the world cell phone subscriberships in 2005 and in 2015.
(e) Use the second derivative to approximate how fast the instantaneous rate of change of world cell phone subscriberships is changing in 2005 and in 2015.
(f) Explain the meaning of $C'(25)$ and $C''(25)$.

Solution

(a) The instantaneous rate of change of $C(t)$ is

$$C'(t) = -0.002184t^2 + 0.0828t - 0.296$$

(b) The instantaneous rate of change of $C'(t)$ is

$$C''(t) = -0.004368t + 0.0828$$

(c) To find the t-value when $C''(t) = -0.004368t + 0.0828 = 0$, we solve as follows.

$$0.0828 = 0.004368t \Rightarrow t = \frac{0.0828}{0.004368} \approx 18.96$$

Thus $C''(t) = 0$ in 2009.
(d) We find these rates of change by evaluating the derivative
In 2005: $C'(15) = -0.002184(15)^2 + 0.0828(15) - 0.296 \approx 0.455$ billion per year
In 2015: $C'(25) = -0.002184(25)^2 + 0.0828(25) - 0.296 \approx 0.409$ billion per year
(e) In 2005: $C''(15) = -0.004368(15) + 0.0828 \approx 0.0173$ billion per year per year
In 2015: $C''(25) = -0.004368(25) + 0.0828 \approx -0.0264$ billion per year per year
(f) In 2015 the number of world cell phone subscriberships increased at a rate of 0.409 billion per year, but the rate of change decreased at a rate of 0.0264 billion per year per year. Thus, in 2015 the number of world cell phone subscriberships was increasing but at a decreasing (slower) rate. See Figure 9.32.   ▇

EXAMPLE   5   Rate of Change of a Derivative

Let $f(x) = 3x^4 + 6x^3 - 3x^2 + 4$.

(a) How fast is $f(x)$ changing at $(1, 10)$?
(b) How fast is $f'(x)$ changing at $(1, 10)$?
(c) Is $f'(x)$ increasing or decreasing at $(1, 10)$?

Solution

(a) Because $f'(x) = 12x^3 + 18x^2 - 6x$, we have

$$f'(1) = 12 + 18 - 6 = 24$$

Thus the rate of change of $f(x)$ at $(1, 10)$ is 24 (y units per x unit).
(b) Because $f''(x) = 36x^2 + 36x - 6$, we have

$$f''(1) = 66$$

Thus the rate of change of $f'(x)$ at $(1, 10)$ is 66 (y units per x unit per x unit).
(c) Because $f''(1) = 66 > 0$, $f'(x)$ is increasing at $(1, 10)$.   ▇

EXAMPLE 6   **Acceleration**

Suppose that a particle travels according to the equation

$$s = 100t - 16t^2 + 200$$

where s is the distance in feet and t is the time in seconds. Then ds/dt is the velocity, and $d^2s/dt^2 = dv/dt$ is the acceleration of the particle. Find the acceleration.

Solution
The velocity is $v = ds/dt = 100 - 32t$ feet per second, and the acceleration is

$$\frac{dv}{dt} = \frac{d^2s}{dt^2} = -32 \text{ (feet/second)/second} = -32 \text{ feet/second}^2$$

✓ **CHECKPOINT**   Suppose that the distance a particle travels is given by

$$s = 4x^3 - 12x^2 + 6$$

where s is in feet and x is in seconds.
  1. Find the function that describes the velocity of this particle.
  2. Find the function that describes the acceleration of this particle.
  3. Is the acceleration always positive?
  4. When does the *velocity* of this particle increase?

Calculator Note   We can use the numerical derivative feature of a graphing calculator to find the second derivative of a function at a point. Figure 9.33 shows how the numerical derivative feature of a graphing calculator can be used to find $f''(2)$ if $f(x) = \sqrt{x^3 - 1}$. See Appendix A, Section 9.8, for details.

```
nDeriv(nDeriv(√(
X^3-1),X,X),X,2)

          .323969225
```

Figure 9.33

The figure shows that $f''(2) = 0.323969225 \approx 0.32397$. We can check this result by calculating $f''(x)$ with formulas.

$$f'(x) = \frac{1}{2}(x^3 - 1)^{-\frac{1}{2}}(3x^2)$$

$$f''(x) = \frac{1}{2}(x^3 - 1)^{-\frac{1}{2}}(6x) + (3x^2)\left[-\frac{1}{4}(x^3 - 1)^{-\frac{3}{2}}(3x^2) \right]$$

$$f''(2) = 0.3239695483 \approx 0.32397$$

Thus we see that the numerical derivative approximation is quite accurate.

✓ **CHECKPOINT**
ANSWERS
  1. The velocity is described by $s'(x) = 12x^2 - 24x$.
  2. The acceleration is described by $s''(x) = 24x - 24$.
  3. No; acceleration is negative when $x < 1$ second, zero when $x = 1$ second, and positive when $x > 1$ second.
  4. The velocity increases when the acceleration is positive, after 1 second.

| EXERCISES | 9.8

In Problems 1–6, find the second derivative.

1. $f(x) = 2x^{10} - 18x^5 - 12x^3 + 4$
2. $y = 6x^5 - 3x^4 + 12x^2$
3. $g(x) = x^3 - \dfrac{1}{x}$
4. $h(x) = x^2 - \dfrac{1}{x^2}$
5. $y = x^3 - \sqrt{x}$
6. $y = 3x^2 - \sqrt[3]{x^2}$

In Problems 7–12, find the third derivative.

7. $y = x^5 - 16x^3 + 12$
8. $y = 6x^3 - 12x^2 + 6x$
9. $f(x) = 2x^9 - 6x^6$
10. $f(x) = 3x^5 - x^6$
11. $y = 1/x$
12. $y = 1/x^2$

In Problems 13–24, find the indicated derivative.

13. If $y = x^5 - x^{1/2}$, find $\dfrac{d^2y}{dx^2}$.

14. If $y = x^4 + x^{1/3}$, find $\dfrac{d^2y}{dx^2}$.

15. If $f(x) = \sqrt{x+1}$, find $f'''(x)$.

16. If $f(x) = \sqrt{x-5}$, find $f'''(x)$.

17. Find $\dfrac{d^4y}{dx^4}$ if $y = 4x^3 - 16x$.

18. Find $y^{(4)}$ if $y = x^6 - 15x^3$.
19. Find $f^{(4)}(x)$ if $f(x) = \sqrt{x}$.
20. Find $f^{(4)}(x)$ if $f(x) = 1/x$.
21. Find $y^{(4)}$ if $y' = \sqrt{4x-1}$.

22. Find $y^{(5)}$ if $\dfrac{d^2y}{dx^2} = \sqrt[3]{3x+2}$.

23. Find $f^{(6)}(x)$ if $f^{(4)}(x) = x(x+1)^{-1}$.

24. Find $f^{(3)}(x)$ if $f'(x) = \dfrac{x^2}{x^2+1}$.

25. If $f(x) = 16x^2 - x^3$, what is the rate of change of $f'(x)$ at $(1, 15)$?
26. If $y = 36x^2 - 6x^3 + x$, what is the rate of change of y' at $(1, 31)$?

In Problems 27–30, use the numerical derivative feature of a graphing calculator to approximate the given second derivatives.

27. $f''(3)$ for $f(x) = x^3 - \dfrac{27}{x}$

28. $f''(-1)$ for $f(x) = \dfrac{x^2}{4} - \dfrac{4}{x^2}$

29. $f''(21)$ for $f(x) = \sqrt{x^2 + 4}$

30. $f''(3)$ for $f(x) = \dfrac{1}{\sqrt{x^2 + 7}}$

In Problems 31 and 32, do the following for each function $f(x)$.

(a) Find $f'(x)$ and $f''(x)$.
(b) Graph $f(x)$, $f'(x)$, and $f''(x)$ with a graphing utility.
(c) Identify x-values where $f''(x) = 0$, $f''(x) > 0$, and $f''(x) < 0$.

(d) Identify x-values where $f'(x)$ has a maximum point or a minimum point, where $f'(x)$ is increasing, and where $f'(x)$ is decreasing.
(e) When $f(x)$ has a maximum point, is $f''(x) > 0$ or $f''(x) < 0$?
(f) When $f(x)$ has a minimum point, is $f''(x) > 0$ or $f''(x) < 0$?

31. $f(x) = x^3 - 3x^2 + 5$ 32. $f(x) = 2 + 3x - x^3$

APPLICATIONS

33. *Acceleration* A particle travels as a function of time according to the formula

$$s = 100 + 10t + 0.01t^3$$

where s is in meters and t is in seconds. Find the acceleration of the particle when $t = 2$.

34. *Acceleration* If the formula describing the distance s (in feet) an object travels as a function of time (in seconds) is

$$s = 100 + 160t - 16t^2$$

what is the acceleration of the object when $t = 4$?

35. *Revenue* The revenue (in dollars) from the sale of x units of a certain product can be described by

$$R(x) = 100x - 0.01x^2$$

Find the instantaneous rate of change of the marginal revenue.

36. *Revenue* Suppose that the revenue (in dollars) from the sale of a product is given by

$$R = 70x + 0.5x^2 - 0.001x^3$$

where x is the number of units sold. How fast is the marginal revenue $\overline{MR}$ changing when $x = 100$?

37. *Sensitivity* When medicine is administered, reaction (measured in change of blood pressure or temperature) can be modeled by

$$R = m^2 \left(\dfrac{c}{2} - \dfrac{m}{3} \right)$$

where c is a positive constant and m is the amount of medicine absorbed into the blood (*Source:* R. M. Thrall et al., *Some Mathematical Models in Biology,* U.S. Department of Commerce, 1967). The sensitivity to the medication is defined to be the rate of change of reaction R with respect to the amount of medicine m absorbed in the blood.

(a) Find the sensitivity.
(b) Find the instantaneous rate of change of sensitivity with respect to the amount of medicine absorbed in the blood.
(c) Which order derivative of reaction gives the rate of change of sensitivity?

38. *Photosynthesis* The amount of photosynthesis that takes place in a certain plant depends on the intensity of light x according to the equation

$$f(x) = 145x^2 - 30x^3$$

   (a) Find the rate of change of photosynthesis with respect to the intensity.
   (b) What is the rate of change when $x = 1$? When $x = 3$?
   (c) How fast is the rate found in part (a) changing when $x = 1$? When $x = 3$?

39. *Revenue* The revenue (in thousands of dollars) from the sale of a product is

$$R = 15x + 30(4x + 1)^{-1} - 30$$

   where x is the number of units sold.
   (a) At what rate is the marginal revenue $\overline{MR}$ changing when the number of units being sold is 25?
   (b) Interpret your result in part (a).

40. *Advertising and sales* The sales of a product S (in thousands of dollars) are given by

$$S = \frac{600x}{x + 40}$$

   where x is the advertising expenditure (in thousands of dollars).
   (a) Find the rate of change of sales with respect to advertising expenditure.
   (b) Use the second derivative to find how this rate is changing at $x = 20$.
   (c) Interpret your result in part (b).

41. *Advertising and sales* The daily sales S (in thousands of dollars) that are attributed to an advertising campaign are given by

$$S = 1 + \frac{3}{t + 3} - \frac{18}{(t + 3)^2}$$

   where t is the number of weeks the campaign runs.
   (a) Find the rate of change of sales at any time t.
   (b) Use the second derivative to find how this rate is changing at $t = 15$.
   (c) Interpret your result in part (b).

42. *Advertising and sales* A product with a large advertising budget has its sales S (in millions of dollars) given by

$$S = \frac{500}{t + 2} - \frac{1000}{(t + 2)^2}$$

   where t is the number of months the product has been on the market.
   (a) Find the rate of change of sales at any time t.
   (b) What is the rate of change of sales at $t = 2$?
   (c) Use the second derivative to find how this rate is changing at $t = 2$.
   (d) Interpret your results from parts (b) and (c).

43. *Average annual wage* By using Social Security Administration data for selected years from 2012 and projected to 2050, the U.S. average annual wage, in thousands of dollars, can be modeled by

$$W(t) = 0.0212t^{2.11}$$

   where t is the number of years past 1975.
   (a) Use the model to find a function that models the instantaneous rate of change of W.
   (b) Find a function that models the rate at which the instantaneous rate from part (a) is changing.
   (c) Find and interpret $W(50)$, $W'(50)$, and $W''(50)$.

44. *Demographics* The makeup of various groups within the U.S. population may reshape the electorate in ways that could change political representation and policies. By using U.S. Department of Labor data from 1980 and projected to 2050 for the civilian non-institutional population ages 16 and older, in millions, the population of those who are non-White or Hispanic can be modeled by the function

$$P(t) = 0.707t^{1.17} + 20.0$$

   where t is the number of years past 1970.
   (a) Find the function that models the instantaneous rate of change of the population of non-Whites or Hispanics.
   (b) Find the function that models how fast the rate found in part (a) changes.
   (c) Find and interpret $P(55)$, $P'(55)$, and $P''(55)$.

45. **Modeling** *Economic dependency ratio* The economic dependency ratio is defined as the number of persons in the total population who are not in the workforce per 100 in the workforce. Since 1960, Baby Boomers in the workforce coupled with a decrease in the birth rate have caused a significant decrease in the economic dependency ratio.

   The table shows the economic dependency ratio for selected years from 1960 and projected to 2050.

Year	Ratio	Year	Ratio
1960	150.4	2015	91.9
1970	140.4	2020	97.4
1980	108.9	2030	106.4
1990	98.3	2040	109.0
2000	93.9	2050	111.4
2010	90.3		

Source: U.S. Bureau of Labor Statistics

   (a) Model these data with a cubic function, $R(x)$, where $R(x)$ is the economic dependency ratio and x is the number of years past 1950. Report the model with three significant digit coefficients.
   (b) Use the reported model from part (a) to find the function that models the rate of change of $R(x)$.

(c) Find the function that gives the rate of change of $R'(x)$.

(d) Find and interpret $R'(90)$ and $R''(90)$.

46. **Modeling *U.S. population*** The table gives the U.S. population to the nearest million for selected years from 1950 and projected to 2050.

Year	Total	Year	Total
1950	160	2010	315
1960	190	2015	328
1970	215	2020	343
1980	235	2030	370
1990	260	2040	391
2000	288	2050	409

Source: Social Security Administration

(a) Find a cubic function $P(t)$ that models these data, where P is the U.S. population in millions and t is the number of years past 1950. Report the model with three significant digit coefficients.

(b) Use the reported model to find the function that models the instantaneous rate of change of the U.S. population.

(c) Use the second derivative to determine how fast this rate was changing in 2000 and 2020.

(d) Write sentences that explain the meanings of $P'(70)$ and $P''(70)$.

47. ***Income by age*** The median income $f(x)$, in thousands of dollars, is a function of the age of workers ages 20–62, x, and can be modeled by

$$f(x) = 0.000864x^3 - 0.128x^2 + 6.61x - 62.6$$

(a) Find the instantaneous rate of change of the median income function.

(b) Find the instantaneous rate of change of the median income for 25-year-old and 55-year-old workers.

(c) Find the second derivative of the median income function.

(d) Use the second derivative to determine how fast the instantaneous rate of change in the median income of workers is changing for 25-year-old and 55-year-old workers.

(e) Explain the meaning $f'(55)$ and $f''(55)$.

- To find the marginal cost and marginal revenue at different levels of production
- To find the marginal profit function, given information about cost and revenue

Applications: Marginals and Derivatives

| APPLICATION PREVIEW |

If the total cost in dollars to produce x kitchen blenders is given by

$$C(x) = 0.001x^3 - 0.3x^2 + 32x + 2500$$

what is the marginal cost function, and what does it tell us about how costs are changing when $x = 80$ and $x = 200$ blenders are produced? See Example 3.

In Section 1.6, "Applications of Functions in Business and Economics," we defined the marginals for linear total cost, total revenue, and profit functions as the rate of change or slope of the respective function. In Section 9.3, we extended the notion of marginal revenue to nonlinear total revenue functions by defining marginal revenue as the derivative of total revenue. In this section, we extend the notion of marginal to nonlinear functions for any total cost or profit function.

Marginal Revenue   As we saw earlier, the instantaneous rate of change (the derivative) of the revenue function gives the **marginal revenue function.**

Marginal Revenue   If $R = R(x)$ is the total revenue function for a commodity, then the **marginal revenue function** is $\overline{MR} = R'(x)$.

Recall that if the demand function for a product in a monopoly market is $p = f(x)$, then the total revenue from the sale of x units is

$$R(x) = px = f(x) \cdot x$$

EXAMPLE 1   Revenue and Marginal Revenue

If the demand for a product in a monopoly market is given by

$$p = 16 - 0.02x$$

where x is the number of units and p is the price per unit, (a) find the total revenue function, and (b) find the marginal revenue for this product at $x = 40$.

Solution

(a) The total revenue function is

$$R(x) = px = (16 - 0.02x)x = 16x - 0.02x^2$$

(b) The marginal revenue function is

$$\overline{MR} = R'(x) = 16 - 0.04x$$

At $x = 40$, $R'(40) = 16 - 1.6 = 14.40$ dollars per unit. Thus the 41st item sold will increase the total revenue by approximately $14.40.

The marginal revenue is an approximation or estimate of the revenue gained from the sale of 1 additional unit. We have used marginal revenue in Example 1 to find that the revenue from the sale of the 41st item will be approximately $14.40. The actual increase in revenue from the sale of the 41st item is

$$R(41) - R(40) = 622.38 - 608 = \$14.38$$

Marginal revenue (and other marginals) can be used to predict for more than one additional unit. For instance, in Example 1, $\overline{MR}$ (40) = $14.40 per unit means that the expected or approximate revenue for the 41st through the 45th items sold would be 5($14.40) = $72.00. The actual revenue for these 5 items is $R(45) - R(40) = \$679.50 - \$608 = \$71.50$.

EXAMPLE 2   Maximum Revenue

Use the graphs in Figure 9.34 to determine the x-value where the revenue function has its maximum. What is happening to the marginal revenue at and near this x-value?

Solution

Figure 9.34(a) shows that the total revenue function has a maximum value at $x = 400$. After that, the total revenue function decreases. This means that the total revenue will be reduced each time a unit is sold if more than 400 are produced and sold. The graph of the marginal revenue function in Figure 9.34(b) shows that the marginal revenue is positive to the left of 400. This indicates that the rate at which the total revenue is changing is positive

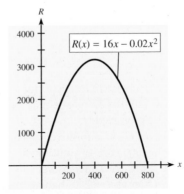

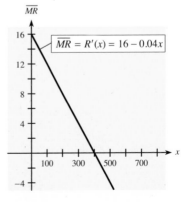

Figure 9.34   (a) Total revenue function   (b) Marginal revenue function

until 400 units are sold; thus the total revenue is increasing. Then, at 400 units, the rate of change is 0. After 400 units are sold, the marginal revenue is negative, which indicates that the total revenue is now decreasing. It is clear from looking at either graph that 400 units should be produced and sold to maximize the total revenue function $R(x)$. That is, the *total revenue* function has its maximum at $x = 400$.

Marginal Cost

As with marginal revenue, the derivative of a total cost function gives the **marginal cost function.**

Marginal Cost	If $C = C(x)$ is a total cost function for a commodity, then its derivative, $\overline{MC} = C'(x)$, is the **marginal cost function.**

Notice that the linear total cost function with equation

$$C(x) = 300 + 6x \quad \text{(in dollars)}$$

has marginal cost \$6 per unit because its slope is 6. Taking the derivative of $C(x)$ also gives

$$\overline{MC} = C'(x) = 6$$

which verifies that the marginal cost is \$6 per unit at all levels of production.

EXAMPLE **3** **Marginal Cost** | **APPLICATION PREVIEW** |

Suppose the daily total cost in dollars for a certain factory to produce x kitchen blenders is

$$C(x) = 0.001x^3 - 0.3x^2 + 32x + 2500$$

(a) Find the marginal cost function for these blenders.
(b) Find and interpret the marginal cost when $x = 80$ and $x = 200$.

Solution
(a) The marginal cost function is the derivative of $C(x)$.

$$\overline{MC} = C'(x) = 0.003x^2 - 0.6x + 32$$

(b) $C'(80) = 0.003(80)^2 - 0.6(80) + 32 = 3.2$ dollars per unit
$C'(200) = 0.003(200)^2 - 0.6(200) + 32 = 32$ dollars per unit

These values for the marginal cost can be used to *estimate* the amount that total cost would change if production were increased by one blender. Thus,

$$C'(80) = 3.2$$

means total cost would increase by *about* \$3.20 if an 81st blender were produced. Note that

$$C(81) - C(80) = 3.141 \approx \$3.14$$

is the actual increase in total cost for an 81st blender.

Also, $C'(200) = 32$ means that total cost would increase by *about* \$32 if a 201st blender were produced.

Whenever $C'(x)$ is positive it means that an additional unit produced adds to or increases the total cost. In addition, the value of the derivative (or marginal cost) measures how fast $C(x)$ is increasing. Thus, our calculations above indicate that $C(x)$ is increasing faster at $x = 200$ than at $x = 80$. The graphs of the total cost function and the marginal cost function in Figure 9.35(a) and (b) also illustrate these facts.

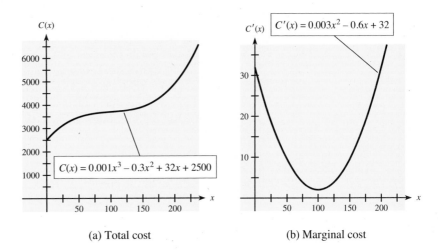

Figure 9.35   (a) Total cost   (b) Marginal cost

The graphs of many marginal cost functions tend to be U-shaped; they eventually will rise, even though there may be an initial interval where they decrease. Looking at the marginal cost graph in Figure 9.35(b), we see that marginal cost reaches its minimum near $x = 100$. We can also see this in Figure 9.35(a) by noting that tangent lines drawn to the total cost graph would have slopes that decrease until about $x = 100$ and then would increase.

Because producing more units can never reduce the total cost of production, the following properties are valid for total cost functions.

1. The total cost can never be negative. If there are fixed costs, the cost of producing 0 units is positive; otherwise, the cost of producing 0 units is 0.
2. The total cost function is always increasing; the more units produced, the higher the total cost. Thus the marginal cost is always positive.
3. There may be limitations on the units produced, such as those imposed by plant space.

EXAMPLE 4   **Total and Marginal Cost**

Suppose the graph in Figure 9.36 shows the monthly total cost for producing a product.

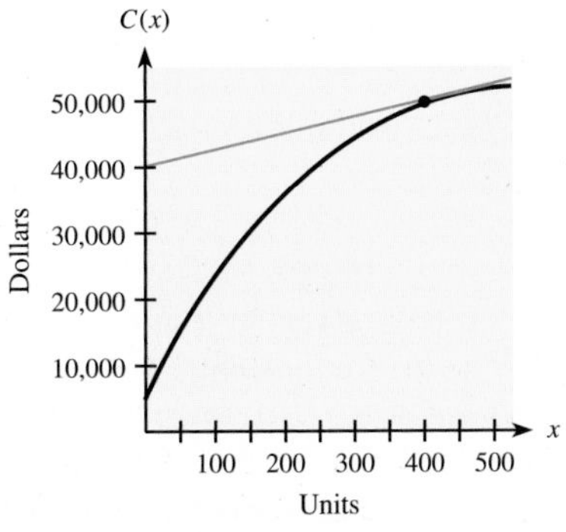

(a) Estimate the total cost of producing 400 items.
(b) Estimate the cost of producing the 401st item.
(c) Will producing the 151st item cost more or less than producing the 401st item?

Solution

(a) The total cost of producing 400 items is the height of the total cost graph when $x = 400$, or about $50,000.

(b) The approximate cost of the 401st item is the marginal cost when $x = 400$, or the slope of the tangent line drawn to the graph at $x = 400$. Figure 9.37 shows the total cost graph with a tangent line at $x = 400$.

Note that the tangent line passes through the point (0, 40,000), so we can find the slope by using the points (0, 40,000) and (400, 50,000).

$$m = \frac{y_2 - y_1}{x_2 - x_1} = \frac{50{,}000 - 40{,}000}{400 - 0} = \frac{10{,}000}{400} = 25$$

Thus the marginal cost at $x = 400$ is 25 dollars per item, so the approximate cost of the 401st item is $25.

(c) From Figure 9.37 we can see that if a tangent line to the graph were drawn where $x = 150$, it would be steeper than the one at $x = 400$. Because the slope of the tangent is the marginal cost and the marginal cost predicts the cost of the next item, this means that it would cost more to produce the 151st item than to produce the 401st.

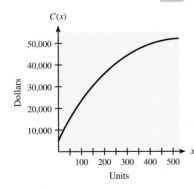

Figure 9.36

Figure 9.37

✓ CHECKPOINT   Suppose the total cost function for a commodity is $C(x) = 0.01x^3 - 0.9x^2 + 33x + 3000$.
1. Find the marginal cost function.
2. What is the marginal cost if $x = 50$ units are produced?
3. Use marginal cost to estimate the cost of producing the 51st unit.
4. Calculate $C(51) - C(50)$ to find the actual cost of producing the 51st unit.
5. True or false: For products that have linear cost functions, the actual cost of producing the $(x + 1)$st unit is equal to the marginal cost at x.

Marginal Profit   As with marginal cost and marginal revenue, the derivative of a profit function for a commodity will give us the **marginal profit function** for the commodity.

Marginal profit   If $P = P(x)$ is the profit function for a commodity, then the **marginal profit function** is $\overline{MP} = P'(x)$.

EXAMPLE 5   **Marginal Profit**

If the total profit, in thousands of dollars, for a product is given by $P(x) = 20\sqrt{x + 1} - 2x - 22$, what is the marginal profit at a production level of 15 units?

Solution
The marginal profit function is

$$\overline{MP} = P'(x) = 20 \cdot \frac{1}{2}(x + 1)^{-1/2} - 2 = \frac{10}{\sqrt{x + 1}} - 2$$

If 15 units are produced, the marginal profit is

$$P'(15) = \frac{10}{\sqrt{15 + 1}} - 2 = \frac{1}{2}$$

This means that the profit from the sale of the 16th unit is approximately $\frac{1}{2}$ (thousand dollars), or $500.

In a **competitive market**, each firm is so small that its actions in the market cannot affect the price of the product. The price of the product is determined in the market by the intersection of the market demand curve (from all consumers) and the market supply curve (from all firms that supply this product). The firm can sell as little or as much as it desires at the given market price, which it cannot change.

Therefore, a firm in a competitive market has a total revenue function given by $R(x) = px$, where p is the market equilibrium price for the product and x is the quantity sold.

EXAMPLE 6   **Profit in a Competitive Market**

A firm in a competitive market must sell its product for $200 per unit. The cost per unit (per month) is $80 + x$, where x represents the number of units sold per month. Find the marginal profit function.

Solution
If the cost per unit is $80 + x$, then the total cost of x units is given by the equation $C(x) = (80 + x)x = 80x + x^2$. The revenue per unit is $200, so the total revenue is given by $R(x) = 200x$. Thus the profit function is

$$P(x) = R(x) - C(x) = 200x - (80x + x^2), \quad \text{or} \quad P(x) = 120x - x^2$$

The marginal profit is $P'(x) = 120 - 2x$.

The marginal profit in Example 6 is not always positive, so producing and selling a certain number of items will maximize profit. Note that the marginal profit will be negative (that is, profit will decrease) if more than 60 items per month are produced. We will discuss methods of maximizing total revenue and profit, and of minimizing average cost, in the next chapter.

✓ **CHECKPOINT**

If the total profit function for a product is $P(x) = 20\sqrt{x+1} - 2x - 22$, then the marginal profit and its derivative are

$$P'(x) = \frac{10}{\sqrt{x+1}} - 2 \quad \text{and} \quad P''(x) = \frac{-5}{\sqrt{(x+1)^3}}$$

6. Is $P''(x) < 0$ for all values of $x \geq 0$?
7. Is the marginal profit decreasing for all $x \geq 0$?

EXAMPLE 7   Marginal Profit

Figure 9.38 shows graphs of a company's total revenue and total cost functions.

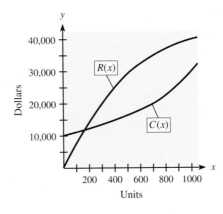

Figure 9.38

(a) If 100 units are being produced and sold, will producing the 101st item increase or decrease the profit?
(b) If 300 units are being produced and sold, will producing the 301st item increase or decrease the profit?
(c) If 1000 units are being produced and sold, will producing the 1001st item increase or decrease the profit?

Solution
At a given x-value, the slope of the tangent line to each function gives the marginal at that x-value.

(a) At $x = 100$, we can see that the graph of $R(x)$ is steeper than the graph of $C(x)$. Thus, the tangent line to $R(x)$ will be steeper than the tangent line to $C(x)$. Hence $\overline{MR}(100) > \overline{MC}(100)$, which means that the revenue from the 101st item will exceed the cost. Therefore, profit will increase when the 101st item is produced. Note that at $x = 100$, total costs are greater than total revenue, so the company is losing money but should still sell the 101st item because it will reduce the amount of loss.

(b) At $x = 300$, we can see that the tangent line to $R(x)$ again will be steeper than the tangent line to $C(x)$. Hence $\overline{MR}(300) > \overline{MC}(300)$, which means that the revenue from the 301st item will exceed the cost. Therefore, profit will increase when the 301st item is produced.

(c) At $x = 1000$, we can see that the tangent line to $C(x)$ will be steeper than the tangent line to $R(x)$. Hence $\overline{MC}(1000) > \overline{MR}(1000)$, which means that the cost of the 1001st item will exceed the revenue. Therefore, profit will decrease when the 1001st item is produced.

EXAMPLE 8   Profit and Marginal Profit

In Example 5, we found that the profit (in thousands of dollars) for a company's products is given by $P(x) = 20\sqrt{x+1} - 2x - 22$ and its marginal profit is given by $P'(x) = \dfrac{10}{\sqrt{x+1}} - 2$.

(a) Use the graphs of $P(x)$ and $P'(x)$ to determine the relationship between the two functions.

(b) When is the marginal profit 0? What is happening to profit at this level of production?

Solution

(a) By comparing the graphs of the two functions (shown in Figure 9.39), we see that for $x > 0$, profit $P(x)$ is increasing over the interval where the marginal profit $P'(x)$ is positive, and profit is decreasing over the interval where the marginal profit $P'(x)$ is negative.

(b) By using ZERO or SOLVER, or by using algebra, we see that $P'(x) = 0$ when $x = 24$. This level of production ($x = 24$) is where profit is maximized, at 30 (thousand dollars). ■

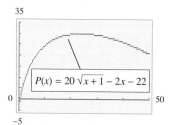

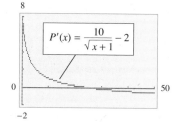

Figure 9.39

✓ **CHECKPOINT**
ANSWERS

1. $\overline{MC} = C'(x) = 0.03x^2 - 1.8x + 33$
2. $C'(50) = 18$
3. The 51st unit will cost approximately $18 to produce.
4. $C(51) - C(50) = 18.61$
5. True   6. Yes   7. Yes, because $P''(x) < 0$ for $x \geq 0$.

| EXERCISES | 9.9

MARGINAL REVENUE, COST, AND PROFIT
In Problems 1–8, total revenue is in dollars and x is the number of units.

1. (a) If the total revenue function for a product is $R(x) = 4x$, what is the marginal revenue function for that product?
    (b) What does this marginal revenue function tell us?

2. If the total revenue function for a product is $R(x) = 32x$, what is the marginal revenue for the product? What does this mean?

3. Suppose that the total revenue function for a commodity is $R = 36x - 0.01x^2$.
    (a) Find $R(100)$ and tell what it represents.
    (b) Find the marginal revenue function.
    (c) Find the marginal revenue at $x = 100$, and tell what it predicts about the sale of the next unit and the next 3 units.

    (d) Find $R(101) - R(100)$ and explain what this value represents.

4. Suppose that the total revenue function for a commodity is $R(x) = 25x - 0.05x^2$.
    (a) Find $R(50)$ and tell what it represents.
    (b) Find the marginal revenue function.
    (c) Find the marginal revenue at $x = 50$, and tell what it predicts about the sale of the next unit and the next 2 units.
    (d) Find $R(51) - R(50)$ and explain what this value represents.

5. Suppose that demand for local cable TV service is given by

    $$p = 80 - 0.4x$$

    where p is the monthly price in dollars and x is the number of subscribers (in hundreds).

(a) Find the total revenue as a function of the number of subscribers.

(b) Find the number of subscribers when the company charges $50 per month for cable service. Then find the total revenue for $p = \$50$.

(c) How could the company attract more subscribers?

(d) Find and interpret the marginal revenue when the price is $50 per month. What does this suggest about the monthly charge to subscribers?

6. Suppose that in a monopoly market, the demand function for a product is given by

$$p = 160 - 0.1x$$

where x is the number of units and p is the price in dollars.

(a) Find the total revenue from the sale of 500 units.

(b) Find and interpret the marginal revenue at 500 units.

(c) Is more revenue expected from the 501st unit sold or from the 701st? Explain.

7. (a) Graph the marginal revenue function from Problem 3.

(b) At what value of x will total revenue be maximized for Problem 3.

(c) What is the maximum revenue?

8. (a) Graph the marginal revenue function from Problem 4.

(b) Determine the number of units that must be sold to maximize total revenue.

(c) What is the maximum revenue?

In Problems 9–16, cost is in dollars and x is the number of units. Find the marginal cost functions for the given cost functions.

9. $C(x) = 40 + 8x$

10. $C(x) = 200 + 16x$

11. $C(x) = 500 + 13x + x^2$

12. $C(x) = 300 + 10x + \frac{1}{100}x^2$

13. $C = x^3 - 6x^2 + 24x + 10$

14. $C = 0.1x^3 - 1.5x^2 + 9x + 15$

15. $C = 400 + 27x + x^3$

16. $C(x) = 50 + 48x + x^3$

17. Suppose that the cost function for a commodity is

$$C(x) = 40 + x^2 \quad \text{dollars}$$

(a) Find the marginal cost at $x = 5$ units and tell what this predicts about the cost of producing 1 additional unit.

(b) Calculate $C(6) - C(5)$ to find the actual cost of producing 1 additional unit.

18. Suppose that the cost function for a commodity is

$$C(x) = 300 + 6x + \frac{1}{20}x^2 \text{ dollars}$$

(a) Find the marginal cost at $x = 8$ units and tell what this predicts about the cost of producing 1 additional unit.

(b) Calculate $C(9) - C(8)$ to find the actual cost of producing 1 additional unit.

19. If the cost function for a commodity is

$$C(x) = x^3 - 4x^2 + 30x + 20 \quad \text{dollars}$$

find the marginal cost at $x = 4$ units and tell what this predicts about the cost of producing 1 additional unit and 3 additional units.

20. If the cost function for a commodity is

$$C(x) = \frac{1}{90}x^3 + 4x^2 + 4x + 10 \quad \text{dollars}$$

find the marginal cost at $x = 3$ units and tell what this predicts about the cost of producing 1 additional unit and 2 additional units.

21. If the cost function for a commodity is

$$C(x) = 300 + 4x + x^2$$

graph the marginal cost function.

22. If the cost function for a commodity is

$$C(x) = x^3 - 12x^2 + 63x + 15$$

graph the marginal cost function.

In each of Problems 23 and 24, the graph of a company's total cost function is shown. For each problem, use the graph to answer the following questions.

(a) Will the 101st item or the 501st item cost more to produce? Explain.

(b) Does this total cost function represent a manufacturing process that is getting more efficient or less efficient? Explain.

23.

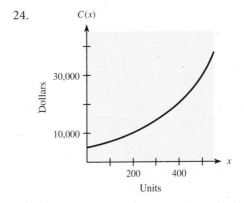

24.

In Problems 25–28, cost, revenue, and profit are in dollars and x is the number of units.

25. If the total profit function is $P(x) = 5x - 25$, find the marginal profit. What does this mean?

26. If the total profit function is $P(x) = 16x - 32$, find the marginal profit. What does this mean?

27. Suppose that the total revenue function for a product is $R(x) = 50x$ and that the total cost function is $C(x) = 1900 + 30x + 0.01x^2$.
    (a) Find the profit from the production and sale of 500 units.
    (b) Find the marginal profit function.
    (c) Find $\overline{MP}$ at $x = 500$ and explain what it predicts.
    (d) Find $P(501) - P(500)$ and explain what this value represents.

28. Suppose that the total revenue function is given by

$$R(x) = 46x$$

and that the total cost function is given by

$$C(x) = 100 + 30x + \tfrac{1}{10}x^2$$

    (a) Find $P(100)$.
    (b) Find the marginal profit function.
    (c) Find $\overline{MP}$ at $x = 100$ and explain what it predicts.
    (d) Find $P(101) - P(100)$ and explain what this value represents.

In each of Problems 29 and 30, the graphs of a company's total revenue function and total cost function are shown. For each problem, use the graph to answer the following questions.

(a) From the sale of 100 items, 400 items, and 700 items, rank from smallest to largest the amount of profit received. Explain your choices and note whether any of these scenarios results in a loss.

(b) From the sale of the 101st item, the 401st item, and the 701st item, rank from smallest to largest the amount of profit received. Explain your choices, and note whether any of these scenarios results in a loss.

29.

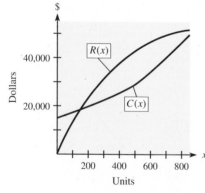

30.

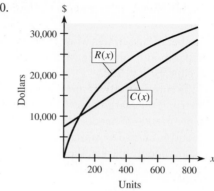

In each of Problems 31 and 32, the graph of a company's profit function is shown. For each problem, use the graph to answer the following questions about points A, B, and C.

(a) Rank from smallest to largest the amounts of profit received at these points. Explain your choices, and note whether any point results in a loss.

(b) Rank from smallest to largest the marginal profit at these points. Explain your choices, and note whether any marginal is negative and what this means.

31.

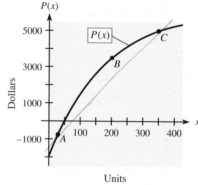

32.

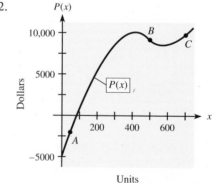

33. (a) Graph the marginal profit function for the profit function $P(x) = 30x - x^2 - 200$, where $P(x)$ is in thousands of dollars and x is hundreds of units.

(b) What level of production and sales will give a 0 marginal profit?

(c) At what level of production and sales will profit be at a maximum?

(d) What is the maximum profit?

34. (a) Graph the marginal profit function for the profit function $P(x) = 16x - 0.1x^2 - 100$, where $P(x)$ is in hundreds of dollars and x is hundreds of units.

(b) What level of production and sales will give a 0 marginal profit?

(c) At what level of production and sales will profit be at a maximum?

(d) What is the maximum profit?

35. The price of a product in a competitive market is $300. If the cost per unit of producing the product is $160 + 0.1x$ dollars, where x is the number of units produced per month, how many units should the firm produce and sell to maximize its profit?

36. The cost per unit of producing a product is $60 + 0.2x$ dollars, where x represents the number of units produced per week. If the equilibrium price determined by a competitive market is $220, how many units should the firm produce and sell each week to maximize its profit?

37. If the daily cost per unit of producing a product by the Ace Company is $10 + 0.1x$ dollars, and if the price on the competitive market is $70, what is the maximum daily profit the Ace Company can expect on this product?

38. The Mary Ellen Candy Company produces chocolate Easter bunnies at a cost per unit of $0.40 + 0.005x$ dollars, where x is the number produced. If the price on the competitive market for a bunny this size is $10.00, how many should the company produce to maximize its profit?

Chapter 9 Summary & Review

KEY TERMS AND FORMULAS

Section 9.1

Limit (p. 536)
One-sided limits (p. 537)
Properties of limits (p. 539)
Polynomial functions (p. 539)

Limits of rational functions (p. 540)
   0/0 indeterminate form
Limits of piecewise defined functions (p. 543)

Section 9.2

Continuous function (p. 549)
Limit at infinity (p. 553)

Vertical asymptote (p. 553)
Horizontal asymptote (p. 554)

Section 9.3

Average rate of change of f over $[a, b]$ (p. 560)
$$\frac{f(b) - f(a)}{b - a}$$
Average velocity (p. 561)
Instantaneous rate of change (p. 561)
Velocity (p. 562)
Derivative of $f(x)$ (p. 562)
$$f'(x) = \lim_{h \to 0} \frac{f(x + h) - f(x)}{h}$$

Derivative notation (p. 563)
$$y', \ f'(x), \ \frac{dy}{dx}, \ \frac{df(x)}{dx}$$
Marginal revenue (p. 563)
$$\overline{MR} = R'(x)$$
Slope of a tangent (p. 565)
Tangent line (p. 566)
Interpretations of the derivative (p. 566)
Differentiability and continuity (p. 567)

Section 9.4

Powers of x Rule (p. 573)
$$\frac{d(x^n)}{dx} = nx^{n-1}$$
Constant Function Rule (p. 575)
$$\frac{d(c)}{dx} = 0 \quad \text{for constant } c$$
Coefficient Rule (p. 575)
$$\frac{d}{dx}[c \cdot f(x)] = c \cdot f'(x)$$

Sum Rule (p. 576)
$$\frac{d}{dx}(u + v) = \frac{du}{dx} + \frac{dv}{dx}$$
Difference Rule (p. 577)
$$\frac{d}{dx}(u - v) = \frac{du}{dx} - \frac{dv}{dx}$$

Section 9.5

Product Rule (p. 584)

$$\frac{d}{dx}(uv) = uv' + vu'$$

Quotient Rule (p. 585)

$$\frac{d}{dx}\left(\frac{u}{v}\right) = \frac{vu' - uv'}{v^2}$$

Section 9.6

Chain Rule (p. 591)

$$\frac{dy}{dx} = \frac{dy}{du} \cdot \frac{du}{dx}$$

Power Rule (p. 592)

$$\frac{d}{dx}(u^n) = nu^{n-1}\frac{du}{dx}$$

Section 9.7

Derivative formulas summary (p. 600)

Section 9.8

Second derivative of $f(x)$ (p. 603)

$$f''(x) = \frac{d}{dx}(f'(x))$$

Higher-order derivatives (p. 603)

Acceleration (p. 605)

Section 9.9

Marginal revenue function (p. 608)
$$\overline{MR} = R'(x)$$
Marginal cost function (p. 610)
$$\overline{MC} = C'(x)$$

Marginal profit function (p. 612)
$$\overline{MP} = P'(x)$$

REVIEW EXERCISES

Section 9.1

In Problems 1–6, use the graph of $y = f(x)$ in Figure 9.40 to find the functional values and limits, if they exist.

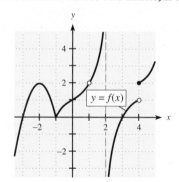

Figure 9.40

1. (a) $f(-2)$ (b) $\lim\limits_{x \to -2} f(x)$
2. (a) $f(-1)$ (b) $\lim\limits_{x \to -1} f(x)$
3. (a) $f(4)$ (b) $\lim\limits_{x \to 4} f(x)$
4. (a) $\lim\limits_{x \to 4^+} f(x)$ (b) $\lim\limits_{x \to 4} f(x)$
5. (a) $f(1)$ (b) $\lim\limits_{x \to 1} f(x)$
6. (a) $f(2)$ (b) $\lim\limits_{x \to 2} f(x)$

In Problems 7–20, find each limit, if it exists.

7. $\lim\limits_{x \to 4} (3x^2 + x + 3)$

8. $\lim\limits_{x \to 4} \dfrac{x^2 - 16}{x + 4}$

9. $\lim\limits_{x \to -1} \dfrac{x^2 - 1}{x + 1}$

10. $\lim\limits_{x \to 5} \dfrac{x^2 - 6x + 5}{x^2 - 5x}$

11. $\lim\limits_{x \to 2} \dfrac{4x^3 - 8x^2}{4x^3 - 16x}$

12. $\lim\limits_{x \to -\frac{1}{2}} \dfrac{x^2 - \frac{1}{4}}{6x^2 + x - 1}$

13. $\lim\limits_{x \to 3} \dfrac{x^2 - 16}{x - 3}$

14. $\lim\limits_{x \to -3} \dfrac{x^2 - x - 12}{2x - 6}$

15. $\lim\limits_{x \to 1} \dfrac{x^2 - 9}{x - 3}$

16. $\lim\limits_{x \to 2} \dfrac{x^2 - 8}{x - 2}$

17. $\lim\limits_{x \to 1} f(x)$ where $f(x) = \begin{cases} 4 - x^2 & \text{if } x < 1 \\ 4 & \text{if } x = 1 \\ 2x + 1 & \text{if } x > 1 \end{cases}$

18. $\lim\limits_{x \to -2} f(x)$ where $f(x) = \begin{cases} x^3 - x & \text{if } x < -2 \\ 2 - x^2 & \text{if } x \geq -2 \end{cases}$

19. $\lim\limits_{h \to 0} \dfrac{3(x + h)^2 - 3x^2}{h}$

20. $\lim\limits_{h \to 0} \dfrac{[(x + h) - 2(x + h)^2] - (x - 2x^2)}{h}$

In Problems 21 and 22, use tables to investigate each limit. Check your result analytically or graphically.

21. $\lim\limits_{x \to 2} \dfrac{x^2 + 10x - 24}{x^2 - 5x + 6}$

22. $\lim\limits_{x \to -\frac{1}{2}} \dfrac{x^2 + \frac{1}{6}x - \frac{1}{6}}{x^2 + \frac{5}{6}x + \frac{1}{6}}$

Section 9.2

Use the graph of $y = f(x)$ in Figure 9.40 to answer the questions in Problems 23 and 24.

23. Is $f(x)$ continuous at
     (a) $x = -1$?   (b) $x = 1$?

24. Is $f(x)$ continuous at
     (a) $x = -2$?   (b) $x = 2$?

In Problems 25–30, suppose that

$$f(x) = \begin{cases} x^2 + 1 & \text{if } x \le 0 \\ x & \text{if } 0 < x < 1 \\ 2x^2 - 1 & \text{if } x \ge 1 \end{cases}$$

25. What is $\lim_{x \to -1} f(x)$?

26. What is $\lim_{x \to 0} f(x)$, if it exists?

27. What is $\lim_{x \to 1} f(x)$, if it exists?

28. Is $f(x)$ continuous at $x = 0$?

29. Is $f(x)$ continuous at $x = 1$?

30. Is $f(x)$ continuous at $x = -1$?

For the functions in Problems 31–34, determine which are continuous. Identify discontinuities for those that are not continuous.

31. $y = \dfrac{x^2 + 25}{x - 5}$

32. $y = \dfrac{x^2 - 3x + 2}{x - 2}$

33. $f(x) = \begin{cases} x + 2 & \text{if } x \le 2 \\ 5x - 6 & \text{if } x > 2 \end{cases}$

34. $y = \begin{cases} x^4 - 3 & \text{if } x \le 1 \\ 2x - 3 & \text{if } x > 1 \end{cases}$

In Problems 35 and 36, use the graphs to find (a) the points of discontinuity, (b) $\lim_{x \to \infty} f(x)$, and (c) $\lim_{x \to -\infty} f(x)$.

35.

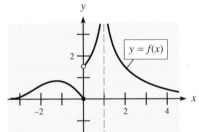

36.

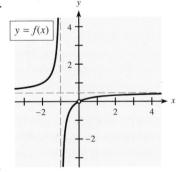

In Problems 37 and 38, evaluate the limits, if they exist. Then state what each limit tells about any horizontal asymptotes.

37. $\lim_{x \to -\infty} \dfrac{2x^2}{1 - x^2}$

38. $\lim_{x \to \infty} \dfrac{3x^{2/3}}{x + 1}$

Section 9.3

39. Find the average rate of change of

$$f(x) = 2x^4 - 3x + 7 \text{ over } [-1, 2]$$

In Problems 40 and 41, decide whether the statements are true or false.

40. $\lim_{h \to 0} \dfrac{f(x + h) - f(x)}{h}$ gives the formula for the slope of the tangent and the instantaneous rate of change of $f(x)$ at any value of x.

41. $\lim_{h \to 0} \dfrac{f(c + h) - f(c)}{h}$ gives the equation of the tangent line to $f(x)$ at $x = c$.

42. Use the definition of derivative to find $f'(x)$ for $f(x) = 3x^2 + 2x - 1$.

43. Use the definition of derivative to find $f'(x)$ if $f(x) = x - x^2$.

Use the graph of $y = f(x)$ in Figure 9.40 to answer the questions in Problems 44–46.

44. Explain which is greater: the average rate of change of f over $[-3, 0]$ or over $[-1, 0]$.

45. Is $f(x)$ differentiable at
  (a) $x = -1$?   (b) $x = 1$?

46. Is $f(x)$ differentiable at
  (a) $x = -2$?   (b) $x = 2$?

47. Let $f(x) = \dfrac{\sqrt[3]{4x}}{(3x^2 - 10)^2}$. Approximate $f'(2)$
  (a) by using the numerical derivative feature of a graphing calculator, and
  (b) by evaluating $\dfrac{f(2 + h) - f(2)}{h}$ with $h = 0.0001$.

48. Use the given table of values for $g(x)$ to
  (a) find the average rate of change of $g(x)$ over $[2, 5]$.
  (b) approximate $g'(4)$ as accurately as possible.

x	2	2.3	3.1	4	4.3	5
$g(x)$	13.2	12.1	9.7	12.2	14.3	18.1

Use the following graph of $f(x)$ to complete Problems 49 and 50.

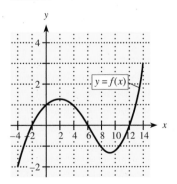

49. Estimate $f'(4)$.

50. Rank the following from smallest to largest and explain.
  A: $f'(2)$ B: $f'(6)$
  C: the average rate of change of $f(x)$ over $[2, 10]$

Section 9.4

51. If $c = 4x^5 - 6x^3$, find c'.
52. If $f(x) = 10x^9 - 5x^6 + 4x - 2^7 + 19$, find $f'(x)$.
53. If $p = q + \sqrt{7}$, find dp/dq.
54. If $y = \sqrt{x}$, find y'.
55. If $f(z) = \sqrt[3]{2^4}$, find $f'(z)$.
56. If $v(x) = 4/\sqrt[3]{x}$, find $v'(x)$.
57. If $y = \dfrac{1}{x} - \dfrac{1}{\sqrt{x}}$, find y'.
58. If $f(x) = \dfrac{3}{2x^2} - \sqrt[3]{x} + 4^5$, find $f'(x)$.
59. Write the equation of the line tangent to the graph of $y = 3x^5 - 6$ at $x = 1$.
60. Write the equation of the line tangent to the curve $y = 3x^3 - 2x$ at the point where $x = 2$.

In Problems 61 and 62, (a) find all x-values where the slope of the tangent equals zero, (b) find points (x, y) where the slope of the tangent equals zero, and (c) use a graphing utility to graph the function and label the points found in part (b).
61. $f(x) = x^3 - 3x^2 + 1$ 62. $f(x) = x^6 - 6x^4 + 8$

Section 9.5

63. If $f(x) = (3x - 1)(x^2 - 4x)$, find $f'(x)$.
64. Find y' if $y = (x^4 + 3)(3x^3 + 1)$.
65. If $p = \dfrac{5q^3}{2q^3 + 1}$, find $\dfrac{dp}{dq}$.
66. Find $\dfrac{ds}{dt}$ if $s = \dfrac{\sqrt{t}}{(3t + 1)}$.
67. Find $\dfrac{dy}{dx}$ for $y = \sqrt{x}\,(3x + 2)$.
68. Find $\dfrac{dC}{dx}$ for $C = \dfrac{5x^4 - 2x^2 + 1}{x^3 + 1}$.

Section 9.6

69. If $y = (x^3 - 4x^2)^3$, find y'.
70. If $y = (5x^6 + 6x^4 + 5)^6$, find y'.
71. If $y = (2x^4 - 9)^9$, find $\dfrac{dy}{dx}$.
72. Find $g'(x)$ if $g(x) = \dfrac{1}{\sqrt{x^3 - 4x}}$.

Section 9.7

73. Find $f'(x)$ if $f(x) = x^2(2x^4 + 5)^8$.
74. Find S' if $S = \dfrac{(3x + 1)^2}{x^2 - 4}$.
75. Find $\dfrac{dy}{dx}$ if $y = [(3x + 1)(2x^3 - 1)]^{12}$.
76. Find y' if $y = \left(\dfrac{x + 1}{1 - x^2}\right)^3$.

77. Find y' if $y = x\sqrt{x^2 - 4}$.
78. Find $\dfrac{dy}{dx}$ if $y = \dfrac{x}{\sqrt[3]{3x - 1}}$.

Section 9.8

In Problems 79 and 80, find the second derivatives.
79. $y = \sqrt{x} - x^2$ 80. $y = x^4 - \dfrac{1}{x}$

In Problems 81 and 82, find the fifth derivatives.
81. $y = (2x + 1)^4$ 82. $y = \dfrac{(1 - x)^6}{24}$

83. If $\dfrac{dy}{dx} = \sqrt{x^2 - 4}$, find $\dfrac{d^3y}{dx^3}$.
84. If $\dfrac{d^2y}{dx^2} = \dfrac{x}{x^2 + 1}$, find $\dfrac{d^4y}{dx^4}$.

APPLICATIONS

Sections 9.1 and 9.2

Cost, revenue, and profit **In Problems 85–88, assume that a company's monthly total revenue and total cost (both in dollars) are given by**

$$R(x) = 140x - 0.01x^2 \quad \text{and} \quad C(x) = 60x + 70{,}000$$

where x is the number of units. (Let $P(x)$ denote the profit function.)
85. Find (a) $\lim\limits_{x \to 4000} R(x)$, (b) $\lim\limits_{x \to 4000} C(x)$, and (c) $\lim\limits_{x \to 4000} P(x)$.
86. Find and interpret (a) $\lim\limits_{x \to 0^+} C(x)$ and (b) $\lim\limits_{x \to 1000} P(x)$.
87. If $\overline{R}(x) = \dfrac{R(x)}{x}$ and $\overline{C}(x) = \dfrac{C(x)}{x}$ are, respectively, the company's average revenue per unit and average cost per unit, find
    (a) $\lim\limits_{x \to 0^+} \overline{R}(x)$.   (b) $\lim\limits_{x \to 0^+} \overline{C}(x)$.
88. Evaluate and explain the meanings of
    (a) $\lim\limits_{x \to \infty} C(x)$.   (b) $\lim\limits_{x \to \infty} \overline{C}(x) = \lim\limits_{x \to \infty} \dfrac{C(x)}{x}$.

Section 9.3

Elderly in the workforce **The graph shows the percent of elderly men and women in the workforce for selected years from 1970 and projected to 2040. Use this graph in Problems 89 and 90.**
89. For the period from 1990 to 2030, find and interpret the annual average rate of change of
    (a) elderly men in the workforce and
    (b) elderly women in the workforce.
90. (a) Find the annual average rate of change of the percent of elderly men in the workforce from 1970 to 1980 and from 2030 to 2040.
    (b) Find the annual average rate of change of the percent of elderly women in the workforce from 1970 to 1980 and from 2030 to 2040.

Elderly in the Workforce, 1970–2040

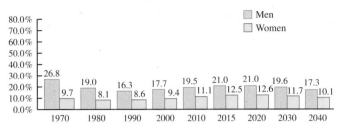

Source: Bureau of the Census, U.S. Department of Commerce

Section 9.4

91. *Demand* Suppose that the demand for x units of a product is given by $x = (100/p) - 1$, where p is the price per unit of the product. Find and interpret the rate of change of demand with respect to price if the price is
    (a) $10.   (b) $20.

92. *Severe weather ice makers* Thunderstorms severe enough to produce hail develop when an upper-level low (a pool of cold air high in the atmosphere) moves through a region where there is warm, moist air at the surface. These storms create an updraft that draws the moist air into subfreezing air above 10,000 feet. Data from the National Weather Service indicates that the strength of the updraft, as measured by its speed s in mph, affects the size of the hail according to

    $$h = 0.000595s^{1.922}$$

    where h is the diameter of the hail (in inches). Find and interpret $h(100)$ and $h'(100)$.

93. *Revenue* The graph shows the revenue function for a commodity. Will the $(A + 1)$st item sold or the $(B + 1)$st item sold produce more revenue? Explain.

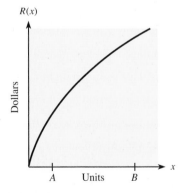

Section 9.5

94. *Revenue* In a 100-unit apartment building, when the price charged per apartment rental is $(830 + 30x)$ dollars, then the number of apartments rented is $100 - x$ and the total revenue for the building is

    $$R(x) = (830 + 30x)(100 - x)$$

    where x is the number of $30 rent increases (and also the resulting number of unrented apartments). Find the marginal revenue when $x = 10$. Does this tell you that the rent should be raised (causing more vacancies) or lowered? Explain.

95. *Productivity* Suppose the productivity of a worker (in units per hour) after x hours of training and time on the job is given by

    $$P(x) = 3 + \frac{70x^2}{x^2 + 1000}$$

    (a) Find and interpret $P(20)$.
    (b) Find and interpret $P'(20)$.

Section 9.6

96. *Demand* The demand q for a product at price p is given by

    $$q = 10{,}000 - 50\sqrt{0.02p^2 + 500}$$

    Find the rate of change of demand with respect to price.

97. *Supply* The number of units x of a product that is supplied at price p is given by

    $$x = \sqrt{p - 1}, \quad p \geq 1$$

    If the price p is $10, what is the rate of change of the supply with respect to the price, and what does it tell us?

Section 9.8

98. *Acceleration* Suppose an object moves so that its distance to a sensor, in feet, is given by

    $$s(t) = 16 + 140t + 8\sqrt{t}$$

    where t is the time in seconds. Find the acceleration at time $t = 4$ seconds.

99. *Profit* Suppose a company's profit (in dollars) is given by

    $$P(x) = 70x - 0.1x^2 - 5500$$

    where x is the number of units. Find and interpret $P'(300)$ and $P''(300)$.

Section 9.9

In Problems 100–107, cost, revenue, and profit are in dollars and x is the number of units.

100. *Cost* If the cost function for a particular good is $C(x) = 3x^2 + 6x + 600$, what is the
     (a) marginal cost function?
     (b) marginal cost if 30 units are produced?
     (c) interpretation of your answer in part (b)?

101. *Cost* If the total cost function for a commodity is $C(x) = 400 + 5x + x^3$, what is the marginal cost when 4 units are produced, and what does it mean?

102. *Revenue* The total revenue function for a commodity is $R = 40x - 0.02x^2$, with x representing the number of units.
    (a) Find the marginal revenue function.
    (b) At what level of production will marginal revenue be 0?

103. *Profit* If the total revenue function for a product is given by $R(x) = 60x$ and the total cost function is given by $C = 200 + 10x + 0.1x^2$, what is the marginal profit at $x = 10$? What does the marginal profit at $x = 10$ predict?

104. *Revenue* The total revenue function for a commodity is given by $R = 80x - 0.04x^2$.
    (a) Find the marginal revenue function.
    (b) What is the marginal revenue at $x = 100$?
    (c) Interpret your answer in part (b).

105. *Revenue* If the revenue function for a product is
$$R(x) = \frac{60x^2}{2x + 1}$$
    find the marginal revenue.

106. *Profit* A firm has monthly costs given by
$$C = 45{,}000 + 100x + x^3$$
    where x is the number of units produced per month. The firm can sell its product in a competitive market for $4600 per unit. Find the marginal profit.

107. *Profit* A small business has weekly costs of
$$C = 100 + 30x + \frac{x^2}{10}$$
    where x is the number of units produced each week. The competitive market price for this business's product is $46 per unit. Find the marginal profit.

108. *Cost, revenue, and profit* The graph shows the total revenue and total cost functions for a company. Use the graph to decide (and justify) at which of points A, B, and C
    (a) the revenue from the next item will be least.
    (b) the profit will be greatest.
    (c) the profit from the sale of the next item will be greatest.
    (d) the next item sold will reduce the profit.

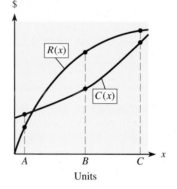

Chapter 9 TEST

1. Evaluate the following limits, if they exist. Use algebraic methods.
    (a) $\lim\limits_{x \to -2} \dfrac{4x - x^2}{4x - 8}$
    (b) $\lim\limits_{x \to \infty} \dfrac{8x^2 - 4x + 1}{2 + x - 5x^2}$
    (c) $\lim\limits_{x \to 7} \dfrac{x^2 - 5x - 14}{x^2 - 6x - 7}$
    (d) $\lim\limits_{x \to -5} \dfrac{5x - 25}{x + 5}$

2. (a) Write the limit definition for $f'(x)$.
    (b) Use the definition from (a) to find $f'(x)$ for $f(x) = 3x^2 - x + 9$.

3. Let $f(x) = \dfrac{4x}{x^2 - 8x}$. Identify all x-values where $f(x)$ is *not* continuous.

4. Use derivative formulas to find the derivative of each of the following. Simplify, except for part (d).
    (a) $B = 0.523W - 5176$
    (b) $p = 9t^{10} - 6t^7 - 17t + 23$
    (c) $y = \dfrac{3x^3}{2x^7 + 11}$
    (d) $f(x) = (3x^5 - 2x + 3)(4x^{10} + 10x^4 - 17)$
    (e) $g(x) = \frac{3}{4}(2x^5 + 7x^3 - 5)^{12}$
    (f) $y = (x^2 + 3)(2x + 5)^6$
    (g) $f(x) = 12\sqrt{x} - \dfrac{10}{x^2} + 17$

5. Find $\dfrac{d^3y}{dx^3}$ for $y = x^3 - x^{-3}$.

6. Let $f(x) = x^3 - 3x^2 - 24x - 10$.
    (a) Write the equation of the line tangent to the graph of $y = f(x)$ at $x = -1$.
    (b) Find all points (both x- and y-coordinates) where $f'(x) = 0$.

7. Find the average rate of change of $f(x) = 4 - x - 2x^2$ over $[1, 6]$.

8. Use the given tables to evaluate the following limits, if they exist.
    (a) $\lim\limits_{x \to 5} f(x)$　　(b) $\lim\limits_{x \to 5} g(x)$　　(c) $\lim\limits_{x \to 5^-} g(x)$

x	4.99	4.999	$\rightarrow 5 \leftarrow$	5.001	5.01
$f(x)$	2.01	2.001	$\rightarrow ? \leftarrow$	1.999	1.99

x	4.99	4.999	$\rightarrow 5 \leftarrow$	5.001	5.01
$g(x)$	-3.99	-3.999	$\rightarrow ? \leftarrow$	6.999	6.99

9. Use the definition of continuity to investigate whether $g(x)$ is continuous at $x = -2$. Show your work.

$$g(x) = \begin{cases} 6 - x & \text{if } x \leq -2 \\ x^3 & \text{if } x > -2 \end{cases}$$

In Problems 10 and 11, suppose a company has its total cost for a product given by $C(x) = 200x + 10,000$ dollars and its total revenue given by $R(x) = 250x - 0.01x^2$ dollars, where x is the number of units produced and sold.

10. (a) Find the marginal revenue function.
    (b) Find $R(72)$ and $R'(72)$ and tell what each represents or predicts.

11. (a) Form the profit function for this product.
    (b) Find the marginal profit function.
    (c) Find the marginal profit when $x = 1000$, and then write a sentence that interprets this result.

12. Suppose that $f(x)$ is a differentiable function. Use the table of values to approximate $f'(3)$ as accurately as possible.

x	2	2.5	2.999	3	3.01	3.1
$f(x)$	0	18.4	44.896	45	46.05	56.18

13. Use the graph to perform the evaluations (a)–(f) and to answer (g)–(i). If no value exists, so indicate.
    (a) $f(1)$
    (b) $\lim\limits_{x \to 6} f(x)$
    (c) $\lim\limits_{x \to 3^-} f(x)$
    (d) $\lim\limits_{x \to -4} f(x)$
    (e) $\lim\limits_{x \to -\infty} f(x)$
    (f) Estimate $f'(4)$.
    (g) Find all x-values where $f'(x)$ does not exist.

(h) Find all x-values where $f(x)$ is not continuous.
(i) Rank from smallest to largest: $f'(-2), f'(2)$, and the average rate of change of $f(x)$ over $[-2, 2]$.

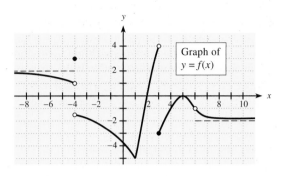

14. Given that the line $y = \frac{2}{3}x - 8$ is tangent to the graph of $y = f(x)$ at $x = 6$, find
    (a) $f'(6)$.
    (b) $f(6)$.
    (c) the instantaneous rate of change of $f(x)$ with respect to x at $x = 6$.

15. The graph shows the total revenue and total cost functions for a company. Use the graph to decide (and justify) at which of points A, B, and C
    (a) profit is the greatest.
    (b) there is a loss.
    (c) producing and selling another item will increase profit.
    (d) the next item sold will decrease profit.

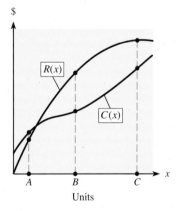

I. Marginal Return to Sales

A tire manufacturer studying the effectiveness of television advertising and other promotions on sales of its GRIPPER-brand tires attempted to fit data it had gathered to the equation

$$S = a_0 + a_1x + a_2x^2 + b_1y$$

where S is sales revenue in millions of dollars, x is millions of dollars spent on television advertising, y is millions of dollars spent on other promotions, and a_0, a_1, a_2, and b_1 are constants. The data, gathered in two different regions of the country where expenditures for other promotions were kept constant (at B_1 and B_2), resulted in the following quadratic equations relating TV advertising and sales.

$$\text{Region 1:} \quad S_1 = 30 + 20x - 0.4x^2 + B_1$$
$$\text{Region 2:} \quad S_2 = 20 + 36x - 1.3x^2 + B_2$$

The company wants to know how to make the best use of its advertising dollars in the regions and whether the current allocation could be improved. Advise management about current advertising effectiveness, allocation of additional expenditures, and reallocation of current advertising expenditures by answering the following questions.

1. In the analysis of sales and advertising, **marginal return to sales** is usually used, and it is given by dS_1/dx for Region 1 and dS_2/dx for Region 2.

    (a) Find $\dfrac{dS_1}{dx}$ and $\dfrac{dS_2}{dx}$.

    (b) If $10 million is being spent on TV advertising in each region, what is the marginal return to sales in each region?

2. Which region would benefit more from additional advertising expenditure, if $10 million is currently being spent in each region?

3. If any additional money is made available for advertising, in which region should it be spent?

4. How could money already being spent be reallocated to produce more sales revenue?

II. Energy from Crude Oil (Modeling)

The table shows the total energy supply from crude oil products, in quadrillion BTUs, for the years from 2010 and projected through 2040.

1. Find the cubic function that models the data, with x equal to the number of years past 2010 and y equal to the number of quadrillion BTUs of energy. Report the model as $y = f(x)$ with three significant digit coefficients.
2. Graph the unrounded model and the data points on the same axes, with x representing the years from 2010 to 2040.
3. Use the reported model $y = f(x)$ for $x = -5$ through $x = 33$.
   (a) Graph $y = f(x)$ for this x-range.
   (b) Find the number of quadrillion BTUs of energy when $x = 9.0$.
   (c) Find the number of quadrillion BTUs of energy when $x = 9.16$.
   (d) Find the number of quadrillion BTUs of energy when $x = 9.3$.
4. What can we conclude about the point (9.16, 15.933) on the graph of $y = f(x)$?
5. Find the rate of change of the reported function $y = f(x)$.
6. Graph $y = f'(x)$ for x-values -5 through 33.
7. (a) What is the value of $y = f'(x)$ at $x = 9.16$?
   (b) Is the value of $y = f'(x)$ positive or negative at $x = 9.0$?
   (c) Is the value of $y = f'(x)$ positive or negative at $x = 9.3$?
8. What can we conclude about the derivative on either side of the x-coordinate of the point (9.16, 15.933)?
9. What can we conclude about the point (25.36, 12.958) on the graph of $y = f(x)$?
10. What can we conclude about the derivative on either side of the x-coordinate of the point (25.36, 12.958)?

Total Energy Supply from Crude Oil Products, in Quadrillion BTUs

Year	BTUs	Year	BTUs	Year	BTUs
2010	11.59	2021	15.75	2032	13.51
2011	12.16	2022	15.42	2033	13.56
2012	13.58	2023	15.09	2034	13.52
2013	14.61	2024	14.74	2035	13.40
2014	15.35	2025	14.50	2036	13.07
2015	15.58	2026	14.11	2037	13.52
2016	16.08	2027	13.91	2038	12.94
2017	16.03	2028	14.71	2039	13.11
2018	16.02	2029	13.54	2040	13.12
2019	16.10	2030	13.47		
2020	15.95	2031	13.52		

Source: http://www.eia.gov/

III. Tangent Lines and Optimization in Business and Economics

In business and economics, common questions of interest include how to maximize profit or revenue, how to minimize average costs, and how to maximize average productivity. For questions such as these, the description of how to obtain the desired maximum or minimum represents an optimal (or best possible) solution to the problem. And the process of finding an optimal solution is called optimization. Answering these questions often involves tangent lines. In this project, we examine how tangent lines can be used to minimize average costs.

Suppose that Wittage, Inc., manufactures paper shredders for home and office use and that its weekly total costs (in dollars) for x shredders are given by

$$C(x) = 0.03x^2 + 12.75x + 6075$$

The average cost per unit for x units [denoted by $\overline{C}(x)$] is the total cost divided by the number of units. Thus, the average cost function for Wittage, Inc., is

$$\overline{C}(x) = \frac{C(x)}{x} = \frac{0.03x^2 + 12.75x + 6075}{x} = \frac{0.03x^2}{x} + \frac{12.75x}{x} + \frac{6075}{x}$$

$$\overline{C}(x) = 0.03x + 12.75 + \frac{6075}{x}$$

Wittage, Inc., would like to know how the company can use marginal costs to gain information about minimizing average costs. To investigate this relationship, answer the following questions.

1. Find Wittage's marginal cost function, then complete the following table.

x	100	200	300	400	500	600
$\overline{C}(x)$						
$\overline{MC}(x)$						

2. (a) For what x-values in the table is $\overline{C}$ getting smaller?
   (b) For these x-values, is $\overline{MC} < \overline{C}, \overline{MC} > \overline{C}$, or $\overline{MC} = \overline{C}$?
3. (a) For what x-values in the table is $\overline{C}$ getting larger?
   (b) For these x-values, is $\overline{MC} < \overline{C}, \overline{MC} > \overline{C}$, or $\overline{MC} = \overline{C}$?
4. Between what pair of consecutive x-values in the table will $\overline{C}$ reach its minimum?
5. Interpret (a) $\overline{C}(200)$ and $\overline{MC}(200)$ and (b) $\overline{C}(400)$ and $\overline{MC}(400)$.
6. On the basis of what $\overline{C}$ and $\overline{MC}$ tell us and the work so far, complete the following statements by filling the blank with *increase, decrease,* or *stay the same.* Explain your reasoning.
   (a) $\overline{MC} < \overline{C}$ means that if Wittage makes one more unit, then $\overline{C}$ will _____.
   (b) $\overline{MC} > \overline{C}$ means that if Wittage makes one more unit, then $\overline{C}$ will _____.
7. (a) On the basis of your answers to Question 6, how will $\overline{C}$ and $\overline{MC}$ be related when $\overline{C}$ is minimized? Write a careful statement.
   (b) Use your idea from part (a) to find x where $\overline{C}$ is minimized. Then check your calculations by graphing $\overline{C}$ and locating the minimum from the graph.

Next, let's examine graphically the relationship that emerged in Question 7 to see whether it is true in general. Figure 9.41 shows a typical total cost function (Wittage's total cost function has a similar shape) and an arbitrary point on the graph.

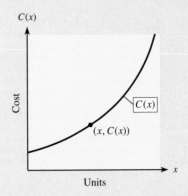

Figure 9.41

8. (a) Measure the slope of the line joining the point $(x, C(x))$ and the origin.
    (b) How is this slope related to the function $\overline{C}$ at $(x, C(x))$?
9. Figure 9.42(a) shows a typical total cost function with several points labeled.
    (a) Explain how you can tell at a glance that $\overline{C}(x)$ is decreasing as the level of production moves from A to B to C to D.
    (b) Explain how you can tell at a glance that $\overline{C}(x)$ is increasing as the level of production moves from Q to R to S to T.

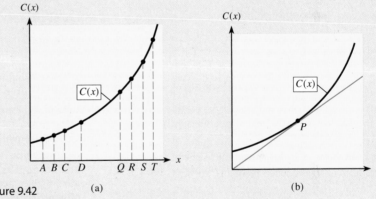

Figure 9.42   (a)   (b)

10. In Figure 9.42(b), a line from $(0, 0)$ is drawn tangent to $C(x)$ and the point of tangency is labeled P. At point P in Figure 9.42(b), what represents
    (a) the average cost, $\overline{C}(x)$?
    (b) the marginal cost, $\overline{MC}(x)$?
    (c) Explain how Figure 9.42 confirms that $\overline{C}(x)$ is minimized when $\overline{MC}(x) = \overline{C}(x)$.

On the basis of the numerical approach for Wittage's total cost function and the general graphical approach, what recommendation would you give Wittage about determining the number of units that will minimize average costs (even if the total cost function changes)?

10

Applications of Derivatives

The derivative can be used to determine where a function has a "turning point" on its graph, so that we can determine where the graph reaches its highest or lowest point within a particular interval. These points are called the relative maxima and relative minima, respectively, and are useful in sketching the graph of the function. The techniques for finding these points are also useful in solving applied problems, such as finding the maximum profit, the minimum average cost, and the maximum productivity. The second derivative can be used to find points of inflection of the graph of a function and to find the point of diminishing returns in certain applications.

The topics and some representative applications discussed in this chapter include the following.

SECTIONS

10.1 **Relative Maxima and Minima: Curve Sketching**

10.2 **Concavity: Points of Inflection**
Second-derivative test

10.3 **Optimization in Business and Economics**
Maximizing revenue
Minimizing average cost
Maximizing profit

10.4 **Applications of Maxima and Minima**

10.5 **Rational Functions: More Curve Sketching**
Asymptotes
More curve sketching

APPLICATIONS

Advertising, aging workers

Water purity, profit, diminishing returns

Maximizing revenue, minimizing average cost, maximizing profit

Company growth, minimizing cost, postal restriction, inventory-cost models, property development

Production costs

© Tyler Panian/Shutterstock.com

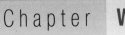

Chapter **Warm-Up**

Prerequisite Problem Type	For Section	Answer	Section for Review
Write $\frac{1}{3}(x^2 - 1)^{-2/3}(2x)$ with positive exponents.	**10.2**	$\dfrac{2x}{3(x^2 - 1)^{2/3}}$	0.3, 0.4 Exponents and radicals
Factor: (a) $x^3 - x^2 - 6x$ (b) $8000 - 80x - 3x^2$	**10.1–10.3**	(a) $x(x - 3)(x + 2)$ (b) $(40 - x)(200 + 3x)$	0.6 Factoring
(a) For what values of x is $\dfrac{2}{3\sqrt[3]{x+2}}$ undefined? (b) For what values of x is $\frac{1}{3}(x^2 - 1)^{-2/3}(2x)$ undefined?	**10.1** **10.2** **10.5**	(a) $x = -2$ (b) $x = -1, x = 1$	1.2 Domains of functions
If $f(x) = \frac{1}{3}x^3 - x^2 - 3x + 2$, and $f'(x) = x^2 - 2x - 3$, (a) find $f(-1)$. (b) find $f'(-2)$.	**10.1**	(a) $\dfrac{11}{3}$ (b) 5	1.2 Function notation
(a) Solve $0 = x^2 - 2x - 3$. (b) If $f'(x) = 3x^2 - 3$, what values of x make $f'(x) = 0$?	**10.1–10.5**	(a) $x = -1, x = 3$ (b) $x = -1, x = 1$	2.1 Solving quadratic equations
Does $\lim\limits_{x \to -2} \dfrac{2x - 4}{3x + 6}$ exist?	**10.5**	No; unbounded	9.1 Limits
(a) Find $f''(x)$ if $f(x) = x^3 - 4x^2 + 3$. (b) Find $P''(x)$ if $P(x) = 48x - 1.2x^2$.	**10.2**	(a) $f''(x) = 6x - 8$ (b) $P''(x) = -2.4$	9.8 Higher-order derivatives
Find the derivatives: (a) $y = \dfrac{1}{3}x^3 - x^2 - 3x + 2$ (b) $f = x + 2\left(\dfrac{80,000}{x}\right)$ (c) $p(t) = 1 + \dfrac{4t}{t^2 + 16}$ (d) $y = (x + 2)^{2/3}$ (e) $y = \sqrt[3]{x^2 - 1}$	**10.1–10.4**	(a) $y' = x^2 - 2x - 3$ (b) $f' = 1 - \dfrac{160,000}{x^2}$ (c) $p'(t) = \dfrac{64 - 4t^2}{(t^2 + 16)^2}$ (d) $y' = \dfrac{2}{3(x + 2)^{1/3}}$ (e) $y' = \dfrac{2x}{3(x^2 - 1)^{2/3}}$	9.4, 9.5, 9.6 Derivatives

OBJECTIVES

10.1

- To find relative maxima and minima and horizontal points of inflection of functions
- To sketch graphs of functions by using information about maxima, minima, and horizontal points of inflection

Relative Maxima and Minima: Curve Sketching

| APPLICATION PREVIEW |

When a company initiates an advertising campaign, there is typically a surge in weekly sales. As the effect of the campaign lessens, sales attributable to it usually decrease. For example, suppose a company models its weekly sales revenue during an advertising campaign by

$$S = \frac{100t}{t^2 + 100}$$

where t is the number of weeks since the beginning of the campaign. The company would like to determine accurately when the revenue function is increasing, when it is decreasing, and when sales revenue is maximized. (See Example 4.)

In this section we will use the derivative of a function to decide whether the function is increasing or decreasing on an interval and to find where the function has relative maximum points and relative minimum points. We will use the information about derivatives of functions to graph the functions and to solve applied problems.

Recall that we can find the maximum value or the minimum value of a quadratic function by finding the vertex of its graph. But for functions of higher degree, the special features of their graphs may be harder to find accurately, even when a graphing utility is used. In addition to intercepts and asymptotes, we can use the first derivative as an aid in graphing. The first derivative identifies the "turning points" of the graph, which help us determine the general shape of the graph and choose a viewing window that includes the interesting points of the graph. Note that for a quadratic function $f(x) = ax^2 + bx + c$, the solution to $f'(x) = 2ax + b = 0$ is $x = -b/2a$. This solution is the x-coordinate of the vertex (turning point) of $f(x)$, found in Section 2.2.

In Figure 10.1(a) we see that the graph of $y = \frac{1}{3}x^3 - x^2 - 3x + 2$ has two "turning points," at $(-1, \frac{11}{3})$ and $(3, -7)$. The curve has a relative maximum at $(-1, \frac{11}{3})$ because this point is higher than any other point "near" it on the curve; the curve has a relative minimum at $(3, -7)$ because this point is lower than any other point "near" it on the curve. A formal definition follows.

Relative Maxima and Minima

The point $(x_1, f(x_1))$ is a **relative maximum point** of the function f if there is an interval around x_1 on which $f(x_1) \geq f(x)$ for all x in the interval. In this case, we say the relative maximum *occurs* at $x = x_1$ and the relative maximum is $f(x_1)$.

The point $(x_2, f(x_2))$ is a **relative minimum point** of the function f if there is an interval around x_2 on which $f(x_2) \leq f(x)$ for all x in the interval. In this case, we say the relative minimum *occurs* at $x = x_2$ and the relative minimum is $f(x_2)$.

In order to determine whether a turning point of a function is a maximum point or a minimum point, it is frequently helpful to know what the graph of the function does in intervals on either side of the turning point. We say a function is **increasing** on an interval if the values of the function increase as the x-values increase (that is, if the graph rises as we move from left to right on the interval). Similarly, a function is **decreasing** on an interval if the values of the function decrease as the x-values increase (that is, if the graph falls as we move from left to right on the interval).

We have seen that if the slope of a line is positive, then the linear function is increasing and its graph is rising. Similarly, if $f(x)$ is differentiable over an interval and if each tangent line to the curve over that interval has positive slope, then the curve is rising over the interval and the function is increasing. Because the derivative of the function gives the slope of the tangent to the curve, we see that if $f'(x) > 0$ on an interval, then $f(x)$ is increasing on that interval. A similar conclusion can be reached when the derivative is negative on the interval.

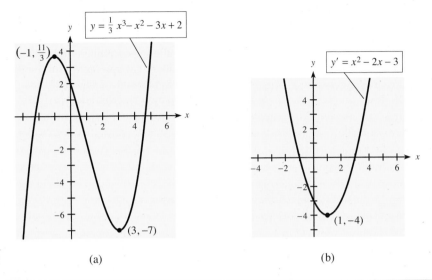

Figure 10.1	(a)	(b)

Increasing and Decreasing Functions

If f is a function that is differentiable on an interval (a, b), then

if $f'(x) > 0$ for all x in (a, b), f is increasing on (a, b).

if $f'(x) < 0$ for all x in (a, b), f is decreasing on (a, b).

Figure 10.1(a) shows the graph of a function, and Figure 10.1(b) shows the graph of its derivative. The figures show that the graph of $y = f(x)$ is increasing for the same x-values for which the graph of $y' = f'(x)$ is above the x-axis (when $f'(x) > 0$). Similarly, the graph of $y = f(x)$ is decreasing for the same x-values $(-1 < x < 3)$ for which the graph of $y' = f'(x)$ is below the x-axis (when $f'(x) < 0$).

The derivative $f'(x)$ can change signs only at values of x at which $f'(x) = 0$ or $f'(x)$ is undefined. We call these values of x **critical values.** The point corresponding to a critical value for x is a **critical point.*** Because a curve changes from increasing to decreasing at a relative maximum and from decreasing to increasing at a relative minimum (see Figure 10.1(a)), we have the following.

Relative Maximum and Minimum

If f has a relative maximum or a relative minimum at $x = x_0$, then $f'(x_0) = 0$ or $f'(x_0)$ is undefined.

Figure 10.2 shows a function with two relative maxima, one at $x = x_1$ and the second at $x = x_3$, and one relative minimum at $x = x_2$. At $x = x_1$ and $x = x_2$, we see that $f'(x) = 0$, and at $x = x_3$ the derivative does not exist.

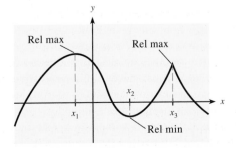

Figure 10.2

*There may be some critical values at which both $f'(x)$ and $f(x)$ are undefined. Critical points do not occur at these values, but studying the derivative on either side of such values may be of interest.

Thus we can find relative maxima and minima for a curve by finding values of x for which the function has critical points. The behavior of the derivative to the left and right of (and near) these points will tell us whether they are relative maxima, relative minima, or neither.

Because the critical values are the only values at which the graph can have turning points, the derivative cannot change sign anywhere except at a critical value. Thus, in an interval between two critical values, the sign of the derivative at any value in the interval will be the sign of the derivative at all values in the interval.

Using the critical values of $f(x)$ and the sign of $f'(x)$ between those critical values, we can create a **sign diagram for $f'(x)$**. The sign diagram for the graph in Figure 10.2 is shown in Figure 10.3. This sign diagram was created from the graph of f, but it is also possible to predict the shape of a graph from a sign diagram.

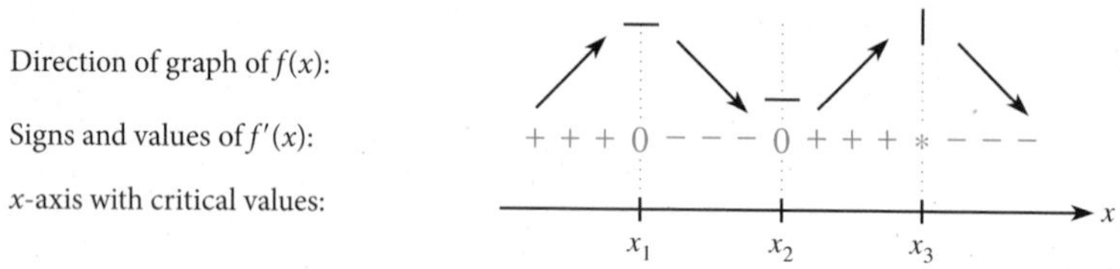

Direction of graph of $f(x)$:

Signs and values of $f'(x)$:

x-axis with critical values:

Figure 10.3   *means $f'(x_3)$ is undefined

Figure 10.4 shows two ways that a function can have a relative maximum at a critical point, and Figure 10.5 shows two ways for a relative minimum.

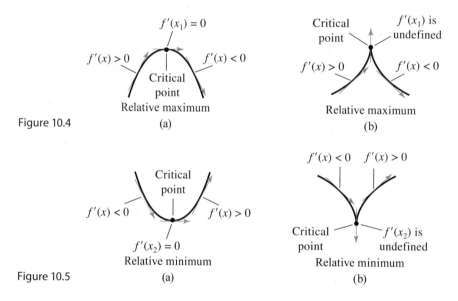

Figure 10.4

Figure 10.5

The preceding discussion suggests the following procedure for finding relative maxima and minima of a function.

First-Derivative Test

Procedure

To find relative maxima and minima of a function:

1. Find the first derivative of the function.
2. Set the derivative equal to 0, and solve for values of x that satisfy $f'(x) = 0$. These are called **critical values.** Values that make $f'(x)$ undefined are also critical values.

Example

Find the relative maxima and minima of $f(x) = \frac{1}{3}x^3 - x^2 - 3x + 2$.

1. $f'(x) = x^2 - 2x - 3$
2. $0 = x^2 - 2x - 3 = (x + 1)(x - 3)$ has solutions $x = -1, x = 3$. No values of x make $f'(x) = x^2 - 2x - 3$ undefined. Critical values are -1 and 3.

(*Continued*)

3. Substitute the critical values into the *original function* to find the **critical points.**
4. Evaluate $f'(x)$ at a value of x to the left and one to the right of each critical point to develop a sign diagram.
   (a) If $f'(x) > 0$ to the left and $f'(x) < 0$ to the right of the critical value, the critical point is a relative maximum point.
   (b) If $f'(x) < 0$ to the left and $f'(x) > 0$ to the right of the critical value, the critical point is a relative minimum point.

3. $f(-1) = \frac{11}{3}$ $f(3) = -7$
   The critical points are $(-1, \frac{11}{3})$ and $(3, -7)$.
4. $f'(-2) = 5 > 0$ and $f'(0) = -3 < 0$
   Thus $(-1, 11/3)$ is a relative maximum point.
   $f'(2) = -3 < 0$ and $f'(4) = 5 > 0$
   Thus $(3, -7)$ is a relative minimum point. The sign diagram for $f'(x)$ is

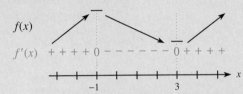

5. Use the information from the sign diagram and selected points to sketch the graph.

5. The information from this sign diagram is shown in Figure 10.6(a). Plotting additional points gives the graph of the function, which is shown in Figure 10.6(b).

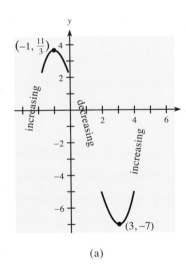

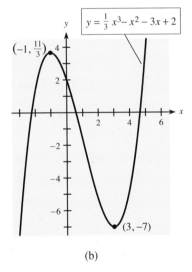

Figure 10.6   (a)   (b)

Because the critical values are the only x-values at which the graph can have turning points, we can test to the left and right of each critical value by testing to the left of the smallest critical value, then testing a value *between* each two successive critical values, and then testing to the right of the largest critical value. The following example illustrates this procedure.

EXAMPLE 1   Maxima and Minima

Find the relative maxima and minima of $f(x) = \frac{1}{4}x^4 - \frac{1}{3}x^3 - 3x^2 + 8$, and sketch its graph.

Solution
1. $f'(x) = x^3 - x^2 - 6x$
2. Setting $f'(x) = 0$ gives $0 = x^3 - x^2 - 6x$. Solving for x gives

$$0 = x(x - 3)(x + 2)$$

$x = 0$	$x - 3 = 0$	$x + 2 = 0$
	$x = 3$	$x = -2$

Thus the critical values are $x = 0$, $x = 3$, and $x = -2$.

3. Substituting the critical values into the original function gives the critical points:

$$f(-2) = \tfrac{8}{3}, \quad \text{so } \left(-2, \tfrac{8}{3}\right) \text{ is a critical point.}$$
$$f(0) = 8, \quad \text{so } (0, 8) \text{ is a critical point.}$$
$$f(3) = -\tfrac{31}{4}, \quad \text{so } \left(3, -\tfrac{31}{4}\right) \text{ is a critical point.}$$

4. Testing $f'(x)$ to the left of the smallest critical value, then between the critical values, and finally to the right of the largest critical value will give the sign diagram. Evaluating $f'(x)$ at the test values $x = -3$, $x = -1$, $x = 1$, and $x = 4$ gives the signs to determine relative maxima and minima.

    The sign diagram for $f'(x)$ is

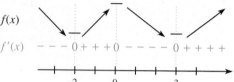

Thus we have

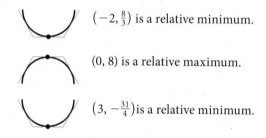

$\left(-2, \tfrac{8}{3}\right)$ is a relative minimum.

$(0, 8)$ is a relative maximum.

$\left(3, -\tfrac{31}{4}\right)$ is a relative minimum.

5. Figure 10.7(a) shows the graph of the function near the critical points, and Figure 10.7(b) shows the graph of the function.

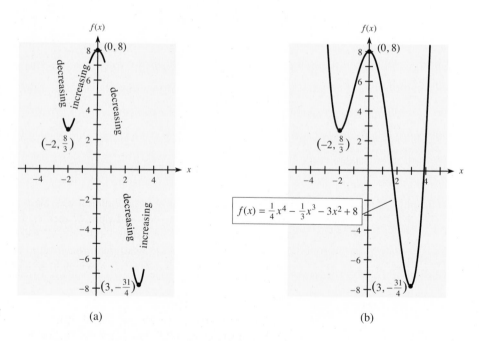

Figure 10.7   (a)   (b)

    Note that we substitute the critical values into the *original function* $f(x)$ to find the y-values of the critical points, but we test for relative maxima and minima by substituting values near the critical values into the *derivative of the function*, $f'(x)$.

    Only four values were needed to test three critical points in Example 1. This method will work *only if* the critical values are tested in order from smallest to largest.

    If the first derivative of f is 0 at x_0 but does not change from positive to negative or from negative to positive as x passes through x_0, then the critical point at x_0 is neither a relative maximum nor a relative minimum. In this case we say that f has a **horizontal point of inflection** (abbreviated HPI) at x_0.

EXAMPLE 2   **Maxima, Minima, and Horizontal Points of Inflection**

Find the relative maxima, relative minima, and horizontal points of inflection of $h(x) = \frac{1}{4}x^4 - \frac{2}{3}x^3 - 2x^2 + 8x + 4$, and sketch its graph.

Solution

1. $h'(x) = x^3 - 2x^2 - 4x + 8$
2. $0 = x^3 - 2x^2 - 4x + 8$ or $0 = x^2(x - 2) - 4(x - 2)$. Therefore, we have $0 = (x - 2)(x^2 - 4)$. Thus $x = -2$ and $x = 2$ are solutions.
3. The critical points are $\left(-2, -\frac{32}{3}\right)$ and $\left(2, \frac{32}{3}\right)$.
4. Using test values, such as $x = -3$, $x = 0$, and $x = 3$, gives the sign diagram for $h'(x)$.

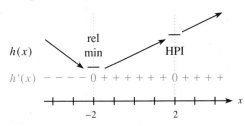

5. Figure 10.8(a) shows the graph of the function near the critical points, and Figure 10.8(b) shows the graph of the function.

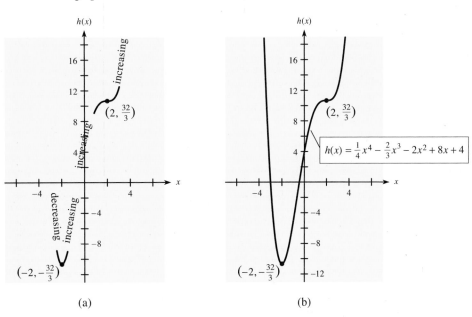

Figure 10.8   (a)   (b)

Technology Note   We can use a graphing calculator as an aid in locating the x-values where $f'(x) = 0$, which are the critical values of $f(x)$, and then making the sign diagram used to classify these critical values, as follows.

1. Enter the derivative of the function in Y1.
2. Graph Y1 and use ZERO to determine the x-values where Y1 = 0; this gives the critical values.
3. Use TABLE to evaluate the derivative to the left and right of each critical value. This creates a sign diagram for $f'(x)$ and can be used to determine whether each critical value is a relative maximum, a relative minimum, or neither.
4. Enter the function in Y2 and use TABLE to evaluate the function at each critical value to find the y-coordinates of the critical points.

Figure 10.9 on the next page illustrates steps (2) and (3) above for $f(x) = \frac{1}{3}x^3 - 4x$, which has derivative $f'(x) = x^2 - 4$. See Appendix A, Section 10.1, for details.

Appendix B, Section 10.1, and the online Excel Guide can be used to find the relative maxima and relative minima of functions with Excel.

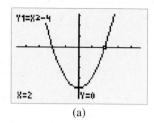

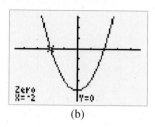

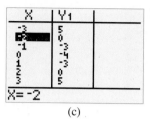

Figure 10.9   (a)   (b)   (c)

EXAMPLE 3   Undefined Derivatives

Find the relative maxima and minima (if any) of the graph of $y = (x + 2)^{2/3}$.

Solution

1. $y' = f'(x) = \dfrac{2}{3}(x + 2)^{-1/3} = \dfrac{2}{3\sqrt[3]{x + 2}}$

2. $0 = \dfrac{2}{3\sqrt[3]{x + 2}}$ has no solutions; $f'(x)$ is undefined at $x = -2$.

3. $f(-2) = 0$, so the critical point is $(-2, 0)$.
4. The sign diagram for $f'(x)$ is

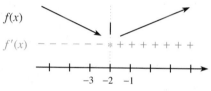

Thus a relative minimum occurs at $(-2, 0)$.   *means $f'(-2)$ is undefined.

5. Figure 10.10(a) shows the graph of the function near the critical points, and Figure 10.10(b) shows the graph of the function.

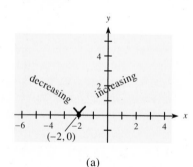

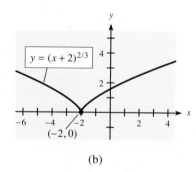

Figure 10.10   (a)   (b)

✓ **CHECKPOINT**

1. The x-values of critical points are found where $f'(x)$ is _____ or _____.
2. Decide whether the following are true or false.
    (a) If $f'(1) = 7$, then $f(x)$ is increasing at $x = 1$.
    (b) If $f'(-2) = 0$, then a relative maximum or a relative minimum occurs at $x = -2$.
    (c) If $f'(-3) = 0$ and $f'(x)$ changes from positive on the left to negative on the right of $x = -3$, then a relative minimum occurs at $x = -3$.
3. If $f(x) = 7 + 3x - x^3$, then $f'(x) = 3 - 3x^2$. Use these functions to decide whether the following statements are true or false.
    (a) The only critical value is $x = 1$.
    (b) The critical points are $(1, 0)$ and $(-1, 0)$.
4. If $f'(x)$ has the following partial sign diagram, make a "stick-figure" sketch of $f(x)$ and label where any maxima and minima occur. Assume that $f(x)$ is defined for all real numbers.

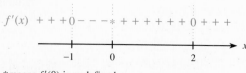

EXAMPLE 4   **Advertising** | **APPLICATION PREVIEW** |

The weekly sales S of a product during an advertising campaign are given by

$$S = \frac{100t}{t^2 + 100}, \qquad 0 \le t \le 20$$

where t is the number of weeks since the beginning of the campaign and S is in thousands of dollars.

(a) Over what interval are sales increasing? Decreasing?
(b) What are the maximum weekly sales?
(c) Sketch the graph for $0 \le t \le 20$.

Solution

(a) To find where S is increasing, we first find $S'(t)$

$$S'(t) = \frac{(t^2 + 100)100 - (100t)2t}{(t^2 + 100)^2}$$

$$= \frac{10,000 - 100t^2}{(t^2 + 100)^2}$$

We see that $S'(t) = 0$ when $10,000 - 100t^2 = 0$, or

$$100(100 - t^2) = 0$$
$$100(10 + t)(10 - t) = 0$$
$$t = -10 \quad \text{or} \quad t = 10$$

Because $S'(t)$ is never undefined ($t^2 + 100 \ne 0$ for any real t) and because $0 \le t \le 20$, our only critical value is $t = 10$. Testing $S'(t)$ to the left and right of $t = 10$ gives the sign diagram.

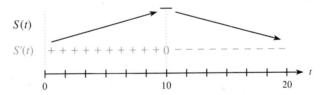

Hence, S is increasing on the interval $[0, 10)$ and decreasing on the interval $(10, 20]$.

(b) Because S is increasing to the left of $t = 10$ and S is decreasing to the right of $t = 10$, the maximum value of S occurs at $t = 10$ and is

$$S = S(10) = \frac{100(10)}{10^2 + 100} = \frac{1000}{200} = 5 \,(\text{thousand dollars})$$

(c) Plotting some additional points gives the graph; see Figure 10.11.

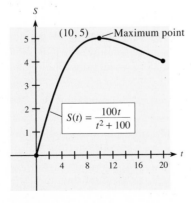

Figure 10.11

Calculator Note   With a graphing calculator, choosing an appropriate viewing window is the key to understanding the graph of a function. To find an appropriate window, set the x-min of the window to the left of the smallest critical value and the x-max to the right of the largest critical value, and set the y-min and y-max to contain the y-coordinates of the critical points. See Example 5 and Appendix A, Section 10.1, for details.

EXAMPLE 5 Critical Points and Viewing Windows

Find the critical values of $f(x) = 0.0001x^3 + 0.003x^2 - 3.6x + 5$. Use them to determine an appropriate viewing window. Then sketch the graph.

Solution

The critical points are helpful in graphing the function. We begin by finding $f'(x)$.

$$f'(x) = 0.0003x^2 + 0.006x - 3.6$$

Now we solve $f'(x) = 0$ to find the critical values.

$$0 = 0.0003x^2 + 0.006x - 3.6$$
$$0 = 0.0003(x^2 + 20x - 12,000)$$
$$0 = 0.0003(x + 120)(x - 100)$$
$$x = -120 \quad \text{or} \quad x = 100$$

We choose a window that includes $x = -120$ and $x = 100$, and use TABLE to find $f(-120) = 307.4$ and $f(100) = -225$. We then graph $y = f(x)$ in a window that includes $y = -225$ and $y = 307.4$ (see Figure 10.12). On this graph we can see that $(-120, 307.4)$ is a relative maximum and that $(100, -225)$ is a relative minimum.

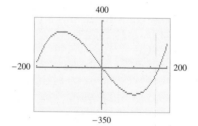

Figure 10.12

EXAMPLE 6 Aging workers

The table shows the millions of Americans who are working full time at selected ages from 27 to 62.

(a) Find a cubic model that gives the number of Americans working full time, y (in millions), as a function of their age, x. Report the model with three significant digits in the coefficients.

(b) Find the relative maximum and minimum of the reported function.

Age	Millions Working Full Time	Age	Millions Working Full Time
27	6.30	47	7.52
32	6.33	52	7.10
37	6.96	57	5.58
42	7.07	62	3.54

Source: Wall Street Journal

Figure 10.13

Solution

(a) A cubic function that models these data is

$$f(x) = -0.000362x^3 + 0.0401x^2 - 1.39x + 21.7$$

Figure 10.13 shows a graph of the model with the data.

(b) We use $f'(x) = -0.001086x^2 + 0.0802x - 1.39$ and solve $f'(x) = 0$ to find the critical values. We can solve

$$-0.001086x^2 + 0.0802x - 1.39 = 0$$

with the quadratic formula or with a graphing calculator. The two solutions are approximately

$$x = 27.8 \quad \text{and} \quad x = 46.1$$

The sign diagram shows that $x = 27.8$ gives a relative minimum and $x = 46.1$ gives a relative maximum.

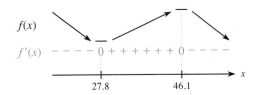

Thus the model indicates a relative minimum of $f(27.8) \approx 6.3$ (million) full-time workers at age 27.8 and a relative maximum of $f(46.1) \approx 7.4$ (million) full-time workers at age 46.1. ■

Spreadsheet Note   Excel can be used to create sign diagrams that determine the maxima and minima of functions. Details are given in Appendix B, Section 10.1, and the Online Excel Guide. ■

✓ CHECKPOINT ANSWERS

1. $f'(x) = 0$ or $f'(x)$ is undefined.
2. (a) True. $f(x)$ is increasing when $f'(x) > 0$.
    (b) False. See Figure 10.8.
    (c) False. A relative maximum occurs at $x = -3$.
3. (a) False. Critical values are $x = 1$ and $x = -1$.
    (b) False. Critical points are $(1, 9)$ and $(-1, 5)$.
4.

Relative maximum at $x = -1$ Relative minimum at $x = 0$ Horizontal point of inflection at $x = 2$

| **EXERCISES** | **10.1**

In Problems 1 and 2, use the indicated points on the graph of $y = f(x)$ to identify points at which $f(x)$ has (a) a relative maximum, (b) a relative minimum, and (c) a horizontal point of inflection.

1.
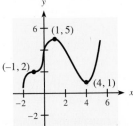

2.

3. Use the graph of $y = f(x)$ in Problem 1 to identify at which of the indicated points the derivative $f'(x)$ (a) changes from positive to negative, (b) changes from negative to positive, and (c) does not change sign.
4. Use the graph of $y = f(x)$ in Problem 2 to identify at which of the indicated points the derivative $f'(x)$ (a) changes from positive to negative, (b) changes from negative to positive, and (c) does not change sign.

In Problems 5 and 6, use the sign diagram for $f'(x)$ to determine (a) the critical values of $f(x)$, (b) intervals on which $f(x)$ increases, (c) intervals on which $f(x)$

decreases, (d) x-values at which relative maxima occur, and (e) x-values at which relative minima occur.

5. $f'(x)$ $\quad --- 0 + + + + + 0 --- \quad x$
$\qquad \qquad \quad 3 \qquad \qquad 7$

6. $f'(x)$ $\quad + + + + 0 + + + + + + +0 ---- \quad x$
$\qquad \qquad \quad -5 \qquad \qquad 8$

In Problems 7–10, (a) find the critical values of the function, and (b) make a sign diagram and determine the relative maxima and minima.

7. $y = 2x^3 - 12x^2 + 6$ 8. $y = x^3 - 3x^2 + 6x + 1$

9. $y = 2x^5 + 5x^4 - 11$ 10. $y = 15x^3 - x^5 + 7$

For each function and graph in Problems 11–14:

(a) **Estimate the coordinates of the relative maxima, relative minima, or horizontal points of inflection by observing the graph.**

(b) **Use $y' = f'(x)$ to find the critical values.**

(c) **Find the critical points.**

(d) **Do the results in part (c) confirm your estimates in part (a)?**

11. $y = x^3 - 3x + 4$ 12. $y = x - \frac{1}{3}x^3$

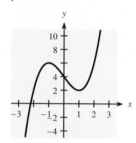 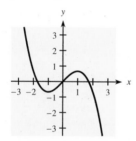

13. $y = x^3 + 3x^2 + 3x - 2$ 14. $y = x^3 - 6x^2 + 12x + 1$

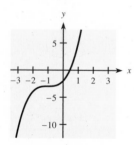 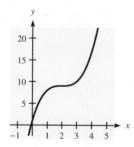

For each function in Problems 15–20:

(a) **Find $y' = f'(x)$.**

(b) **Find the critical values.**

(c) **Find the critical points.**

(d) **Find intervals of x-values where the function is increasing and where it is decreasing.**

(e) **Classify the critical points as relative maxima, relative minima, or horizontal points of inflection. In each case, check your conclusions with a graphing utility.**

15. $y = \frac{1}{2}x^2 - x$ 16. $y = x^2 + 4x$

17. $y = \frac{x^3}{3} + \frac{x^2}{2} - 2x + 1$ 18. $y = \frac{x^4}{4} - \frac{x^3}{3} - 2$

19. $y = x^{2/3}$ 20. $y = -(x - 3)^{2/3}$

For each function and graph in Problems 21–24:

(a) **Use the graph to identify x-values for which $y' > 0$, $y' < 0$, $y' = 0$, and y' does not exist.**

(b) **Use the derivative to check your conclusions.**

21. $y = 6 - x - x^2$ 22. $y = \frac{1}{2}x^2 - 4x + 1$

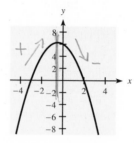 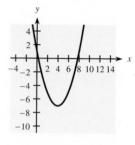

23. $y = 6 + x^3 - \frac{1}{15}x^5$ 24. $y = x^4 - 2x^2 - 1$

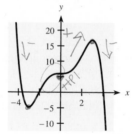 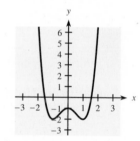

For each function in Problems 25–30, find the relative maxima, relative minima, horizontal points of inflection, and sketch the graph. Check your graph with a graphing utility.

25. $y = \frac{1}{3}x^3 - x^2 + x + 1$

26. $y = \frac{1}{4}x^4 - \frac{2}{3}x^3 + \frac{1}{2}x^2 - 2$

27. $y = \frac{1}{3}x^3 + x^2 - 24x + 20$

28. $C(x) = x^3 - \frac{3}{2}x^2 - 18x + 5$

29. $y = 3x^5 - 5x^3 + 1$

30. $y = \frac{1}{6}x^6 - x^4 + 7$

In Problems 31–36, both a function and its derivative are given. Use them to find critical values, critical points, intervals on which the function is increasing and decreasing, relative maxima, relative minima, and horizontal points of inflection; sketch the graph of each function.

31. $y = (x^2 - 2x)^2$ $\dfrac{dy}{dx} = 4x(x - 1)(x - 2)$

32. $f(x) = (x^2 - 4)^2$ $f'(x) = 4x(x + 2)(x - 2)$

33. $y = \dfrac{x^3(x - 5)^2}{27}$ $\dfrac{dy}{dx} = \dfrac{5x^2(x - 3)(x - 5)}{27}$

34. $y = \dfrac{x^2(x - 5)^3}{27}$ $\dfrac{dy}{dx} = \dfrac{5x(x - 2)(x - 5)^2}{27}$

35. $f(x) = x^{2/3}(x - 5)$ $f'(x) = \dfrac{5(x - 2)}{3x^{1/3}}$

36. $f(x) = x - 3x^{2/3}$ $f'(x) = \dfrac{x^{1/3} - 2}{x^{1/3}}$

In Problems 37–42, use the derivative to locate critical points and determine a viewing window that shows all features of the graph. Use a graphing utility to sketch a complete graph.

37. $f(x) = x^3 - 225x^2 + 15{,}000x - 12{,}000$
38. $f(x) = x^3 - 15x^2 - 16{,}800x + 80{,}000$
39. $f(x) = x^4 - 160x^3 + 7200x^2 - 40{,}000$
40. $f(x) = x^4 - 240x^3 + 16{,}200x^2 - 60{,}000$
41. $y = 7.5x^4 - x^3 + 2$
42. $y = 2 - x^3 - 7.5x^4$

In each of Problems 43–46, a graph of $f'(x)$ is given. Use the graph to determine the critical values of $f(x)$, where $f(x)$ is increasing, where it is decreasing, and where it has relative maxima, relative minima, and horizontal points of inflection. In each case sketch a possible graph for $f(x)$ that passes through $(0, 0)$.

43. $f'(x) = x^2 - x - 2$ 44. $f'(x) = 4x - x^2$

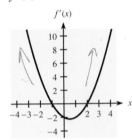
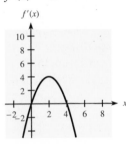

45. $f'(x) = x^3 - 3x^2$ 46. $f'(x) = x(x - 2)^2$

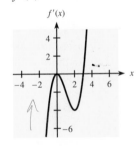
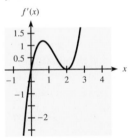

In Problems 47 and 48, two graphs are given. One is the graph of f and the other is the graph of f'. Decide which is which and explain your reasoning.

47.
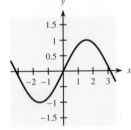

48.
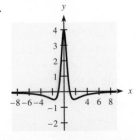
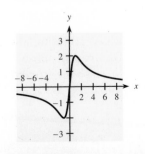

APPLICATIONS

49. *Advertising and sales* Suppose that the daily sales (in dollars) t days after the end of an advertising campaign are given by

$$S = 1000 + \frac{400}{t + 1}, \quad t \ge 0$$

Does S increase for all $t \ge 0$, decrease for all $t \ge 0$, or change direction at some point?

50. *Pricing and sales* Suppose that a chain of auto service stations, Quick-Oil, Inc., has found that its monthly sales volume y (in thousands of dollars) is related to the price p (in dollars) of an oil change by

$$y = \frac{90}{\sqrt{p + 5}}, \quad p > 10$$

Is y increasing or decreasing for all values of $p > 10$?

51. *Productivity* A time study showed that, on average, the productivity of a worker after t hours on the job can be modeled by

$$P(t) = 27t + 6t^2 - t^3, \quad 0 \le t \le 8$$

where P is the number of units produced per hour.
   (a) Find the critical values of this function.
   (b) Which critical value makes sense in this model?
   (c) For what values of t is P increasing?
   (d) Graph the function for $0 \le t \le 8$.

52. *Production* Analysis of daily output of a factory shows that, on average, the number of units per hour y produced after t hours of production is

$$y = 70t + \frac{1}{2}t^2 - t^3, \quad 0 \le t \le 8$$

   (a) Find the critical values of this function.
   (b) Which critical values make sense in this particular problem?
   (c) For which values of t, for $0 \le t \le 8$, is y increasing?
   (d) Graph this function.

53. *Production costs* Suppose that the average cost, in dollars, of producing a shipment of a certain product is

$$\overline{C} = 5000x + \frac{125{,}000}{x}, \quad x > 0$$

where x is the number of machines used in the production process.
   (a) Find the critical values of this function.
   (b) Over what interval does the average cost decrease?
   (c) Over what interval does the average cost increase?

54. *Average costs* Suppose the average costs of a mining operation depend on the number of machines used, and average costs, in dollars, are given by

$$\overline{C}(x) = 2900x + \frac{1{,}278{,}900}{x}, \quad x > 0$$

where x is the number of machines used.

(a) Find the critical values of $\overline{C}(x)$ that lie in the domain of the problem.
(b) Over what interval in the domain do average costs decrease?
(c) Over what interval in the domain do average costs increase?
(d) How many machines give minimum average costs?
(e) What is the minimum average cost?

55. **Marginal revenue** Suppose the weekly marginal revenue function for selling x units of a product is given by the graph in the figure.

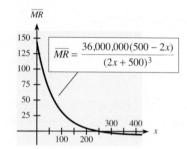

(a) At each of $x = 150$, $x = 250$, and $x = 350$, what is happening to revenue?
(b) Over what interval is revenue increasing?
(c) How many units must be sold to maximize revenue?

56. **Earnings** Suppose that the rate of change $f'(x)$ of the average annual earnings of new car salespersons is shown in the figure.

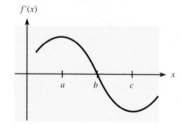

(a) If a, b, and c represent certain years, what is happening to $f(x)$, the average annual earnings of the salespersons, at a, b, and c?
(b) Over what interval (involving a, b, or c) is there an increase in $f(x)$, the average annual earnings of the salespersons?

57. **Revenue** The weekly revenue of a certain recently released film is given by

$$R(t) = \frac{50t}{t^2 + 36}, \quad t \geq 0$$

where R is in millions of dollars and t is in weeks.
(a) Find the critical values.
(b) For how many weeks will weekly revenue increase?

58. **Medication** Suppose that the concentration C of a medication in the bloodstream t hours after an injection is given by

$$C(t) = \frac{0.2t}{t^2 + 1}$$

(a) Determine the number of hours before C attains its maximum.
(b) Find the maximum concentration.

59. **Candidate recognition** Suppose that the proportion P of voters who recognize a candidate's name t months after the start of the campaign is given by

$$P(t) = \frac{13t}{t^2 + 100} + 0.18$$

(a) How many months after the start of the campaign is recognition at its maximum?
(b) To have greatest recognition on November 1, when should a campaign be launched?

60. **Medication** The number of milligrams x of a medication in the bloodstream t hours after a dose is taken can be modeled by

$$x(t) = \frac{2000t}{t^2 + 16}$$

(a) For what t-values is x increasing?
(b) Find the t-value at which x is maximum.
(c) Find the maximum value of x.

61. **Worldwide cell phone subscriberships** In 2013, worldwide cell phone subscriberships surpassed the world's total population of 6.8 billion. Using data from 2009 and projected to 2020, the billions of subscriberships can be modeled by

$$C(t) = 0.000286t^3 - 0.0443t^2 + 1.49t - 5.36$$

where t is equal to the number of years after 2000. When does this model estimate that the number of subscribers reached its maximum, and what maximum does it estimate? *Source: portioresearch.com*

62. **Economic dependency ratio** The economic dependency ratio is defined as the number of persons in the total population who are not in the workforce per 100 in the workforce. Since 1960, Baby Boomers in the workforce and a decrease in the birth rate have caused a significant decrease in the economic dependency ratio. With data for selected years from 1960 and projected to 2050, the economic dependency ratio R can be modeled by the function

$$R(x) = -0.0002x^3 + 0.052x^2 - 4.06x + 192$$

where x is the number of years past 1950 (*Source:* U.S. Department of Labor). Use this model to find the year in which the economic dependency ratio reached its minimum. What was happening in the United States around this time that helps explain why the minimum occurred in this year?

63. **Modeling *China's labor pool*** The following table shows the millions of individuals, ages 15 to 59, in China's labor pool for selected years from 1975 and projected to 2050.
    (a) Find the cubic function $L(t)$ that best models the size of this labor pool, where t is the number of years after 1970. Report the model with three significant digit coefficients.
    (b) Use the reported model to estimate the maximum size of this labor pool before it begins to shrink and the year when the maximum occurs.

Year	Labor Pool (millions)	Year	Labor Pool (millions)
1975	490	2015	920
1980	560	2020	920
1985	650	2025	905
1990	730	2030	875
1995	760	2035	830
2000	800	2040	820
2005	875	2045	800
2010	910	2050	670

Source: United Nations

64. **Modeling *Energy from crude oil*** The table shows the total energy supply from crude oil products, in quadrillion BTUs, for selected years from 2010 and projected to 2040.
    (a) Find the cubic function that is the best model for the data. Use x as the number of years after 2010 and $C(x)$ as the quadrillion BTUs of energy from crude oil products. Report the model with three significant digit coefficients.
    (b) Find the critical values of the reported model.

(c) Find the reported model's critical points and classify them.
(d) Do you think this model will be valid for years past 2040? Explain.

Total Energy Supply from Crude Oil Products (in quadrillion BTUs)

Year	BTUs	Year	BTUs
2010	11.6	2030	13.5
2015	15.6	2035	13.4
2020	16.0	2040	13.1
2025	14.5		

Source: U.S. Energy Information Administration

65. **Modeling *Employment in manufacturing*** The table shows the total employment in U.S. manufacturing, in millions, for selected years from 2010 and projected to 2040.
    (a) Find the cubic function that is the best model for the data. Use t as the number of years after 2010 and $M(t)$ as the millions employed in U.S. manufacturing. Report the model with three significant digit coefficients.
    (b) Find the reported model's critical points and classify them.
    (c) Do you think this model will be accurate for years between the critical points? Explain.

U.S. Employment in Manufacturing (in millions)

Year	Employment	Year	Employment
2010	11.5	2030	12.5
2015	12.4	2035	11.8
2020	12.8	2040	11.0
2025	12.9		

Source: U.S. Department of Energy

OBJECTIVES   **10.2**

- To determine the concavity of the graph of a function at a point
- To find points of inflection of graphs of functions
- To use the second-derivative test to graph functions

Concavity; Points of Inflection

■ | APPLICATION PREVIEW | ■

Suppose that a retailer wishes to sell his store and uses the graph in Figure 10.14 on the next page to show how profits have increased since he opened the store and the potential for profit in the future. Can we conclude that profits will continue to grow, or should we be concerned about future earnings?

Note that although profits are still increasing in 2017, they seem to be increasing more slowly than in previous years. Indeed, they appear to have been growing at a decreasing rate since about 2015. We can analyze this situation more carefully by identifying the point at which profit begins to grow more slowly, that is, the *point of diminishing returns*. (See Example 3.)

(Continued)

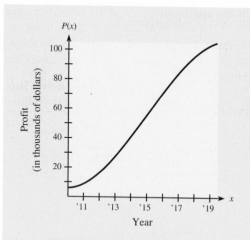

Figure 10.14

Just as we used the first derivative to determine whether a curve was increasing or decreasing on a given interval, we can use the second derivative to determine whether the curve is concave up or concave down on an interval.

Concavity   A curve is said to be **concave up** on an interval (a, b) if at each point on the interval the curve is above its tangent at the point (see Figure 10.15(a)). If the curve is below all its tangents on a given interval, it is **concave down** on the interval (Figure 10.15(b)).

Looking at Figure 10.15(a), we see that the *slopes* of the tangent lines increase over the interval where the graph is concave up. Because $f'(x)$ gives the slopes of those tangents, it follows that $f'(x)$ is increasing over the interval where $f(x)$ is concave up. However, we know that $f'(x)$ is increasing when its derivative, $f''(x)$, is positive. That is, if the second derivative is positive, the curve is concave up.

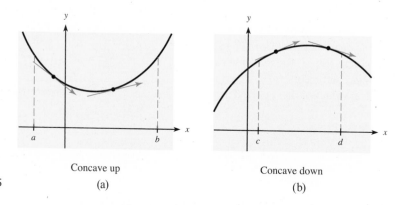

Concave up	Concave down
(a)	(b)

Figure 10.15

Similarly, if the second derivative of a function is negative over an interval, the slopes of the tangents to the graph decrease over that interval. This happens when the tangent lines are above the graph, as in Figure 10.15(b), so the graph must be concave down on this interval.

Thus we see that the second derivative can be used to determine the concavity of a curve.

Concave Up and Concave Down

Assume that the first and second derivatives of function f exist.

The function f is **concave up** on an interval I, if $f''(x) > 0$ on I, and **concave up** at the point $(a, f(a))$, if $f''(a) > 0$.

The function f is **concave down** on an interval I, if $f''(x) < 0$ on I, and **concave down** at the point $(a, f(a))$, if $f''(a) < 0$.

EXAMPLE 1 Concavity at a Point

Is the graph of $f(x) = x^3 - 4x^2 + 3$ concave up or down at the point
(a) $(1, 0)$? (b) $(2, -5)$?

Solution

(a) We must find $f''(x)$ before we can answer this question.

$$f'(x) = 3x^2 - 8x \qquad f''(x) = 6x - 8$$

Then $f''(1) = 6(1) - 8 = -2$, so the graph is concave down at $(1, 0)$.

(b) Because $f''(2) = 6(2) - 8 = 4$, the graph is concave up at $(2, -5)$. The graph of $f(x) = x^3 - 4x^2 + 3$ is shown in Figure 10.16(a). ■

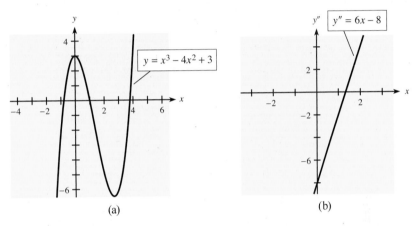

Figure 10.16 (a) (b)

Points of Inflection   Looking at the graph of $y = x^3 - 4x^2 + 3$ [Figure 10.16(a)], we see that the curve is concave down on the left and concave up on the right. Thus it has changed from concave down to concave up, and from Example 1 we would expect the concavity to change somewhere between $x = 1$ and $x = 2$. Figure 10.16(b) shows the graph of $y'' = f''(x) = 6x - 8$, and we can see that $y'' = 0$ when $x = \frac{4}{3}$ and that $y'' < 0$ for $x < \frac{4}{3}$ and $y'' > 0$ for $x > \frac{4}{3}$. Thus the second derivative changes sign at $x = \frac{4}{3}$, so the concavity of the graph of $y = f(x)$ changes at $x = \frac{4}{3}$, $y = -\frac{47}{27}$. The point where concavity changes is called a **point of inflection.**

Point of Inflection

A point (x_0, y_0) on the graph of a function f is called a **point of inflection** if the curve is concave up on one side of the point and concave down on the other side. The second derivative at this point, $f''(x_0)$, will be 0 or undefined.

In general, we can find points of inflection and information about concavity as follows.

Finding Points of Inflection and Concavity

Procedure	Example
To find the point(s) of inflection of a curve and intervals where it is concave up and where it is concave down:	Find the points of inflection and concavity of the graph of $y = \dfrac{x^4}{2} - x^3 + 5$.
1. Find the second derivative of the function.	1. $y' = f'(x) = 2x^3 - 3x^2$ $y'' = f''(x) = 6x^2 - 6x$
2. Set the second derivative equal to 0, and solve for x. Potential points of inflection occur at these values of x or at values of x where $f(x)$ is defined and $f''(x)$ is undefined.	2. $0 = 6x^2 - 6x = 6x(x - 1)$ has solutions $x = 0$, $x = 1$. $f''(x)$ is defined everywhere.

(Continued)

Finding Points of Inflection and Concavity (*Continued*)

Procedure	Example

3. Find the potential points of inflection.
4. If the second derivative has opposite signs on the two sides of one of these values of x, a point of inflection occurs.
   The curve is concave up where $f''(x) > 0$ and concave down where $f''(x) < 0$.
   The changes in the sign of $f''(x)$ correspond to changes in concavity and occur at points of inflection.

3. $(0, 5)$ and $\left(1, \frac{9}{2}\right)$ are potential points of inflection.
4. A **sign diagram for $f''(x)$** is

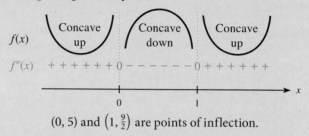

$(0, 5)$ and $\left(1, \frac{9}{2}\right)$ are points of inflection.

See the graph in Figure 10.17.

The graph of $y = \frac{1}{2}x^4 - x^3 + 5$ is shown in Figure 10.17. Note the points of inflection at $(0, 5)$ and $\left(1, \frac{9}{2}\right)$. The point of inflection at $(0, 5)$ is a horizontal point of inflection because $f'(x)$ is also 0 at $x = 0$.

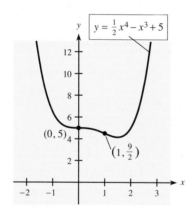

Figure 10.17

EXAMPLE 2 Water Purity

Suppose that a real estate developer wishes to remove pollution from a small lake so that she can sell lakefront homes on a "crystal clear" lake. The graph in Figure 10.18 shows the relation between dollars spent on cleaning the lake and the purity of the water. The point of inflection on the graph is called the **point of diminishing returns** on her investment because it is where the *rate* of return on her investment changes from increasing to decreasing. Show that the rate of change in the purity of the lake, $f'(x)$, is maximized at this point, $x = c$. Assume that $f(c)$, $f'(c)$, and $f''(c)$ are defined.

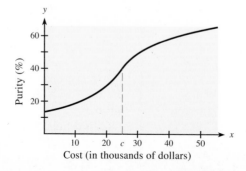

Figure 10.18

Solution

Because $x = c$ is a point of inflection for $f(x)$, we know that the concavity must change at $x = c$. From the figure we see the following.

$$x < c: \quad f(x) \text{ is concave up, so } f''(x) > 0.$$
$$f''(x) > 0 \text{ means that } f'(x) \text{ is increasing.}$$
$$x > c: \quad f(x) \text{ is concave down, so } f''(x) < 0.$$
$$f''(x) < 0 \text{ means that } f'(x) \text{ is decreasing.}$$

Thus $f'(x)$ has $f'(c)$ as its relative maximum.

EXAMPLE   3   **Diminishing Returns | APPLICATION PREVIEW |**

Suppose the annual profit for a store (in thousands of dollars) is given by

$$P(x) = -0.2x^3 + 3x^2 + 6$$

where x is the number of years past 2010. If this model is accurate, find the point of diminishing returns for the profit.

Solution

The point of diminishing returns occurs at the point of inflection. Thus we seek the point where the graph of this function changes from concave up to concave down, if such a point exists.

$$P'(x) = -0.6x^2 + 6x$$
$$P''(x) = -1.2x + 6$$
$$P''(x) = 0 \quad \text{when} \quad 0 = -1.2x + 6 \quad \text{or} \quad \text{when } x = 5$$

Thus $x = 5$ is a possible point of inflection. We test $P''(x)$ to the left and the right of $x = 5$.

$$P''(4) = 1.2 > 0 \Rightarrow \text{concave up to the left of } x = 5$$
$$P''(6) = -1.2 < 0 \Rightarrow \text{concave down to the right of } x = 5$$

Thus $(5, 56)$ is the point of inflection for the graph, and the point of diminishing returns for the profit is when $x = 5$ (in the year 2015) and is $P(5) = 56$ thousand dollars. Figure 10.19 shows the graphs of $P(x)$, $P'(x)$, and $P''(x)$. At $x = 5$, we see that the point of diminishing returns on the graph of $P(x)$ corresponds to the maximum point of the graph of $P'(x)$ and the zero (or x-intercept) of the graph of $P''(x)$.

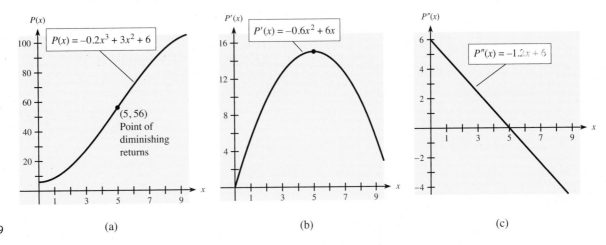

Figure 10.19   (a)   (b)   (c)

✓ CHECKPOINT

1. If $f''(x) > 0$, then $f(x)$ is concave _____.
2. At what value of x does the graph $y = \frac{1}{3}x^3 - 2x^2 + 2x$ have a point of inflection?
3. On the graph below, locate any points of inflection (approximately) and label where the curve satisfies $f''(x) > 0$ and $f''(x) < 0$.

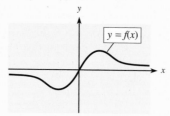

4. Determine whether the following is true or false. If $f''(0) = 0$, then $f(x)$ has a point of inflection at $x = 0$.

Second-Derivative Test   We can use information about points of inflection and concavity to help sketch graphs. For example, if we know that the curve is concave up at a critical point where $f'(x) = 0$, then the point must be a relative minimum because the tangent to the curve is horizontal at the critical point, and only a point at the bottom of a "concave up" curve could have a horizontal tangent [see Figure 10.20(a)].

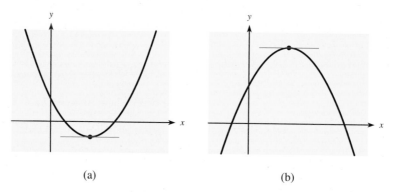

Figure 10.20   (a)   (b)

On the other hand, if the curve is concave down at a critical point where $f'(x) = 0$, then the point is a relative maximum [see Figure 10.20(b)].

Thus we can use the **second-derivative test** to determine whether a critical point where $f'(x) = 0$ is a relative maximum or minimum.

Second-Derivative Test

Procedure	Example
To find relative maxima and minima of a function:	Find the relative maxima and minima of $y = f(x) = \frac{1}{3}x^3 - x^2 - 3x + 2$.
1. Find the critical values of the function.	1. $f'(x) = x^2 - 2x - 3$ $0 = (x - 3)(x + 1)$ has solutions $x = -1$ and $x = 3$. No values of x make $f'(x)$ undefined.
2. Substitute the critical values into $f(x)$ to find the critical points.	2. $f(-1) = \frac{11}{3}$ $f(3) = -7$ The critical points are $\left(-1, \frac{11}{3}\right)$ and $(3, -7)$.
3. Evaluate $f''(x)$ at each critical value for which $f'(x) = 0$. (a) If $f''(x_0) < 0$, a relative maximum occurs at x_0. (b) If $f''(x_0) > 0$, a relative minimum occurs at x_0. (c) If $f''(x_0) = 0$, or $f''(x_0)$ is undefined, the second-derivative test fails; use the first-derivative test.	3. $f''(x) = 2x - 2$ $f''(-1) = 2(-1) - 2 = -4 < 0$, so $\left(-1, \frac{11}{3}\right)$ is a relative maximum point. $f''(3) = 2(3) - 2 = 4 > 0$, so $(3, -7)$ is a relative minimum point. (The graph is shown in Figure 10.21.)

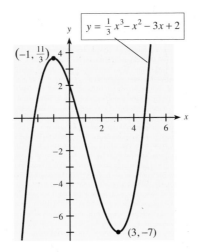

Figure 10.21

Because the second-derivative test (just shown) and the first-derivative test (in Section 10.1) are both methods for classifying critical values, let's compare the advantages and disadvantages of the second derivative test.

Advantage of the Second-Derivative Test	Disadvantages of the Second-Derivative Test
It is quick and easy for many functions.	The second derivative is difficult to find for some functions. The second-derivative test sometimes fails to give results.

EXAMPLE 4 Maxima, Minima, and Points of Inflection

Find the relative maxima and minima and points of inflection of $y = 3x^4 - 4x^3 - 2$.

Solution

$$y' = f'(x) = 12x^3 - 12x^2$$

Solving $0 = 12x^3 - 12x^2 = 12x^2(x - 1)$ gives $x = 1$ and $x = 0$. Thus the critical points are $(1, -3)$ and $(0, -2)$.

$$y'' = f''(x) = 36x^2 - 24x$$
$$f''(1) = 12 > 0 \Rightarrow (1, -3) \text{ is a relative minimum point.}$$
$$f''(0) = 0 \Rightarrow \text{the second-derivative test fails.}$$

Because the second-derivative test fails, we must use the first-derivative test at the critical point $(0, -2)$. A sign diagram for $f'(x)$ shows that $(0, -2)$ is a horizontal point of inflection.

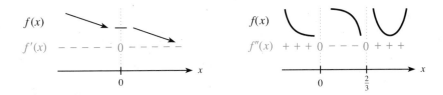

We look for points of inflection by setting $f''(x) = 0$ and solving for x. We find that $0 = 36x^2 - 24x$ has solutions $x = 0$ and $x = \frac{2}{3}$. The sign diagram for $f''(x)$ shows points of inflection at both of these x-values, that is at the points $(0, -2)$ and $\left(\frac{2}{3}, -\frac{70}{27}\right)$. The point $(0, -2)$ is a special point, where the curve changes concavity *and* has a horizontal tangent (see Figure 10.22 on the next page).

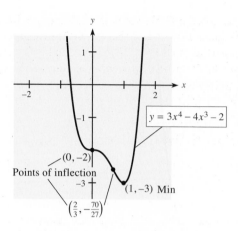

Figure 10.22

Calculator Note

We can use a graphing calculator to explore the relationships among f, f', and f'', as we did in the previous section for f and f'. See Appendix A, Section 10.2, for details. ■

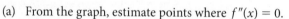

EXAMPLE 5

Concavity from the Graph of $y = f(x)$

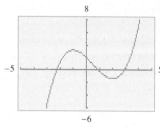

Figure 10.23

Figure 10.23 shows the graph of $f(x) = \frac{1}{6}(2x^3 - 3x^2 - 12x + 12)$.

(a) From the graph, estimate points where $f''(x) = 0$.
(b) From the graph, observe intervals where $f''(x) > 0$ and where $f''(x) < 0$.
(c) Check the conclusions from parts (a) and (b) by calculating $f''(x)$ and graphing it.

Solution

(a) From Figure 10.23, the point of inflection appears to be near $x = \frac{1}{2}$, so we expect $f''(x) = 0$ at (or very near) $x = \frac{1}{2}$.
(b) We see that the graph is concave down (so $f''(x) < 0$) to the left of the point of inflection. That is, $f''(x) < 0$ when $x < \frac{1}{2}$. Similarly, $f''(x) > 0$ when $x > \frac{1}{2}$.

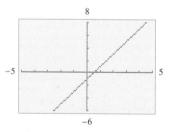

Figure 10.24

(c) $f(x) = \frac{1}{6}(2x^3 - 3x^2 - 12x + 12)$
$f'(x) = \frac{1}{6}(6x^2 - 6x - 12) = x^2 - x - 2$
$f''(x) = 2x - 1$
Thus $f''(x) = 0$ when $x = \frac{1}{2}$.
Figure 10.24 shows the graph of $f''(x) = 2x - 1$. We see that the graph crosses the x-axis ($f''(x) = 0$) when $x = \frac{1}{2}$, is below the x-axis ($f''(x) < 0$) when $x < \frac{1}{2}$, and is above the x-axis ($f''(x) > 0$) when $x > \frac{1}{2}$. This verifies our conclusions from parts (a) and (b). ■

Spreadsheet Note

We can investigate the function in Example 5 by using Excel to find the outputs of $f(x) = \frac{1}{6}(2x^3 - 3x^2 - 12x + 12)$, $f'(x)$, and $f''(x)$ for values of x at and near the critical points and including values where $f''(x) = 0$. The following Excel Spreadsheet illustrates how we extend the Appendix B, Section 10.1, technique for relative maxima and minima by adding a column for $f''(x)$.

	A	B	C	D
	x	$f(x)$	$f'(x)$	$f''(x)$
1				
2	−2	1.3333	4	−5
3	−1	3.1667	0	−3
4	0	2	−2	−1
5	1/2	0.9167	−2.25	0
6	1	−0.1667	−2	1
7	2	−1.3333	0	3
8	3	0.5	4	5

By looking at column C, we see that $f'(x)$ changes from positive to negative as x passes through the value $x = -1$ and that the derivative changes from negative to positive as x passes through the value $x = 2$. Thus the graph has a relative maximum at $x = -1$ and a relative minimum at $x = 2$. By looking at column D, we can also see that $f''(x)$ changes from negative to positive as x passes through the value $x = 1/2$, so the graph has a point of inflection at $x = 1/2$. ■

EXAMPLE 6

Concavity from the Graph of $y = f'(x)$

Figure 10.25 shows the graph of $f'(x) = -x^2 - 2x$. Use the graph of $f'(x)$ to do the following.

(a) Find intervals on which $f(x)$ is concave down and concave up.
(b) Find x-values at which $f(x)$ has a point of inflection.
(c) Check the conclusions from parts (a) and (b) by finding $f''(x)$ and graphing it.
(d) For $f(x) = \frac{1}{3}(9 - x^3 - 3x^2)$, calculate $f'(x)$ to verify that this could be $f(x)$.

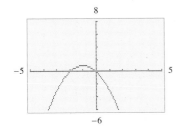

Figure 10.25

Solution
(a) Concavity for $f(x)$ can be found from the sign of $f''(x)$. Because $f''(x)$ is the first derivative of $f'(x)$, wherever the graph of $f'(x)$ is increasing, it follows that $f''(x) > 0$. Thus $f''(x) > 0$ and $f(x)$ is concave up when $x < -1$. Similarly, $f''(x) < 0$, and $f(x)$ is concave down, when $f'(x)$ is decreasing—that is, when $x > -1$.
(b) From (a) we know that $f''(x)$ changes sign at $x = -1$, so $f(x)$ has a point of inflection at $x = -1$. Note that $f'(x)$ has its maximum at the x-value where $f(x)$ has a point of inflection. In fact, points of inflection for $f(x)$ will correspond to relative extrema for $f'(x)$.
(c) For $f'(x) = -x^2 - 2x$, we have $f''(x) = -2x - 2$. Figure 10.26 shows the graph of $y = f''(x)$ and verifies our conclusions from (a) and (b).
(d) If $f(x) = \frac{1}{3}(9 - x^3 - 3x^2)$, then $f'(x) = \frac{1}{3}(-3x^2 - 6x) = -x^2 - 2x$. Figure 10.27 shows the graph of $f(x) = \frac{1}{3}(9 - x^3 - 3x^2)$. Note that the point of inflection and the concavity correspond to what we discovered in parts (a) and (b). ■

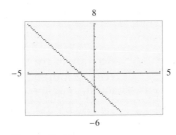

Figure 10.26

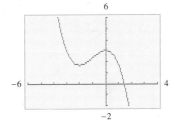

Figure 10.27

The relationships among $f(x)$, $f'(x)$, and $f''(x)$ that we explored in Example 6 can be summarized as follows.

$f(x)$	Concave Up	Concave Down	Point of Inflection	
$f'(x)$	increasing	decreasing	maximum	minimum
$f''(x)$	positive (+)	negative (−)	(+) to (−)	(−) to (+)

1. Up
2. At $x = 2$
3. Points of inflection at A, B, and C
    $f''(x) < 0$ to the left of A and between B and C
    $f''(x) > 0$ between A and B and to the right of C

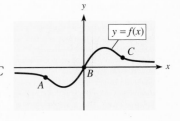

4. (a) False. Consider $f(x) = x^4$.

| EXERCISES | 10.2

In Problems 1 and 2, determine whether each function is concave up or concave down at the indicated points.

1. $f(x) = x^3 - 3x^2 + 1$ at (a) $x = -2$ (b) $x = 3$
2. $f(x) = x^3 + 6x - 4$ at (a) $x = -5$ (b) $x = 7$

In Problems 3–8, use the indicated x-values on the graph of $y = f(x)$ to find the following.

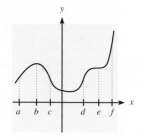

3. Find intervals over which the graph is concave down.
4. Find intervals over which the graph is concave up.
5. Find intervals where $f''(x) > 0$.
6. Find intervals where $f''(x) < 0$.
7. Find the x-coordinates of three points of inflection.
8. Find the x-coordinate of a horizontal point of inflection.

In Problems 9–12, a function and its graph are given. Use the second derivative to determine intervals on which the function is concave up, to determine intervals on which it is concave down, and to locate points of inflection. Check these results against the graph shown.

9. $f(x) = x^3 - 6x^2 + 5x + 6$

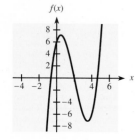

10. $y = x^3 - 9x^2$

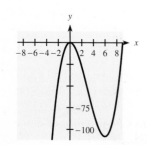

11. $f(x) = \frac{1}{4}x^4 + \frac{1}{2}x^3 - 3x^2 + 3$

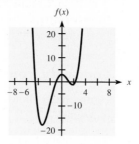

12. $y = 2x^4 - 6x^2 + 4$

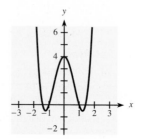

Find the relative maxima, relative minima, and points of inflection, and sketch the graphs of the functions, in Problems 13–18.

13. $y = x^2 - 4x + 2$ 14. $y = x^3 - x^2$
15. $y = \frac{1}{3}x^3 - 2x^2 + 3x + 2$
16. $y = x^3 - 3x^2 + 6$
17. $y = x^4 - 16x^2$
18. $y = x^4 - 8x^3 + 16x^2$

In Problems 19–22, a function and its first and second derivatives are given. Use these to find relative maxima, relative minima, and points of inflection; sketch the graph of each function.

19. $f(x) = 3x^5 - 20x^3$
    $f'(x) = 15x^2(x - 2)(x + 2)$
    $f''(x) = 60x(x^2 - 2)$
20. $f(x) = x^5 - 5x^4$
    $f'(x) = 5x^3(x - 4)$
    $f''(x) = 20x^2(x - 3)$
21. $y = x^{1/3}(x - 4)$ 22. $y = x^{4/3}(x - 7)$
    $y' = \dfrac{4(x - 1)}{3x^{2/3}}$ $y' = \dfrac{7x^{1/3}(x - 4)}{3}$
    $y'' = \dfrac{4(x + 2)}{9x^{5/3}}$ $y'' = \dfrac{28(x - 1)}{9x^{2/3}}$

In Problems 23 and 24, a function and its graph are given.

(a) From the graph, estimate where $f''(x) > 0$, where $f''(x) < 0$, and where $f''(x) = 0$.
(b) Use (a) to decide where $f'(x)$ has its relative maxima and relative minima.
(c) Verify your results in parts (a) and (b) by finding $f'(x)$ and $f''(x)$ and then graphing each with a graphing utility.

23. $f(x) = -\frac{1}{3}x^3 + x^2 + 8x - 12$

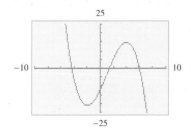

24. $f(x) = \frac{1}{3}x^3 + 2x^2 - 12x - 20$

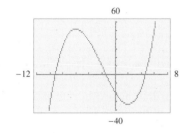

 In Problems 25 and 26, $f'(x)$ and its graph are given. Use the graph of $f'(x)$ to determine the following.

(a) Where is the graph of $f(x)$ concave up and where is it concave down?

(b) Where does $f(x)$ have any points of inflection?

(c) Find $f''(x)$ and graph it. Then use that graph to check your conclusions from parts (a) and (b).

(d) Sketch a possible graph of $f(x)$.

25. $f'(x) = 4x - x^2$

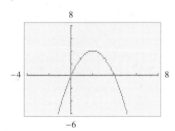

26. $f'(x) = x^2 - x - 2$

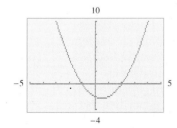

In Problems 27 and 28, use the graph shown in the figure and identify points from A through I that satisfy the given conditions.

27. (a) $f'(x) > 0$ and $f''(x) > 0$
    (b) $f'(x) < 0$ and $f''(x) < 0$
    (c) $f'(x) = 0$ and $f''(x) > 0$
    (d) $f'(x) > 0$ and $f''(x) = 0$
    (e) $f'(x) = 0$ and $f''(x) = 0$

28. (a) $f'(x) > 0$ and $f''(x) < 0$
    (b) $f'(x) < 0$ and $f''(x) > 0$
    (c) $f'(x) = 0$ and $f''(x) < 0$
    (d) $f'(x) < 0$ and $f''(x) = 0$

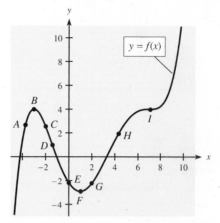

In Problems 29 and 30, a graph is given. Tell where $f(x)$ is concave up, where it is concave down, and where it has points of inflection on the interval $-2 < x < 2$, if the given graph is the graph of

(a) $f(x)$.   (b) $f'(x)$.   (c) $f''(x)$.

29.

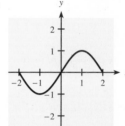

30.

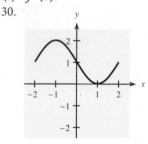

APPLICATIONS

31. *Productivity—diminishing returns* The figure is a typical graph of worker productivity as a function of time on the job.

(a) If P represents the productivity and t represents the time, write a mathematical symbol that represents the rate of change of productivity with respect to time.

(b) Which of A, B, and C is the critical point for the rate of change found in part (a)? This point actually corresponds to the point at which the rate of production is maximized, or the point for maximum worker efficiency. In economics, this is called the point of diminishing returns.

(c) Which of A, B, and C corresponds to the upper limit of worker productivity?

32. *Population growth* The figure shows the growth of a population as a function of time.

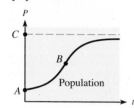

(a) If P represents the population and t represents the time, write a mathematical symbol that represents the rate of change (growth *rate*) of the population with respect to time.

(b) Which of A, B, and C corresponds to the point at which the growth *rate* attains its maximum?

(c) Which of A, B, and C corresponds to the upper limit of population?

33. *Advertising and sales* The figure shows the daily sales volume S as a function of time t since an ad campaign began.

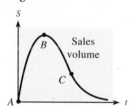

(a) Which of A, B, and C is the point of inflection for the graph?

(b) On which side of C is $d^2S/dt^2 > 0$?

(c) Does the *rate of change* of sales volume attain its minimum at C?

34. *Oxygen purity* The figure shows the oxygen level P (for purity) in a lake t months after an oil spill.

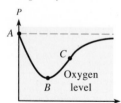

(a) Which of A, B, and C is the point of inflection for the graph?

(b) On which side of C is $d^2P/dt^2 < 0$?

(c) Does the *rate of change* of purity attain its maximum at C?

35. *Production* Suppose that the total number of units produced by a worker in t hours of an 8-hour shift can be modeled by the production function $P(t)$:

$$P(t) = 27t + 12t^2 - t^3$$

(a) Find the number of hours before production is maximized.

(b) Find the number of hours before the rate of production is maximized. That is, find the point of diminishing returns.

36. *Poiseuille's law—velocity of blood* ccording to Poiseuille's law, the speed S of blood through an artery of radius r at a distance x from the artery wall is given by

$$S = k[r^2 - (r - x)^2]$$

where k is a constant. Find the distance x that maximizes the speed.

37. *Advertising and sales—diminishing returns* Suppose that a company's daily sales volume attributed to an advertising campaign is given by

$$S(t) = \frac{3}{t + 3} - \frac{18}{(t + 3)^2} + 1$$

(a) Find how long it will be before sales volume is maximized.

(b) Find how long it will be before the rate of change of sales volume is minimized. That is, find the point of diminishing returns.

38. *Oxygen purity—diminishing returns* Suppose that the oxygen level P (for purity) in a body of water t months after an oil spill is given by

$$P(t) = 500\left[1 - \frac{4}{t + 4} + \frac{16}{(t + 4)^2}\right]$$

(a) Find how long it will be before the oxygen level reaches its minimum.

(b) Find how long it will be before the rate of change of P is maximized. That is, find the point of diminishing returns.

39. *Energy use* Energy use per capita indexed to 1995 means that per capita energy use for any year is viewed as a percent of per capita use in 1995. Using U.S. Department of Energy data for selected years from 1985 and projected to 2035, the per capita energy use, as a percent of the use in 1995, can be modeled by the function

$$E(t) = 0.000772t^3 - 0.0760t^2 + 1.83t + 87.3$$

where t is the number of years past 1980.

(a) Find the critical points for this function.

(b) Use the second-derivative test to classify and interpret these critical points.

40. *Modeling Foreign-born population* The figure gives the percent of the U.S. population that was foreign-born for selected years from 1910 and projected to 2020.

(a) Find the cubic function that is the best fit for the data. Use $x = 0$ to represent 1900, and report the model to three significant digits.

(b) Find the critical point of the reported model when $x > 50$.

(c) Interpret this point in terms of the percent of foreign-born people in the U.S. population.

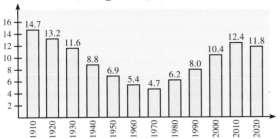

Percent of Population That Was Foreign-Born, 1910–2020

Source: U.S. Bureau of the Census

41. **Modeling *Civilian labor force*** The table gives the size of the U.S. civilian labor force (in millions) for selected years from 1950 and projected to 2050.
    (a) Find a cubic function that models these data, with x equal to the number of years after 1950 and y equal to the labor force in millions. Report the model with three significant digit coefficients.
    (b) Use the reported model to find the year when the second derivative predicts that the rate of change of the civilian labor force begins to decrease.

Year	Civilian Labor Force (in millions)	Year	Civilian Labor Force (in millions)
1950	62.2	2010	157.7
1960	69.6	2015	162.8
1970	82.8	2020	164.7
1980	106.9	2030	170.1
1990	125.8	2040	180.5
2000	140.9	2050	191.8

Source: U.S. Bureau of Labor Statistics

42. **Modeling *Elderly men in the workforce*** The table gives the percent of men 65 years or older in the workforce for selected years from 1920 and projected to 2030.
    (a) With $x = 0$ representing 1900, find the cubic function that models these data. Report the model with three significant digits.

(b) Use the reported model to determine when the rate of change of the percent of elderly men in the workforce reached its minimum.
(c) On the graph of this model, to what does the result in part (b) correspond?

Year	Percent	Year	Percent
1920	55.6	1980	19.0
1930	54.0	1990	16.3
1940	41.8	2000	17.7
1950	45.8	2010	22.6
1960	33.1	2020	27.2
1970	21.8	2030	27.6

Source: U.S. Bureau of the Census

43. **Modeling *Home health care*** The table gives the annual cost per person age 65 and older for home health care from 2006 and projected through 2021. While these annual costs per person may seem modest, the majority of older Americans live healthy, active lives and require no special care.
    (a) With t equal to the number of years past 2005, find a cubic function $C(t)$ that models the annual cost per person age 65 and older for home health care. Report the model with three significant digits.
    (b) Find the point of inflection of the reported model. Examine the function on either side of the point of inflection and interpret your results.

Year	Cost (in $)	Year	Cost (in $)
2006	1418	2014	1967
2007	1529	2015	2045
2008	1589	2016	2122
2009	1673	2017	2203
2010	1764	2018	2305
2011	1787	2019	2416
2012	1841	2020	2532
2013	1883	2021	2653

Source: Centers for Medicare and Medicaid Services

OBJECTIVES

10.3

- To find absolute maxima and minima
- To maximize revenue, given the total revenue function
- To minimize the average cost, given the total cost function
- To find the maximum profit from total cost and total revenue functions, or from a profit function

Optimization in Business and Economics

| APPLICATION PREVIEW |

Suppose a travel agency charges $300 per person for a certain tour when the tour group has 25 people, but will reduce the price by $10 per person for each additional person above the 25. In order to determine the group size that will maximize the revenue, the agency must create a model for the revenue. With such a model, the techniques of calculus can be useful in determining the maximum revenue. (See Example 2.)

Most companies (such as the travel agency) are interested in obtaining the greatest possible revenue or profit. Similarly, manufacturers of products

(Continued)

are concerned about producing their products for the lowest possible average cost per unit. Therefore, rather than just finding the relative maxima or relative minima of a function, we will consider where the absolute maximum or absolute minimum of a function occurs in a given interval.

In this section we will discuss how to find the absolute extrema of a function and then use these techniques to solve applications involving revenue, cost, and profit.

As the name implies, **absolute extrema** are the functional values that are the largest or smallest values over the entire domain of the function (or over the interval of interest).

Absolute Extrema	The value $f(a)$ is the **absolute maximum** of f if $f(a) \geq f(x)$ for all x in the domain of f (or over the interval of interest). The value $f(b)$ is the **absolute minimum** of f if $f(b) \leq f(x)$ for all x in the domain of f (or over the interval of interest).

Let us begin by considering the graph of $y = (x - 1)^2$, shown in Figure 10.28(a). This graph has a relative minimum at $(1, 0)$. Note that the relative minimum is the lowest point on the graph. In this case, the point $(1, 0)$ is an **absolute minimum point**, and 0 is the absolute minimum for the function. Similarly, when there is a point that is the highest point on the graph over the domain of the function, we call the point an **absolute maximum point** of the graph of the function.

In Figure 10.28(a), we see that there is no relative maximum. However, if the domain of the function is restricted to the interval $\left[\frac{1}{2}, 2\right]$, then we get the graph shown in Figure 10.28(b). In this case, there is an absolute maximum of 1 at the point $(2, 1)$ and the absolute minimum of 0 is still at $(1, 0)$.

If the domain of $y = (x - 1)^2$ is restricted to the interval $[2, 3]$, the resulting graph is that shown in Figure 10.28(c). In this case, the absolute minimum is 1 and occurs at the point $(2, 1)$, and the absolute maximum is 4 and occurs at $(3, 4)$.

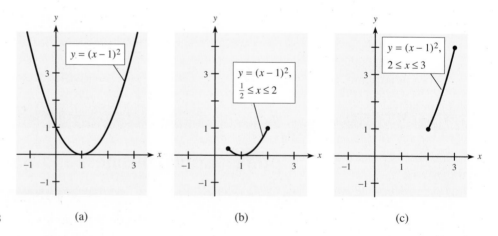

Figure 10.28   (a)   (b)   (c)

As the preceding discussion indicates, if the domain of a continuous function is limited to a closed interval, the absolute maximum or minimum may occur at an endpoint of the domain. In testing functions with limited domains for absolute maxima and minima, we must compare the function values at the endpoints of the domain with the function values at the critical values found by taking derivatives. In the management, life, and social sciences, a limited domain occurs very often, because many quantities are required to be positive, or at least nonnegative.

Maximizing Revenue Because the marginal revenue is the first derivative of the total revenue, the total revenue function will have a critical point at the point where the marginal revenue equals 0. With the total revenue function $R(x) = 16x - 0.02x^2$, the point where $R'(x) = 0$ is clearly a maximum because $R(x)$ is a parabola that opens downward. But the domain may be limited, the revenue function may not always be a parabola, or the critical point may not always be a maximum, so it is important to verify where the maximum value occurs.

EXAMPLE 1 Revenue

The total revenue in dollars for a firm is given by

$$R(x) = 8000x - 40x^2 - x^3$$

where x is the number of units sold per day. If only 50 units can be sold per day, find the number of units that must be sold to maximize revenue. Find the maximum revenue.

Solution
This revenue function is limited to x in the interval $[0, 50]$. Thus, the maximum revenue will occur at a critical value in this interval or at an endpoint. $R'(x) = 8000 - 80x - 3x^2$, so we must solve $8000 - 80x - 3x^2 = 0$ for x:

$$(40 - x)(200 + 3x) = 0 \qquad \text{means} \qquad 40 - x = 0 \qquad \text{or} \qquad 200 + 3x = 0$$

Thus $\qquad\qquad\qquad\qquad\qquad x = 40 \quad \text{or} \quad x = -\dfrac{200}{3}$

The negative value of x is not relevant, so for $x = 40$ we use either the second-derivative test or the first-derivative test with a sign diagram.

Second-Derivative Test	First-Derivative Test
$R''(x) = -80 - 6x$, $R''(40) = -320 < 0$, so $x = 40$ gives a relative maximum.	

These tests show that a relative maximum occurs at $x = 40$, giving revenue $R(40) = \$192,000$. Checking the endpoints $x = 0$ and $x = 50$ gives $R(0) = \$0$ and $R(50) = \$175,000$. Thus $R = \$192,000$ at $x = 40$ is the (absolute) maximum revenue. ■

✓ CHECKPOINT   1. True or false: If $R(x)$ is the revenue function, we find all possible points where $R(x)$ could be maximized by solving $\overline{MR} = 0$ for x.

EXAMPLE 2 Revenue | APPLICATION PREVIEW |

A travel agency will plan tours for groups of 25 or larger. If the group contains exactly 25 people, the price is \$300 per person. However, the price per person is reduced by \$10 for each additional person above the 25. What size group will produce the largest revenue for the agency?

Solution
The total revenue is

$$R = (\text{number of people})(\text{price per person})$$

The table shows how the revenue is changed by increases in the size of the group.

Number in Group	Price per Person	Revenue
25	300	7500
25 + 1	300 − 10	7540
25 + 2	300 − 20	7560
⋮	⋮	⋮
25 + x	300 − 10x	$(25 + x)(300 - 10x)$

Thus when x is the number of people added to the 25, the total revenue will be

$$R = R(x) = (25 + x)(300 - 10x) \qquad \text{or} \qquad R(x) = 7500 + 50x - 10x^2$$

This is a quadratic function, so its graph is a parabola that is concave down. Thus, a maximum will occur at its vertex, where $R'(x) = 0$.

$R'(x) = 50 - 20x$, and the solution to $0 = 50 - 20x$ is $x = 2.5$. Thus adding 2.5 people to the group should maximize the total revenue. But we cannot add half a person, so we will test the total revenue function for 27 people and 28 people. This will determine the optimal group size.

For $x = 2$ (giving 27 people) we get $R(2) = 7500 + 50(2) - 10(2)^2 = 7560$. For $x = 3$ (giving 28 people) we get $R(3) = 7500 + 50(3) - 10(3)^2 = 7560$. Note that both 27 and 28 people give the same total revenue and that this revenue is greater than the revenue for 25 people. Thus the revenue is maximized at either 27 or 28 people in the group. (See Figure 10.29.)

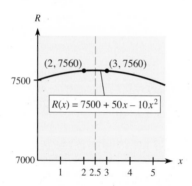

Figure 10.29

Minimizing Average Cost   Because the total cost function is always increasing for $x \geq 0$, the number of units that will make the total cost a minimum is always $x = 0$ units, which gives an absolute minimum. However, it is more useful to find the number of units that will make the **average cost** per unit a minimum.

Average Cost

If the total cost function is $C = C(x)$, then the per unit **average cost function** is

$$\overline{C} = \frac{C(x)}{x}$$

Note that the average cost per unit is undefined if no units are produced.

We can use derivatives to find the minimum of the average cost function, as the next example shows.

EXAMPLE   3   Average Cost

If the total cost function for a commodity is given by $C = \frac{1}{4}x^2 + 4x + 100$ dollars, where x represents the number of units produced, producing how many units will result in a minimum *average cost* per unit? Find the minimum average cost.

Solution

Begin by finding the average cost function and its derivative:

$$\overline{C} = \frac{\frac{1}{4}x^2 + 4x + 100}{x} = \frac{1}{4}x + 4 + \frac{100}{x}$$

$$\overline{C}' = \overline{C}'(x) = \frac{1}{4} - \frac{100}{x^2}$$

Setting $\overline{C}' = 0$ gives

$$0 = \frac{1}{4} - \frac{100}{x^2} \quad \text{so} \quad 0 = x^2 - 400 \quad \text{and} \quad x = \pm 20$$

Because the quantity produced must be positive, 20 units should minimize the average cost per unit. We show that it is an absolute minimum by using the second derivative to show $\overline{C}$ is concave up for all positive x.

$$\overline{C}''(x) = \frac{200}{x^3} \quad \text{so} \quad \overline{C}''(x) > 0 \quad \text{when } x > 0$$

Thus the minimum average cost per unit occurs if 20 units are produced. The graph of the average cost per unit is shown in Figure 10.30. The minimum average cost per unit is $\overline{C}(20) = \$14$. ■

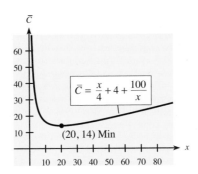

Figure 10.30

✓ **CHECKPOINT**   2. If $C(x) = 0.01x^2 + 20x + 2500$, form $\overline{C}(x)$, the average cost function, and find the minimum average cost.

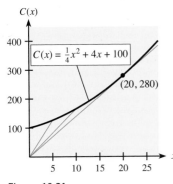

Figure 10.31

The graph of the cost function for the commodity in Example 3 is shown in Figure 10.31, along with several lines that join $(0, 0)$ to a point of the form $(x, C(x))$ on the total cost graph. Note that the slope of each of these lines has the form

$$\frac{C(x) - 0}{x - 0} = \frac{C(x)}{x}$$

so the slope of each line is the *average cost* for the given number of units at the point on $C(x)$. Note that the line from the origin to the point on the curve where $x = 20$ is tangent to the total cost curve. Hence, when $x = 20$, the slope of this line represents both the derivative of the cost function (the *marginal cost*) and the *average cost*. All lines from the origin to points with x-values larger than 20 or smaller than 20 are steeper (and therefore have greater slopes) than the line to the point where $x = 20$. Thus the minimum average cost occurs where the average cost equals the marginal cost. You will be asked to show this analytically in Problems 21 and 22 of the 10.3 Exercises.

Maximizing Profit   We have defined the marginal profit function as the derivative of the profit function. That is,

$$\overline{MP} = P'(x)$$

In this chapter we have seen how to use the derivative to find maxima and minima for various functions. Now we can apply those same techniques in the context of **profit maximization.** We can use marginal profit to maximize profit functions.

If there is a physical limitation on the number of units that can be produced in a given period of time, then the endpoints of the interval caused by these limitations should also be checked.

EXAMPLE 4   Profit

Suppose that the production capacity for a certain commodity cannot exceed 30. If the total profit for this commodity is

$$P(x) = 4x^3 - 210x^2 + 3600x - 200 \quad \text{dollars}$$

where x is the number of units sold, find the number of items that will maximize profit.

Solution

The restrictions on capacity mean that $P(x)$ is restricted by $0 \le x \le 30$. The marginal profit function is

$$P'(x) = 12x^2 - 420x + 3600$$

Setting $P'(x)$ equal to 0, we get

$$0 = 12(x - 15)(x - 20)$$

so $P'(x) = 0$ at $x = 15$ *and* $x = 20$. A sign diagram for $P'(x)$ tests these critical values.

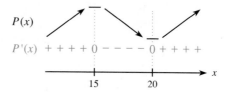

Thus, at (15, 20,050) the total profit function has a *relative* maximum, but we must check the endpoints (0 and 30) before deciding whether it is the absolute maximum.

$$P(0) = -200 \quad \text{and} \quad P(30) = 26,800$$

Thus the absolute maximum profit is $26,800, and it occurs at the endpoint, $x = 30$. Figure 10.32 shows the graph of the profit function. ■

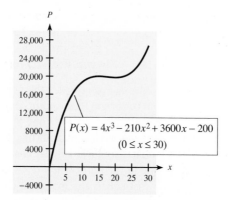

Figure 10.32

In a **monopolistic market,** the seller who has a monopoly controls the supply of a product and can force the price higher by limiting supply.

If the demand function for the product is $p = f(x)$, total revenue for the sale of x units is $R(x) = px = f(x) \cdot x$. Note that the price p is fixed by the market in a competitive market but varies with output for the monopolist.

If $\overline{C} = \overline{C}(x)$ represents the average cost per unit sold, then the total cost for the x units sold is $C = \overline{C} \cdot x = \overline{C}x$. Because we have both total cost and total revenue as a function of the quantity, x, we can maximize the profit function, $P(x) = px - \overline{C}x$, where p represents the demand function $p = f(x)$ and $\overline{C}$ represents the average cost function $\overline{C} = \overline{C}(x)$.

EXAMPLE 5 Profit in a Monopoly Market

MONOPOLY

One big player controls supply and dictates pricing.

© Eric Isselee/Shutterstock.com

The price of a product in dollars is related to the number of units x demanded daily by

$$p = 168 - 0.2x$$

A monopolist finds that the daily average cost for this product is

$$\overline{C} = 120 + x \quad \text{dollars}$$

(a) How many units must be sold daily to maximize profit?
(b) What is the selling price at this "optimal" level of production?
(c) What is the maximum possible daily profit?

Solution

(a) The total revenue function for the product is

$$R(x) = px = (168 - 0.2x)x = 168x - 0.2x^2$$

and the total cost function is

$$C(x) = \overline{C} \cdot x = (120 + x)x = 120x + x^2$$

Thus the profit function is

$$P(x) = R(x) - C(x) = 168x - 0.2x^2 - (120x + x^2) = 48x - 1.2x^2$$

Then $P'(x) = 48 - 2.4x$, so $P'(x) = 0$ when $x = 20$. We see that $P''(20) = -2.4$, so by the second-derivative test, $P(x)$ has a maximum at $x = 20$. That is, selling 20 units will maximize profit.

(b) The selling price is determined by $p = 168 - 0.2x$, so the price that will result from supplying 20 units per day is $p = 168 - 0.2(20) = 164$. That is, the "optimal" selling price is \$164 per unit.

(c) The profit at $x = 20$ is $P(20) = 48(20) - 1.2(20)^2 = 960 - 480 = 480$. Thus the maximum possible profit is \$480 per day. ■

COMPETITION

Several players are so small that individual actions cannot affect product

© Stephen Coburn/Shutterstock.com

In a **competitive market,** each firm is so small that its actions in the market cannot affect the price of the product. The price of the product is determined in the market by the intersection of the market demand curve (from all consumers) and market supply curve (from all firms that supply this product). The firm can sell as little or as much as it desires at the market equilibrium price.

Therefore, a firm in a competitive market has a total revenue function given by $R(x) = px$, where p is the market equilibrium price for the product and x is the quantity sold.

EXAMPLE 6 Profit in a Competitive Market

A firm in a competitive market must sell its product for \$200 per unit. The average cost per unit (per month) is $\overline{C} = 80 + x$, where x is the number of units sold per month. How many units should be sold to maximize profit?

Solution

If the average cost per unit is $\overline{C} = 80 + x$, then the total cost of x units is given by $C(x) = (80 + x)x = 80x + x^2$. The revenue per unit is \$200, so the total revenue is given by $R(x) = 200x$. Thus the profit function is

$$P(x) = R(x) - C(x) = 200x - (80x + x^2), \quad \text{or} \quad P(x) = 120x - x^2$$

Then $P'(x) = 120 - 2x$. Setting $P'(x) = 0$ and solving for x gives $x = 60$. Because $P''(60) = -2$, the profit is maximized when the firm sells 60 units per month. ■

✓ CHECKPOINT
3. (a) If $p = 5000 - x$ gives the demand function in a monopoly market, find $R(x)$, if it is possible with this information.
   (b) If $p = 5000 - x$ gives the demand function in a competitive market, find $R(x)$, if it is possible with this information.
4. If $R(x) = 400x - 0.25x^2$ and $C(x) = 150x + 0.25x^2 + 8500$ are the total revenue and total cost (both in dollars) for x units of a product, find the number of units that gives maximum profit and find the maximum profit.

Calculator Note

As we have seen, graphing calculators can be used to locate maximum values. In addition, if it is difficult to determine critical values algebraically, we may be able to approximate them graphically. For example, if

$$P(x) = 2500 - \frac{3000}{x + 1} - 12x - x^2 \qquad \text{then} \qquad P'(x) = \frac{3000}{(x + 1)^2} - 12 - 2x$$

Finding the critical values by solving $P'(x) = 0$ is difficult unless we use a graphing approach. Figure 10.33(a) shows the graph of $P(x)$, and Figure 10.33(b) shows the graph of $P'(x)$. These figures indicate that the maximum occurs near $x = 10$.

By adjusting the viewing window for $P'(x)$, we obtain the graph in Figure 10.34. This shows that $P'(x) = 0$ when $x = 9$. The maximum is $P(9) = 2500 - 300 - 108 - 81 = 2011$. See Appendix A, Section 10.3, for additional details.

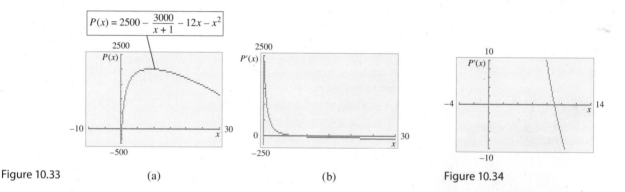

Figure 10.33   (a)   (b)   Figure 10.34

Spreadsheet Note

The solver feature of Excel can also be used to find optimal values of functions. See Appendix B, Section 10.3, and the Online Excel Guide for complete details.

✓ CHECKPOINT
ANSWERS

1. False. $\overline{MR} = R'(x)$, but there may also be critical points where $R'(x)$ is undefined, or $R(x)$ may be maximized at endpoints of a restricted domain.
2. $\overline{C}(x) = \dfrac{C(x)}{x} = 0.01x + 20 + \dfrac{2500}{x}$

   The minimum average cost per unit is $\overline{C}(500) = 30$ dollars per unit.
3. (a) $R(x) = p \cdot x = 5000x - x^2$
   (b) In a competitive market, we need to know the supply function and find the equilibrium price before we can form $R(x)$.
4. The (absolute) maximum profit occurs when $x = 250$ and is $P(250) = \$22{,}750$.

| EXERCISES | 10.3

In Problems 1–4, find the absolute maxima and minima for $f(x)$ on the interval $[a, b]$.

1. $f(x) = x^3 - 2x^2 - 4x + 2, [-1, 3]$
2. $f(x) = x^3 - 3x + 3, [-3, 1.5]$
3. $f(x) = x^3 + x^2 - x + 1, [-2, 0]$
4. $f(x) = x^3 - x^2 - x, [-0.5, 2]$

MAXIMIZING REVENUE

5. (a) If the total revenue function for a hammer is $R = 36x - 0.01x^2$, then sale of how many hammers, x, will maximize the total revenue in dollars? Find the maximum revenue?
   (b) Find the maximum revenue if production is limited to at most 1500 hammers.

6. (a) If the total revenue function for a blender is $R(x) = 25x - 0.05x^2$, sale of how many units, x, will provide the maximum total revenue in dollars? Find the maximum revenue.
   (b) Find the maximum revenue if production is limited to at most 200 blenders.

7. If the total revenue function for a computer is $R(x) = 2000x - 20x^2 - x^3$, find the level of sales, x, that maximizes revenue and find the maximum revenue in dollars.

8. A firm has total revenues given by

   $$R(x) = 2800x - 8x^2 - x^3 \text{ dollars}$$

   for x units of a product. Find the maximum revenue from sales of that product.

9. An agency charges $100 per person for a trip to a concert if 70 people travel in a group. But for each person above the 70, the charge will be reduced by $1.00. How many people will maximize the total revenue for the agency if the trip is limited to at most 90 people?

10. A company handles an apartment building with 70 units. Experience has shown that if the rent for each of the units is $1080 per month, all the units will be filled, but 1 unit will become vacant for each $20 increase in the monthly rate. What rent should be charged to maximize the total revenue from the building if the upper limit on the rent is $1300 per month?

11. A cable TV company has 4000 customers paying $110 each month. If each $1 reduction in price attracts 50 new customers, find the price that yields maximum revenue. Find the maximum revenue.

12. If club members charge $5 admission to a classic car show, 1000 people will attend, and for each $1 increase in price, 100 fewer people will attend. What price will give the maximum revenue for the show? Find the maximum revenue.

13. The function $\overline{R}(x) = R(x)/x$ defines the average revenue for selling x units. For

    $$R(x) = 2000x + 20x^2 - x^3$$

    (a) find the maximum average revenue.
    (b) show that $\overline{R}(x)$ attains its maximum at an x-value where $\overline{R}(x) = \overline{MR}$.

14. For the revenue function given by

    $$R(x) = 2800x + 8x^2 - x^3$$

    (a) find the maximum average revenue.
    (b) show that $\overline{R}(x)$ attains its maximum at an x-value where $\overline{R}(x) = \overline{MR}$.

MINIMIZING AVERAGE COST

15. If the total cost function for a lamp is $C(x) = 250 + 33x + 0.1x^2$ dollars, producing how many units, x, will result in a minimum average cost per unit? Find the minimum average cost.

16. If the total cost function for a product is $C(x) = 300 + 10x + 0.03x^2$ dollars, producing how many units, x, will result in a minimum average cost per unit? Find the minimum average cost.

17. If the total cost function for a product is $C(x) = 810 + 0.1x^2$ dollars, producing how many units, x, will result in a minimum average cost per unit? Find the minimum average cost.

18. If the total cost function for a product is $C(x) = 250 + 6x + 0.1x^2$ dollars, producing how many units, x, will minimize the average cost? Find the minimum average cost.

19. If the total cost function for a product is $C(x) = 100(0.02x + 4)^3$ dollars, where x represents the number of hundreds of units produced, producing how many units will minimize average cost? Find the minimum average cost.

20. If the total cost function for a product is $C(x) = (x + 5)^3$ dollars, where x represents the number of hundreds of units produced, producing how many units will minimize average cost? Find the minimum average cost.

21. For the cost function $C(x) = 25 + 13x + x^2$, show that average costs are minimized at the x-value where

    $$\overline{C}(x) = \overline{MC}$$

22. For the cost function $C(x) = 300 + 10x + 0.03x^2$, show that average costs are minimized at the x-value where

    $$\overline{C}(x) = \overline{MC}$$

The graphs in Problems 23 and 24 show total cost functions. For each problem:

(a) **Explain how to use the total cost graph to determine the level of production at which average cost is minimized.**

(b) **Determine that level of production.**

23.

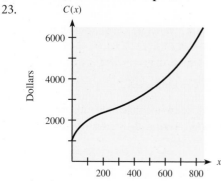

24.

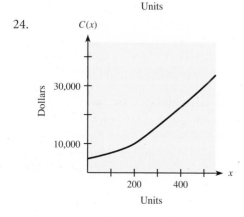

MAXIMIZING PROFIT

25. If the profit function for a product is $P(x) = 5600x + 85x^2 - x^3 - 200{,}000$ dollars, selling how many items, x, will produce a maximum profit? Find the maximum profit.

26. If the profit function for a commodity is $P = 6400x - 18x^2 - \frac{1}{3}x^3 - 40{,}000$ dollars, selling how many units, x, will result in a maximum profit? Find the maximum profit.

27. A manufacturer estimates that its product can be produced at a total cost of $C(x) = 45{,}000 + 100x + x^3$ dollars. If the manufacturer's total revenue from the sale of x units is $R(x) = 4600x$ dollars, determine the level of production x that will maximize the profit. Find the maximum profit.

28. A product can be produced at a total cost $C(x) = 800 + 100x^2 + x^3$ dollars, where x is the number produced. If the total revenue is given by $R(x) = 60{,}000x - 50x^2$ dollars, determine the level of production, x, that will maximize the profit. Find the maximum profit.

29. A firm can produce only 1000 units per month. The monthly total cost is given by $C(x) = 300 + 200x$ dollars, where x is the number produced. If the total revenue is given by $R(x) = 250x - \frac{1}{100}x^2$ dollars, how

many items, x, should the firm produce for maximum profit? Find the maximum profit.

30. A firm can produce 100 units per week. If its total cost function is $C = 500 + 1500x$ dollars and its total revenue function is $R = 1600x - x^2$ dollars, how many units, x, should it produce to maximize its profit? Find the maximum profit.

31. *Marginal revenue and marginal cost* The figure shows the graph of a quadratic revenue function and a linear cost function.

    (a) At which of the four x-values shown is the distance between the revenue and the cost greatest?

    (b) At which of the four x-values shown is the profit largest?

    (c) At which of the four x-values shown is the slope of the tangent to the revenue curve equal to the slope of the cost line?

    (d) What is the relationship between marginal cost and marginal revenue when profit is at its maximum value?

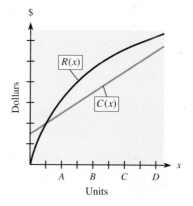

32. *Marginal revenue and marginal cost* The figure shows the graph of revenue function $y = R(x)$ and cost function $y = C(x)$.

    (a) At which of the four x-values shown is the profit largest?

    (b) At which of the four x-values shown is the slope of the tangent to the revenue curve equal to the slope of the tangent to the cost curve?

    (c) What is the relationship between marginal cost and marginal revenue when profit is at its maximum value?

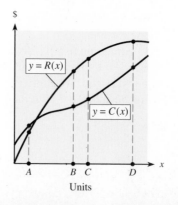

33. A company handles an apartment building with 50 units. Experience has shown that if the rent for each of the units is $720 per month, all of the units will be filled, but 1 unit will become vacant for each $20 increase in this monthly rate. If the monthly cost of maintaining the apartment building is $12 per rented unit, what rent should be charged per month to maximize the profit?

34. A travel agency will plan a tour for groups of size 25 or larger. If the group contains exactly 25 people, the cost is $500 per person. However, each person's cost is reduced by $10 for each additional person above the 25. If the travel agency incurs a cost of $125 per person for the tour, what size group will give the agency the maximum profit?

35. A firm has monthly average costs, in dollars, given by

$$\overline{C} = \frac{45,000}{x} + 100 + x$$

where x is the number of units produced per month. The firm can sell its product in a competitive market for $1600 per unit. If production is limited to 600 units per month, find the number of units that gives maximum profit, and find the maximum profit.

36. A small business has weekly average costs, in dollars, of

$$\overline{C} = \frac{100}{x} + 30 + \frac{x}{10}$$

where x is the number of units produced each week. The competitive market price for this business's product is $46 per unit. If production is limited to 150 units per week, find the level of production that yields maximum profit, and find the maximum profit.

37. The weekly demand function for x units of a product sold by only one firm is $p = 600 - \frac{1}{2}x$ dollars, and the average cost of production and sale is $\overline{C} = 300 + 2x$ dollars.
    (a) Find the quantity that will maximize profit.
    (b) Find the selling price at this optimal quantity.
    (c) What is the maximum profit?

38. The monthly demand function for x units of a product sold by a monopoly is $p = 8000 - x$ dollars, and its average cost is $\overline{C} = 4000 + 5x$ dollars.
    (a) Determine the quantity that will maximize profit.
    (b) Determine the selling price at the optimal quantity.
    (c) Determine the maximum profit.

39. The monthly demand function for a product sold by a monopoly is $p = 1960 - \frac{1}{3}x^2$ dollars, and the average cost is $\overline{C} = 1000 + 2x + x^2$ dollars. Production is limited to 1000 units and x is in hundreds of units.
    (a) Find the quantity that will give maximum profit.
    (b) Find the maximum profit.

40. The monthly demand function for x units of a product sold by a monopoly is $p = 5900 - \frac{1}{2}x^2$ dollars, and its average cost is $\overline{C} = 3020 + 2x$ dollars. If production is limited to 100 units, find the number of units that

maximizes profit. Will the maximum profit result in a profit or loss?

41. An industry with a monopoly on a product has its average weekly costs, in dollars, given by

$$\overline{C} = \frac{10,000}{x} + 60 - 0.03x + 0.00001x^2$$

The weekly demand for x units of the product is given by $p = 120 - 0.015x$ dollars. Find the price the industry should set and the number of units it should produce to obtain maximum profit. Find the maximum profit.

42. A large corporation with monopolistic control in the marketplace has its average daily costs, in dollars, given by

$$\overline{C} = \frac{800}{x} + 100x + x^2$$

The daily demand for x units of its product is given by $p = 60,000 - 50x$ dollars. Find the quantity that gives maximum profit, and find the maximum profit. What selling price should the corporation set for its product?

43. Coastal Soda Sales has been granted exclusive market rights to the upcoming Beaufort Seafood Festival. This means that during the festival Coastal will have a monopoly, and it is anxious to take advantage of this position in its pricing strategy. The daily demand function is

$$p = 2 - 0.0004x$$

and the daily total cost function is

$$C(x) = 800 + 0.2x + 0.0001x^2$$

where x is the number of units.
    (a) Determine Coastal's total revenue and profit functions.
    (b) What profit-maximizing price per soda should Coastal charge, how many sodas per day would it expect to sell at this price, and what would be the daily profits?
    (c) If the festival organizers wanted to set an economically efficient price of $1.25 per soda, how would this change the results from part (b)? Would Coastal be willing to provide sodas for the festival at this regulated price? Why or why not?

44. A retiree from a large Atlanta financial services firm decides to keep busy and supplement her retirement income by opening a small upscale folk art company near Charleston, South Carolina. The company, Sand Dollar Art, manufactures and sells in a purely competitive market, and the following monthly market information for x units at p per unit applies:

$$\begin{aligned} \text{Demand:} \quad & p = 2000 - 4.5x \\ \text{Supply:} \quad & p = 100 + 0.25x \end{aligned}$$

(a) Find the market equilibrium quantity and price for this market.

(b) If Sand Dollar Art's monthly cost function is

$$C(x) = 400 + 100x + x^2$$

find the profit-maximizing monthly quantity. What are the total monthly revenues and total monthly costs? What monthly profit does Sand Dollar Art earn?

(c) Assuming that Sand Dollar Art is representative of firms in this competitive market, what is its market share?

MISCELLANEOUS APPLICATIONS

45. Modeling *Social Security beneficiaries* The numbers of millions of Social Security beneficiaries for selected years and projected into the future are given in the table.

(a) Find the cubic function that models these data, with x equal to the number of years past 1950. Report the model with three significant digits.

(b) Find the point of inflection of the graph of the reported model for $x > 0$.

(c) Graph this function and discuss what the point of inflection indicates.

Year	Number of Beneficiaries (millions)	Year	Number of Beneficiaries (millions)
1950	2.9	2002	44.8
1960	14.3	2010	53.3
1970	25.2	2020	68.8
1980	35.1	2030	82.7
1990	39.5		

Source: Social Security Trustees Report

46. Modeling *Workforce participation: Women* For women age 16 and older, the table gives the percent of this group that participates in the U.S. workforce for selected years from 1950 and projected to 2050.

Year	Percent	Year	Percent
1950	33.9	2010	62.2
1960	37.7	2015	62.1
1970	43.3	2020	60.3
1980	51.5	2030	57.4
1990	57.5	2040	56.7
2000	60.2	2050	56.6

Source: U.S. Bureau of the Census

(a) With x as the number of years past 1940, find a quartic function that models the data. Report the model with three significant digit coefficients.

(b) For the years from 1950 to 2050, determine all critical points of the reported model.

(c) Find the absolute maximum and absolute minimum of the reported model. Interpret the coordinates of each point.

(d) Find the absolute maximum and absolute minimum of the data set.

47. *Dow Jones Industrial Average* The figure shows the Dow Jones Industrial Average for all of 2001, the year of the terrorist attacks on New York City and Washington, D.C.

(a) Approximate when during 2001 the Dow reached its absolute maximum for that year.

(b) When do you think the Dow reached its absolute minimum for this period? What happened to trigger this?

Dow Jones Industrial Average

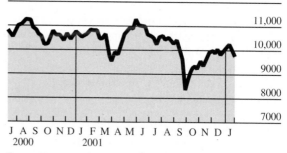

Source: From *The Wall Street Journal*, January 17, 2002. Copyright © 2002 by Dow Jones & Co. Reprinted by permission of Dow Jones & Co. via Copyright Clearance Center.

48. *Dow Jones averages* The figure shows the daily Dow Jones Industrial Average (DJIA) and its 30-day moving average from late July to early November. Use the figure to complete the following.

(a) Approximate the absolute maximum point and absolute minimum point for the daily DJIA.

(b) Approximate the absolute maximum point and absolute minimum point for the DJIA 30-day moving average.

$INDU (Dow Jones Industrial Average) INDX

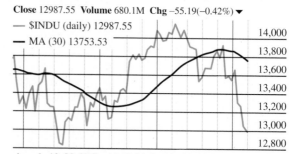

Source: StockCharts.com

49. *Social Security support* The graph shows the number of workers, $W = f(t)$, still in the workforce per Social Security beneficiary (historically and projected into the future) as a function of time t, in calendar years with $1950 \leq t \leq 2050$. Use the graph to answer the following.
    (a) What is the absolute maximum of $f(t)$?
    (b) What is the absolute minimum of $f(t)$?
    (c) Does this graph suggest that Social Security taxes will rise or will fall in the early 21st century? Explain.

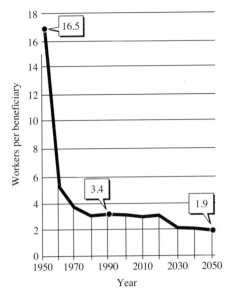

Source: Social Security Administration

OBJECTIVES

10.4

- To apply the procedures for finding maxima and minima to solve problems from the management, life, and social sciences

Applications of Maxima and Minima

| APPLICATION PREVIEW |

Suppose that a company needs 1,000,000 items during a year and that preparation costs are $800 for each production run. Suppose further that it costs the company $6 to produce each item and $1 to store each item for up to a year. Find the number of units that should be produced in each production run so that the total costs of production and storage are minimized. This question is answered using an inventory cost model (see Example 4).

This inventory-cost determination is a typical example of the kinds of questions and important business applications that require the use of the derivative for finding maxima and minima. As managers, workers, or consumers, we may be interested in such things as maximum revenue, maximum profit, minimum cost, maximum medical dosage, maximum utilization of resources, and so on.

If we have functions that model cost, revenue, or population growth, we can apply the methods of this chapter to find the maxima and minima of those functions.

EXAMPLE 1   Company Growth

Suppose that a new company begins production in 2015 with eight employees and the growth of the company over the next 10 years is predicted by

$$N = N(t) = 8\left(1 + \frac{160t}{t^2 + 16}\right), \quad 0 \leq t \leq 10$$

where N is the number of employees t years after 2015.

Determine in what year the number of employees will be maximized and the maximum number of employees.

Solution

This function will have a relative maximum when $N'(t) = 0$.

$$N'(t) = 8\left[\frac{(t^2 + 16)(160) - (160t)(2t)}{(t^2 + 16)^2}\right]$$

$$= 8\left[\frac{160t^2 + 2560 - 320t^2}{(t^2 + 16)^2}\right]$$

$$= 8\left[\frac{2560 - 160t^2}{(t^2 + 16)^2}\right]$$

Because $N'(t) = 0$ when its numerator is 0 (note that this denominator is never 0), we must solve

$$2560 - 160t^2 = 0$$
$$160(4 + t)(4 - t) = 0$$

so
$$t = -4 \text{ or } t = 4$$

We are interested only in positive t-values, so we test $t = 4$.

$$\left.\begin{array}{l} N'(0) = 8\left[\dfrac{2560}{256}\right] > 0 \\[4mm] N'(10) = 8\left[\dfrac{-13,440}{(116)^2}\right] < 0 \end{array}\right\} \Rightarrow \textit{relative maximum}$$

The relative maximum is

$$N(4) = 8\left(1 + \frac{640}{32}\right) = 168$$

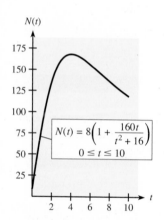

$$N(t) = 8\left(1 + \frac{160t}{t^2 + 16}\right)$$
$$0 \le t \le 10$$

Figure 10.35

At $t = 0$, the number of employees is $N(0) = 8$, and it increases to $N(4) = 168$. After $t = 4$ (in 2019), $N(t)$ decreases to $N(10) = 118$ (approximately), so $N(4) = 168$ is the maximum number of employees. Figure 10.35 verifies these conclusions. ∎

Sometimes we must develop the function we need from the statement of the problem. In this case, it is important to understand what is to be maximized or minimized and to express that quantity as a function of *one* variable.

EXAMPLE 2   Minimizing Cost

A farmer needs to enclose a rectangular pasture containing 1,600,000 square feet. Suppose that along the road adjoining his property he wants to use a more expensive fence and that he needs no fence on one side perpendicular to the road because a river bounds his property on that side. If the fence costs $15 per foot along the road and $10 per foot along the two remaining sides that must be fenced, what dimensions of his rectangular field will minimize his cost?

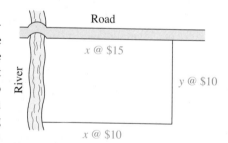

Figure 10.36

Solution

In Figure 10.36, x represents the length of the pasture along the road (and parallel to the road) and y represents the width. The cost function for the fence used is

$$C = 15x + 10y + 10x = 25x + 10y$$

We cannot use a derivative to find where C is minimized unless we write C as a function of x or y only. Because the area of the rectangular field must be 1,600,000 square feet, we have

$$A = xy = 1,600,000$$

Solving for y in terms of x and substituting give C as a function of x.

$$y = \frac{1,600,000}{x}$$

$$C = 25x + 10\left(\frac{1,600,000}{x}\right) = 25x + \frac{16,000,000}{x}$$

To find $C'(x)$, we first rewrite: $C = 25x + 16,000,000x^{-1}$. Then

$$C'(x) = 25 - 16,000,000x^{-2} = 25 - \frac{16,000,000}{x^2}$$

We find the critical values of C by solving $C'(x) = 0$ as follows:

$$0 = 25 - \frac{16,000,000}{x^2}$$
$$0 = 25x^2 - 16,000,000 \quad \text{(multiply both sides by } x^2\text{)}$$
$$25x^2 = 16,000,000$$
$$x^2 = 640,000 \Rightarrow x = \pm\sqrt{640,000} = \pm 800$$

We use $x = 800$ feet because $x = -800$ is meaningless in this application. Testing to see whether $x = 800$ gives the minimum cost, we find

$$C''(x) = \frac{32,000,000}{x^3}$$

$C''(x) > 0$ for $x > 0$, so $C(x)$ is concave up for all positive x. Thus $x = 800$ gives the absolute minimum, and $C(800) = 40,000$ is the minimum cost. The other dimension of the rectangular field is $y = 1,600,000/800 = 2000$ feet. Figure 10.37 verifies that $C(x)$ reaches its minimum (of 40,000) at $x = 800$. ◼

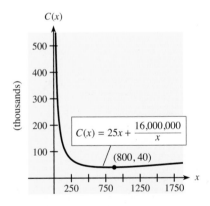

Figure 10.37

$x = 800$ ft @ \$15/ft = \$12,000 for the road side

© Melissa Brandes/Shutterstock.com

EXAMPLE 3   Postal Restrictions

Postal restrictions limit the size of packages sent through the mail. If the restrictions are that the length plus the girth may not exceed 108 inches, find the volume of the largest box with square cross section that can be mailed.

Solution
Let l equal the length of the box, and let s equal a side of the square end. See Figure 10.38. The volume we seek to maximize is given by

$$V = s^2 l$$

We can use the restriction that girth plus length equals 108,

$$4s + l = 108$$

to express V as a function of s or l. Because $l = 108 - 4s$, the equation for V becomes

$$V = s^2(108 - 4s) = 108s^2 - 4s^3$$

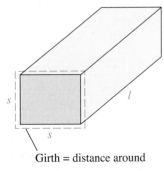

Girth = distance around

Figure 10.38

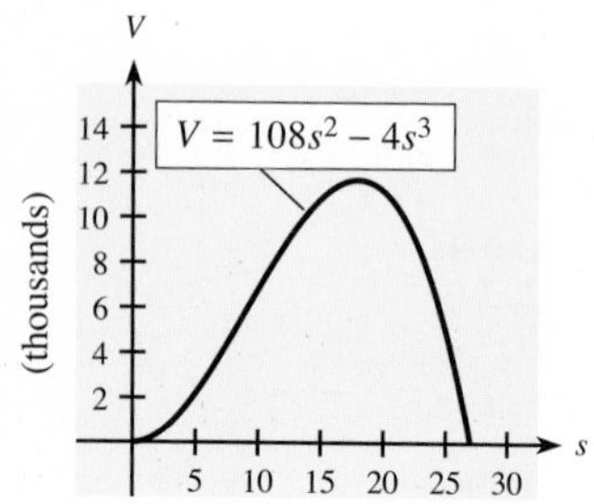

V = 108s² − 4s³

(thousands)

Figure 10.39

We solve $dV/ds = 0$ to find the critical values.

$$\frac{dV}{ds} = 216s - 12s^2$$

$$0 = s(216 - 12s)$$

The critical values are $s = 0$, $s = \frac{216}{12} = 18$. The critical value $s = 0$ will not maximize the volume, for in this case, $V = 0$. Testing to the left and right of $s = 18$ gives

$$V'(17) > 0 \quad \text{and} \quad V'(19) < 0$$

Thus $s = 18$ inches and $l = 108 - 4(18) = 36$ inches yield a maximum volume of 11,664 cubic inches. Once again we can verify our results graphically. Figure 10.39 shows that $V = 108s^2 - 4s^3$ achieves its maximum when $s = 18$. ■

✓ **CHECKPOINT**

Suppose we want to find the minimum value of $C = 5x + 2y$ and we know that x and y must be positive and that $xy = 1000$.
1. What equation do we differentiate to solve this problem?
2. Find the critical values.
3. Find the minimum value of C.

We next consider **inventory cost models,** in which x items are produced in each production run and items are removed from inventory at a fixed constant rate. Because items are removed at a constant rate, the average number stored at any time is $x/2$. Also, when $x = 0$, new items must be added to inventory from a production run. Thus the number of units in storage changes with time and is illustrated in Figure 10.40. In these models there are costs associated with both production and storage, but lowering one of these costs means increasing the other. To see how inventory cost models work, consider the following example.

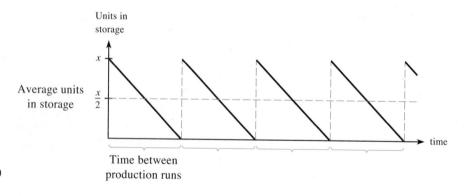

Figure 10.40

EXAMPLE 4 **Inventory Cost Model** | **APPLICATION PREVIEW** |

Suppose that a company needs 1,000,000 items during a year and that preparation costs are $800 for each production run. Suppose further that it costs the company $6 to produce each item and $1 to store an item for up to a year. If each production run consists of x items, find x so that the total costs of production and storage are minimized.

Solution
The total production costs are given by

$$\left(\begin{array}{c}\text{No. of}\\\text{runs}\end{array}\right)\left(\begin{array}{c}\text{cost}\\\text{per run}\end{array}\right) + \left(\begin{array}{c}\text{no. of}\\\text{items}\end{array}\right)\left(\begin{array}{c}\text{cost}\\\text{per item}\end{array}\right) = \left(\frac{1,000,000}{x}\right)(\$800) + (1,000,000)(\$6)$$

The total storage costs are

$$\left(\begin{array}{c}\text{Average}\\\text{no. stored}\end{array}\right)\left(\begin{array}{c}\text{storage cost}\\\text{per item}\end{array}\right) = \left(\frac{x}{2}\right)(\$1)$$

Thus the total costs of production and storage are

$$C = \left(\frac{1,000,000}{x}\right)(800) + 6,000,000 + \frac{x}{2} = \frac{800,000,000}{x} + 6,000,000 + \frac{x}{2}$$

We wish to find x so that C is minimized.

$$C' = \frac{-800,000,000}{x^2} + \frac{1}{2}$$

If $x > 0$, critical values occur when $C' = 0$.

$$0 = \frac{-800,000,000}{x^2} + \frac{1}{2}$$

$$\frac{800,000,000}{x^2} = \frac{1}{2}$$

$$1,600,000,000 = x^2$$

$$x = \pm 40,000$$

Because x must be positive, we test $x = 40,000$ with the second derivative.

$$C''(x) = \frac{1,600,000,000}{x^3}, \quad \text{so} \quad C''(40,000) > 0$$

Note that $x = 40,000$ yields an absolute minimum value of C, because $C'' > 0$ for all $x > 0$. That is, production runs of 40,000 items yield minimum total costs for production and storage. ▪

Technology Note   Problems of the types we've studied in this section could also be solved (at least approximately) with a graphing calculator or Excel. With this approach, our first goal is still to express the quantity to be maximized or minimized as a function of one variable. Then that function can be graphed, and the (at least approximate) optimal value can be obtained from the graph. This method is especially useful with problems that have critical values that are difficult to find algebraically, like the one in Example 5. ▪

EXAMPLE 5   Property Development

A developer of a campground has to pay for utility line installation to the community center from a transformer on the street at the corner of her property. Because of local restrictions, the lines must be underground on her property. Suppose that the costs are $50 per meter along the street and $100 per meter underground. How far from the transformer should the line enter the property to minimize installation costs? See Figure 10.41.

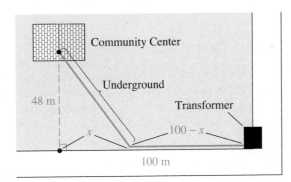

Figure 10.41

Solution

If the developer had the cable placed underground from the community center perpendicular to the street and then to the transformer, then $x = 0$ in Figure 10.41 and the cost would be

$$\$100(48) + \$50(100) = \$9800$$

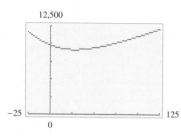

Figure 10.42

It may be possible to save some money by placing the cable on a diagonal to the street, but then only $x \le 100$ makes sense. By using the Pythagorean Theorem, we find that the length of the underground cable that meets the street x meters closer to the transformer is

$$\sqrt{48^2 + x^2} = \sqrt{2304 + x^2} \quad \text{meters}$$

Thus the cost C of installation is given by

$$C = 100\sqrt{2304 + x^2} + 50(100 - x) \text{ dollars}$$

Figure 10.42 shows the graph of this function over an interval for x that contains $0 \le x \le 100$. Because any extrema must occur in this interval, we can find the minimum by using MIN on a graphing calculator or Excel. See Appendices A and B, Section 10.4, and the Online Excel Guide for the steps. The minimum cost is $9156.92, when $x = 27.7$ meters (that is, when the cable meets the street 72.3 meters from the transformer). ∎

✓ **CHECKPOINT ANSWERS**

1. We must differentiate C, but first C must be expressed as a function of one variable:

$$C(x) = 5x + \frac{2000}{x}$$

2. $C'(x) = 5 - 2000/x^2$, so $C'(x) = 0$ when $x = \pm 20$. The only relevant critical value is $x = 20$.

3. The minimum value occurs when $x = 20$ and $y = 50$ and is $C = 200$.

| EXERCISES | 10.4

APPLICATIONS

1. **Return to sales** The manufacturer of GRIPPER tires modeled its return to sales from television advertising expenditures in two regions, as follows:

   Region 1: $S_1 = 30 + 20x_1 - 0.4x_1^2$
   Region 2: $S_2 = 20 + 36x_2 - 1.3x_2^2$

   where S_1 and S_2 are the sales revenue in millions of dollars and x_1 and x_2 are millions of dollars of expenditures for television advertising.
   (a) What advertising expenditures would maximize sales revenue in each district?
   (b) How much money will be needed to maximize sales revenue in both districts?

2. **Projectiles** A ball thrown into the air from a building 100 ft high travels along a path described by

   $$y = \frac{-x^2}{110} + x + 100$$

   where y is its height in feet and x is the horizontal distance from the building in feet. What is the maximum height the ball will reach?

3. **Profit** The profit from a grove of orange trees is given by $x(200 - x)$ dollars, where x is the number of orange trees per acre. How many trees per acre will maximize the profit?

4. **Reaction rates** The velocity v of an autocatalytic reaction can be represented by the equation

   $$v = x(a - x)$$

   where a is the amount of material originally present and x is the amount that has been decomposed at any given time. Find the maximum velocity of the reaction.

5. **Productivity** Analysis of daily output of a factory during an 8-hour shift shows that the hourly number of units y produced after t hours of production is

   $$y = 70t + \tfrac{1}{2}t^2 - t^3, \quad 0 \le t \le 8$$

   (a) After how many hours will the hourly number of units be maximized?
   (b) What is the maximum hourly output?

6. **Productivity** A time study showed that, on average, the productivity of a worker after t hours on the job can be modeled by

   $$P = 27t + 6t^2 - t^3, \quad 0 \le t \le 8$$

   where P is the number of units produced per hour. After how many hours will productivity be maximized? What is the maximum productivity?

7. **Consumer expenditure** Suppose that the demand x (in units) for a product is $x = 10,000 - 100p$, where p dollars is the market price per unit. Then the consumer expenditure for the product is

   $$E = px = 10,000p - 100p^2$$

   For what market price will expenditure be greatest?

8. **Production costs** Suppose that the monthly cost in dollars of mining a certain ore is related to the number of pieces of equipment used, according to

   $$C = 25,000x + \frac{870,000}{x}, \quad x > 0$$

where *x* is the number of pieces of equipment used. Using how many pieces of equipment will minimize the cost?

Medication For Problems 9 and 10, consider that when medicine is administered, reaction (measured in change of blood pressure or temperature) can be modeled by

$$R = m^2 \left(\frac{c}{2} - \frac{m}{3} \right)$$

where *c* is a positive constant and *m* is the amount of medicine absorbed into the blood (*Source*: R. M. Thrall et al., *Some Mathematical Models in Biology*, U.S. Department of Commerce, 1967).

9. Find the amount of medicine that is being absorbed into the blood when the reaction is maximum.

10. The rate of change of reaction *R* with respect to the amount of medicine *m* is defined to be the sensitivity.
 (a) Find the sensitivity, *S*.
 (b) Find the amount of medicine that is being absorbed into the blood when the sensitivity is maximum.

11. *Advertising and sales* An inferior product with a large advertising budget sells well when it is introduced, but sales fall as people discontinue use of the product. Suppose that the weekly sales *S* are given by

$$S = \frac{200t}{(t + 1)^2}, \quad t \geq 0$$

where *S* is in millions of dollars and *t* is in weeks. After how many weeks will sales be maximized?

12. *Revenue* A newly released film has its weekly revenue given by

$$R(t) = \frac{50t}{t^2 + 36}, \quad t \geq 0$$

where *R* is in millions of dollars and *t* is in weeks.
 (a) After how many weeks will the weekly revenue be maximized?
 (b) What is the maximum weekly revenue?

13. *News impact* Suppose that the percent *p* (as a decimal) of people who could correctly identify two of eight defendants in a drug case *t* days after their trial began is given by

$$p(t) = \frac{6.4t}{t^2 + 64} + 0.05$$

Find the number of days before the percent is maximized, and find the maximum percent.

14. *Candidate recognition* Suppose that in an election year the proportion *p* of voters who recognize a certain candidate's name *t* months after the campaign started is given by

$$p(t) = \frac{7.2t}{t^2 + 36} + 0.2$$

After how many months is the proportion maximized?

15. *Minimum fence* Two equal rectangular lots are to be enclosed by fencing the perimeter of a rectangular lot and then putting a fence across its middle. If each lot is to contain 1200 square feet, what is the minimum amount of fence needed to enclose the lots (include the fence across the middle)?

16. *Minimum fence* The running yard for a dog kennel must contain at least 900 square feet. If a 20-foot side of the kennel is used as part of one side of a rectangular yard with 900 square feet, what dimensions will require the least amount of fencing?

17. *Minimum cost* A rectangular field with one side along a river is to be fenced. Suppose that no fence is needed along the river, the fence on the side opposite the river costs $20 per foot, and the fence on the other sides costs $5 per foot. If the field must contain 45,000 square feet, what dimensions will minimize costs?

18. *Minimum cost* From a tract of land a developer plans to fence a rectangular region and then divide it into two identical rectangular lots by putting a fence down the middle. Suppose that the fence for the outside boundary costs $5 per foot and the fence for the middle costs $2 per foot. If each lot contains 13,500 square feet, find the dimensions of each lot that yield the minimum cost for the fence.

19. *Optimization at a fixed cost* A rectangular area is to be enclosed and divided into thirds. The family has $800 to spend for the fencing material. The outside fence costs $10 per running foot installed, and the dividers cost $20 per running foot installed. What are the dimensions that will maximize the area enclosed? (The answer contains a fraction.)

20. *Minimum cost* A kennel of 640 square feet is to be constructed as shown. The cost is $4 per running foot for the sides and $1 per running foot for the ends and dividers. What are the dimensions of the kennel that will minimize the cost?

21. *Minimum cost* The base of a rectangular box is to be twice as long as it is wide. The volume of the box is 256 cubic inches. The material for the top costs $0.10 per square inch and the material for the sides and bottom costs $0.05 per square inch. Find the dimensions that will make the cost a minimum.

22. *Velocity of air during a cough* According to B. F. Visser, the velocity *v* of air in the trachea during a cough is related to the radius *r* of the trachea according to

$$v = ar^2(r_0 - r)$$

where *a* is a constant and r_0 is the radius of the trachea in a relaxed state. Find the radius *r* that produces the maximum velocity of air in the trachea during a cough.

23. *Inventory cost model* Suppose that a company needs 1,500,000 items during a year and that preparation

for each production run costs $600. Suppose also that it costs $15 to produce each item and $2 per year to store an item. Use the inventory cost model to find the number of items in each production run so that the total costs of production and storage are minimized.

24. *Inventory cost model* Suppose that a company needs 60,000 items during a year and that preparation for each production run costs $400. Suppose further that it costs $4 to produce each item and $0.75 to store an item for one year. Use the inventory cost model to find the number of items in each production run that will minimize the total costs of production and storage.

25. *Inventory cost model* A company needs 150,000 items per year. It costs the company $360 to prepare a production run of these items and $7 to produce each item. If it also costs the company $0.75 per year for each item stored, find the number of items that should be produced in each run so that total costs of production and storage are minimized.

26. *Inventory cost model* A company needs 450,000 items per year. Production costs are $500 to prepare for a production run and $10 for each item produced. Inventory costs are $2 per item per year. Find the number of items that should be produced in each run so that the total costs of production and storage are minimized.

27. *Volume* A rectangular box with a square base is to be formed from a square piece of metal with 12-inch sides. If a square piece with side x is cut from each corner of the metal and the sides are folded up to form an open box, the volume of the box is $V = (12 - 2x)^2 x$. What value of x will maximize the volume of the box?

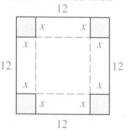

28. *Volume*
   (a) A square piece of cardboard 36 centimeters on a side is to be formed into a rectangular box by cutting squares with length x from each corner and folding up the sides. What is the maximum volume possible for the box?
   (b) Show that if the piece of cardboard is k centimeters on each side, cutting squares of size $k/6$ and folding up the sides gives the maximum volume.

29. *Revenue* The owner of an orange grove must decide when to pick one variety of oranges. She can sell them for $24 a bushel if she sells them now, with each tree yielding an average of 5 bushels. The yield increases by half a bushel per week for the next 5 weeks, but the price per bushel decreases by $1.50 per bushel each week. When should the oranges be picked for maximum return?

30. *Minimum material*
   (a) A box with an open top and a square base is to be constructed to contain 4000 cubic inches. Find

the dimensions that will require the minimum amount of material to construct the box.
   (b) A box with an open top and a square base is to be constructed to contain k cubic inches. Show that the minimum amount of material is used to construct the box when each side of the base is $x = (2k)^{1/3}$ and the height is $y = (k/4)^{1/3}$

31. *Minimum cost* A printer has a contract to print 100,000 posters for a political candidate. He can run the posters by using any number of plates from 1 to 30 on his press. If he uses x metal plates, they will produce x copies of the poster with each impression of the press. The metal plates cost $20.00 to prepare, and it costs $125.00 per hour to run the press. If the press can make 1000 impressions per hour, how many metal plates should the printer make to minimize costs?

32. *Shortest time* A vacationer on an island 8 miles offshore from a point that is 48 miles from town must travel to town occasionally. (See the figure.) The vacationer has a boat capable of traveling 30 mph and can go by auto along the coast at 55 mph. At what point should the car be left to minimize the time it takes to get to town?

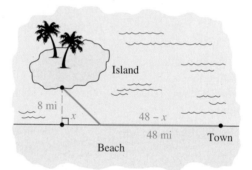

33. *U.S. oil reserves* By using U.S. Department of Energy data for selected years from 2011 and projected to 2040, the U.S. onshore oil reserves in the lower 48 states (in billions of barrels) can be modeled by the function

$$R(t) = -0.00044t^3 + 0.0042t^2 + 0.52t + 19$$

where t is the number of years past 2010.
   (a) In what year does the model predict the onshore oil reserves in the lower 48 states will reach a maximum? Find the predicted maximum. Is this the absolute maximum for the period from 2011 to 2040?
   (b) Find the absolute minimum point for the period from 2011 to 2040.
   (c) Find the t-value of the point of inflection for $R(t)$. At that t-value, is the rate of change of these oil reserves maximum or minimum? Explain.

OBJECTIVES

10.5

- To locate horizontal asymptotes
- To locate vertical asymptotes
- To sketch graphs of functions that have vertical and/or horizontal asymptotes

Rational Functions: More Curve Sketching

■ | APPLICATION PREVIEW |

Suppose that the total cost of producing a shipment of a product is

$$C(x) = 5000x + \frac{125{,}000}{x}, \quad x > 0$$

where x is the number of machines used in the production process. To find the number of machines that will minimize the total cost, we find the minimum value of this rational function. (See Example 3.) The graph of this function contains a vertical asymptote at $x = 0$. We will discuss graphs and applications involving asymptotes in this section.

The procedures for using the first-derivative test and the second-derivative test are given in previous sections, but none of the graphs discussed in those sections contains vertical asymptotes or horizontal asymptotes. In this section, we consider how to use information about asymptotes along with the first and second derivatives, and we present a unified approach to curve sketching.

Asymptotes   In Section 2.4, "Special Functions and Their Graphs," we first discussed asymptotes and saw that they are important features of the graphs that have them. Then, in our discussion of limits in Sections 9.1 and 9.2, we discovered the relationship between certain limits and asymptotes. The formal definition of **vertical asymptotes** uses limits.

Vertical Asymptote

The line $x = x_0$ is a **vertical asymptote** of the graph of $y = f(x)$ if the values of $f(x)$ approach ∞ or $-\infty$ as x approaches x_0 (from the left or the right).

From our work with limits, recall that a vertical asymptote will occur on the graph of a function at an x-value at which the denominator (but not the numerator) of the function is equal to zero. These observations allow us to determine where vertical asymptotes occur.

Vertical Asymptote of a Rational Function

The graph of the rational function

$$h(x) = \frac{f(x)}{g(x)}$$

has a vertical asymptote at $x = c$ if $g(c) = 0$ and $f(c) \neq 0$.

Because a **horizontal asymptote** tells us the behavior of the values of the function (y-coordinates) when x increases or decreases without bound, we use limits at infinity to determine the existence of horizontal asymptotes.

Horizontal Asymptote

The graph of a rational function $y = f(x)$ will have a **horizontal asymptote** at $y = b$, for a constant b, if

$$\lim_{x \to \infty} f(x) = b \quad \text{or} \quad \lim_{x \to -\infty} f(x) = b$$

Otherwise, the graph has no horizontal asymptote.

For a rational function f, $\lim\limits_{x \to \infty} f(x) = b$ if and only if $\lim\limits_{x \to -\infty} f(x) = b$, so we only need to find one of these limits to locate a horizontal asymptote. In Problems 37 and 38 in the 9.2 Exercises, the following statements regarding horizontal asymptotes of the graphs of rational functions were proved.

Horizontal Asymptotes of Rational Functions

Consider the rational function $y = \dfrac{f(x)}{g(x)} = \dfrac{a_n x^n + \cdots + a_1 x + a_0}{b_m x^m + \cdots + b_1 x + b_0}$.

1. If $n < m$ (that is, if the degree of the numerator is less than that of the denominator), a horizontal asymptote occurs at $y = 0$ (the x-axis).
2. If $n = m$ (that is, if the degree of the numerator equals that of the denominator), a horizontal asymptote occurs at $y = \dfrac{a_n}{b_m}$ (the ratio of the leading coefficients).
3. If $n > m$ (that is, if the degree of the numerator is greater than that of the denominator), there is no horizontal asymptote.

EXAMPLE 1   Vertical and Horizontal Asymptotes

Find any vertical and horizontal asymptotes for

(a) $f(x) = \dfrac{2x - 1}{x + 2}$ (b) $f(x) = \dfrac{x^2 + 3}{1 - x}$ (c) $g(x) = \dfrac{10x}{x^2 + 9}$

Solution

(a) The denominator of this function is 0 at $x = -2$, and because this value does not make the numerator 0, there is a vertical asymptote at $x = -2$. Because the function is rational, with the degree of the numerator equal to that of the denominator and with the ratio of the leading coefficients equal to 2, the graph of the function has a horizontal asymptote at $y = 2$. The graph is shown in Figure 10.43(a).

(b) At $x = 1$, the denominator of $f(x)$ is 0 and the numerator is not, so a vertical asymptote occurs at $x = 1$. The function is rational with the degree of the numerator greater than that of the denominator, so there is no horizontal asymptote. The graph is shown in Figure 10.43(b).

(c) The denominator of this function is never zero ($x^2 + 9 \geq 9$ for every real number). Therefore, there is no vertical asymptote. This is a rational function with the degree of the numerator less than that of the denominator; thus $y = 0$ (the x-axis) is a horizontal asymptote. The graph is shown in Figure 10.43(c). ■

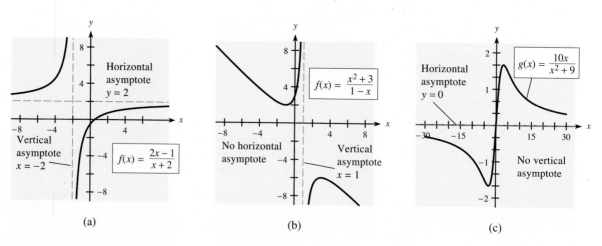

Figure 10.43   (a)   (b)   (c)

More Curve Sketching   We now extend our first- and second-derivative techniques of curve sketching to include functions that have asymptotes.

In general, the following steps are helpful when we sketch the graph of a function.

Graphing Guidelines

1. Determine the domain of the function. The domain may be restricted by the nature of the problem or by the equation.
2. Look for vertical asymptotes, especially if the function is a rational function.
3. Look for horizontal asymptotes, especially if the function is a rational function.
4. Find the relative maxima and minima by using the first-derivative test or the second-derivative test.
5. Use the second derivative to find the points of inflection if this derivative is easily found.
6. Use other information (intercepts, for example) and plot additional points to complete the sketch of the graph.

EXAMPLE 2 **Graphing with Asymptotes**

Sketch the graph of the function $f(x) = \dfrac{x^2}{(x + 1)^2}$.

Solution

1. The domain is the set of all real numbers except $x = -1$.
2. Because $x = -1$ makes the denominator 0 and does not make the numerator 0, there is a vertical asymptote at $x = -1$.
3. Because $\dfrac{x^2}{(x + 1)^2} = \dfrac{x^2}{x^2 + 2x + 1}$

   the function is rational with the degree of the numerator equal to that of the denominator and with the ratio of the leading coefficients equal to 1. Hence, the graph of the function has a horizontal asymptote at $y = 1$.
4. To find any maxima and minima, we first find $f'(x)$.

$$f'(x) = \frac{(x + 1)^2 (2x) - x^2 [2(x + 1)]}{(x + 1)^4} = \frac{2x(x + 1)[(x + 1) - x]}{(x + 1)^4} = \frac{2x}{(x + 1)^3}$$

Thus $f'(x) = 0$ when $x = 0$ (and $y = 0$), and $f'(x)$ is undefined at $x = -1$ (where the vertical asymptote occurs). Using $x = 0$ and $x = -1$ gives the following sign diagram for $f'(x)$. The sign diagram shows that the critical point $(0, 0)$ is a relative minimum and shows how the graph approaches the vertical asymptote at $x = -1$.

$*x = -1$ is a vertical asymptote.

5. The second derivative is

$$f''(x) = \frac{(x + 1)^3 (2) - 2x[3(x + 1)^2]}{(x + 1)^6}$$

Factoring $(x + 1)^2$ from the numerator and simplifying give

$$f''(x) = \frac{2 - 4x}{(x + 1)^4}$$

We can see that $f''(0) = 2 > 0$, so the second-derivative test also shows that $(0, 0)$ is a relative minimum. We see that $f''(x) = 0$ when $x = \frac{1}{2}$. Checking $f''(x)$ between $x = -1$ (where it is undefined) and $x = \frac{1}{2}$ shows that the graph is concave up on this interval.

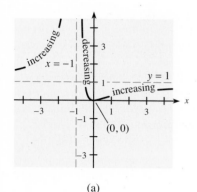

(a)

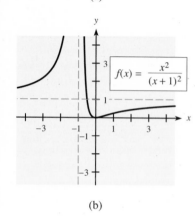

(b)

Figure 10.44

Note also that $f''(x) < 0$ for $x > \frac{1}{2}$, so the point $\left(\frac{1}{2}, \frac{1}{9}\right)$ is a point of inflection. Also see the sign diagram for $f''(x)$.

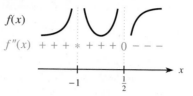

*$x = -1$ is a vertical asymptote.

6. To see how the graph approaches the horizontal asymptote, we check $f(x)$ for large values of $|x|$.

$$f(-100) = \frac{(-100)^2}{(-99)^2} = \frac{10,000}{9,801} > 1, \quad f(100) = \frac{100^2}{101^2} = \frac{10,000}{10,201} < 1$$

Thus the graph has the characteristics shown in Figure 10.44(a). The graph is shown in Figure 10.44(b).

When we wish to learn about a function $f(x)$ or sketch its graph, it is important to understand what information we obtain from $f(x)$, from $f'(x)$, and from $f''(x)$. The following summary may be helpful.

■ | DERIVATIVES AND GRAPHS |

Source	Information Provided
$f(x)$	y-coordinates; horizontal asymptotes, vertical asymptotes; domain restrictions
$f'(x)$	Increasing [$f'(x) > 0$]; decreasing [$f'(x) < 0$]; critical points [$f'(x) = 0$ or $f'(x)$ undefined]; sign-diagram tests for maxima and minima
$f''(x)$	Concave up [$f''(x) > 0$]; concave down [$f''(x) < 0$]; possible points of inflection [$f''(x) = 0$ or $f''(x)$ undefined]; sign-diagram tests for points of inflection; second-derivative test for maxima and minima

✓ CHECKPOINT

1. Let $f(x) = \dfrac{2x + 10}{x - 1}$ and decide whether the following are true or false.

   (a) $f(x)$ has a vertical asymptote at $x = 1$.
   (b) $f(x)$ has $y = 2$ as its horizontal asymptote.

2. Let $f(x) = \dfrac{x^3 - 16}{x} + 1$; then $f'(x) = \dfrac{2x^3 + 16}{x^2}$ and $f''(x) = \dfrac{2x^3 - 32}{x^3}$.

   Use these to determine whether the following are true or false.

   (a) There are no asymptotes.
   (b) $f'(x) = 0$ when $x = -2$
   (c) A partial sign diagram for $f'(x)$ is

   $f'(x)$ $- - - - 0 + + + + + * + + + +$

   $\qquad\qquad\quad -2 \qquad\quad 0$

   *means $f'(0)$ is undefined.

   (d) There is a relative minimum at $x = -2$.
   (e) A partial sign diagram for $f''(x)$ is

   $f''(x) + + + + * - - - - - 0 + + + +$

   $\qquad\qquad\quad 0 \qquad\quad \sqrt[3]{16}$

   *means $f''(0)$ is undefined.

   (f) There are points of inflection at $x = 0$ and $x = \sqrt[3]{16}$.

EXAMPLE 3 **Production Costs** **| APPLICATION PREVIEW |**

Suppose that the total cost of producing a shipment of a certain product is

$$C(x) = 5000x + \frac{125{,}000}{x}, \quad x > 0$$

where x is the number of machines used in the production process.

(a) Determine any asymptotes for $C(x)$.
(b) How many machines should be used to minimize the total cost?
(c) Graph this total cost function.

Solution

(a) Writing this function with all terms over a common denominator gives

$$C(x) = 5000x + \frac{125{,}000}{x} = \frac{5000x^2 + 125{,}000}{x}$$

The domain of $C(x)$ does not include 0, and $C \to \infty$ as $x \to 0^+$, so there is a vertical asymptote at $x = 0$. Thus the cost increases without bound as the number of machines used in the process approaches zero. Because the numerator has a higher degree than the denominator, there is no horizontal asymptote.

(b) Finding the derivative of $C(x)$ gives

$$C'(x) = 5000 - \frac{125{,}000}{x^2} = \frac{5000x^2 - 125{,}000}{x^2}$$

Setting $C'(x) = 0$ and solving for x gives the critical values of x.

$$0 = \frac{5000(x + 5)(x - 5)}{x^2}$$

$$x = 5 \quad \text{or} \quad x = -5$$

Because $C''(x) = 250{,}000x^{-3} = \dfrac{250{,}000}{x^3}$ is positive for all $x > 0$, using 5 machines minimizes the cost at $C(5) = 50{,}000$.

(c) The graph is shown in Figure 10.45.

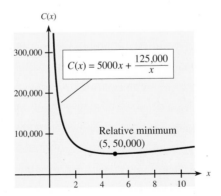

$C(x)$

$C(x) = 5000x + \dfrac{125{,}000}{x}$

Relative minimum
(5, 50,000)

Figure 10.45

EXAMPLE 4 **Horizontal and Vertical Asymptotes**

Figure 10.46 on the next page shows the graph of $f(x) = \dfrac{71x^2}{28(3 - 2x^2)}$.

(a) Determine whether the function has horizontal or vertical asymptotes, and estimate where they occur.
(b) Check your conclusions to part (a) analytically.

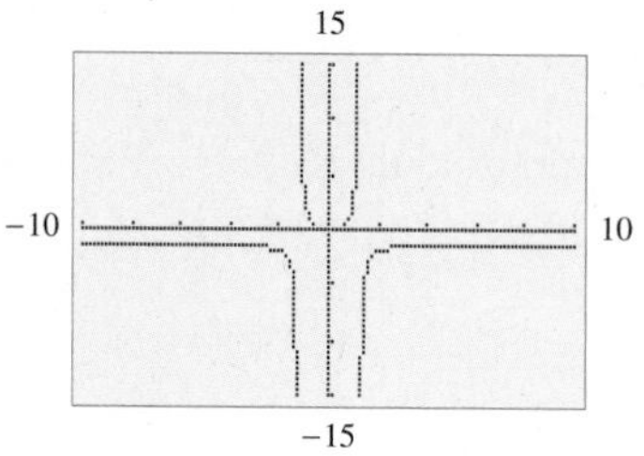

Figure 10.46

Solution

(a) The graph appears to have a horizontal asymptote somewhere between $y = -1$ and $y = -2$, perhaps near $y = -1.5$. Also, there are two vertical asymptotes located approximately at $x = 1.25$ and $x = -1.25$.

(b) The function

$$f(x) = \frac{71x^2}{28(3 - 2x^2)} = \frac{71x^2}{84 - 56x^2}$$

is a rational function with the degree of the numerator equal to that of the denominator and with the ratio of the leading coefficients equal to $-71/56$. Thus the graph of the function has a horizontal asymptote at $y = -71/56 \approx -1.27$.

Vertical asymptotes occur at x-values where $28(3 - 2x^2) = 0$. Solving gives

$$3 - 2x^2 = 0 \quad \text{or} \quad 3 = 2x^2 \quad \text{so} \quad \frac{3}{2} = x^2$$

$$\text{Thus} \qquad \pm\sqrt{\frac{3}{2}} = x \quad \text{or} \quad x \approx \pm 1.225$$

Calculator Note The procedures previously outlined in this section are necessary to generate a complete and accurate graph. With a graphing calculator, the graph of a function is easily generated as long as the viewing window dimensions are appropriate. We frequently need information provided by derivatives to obtain a window that shows all features of a graph. See Example 5 and Appendix A, Section 10.5, for details.

EXAMPLE 5

Graphing with Technology

The standard viewing window of the graph of $f(x) = \dfrac{x + 10}{x^2 + 300}$ appears blank (check and see). Find any asymptotes, maxima, and minima, and determine an appropriate viewing window. Sketch the graph.

Solution

Because $x^2 + 300 = 0$ has no real solution, there are no vertical asymptotes. The function is rational with the degree of the numerator less than that of the denominator, so the horizontal asymptote is $y = 0$, which is the x-axis.

We then find an appropriate viewing window by locating the critical points.

$$f'(x) = \frac{(x^2 + 300)(1) - (x + 10)(2x)}{(x^2 + 300)^2}$$

$$= \frac{x^2 + 300 - 2x^2 - 20x}{(x^2 + 300)^2} = \frac{300 - 20x - x^2}{(x^2 + 300)^2}$$

$f'(x) = 0$ when the numerator is zero. Thus

$$300 - 20x - x^2 = 0$$
$$0 = x^2 + 20x - 300$$
$$0 = (x + 30)(x - 10)$$
$$x + 30 = 0 \quad x - 10 = 0$$
$$x = -30 \quad\quad x = 10$$

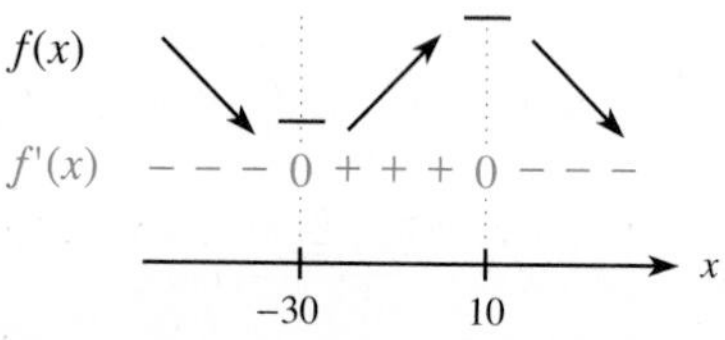

The critical points are $x = -30$, $y = -\frac{1}{60} \approx -0.01666667$ and $x = 10$, $y = \frac{1}{20} = 0.05$. See the sign diagram for $f'(x)$.

Without using the information above, a graphing calculator may not give a useful graph. An x-range that includes -30 and 10 is needed. Because $y = 0$ is a horizontal asymptote, these relative extrema are absolute, and the y-range must be quite small for the shape of the graph to be seen clearly. Figure 10.47 shows the graph.

$f(x)$ ↘ ↘ ↗ ↗ ↘ ↘

$f'(x)$ − − − 0 + + + 0 − − −

$-30 \qquad 10 \qquad x$

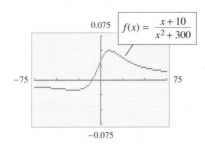

0.075

$$f(x) = \frac{x + 10}{x^2 + 300}$$

−75   75

Figure 10.47   −0.075

1. (a) True. (b) True.
2. (a) False. There are no horizontal asymptotes, but $x = 0$ is a vertical asymptote.
   (b) True
   (c) True
   (d) True
   (e) True
   (f) False, only at $(\sqrt[3]{16}, 1)$. At $x = 0$ there is a vertical asymptote.

| EXERCISES | 10.5

In Problems 1–4, a function and its graph are given. Use the graph to find each of the following, if they exist. Then confirm your results analytically.

(a) vertical asymptotes   (b) $\lim\limits_{x \to \infty} f(x)$

(c) $\lim\limits_{x \to -\infty} f(x)$ (d) horizontal asymptotes

1. $f(x) = \dfrac{x - 4}{x - 2}$ 2. $f(x) = \dfrac{8}{x + 2}$

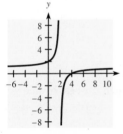

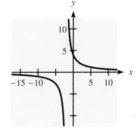

3. $f(x) = \dfrac{3(x^4 + 2x^3 + 6x^2 + 2x + 5)}{(x^2 - 4)^2}$

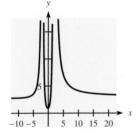

4. $f(x) = \dfrac{x^2}{(x - 2)^2}$

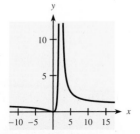

In Problems 5–10, find any horizontal and vertical asymptotes for each function.

5. $y = \dfrac{2x}{x - 3}$ 6. $y = \dfrac{3x - 1}{x + 5}$

7. $y = \dfrac{x + 1}{x^2 - 4}$ 8. $y = \dfrac{4x}{9 - x^2}$

9. $y = \dfrac{3x^3 - 6}{x^2 + 4}$ 10. $y = \dfrac{6x^3}{4x^2 + 9}$

For each function in Problems 11–18, find any horizontal and vertical asymptotes, and use information from the first derivative to sketch the graph.

11. $f(x) = \dfrac{2x + 2}{x - 3}$ 12. $f(x) = \dfrac{5x - 15}{x + 2}$

13. $y = \dfrac{x^2 + 4}{x}$ 14. $y = \dfrac{x^2 + 4}{x^2}$

15. $y = \dfrac{27x^2}{(x + 1)^3}$ 16. $y = \left(\dfrac{x + 2}{x - 3}\right)^2$

17. $f(x) = \dfrac{16x}{x^2 + 1}$ 18. $f(x) = \dfrac{4x^2}{x^4 + 1}$

In Problems 19–24, a function and its first and second derivatives are given. Use these to find any horizontal and vertical asymptotes, critical points, relative maxima, relative minima, and points of inflection. Then sketch the graph of each function.

19. $y = \dfrac{x}{(x - 1)^2}$ 20. $y = \dfrac{(x - 1)^2}{x^2}$

   $y' = -\dfrac{x + 1}{(x - 1)^3}$ $y' = \dfrac{2(x - 1)}{x^3}$

   $y'' = \dfrac{2x + 4}{(x - 1)^4}$ $y'' = \dfrac{6 - 4x}{x^4}$

21. $y = x + \dfrac{3}{\sqrt[3]{x-3}}$

$y' = 1 - \dfrac{1}{(x-3)^{4/3}}$

$y'' = \dfrac{4}{3(x-3)^{7/3}}$

22. $y = 3\sqrt[3]{x} + \dfrac{1}{x}$

$y' = \dfrac{x^{4/3} - 1}{x^2}$

$y'' = \dfrac{6 - 2x^{4/3}}{3x^3}$

23. $f(x) = \dfrac{9(x-2)^{2/3}}{x^2}$

$f'(x) = \dfrac{12(3-x)}{x^3(x-2)^{1/3}}$

$f''(x) = \dfrac{4(7x^2 - 42x + 54)}{x^4(x-2)^{4/3}}$

24. $f(x) = \dfrac{3x^{2/3}}{x+1}$

$f'(x) = \dfrac{2-x}{x^{1/3}(x+1)^2}$

$f''(x) = \dfrac{2(2x^2 - 8x - 1)}{3x^{4/3}(x+1)^3}$

In Problems 25–28, a function and its graph are given.
(a) **Use the graph to estimate the locations of any horizontal or vertical asymptotes.**
(b) **Use the function to determine precisely the locations of any asymptotes.**

25. $f(x) = \dfrac{9x}{17 - 4x}$

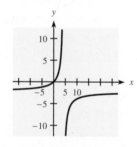

26. $f(x) = \dfrac{5 - 13x}{3x + 20}$

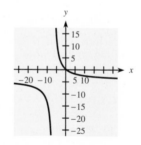

27. $f(x) = \dfrac{20x^2 + 98}{9x^2 - 49}$

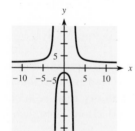

28. $f(x) = \dfrac{15x^2 - x}{7x^2 - 35}$

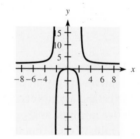

For each function in Problems 29–34, complete the following steps.
(a) **Use a graphing calculator to graph the function in the standard viewing window.**
(b) **Analytically determine the location of any asymptotes and extrema.**

(c) **Graph the function in a viewing window that shows all features of the graph. State the ranges for x-values and y-values for your viewing window.**

29. $f(x) = \dfrac{x + 25}{x^2 + 1400}$

30. $f(x) = \dfrac{x - 50}{x^2 + 1100}$

31. $f(x) = \dfrac{100(9 - x^2)}{x^2 + 100}$

32. $f(x) = \dfrac{200x^2}{x^2 + 100}$

33. $f(x) = \dfrac{1000x - 4000}{x^2 - 10x - 2000}$

34. $f(x) = \dfrac{900x + 5400}{x^2 - 30x - 1800}$

APPLICATIONS

35. *Cost-benefit* The percent p of particulate pollution that can be removed from the smokestacks of an industrial plant by spending C dollars is given by

$$p = \frac{100C}{7300 + C}$$

(a) Find any C-values at which the rate of change of p with respect to C does not exist. Make sure that these make sense in the problem.
(b) Find C-values for which p is increasing.
(c) If there is a horizontal asymptote, find it.
(d) Can 100% of the pollution be removed?

36. *Cost-benefit* The percent p of impurities that can be removed from the waste water of a manufacturing process at a cost of C dollars is given by

$$p = \frac{100C}{8100 + C}$$

(a) Find any C-values at which the rate of change of p with respect to C does not exist. Make sure that these make sense in the problem.
(b) Find C-values for which p is increasing.
(c) Find any horizontal asymptotes.
(d) Can 100% of the pollution be removed?

37. *Revenue* A recently released film has its weekly revenue given by

$$R(t) = \frac{50t}{t^2 + 36}, \quad t \geq 0$$

where $R(t)$ is in millions of dollars and t is in weeks.
(a) Graph $R(t)$.
(b) When will revenue be maximized?
(c) Suppose that if revenue decreases for 4 consecutive weeks, the film will be removed from theaters and will be released as a video 12 weeks later. When will the video come out?

38. *Minimizing average cost* If the total daily cost, in dollars, of producing plastic rafts for swimming pools is given by

$$C(x) = 500 + 8x + 0.05x^2$$

where x is the number of rafts produced per day, then the average cost per raft produced is given by $\overline{C}(x) = C(x)/x$, for $x > 0$.

(a) Graph this function.

(b) Discuss what happens to the average cost as the number of rafts decreases, approaching 0.

(c) Find the level of production that minimizes average cost.

39. *Wind chill* If x is the wind speed in miles per hour and is greater than or equal to 5, then the wind chill (in degrees Fahrenheit) for an air temperature of 0°F can be approximated by the function

$$f(x) = \frac{289.173 - 58.5731x}{x + 1}, \quad x \geq 5$$

(a) Ignoring the restriction $x \geq 5$, does $f(x)$ have a vertical asymptote? If so, what is it?

(b) Does $f(x)$ have a vertical asymptote within its domain?

(c) Does $f(x)$ have a horizontal asymptote? If so, what is it?

(d) In the context of wind chill, does $\lim\limits_{x \to \infty} f(x)$ have a physical interpretation? If so, what is it, and is it meaningful?

40. *Profit* An entrepreneur starts new companies and sells them when their growth is maximized. Suppose that the annual profit for a new company is given by

$$P(x) = 22 - \frac{1}{2}x - \frac{18}{x + 1}$$

where P is in thousands of dollars and x is the number of years after the company is formed. If she wants to sell the company before profits begin to decline, after how many years should she sell it?

41. *Productivity* The figure is a typical graph of worker productivity per hour P as a function of time t on the job.

(a) What is the horizontal asymptote?

(b) What is $\lim\limits_{x \to \infty} P(t)$?

(c) What is the horizontal asymptote for $P'(t)$?

(d) What is $\lim\limits_{x \to \infty} P'(t)$?

42. *Sales volume* The figure shows a typical curve that gives the volume of sales S as a function of time t after an ad campaign.

(a) What is the horizontal asymptote?

(b) What is $\lim\limits_{t \to \infty} S(t)$?

(c) What is the horizontal asymptote for $S'(t)$?

(d) What is $\lim\limits_{t \to \infty} S'(t)$?

43. *Females in the workforce* For selected years from 1950 and projected to 2050, the table shows the percent of total U.S. workers who were female.

Year	% Female	Year	% Female
1950	29.6	2010	47.9
1960	33.4	2015	48.3
1970	38.1	2020	48.1
1980	42.5	2030	48.0
1990	45.2	2040	47.9
2000	46.6	2050	47.7

Source: U.S. Bureau of Labor Statistics

Assume these data can be modeled with the function

$$p(t) = \frac{78.6t + 2090}{1.38t + 64.1}$$

where $p(t)$ is the percent of the U.S. workforce that is female and t is the number of years past 1950.

(a) Find $\lim\limits_{t \to \infty} p(t)$.

(b) Interpret your answer to part (a).

(c) Does $p(t)$ have any vertical asymptotes within its domain $t \geq 0$?

(d) Whenever $p(t) < 0$ or $p(t) > 100$, the model would be inappropriate. Determine whether the model is ever inappropriate for $t \geq 0$.

44. *Modeling Obesity* Obesity (BMI $\geq$ 30) is a serious problem in the United States and expected to get worse. Being overweight increases the risk of diabetes, heart disease, and many other ailments, but the severely obese (BMI $\geq$ 40) are most at risk and are the most expensive to treat. The percent of Americans who are obese and severely obese from 1990 projected to 2030 are shown in the table below. Report the following models with three significant digits.

(a) Find the linear function, $O(x)$, that models the percent of Americans who are obese, with x equal to the number of years past 1980.

(b) Find the linear function, $S(x)$, that models the percent of Americans who are severely obese, with x equal to the number of years past 1980.

(c) Form the rational function $F(x)$ that gives the fraction of obese Americans who are severely obese.

(d) Find $\lim\limits_{x \to \infty} F(x)$ and tell what it means in terms of these data.

(e) What does part (d) tell us about the graph of $y = F(x)$?

(f) Does $F(x)$ have any vertical asymptotes for $x > 0$?

Year	1990	2000	2010	2015	2020	2025	2030
% Obese	12.7	22.1	30.9	34.5	37.4	39.9	42.2
% Severely Obese	0.8	2.2	4.9	6.4	7.9	9.5	11.1

Source: American Journal of Preventive Medicine 42 (June 2012) 563–70, ajpmonline.org

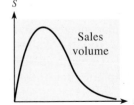

45. *Barometric pressure* The figure shows a barograph readout of the barometric pressure as recorded by Georgia Southern University's meteorological equipment. The figure shows a tremendous drop in barometric pressure on Saturday morning, March 13, 1993.

(a) If $B(t)$ is barometric pressure expressed as a function of time, as shown in the figure, does $B(t)$ have a vertical asymptote sometime after 8 A.M. on Saturday, March 13, 1993? Explain why or why not.

(b) Consult your library or some other resource to find out what happened in Georgia (and in the eastern United States) on March 13, 1993, to cause such a dramatic drop in barometric pressure.

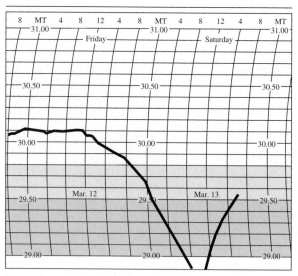

Source: Statesboro Herald, March 14, 1993.

Chapter 10 Summary & Review

KEY TERMS AND FORMULAS

Section 10.1

Relative maxima and minima (p. 630)
Increasing (p. 630)
  $f'(x) > 0$
Decreasing (p. 630)
  $f'(x) < 0$

Critical points (p. 631)
  $f'(x) = 0$ or $f'(x)$ undefined
Sign diagram for $f'(x)$ (p. 632)
First-derivative test (p. 632)
Horizontal point of inflection (p. 634)

Section 10.2

Concave up (p. 644)
  $f''(x) > 0$
Concave down (p. 644)
  $f''(x) < 0$

Point of inflection (p. 645)
  May occur where $f''(x) = 0$ or $f''(x)$ undefined
Sign diagram for $f''(x)$ (p. 646)
Second-derivative test (p. 648)

Section 10.3

Absolute extrema (p. 656)
Maximizing revenue (p. 657)
Average cost (p. 658)
  $\overline{C}(x) = C(x)/x$

Profit maximization (p. 659)
Monopolistic market (p. 660)
  $R(x) = p \cdot x$ where $p = f(x)$ is the demand function
Competitive market (p. 661)
  $R(x) = p \cdot x$ where $p = $ equilibrium price

Section 10.4

Inventory cost models (p. 670)

Section 10.5

Asymptotes for $f(x)/g(x)$ (p. 675)
  Vertical: $x = c$ if $g(c) = 0$ and $f(c) \neq 0$
    y unbounded near $x = c$

Horizontal: $y = b$
  $f(x) \to b$ as $x \to \infty$ or $x \to -\infty$
  Use the highest power terms of $f(x)$ and $g(x)$
Graphing guidelines (p. 677)
Derivatives and graphs (p. 678)

REVIEW EXERCISES

Section 10.1

In Problems 1–4, find all critical points and determine whether they are relative maxima, relative minima, or horizontal points of inflection.

1. $y = -x^2$
2. $p = q^2 - 4q - 5$
3. $f(x) = 1 - 3x + 3x^2 - x^3$
4. $f(x) = \dfrac{3x}{x^2 + 1}$

In Problems 5–10:
(a) Find all critical values, including those at which $f'(x)$ is undefined.
(b) Find the relative maxima and minima, if any exist.
(c) Find the horizontal points of inflection, if any exist.
(d) Sketch the graph.

5. $y = x^3 + x^2 - x - 1$
6. $f(x) = 4x^3 - x^4$
7. $f(x) = x^3 - \dfrac{15}{2}x^2 - 18x + \dfrac{3}{2}$
8. $y = 5x^7 - 7x^5 - 1$
9. $y = x^{2/3} - 1$
10. $y = x^{2/3}(x - 4)^2$

Section 10.2

11. Is the graph of $y = x^4 - 3x^3 + 2x - 1$ concave up or concave down at $x = 2$?
12. Find intervals on which the graph of $y = x^4 - 2x^3 - 12x^2 + 6$ is concave up and intervals on which it is concave down, and find points of inflection.
13. Find the relative maxima, relative minima, and points of inflection of the graph of $y = x^3 - 3x^2 - 9x + 10$.

In Problems 14 and 15, find any relative maxima, relative minima, and points of inflection, and sketch each graph.

14. $y = x^3 - 12x$
15. $y = 2 + 5x^3 - 3x^5$

Section 10.3

16. Given $R = 280x - x^2$, find the absolute maximum and minimum for R when (a) $0 \le x \le 200$ and (b) $0 \le x \le 100$.

17. Given $y = 6400x - 18x^2 - \dfrac{x^3}{3}$, find the absolute maximum and minimum for y when (a) $0 \le x \le 50$ and (b) $0 \le x \le 100$.

Section 10.5

In Problems 18 and 19, use the graphs to find the following items.
(a) vertical asymptotes
(b) horizontal asymptotes
(c) $\lim\limits_{x \to \infty} f(x)$
(d) $\lim\limits_{x \to -\infty} f(x)$

18.

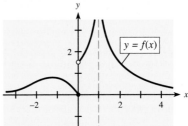

19.

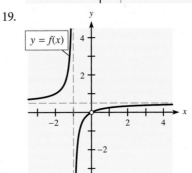

In Problems 20 and 21, find any horizontal asymptotes and any vertical asymptotes.

20. $y = \dfrac{3x + 2}{2x - 4}$

21. $y = \dfrac{x^2}{1 - x^2}$

In Problems 22–24:
(a) Find any horizontal and vertical asymptotes.
(b) Find any relative maxima and minima.
(c) Sketch each graph.

22. $y = \dfrac{3x}{x + 2}$

23. $y = \dfrac{8(x - 2)}{x^2}$

24. $y = \dfrac{x^2}{x - 1}$

Sections 10.1 and 10.2

In Problems 25 and 26, a function and its graph are given.
(a) Use the graph to determine (estimate) x-values where $f'(x) > 0$, where $f'(x) < 0$, and where $f'(x) = 0$.
(b) Use the graph to determine x-values where $f''(x) > 0$, where $f''(x) < 0$, and where $f''(x) = 0$.
(c) Check your conclusions to (a) by finding $f'(x)$ and graphing it with a graphing calculator.
(d) Check your conclusions to (b) by finding $f''(x)$ and graphing it with a graphing calculator.

25. $f(x) = x^3 - 4x^2 + 4x$

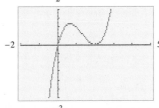

26. $f(x) = 0.0025x^4 + 0.02x^3 - 0.48x^2 + 0.08x + 4$

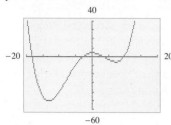

In Problems 27 and 28, $f'(x)$ and its graph are given.
(a) Use the graph of $f'(x)$ to determine (estimate) where the graph of $f(x)$ is increasing, where it is decreasing, and where it has relative extrema.
(b) Use the graph of $f'(x)$ to determine where $f''(x) > 0$, where $f''(x) < 0$, and where $f''(x) = 0$.
(c) Verify that the given $f(x)$ has $f'(x)$ as its derivative, and graph $f(x)$ to check your conclusions in part (a).
(d) Calculate $f''(x)$ and graph it to check your conclusions in part (b).

27. $f'(x) = x^2 + 4x - 5$ $\left(\text{for } f(x) = \dfrac{x^3}{3} + 2x^2 - 5x\right)$

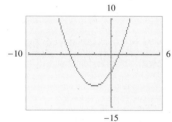

28. $f'(x) = 6x^2 - x^3$ $\left(\text{for } f(x) = 2x^3 - \dfrac{x^4}{4}\right)$

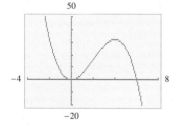

In Problems 29 and 30, $f''(x)$ and its graph are given.
(a) Use the graph to determine (estimate) where the graph of $f(x)$ is concave up, where it is concave down, and where it has points of inflection.
(b) Verify that the given $f(x)$ has $f''(x)$ as its second derivative, and graph $f(x)$ to check your conclusions in part (a).

29. $f''(x) = 4 - x$ $\left(\text{for } f(x) = 2x^2 - \dfrac{x^3}{6}\right)$

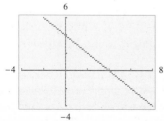

30. $f''(x) = 6 - x - x^2$ $\left(\text{for } f(x) = 3x^2 - \dfrac{x^3}{6} - \dfrac{x^4}{12}\right)$

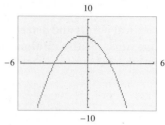

APPLICATIONS

Sections 10.1–10.3

In Problems 31–36, cost, revenue, and profit are in dollars and x is the number of units.

31. *Cost* Suppose the total cost function for a product is

$$C(x) = 3x^2 + 15x + 75$$

How many units will minimize the average cost? Find the minimum average cost.

32. *Revenue* Suppose the total revenue function for a product is given by

$$R(x) = 32x - 0.01x^2$$

(a) How many units will maximize the total revenue? Find the maximum revenue.
(b) If production is limited to 1500 units, how many units will maximize the total revenue? Find the maximum revenue.

33. *Profit* Suppose the profit function for a product is

$$P(x) = 1080x + 9.6x^2 - 0.1x^3 - 50,000$$

Find the maximum profit.

34. *Profit* How many units (x) will maximize profit if $R(x) = 46x - 0.01x^2$ and $C(x) = 0.05x^2 + 10x + 1100$?

35. *Profit* A product can be produced at a total cost of $C(x) = 800 + 4x$, where x is the number produced and is limited to at most 150 units. If the total revenue is given by $R(x) = 80x - \frac{1}{4}x^2$, determine the level of production that will maximize the profit.

36. *Average cost* The total cost function for a product is $C = 2x^2 + 54x + 98$. How many units must be produced to minimize average cost?

37. *Marginal profit* The figure shows the graph of a marginal profit function for a company. At what level of sales will profit be maximized? Explain.

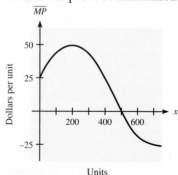

38. *Productivity—diminishing returns* Suppose the productivity P of an individual worker (in number of items produced per hour) is a function of the number of hours of training t according to

$$P(t) = 5 + \frac{95t^2}{t^2 + 2700}$$

Find the number of hours of training at which the rate of change of productivity is maximized. (That is, find the point of diminishing returns.)

39. *Output* The figure shows a typical graph of output y (in thousands of dollars) as a function of capital investment I (also in thousands of dollars).
   (a) Is the point of diminishing returns closest to the point at which $I = 20$, $I = 60$, or $I = 120$? Explain.
   (b) The average output per dollar of capital investment is defined as the total output divided by the amount of capital investment; that is,

$$\text{Average output} = \frac{f(I)}{I}$$

   Calculate the slope of a line from $(0, 0)$ to an arbitrary point $(I, f(I))$ on the output graph. How is this slope related to the average output?
   (c) Is the maximum average output attained when the capital investment is closest to $I = 40$, to $I = 70$, or to $I = 140$? Explain.

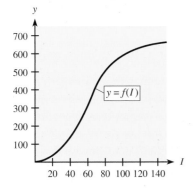

40. *Revenue* MMR II Extreme Bike Shop sells 54 of its most popular mountain bikes per month at a price of $1540 each. Market research indicates that MMR II could sell one more of these bikes if the price were $10 lower. At what selling price will MMR II maximize the revenue from these bikes?

41. *Profit* If in Problem 40 the mountain bikes cost the shop $680 each, at what selling price will MMR II's profit be a maximum?

42. *Profit* Suppose that for a product in a competitive market, the demand function is $p = 1200 - 2x$ and the supply function is $p = 200 + 2x$, where x is the number of units and p is in dollars. A firm's average cost function for this product is

$$\overline{C}(x) = \frac{12,000}{x} + 50 + x$$

Find the maximum profit. (*Hint:* First find the equilibrium price.)

43. *Profit* The monthly demand function for x units of a product sold at $\$p$ per unit by a monopoly is $p = 800 - x$, and its average cost is $\overline{C} = 200 + x$.
   (a) Determine the quantity that will maximize profit.
   (b) Find the selling price at the optimal quantity.

44. *Profit* Suppose that in a monopolistic market, the demand function for a commodity is

$$p = 7000 - 10x - \frac{x^2}{3}$$

where x is the number of units and p is in dollars. If a company's average cost function for this commodity is

$$\overline{C}(x) = \frac{40,000}{x} + 600 + 8x$$

find the maximum profit.

Section 10.4

45. *Reaction to a drug* The reaction R to an injection of a drug is related to the dose x (in milligrams) according to

$$R(x) = x^2\left(500 - \frac{x}{3}\right)$$

Find the dose that yields the maximum reaction.

46. *Productivity* The number of parts produced per hour by a worker is given by

$$N = 4 + 3t^2 - t^3$$

where t is the number of hours on the job without a break. If the worker starts at 8 A.M., when will she be at maximum productivity during the morning?

47. *Population* Population estimates show that the equation $P = 300 + 10t - t^2$ represents the size of the graduating class of a high school, where t represents the number of years after 2015, $0 \le t \le 10$. What will be the largest graduating class in the next 10 years?

48. *Night brightness* Suppose that an observatory is to be built between cities A and B, which are 30 miles apart. For the best viewing, the observatory should be located where the night brightness from these cities is minimum. If the night brightness of city A is 8 times that of city B, then the night brightness b between the two cities and x miles from A is given by

$$b = \frac{8k}{x^2} + \frac{k}{(30 - x)^2}$$

where k is a constant. Find the best location for the observatory; that is, find x that minimizes b.

49. *Product design* A playpen manufacturer wants to make a rectangular enclosure with maximum play area. To remain competitive, he wants the perimeter of the base to be only 16 feet. What dimensions should the playpen have?

50. *Printing design* A page is to contain 56 square inches of print and have a $\frac{3}{4}$-inch margin at the bottom and 1-inch margins at the top and on both sides. Find the

dimensions that minimize the size of the page (and hence the costs for paper).

51. **Drug sensitivity** The reaction R to an injection of a drug is related to the dose x, in milligrams, according to

$$R(x) = x^2 \left(500 - \frac{x}{3} \right)$$

The sensitivity to the drug is defined by dR/dx. Find the dose that maximizes sensitivity.

52. **Per capita health care costs** For the years from 2000 and projected to 2018, the U.S. per capita out-of-pocket cost for health care C (in dollars) can be modeled by the function

$$C(t) = 0.118t^3 - 2.51t^2 + 40.2t + 677$$

where t is the number of years past 2000 (*Source:* U.S. Centers for Medicare and Medicaid Services).
   (a) When does the rate of change of health care costs per capita reach its minimum?
   (b) On a graph of $C(t)$, what feature occurs at the t-value found in part (a)?

53. **Inventory cost model** A company needs to produce 288,000 items per year. Production costs are $1500 to prepare for a production run and $30 for each item produced. Inventory costs are $1.50 per year for each item stored. Find the number of items that should be produced in each run so that the total costs of production and storage are minimum.

Section 10.5

54. **Average cost** Suppose the total cost of producing x units of a product is given by

$$C(x) = 4500 + 120x + 0.05x^2 \quad \text{dollars}$$

   (a) Find any asymptotes of the average cost function $\overline{C}(x) = C(x)/x$.
   (b) Graph the average cost function.

55. **Market share** Suppose a company's percent share of the market (actual and projected) for a new product t quarters after its introduction is given by

$$M(t) = \frac{3.8t^2 + 3}{0.1t^2 + 1}$$

   (a) Find the company's market share when the product is introduced.
   (b) Find any horizontal asymptote of the graph of $M(t)$, and write a sentence that explains the meaning of this asymptote.

Chapter 10 TEST

Find the local maxima, local minima, points of inflection, and asymptotes, if they exist, for each of the functions in Problems 1–3. Graph each function.

1. $f(x) = x^3 + 6x^2 + 9x + 3$

2. $y = 4x^3 - x^4 - 10$

3. $y = \dfrac{x^2 - 3x + 6}{x - 2}$

In Problems 4–6, use the function $y = 3x^5 - 5x^3 + 2$.

4. Over what intervals is the graph of this function concave up?

5. Find the points of inflection of this function.

6. Find the relative maxima and minima of this function.

7. Find the absolute maximum and minimum for $f(x) = 2x^3 - 15x^2 + 3$ on the interval $[-2, 8]$.

8. Find all horizontal and vertical asymptotes of the function

$$f(x) = \frac{200x - 500}{x + 300}.$$

9. Use the graph of $y = f(x)$ and the indicated points to complete the chart. Enter $+$, $-$, or 0, according to whether f, f', and f'' are positive, negative, or zero at each point.

Point	f	f'	f''
A			
B			
C			

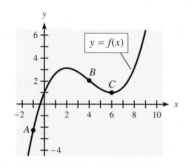

10. Use the figure to complete the following.
   (a) $\lim\limits_{x \to -\infty} f(x) = ?$
   (b) What is the vertical asymptote?

(c) Find the horizontal asymptote.

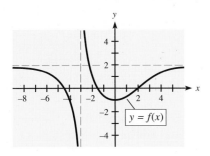

11. If $f(6) = 10, f'(6) = 0$, and $f''(6) = -3$, what can we conclude about the point on the graph of $y = f(x)$ where $x = 6$? Explain.

12. The aged dependency ratio is defined as the number of individuals age 65 and older per 100 individuals ages 20 to 64. The aging of the Baby Boomer generation along with medical advancements and lifestyle changes for all individuals have caused this ratio to rise, shaping society's plans for the needs of a greater number of older individuals. By using Social Security Administration data for selected years from 1990 and projected to 2045, the aged dependency ratio can be modeled by the function

$$A(t) = -0.000497t^3 + 0.0449t^2 - 0.669t + 22.3$$

where t is the number of years past 1990.
(a) Find the critical points for this model, and classify them as relative maxima or relative minima.
(b) Interpret the points found in part (a)

13. The revenue function for a product is $R(x) = 164x$ dollars and the cost function for the product is

$$C(x) = 0.01x^2 + 20x + 300 \text{ dollars}$$

where x is the number of units produced and sold.
(a) How many units of the product should be sold to obtain maximum profit?
(b) What is the maximum possible profit?

14. The cost of producing x units of a product is given by

$$C(x) = 100 + 20x + 0.01x^2 \text{ dollars}$$

How many units should be produced to minimize average cost?

15. A firm sells 100 TV sets per month at $300 each, but market research indicates that it can sell 1 more set per month for each $2 reduction of the price. At what price will the revenue be maximized?

16. An open-top box is made by cutting squares from the corners of a piece of tin and folding up the sides. If the piece of tin was originally 20 centimeters on a side, how long should the sides of the removed squares be to maximize the resulting volume?

17. A company estimates that it will need 784,000 items during the coming year. It costs $420 to manufacture each item, $2500 to prepare for each production run, and $5 per year for each item stored. How many units should be in each production run so that the total costs of production and storage are minimized?

18. **Modeling** The table gives the number of women age 16 years and older (in millions) in the U.S. civilian workforce for selected years from 1950 and projected to 2050.
(a) Use x as the number of years past 1950 to create a cubic model using these data. Report the model with three significant digit coefficients.
(b) During what year does the reported model indicate that the rate of change of the number of women in the workforce reached its maximum?
(c) What feature of the graph of the function found in part (a) is the result found in part (b)?

Year	Millions of Women in the Workforce
1950	18.4
1960	23.2
1970	31.5
1980	45.5
1990	56.8
2000	65.6
2010	75.5
2015	78.6
2020	79.2
2030	81.6
2040	86.5
2050	91.5

Source: U.S. Bureau of Labor Statistics

Extended Applications & Group Projects

I. Production Management

Metal Containers, Inc. is reviewing the way it submits bids on U.S. Army contracts. The army often requests open-top boxes, with square bases and of specified volumes. The army also specifies the materials for the boxes, and the base is usually made of a different material than the sides. The box is assembled by riveting a bracket at each of the eight corners. For Metal Containers, the total cost of producing a box is the sum of the cost of the materials for the box and the labor costs associated with affixing each bracket.

Instead of estimating each job separately, the company wants to develop an overall approach that will allow it to cost out proposals more easily. To accomplish this, company managers need you to devise a formula for the total cost of producing each box and determine the dimensions that allow a box of specified volume to be produced at minimum cost. Use the following notation to help you solve this problem.

Cost of the material for the base = A per square unit

Cost of the material for the sides = B per square unit

Cost of each bracket = C

Cost to affix each bracket = D

Length of the sides of the base = x

Height of the box = h

Volume specified by the army = V

1. Write an expression for the company's total cost in terms of these quantities.
2. At the time an order is received for boxes of a specified volume, the costs of the materials and labor will be fixed and only the dimensions will vary. Find a formula for each dimension of the box so that the total cost is a minimum.
3. The army requests bids on boxes of 48 cubic feet with base material costing the container company $12 per square foot and side material costing $8 per square foot. Each bracket costs $5, and the associated labor cost is $1 per bracket. Use your formulas to find the dimensions of the box that meet the army's requirements at a minimum cost. What is this cost?

Metal Containers asks you to determine how best to order the brackets it uses on its boxes. You are able to obtain the following information: The company uses approximately 100,000 brackets a year, and the purchase price of each is $5. It buys the same number of brackets (say, n) each time it places an order with the supplier, and it costs $60 to process each order. Metal Containers also has additional costs associated with storing, insuring, and financing its inventory of brackets. These carrying costs amount to 15% of the average value of inventory annually. The brackets are used steadily and deliveries are made just as inventory reaches zero, so that inventory fluctuates between zero and n brackets.

4. If the total annual cost associated with the bracket supply is the sum of the annual purchasing cost and the annual carrying costs, what order size n would minimize the total cost?
5. In the general case of the bracket-ordering problem, the order size n that minimizes the total cost of the bracket supply is called the economic order quantity, or EOQ. Use the following notations to determine a general formula for the EOQ.

Fixed cost per order = F

Unit cost = C

Quantity purchased per year = P

Carrying cost (as a decimal rate) = r

II. Room Pricing in the Off-Season (Modeling)

The data in the table, from a survey of resort hotels with comparable rates on Hilton Head Island, show that room occupancy during the off-season (November through February) is related to the price charged for a basic room.

Price per Day	Occupancy Rate, %
$104	53
134	47
143	46
149	45
164	40
194	32

The goal is to use these data to help answer the following questions.

A. What price per day will maximize the daily off-season revenue for a typical hotel in this group if it has 200 rooms available?
B. Suppose that for this typical hotel the daily cost is $5510 plus $30 per occupied room. What price will maximize the profit for this hotel in the off-season?

The price per day that will maximize the off-season profit for this typical hotel applies to this group of hotels. To find the room price per day that will maximize the daily revenue and the room price per day that will maximize the profit for this hotel (and thus the group of hotels) in the off-season, complete the following.

1. Multiply each occupancy rate by 200 to get the hypothetical room occupancy. Create the revenue data points that compare the price with the revenue, R, which is equal to price times the room occupancy.
2. Use technology to create an equation that models the revenue, R, as a function of the price per day, x.
3. Use maximization techniques to find the price that these hotels should charge to maximize the daily revenue.
4. Use technology to get the occupancy as a function of the price, and use the occupancy function to create a daily cost function.
5. Form the profit function.
6. Use maximization techniques to find the price that will maximize the profit.

© Kenneth Sponsler/Shutterstock.com

11

Derivatives Continued

In this chapter we will develop derivative formulas for logarithmic and exponential functions, focusing primarily on base *e* exponentials and logarithms. We will apply logarithmic and exponential functions and use their derivatives to solve maximization and minimization problems in the management and life sciences.

We will also develop methods for finding the derivative of one variable with respect to another even when the first variable is not a function of the other. This method is called implicit differentiation. We will use implicit differentiation with respect to time to solve problems involving rates of change of two or more variables. These problems are called related-rates problems.

The topics and some representative applications discussed in this chapter include the following.

SECTIONS

11.1 **Derivatives of Logarithmic Functions**

11.2 **Derivatives of Exponential Functions**

11.3 **Implicit Differentiation**

11.4 **Related Rates**

11.5 **Applications in Business and Economics**
Elasticity of demand
Taxation in a competitive market

APPLICATIONS

Life span, cost

Future value, revenue

Production, demand

Allometric relationships, blood flow, flight, oil slicks

Elasticity and revenue, maximizing tax revenue

Prerequisite Problem Type	For Section	Answer	Section for Review
(a) Simplify: $\dfrac{1}{(3x)^{1/2}} \cdot \dfrac{1}{2}(3x)^{-1/2} \cdot 3$ (b) Write with a positive exponent: $\sqrt{x^2 - 1}$	**11.1**	(a) $\dfrac{1}{2x}$ (b) $(x^2 - 1)^{1/2}$	0.4 Rational exponents
(a) Vertical lines have _____ slopes. (b) Horizontal lines have _____ slopes.	**11.3**	(a) Undefined (b) 0	1.3 Slopes
Write the equation of the line passing through $(-2, -2)$ with slope 5.	**11.3**	$y = 5x + 8$	1.3 Equations of lines
Solve: $x^2 + y^2 - 9 = 0$, for y.	**11.3**	$y = \pm\sqrt{9 - x^2}$	2.1 Quadratic equations
(a) Write $\log_a (x + h) - \log_a x$ as an expression involving one logarithm. (b) Does $\ln x^4 = 4 \ln x$? (c) Does $\dfrac{x}{h} \log_a\left(\dfrac{x+h}{x}\right) = \log_a\left(1 + \dfrac{h}{x}\right)^{x/h}$? (d) Expand $\ln (xy)$ to separate x and y. (e) If $y = a^x$, then $x = $ _____. (f) Simplify $\ln e^x$.	**11.1–11.2**	(a) $\log_a\left(\dfrac{x+h}{x}\right)$ (b) Yes (c) Yes (d) $\ln x + \ln y$ (e) $\log_a y$ (f) x	5.2 Logarithms
Find the derivative of (a) $y = x^2 - 2x - 2$ (b) $T(q) = 400q - \dfrac{4}{3}q^2$ (c) $y = \sqrt{9 - x^2}$	**11.1–11.5**	(a) $y' = 2x - 2$ (b) $T'(q) = 400 - \dfrac{8}{3}q$ (c) $y' = -x(9 - x^2)^{-1/2}$	9.4–9.6 Derivatives
If $\dfrac{dy}{dx} = \dfrac{-1}{2\sqrt{x}}$, find the slope of the tangent to $y = f(x)$ at $x = 4$.	**11.3**	Slope $= -\dfrac{1}{4}$	9.3–9.7 Derivatives

OBJECTIVE 11.1

- To find derivatives of logarithmic functions

Derivatives of Logarithmic Functions

| APPLICATION PREVIEW |

The table shows the expected life spans at birth for people born in certain years in the United States, with projections to 2020. These data can be modeled by the function $l(x) = 11.249 + 14.244 \ln x$, where x is the number of years past 1900. The graph of this function is shown in Figure 11.1. If we wanted to use this model to find the rate of change of life span with respect to the number of years past 1900, we would need the derivative of this function and hence the derivative of the logarithmic function $\ln x$. (See Example 4.)

Year	Life Span (years)	Year	Life Span (years)
1920	54.1	1985	74.7
1930	59.7	1990	75.4
1940	62.9	1995	75.8
1950	68.2	2000	77.0
1960	69.7	2005	77.9
1970	70.8	2010	78.1
1975	72.6	2015	78.9
1980	73.7	2020	79.5

Source: National Center for Health Statistics

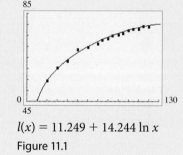

$l(x) = 11.249 + 14.244 \ln x$

Figure 11.1

Logarithmic Functions   Recall that we define the **logarithmic function** $y = \log_a x$ as follows.

Logarithmic Functions

For $a > 0$ and $a \neq 1$, the **logarithmic function**

$$y = \log_a x \quad \text{(logarithmic form)}$$

has domain $x > 0$, base a, and is defined by

$$a^y = x \quad \text{(exponential form)}$$

The a is called the **base** in both $\log_a x = y$ and $a^y = x$, and y is the *logarithm* in $\log_a x = y$ and the *exponent* in $a^y = x$. Thus **a logarithm is an exponent.** Although logarithmic functions can have any base a, where $a > 0$ and $a \neq 1$, most problems in calculus and many of the applications to the management, life, and social sciences involve logarithms with base e, called **natural logarithms.** In this section we'll see why this base is so important. Recall that the natural logarithmic function ($y = \log_e(x)$) is written $y = \ln x$; see Figure 11.2 for the graph.

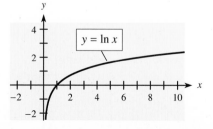

Figure 11.2

Derivative of $y = \ln x$ From Figure 11.2 we see that for $x > 0$, the graph of $y = \ln x$ is always increasing, so the slope of the tangent line to any point must be positive. This means that the derivative of $y = \ln x$ is always positive. Figure 11.3 shows the graph of $y = \ln x$ with tangent lines drawn at several points and with their slopes indicated.

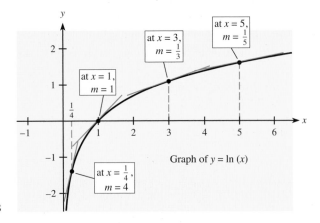

Figure 11.3

Graph of $y = \ln (x)$

Note in Figure 11.3 that at each point where a tangent line is drawn, the slope of the tangent line is the reciprocal of the x-coordinate. In fact, this is true for every point on $y = \ln x$, so the slope of the tangent at any point is given by $1/x$. Thus we have the following:

Derivative of $y = \ln x$

If $y = \ln x$, then $\dfrac{dy}{dx} = \dfrac{1}{x}$.

This formula for the derivative of $y = \ln x$ is proved at the end of this section.

EXAMPLE 1   **Derivatives Involving $\ln x$**

(a) If $y = x^3 + 3 \ln x$, find dy/dx.   (b) If $y = x^2 \ln x$, find y'.

Solution

(a) $\dfrac{dy}{dx} = 3x^2 + 3\left(\dfrac{1}{x}\right) = 3x^2 + \dfrac{3}{x}$

(b) By the Product Rule,

$$y' = x^2 \cdot \dfrac{1}{x} + (\ln x)(2x) = x + 2x \ln x$$

We can use the Chain Rule to find the formula for the derivative of $y = \ln u$, where $u = f(x)$.

Derivatives of Natural Logarithmic Functions

If $y = \ln u$, where u is a differentiable function of x, then

$$\dfrac{dy}{dx} = \dfrac{1}{u} \cdot \dfrac{du}{dx}$$

EXAMPLE 2   **Derivatives of $y = \ln (u)$**

Find the derivative for each of the following.

(a) $f(x) = \ln (x^4 - 3x + 7)$ (b) $f(x) = \tfrac{1}{3} \ln (2x^6 - 3x + 2)$

(c) $g(x) = \dfrac{\ln (2x + 1)}{2x + 1}$

Solution

(a) $f'(x) = \dfrac{1}{x^4 - 3x + 7}(4x^3 - 3) = \dfrac{4x^3 - 3}{x^4 - 3x + 7}$

(b) $f'(x)$ is $\frac{1}{3}$ of the derivative of $\ln(2x^6 - 3x + 2)$.

$$f'(x) = \frac{1}{3} \cdot \frac{1}{2x^6 - 3x + 2}(12x^5 - 3)$$

$$= \frac{4x^5 - 1}{2x^6 - 3x + 2}$$

(c) We begin with the Quotient Rule.

$$g'(x) = \frac{(2x + 1)\dfrac{1}{2x + 1}(2) - [\ln(2x + 1)]2}{(2x + 1)^2}$$

$$= \frac{2 - 2\ln(2x + 1)}{(2x + 1)^2}$$

✓ CHECKPOINT

1. If $y = \ln(3x^2 + 2)$, find y'.
2. If $y = \ln x^6$, find y'.

Using Properties of Logarithms   A logarithmic function of products, quotients, or powers, such as $y = \ln[x(x^5 - 2)^{10}]$, can be rewritten with properties of logarithms so that finding the derivative is much easier. The properties of logarithms, which were introduced in Section 5.2, are stated here for logarithms with an arbitrary base a (with $a > 0$ and $a \neq 1$) and for natural logarithms.

Properties of Logarithms

Let M, N, p, and a be real numbers with $M > 0$, $N > 0$, $a > 0$, and $a \neq 1$.

Base a Logarithms

I. $\log_a(a^x) = x$ (for any real x)

II. $a^{\log_a(x)} = x$ (for $x > 0$)

III. $\log_a(MN) = \log_a M + \log_a N$

IV. $\log_a\left(\dfrac{M}{N}\right) = \log_a M - \log_a N$

V. $\log_a(M^p) = p \log_a M$

Natural Logarithms

I. $\ln(e^x) = x$

II. $e^{\ln x} = x$

III. $\ln(MN) = \ln M + \ln N$

IV. $\ln\left(\dfrac{M}{N}\right) = \ln M - \ln N$

V. $\ln(M^p) = p \ln M$

For example, to find the derivative of $f(x) = \ln \sqrt[3]{2x^6 - 3x + 2}$, it is easier to rewrite this as follows:

$$f(x) = \ln[(2x^6 - 3x + 2)^{1/3}] = \tfrac{1}{3}\ln(2x^6 - 3x + 2)$$

Then take the derivative, as in Example 2(b).

EXAMPLE 3   **Logarithm Properties and Derivatives**

Use logarithm properties to find the derivatives for

(a) $y = \ln[x(x^5 - 2)^{10}]$.

(b) $f(x) = \ln\left(\dfrac{\sqrt[3]{3x + 5}}{x^2 + 11}\right)^4$.

Solution

(a) We use logarithm Properties III and V to rewrite the function.

$$y = \ln x + \ln (x^5 - 2)^{10} \qquad \text{Property III}$$
$$y = \ln x + 10 \ln (x^5 - 2) \qquad \text{Property V}$$

We now take the derivative.

$$\frac{dy}{dx} = \frac{1}{x} + 10 \cdot \frac{1}{x^5 - 2} \cdot 5x^4$$
$$= \frac{1}{x} + \frac{50x^4}{x^5 - 2}$$

(b) Again we begin by using logarithm properties.

$$f(x) = \ln \left(\frac{\sqrt[3]{3x + 5}}{x^2 + 11} \right)^4 = 4 \ln \left(\frac{\sqrt[3]{3x + 5}}{x^2 + 11} \right) \qquad \text{Property V}$$
$$f(x) = 4 \left[\tfrac{1}{3} \ln (3x + 5) - \ln (x^2 + 11) \right] \qquad \text{Properties IV and V}$$

We now take the derivative.

$$f'(x) = 4 \left(\frac{1}{3} \cdot \frac{1}{3x + 5} \cdot 3 - \frac{1}{x^2 + 11} \cdot 2x \right)$$
$$= 4 \left(\frac{1}{3x + 5} - \frac{2x}{x^2 + 11} \right) = \frac{4}{3x + 5} - \frac{8x}{x^2 + 11} \qquad \blacksquare$$

✓ **CHECKPOINT**

3. If $y = \ln \sqrt[3]{x^2 + 1}$, find y'.

4. Find $f'(x)$ for $f(x) = \ln \left[\dfrac{2x^4}{(5x + 7)^5} \right]$.

EXAMPLE 4 **Life Span** **| APPLICATION PREVIEW |**

Assume that the average life span (in years) for people born from 1920 and projected to 2020 can be modeled by

$$l(x) = 11.249 + 14.244 \ln x$$

where x is the number of years past 1900.
(a) Find the function that models the rate of change of life span.
(b) Does $l(x)$ have a maximum value for $x > 0$?
(c) Evaluate $\lim\limits_{x \to \infty} l'(x)$.
(d) What do the results of parts (b) and (c) tell us about the average life span?

Solution

(a) The rate of change of life span is given by the derivative.

$$l'(x) = 0 + 14.244 \left(\frac{1}{x} \right) = \frac{14.244}{x}$$

(b) For $x > 0$, we see that $l'(x) > 0$. Hence $l(x)$ is increasing for all values of $x > 0$, so $l(x)$ never achieves a maximum value. That is, there is no maximum life span.

(c) $\lim\limits_{x \to \infty} l'(x) = \lim\limits_{x \to \infty} \dfrac{14.244}{x} = 0$

(d) If this model is accurate, life span will continue to increase, but at an ever slower rate. $\blacksquare$

EXAMPLE 5 **Cost**

Suppose the cost function for x skateboards is given by

$$C(x) = 18,250 + 615 \ln (4x + 10)$$

where $C(x)$ is in dollars. Find the marginal cost when 100 units are produced, and explain what it means.

Solution
Marginal cost is given by $C'(x)$.

$$\overline{MC} = C'(x) = 615\left(\frac{1}{4x+10}\right)(4) = \frac{2460}{4x+10}$$

$$\overline{MC}(100) = \frac{2460}{4(100)+10} = \frac{2460}{410} = 6$$

When 100 units are produced, the marginal cost is 6. This means that the approximate cost of producing the 101st skateboard is $6.

Derivative of $y = \log_a (x)$ So far we have found derivatives of natural logarithmic functions. If we have a logarithmic function with a base other than e, then we can use the **change-of-base formula.**

Change-of-Base Formula

To express a logarithm base a as a natural logarithm, use

$$\log_a x = \frac{\ln x}{\ln a}$$

We can apply this change-of-base formula to find the derivative of a logarithm with any base, as the following example illustrates.

EXAMPLE 6   **Derivative of $y = \log_a (u)$**

If $y = \log_4 (x^3 + 1)$, find dy/dx.

Solution
By using the change-of-base formula, we have

$$y = \log_4 (x^3 + 1) = \frac{\ln (x^3 + 1)}{\ln 4} = \frac{1}{\ln 4} \cdot \ln (x^3 + 1)$$

Thus

$$\frac{dy}{dx} = \frac{1}{\ln 4} \cdot \frac{1}{x^3 + 1} \cdot 3x^2 = \frac{3x^2}{(x^3 + 1)\ln 4}$$

Note that this formula means that logarithms with bases other than e will have the *constant* $1/\ln a$ as a factor in their derivatives (as Example 6 had $1/\ln 4$ as a factor). This means that derivatives involving natural logarithms have a simpler form, and we see why base e logarithms are used more frequently in calculus.

EXAMPLE 7   **Critical Values**

Let $f(x) = x \ln x - x$. Use the graph of the derivative of $f(x)$ for $x > 0$ to answer the following questions.

(a) At what value a does the graph of $f'(x)$ cross the x-axis (that is, where is $f'(x) = 0$)?
(b) What value a is a critical value for $y = f(x)$?
(c) Does $f(x)$ have a relative maximum or a relative minimum at $x = a$?

Solution

$$f'(x) = x \cdot \frac{1}{x} + \ln x - 1 = \ln x. \text{ The graph of } f'(x) = \ln x \text{ is shown in Figure 11.4(a).}$$

(a) The graph crosses the x-axis at $x = 1$, so $f'(a) = 0$ if $a = 1$.

(b) $a = 1$ is a critical value of $f(x)$.

(c) Because $f'(x)$ is negative for $x < 1$, $f(x)$ is decreasing for $x < 1$.
Because $f'(x)$ is positive for $x > 1$, $f(x)$ is increasing for $x > 1$.
Therefore, $f(x)$ has a relative minimum at $x = 1$.
The graph of $f(x) = x \ln x - x$ is shown in Figure 11.4(b). It has a relative minimum at the point $(1, -1)$. ∎

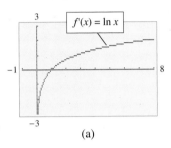

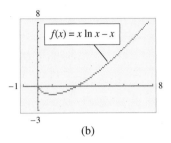

Figure 11.4   (a)   (b)

Calculator Note   As with polynomial functions, critical values involving logarithmic functions can be found by using the SOLVER feature of a graphing calculator. Figure 11.5 shows the steps used to find the two critical values of $y = x^2 - 8 \ln x$ by solving $0 = 2x - \dfrac{8}{x}$ with SOLVER. The critical values are $x = 2$ and $x = -2$. See Appendix A, Section 11.1, for details. ∎

Figure 11.5

 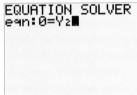 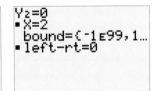

Proof That $\dfrac{d}{dx}(\ln x) = \dfrac{1}{x}$ For completeness, we now include the formal proof that if $y = \ln x$, then $dy/dx = 1/x$.

$$\frac{dy}{dx} = \lim_{h \to 0} \frac{f(x + h) - f(x)}{h}$$

$$= \lim_{h \to 0} \frac{\ln (x + h) - \ln x}{h}$$

$$= \lim_{h \to 0} \frac{\ln \left(\dfrac{x + h}{x} \right)}{h} \qquad \text{Property IV}$$

$$= \lim_{h \to 0} \frac{x}{x} \cdot \frac{1}{h} \ln \left(\frac{x + h}{x} \right) \qquad \text{Introduce } \frac{x}{x}$$

$$= \lim_{h \to 0} \frac{1}{x} \cdot \frac{x}{h} \ln \left(1 + \frac{h}{x} \right)$$

$$= \lim_{h \to 0} \frac{1}{x} \ln \left(1 + \frac{h}{x} \right)^{x/h} \qquad \text{Property V}$$

$$= \frac{1}{x} \lim_{h \to 0} \left[\ln \left(1 + \frac{h}{x} \right)^{x/h} \right] \qquad *$$

$$= \frac{1}{x} \ln \left[\lim_{h \to 0} \left(1 + \frac{h}{x} \right)^{x/h} \right] .$$

If we let $a = \dfrac{h}{x}$, then $h \to 0$ means $a \to 0$, and we have

$$\frac{dy}{dx} = \frac{1}{x} \ln \left[\lim_{a \to 0} (1 + a)^{1/a} \right]$$

---

*The next step uses a new limit property for continuous composite functions. In particular, $\lim_{x \to c} \ln (f(x)) = \ln [\lim_{x \to c} f(x)]$ when $\lim_{x \to c} f(x)$ exists and is positive.

In Problem 47 in the 9.1 Exercises, we saw that $\lim_{a \to 0} (1 + a)^{1/a} = e$. Hence,

$$\frac{dy}{dx} = \frac{1}{x} \ln e = \frac{1}{x}$$

✓ CHECKPOINT ANSWERS

1. $y' = \dfrac{6x}{3x^2 + 2}$

2. $y' = \dfrac{6}{x}$

3. $y' = \dfrac{2x}{3(x^2 + 1)}$

4. $f'(x) = \dfrac{4}{x} - \dfrac{25}{5x + 7}$

| EXERCISES | 11.1

Find the derivatives of the functions in Problems 1–10.

1. $f(x) = 4 \ln x$
2. $y = 3 \ln x$
3. $y = \ln 8x$
4. $y = \ln 5x$
5. $y = \ln x^4$
6. $f(x) = \ln x^3$
7. $f(x) = \ln (4x + 9)$
8. $y = \ln (6x + 1)$
9. $y = \ln (2x^2 - x) + 3x$
10. $y = \ln (8x^3 - 2x) - 2x$
11. Find dp/dq if $p = \ln (q^2 + 1)$.
12. Find $\dfrac{ds}{dq}$ if $s = \ln \left(\dfrac{q^2}{4} + 1 \right)$.

In each of Problems 13–18, find the derivative of the function in part (a). Then find the derivative of the function in part (b) or show that the function in part (b) is the same function as that in part (a).

13. (a) $y = \ln x - \ln (x - 1)$
    (b) $y = \ln \dfrac{x}{x - 1}$
14. (a) $y = \ln (x - 1) + \ln (2x + 1)$
    (b) $y = \ln [(x - 1)(2x + 1)]$
15. (a) $y = \frac{1}{3} \ln (x^2 - 1)$
    (b) $y = \ln \sqrt[3]{x^2 - 1}$
16. (a) $y = 3 \ln (x^4 - 1)$
    (b) $y = \ln (x^4 - 1)^3$
17. (a) $y = \ln (4x - 1) - 3 \ln x$
    (b) $y = \ln \left(\dfrac{4x - 1}{x^3} \right)$
18. (a) $y = 3 \ln x - \ln (x + 1)$
    (b) $y = \ln \left(\dfrac{x^3}{x + 1} \right)$
19. Find $\dfrac{dp}{dq}$ if $p = \ln \left(\dfrac{q^2 - 1}{q} \right)$.
20. Find $\dfrac{ds}{dt}$ if $s = \ln [t^3(t^2 - 1)]$.
21. Find $\dfrac{dy}{dt}$ if $y = \ln \left(\dfrac{t^2 + 3}{\sqrt{1 - t}} \right)$.
22. Find $\dfrac{dy}{dx}$ if $y = \ln \left(\dfrac{3x + 2}{x^2 - 5} \right)^{1/4}$.

23. Find $\dfrac{dy}{dx}$ if $y = \ln (x^3 \sqrt{x + 1})$.

24. Find $\dfrac{dy}{dx}$ if $y = \ln [x^2(x^4 - x + 1)]$.

In Problems 25–38, find y'.

25. $y = x - \ln x$
26. $y = x^2 \ln (2x + 3)$
27. $y = \dfrac{\ln x}{x}$
28. $y = \dfrac{1 + \ln x}{x^2}$
29. $y = \ln (x^4 + 3)^2$
30. $y = \ln (3x + 1)^{1/2}$
31. $y = (\ln x)^4$
32. $y = (\ln x)^{-1}$
33. $y = [\ln (x^4 + 3)]^2$
34. $y = \sqrt{\ln (3x + 1)}$
35. $y = \log_4 x$
36. $y = \log_5 x$
37. $y = \log_6 (x^4 - 4x^3 + 1)$
38. $y = \log_2 (1 - x - x^2)$

In Problems 39–42, find the relative maxima and relative minima, and sketch the graph with a graphing utility to check your results.

39. $y = x \ln x$
40. $y = x^2 \ln x$
41. $y = x^2 - 8 \ln x$
42. $y = \ln x - x$

APPLICATIONS

43. *Marginal cost* Suppose that the total cost (in dollars) for a product is given by

    $$C(x) = 1500 + 200 \ln (2x + 1)$$

    where x is the number of units produced.
    (a) Find the marginal cost function.
    (b) Find the marginal cost when 200 units are produced, and interpret your result.
    (c) Total cost functions always increase because producing more items costs more. What then must be true of the marginal cost function? Does it apply in this problem?

44. *Investment* If money is invested at the constant rate r, the time to increase the investment by a factor x is

    $$t = \frac{\ln x}{r}$$

(a) At what rate $\dfrac{dt}{dx}$ is the time changing at $x = 2$?

(b) What happens to $\dfrac{dt}{dx}$ as x gets very large? Interpret this result.

45. *Marginal revenue* The total revenue, in dollars, from the sale of x units of a product is given by

$$R(x) = \frac{2500x}{\ln(10x + 10)}$$

(a) Find the marginal revenue function.

(b) Find the marginal revenue when 100 units are sold, and interpret your result.

46. *Supply* Suppose that the supply of x units of a product at price p dollars per unit is given by

$$p = 10 + 50 \ln(3x + 1)$$

(a) Find the rate of change of supply price with respect to the number of units supplied.

(b) Find the rate of change of supply price when the number of units is 33.

(c) Approximate the price increase associated with the number of units supplied changing from 33 to 34.

47. *Demand* The demand function for a product is given by $p = 4000/\ln(x + 10)$, where p is the price per unit in dollars when x units are demanded.

(a) Find the rate of change of price with respect to the number of units sold when 40 units are sold.

(b) Find the rate of change of price with respect to the number of units sold when 90 units are sold.

(c) Find the second derivative to see whether the rate at which the price is changing at 40 units is increasing or decreasing.

48. *pH level* The pH of a solution is given by

$$pH = -\log[H^+]$$

where $[H^+]$ is the concentration of hydrogen ions (in gram atoms per liter). What is the rate of change of pH with respect to $[H^+]$?

49. *Drug concentration* Concentration (in mg/ml) in the bloodstream of a certain drug is related to the time t (in minutes) after an injection and can be calculated using y in the equation

$$y = A \ln(t) - Bt + C$$

where A, B, and C are positive constants. In terms of A and B, find t at which y (and hence the drug concentration) reaches its maximum.

50. *Decibels* The loudness of sound (L, measured in decibels) perceived by the human ear depends on intensity levels (I) according to

$$L = 10 \log(I/I_0)$$

where I_0 is the standard threshold of audibility. If $x = I/I_0$, then using the change-of-base formula, we get

$$L = \frac{10 \ln(x)}{\ln 10}$$

At what rate is the loudness changing with respect to x when the intensity is 100 times the standard threshold of audibility (that is, when $x = 100$)?

51. *Richter scale* The Richter scale reading, R, used for measuring the magnitude of an earthquake with intensity I is determined by

$$R = \frac{\ln(I/I_0)}{\ln 10}$$

where I_0 is a standard minimum threshold of intensity. If $I_0 = 1$, what is the rate of change of the Richter scale reading with respect to intensity?

52. *Women in the workforce* From 1950 and projected to 2050, the percent of women in the workforce can be modeled by

$$w(x) = 9.42 + 8.70 \ln x$$

where x is the number of years past 1940 (*Source:* U.S. Bureau of Labor Statistics). If this model is accurate, at what rate will the percent be changing in 2020?

53. *Modeling Obesity* Obesity (BMI ≥ 30) is a serious problem in the United States and expected to get worse. Being overweight increases the risk of diabetes, heart disease, and many other ailments. The table gives the percent of Americans who are obese for selected years from 1990 and projected to 2030.

Year	1990	2000	2010	2015	2020	2025	2030
% Obese	12.7	22.1	30.9	34.5	37.4	39.9	42.2

Source: American Journal of Preventive Medicine 42 (June 2012).

(a) Find a logarithmic function that models the data, with x equal to the number of years after 1980 and y equal to the percent. Report the model with three significant digits.

(b) Use the reported model to find the instantaneous rate of change of obesity.

(c) At what rate is obesity projected to increase in 2020?

54. *Modeling Diabetes* The following table gives the percent of adult Americans with diabetes for selected years from 2010 and projected to 2050.

(a) Find a logarithmic function that models the data, with x equal to the number of years after 2000 and y equal to the percent of the adult American population with diabetes. Report the model with three significant digits.

(b) Use the reported model to find the instantaneous rate of change of the percent of Americans with diabetes.

(c) At what rate is the percent of Americans with diabetes projected to increase in 2030?

Year	Percent	Year	Percent
2010	15.7	2035	29.0
2015	18.9	2040	31.4
2020	21.1	2045	32.1
2025	24.2	2050	34.3
2030	27.2		

OBJECTIVE

11.2

- To find derivatives of exponential functions

Derivatives of Exponential Functions

| APPLICATION PREVIEW |

We saw in Chapter 6, "Mathematics of Finance," that the amount that accrues when $100 is invested at 8%, compounded continuously, is

$$S(t) = 100e^{0.08t}$$

where t is the number of years. If we want to find the rate at which the money in this account is growing at the end of 1 year, then we need to find the derivative of this exponential function. (See Example 4.)

Derivative of $y = e^x$ Just as base e logarithms are most convenient, we begin by focusing on base e exponentials. Figure 11.6(a) shows the graph of $f(x) = e^x$, and Figure 11.6(b) shows the same graph with tangent lines drawn to several points.

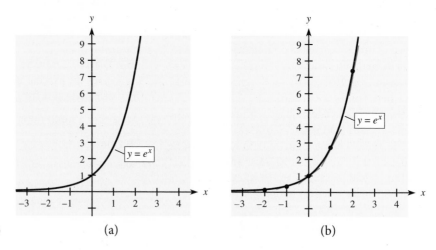

Figure 11.6   (a)   (b)

Note in Figure 11.6(b) that when $x < 0$, tangent lines have slopes near 0, much like the y-coordinates of $f(x) = e^x$. Furthermore, as x increases to $x = 0$ and for $x > 0$, the slopes of the tangents increase just as the values of the function do. This suggests that the function that gives the slope of the tangent to the graph of $f(x) = e^x$ (that is, the derivative) is similar to the function itself. In fact, the derivative of $f(x) = e^x$ is exactly the function itself, which we prove as follows:

From logarithms Property I, we know that

$$\ln e^x = x$$

Taking the derivative, with respect to x, of both sides of this equation, we have

$$\frac{d}{dx}(\ln e^x) = \frac{d}{dx}(x)$$

Using the Chain Rule for logarithms gives

$$\frac{1}{e^x} \cdot \frac{d}{dx}(e^x) = 1$$

and solving for $\frac{d}{dx}(e^x)$ yields

$$\frac{d}{dx}(e^x) = e^x$$

Thus we can conclude the following.

Derivative of $y = e^x$ | If $y = e^x$, then $\dfrac{dy}{dx} = e^x$.

EXAMPLE 1   **Derivative of an Exponential Function**

If $y = 3e^x + 4x - 11$, find $\dfrac{dy}{dx}$.

Solution

$$\frac{dy}{dx} = 3e^x + 4$$

Derivative of $y = e^u$ As with logarithmic functions, the Chain Rule permits us to expand our derivative formulas.

Derivatives of Exponential Functions | If $y = e^u$, where u is a differentiable function of x, then

$$\frac{dy}{dx} = e^u \cdot \frac{du}{dx}$$

EXAMPLE 2   **Derivatives of $y = e^u$**

(a) If $f(x) = e^{4x^3}$ find $f'(x)$.
(b) If $s = 3te^{3t^2 + 5t}$, find ds/dt.
(c) If $u = w/e^{3w}$, find u'.

Solution
(a) $f'(x) = e^{4x^3} \cdot (12x^2) = 12x^2\, e^{4x^3}$
(b) We begin with the Product Rule.

$$\frac{ds}{dt} = 3t \cdot e^{3t^2 + 5t}(6t + 5) + e^{3t^2 + 5t} \cdot 3$$

$$= (18t^2 + 15t)e^{3t^2 + 5t} + 3e^{3t^2 + 5t}$$

$$= 3e^{3t^2 + 5t}(6t^2 + 5t + 1)$$

(c) The function is a quotient. Using the Quotient Rule gives

$$u' = \frac{(e^{3w})(1) - (w)(e^{3w} \cdot 3)}{(e^{3w})^2} = \frac{e^{3w} - 3we^{3w}}{e^{6w}} = \frac{1 - 3w}{e^{3w}}$$

EXAMPLE 3   **Derivatives and Logarithmic Properties**

If $y = e^{\ln x^2}$, find y'.

Solution

$$y' = e^{\ln x^2} \cdot \frac{1}{x^2} \cdot 2x = \frac{2}{x} e^{\ln x^2}$$

By logarithm Property II (see the previous section), $e^{\ln u} = u$, and we can simplify the derivative to

$$y' = \frac{2}{x} \cdot x^2 = 2x$$

Note that if we had used this property *before* taking the derivative, we would have had

$$y = e^{\ln x^2} = x^2$$

Then the derivative is $y' = 2x$.

✓ CHECKPOINT   1. If $y = 2e^{4x}$, find y'.   2. If $y = e^{x^2 + 6x}$, find y'.   3. If $s = te^{t^2}$, find ds/dt.

EXAMPLE 4   Future Value | APPLICATION PREVIEW |

When \$100 is invested at 8% compounded continuously, the amount that accrues after t years, which is called the future value, is $S(t) = 100e^{0.08t}$. At what rate is the money in this account growing

(a) at the end of 1 year?   (b) at the end of 10 years?

Solution

The rate of growth of the money is given by

$$S'(t) = 100e^{0.08t}(0.08) = 8e^{0.08t}$$

(a) The rate of growth of the money at the end of 1 year is

$$S'(1) = 8e^{0.08} \approx 8.666$$

Thus the future value will change by about \$8.67 during the second year.

(b) The rate of growth of the money at the end of 10 years is

$$S'(10) = 8e^{0.08(10)} \approx 17.804$$

Thus the future value will change by about \$17.80 during the eleventh year.

EXAMPLE 5   Revenue

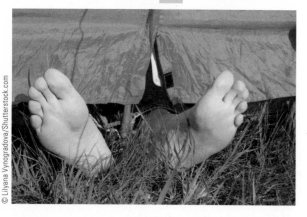

© Lilyana Vynogradova/Shutterstock.com

North Forty, Inc. is a manufacturer of wilderness camping equipment. The revenue function for its best-selling tent, the Sierra, can be modeled by the function

$$R(x) = 250xe^{(1 - 0.01x)}$$

where $R(x)$ is the revenue in thousands of dollars from the sale of x thousand Sierra tents. Find the marginal revenue when 75,000 tents are sold, and explain what it means.

Solution

The marginal revenue function is given by $R'(x)$, and to find this derivative we use the Product Rule.

$$R'(x) = \overline{MR} = 250x[e^{(1 - 0.01x)} \cdot (-0.01)] + e^{(1 - 0.01x)}(250)$$
$$\overline{MR} = 250e^{(1 - 0.01x)}(1 - 0.01x)$$

To find the marginal revenue when 75,000 tents are sold, we use $x = 75$.

$$\overline{MR}(75) = 250e^{(1 - 0.75)}(1 - 0.75) \approx 80.25$$

This means that the sale of one (thousand) more Sierra tents will yield approximately \$80.25 (thousand) in additional revenue.

✓ CHECKPOINT   4. If the sales of a product are given by $S = 1000e^{-0.2x}$, where x is the number of days after the end of an advertising campaign, what is the rate of decline in sales 20 days after the end of the campaign?

Derivative of $y = a^u$ In a manner similar to that used to find the derivative of $y = e^x$, we can develop a formula for the derivative of $y = a^x$ for any base $a > 0$ and $a \neq 1$.

Derivative of $y = a^u$

If $y = a^x$, with $a > 0$, $a \neq 1$, then

$$\frac{dy}{dx} = a^x \ln a$$

If $y = a^u$, with $a > 0$, $a \neq 1$, where u is a differentiable function of x, then

$$\frac{dy}{dx} = a^u \frac{du}{dx} \ln a$$

EXAMPLE 6 Derivatives of $y = a^u$

(a) If $y = 4^x$, find dy/dx.   (b) If $y = 5^{x^2+x}$, find y'.

Solution

(a) $\dfrac{dy}{dx} = 4^x \ln 4$

(b) $y' = 5^{x^2+x}(2x + 1) \ln 5$

Calculator Note   As with other functions, we can make use of a graphing calculator to check derivatives of functions involving exponentials. See Appendix A, Section 11.2, for details.

  ### EXAMPLE 7 Technology and Critical Values

For the function $y = e^x - 3x^2$, complete the following.

(a) Approximate the critical values of the function to four decimal places.
(b) Determine whether relative maxima or relative minima occur at the critical values.

Solution

(a) The derivative is $y' = e^x - 6x$. Using the ZERO feature of a graphing calculator, we find that $y' = 0$ at $x \approx 0.2045$ (see Figure 11.7(a)) and at $x \approx 2.8331$.

(b) From the graph of $y' = e^x - 6x$ in Figure 11.7(a), we can observe where $y' > 0$ and where $y' < 0$. From this we can make a sign diagram to determine relative maxima and relative minima.

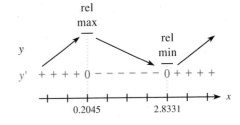

The graph of $y = e^x - 3x^2$ in Figure 11.7(b) shows that the relative maximum point is $(0.2045, 1.1015)$ and that the relative minimum point is $(2.8331, -7.0813)$.

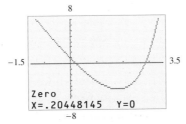

Graph of $y' = e^x - 6x$

(a)

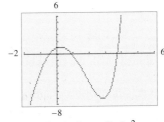

Graph of $y = e^x - 3x^2$

(b)

Figure 11.7

Calculator Note As with other functions, the SOLVER feature of a graphing calculator can also be used to find maxima and minima of functions involving exponentials. See Appendix A, Section 11.2, for details. ∎

✓ CHECKPOINT
ANSWERS

1. $y' = 8e^{4x}$
2. $y' = (2x + 6)e^{x^2 + 6x}$
3. $\dfrac{ds}{dt} = e^{t^2} + 2t^2 e^{t^2}$.
4. The rate of decline is given by $S'(20) = -200e^{-4} \approx -3.663$ sales/day.

| EXERCISES | 11.2

Find the derivatives of the functions in Problems 1–34.

1. $y = 5e^x - x$
2. $y = x^2 - 3e^x$
3. $f(x) = e^x - x^e$
4. $f(x) = 4e^x - \ln x$
5. $g(x) = 500(1 - e^{-0.1x})$
6. $h(x) = 750e^{0.04x}$
7. $y = e^{x^3}$
8. $y = e^{x^2 - 1}$
9. $y = 6e^{3x^2}$
10. $y = 1 - 2e^{-x^3}$
11. $y = 2e^{(x^2 + 1)^3}$
12. $y = e^{\sqrt{x^2 - 9}}$
13. $y = e^{\ln x^3}$
14. $y = e^3 + e^{\ln x}$
15. $y = e^{-1/x}$
16. $y = 2e^{\sqrt{x}}$
17. $y = e^{-1/x^2} + e^{-x^2}$
18. $y = \dfrac{2}{e^{2x}} + \dfrac{e^{2x}}{2}$
19. $s = t^2 e^t$
20. $p = 4qe^{q^3}$
21. $y = e^{x^4} - (e^x)^4$
22. $y = 4(e^x)^3 - 4e^{x^3}$
23. $y = \ln(e^{4x} + 2)$
24. $y = \ln(e^{2x} + 1)$
25. $y = e^{-3x} \ln(2x)$
26. $y = e^{2x^2} \ln(4x)$
27. $y = \dfrac{1 + e^{5x}}{e^{3x}}$
28. $y = \dfrac{x}{1 + e^{2x}}$
29. $y = (e^{3x} + 4)^{10}$
30. $y = \dfrac{e^x - e^{-x}}{e^x + e^{-x}}$
31. $y = 6^x$
32. $y = 3^x$
33. $y = 4^{x^2}$
34. $y = 5^{2x - 1}$

35. (a) What is the slope of the line tangent to $y = xe^{-x}$ at $x = 1$?
    (b) Write the equation of the line tangent to the graph of $y = xe^{-x}$ at $x = 1$.

36. (a) What is the slope of the line tangent to $y = e^{-x}/(1 + e^{-x})$ at $x = 0$?
    (b) Write the equation of the line tangent to the graph of $y = e^{-x}/(1 + e^{-x})$ at $x = 0$.

37. The equation for the standard normal probability distribution is

$$y = \frac{1}{\sqrt{2\pi}} e^{-z^2/2}$$

(a) At what value of z will the curve be at its highest point?
(b) Graph this function with a graphing utility to verify your answer.

38. (a) Find the mode of the normal distribution* given by

$$y = \frac{1}{\sqrt{2\pi}} e^{-(x - 10)^2/2}$$

(b) What is the mean of this normal distribution?
(c) Use a graphing utility to verify your answer.

 In Problems 39–42, find any relative maxima and minima. Use a graphing utility to check your results.

39. $y = \dfrac{e^x}{x}$
40. $y = \dfrac{x}{e^x}$
41. $y = x - e^x$
42. $y = \dfrac{x^2}{e^x}$

APPLICATIONS

43. *Future value* If $\$P$ is invested for n years at 10% compounded continuously, the future value that results after n years is given by the function

$$S = Pe^{0.1n}$$

(a) At what rate is the future value growing at any time (for any nonnegative n)?
(b) At what rate is the future value growing after 1 year $(n = 1)$?
(c) Is the rate of growth of the future value after 1 year greater than 10%? Why?

44. *Future value* The future value that accrues when $\$700$ is invested at 9%, compounded continuously, is

$$S(t) = 700e^{0.09t}$$

where t is the number of years.
(a) At what rate is the money in this account growing when $t = 4$?
(b) At what rate is it growing when $t = 10$?

45. *Sales decay* After the end of an advertising campaign, the sales of a product are given by

$$S = 100{,}000e^{-0.5t}$$

where S is weekly sales in dollars and t is the number of weeks since the end of the campaign.

---

*The mode occurs at the highest point on normal curves and equals the mean.

(a) Find the rate of change of S (that is, the rate of *sales decay*).

(b) Give a reason from looking at the function and another reason from looking at the derivative that explain how you know sales are decreasing.

46. **Sales decay** The sales decay for a product is given by

$$S = 50{,}000e^{-0.8t}$$

where S is the daily sales in dollars and t is the number of days since the end of a promotional campaign. Find the rate of sales decay.

47. **Marginal cost** Suppose that the total cost in dollars of producing x units of a product is given by

$$C(x) = 10{,}000 + 20xe^{x/600}$$

Find the marginal cost when 600 units are produced.

48. **Marginal revenue** Suppose that the revenue in dollars from the sale of x units of a product is given by

$$R(x) = 1000xe^{-x/50}$$

Find the marginal revenue function.

49. **Drugs in a bloodstream** The percent concentration y of a certain drug in the bloodstream at any time t (in hours) is given by

$$y = 100(1 - e^{-0.462t})$$

(a) What function gives the instantaneous rate of change of the concentration of the drug in the bloodstream?

(b) Find the rate of change of the concentration after 1 hour. Give your answer to three decimal places.

50. **Radioactive decay** The amount of the radioactive isotope thorium-234 present at time t in years is given by

$$Q(t) = 100e^{-0.02828t}$$

(a) Find the function that describes how rapidly the isotope is decaying.

(b) Find the rate of radioactive decay of the isotope after 10 years.

51. **Pollution** Pollution levels in Lake Sagamore have been modeled by the equation

$$x = 0.05 + 0.18e^{-0.38t}$$

where x is the volume of pollutants (in cubic kilometers) and t is the time (in years). What is the rate of change of x with respect to time?

52. **Drug concentration** Suppose the concentration $C(t)$, in mg/ml, of a drug in the bloodstream t minutes after an injection is given by

$$C(t) = 20te^{-0.04t}$$

(a) Find the instantaneous rate of change of the concentration after 10 minutes.

(b) Find the maximum concentration and when it occurs.

53. **National health care** With U.S. Department of Health and Human Services data from 2000 and projected to 2018, the total public expenditures for health care H can be modeled by

$$H = 624e^{0.07t}$$

where t is the number of years past 2000 and H is in billions of dollars. If this model is accurate, at what rate will health care expenditures change in 2020?

54. **Personal consumption** By using U.S. Bureau of Labor Statistics data for selected years from 1988 and projected to 2018, the billions of dollars spent for personal consumption in the United States can be modeled by

$$P = 2969e^{0.051t}$$

where t is the number of years past 1985. If this model is accurate, find and interpret the rate of change of personal consumption in 2015.

55. **Richter scale** The intensity of an earthquake is related to the Richter scale reading R by

$$\frac{I}{I_0} = 10^R$$

where I_0 is a standard minimum intensity. If $I_0 = 1$, what is the rate of change of the intensity I with respect to the Richter scale reading?

56. **Decibel readings** The intensity level of sound, I, is given by

$$\frac{I}{I_0} = 10^{L/10}$$

where L is the decibel reading and I_0 is the standard threshold of audibility. If $I/I_0 = y$, at what rate is y changing with respect to L when $L = 20$?

57. **U.S. debt** For selected years from 1900 to 2014, the national debt d, in billions of dollars, can be modeled by

$$d = 1.60e^{0.083t}$$

where t is the number of years past 1900 (*Source:* Bureau of Public Debt, U.S. Treasury).

(a) What function describes how fast the national debt is changing?

(b) Find the instantaneous rate of change of the national debt model $d(t)$ in 1950 and 2025.

58. **Blood pressure** Medical research has shown that between heartbeats, the pressure in the aorta of a normal adult is a function of time in seconds and can be modeled by the equation

$$P = 95e^{-0.491t}$$

(a) Use the derivative to find the rate at which the pressure is changing at any time t.

(b) Use the derivative to find the rate at which the pressure is changing after 0.1 second.

(c) Is the pressure increasing or decreasing?

59. *Spread of disease* Suppose that the spread of a disease through the student body at an isolated college campus can be modeled by

$$y = \frac{10{,}000}{1 + 9999e^{-0.99t}}$$

where y is the total number affected at time t (in days). Find the rate of change of y.

60. *Spread of a rumor* The number of people $N(t)$ in a community who are reached by a particular rumor at time t (in days) is given by

$$N(t) = \frac{50{,}500}{1 + 100e^{-0.7t}}$$

Find the rate of change of $N(t)$.

61. *Population* By using Social Security Administration data for selected years from 1950 and projected to 2050, the population of Americans ages 20 to 64 can be modeled by the function

$$P(t) = \frac{250}{1 + 1.91e^{-0.0280t}}$$

where t is the number of years after 1950 and $P(t)$ is in millions.

(a) Find and interpret $P'(45)$.

(b) At what rate is this population projected to increase in 2040?

(c) What does this tell us about this population after 2040?

62. *Carbon dioxide emissions* By using U.S. Department of Energy data for selected years from 2010 and projected to 2032, the millions of metric tons of carbon dioxide (CO_2) emissions from biomass energy combustion in the United States can be modeled with the function

$$E(t) = \frac{1310}{1 + 2.95e^{-0.0619t}}$$

where t is the number of years past 2010.

(a) Find the function that models the rate of change of $E(t)$. Report this function as $E'(t)$ with three significant digits.

(b) Find and interpret $E'(15)$.

63. **Modeling *Severe obesity*** Obesity (BMI $\geq$ 30) increases the risk of diabetes, heart disease, and many other ailments, but the severely obese (BMI $\geq$ 40) are most at risk and are the most expensive to treat. The table gives the percent of the Americans who are severely obese for the years from 1990 and projected to 2030.

(a) Find an exponential function that models the data, with x equal to the number of years after 1980 and y equal to the percent. Report the model with three significant digits.

(b) Use the reported model to find the instantaneous rate of change of severe obesity.

(c) At what rate is severe obesity projected to increase in 2020?

Year	1990	2000	2010	2015	2020	2025	2030
% Severely Obese	0.8	2.2	4.9	6.4	7.9	9.5	11.1

Source: *American Journal of Preventive Medicine* 42 (June 2012)

64. **Modeling *Centenarians*** The following table gives the projected population, in thousands, of Americans over 100 years of age.

(a) Create an exponential function that models the projected population y, in thousands, as a function of the number of years past 2010. Report the model with three significant digits.

(b) Use the reported model to find a function that gives the rate of change in the projected number of centenarians.

(c) What is the projected rate of change in the number of centenarians in 2040?

Year	Centenarians (thousands)	Year	Centenarians (thousands)
2015	78	2040	230
2020	106	2045	310
2025	143	2050	442
2030	168	2055	564
2035	188	2060	690

Source: U.S. Census Bureau

65. *World tourism* Using data from 2000 and projected to 2050, the receipts (in billions of dollars) for world tourism can be modeled by the function

$$y = 165.55(1.055^x)$$

where x is the number of years past 1980.

(a) Write the function that models the rate of change in the tourism receipts. Use four significant digits.

(b) Predict the rate of change of tourism receipts in 2015.

66. *Energy use* When U.S. energy use per dollar of GDP is indexed to 1980, that energy use for any year is viewed as a percent of the use per dollar of GDP in 1980. By using U.S. Department of Energy data for selected years from 1985 and projected to 2035, energy use per dollar of GDP can be modeled by the function

$$E(t) = 96e^{-0.019t}$$

where t is the number of years past 1980.

(a) Find the function that describes how rapidly the energy use per dollar of GDP is changing.

(b) Find and interpret $E(50)$ and $E'(50)$.

67. *Modeling Annual wages* The table gives the average annual wage, in thousands of dollars, for selected years from 2012 and projected to 2050.

(a) Find an exponential function that models the data, with x equal to the number of years after 2010 and y equal to the average annual wage in thousands of dollars. Report the model with three significant digits.

(b) Use the reported model to find a function that gives the instantaneous rate of change of the average annual wage. Report this function with three significant digits.

(c) Use the result in part (b) to predict the rate of growth of the average annual wage in 2040.

Year	Annual Wage ($ thousands)	Year	Annual Wage ($ thousands)
2012	44.6	2030	93.2
2014	48.6	2035	113.2
2016	53.3	2040	137.6
2018	58.7	2045	167.1
2020	63.7	2050	202.5
2025	76.8		

Source: Social Security Administration

OBJECTIVES

11.3

- To find derivatives by using implicit differentiation
- To find slopes of tangents by using implicit differentiation

Implicit Differentiation

| APPLICATION PREVIEW |

In the retail electronics industry, suppose the monthly demand for Precision, Inc., headphones is given by

$$p = \frac{10{,}000}{(x+1)^2}$$

where p is the price in dollars per set of headphones and x is demand in hundreds of sets of headphones. If we want to find the rate of change of the quantity demanded with respect to the price, then we need to find dx/dp. (See Example 8.) Although we can solve this equation for x so that dx/dp can be found, the resulting equation does not define x as a function of p. In this case, and in other cases where we cannot solve equations for the variables we need, we can find derivatives with a technique called implicit differentiation.

Up to this point, functions involving x and y have been written in the form $y = f(x)$, defining y as an *explicit function* of x. However, not all equations involving x and y can be written in the form $y = f(x)$, and we need a new technique for taking their derivatives. For example, solving $y^2 = x$ for y gives $y = \pm\sqrt{x}$ so that y is not a function of x. We can write $y = \sqrt{x}$ and $y = -\sqrt{x}$, but then finding the derivative $\dfrac{dy}{dx}$ at a point on the graph of $y^2 = x$ would require determining which of these functions applies before taking the derivative. Alternatively, we can *imply* that y is a function of x and use a technique called **implicit differentiation**. When y is an implied function of x, we find $\dfrac{dy}{dx}$ by differentiating both sides of the equation with respect to x and then algebraically solving for $\dfrac{dy}{dx}$.

EXAMPLE   1   **Implicit Differentiation**

(a) Use implicit differentiation to find $\dfrac{dy}{dx}$ for $y^2 = x$.

(b) Find the slopes of the tangents to the graph of $y^2 = x$ at the points $(4, 2)$ and $(4, -2)$.

Solution

(a) First take the derivative of both sides of the equation with respect to x.

$$\frac{d}{dx}(y^2) = \frac{d}{dx}(x)$$

Because y is an implied, or implicit, function of x, we can think of $y^2 = x$ as meaning $[u(x)]^2 = x$ for some function u. We use the Chain Rule to take the derivative of y^2 in the same way we would for $[u(x)]^2$. Thus

$$\frac{d}{dx}(y^2) = \frac{d}{dx}(x) \quad \text{gives} \quad 2y\frac{dy}{dx} = 1$$

Solving for $\frac{dy}{dx}$ gives $\frac{dy}{dx} = \frac{1}{2y}$.

(b) To find the slopes of the tangents at the points $(4, 2)$ and $(4, -2)$, we use the coordinates of the points to evaluate $\frac{dy}{dx}$ at those points.

$$\left.\frac{dy}{dx}\right|_{(4,\,2)} = \frac{1}{2(2)} = \frac{1}{4} \quad \text{and} \quad \left.\frac{dy}{dx}\right|_{(4,\,-2)} = \frac{1}{2(-2)} = \frac{-1}{4}$$

Figure 11.8 shows the graph of $y^2 = x$ with tangent lines drawn at $(4, 2)$ and $(4, -2)$. ∎

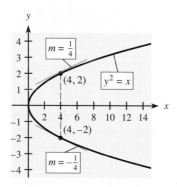

Figure 11.8

As noted previously, solving $y^2 = x$ for y gives the two functions $y = \sqrt{x} = x^{1/2}$ and $y = -\sqrt{x} = -x^{1/2}$; see Figures 11.9(a) and (b).

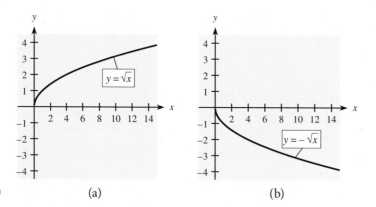

Figure 11.9   (a)   (b)

Let us now compare the results obtained in Example 1 with the results from the derivatives of the two functions $y = \sqrt{x}$ and $y = -\sqrt{x}$. The derivatives of these two functions are as follows:

$$\text{For } y = x^{1/2}, \text{ then } \frac{dy}{dx} = \frac{1}{2}x^{-1/2} = \frac{1}{2\sqrt{x}}$$

$$\text{For } y = -x^{1/2}, \text{ then } \frac{dy}{dx} = -\frac{1}{2}x^{-1/2} = \frac{-1}{2\sqrt{x}}$$

We cannot find the slope of a tangent line at $x = 4$ unless we also know the y-coordinate and hence which function and which derivative to use.

$$\text{At } (4, 2) \text{ use } y = \sqrt{x} \text{ so } \left.\frac{dy}{dx}\right|_{(4,\,2)} = \frac{1}{2\sqrt{4}} = \frac{1}{4}$$

$$\text{At } (4, -2) \text{ use } y = -\sqrt{x} \text{ so } \left.\frac{dy}{dx}\right|_{(4,\,-2)} = \frac{-1}{2\sqrt{4}} = \frac{-1}{4}$$

These are the same results that we obtained directly after implicit differentiation. In this example, we could easily solve for y in terms of x and work the problem two different ways. However, sometimes solving explicitly for y is difficult or impossible. For these functions, implicit differentiation is necessary and extends our ability to find derivatives.

EXAMPLE 2   Slope of a Tangent

Find the slope of the tangent to the graph of $x^2 + y^2 - 9 = 0$ at $(\sqrt{5}, 2)$.

Solution
We find the derivative dy/dx from $x^2 + y^2 - 9 = 0$ by taking the derivative term by term on both sides of the equation.

$$\frac{d}{dx}(x^2) + \frac{d}{dx}(y^2) + \frac{d}{dx}(-9) = \frac{d}{dx}(0)$$

$$2x + 2y \cdot \frac{dy}{dx} + 0 = 0$$

Solving for dy/dx gives

$$\frac{dy}{dx} = -\frac{2x}{2y} = -\frac{x}{y}$$

The slope of the tangent to the curve at $(\sqrt{5}, 2)$ is the value of the derivative at this point. Evaluating $dy/dx = -x/y$ at $(\sqrt{5}, 2)$ gives the slope of the tangent as $-\sqrt{5}/2$. ∎

As we have seen in Examples 1 and 2, the Chain Rule yields $\frac{d}{dx}(y^n) = ny^{n-1}\frac{dy}{dx}$. Just as we needed this result to find derivatives of powers of y, the next example discusses how to find the derivative of xy.

EXAMPLE 3   Equation of a Tangent Line

Write the equation of the tangent to the graph of $x^3 + xy + 4 = 0$ at the point $(2, -6)$.

Solution
Note that the second term is the *product* of x and y. Because we are assuming that y is a function of x, and because x is a function of x, we must use the Product Rule to find $\frac{d}{dx}(xy)$.

$$\frac{d}{dx}(xy) = x \cdot 1\frac{dy}{dx} + y \cdot 1 = x\frac{dy}{dx} + y$$

Thus we have

$$\frac{d}{dx}(x^3) + \frac{d}{dx}(xy) + \frac{d}{dx}(4) = \frac{d}{dx}(0)$$

$$3x^2 + \left(x\frac{dy}{dx} + y\right) + 0 = 0$$

Solving for dy/dx gives
$$\frac{dy}{dx} = \frac{-3x^2 - y}{x}$$

The slope of the tangent to the curve at $x = 2$, $y = -6$ is

$$m = \frac{-3(2)^2 - (-6)}{2} = -3$$

The equation of the tangent line is

$$y - (-6) = -3[x - (2)], \quad \text{or} \quad y = -3x$$

Technology Note   A graphing utility can be used to graph the function of Example 3 and the line that is tangent to the curve at $(2, -6)$. To graph the equation, we solve the equation for y, getting

$$y = \frac{-x^3 - 4}{x}$$

The graph of the equation and the line that is tangent to the curve at $(2, -6)$ are shown in Figure 11.10.

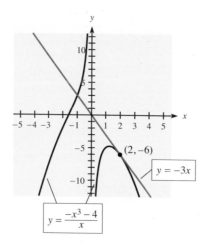

Figure 11.10

EXAMPLE 4   Implicit Differentiation

Find dy/dx if $x^4 + 5xy^4 = 2y^2 + x - 1$.

Solution

By viewing $5xy^4$ as the product $(5x)(y^4)$ and differentiating implicitly, we get

$$\frac{d}{dx}(x^4) + \frac{d}{dx}(5xy^4) = \frac{d}{dx}(2y^2) + \frac{d}{dx}(x) - \frac{d}{dx}(1)$$

$$4x^3 + 5x\left(4y^3\frac{dy}{dx}\right) + y^4(5) = 4y\frac{dy}{dx} + 1$$

To solve for $\dfrac{dy}{dx}$, we first rewrite the equation with terms containing $\dfrac{dy}{dx}$ on one side and the other terms on the other side.

$$20xy^3\frac{dy}{dx} - 4y\frac{dy}{dx} = 1 - 4x^3 - 5y^4$$

Now $\dfrac{dy}{dx}$ is a factor of one side. To complete the solution, we factor out $\dfrac{dy}{dx}$ and divide both sides by its coefficient.

$$(20xy^3 - 4y)\frac{dy}{dx} = 1 - 4x^3 - 5y^4$$

$$\frac{dy}{dx} = \frac{1 - 4x^3 - 5y^4}{20xy^3 - 4y}$$

✓ **CHECKPOINT**

1. Find the following:

   (a) $\dfrac{d}{dx}(x^3)$ (b) $\dfrac{d}{dx}(y^4)$ (c) $\dfrac{d}{dx}(x^2y^5)$

2. Find $\dfrac{dy}{dx}$ for $x^3 + y^4 = x^2y^5$.

EXAMPLE 5   Production

Suppose that a company's weekly production output is \$384,000 and that this output is related to hours of labor x and dollars of capital investment y by

$$384{,}000 = 30x^{1/3}\,y^{2/3}$$

(This relationship is an example of a Cobb-Douglas production function, studied in more detail in Chapter 14.) Find and interpret the rate of change of capital investment with respect to labor hours when labor hours are 512 and capital investment is \$64,000.

Solution

The desired rate of change is given by the value of dy/dx when $x = 512$ and $y = 64{,}000$. Taking the derivative implicitly gives

$$\frac{d}{dx}(384{,}000) = \frac{d}{dx}(30x^{1/3}\,y^{2/3})$$

$$0 = 30x^{1/3}\left(\frac{2}{3}y^{-1/3}\frac{dy}{dx}\right) + y^{2/3}(10x^{-2/3})$$

$$0 = \frac{20x^{1/3}}{y^{1/3}}\cdot\frac{dy}{dx} + \frac{10y^{2/3}}{x^{2/3}}$$

$$\frac{-20x^{1/3}}{y^{1/3}}\cdot\frac{dy}{dx} = \frac{10y^{2/3}}{x^{2/3}}$$

Multiplying both sides by $\dfrac{-y^{1/3}}{20x^{1/3}}$ gives

$$\frac{dy}{dx} = \left(\frac{10y^{2/3}}{x^{2/3}}\right)\left(\frac{-y^{1/3}}{20x^{1/3}}\right) = \frac{-y}{2x}$$

When $x = 512$ and $y = 64{,}000$, we obtain

$$\frac{dy}{dx} = \frac{-64{,}000}{2(512)} = -62.5$$

This means that when labor hours are 512 and capital investment is \$64,000, if labor hours change by 1 hour, then capital investment could decrease by about \$62.50. ■

EXAMPLE 6   Horizontal and Vertical Tangents

(a) At what point(s) does $x^2 + 4y^2 - 2x + 4y - 2 = 0$ have a horizontal tangent?
(b) At what point(s) does it have a vertical tangent?

Solution

(a) First we find the derivative implicitly, with y' representing $\dfrac{dy}{dx}$.

$$2x + 8y \cdot y' - 2 + 4y' - 0 = 0$$

We isolate the y' terms, factor out y', and solve for y'.

$$8yy' + 4y' = 2 - 2x$$
$$(8y + 4)y' = 2 - 2x$$
$$y' = \frac{2 - 2x}{8y + 4} = \frac{1 - x}{4y + 2}$$

Horizontal tangents will occur where $y' = 0$; that is, where $1 - x = 0$, or $x = 1$. We can now find the corresponding y-value(s) by substituting 1 for x in the original equation and solving.

$$1 + 4y^2 - 2 + 4y - 2 = 0$$
$$4y^2 + 4y - 3 = 0$$
$$(2y - 1)(2y + 3) = 0$$
$$y = \frac{1}{2} \quad \text{or} \quad y = -\frac{3}{2}$$

Thus horizontal tangents occur at $\left(1, \frac{1}{2}\right)$ and $\left(1, -\frac{3}{2}\right)$; see Figure 11.11.

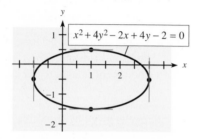

Figure 11.11

(b) Vertical tangents will occur where the derivative is undefined—that is, where $4y + 2 = 0$, or $y = -\frac{1}{2}$. To find the corresponding x-value(s), we substitute $-\frac{1}{2}$ in the equation for y and solve for x.

$$x^2 + 4\left(-\frac{1}{2}\right)^2 - 2x + 4\left(-\frac{1}{2}\right) - 2 = 0$$
$$x^2 - 2x - 3 = 0$$
$$(x - 3)(x + 1) = 0$$
$$x = 3 \quad \text{or} \quad x = -1$$

Thus vertical tangents occur at $\left(3, -\frac{1}{2}\right)$ and $\left(-1, -\frac{1}{2}\right)$; see Figure 11.11.

EXAMPLE 7   Implicit Derivatives with Logarithms and Exponentials

Find dy/dx for each of the following.
(a) $\ln xy = 6$ (b) $4x^2 + e^{xy} = 6y$

Solution
(a) Using the properties of logarithms, we have

$$\ln x + \ln y = 6$$

which leads to the implicit derivative

$$\frac{1}{x} + \frac{1}{y}\frac{dy}{dx} = 0$$

Solving gives $\dfrac{1}{y}\dfrac{dy}{dx} = -\dfrac{1}{x}$ so $\dfrac{dy}{dx} = -\dfrac{y}{x}$

(b) We take the derivative of both sides of $4x^2 + e^{xy} = 6y$ and obtain

$$8x + e^{xy}\left(x\frac{dy}{dx} + y\right) = 6\frac{dy}{dx}$$
$$8x + xe^{xy}\frac{dy}{dx} + ye^{xy} = 6\frac{dy}{dx}$$
$$8x + ye^{xy} = 6\frac{dy}{dx} - xe^{xy}\frac{dy}{dx}$$

$$8x + ye^{xy} = (6 - xe^{xy})\frac{dy}{dx} \quad \text{so} \quad \frac{8x + ye^{xy}}{6 - xe^{xy}} = \frac{dy}{dx}$$

EXAMPLE 8 **Demand | APPLICATION PREVIEW |**

Suppose the demand for Precision, Inc. headphones is given by

$$p = \frac{10,000}{(x + 1)^2}$$

where p is the price per set in dollars and x is in hundreds of headphone sets demanded. Find the rate of change of demand with respect to price when 19 (hundred) sets are demanded.

Solution

The rate of change of demand with respect to price is dx/dp. Using implicit differentiation, we get

$$\frac{d}{dp}(p) = \frac{d}{dp}\left[\frac{10,000}{(x + 1)^2}\right] = \frac{d}{dp}[10,000(x + 1)^{-2}]$$

$$1 = 10,000\left[-2(x + 1)^{-3}\frac{dx}{dp}\right]$$

$$1 = \frac{-20,000}{(x + 1)^3} \cdot \frac{dx}{dp}$$

$$\frac{(x + 1)^3}{-20,000} = \frac{dx}{dp}$$

When 19 (hundred) headphone sets are demanded we use $x = 19$, and the rate of change of demand with respect to price is

$$\frac{dx}{dp}\bigg|_{x=19} = \frac{(19 + 1)^3}{-20,000} = \frac{8000}{-20,000} = -0.4$$

This result means that when 19 (hundred) headphone sets are demanded, if the price per set is increased by $1, then the expected change in demand is a decrease of about 0.4 hundred, or 40, headphone sets. ∎

Calculator Note To graph a function with a graphing calculator, we need to write y as an *explicit* function of x (such as $y = \sqrt{4 - x^2}$). If an equation defines y as an *implicit* function of x, we have to solve for y in terms of x before we can use the graphing calculator. Sometimes we cannot solve for y, and other times, such as for

$$x^{2/3} + y^{2/3} = 8^{2/3}$$

y cannot be written as a single function of x. If this equation is solved for y as a single function and a graphing calculator is used to graph that function, the resulting graph usually shows only the portion of the graph that lies in Quadrants I and II. To graph $x^{2/3} + y^{2/3} = 8^{2/3}$, we solve for y.

$$y^{2/3} = 8^{2/3} - x^{2/3} \Rightarrow y^2 = (8^{2/3} - x^{2/3})^3 \Rightarrow y = \pm\sqrt{(8^{2/3} - x^{2/3})^3}$$

Graphing $Y_1 = \sqrt{(8^{2/3} - x^{2/3})^3}$ and $Y_2 = -\sqrt{(8^{2/3} - x^{2/3})^3}$ on a window such as 5:ZSquare gives a complete graph like the one shown in Figure 11.12. In general, for the graph of an implicitly defined function, a graphing calculator must be used carefully (and sometimes cannot be used at all). ∎

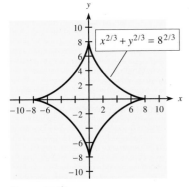

Figure 11.12

1. (a) $\dfrac{d}{dx}(x^3) = 3x^2$ (b) $\dfrac{d}{dx}(y^4) = 4y^3\dfrac{dy}{dx}$

   (c) $\dfrac{d}{dx}(x^2y^5) = x^2\left(5y^4\dfrac{dy}{dx}\right) + y^5(2x)$

2. $\dfrac{dy}{dx} = \dfrac{3x^2 - 2xy^5}{5x^2y^4 - 4y^3}$

| EXERCISES | 11.3

In Problems 1–6, find dy/dx at the given point without first solving for y.

1. $x^2 - 4y - 17 = 0$ at $(1, -4)$
2. $3x^2 - 10y + 400 = 0$ at $(10, 70)$
3. $xy^2 = 8$ at $(2, 2)$
4. $e^y = x$ at $(1, 0)$
5. $x^2 + 3xy - 4 = 0$ at $(1, 1)$
6. $x^2 + 5xy + 4 = 0$ at $(1, -1)$

Find dy/dx for the functions in Problems 7–10.

7. $x^2 + 2y^2 - 4 = 0$
8. $x + y^2 - 4y + 6 = 0$
9. $x^2 + 4x + y^2 - 3y + 1 = 0$
10. $x^2 - 5x + y^3 - 3y - 3 = 0$
11. If $x^2 + y^2 = 4$, find y'.
12. If $p^2 + 4p - q = 4$, find dp/dq.
13. If $xy^2 - y^3 = 1$, find y'.
14. If $p^2 - q = 4$, find dp/dq.
15. If $p^2q = 4p - 2$, find dp/dq.
16. If $x^2 - 3y^4 = 2x^5 + 7y^3 - 5$, find dy/dx.
17. If $3x^5 - 5y^3 = 5x^2 + 3y^5$, find dy/dx.
18. If $x^2 + 3x^2y^4 = y + 8$, find dy/dx.
19. If $x^4 + 2x^3y^2 = x - y^3$, find dy/dx.
20. If $(x + y)^2 = 5x^4y^3$, find dy/dx.
21. Find dy/dx for $x^4 + 3x^3y^2 - 2y^5 = (2x + 3y)^2$.
22. Find y' for $2x + 2y = \sqrt{x^2 + y^2}$.

For Problems 23–26, find the slope of the tangent to the curve.

23. $x^2 + 4x + y^2 + 2y - 4 = 0$ at $(1, -1)$
24. $x^2 - 4x + 2y^2 - 4 = 0$ at $(2, 2)$
25. $x^2 + 2xy + 3 = 0$ at $(-1, 2)$
26. $y + x^2 + xy = 13$ at $(2, 3)$
27. Write the equation of the line tangent to the curve $x^2 - 2y^2 + 4 = 0$ at $(2, 2)$.
28. Write the equation of the line tangent to the curve $x^2 + y^2 + 2x - 3 = 0$ at $(-1, 2)$.
29. Write the equation of the line tangent to the curve $4x^2 + 3y^2 - 4y - 3 = 0$ at $(-1, 1)$.
30. Write the equation of the line tangent to the curve $xy + y^2 = 0$ at $(3, 0)$.
31. If $\ln x = y^2$, find dy/dx.
32. If $\ln (x + y) = y^2$, find dy/dx.
33. If $y^2 \ln x = 4$, find dy/dx.
34. If $\ln (xy - 1) = x + 2$, find dy/dx.
35. Find the slope of the tangent to the curve $y^2 \ln x + x^2y = 3$ at the point $(1, 3)$.
36. Write the equation of the line tangent to the curve $x \ln y + 2xy = 2$ at the point $(1, 1)$.
37. If $xe^y = 6$, find dy/dx.
38. If $x + e^{xy} = 10$, find dy/dx.
39. If $xe^{xy} = 4$, find dy/dx.
40. If $x - xe^y = 3$, find dy/dx.
41. If $ye^x - y = 3$, find dy/dx.

42. If $x^2y = e^{x+y}$, find dy/dx.
43. Find the slope of the line tangent to the graph of $ye^x = y^2 + x - 2$ at $(0, 2)$.
44. Find the slope of the line tangent to the curve $xe^y = 3x^2 + y - 24$ at $(3, 0)$.
45. Write the equation of the line tangent to the curve $xe^y = 2y + 3$ at $(3, 0)$.
46. Write the equation of the line tangent to the curve $ye^x = 2y + 1$ at $(0, -1)$.
47. At what points does the curve defined by $x^2 + 4y^2 - 4x - 4 = 0$ have
    (a) horizontal tangents?
    (b) vertical tangents?
48. At what points does the curve defined by $x^2 + 4y^2 - 4 = 0$ have
    (a) horizontal tangents?
    (b) vertical tangents?
49. In Problem 11, the derivative y' was found to be

$$y' = \frac{-x}{y}$$

when $x^2 + y^2 = 4$.
    (a) Take the implicit derivative of the equation for y' to show that

$$y'' = \frac{-y + xy'}{y^2}$$

    (b) Substitute $-x/y$ for y' in the expression for y'' in part (a) and simplify to show that

$$y'' = -\frac{(x^2 + y^2)}{y^3}$$

    (c) Does $y'' = -4/y^3$? Why or why not?
50. (a) Find y' implicitly for $x^3 - y^3 = 8$.
    (b) Then, by taking derivatives implicitly, use part (a) to show that

$$y'' = \frac{2x(y - xy')}{y^3}$$

    (c) Substitute x^2/y^2 for y' in the expression for y'' and simplify to show that

$$y'' = \frac{2x(y^3 - x^3)}{y^5}$$

    (d) Does $y'' = -16x/y^5$? Why or why not?
51. Find y'' for $\sqrt{x} + \sqrt{y} = 1$ and simplify.
52. Find y'' for $\dfrac{1}{x} - \dfrac{1}{y} = 1$.

In Problems 53 and 54, find the maximum and minimum values of y. Use a graphing utility to verify your conclusion.

53. $x^2 + y^2 - 9 = 0$ 54. $4x^2 + y^2 - 8x = 0$

APPLICATIONS

55. *Advertising and sales* Suppose that a company's sales volume y (in thousands of units) is related to its advertising expenditures x (in thousands of dollars) according to

$$xy - 20x + 10y = 0$$

Find the rate of change of sales volume with respect to advertising expenditures when $x = 10$ (thousand dollars).

56. *Insect control* Suppose that the number of mosquitoes N (in thousands) in a certain swampy area near a community is related to the number of pounds of insecticide x sprayed on the nesting areas according to

$$Nx - 10x + N = 300$$

Find the rate of change of N with respect to x when 49 pounds of insecticide is used.

57. *Production* Suppose that a company can produce 12,000 units when the number of hours of skilled labor y and unskilled labor x satisfy

$$384 = (x + 1)^{3/4}(y + 2)^{1/3}$$

Find the rate of change of skilled-labor hours with respect to unskilled-labor hours when $x = 255$ and $y = 214$. This can be used to approximate the change in skilled-labor hours required to maintain the same production level when unskilled-labor hours are increased by 1 hour.

58. *Production* Suppose that production of 10,000 units of a certain agricultural crop is related to the number of hours of labor x and the number of acres of the crop y according to

$$300x + 30{,}000y = 11xy - 0.0002x^2 - 5y$$

Find the rate of change of the number of hours with respect to the number of acres.

59. *Demand* If the demand function for q units of a product at $\$p$ per unit is given by

$$p(q + 1)^2 = 200{,}000$$

find the rate of change of quantity with respect to price when $p = \$80$. Interpret this result.

60. *Demand* If the demand function for q units of a commodity at $\$p$ per unit is given by

$$p^2(2q + 1) = 100{,}000$$

find the rate of change of quantity with respect to price when $p = \$50$. Interpret this result.

61. *Radioactive decay* The number of grams of radium, y, that will remain after t years if 100 grams existed originally can be found by using the equation

$$-0.000436t = \ln\left(\frac{y}{100}\right)$$

Use implicit differentiation to find the rate of change of y with respect to t—that is, the rate at which the radium will decay.

62. *Disease control* Suppose the proportion P of people affected by a certain disease is described by

$$\ln\left(\frac{P}{1 - P}\right) = 0.5t$$

where t is the time in months. Find dP/dt, the rate at which P grows.

63. *Temperature-humidity index* The temperature-humidity index (THI) is given by

$$\text{THI} = t - 0.55(1 - h)(t - 58)$$

where t is the air temperature in degrees Fahrenheit and h is the relative humidity. If the THI remains constant, find the rate of change of humidity with respect to temperature if the temperature is 70°F (*Source:* "Temperature-Humidity Indices," *UMAP Journal*, Fall 1989).

OBJECTIVE   **11.4**

- To use implicit differentiation to solve problems that involve related rates

Related Rates

▮ | APPLICATION PREVIEW |

According to Poiseuille's law, the flow of blood F is related to the radius r of the vessel according to

$$F = kr^4$$

where k is a constant. When the radius of a blood vessel is reduced, such as by cholesterol deposits, the flow of blood is also restricted. Drugs can be administered that increase the radius of the blood vessel and, hence, the flow of blood. The rate of change of the blood flow and the rate of change of the radius of the blood vessel are time rates of change that are related to each other, so they are called related rates. We can use these related rates to find the percent rate of change in the blood flow that corresponds to the percent rate of change in the radius of the blood vessel caused by the drug. (See Example 2.)

Related Rates   We have seen that the derivative represents the instantaneous rate of change of one variable with respect to another. When the derivative is taken with respect to time, it represents the rate at which that variable is changing with respect to time. For example, if distance x is measured in miles and time t in hours, then dx/dt is measured in miles per hour and indicates how fast x is changing. Similarly, if V represents the volume (in cubic feet) of water in a swimming pool and t is time (in minutes), then dV/dt is measured in cubic feet per minute (ft³/min) and might measure the rate at which the pool is being filled with water or being emptied.

Sometimes, two (or more) quantities that depend on time are also related to each other. For example, the height of a tree h (in feet) is related to the radius r (in inches) of its trunk, and this relationship can be modeled by

$$h = kr^{2/3}$$

where k is a constant.* Of course, both h and r are also related to time, so the rates of change dh/dt and dr/dt are related to each other. Thus they are called **related rates.**

The specific relationship between dh/dt and dr/dt can be found by differentiating $h = kr^{2/3}$ implicitly with respect to time t.

EXAMPLE 1   Tree Height and Trunk Radius

Suppose that for a certain type of tree, the height of the tree (in feet) is related to the radius of its trunk (in inches) by

$$h = 15r^{2/3}$$

Suppose that the rate of change of r is $\frac{3}{4}$ inch per year. Find how fast the height is changing when the radius is 8 inches.

Solution
To find how the rates dh/dt and dr/dt are related, we differentiate $h = 15r^{2/3}$ implicitly with respect to time t.

$$\frac{dh}{dt} = 10r^{-1/3}\frac{dr}{dt}$$

Using $r = 8$ inches and $dr/dt = \frac{3}{4}$ inch per year gives

$$\frac{dh}{dt} = 10(8)^{-1/3}(3/4) = \frac{15}{4} = 3\tfrac{3}{4} \text{ feet per year}$$

Percent Rates of Change   The work in Example 1 shows how to obtain related rates, but the different units (feet per year and inches per year) may be somewhat difficult to interpret. For this reason, many applications in the life sciences deal with **percent rates of change.** The percent rate of change of a quantity is the rate of change of the quantity divided by the quantity.

EXAMPLE 2   Blood Flow | APPLICATION PREVIEW |

Poiseuille's law expresses the flow of blood F as a function of the radius r of the vessel according to

$$F = kr^4$$

where k is a constant. When the radius of a blood vessel is restricted, such as by cholesterol deposits, drugs can be administered that will increase the radius of the blood vessel (and hence the blood flow). Find the percent rate of change of the flow of blood that corresponds to the percent rate of change of the radius of a blood vessel caused by the drug.

*T. McMahon, "Size and Shape in Biology," *Science* 179 (1979): 1201.

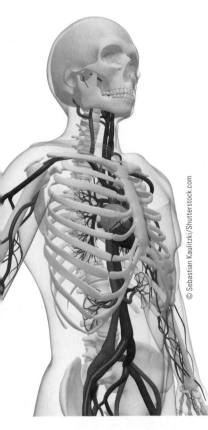

© Sebastian Kaulitzki/Shutterstock.com

Solution

We seek the percent rate of change of flow, $(dF/dt)/F$, that results from a given percent rate of change of the radius $(dr/dt)/r$. We first find the related rates of change by differentiating

$$F = kr^4$$

implicitly with respect to time.

$$\frac{dF}{dt} = k\left(4r^3\frac{dr}{dt}\right)$$

Then the percent rate of change of flow can be found by dividing both sides of the equation by F.

$$\frac{\frac{dF}{dt}}{F} = \frac{4kr^3\frac{dr}{dt}}{F}$$

If we replace F on the right side of the equation with kr^4 and reduce, we get

$$\frac{\frac{dF}{dt}}{F} = \frac{4kr^3\frac{dr}{dt}}{kr^4} = 4\left(\frac{\frac{dr}{dt}}{r}\right)$$

Thus we see that the percent rate of change of the flow of blood is 4 times the corresponding percent rate of change of the radius of the blood vessel. This means that a drug that would cause a 12% increase in the radius of a blood vessel at a certain time would produce a corresponding 48% increase in blood flow through that vessel at that time. ■

Solving Related-Rates Problems

In the examples above, the equation relating the time-dependent variables has been given. For some problems, the original equation relating the variables must first be developed from the statement of the problem. These problems can be solved with the aid of the following procedure.

Solving a Related-Rates Problem

Procedure	**Example**
To solve a related-rates problem:	Sand falls at a rate of 5 ft³/min on a conical pile, with the diameter always equal to the height of the pile. At what rate is the height increasing when the pile is 10 ft high?
1. Use geometric and/or physical conditions to write an equation that relates the time-dependent variables.	1. The conical pile has its volume given by $$V = \frac{1}{3}\pi r^2 h$$
2. Substitute into the equation values or relationships that are true at *all times*.	2. The radius $r = \frac{1}{2}h$ at all times, so $$V = \frac{1}{3}\pi\left(\frac{1}{4}h^2\right)h = \frac{\pi}{12}h^3$$
3. Differentiate both sides of the equation implicitly with respect to time. This equation is valid for all times.	3. $\dfrac{dV}{dt} = \dfrac{\pi}{12}\left(3h^2\dfrac{dh}{dt}\right) = \dfrac{\pi}{4}h^2\dfrac{dh}{dt}$
4. Substitute the values that are known at the instant specified, and solve the equation.	4. $\dfrac{dV}{dt} = 5$ at all times, so when $h = 10$, $$5 = \frac{\pi}{4}(10^2)\frac{dh}{dt}$$
5. Solve for the specified quantity at the given time.	5. $\dfrac{dh}{dt} = \dfrac{20}{100\pi} = \dfrac{1}{5\pi}$ (feet/minute)

Note that you should *not* substitute numerical values for any quantity that varies with time until after the derivative is taken. If values are substituted before the derivative is taken, that quantity will have the constant value resulting from the substitution and hence will have a derivative equal to zero.

EXAMPLE 3   Hot Air Balloon

A hot air balloon has a velocity of 50 feet per minute and is flying at a constant height of 500 feet. An observer on the ground is watching the balloon approach. How fast is the distance between the balloon and the observer changing when the balloon is 1000 feet from the observer?

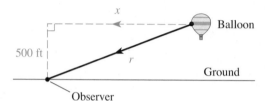

Figure 11.13

Solution
If we let r be the distance between the balloon and the observer and let x be the horizontal distance from the balloon to a point directly above the observer, then we see that these quantities are related by the equation

$$x^2 + 500^2 = r^2 \qquad \text{(See Figure 11.13)}$$

Because the distance x is decreasing, we know that dx/dt must be negative. Thus we are given that $dx/dt = -50$ at all times, and we need to find dr/dt when $r = 1000$. Taking the derivative with respect to t of both sides of the equation $x^2 + 500^2 = r^2$ gives

$$2x\frac{dx}{dt} + 0 = 2r\frac{dr}{dt}$$

Using $dx/dt = -50$ and $r = 1000$, we get

$$2x(-50) = 2000\frac{dr}{dt}$$
$$\frac{dr}{dt} = \frac{-100x}{2000} = \frac{-x}{20}$$

Using $r = 1000$ in $x^2 + 500^2 = r^2$ gives $x^2 = 750{,}000$. Thus $x = 500\sqrt{3}$, and

$$\frac{dr}{dt} = \frac{-500\sqrt{3}}{20} = -25\sqrt{3} \approx -43.3$$

The distance is decreasing at 43.3 feet per minute.

✓ CHECKPOINT

1. If V represents volume, write a mathematical symbol that represents "the rate of change of volume with respect to time."
2. (a) Differentiate $x^2 + 64 = y^2$ implicitly with respect to time.
   (b) Suppose that we know that y is increasing at 2 units per minute. Use part (a) to find the rate of change of x at the instant when $x = 6$ and $y = 10$.
3. True or false: In solving a related-rates problem, we substitute all numerical values into the equation before we take derivatives.

EXAMPLE 4   Spread of an Oil Slick

Suppose that oil is spreading in a circular pattern from a leak at an offshore rig. If the rate at which the radius of the oil slick is growing is 1 foot per minute at what rate is the area of the oil slick growing when the radius is 600 feet?

Solution

The area of the circular oil slick is given by

$$A = \pi r^2$$

where r is the radius. The rate at which the area is changing is

$$\frac{dA}{dt} = 2\pi r \frac{dr}{dt}$$

Using $r = 600$ feet and $dr/dt = 1$ foot per minute gives

$$\frac{dA}{dt} = 2\pi\,(600\text{ ft})\,(1\text{ ft/min}) = 1200\pi \text{ ft}^2/\text{min}$$

Thus when the radius of the oil slick is 600 feet, the area is growing at the rate of 1200π square feet per minute, or approximately 3770 square feet per minute. ■

✓ CHECKPOINT ANSWERS

1. dV/dt

2. (a) $x\dfrac{dx}{dt} = y\dfrac{dy}{dt}$ (b) $\dfrac{10}{3}$ units per minute

3. False; the numerical value for any variable that is changing with time should not be substituted until after the derivative is taken.

| EXERCISES | 11.4

In Problems 1–4, find dy/dt using the given values.

1. $y = x^3 - 3x$ for $x = 2$, $dx/dt = 4$
2. $y = 3x^3 + 5x^2 - x$ for $x = 4$, $dx/dt = 3$
3. $xy = 4$ for $x = 8$, $dx/dt = -2$
4. $xy = x + 3$ for $x = 3$, $dx/dt = -1$

In Problems 5–8, assume that x and y are differentiable functions of t. In each case, find dx/dt given that $x = 5$, $y = 12$, and $dy/dt = 2$.

5. $x^2 + y^2 = 169$
6. $y^2 - x^2 = 119$
7. $y^2 = 2xy + 24$
8. $x^2(y - 6) = 12y + 6$
9. If $x^2 + y^2 = z^2$, find dy/dt when $x = 3$, $y = 4$, $dx/dt = 10$, and $dz/dt = 2$.
10. If $s = 2\pi r(r + h)$, find dr/dt when $r = 2$, $h = 8$, $dh/dt = 3$, and $ds/dt = 10\pi$.
11. A point is moving along the graph of the equation $y = -4x^2$. At what rate is y changing when $x = 5$ and is changing at a rate of 2 units/sec?
12. A point is moving along the graph of the equation $y = 5x^3 - 2x$. At what rate is y changing when $x = 4$ and is changing at a rate of 3 units/sec?
13. The radius of a circle is increasing at a rate of 2 ft/min. At what rate is its area changing when the radius is 3 ft? (Recall that for a circle, $A = \pi r^2$.)
14. The area of a circle is changing at a rate of 1 in²/sec. At what rate is its radius changing when the radius is 2 in.?

15. The volume of a cube is increasing at a rate of 64 in³/sec. At what rate is the length of each edge of the cube changing when the edges are 6 in. long? (Recall that for a cube, $V = x^3$.)
16. The lengths of the edges of a cube are increasing at a rate of 8 ft/min. At what rate is the surface area changing when the edges are 24 ft long? (Recall that for a cube, $S = 6x^2$.)

APPLICATIONS

17. *Profit* Suppose that the daily profit (in dollars) from the production and sale of x units of a product is given by

$$P = 180x - \frac{x^2}{1000} - 2000$$

At what rate per day is the profit changing when the number of units produced and sold is 100 and is increasing at a rate of 10 units per day?

18. *Profit* Suppose that the monthly revenue and cost (in dollars) for x units of a product are

$$R = 400x - \frac{x^2}{20} \quad \text{and} \quad C = 5000 + 70x$$

At what rate per month is the profit changing if the number of units produced and sold is 200 and is increasing at a rate of 5 units per month?

19. **Demand** Suppose that the price p (in dollars) of a product is given by the demand function

$$p = \frac{1000 - 10x}{400 - x}$$

where x represents the quantity demanded. If the daily demand is *decreasing* at a rate of 20 units per day, at what rate is the price changing when the demand is 20 units?

20. **Supply** The supply function for a product is given by $p = 40 + 100\sqrt{2x + 9}$, where x is the number of units supplied and p is the price in dollars. If the price is increasing at a rate of $1 per month, at what rate is the supply changing when $x = 20$?

21. **Capital investment and production** Suppose that for a particular product, the number of units x produced per month depends on the number of thousands of dollars y invested, with $x = 30y + 20y^2$. At what rate will production increase if $10,000 is invested and if the investment capital is increasing at a rate of $1000 per month?

22. **Boyle's law** Boyle's law for enclosed gases states that at a constant temperature, the pressure is related to the volume by the equation

$$P = \frac{k}{V}$$

where k is a constant. If the volume is increasing at a rate of 5 cubic inches per hour, at what rate is the pressure changing when the volume is 30 cubic inches and $k = 2$ inch-pounds?

Tumor growth For Problems 23 and 24, suppose that a tumor in a person's body has a spherical shape and that treatment is causing the radius of the tumor to decrease at a rate of 1 millimeter per month.

23. At what rate is the volume decreasing when the radius is 3 mm? (Recall that $V = \frac{4}{3}\pi r^3$.)

24. At what rate is the surface area of the tumor decreasing when the radius is 3 mm? (Recall that for a sphere, $S = 4\pi r^2$.)

25. **Allometric relationships—fish** For many species of fish, the allometric relationship between the weight W and the length L is approximately $W = kL^3$, where k is a constant. Find the percent rate of change of the weight as a corresponding percent rate of change of the length.

26. **Blood flow** The resistance R of a blood vessel to the flow of blood is a function of the radius r of the blood vessel and is given by

$$R = \frac{k}{r^4}$$

where k is a constant. Find the percent rate of change of the resistance of a blood vessel in terms of the percent rate of change in the radius of the blood vessel.

27. **Allometric relationships—crabs** For fiddler crabs, data gathered by Thompson* show that the allometric relationship between the weight C of the claw and the weight W of the body is given by

$$C = 0.11W^{1.54}$$

Find the percent rate of change of the claw weight in terms of the percent rate of change of the body weight for fiddler crabs.

28. **Body weight and surface area** For human beings, the surface area S of the body is related to the body's weight W according to

$$S = kW^{2/3}$$

where k is a constant. Find the percent rate of change of the body's surface area in terms of the percent rate of change of the body's weight.

29. **Cell growth** A bacterial cell has a spherical shape. If the volume of the cell is increasing at a rate of 4 cubic micrometers per day, at what rate is the radius of the cell increasing when it is 2 micrometers? (Recall that for a sphere, $V = \frac{4}{3}\pi r^3$.)

30. **Water purification** Assume that water is being purified by causing it to flow through a conical filter that has a height of 15 inches and a radius of 5 inches. If the depth of the water is decreasing at a rate of 1 inch per minute when the depth is 6 inches, at what rate is the volume of water flowing out of the filter at this instant?

31. **Volume and radius** Suppose that air is being pumped into a spherical balloon at a rate of 5 in³/min. At what rate is the radius of the balloon increasing when the radius is 5 in.?

32. **Boat docking** Suppose that a boat is being pulled toward a dock by a winch that is 5 ft above the level of the boat deck, as shown in the figure. If the winch is pulling the cable at a rate of 3 ft/min, at what rate is the boat approaching the dock when it is 12 ft from the dock?

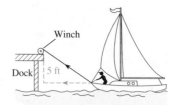

33. **Ladder safety** A 30-ft ladder is leaning against a wall. If the bottom is pulled away from the wall at a rate of 1 ft/sec, at what rate is the top of the ladder sliding down the wall when the bottom is 18 ft from the wall?

34. **Flight** A kite is 30 ft high and is moving horizontally at a rate of 10 ft/min. If the kite string is taut, at what rate is the string being played out when 50 ft of string is out?

*d'Arcy Thompson, *On Growth and Form* (Cambridge, England: Cambridge University Press, 1961).

35. *Flight* A plane is flying at a constant altitude of 1 mile and a speed of 300 mph. If it is flying toward an observer on the ground, how fast is the plane approaching the observer when it is 5 miles from the observer?

36. *Distance* Two boats leave the same port at the same time, with boat A traveling north at 15 knots (nautical miles per hour) and boat B traveling east at 20 knots. How fast is the distance between them changing when boat A is 30 nautical miles from port?

37. *Distance* Two cars are approaching an intersection on roads that are perpendicular to each other. Car A is north of the intersection and traveling south at

40 mph. Car B is east of the intersection and traveling west at 55 mph. How fast is the distance between the cars changing when car A is 15 miles from the intersection and car B is 8 miles from the intersection?

38. *Water depth* Water is flowing into a barrel in the shape of a right circular cylinder at the rate of 200 in³/min. If the radius of the barrel is 18 in., at what rate is the depth of the water changing when the water is 30 in. deep?

39. *Water depth* Suppose that water is being pumped into a rectangular swimming pool of uniform depth at 10 ft³/hr. If the pool is 10 ft wide and 25 ft long, at what rate is the water rising when it is 4 ft deep?

OBJECTIVES

11.5

- To find the elasticity of demand
- To find the tax per unit that will maximize tax revenue

Applications in Business and Economics

| APPLICATION PREVIEW |

Suppose that the demand for a product is given by

$$p = \frac{1000}{(q + 1)^2}$$

We can measure how sensitive the demand for this product is to price changes by finding the elasticity of demand. (See Example 2.) This elasticity can be used to measure the effects that price changes have on total revenue. We also consider taxation in a competitive market, which examines how a tax levied on goods shifts market equilibrium. We will find the tax per unit that, despite changes in market equilibrium, maximizes tax revenue.

Elasticity of Demand   We know from the law of demand that consumers will respond to changes in prices; if prices increase, the quantities demanded will decrease. But the degree of responsiveness of the consumers to price changes will vary widely for different products. For example, a price increase in insulin will not greatly decrease the demand for it by diabetics, but a price increase in clothes may cause consumers to buy less and wear their old clothes longer. When the response to price changes is considerable, we say the demand is *elastic*. When price changes cause relatively small changes in demand for a product, the demand is said to be *inelastic* for that product. Economists measure the **elasticity of demand** as follows.

Elasticity

The **elasticity of demand** at the point (q_A, p_A) is

$$\eta = -\frac{p}{q} \cdot \frac{dq}{dp}\Big|_{(q_A, p_A)}$$

EXAMPLE   1   Elasticity

Find the elasticity of the demand function $p + 5q = 100$ when
(a)   the price is $40.   (b)   the price is $60.   (c)   the price is $50.

Solution
Solving the demand function for q gives $q = 20 - \frac{1}{5}p$. Then $dq/dp = -\frac{1}{5}$ and

$$\eta = -\frac{p}{q}\left(-\frac{1}{5}\right)$$

(a) When $p = 40, q = 12$ and $\eta = -\dfrac{p}{q}\left(-\dfrac{1}{5}\right)\Big|_{(12,\,40)} = -\dfrac{40}{12}\left(-\dfrac{1}{5}\right) = \dfrac{2}{3}$.

(b) When $p = 60, q = 8$ and $\eta = -\dfrac{p}{q}\left(-\dfrac{1}{5}\right)\Big|_{(8,\,60)} = -\dfrac{60}{8}\left(-\dfrac{1}{5}\right) = \dfrac{3}{2}$.

(c) When $p = 50, q = 10$ and $\eta = -\dfrac{p}{q}\left(-\dfrac{1}{5}\right)\Big|_{(10,\,50)} = -\dfrac{50}{10}\left(-\dfrac{1}{5}\right) = 1$.

Note that in Example 1 the demand equation was $p + 5q = 100$, so the demand "curve" is a straight line, with slope $m = -5$. But the elasticity was $\eta = \frac{2}{3}$ at $(12, 40)$, $\eta = \frac{3}{2}$ at $(8, 60)$, and $\eta = 1$ at $(10, 50)$. This illustrates that the elasticity of demand may be different at different points on the demand curve, even though the slope of the demand "curve" is constant. (See Figure 11.14.)

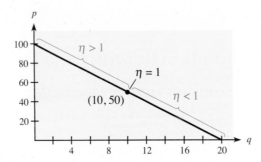

Figure 11.14

This example shows that the elasticity of demand is more than just the slope of the demand curve, which is the rate at which the demand is changing. Recall that the elasticity measures the consumers' degree of responsiveness to a price change.

Economists use η to measure how responsive demand is to price at different points on the demand curve for a product.

Elasticity of Demand

- If $\eta > 1$, the demand is **elastic,** and the percent decrease in demand is greater than the corresponding percent increase in price.
- If $\eta < 1$, the demand is **inelastic,** and the percent decrease in demand is less than the corresponding percent increase in price.
- If $\eta = 1$, the demand is **unitary elastic,** and the percent decrease in demand is approximately equal to the corresponding percent increase in price.

We can also use implicit differentiation to find dq/dp in evaluating the point elasticity of demand.

EXAMPLE 2   Elasticity | APPLICATION PREVIEW |

The demand for a certain product is given by

$$p = \frac{1000}{(q + 1)^2}$$

where p is the price per unit in dollars and q is demand in units of the product. Find the elasticity of demand with respect to price when $q = 19$.

Solution

To find the elasticity, we need to find dq/dp. Using implicit differentiation, we get the following:

$$\frac{d}{dp}(p) = \frac{d}{dp}[1000(q+1)^{-2}]$$

$$1 = 1000\left[-2(q+1)^{-3}\frac{dq}{dp}\right]$$

$$1 = \frac{-2000}{(q+1)^3}\frac{dq}{dp}$$

$$\frac{(q+1)^3}{-2000} = \frac{dq}{dp}$$

When $q = 19$, we have $p = 1000/(19+1)^2 = 1000/400 = 5/2$ and

$$\frac{dq}{dp}\bigg|_{(q=19)} = \frac{(19+1)^3}{-2000} = \frac{8000}{-2000} = -4$$

The elasticity of demand when $q = 19$ is

$$\eta = \frac{-p}{q}\cdot\frac{dq}{dp} = -\frac{(5/2)}{19}\cdot(-4) = \frac{10}{19} < 1$$

Thus the demand for this product is inelastic. ◼

Elasticity and Revenue

Elasticity is related to revenue in a special way. We can see how by computing the derivative with respect to p of the revenue function

$$R = pq$$

$$\frac{dR}{dp} = p\cdot\frac{dq}{dp} + q\cdot 1$$

$$= \frac{q}{q}\cdot p\cdot\frac{dq}{dp} + q = q\cdot\frac{p}{q}\cdot\frac{dq}{dp} + q$$

$$= q(-\eta) + q$$

$$= q(1 - \eta)$$

From this we can summarize the relationship of elasticity and revenue.

Elasticity and Revenue

The rate of change of revenue R with respect to price p is related to elasticity in the following way.

- Elastic ($\eta > 1$) means $\dfrac{dR}{dp} < 0.$ $\begin{cases} \text{Hence if price increases, revenue decreases,} \\ \text{and if price decreases, revenue increases.} \end{cases}$

- Inelastic ($\eta < 1$) means $\dfrac{dR}{dp} > 0.$ $\begin{cases} \text{Hence if price increases, revenue increases,} \\ \text{and if price decreases, revenue decreases.} \end{cases}$

- Unitary elastic ($\eta = 1$) means $\dfrac{dR}{dp} = 0.$ Hence an increase or decrease in price will not change revenue. Revenue is optimized at this point.

EXAMPLE 3 **Elasticity and Revenue**

The demand for a product is given by

$$p = 10\sqrt{100 - q}, \qquad 0 \le q \le 100$$

(a) Find the point at which demand is of unitary elasticity, and find intervals in which the demand is inelastic and in which it is elastic.
(b) Find q where revenue is increasing, where it is decreasing, and where it is maximized.
(c) Use a graphing utility to show the graph of the revenue function $R = pq$, with $0 \le q \le 100$, and confirm the results from part (b).

Solution

The elasticity is

$$\eta = \frac{-p}{q} \cdot \frac{dq}{dp} = -\frac{10\sqrt{100 - q}}{q} \cdot \frac{dq}{dp}$$

Finding dq/dp implicitly, we have

$$1 = 10\left[\frac{1}{2}(100 - q)^{-1/2}\left(-\frac{dq}{dp}\right)\right] = \frac{-5}{\sqrt{100 - q}} \cdot \frac{dq}{dp}$$

so

$$\frac{dq}{dp} = \frac{-\sqrt{100 - q}}{5}$$

Thus

$$\eta = \left(-\frac{10\sqrt{100 - q}}{q}\right)\left(\frac{-\sqrt{100 - q}}{5}\right) = \frac{200 - 2q}{q}$$

(a) Unitary elasticity occurs where $\eta = 1$.

$$1 = \frac{200 - 2q}{q}$$
$$q = 200 - 2q$$
$$3q = 200$$
$$q = 66\frac{2}{3}$$

so unitary elasticity occurs when $66\frac{2}{3}$ units are sold, at a price of \$57.74. For values of q between 0 and $66\frac{2}{3}$, $\eta > 1$ and demand is elastic. For values of q between $66\frac{2}{3}$ and 100, $\eta < 1$ and demand is inelastic.

(b) Because $\eta > 1$ and p decreases when q increases over $0 < q < 66\frac{2}{3}$, R increases. Similarly, because $\eta < 1$ and p decreases when q increases over $66\frac{2}{3} < q < 100$, R decreases. Thus, revenue is maximized where $\eta = 1$, at $q = 66\frac{2}{3}$ and $p = 57.74$.

(c) The graph of this revenue function,

$$R = 10q\sqrt{100 - q}$$

is shown in Figure 11.15. The maximum revenue is \$3849 at $q = 66\frac{2}{3}$. ∎

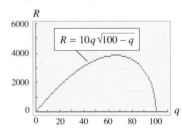

Figure 11.15

✓ **CHECKPOINT**	1. Write the formula for point elasticity, η.
	 2. (a) If $\eta > 1$, the demand is called _____.
	   (b) If $\eta < 1$, the demand is called _____.
	   (c) If $\eta = 1$, the demand is called _____.
	 3. Find the elasticity of demand for $q = \dfrac{100}{p} - 1$ when $p = 10$ and $q = 9$.

Taxation in a Competitive Market

Many taxes imposed by governments are "hidden." That is, the tax is levied on goods produced, and the producers must pay the tax. Of course, the tax becomes a cost to the producers, and they pass that cost on to the consumer in the form of higher prices for goods.

Suppose the government imposes a tax of t dollars on each unit produced and sold by producers. If we are in pure competition in which the consumers' demand depends only on price, the *demand function* will not change. The tax will change the supply function, of course, because at each level of output q, the firm will want to charge a price $p + t$ per unit, where p is the original price per unit and t is the tax per unit.

The graphs of the market demand function, the original market supply function, and the market supply function after taxes are shown in Figure 11.16. Because the tax added to each item is constant, the graph of the new supply function is t units above the original supply function. If $p = f(q)$ defines the original supply function, then $p = f(q) + t$ defines the supply function after taxation.

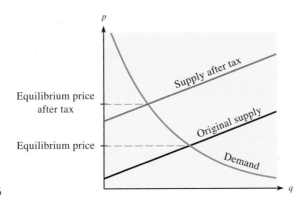

Figure 11.16

Note that after the taxes are imposed, *no* items are supplied at the price that was the equilibrium price before taxation. After the taxes are imposed, the consumers simply have to pay more for the product. Because taxation does not change the demand curve, the quantity purchased at market equilibrium will be less than it was before taxation. Thus governments planning taxes should recognize that they will not collect taxes on the original equilibrium quantity. They will collect on the *new* equilibrium quantity, a quantity reduced by their taxation. Thus a large tax on each item may reduce the quantity demanded at the new market equilibrium so much that very little revenue results from the tax!

If the tax revenue is represented by $T = tq$, where t is the tax per unit and q is the equilibrium quantity of the supply and demand functions after taxation, we can use the following procedure for maximizing the total tax revenue in a competitive market.

Maximizing Total Tax Revenue

Procedure	Example
To find the tax per item (under pure competition) that will maximize total tax revenue:	Suppose the demand and supply functions are given by $p = 600 - q$ and $p = 200 + \frac{1}{3}q$, respectively, where $\$p$ is the price per unit and q is the number of units. Find the tax t that will maximize the total tax revenue T.
1. Write the supply function after taxation.	1. $p = 200 + \frac{1}{3}q + t$
2. Set the demand function and the new supply function equal, and solve for t in terms of q.	2. $600 - q = 200 + \frac{1}{3}q + t$ $400 - \frac{4}{3}q = t$
3. Form the total tax revenue function, $T = tq$, by multiplying the expression for t by q, and then take its derivative with respect to q.	3. $T = tq = 400q - \frac{4}{3}q^2$ $T'(q) = \dfrac{dT}{dq} = 400 - \frac{8}{3}q$
4. Set $T' = 0$, and solve for q. This is the q that should maximize T. Use the second-derivative test to verify it.	4. $0 = 400 - \frac{8}{3}q$ $q = 150$ $T''(q) = -\frac{8}{3}$. Thus T is maximized at $q = 150$.
5. Substitute the value of q in the equation for t (in Step 2). This is the value of t that will maximize T.	5. $t = 400 - \frac{4}{3}(150) = 200$ A tax of $200 per item will maximize the total tax revenue. The total tax revenue for the period would be $\$200 \cdot (150) = \$30,000$.

Note that in the example just given, if a tax of $300 were imposed, market equilibrium would occur at $q = 75$, and the total tax revenue the government would receive would be

$$(\$300)(75) = \$22,500$$

Thus, with a tax of $300 per unit rather than $200, the government's tax revenue would be $22,500 rather than $30,000. In addition, with a $300 tax, suppliers would sell only 75 units rather than 150, and consumers would pay $p = 200 + 25 + 300 = \$525$ per item, which is $75 more than the price with a $200 tax. Thus everyone would suffer if the tax rate were raised to $300.

✓ CHECKPOINT   4. For problems involving taxation in a competitive market, if supply is $p = f(q)$ and demand is $p = g(q)$, is the tax t added to $f(q)$ or to $g(q)$?

EXAMPLE 4   Maximizing Tax Revenue

The demand and supply functions for a product are $p = 900 - 20q - \frac{1}{3}q^2$ and $p = 200 + 10q$, respectively, where p is in dollars and q is the number of units. Find the tax per unit that will maximize the tax revenue T.

Solution

After taxation, the supply function is $p = 200 + 10q + t$, where t is the tax per unit. The demand function will meet the new supply function where

$$900 - 20q - \frac{1}{3}q^2 = 200 + 10q + t$$

so

$$t = 700 - 30q - \frac{1}{3}q^2$$

Then the total tax T is $T = tq = 700q - 30q^2 - \frac{1}{3}q^3$, and we maximize T as follows:

$$T'(q) = 700 - 60q - q^2$$
$$0 = -(q + 70)(q - 10)$$
$$q = 10 \quad \text{or} \quad q = -70$$
$$T''(q) = -60 - 2q, \text{ so } T''(q) < 0 \text{ for } q \geq 0$$

Thus the curve is concave down for $q \geq 0$, and $q = 10$ gives an absolute maximum for the tax revenue T. The maximum possible tax revenue is

$$T(10) \approx \$3666.67$$

The tax per unit that maximizes T is

$$t = 700 - 30(10) - \frac{1}{3}(10)^2 \approx \$366.67$$

An infamous example of a tax increase that resulted in decreased tax revenue and economic disaster is the "luxury tax" enacted in 1991. This was a 10% excise tax on the sale of expensive jewelry, furs, airplanes, certain expensive boats, and luxury automobiles. The Congressional Joint Tax Committee had estimated that the luxury tax would raise $6 million from airplanes alone, but it raised only $53,000 while it destroyed the small-airplane market (one company lost $130 million and 480 jobs in a single year). It also capsized the boat market. The luxury tax was repealed at the end of 1993 (except for automobiles).*

✓ **CHECKPOINT ANSWERS**

1. $\eta = \dfrac{-p}{q} \cdot \dfrac{dq}{dp}$

2. (a) elastic   (b) inelastic   (c) unitary elastic

3. $\eta = \dfrac{10}{9}$ (elastic)

4. Tax t is added to supply: $p = f(q) + t$.

Fortune, Sept. 6, 1993; Motor Trend, December 1993.

| EXERCISES | 11.5

ELASTICITY OF DEMAND

In Problems 1–8, p is in dollars and q is the number of units.

1. (a) Find the elasticity of the demand function
      $p + 4q = 80$ at $(10, 40)$.
   (b) How will a price increase affect total revenue?
2. (a) Find the elasticity of the demand function
      $2p + 3q = 150$ at the price $p = 15$.
   (b) How will a price increase affect total revenue?
3. (a) Find the elasticity of the demand function
      $p^2 + 2p + q = 49$ at $p = 6$.
   (b) How will a price increase affect total revenue?
4. (a) Find the elasticity of the demand function
      $pq = 81$ at $p = 3$.
   (b) How will a price increase affect total revenue?
5. Suppose that the demand for a product is given by
   $pq + p = 5000$.
   (a) Find the elasticity when $p = \$50$ and $q = 99$.
   (b) Tell what type of elasticity this is: unitary, elastic, or inelastic.
   (c) How would revenue be affected by a price increase?

6. Suppose that the demand for a product is given by $2p^2q = 10{,}000 + 9000p^2$.
   (a) Find the elasticity when $p = \$50$ and $q = 4502$.
   (b) Tell what type of elasticity this is: unitary, elastic, or inelastic.
   (c) How would revenue be affected by a price increase?

7. Suppose that the demand for a product is given by $pq + p + 100q = 50{,}000$.
   (a) Find the elasticity when $p = \$401$.
   (b) Tell what type of elasticity this is.
   (c) How would a price increase affect revenue?

8. Suppose that the demand for a product is given by
   $$(p + 1)\sqrt{q + 1} = 1000$$
   (a) Find the elasticity when $p = \$39$.
   (b) Tell what type of elasticity this is.
   (c) How would a price increase affect revenue?

9. Suppose the demand function for a product is given by
   $$p = \frac{1}{2}\ln\left(\frac{5000 - q}{q + 1}\right)$$
   where p is in hundreds of dollars and q is the number of tons.
   (a) What is the elasticity of demand when the quantity demanded is 2 tons and the price is \$371?
   (b) Is the demand elastic or inelastic?

10. Suppose the weekly demand function for a product is
    $$q = \frac{5000}{1 + e^{2p}} - 1$$
    where p is the price in thousands of dollars and q is the number of units demanded. What is the elasticity of demand when the price is \$1000 and the quantity demanded is 595?

In Problems 11 and 12, the demand functions for specialty steel products are given, where p is in dollars and q is the number of units. For both problems:
(a) **Find the elasticity of demand as a function of the quantity demanded, q.**
(b) **Find the point at which the demand is of unitary elasticity and find intervals in which the demand is inelastic and in which it is elastic.**
(c) **Use information about elasticity in part (b) to decide where the revenue is increasing, where it is decreasing, and where it is maximized.**
(d) **Graph the revenue function $R = pq$, and use it to find where revenue is maximized. Is it at the same quantity as that determined in part (c)?**

11. $p = 120\sqrt[3]{125 - q}$
12. $p = 30\sqrt{49 - q}$
13. South West Electronics Corporation (SWEC) designs high-tech business and residential security systems. The company's marketing analyst has been assigned to analyze market demand for SWEC's top-selling

system, The Terminator. Monthly demand for The Terminator has been estimated as follows:
$$q = 445 - 8p + 25A + 4.5C + 6Y$$
where $q =$ Expected number of system sales per month
$p =$ Selling price for The Terminator (in dollars)
$A =$ Advertising (in thousands of dollars)
$C =$ SWEC's only competitor's average price (in dollars)
$Y =$ Disposable annual per capita income (in thousands of dollars)

A recent survey of the potential customer market indicates that monthly advertising is \$25,000, per capita disposable income is \$80,000 per year, and the average price of the only competitor is \$100.
(a) Based on this information, what is the monthly demand function $p = f(q)$ for The Terminator?
(b) Find the elasticity of demand for The Terminator.
(c) If SWEC's current price for The Terminator is \$175, is demand elastic, inelastic, or unitary elastic? Is SWEC's revenue for The Terminator maximized at the current price?
(d) Use the elasticity of demand found in part (b) to determine the price for The Terminator that would maximize SWEC's revenue. Find the maximum revenue.

14. The owner and manager of Pleasantville Deli is considering the expansion of current menu offerings to include a new line of take-out sandwiches. The deli serves primarily the Pleasantville business districts and students from a nearby college, yet it is unclear exactly what level of demand to anticipate for this new product offering and how to price the product in order to maximize sales.
    Over the past few months the information shown in the table was collected from existing customers and from mailings to businesses in the area.

Sandwich Price	\$12	10	8	6	4	2
Weekly Demand	0	400	800	1200	1600	2000

(a) Develop a function $p = f(q)$ that represents this demand schedule.
(b) Compute the elasticity of demand for the new sandwich.
(c) Find the elasticity at the possible prices of \$4 and \$10. Classify these prices as elastic, inelastic, or unitary elastic.
(d) Determine the price and quantity that would maximize weekly revenues for the new sandwich. Find the maximum weekly revenue.

TAXATION IN A COMPETITIVE MARKET

In Problems 15–24, p is the price per unit in dollars and q is the number of units.
15. If the weekly demand function is $p = 30 - q$ and the supply function before taxation is $p = 6 + 2q$, what tax per item will maximize the total tax revenue?

16. If the demand function for a fixed period of time is given by $p = 38 - 2q$ and the supply function before taxation is $p = 8 + 3q$, what tax per item will maximize the total tax revenue?

17. If the demand and supply functions for a product are $p = 800 - 2q$ and $p = 100 + 0.5q$, respectively, find the tax per unit t that will maximize the tax revenue T.

18. If the demand and supply functions for a product are $p = 2100 - 3q$ and $p = 300 + 1.5q$, respectively, find the tax per unit t that will maximize the tax revenue T.

19. If the weekly demand function is $p = 200 - 2q^2$ and the supply function before taxation is $p = 20 + 3q$, what tax per item will maximize the total tax revenue?

20. If the monthly demand function is $p = 7230 - 5q^2$ and the supply function before taxation is $p = 30 + 30q^2$, what tax per item will maximize the total revenue?

21. Suppose the weekly demand for a product is given by $p + 2q = 840$ and the weekly supply before taxation is given by $p = 0.02q^2 + 0.55q + 7.4$. Find the tax per item that produces maximum tax revenue. Find the tax revenue.

22. If the daily demand for a product is given by the function $p + q = 1000$ and the daily supply before taxation is $p = q^2/30 + 2.5q + 920$, find the tax per item that maximizes tax revenue. Find the tax revenue.

23. If the demand and supply functions for a product are $p = 2100 - 10q - 0.5q^2$ and $p = 300 + 5q + 0.5q^2$, respectively, find the tax per unit t that will maximize the tax revenue T.

24. If the demand and supply functions for a product are $p = 5000 - 20q - 0.7q^2$ and $p = 500 + 10q + 0.3q^2$, respectively, find the tax per unit t that will maximize the tax revenue T.

Chapter 11   Summary & Review

KEY TERMS AND FORMULAS

Section 11.1

Logarithmic function (p. 694)
$y = \log_a x$, defined by $x = a^y$
Natural logarithm (p. 694)
$\ln x = \log_e x$; $y = \ln (x)$ means $e^y = x$
Derivatives of logarithmic functions (p. 695)
$$\frac{d}{dx}(\ln x) = \frac{1}{x}$$
$$\frac{d}{dx}(\ln u) = \frac{1}{u} \cdot \frac{du}{dx}$$

Logarithmic Properties I–V for natural logarithms (p. 696)
$\ln e^x = x$; $e^{\ln x} = x$
$\ln (MN) = \ln M + \ln N$
$\ln (M/N) = \ln M - \ln N$
$\ln (M^P) = p (\ln M)$
Change-of-base formula (p. 698)
$$\log_a x = \frac{\ln x}{\ln a}$$

Section 11.2

Derivatives of exponential functions (p. 703)
$$\frac{d}{dx}(e^x) = e^x$$

$$\frac{d}{dx}e^u = e^u \frac{du}{dx}$$
$$\frac{d}{dx}a^u = a^u \frac{du}{dx} \ln a$$

Section 11.3

Implicit differentiation (p. 709)
$$\frac{d}{dx}(y^n) = ny^{n-1}\frac{dy}{dx}$$

Section 11.4

Related rates (p. 718)
Differentiate implicitly with respect to time, t

Percent rates of change (p. 718)
Solving related rates problems (p. 719)

Section 11.5

Elasticity of demand (p. 723)
$$\eta = \frac{-p}{q} \cdot \frac{dq}{dp}$$
Elastic
$\eta > 1$
Inelastic
$\eta < 1$

Unitary elastic
$\eta = 1$
Elasticity and revenue (p. 725)
Taxation in competitive market (p. 727)
Supply function after taxation
$p = f(q) + t$

REVIEW EXERCISES

Sections 11.1 and 11.2

In Problems 1–12, find the derivative of each function.

1. $y = 10e^{3x^2 - x}$

2. $y = 3\ln(4x + 11)$

3. $p = \ln\left(\dfrac{q}{q^2 - 1}\right)$

4. $y = xe^{x^2}$

5. $f(x) = 5e^{2x} - 40e^{-0.1x} + 11$

6. $g(x) = (2e^{3x+1} - 5)^3$

7. $s = \dfrac{3}{4}\ln(x^{12} - 2x^4 + 5)$

8. $w = (t^2 + 1)\ln(t^2 + 1) - t^2$

9. $y = 3^{3x-4}$

10. $y = 1 + \log_8(x^{10})$

11. $y = \dfrac{\ln x}{x}$

12. $y = \dfrac{1 + e^{-x}}{1 - e^{-x}}$

13. Write the equation of the line tangent to $y = 4e^{x^3}$ at $x = 1$.

14. At $x = 1$, write the equation of the line tangent to $y = 8 + 3x^2 \ln x$

Section 11.3

In Problems 15–20, find the indicated derivative.

15. If $y \ln x = 5y^2 + 11$, find dy/dx.

16. Find dy/dx for $e^{xy} = y$.

17. Find dy/dx for $y^2 = 4x - 1$.

18. Find dy/dx for $x^2 + 3y^2 + 2x - 3y + 2 = 0$.

19. Find dy/dx for $3x^2 + 2x^3y^2 - y^5 = 7$.

20. Find the second derivative y'' if $x^2 + y^2 = 1$.

21. Find the slope of the tangent to the curve $x^2 + 4x - 3y^2 + 6 = 0$ at $(3, 3)$.

22. Find the points where tangents to the graph of the equation in Problem 21 are horizontal.

Section 11.4

23. Suppose $3x^2 - 2y^3 = 10y$, where x and y are differentiable functions of t. If $dx/dt = 2$, find dy/dt when $x = 10$ and $y = 5$.

24. A right triangle with legs of lengths x and y has its area given by

$$A = \frac{1}{2}xy$$

If the rate of change of x is 2 units per minute and the rate of change of y is 5 units per minute, find the rate of change of the area when $x = 4$ and $y = 1$.

APPLICATIONS

Section 11.1

25. *Demographics* By using U.S. Census Bureau data for selected years from 1980 and projected to 2040, the number of White non-Hispanic individuals in the U.S.

civilian non-institutional population age 16 years and older, in millions, can be modeled by the function

$$P(t) = 96.1 + 17.4 \ln t$$

where t is the number of years past 1970.
(a) Find $P'(t)$.
(b) Find and interpret $P(58)$ and $P'(58)$.

26. *Life expectancy at age 65* By using Social Security Administration data for selected years from 1950 and projected to 2050, the additional years of life expectancy at age 65 can be modeled by

$$L(y) = -14.6 + 7.10 \ln y$$

where y equals the number of years past 1950.
(a) Find the function that models the rate of change of the years of life expectancy past age 65.
(b) Find and interpret $L(50)$ and $L(80)$.
(c) Find and interpret $L'(50)$ and $L'(80)$.

Section 11.2

27. *Compound interest* If $1000 is invested for n years at 12% compounded continuously, the future value of the investment is given by

$$S = 1000e^{0.12n}$$

(a) Find the function that gives the rate of change of this investment.
(b) Compare the rate at which the future value is growing after 1 year and after 10 years.

28. *Disposable income* Disposable income is the amount available for spending and saving after taxes have been paid and is one gauge for the state of the economy. By using U.S. Energy Administration data for selected years from 2010 and projected to 2040, the total U.S. disposable income, in billions, can be modeled by

$$D(t) = 10{,}020e^{0.02292t}$$

where t is the number of years past 2010.
(a) Find the function that models the rate of change of disposable income.
(b) Find and interpret $D(20)$ and $D'(20)$.

29. *Radioactive decay* A breeder reactor converts stable uranium-238 into the isotope plutonium-239. The decay of this isotope is given by

$$A(t) = A_0 e^{-0.00002876t}$$

where $A(t)$ is the amount of isotope at time t, in years, and A_0 is the original amount. This isotope has a half-life of 24,101 years (that is, half of it will decay away in 24,101 years).
(a) At what rate is $A(t)$ decaying at $t = 24{,}101$ years?
(b) At what rate is $A(t)$ decaying after 1 year?
(c) Is the rate of decay at its half-life greater or less than after 1 year?

30. **Marginal cost** The average cost of producing x units of a product is $\overline{C} = 600e^{x/600}$ dollars per unit. What is the marginal cost when 600 units are produced?

31. **Inflation** The impact of inflation on a $60,000 pension can be measured by the purchasing power P of $60,000 after t years. For an inflation rate of 5% per year, compounded annually, P is given by

$$P = 60{,}000e^{-0.0488t}$$

At what rate is purchasing power changing when $t = 10$? (*Source: Viewpoints*, VALIC)

Section 11.4

32. **Evaporation** A spherical droplet of water evaporates at a rate of 1 mm³/min. Find the rate of change of the radius when the droplet has a radius of 2.5 mm.

33. **Worker safety** A sign is being lowered over the side of a building at the rate of 2 ft/min. A worker handling a guide line is 7 ft away from a spot directly below the sign. How fast is the worker taking in the guide line at the instant the sign is 25 ft from the worker's hands? See the figure.

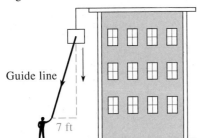

Guide line

7 ft

34. **Environment** Suppose that in a study of water birds, the relationship between the area A of wetlands (in square miles) and the number of different species S of birds found in the area was determined to be

$$S = kA^{1/3}$$

where k is constant. Find the percent rate of change of the number of species in terms of the percent rate of change of the area.

Section 11.5

35. **Taxes** Can increasing the tax per unit sold actually lead to a decrease in tax revenues?

36. **Taxes** Suppose the demand and supply functions for a product are

$$p = 2800 - 8q - \frac{q^2}{3} \quad \text{and} \quad p = 400 + 2q$$

respectively, where p is in dollars and q is the number of units. Find the tax per unit t that will maximize the tax revenue T, and find the maximum tax revenue.

37. **Taxes** Suppose the supply and demand functions for a product are

$$p = 40 + 20q \quad \text{and} \quad p = \frac{5000}{q + 1}$$

respectively, where p is in dollars and q is the number of units. Find the tax t that maximizes the tax revenue T, and find the maximum tax revenue.

38. **Elasticity** A demand function is given by

$$pq = 27$$

where p is in dollars and q is the number of units.
(a) Find the elasticity of demand at $(9, 3)$.
(b) How will a price increase affect total revenue?

39. **Elasticity** Suppose the demand for a product is given by

$$p^2(2q + 1) = 10{,}000$$

where p is in dollars and q is the number of units.
(a) Find the elasticity of demand when $p = \$20$.
(b) How will a price increase affect total revenue?

40. **Elasticity** Suppose the weekly demand function for a product is given by

$$p = 100e^{-0.1q}$$

where p is the price in dollars and q is the number of tons demanded.
(a) What is the elasticity of demand when the price is $36.79 and the quantity demanded is 10?
(b) How will a price increase affect total revenue?

41. **Revenue** A product has the demand function

$$p = 100 - 0.5q$$

where p is in dollars and q is the number of units.
(a) Find the elasticity $\eta(q)$ as a function of q, and graph the function

$$f(q) = \eta(q)$$

(b) Find where $f(q) = 1$, which gives the quantity for which the product has unitary elasticity.
(c) The revenue function for this product is

$$R(q) = pq = (100 - 0.5q)q$$

Graph $R(q)$ and find the q-value for which the maximum revenue occurs.
(d) What is the relationship between elasticity and maximum revenue?

Chapter 11   TEST

In Problems 1–8, find the derivative of each function.

1. $y = 5e^{x^3} + x^2$

2. $y = 4\ln(x^3 + 1)$

3. $y = \ln(x^4 + 1)^3$

4. $f(x) = 10(3^{2x})$

5. $S = te^{t^4}$

6. $y = \dfrac{e^{x^3+1}}{x}$

7. $y = \dfrac{3\ln x}{x^4}$

8. $g(x) = 2\log_5(4x + 7)$

9. Find y' if $3x^4 + 2y^2 + 10 = 0$.

10. Let $x^2 + y^2 = 100$.

   If $\dfrac{dx}{dt} = 2$, find $\dfrac{dy}{dt}$ when $x = 6$ and $y = 8$.

11. Find y' if $xe^y = 10y$.

12. Suppose the weekly revenue and weekly cost (both in dollars) for a product are given by $R(x) = 300x - 0.001x^2$ and $C(x) = 4000 + 30x$, respectively, where x is the number of units produced and sold. Find the rate at which profit is changing with respect to time when the number of units produced and sold is 50 and is increasing at a rate of 5 units per week.

13. Suppose the demand for a product is $p^2 + 3p + q = 1500$, where p is in dollars and q is the number of units. Find the elasticity of demand at $p = 30$. If the price is raised to \$31, does revenue increase or decrease?

14. Suppose the demand function for a product is given by $(p + 1)q^2 = 10,000$, where p is the price and q is the quantity. Find the rate of change of quantity with respect to price when $p = \$99$.

15. The sales of a product are given by $S = 80,000e^{-0.4t}$, where S is the daily sales and t is the number of days after the end of an advertising campaign. Find the rate of sales decay 10 days after the end of the ad campaign.

16. By using U.S. Centers for Medicare and Medicaid Services data from 2000 and projected to 2018, the total U.S. expenditures (in billions of dollars) for health services and supplies can be modeled by

$$y = 1319e^{0.062t}$$

where $t = 0$ in 2000.

(a) Find the function that models the rate of change.

(b) Find the model's value for the rate of change of U.S. expenditures for health services and supplies in 2005 and in 2020.

17. Suppose the demand and supply functions for a product are $p = 1100 - 5q$ and $p = 20 + 0.4q$, respectively, where p is in dollars and q is the number of units. Find the tax per unit t that will maximize the tax revenue $T = tq$.

18. *Modeling* Projections indicate that the percent of U.S. adults with diabetes could dramatically increase, and already in 2007 this disease had cost the country almost \$175 billion. The table gives the percent of U.S. adults with diabetes for selected years from 2010 and projected to 2050.

(a) Find a logarithmic model, $y = f(x)$, for these data. Use $x = 0$ to represent 2000.

(b) Find the function that models the rate of change of the percent of U.S. adults with diabetes.

(c) Find and interpret $f(25)$ and $f'(25)$.

Year	Percent	Year	Percent
2010	15.7	2035	29.0
2015	18.9	2040	31.4
2020	21.1	2045	32.1
2025	24.2	2050	34.3
2030	27.2		

Source: Centers for Disease Control and Prevention

19. Prices for goods and services tend to rise over time, and this results in the erosion of purchasing power. With the 2012 dollar as a reference, a purchasing power of 0.921 for a certain year means that in that year a dollar will purchase 92.1% of the goods and services that could be purchased for \$1.00 in 2012. By using Social Security Administration data for selected years from 2012 and projected to 2050, the purchasing power of a 2012 dollar can be modeled by the function

$$P(t) = 1.078(0.9732^t)$$

where t is the number of years past 2010.

(a) Find the function that models the rate of change of the purchasing power of a 2012 dollar.

(b) Find and interpret $P(18)$ and $P'(18)$.

Extended Applications & Group Projects

I. Inflation

Hollingsworth Pharmaceuticals specializes in manufacturing generic medicines. Recently it developed an antibiotic with outstanding profit potential. The new antibiotic's total costs, sales, and sales growth, as well as projected inflation, are described as follows.

Total monthly costs, in dollars, to produce x units (1 unit is 100 capsules):

$$C(x) = \begin{cases} 15,000 + 10x & 0 \le x \le 11,000 \\ 15,000 + 10x + 0.001(x - 11,000)^2 & x \ge 11,000 \end{cases}$$

Sales: 10,000 units per month and growing at 1.25% per month, compounded continuously
Selling price: $34 per unit
Inflation: Approximately 0.25% per month, compounded continuously, affecting both total costs and selling price

Company owners are pleased with the sales growth but are concerned about the projected increase in variable costs when production levels exceed 11,000 units per month. The consensus is that improvements eventually can be made that will reduce costs at higher production levels, thus altering the current cost function model. To plan properly for these changes, Hollingsworth Pharmaceuticals would like you to determine when the company's profits will begin to decrease. To help you determine this, answer the following.

1. If inflation is assumed to be compounded continuously, the selling price and total costs must be multiplied by the factor $e^{0.0025t}$. In addition, if sales growth is assumed to be compounded continuously, then sales must be multiplied by a factor of the form e^{rt}, where r is the monthly sales growth rate (expressed as a decimal) and t is time in months. Use these factors to write each of the following as a function of time t:
   (a) selling price p per unit (including inflation).
   (b) number of units x sold per month (including sales growth).
   (c) total revenue. (Recall that $R = px$.)
2. Determine how many months it will be before monthly sales exceed 11,000 units.
3. If you restrict your attention to total costs when $x \ge 11,000$, then, after expanding and collecting like terms, $C(x)$ can be written as follows:

   $$C(x) = 136,000 - 12x + 0.001x^2 \quad \text{for } x \ge 11,000$$

   Use this form for $C(x)$ with your result from Question 1(b) and with the inflationary factor $e^{0.0025t}$ to express these total costs as a function of time.
4. Form the profit function that would be used when monthly sales exceed 11,000 units by using the total revenue function from Question 1(c) and the total cost function from Question 3. This profit function should be a function of time t.
5. Find how long it will be before the profit is maximized. You may have to solve $P'(t) = 0$ by using a graphing calculator or computer to find the t-intercept of the graph of $P'(t)$. In addition, because $P'(t)$ has large numerical coefficients, you may want to divide both sides of $P'(t) = 0$ by 1000 before solving or graphing.

II. Renewable Electric Power (Modeling)

In the United States, consumers of electric power include all sectors of society—residential, commercial, industrial, and transportation related. Because of this broad dependency, reliable and uninterruptible sources of electric power (and those with the lowest environmental impact) are important. The United States has many sources of electrical power generation—predominately coal, natural gas, nuclear power, and renewable sources. The focus of this application is renewable sources because these will not be depleted, even though some renewables can be significant polluters (such as energy from combustion of wood or municipal waste).

The following table shows data for billions of kilowatt-hours of U.S. electrical power generation for selected years from 2016 and projected to 2040.

Power Generation (Billions of kilowatt-hours)

Year	Renewable	Total	Year	Renewable	Total
2016	629	4233	2030	748	4815
2018	651	4342	2032	764	4880
2020	667	4402	2034	778	4962
2022	683	4483	2036	800	5044
2024	700	4583	2038	825	5132
2026	716	4663	2040	851	5219
2028	730	4745			

Source: U.S. Department of Energy

1. Find an exponential function that models the billions of kilowatt-hours of renewable electric power generation, with x as the number of years after 2010. Store the unrounded model as $r(x)$ in Y_1 on your calculator, and report the rounded model as $R(x)$, with 3 significant digits and with base e.
2. Find a linear model for the billions of kilowatt-hours of total electric power generation, with x as the number of years after 2010. Store the unrounded model as $t(x)$ in Y_2 on your calculator, and report the rounded model as $T(x)$, with 3 significant digits.
3. Use the results of Questions 1 and 2 to create a function that models the percent of total electrical power generation that is renewable. Use the unrounded models in Y_1 and Y_2 to create the unrounded model as $p(x)$ and stored in Y_3, and use the rounded models $R(x)$ and $T(x)$ to report the rounded model as $P(x)$.
4. (a) Use the given data sets to create a data set that corresponds to $p(x)$.
   (b) Graph the data set and the model in Y_3 on the same set of axes.
   (c) How well does the model fit these data points?

In Questions 5–9, use the rounded models $R(x)$ and $P(x)$.

5. Find the function that models the rate of change of
   (a) $R(x)$ (b) $P(x)$
6. Find and interpret
   (a) $R(6)$ and $R'(6)$ (b) $R(25)$ and $R'(25)$
7. Find and interpret
   (a) $P(6)$ and $P'(6)$ (b) $P(25)$ and $P'(25)$
8. Find and interpret $P''(6)$ and $P''(25)$.
9. What does your analysis indicate about how the percent of total U.S. electrical power generation that is renewable is changing?

© Vasily Smirnov/Shutterstock.com

Indefinite Integrals

When we know the derivative of a function, it is often useful to determine the function itself. For example, accountants can use linear regression to translate information about marginal cost into a linear equation defining (approximately) the marginal cost function and then use the process of antidifferentiation (or integration) as part of finding the (approximate) total cost function. We can also use integration to find total revenue functions from marginal revenue functions, to optimize profit from information about marginal cost and marginal revenue, and to find national consumption from information about marginal propensity to consume.

Integration can also be used in the social and life sciences to predict growth or decay from expressions giving rates of change. For example, we can find equations for population size from the rate of change of growth, we can write equations for the number of radioactive atoms remaining in a substance if we know the rate of the decay of the substance, and we can determine the volume of blood flow from information about rate of flow.

The topics and some representative applications discussed in this chapter include the following.

SECTIONS

12.1 **Indefinite Integrals**

12.2 **The Power Rule**

12.3 **Integrals Involving Exponential and Logarithmic Functions**

12.4 **Applications of the Indefinite Integral in Business and Economics**
Total cost and profit
National consumption and savings

12.5 **Differential Equations**
Solution of differential equations
Separable differential equations
Applications of differential equations

APPLICATIONS

Revenue, population growth

Revenue, productivity

Real estate inflation, population growth

Cost, revenue, maximum profit, national consumption and savings

Carbon-14 dating, drug in an organ

Prerequisite Problem Type	For Section	Answer	Section for Review
Write as a power: (a) $\sqrt{x}$ (b) $\sqrt{x^2 - 9}$	**12.1–12.4**	(a) $x^{1/2}$ (b) $(x^2 - 9)^{1/2}$	0.4 Radicals
Expand $(x^2 + 4)^2$.	**12.2**	$x^4 + 8x^2 + 16$	0.5 Special powers
Divide $x^4 - 2x^3 + 4x^2 - 7x - 1$ by $x^2 - 2x$.	**12.3**	$x^2 + 4 + \dfrac{x - 1}{x^2 - 2x}$	0.5 Division
Find the derivative of (a) $f(x) = 2x^{1/2}$ (b) $u = x^3 - 3x$	**12.1–12.3**	(a) $f'(x) = x^{-1/2}$ (b) $u' = 3x^2 - 3$	9.4 Derivatives
If $y = \dfrac{(x^2 + 4)^6}{6}$, what is y'?	**12.2**	$y' = (x^2 + 4)^5 2x$	9.6 Derivatives
(a) If $y = \ln u$, what is y'? (b) If $y = e^u$, what is y'?	**12.3**	(a) $y' = \dfrac{1}{u} \cdot u'$ (b) $y' = e^u \cdot u'$	11.1, 11.2 Derivatives
Solve for y: $\ln y = kt + C$	**12.5**	$y = e^{kt + C}$	5.2 Logarithmic functions
Solve for k: $0.5 = e^{5730k}$	**12.5**	$k \approx -0.00012097$	5.3 Exponential equations

- To find certain indefinite integrals

Indefinite Integrals

■ | APPLICATION PREVIEW | ■

In our study of the theory of the firm, we have worked with total cost, total revenue, and profit functions and have found their marginal functions. In practice, it is often easier for a company to measure marginal cost, revenue, and profit and use these data to form marginal functions from which it can find total cost, revenue, and profit functions. For example, Jarus Technologies manufactures motherboards, and the company's sales records show that the marginal revenue (in dollars per unit) for its motherboards is given by

$$\overline{MR} = 300 - 0.2x$$

where x is the number of units sold. If we want to use this function to find the total revenue function for Jarus Technologies' motherboards, we need to find $R(x)$ from $\overline{MR} = R'(x)$. (See Example 7.) In this situation, we need to be able to reverse the process of differentiation. This reverse process is called antidifferentiation, or integration.

Indefinite Integrals   We have studied procedures for and applications of finding derivatives of a given function. We now turn our attention to reversing this process of differentiation. When we know the derivative of a function, the process of finding the function itself is called **antidifferentiation.** For example, if the derivative of a function is $2x$, we know that the function could be $f(x) = x^2$ because $\dfrac{d}{dx}(x^2) = 2x$. But the function could also be $f(x) = x^2 + 4$ because $\dfrac{d}{dx}(x^2 + 4) = 2x$. It is clear that any function of the form $f(x) = x^2 + C$, where C is an arbitrary constant, will have $f'(x) = 2x$ as its derivative. Thus we say that the **general antiderivative** of $f'(x) = 2x$ is $f(x) = x^2 + C$, where C is an arbitrary constant.

The process of finding an antiderivative is also called **integration.** The function that results when integration takes place is called an **indefinite integral** or, more simply, an **integral.** We can denote the indefinite integral (that is, the general antiderivative) of a function $f(x)$ by $\int f(x)\,dx$. Thus we can write $\int 2x\,dx$ to indicate the general antiderivative of the function $f(x) = 2x$. The expression is read as "the integral of $2x$ with respect to x." In this case, $2x$ is called the **integrand.** The **integral sign,** $\int$, indicates the process of integration, and the dx indicates that the integral is to be taken with respect to x. Because the antiderivative of $2x$ is $x^2 + C$, we can write

$$\int 2x\,dx = x^2 + C$$

EXAMPLE   1   **Antidifferentiation**

If $f'(x) = 3x^2$, what is $f(x)$?

Solution

The derivative of the function $f(x) = x^3$ is $f'(x) = 3x^2$. But other functions also have this derivative. They will all be of the form $f(x) = x^3 + C$, where C is a constant. Thus we write

$$\int 3x^2\,dx = x^3 + C$$

EXAMPLE 2 | Integration

If $f'(x) = x^3$, what is $f(x)$?

Solution

We know that $\dfrac{d}{dx}(x^4) = 4x^3$, so the derivative of $f(x) = \frac{1}{4}x^4$ is $f'(x) = x^3$. Thus

$$f(x) = \int f'(x)dx = \int x^3 dx = \frac{1}{4}x^4 + C$$

Powers of x Formula   It is easily seen that

$$\int x^4\, dx = \frac{x^5}{5} + C \text{ because } \frac{d}{dx}\left(\frac{x^5}{5} + C\right) = x^4$$

$$\int x^5\, dx = \frac{x^6}{6} + C \text{ because } \frac{d}{dx}\left(\frac{x^6}{6} + C\right) = x^5$$

In general, we have the following.

Powers of x Formula

$$\int x^n\, dx = \frac{x^{n+1}}{n+1} + C \quad (\text{for } n \neq -1)$$

In the Powers of x Formula, we see that $n \neq -1$ is essential, because if $n = -1$, then the denominator $n + 1 = 0$. We will discuss the case when $n = -1$ later. (Can you think what function has $1/x$ as its derivative?)

In addition, we can see that this Powers of x Formula applies for any $n \neq -1$ by noting that

$$\frac{d}{dx}\left(\frac{x^{n+1}}{n+1} + C\right) = \frac{d}{dx}\left(\frac{1}{n+1}x^{n+1} + C\right) = \frac{n+1}{n+1}x^n = x^n$$

EXAMPLE 3 | Powers of x Formula

Evaluate $\int x^{-1/2}\, dx$.

Solution

Using the formula, we get

$$\int x^{-1/2}dx = \frac{x^{-1/2+1}}{-1/2+1} + C = \frac{x^{1/2}}{1/2} + C = 2x^{1/2} + C$$

We can check by noting that the derivative of $2x^{1/2} + C$ is $x^{-1/2}$. ✔

Note that the indefinite integral in Example 3 is a function (actually a number of functions, one for each value of C). Graphs of several members of this family of functions are shown in Figure 12.1. Note that at any given x-value, the tangent line to each curve has the same slope, indicating that all family members have the same derivative.

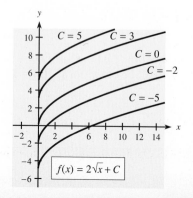

Figure 12.1

EXAMPLE 4 **Powers of x Formula**

Find (a) $\int \sqrt[3]{x}\, dx$ and (b) $\int \frac{1}{x^2}\, dx$.

Solution

(a) $\displaystyle \int \sqrt[3]{x}\, dx = \int x^{1/3}\, dx = \frac{x^{4/3}}{4/3} + C$

$\displaystyle \qquad = \frac{3}{4} x^{4/3} + C = \frac{3}{4} \sqrt[3]{x^4} + C$

(b) We write the power of x in the numerator so that the integral has the form in the formula above.

$$\int \frac{1}{x^2}\, dx = \int x^{-2}\, dx = \frac{x^{-2+1}}{-2+1} + C = \frac{x^{-1}}{-1} + C = \frac{-1}{x} + C$$

Other Formulas and Properties

Other formulas will be useful in evaluating integrals. The following table shows how some new integration formulas result from differentiation formulas.

Integration Formulas

Derivative	Resulting Integral
$\dfrac{d}{dx}(x) = 1$	$\displaystyle \int 1\, dx = \int dx = x + C$
$\dfrac{d}{dx}[c \cdot u(x)] = c \cdot \dfrac{d}{dx} u(x)$	$\displaystyle \int c\, u(x)\, dx = c \int u(x)\, dx$
$\dfrac{d}{dx}[u(x) \pm v(x)] = \dfrac{d}{dx} u(x) \pm \dfrac{d}{dx} v(x)$	$\displaystyle \int [u(x) \pm v(x)]\, dx = \int u(x)\, dx \pm \int v(x)\, dx$

These formulas indicate that we can integrate functions term by term just as we were able to take derivatives term by term.

EXAMPLE 5 **Using Integration Formulas**

Evaluate: (a) $\int 4\, dx$ (b) $\int 8x^5 dx$ (c) $\int (x^3 + 4x)\, dx$

Solution

(a) $\int 4\, dx = 4\int dx = 4(x + C_1) = 4x + C$

(Because C_1 is an unknown constant, we can write $4C_1$ as the unknown constant C.)

(b) $\displaystyle \int 8x^5 dx = 8 \int x^5 dx = 8 \left(\frac{x^6}{6} + C_1 \right) = \frac{4x^6}{3} + C$

(c) $\displaystyle \int (x^3 + 4x)\, dx = \int x^3\, dx + \int 4x\, dx$

$$= \left(\frac{x^4}{4} + C_1 \right) + \left(4 \cdot \frac{x^2}{2} + C_2 \right)$$

$$= \frac{x^4}{4} + 2x^2 + C_1 + C_2$$

$$= \frac{x^4}{4} + 2x^2 + C$$

Note that we need only one constant because the sum of C_1 and C_2 is just a new constant.

EXAMPLE 6

Integral of a Polynomial

Evaluate $\int (x^2 - 4)^2 \, dx$.

Solution

We expand $(x^2 - 4)^2$ so that the integrand is in a form that fits the basic integration formulas.

$$\int (x^2 - 4)^2 \, dx = \int (x^4 - 8x^2 + 16) \, dx = \frac{x^5}{5} - \frac{8x^3}{3} + 16x + C$$

✓ CHECKPOINT

1. True or false:
   (a) $\int (4x^3 - 2x) \, dx = \int 4x^3 \, dx - \int 2x \, dx$
   $$= (x^4 + C) - (x^2 + C) = x^4 - x^2$$
   (b) $\int \dfrac{1}{3x^2} \, dx = \dfrac{1}{3(x^3/3)} + C = \dfrac{1}{x^3} + C$

2. Evaluate $\int (2x^3 + x^{-1/2} - 4x^{-5}) \, dx$.

EXAMPLE 7

Revenue | APPLICATION PREVIEW |

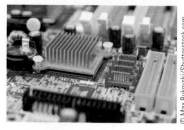
© Max Bukovski/Shutterstock.com

Sales records at Jarus Technologies show that the rate of change of the revenue (that is, the marginal revenue) in dollars per unit for a motherboard is $\overline{MR} = 300 - 0.2x$, where x represents the quantity sold. Find the total revenue function for the product. Then find the total revenue from the sale of 1000 motherboards.

Solution

We know that the marginal revenue can be found by differentiating the total revenue function. That is,

$$R'(x) = 300 - 0.2x$$

Thus integrating the marginal revenue function gives the total revenue function.

$$R(x) = \int (300 - 0.2x) \, dx = 300x - 0.1x^2 + K^*$$

We can use the fact that there is no revenue when no units are sold to evaluate K. Setting $x = 0$ and $R = 0$ gives $0 = 300(0) - 0.1(0)^2 + K$, so $K = 0$. Thus the total revenue function is

$$R(x) = 300x - 0.1x^2$$

The total revenue from the sale of 1000 motherboards is

$$R(1000) = 300(1000) - 0.1(1000^2) = \$200,000$$

We can check that the $R(x)$ we found in Example 7 is correct by verifying that $R'(x) = 300 - 0.2x$ and $R(0) = 0$. Also, graphs can help us check the reasonableness of our result. Figure 12.2 shows the graphs of $\overline{MR} = 300 - 0.2x$ and of the $R(x)$ we found. Note that $R(x)$ passes through the origin, indicating $R(0) = 0$. Also, reading both graphs from left to right, we see that $R(x)$ increases when $\overline{MR} > 0$, attains its maximum when $\overline{MR} = 0$, and decreases when $\overline{MR} < 0$.

---

*Here we are using K rather than C to represent the constant of integration to avoid confusion between the constant C and the cost function $C = C(x)$.

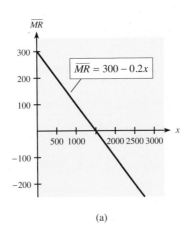

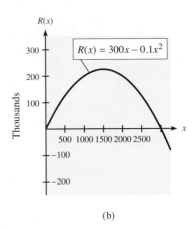

Figure 12.2   (a)   (b)

Calculator Note   We mentioned in Chapter 9, "Derivatives," that graphing calculators have a numerical derivative feature that can be used to check graphically the derivative of a function that has been calculated with a formula. We can also use the numerical integration feature on graphing calculators to check our integration (if we assume temporarily that the constant of integration is 0). We do this by graphing the integral calculated with a formula and the numerical integral from the graphing calculator on the same set of axes. If the graphs lie on top of one another, the integrals agree. Figure 12.3 illustrates this for the function $f(x) = 3x^2 - 2x + 1$. Its integral, with the constant of integration set equal to 0, is shown as $y_1 = x^3 - x^2 + x$ in Figure 12.3(a), and the graphs of the two integrals are shown in Figure 12.3(b). Of course, it is often easier to use the derivative to check integration. See Appendix A, Section 12.1, for details.

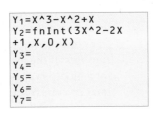

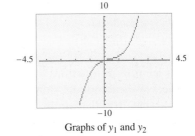

Figure 12.3   (a)   (b)

1. (a) False; $\int (4x^3 - 2x)\, dx = x^4 - x^2 + C$

   (b) False; $\int \dfrac{1}{3x^2}\, dx = \dfrac{-1}{3x} + C$

2. $\int (2x^3 + x^{-1/2} - 4x^{-5})\, dx = \dfrac{x^4}{2} + 2x^{1/2} + x^{-4} + C$

| EXERCISES | 12.1

1. If $f'(x) = 4x^3$, what is $f(x)$?
2. If $f'(x) = 5x^4$, what is $f(x)$?
3. If $f'(x) = x^6$, what is $f(x)$?
4. If $g'(x) = x^4$, what is $g(x)$?

Evaluate the integrals in Problems 5–28. Check your answers by differentiating.

5. $\int x^7 dx$
6. $\int x^5\, dx$
7. $\int 8x^3 dx$
8. $\int 16x^9\, dx$

9. $\int (3^3 + x^{13})\, dx$
10. $\int (5^2 + x^{10})\, dx$
11. $\int (3 - x^{3/2})\, dx$
12. $\int (8 + x^{2/3})\, dx$
13. $\int (x^4 - 9x^2 + 3)\, dx$
14. $\int (3x^2 - 4x - 4)\, dx$
15. $\int (13 - 6x + 21x^6)\, dx$
16. $\int (12x^5 + 12x^3 - 7)\, dx$
17. $\int (2 + 2\sqrt{x})\, dx$
18. $\int (17 + \sqrt{x^3})\, dx$
19. $\int 6\sqrt[4]{x}\, dx$
20. $\int 3\sqrt[3]{x^2}\, dx$
21. $\int \dfrac{5}{x^4}\, dx$
22. $\int \dfrac{6}{x^5}\, dx$

23. $\int \dfrac{dx}{2\sqrt[3]{x^2}}$

24. $\int \dfrac{2\,dx}{5\sqrt{x^3}}$

25. $\int \left(x^3 - 4 + \dfrac{5}{x^6}\right) dx$

26. $\int \left(x^3 - 7 - \dfrac{3}{x^4}\right) dx$

27. $\int \left(x^9 - \dfrac{1}{x^3} + \dfrac{2}{\sqrt[3]{x}}\right) dx$

28. $\int \left(3x^8 + \dfrac{4}{x^8} - \dfrac{5}{\sqrt[5]{x}}\right) dx$

In Problems 29–32, use algebra to rewrite the integrands; then integrate and simplify.

29. $\int (4x^2 - 1)^2 x^3\,dx$

30. $\int (x^3 + 1)^2 x\,dx$

31. $\int \dfrac{x + 1}{x^3}\,dx$

32. $\int \dfrac{x - 3}{\sqrt{x}}\,dx$

In Problems 33 and 34, find the antiderivatives and graph the resulting family members that correspond to $C = 0$, $C = 4$, $C = -4$, $C = 8$, and $C = -8$.

33. $\int (2x + 3)\,dx$

34. $\int (4 - x)\,dx$

35. If $\int f(x)\,dx = 2x^9 - 7x^5 + C$, find $f(x)$.

36. If $\int g(x)\,dx = 11x^{10} - 4x^3 + C$, find $g(x)$.

In each of Problems 37–40, a family of functions is given and graphs of some members of the family are shown. Write the indefinite integral that gives the family.

37. $F(x) = 5x - \dfrac{x^2}{4} + C$

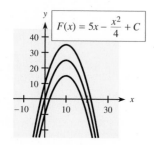

38. $F(x) = \dfrac{x^2}{2} + 3x + C$

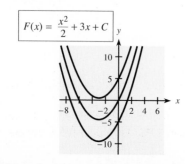

39. $F(x) = x^3 - 3x^2 + C$

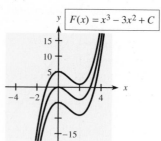

40. $F(x) = 12x - x^3 + C$

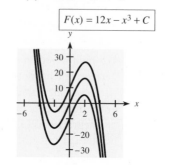

APPLICATIONS

41. **Revenue** If the marginal revenue (in dollars per unit) for a month for a commodity is $\overline{MR} = -0.4x + 30$, find the total revenue function.

42. **Revenue** If the marginal revenue (in dollars per unit) for a month for a commodity is $\overline{MR} = -0.05x + 25$, find the total revenue function.

43. **Revenue** If the marginal revenue (in dollars per unit) for a month is given by $\overline{MR} = -0.3x + 450$, what is the total revenue from the production and sale of 50 units?

44. **Revenue** If the marginal revenue (in dollars per unit) for a month is given by $\overline{MR} = -0.006x + 36$, find the total revenue from the sale of 75 units.

45. **Stimulus-response** Suppose that when a sense organ receives a stimulus at time t, the total number of action potentials is $P(t)$. If the rate at which action potentials are produced is $t^3 + 4t^2 + 6$, and if there are 0 action potentials when $t = 0$, find the formula for $P(t)$.

46. **Projectiles** Suppose that a particle has been shot into the air in such a way that the rate at which its height is changing is $v = 320 - 32t$, in feet per second, and suppose that it is 1600 feet high when $t = 10$ seconds. Write the equation that describes the height of the particle at any time t.

47. **Pollution** A factory is dumping pollutants into a river at a rate given by $dx/dt = t^{3/4}/600$ tons per week, where t is the time in weeks since the dumping began and x is the number of tons of pollutants.
    (a) Find the equation for total tons of pollutants dumped.
    (b) How many tons were dumped during the first year?

48. **Population growth** The rate of growth of the population of a city is predicted to be
$$\dfrac{dp}{dt} = 1000t^{1.08}$$

where p is the population at time t and t is measured in years from the present. Suppose that the current population is 100,000. What is the predicted
(a) rate of growth 5 years from the present?
(b) population 5 years from the present?

49. *Average cost* The DeWitt Company has found that the rate of change of its average cost for a product is

$$\overline{C}'(x) = \frac{1}{4} - \frac{100}{x^2}$$

where x is the number of units and cost is in dollars. The average cost of producing 20 units is $40.00.
(a) Find the average cost function for the product.
(b) Find the average cost of 100 units of the product.

50. *Oil leakage* An oil tanker hits a reef and begins to leak. The efforts of the workers repairing the leak cause the rate at which the oil is leaking to decrease. The oil was leaking at a rate of 31 barrels per hour at the end of the first hour after the accident, and the rate is decreasing at a rate of one barrel per hour.
(a) What function describes the rate of loss?
(b) How many barrels of oil will leak in the first 6 hours?
(c) When will the oil leak be stopped? How much will have leaked altogether?

51. *Health expenditures per capita* National health expenditures per capita E (in dollars) have risen dramatically since 2000. By using data from the Centers for Medicare and Medicaid Services from 2006 and projected to 2021, the rate of change of health expenditures per capita can be modeled by

$$\frac{dE}{dt} = 41.22t - 116.4$$

dollars per year, where t is the number of years past 2000.
(a) Find the function that models the national health expenditures per capita if these expenditures were $9808 in 2014. Use four significant digits.
(b) Use the model from part (a) to predict the national health expenditures per capita for 2020.

52. *Total taxable payroll* By using Social Security Administration data for selected years from 2012 and projected to 2050, the rate of change of the total U.S. taxable payroll can be modeled by

$$\frac{dP}{dt} = 27.0t + 112$$

billions of dollars per year, where t is the number of years past 2010.
(a) If total U.S. taxable payroll is $7088 billion in 2016, find the function $P(t)$. Use three significant digits.
(b) What does $P(t)$ predict for the total U.S. taxable payroll in 2025?

53. *Wind chill* When the air temperature is 20° F, the rate of change of the wind chill temperature t (°F) is given by

$$\frac{dt}{dw} = -4.352w^{-0.84}$$

where w is wind speed in miles per hour.
(a) Will both the rate of change of wind chill temperature and wind chill temperature decrease as the windspeed increases? Explain.
(b) If the wind chill temperature is 1°F when the wind speed is 31 mph and the air temperature is 20°F, find the function that models the wind chill temperature.

54. *Severe weather ice makers* Hail is produced in severe thunderstorms when an updraft draws moist surface air into subfreezing air above 10,000 feet. The speed of the updraft s, in mph, affects the diameter of hail (in inches). According to National Weather Service data, the rate of change of hail size with respect to updraft speed is

$$\frac{dh}{ds} = 0.001144s^{0.922}$$

inches of diameter per mph of updraft.
(a) When updraft speeds approach 60 mph, hail is golf-ball-sized. Use $h = 1.5$ and $s = 60$ to find a model for $h(s)$.
(b) Use the model to find the hail size for an updraft speed of 100 mph.

55. *U.S. population* With U.S. Census Bureau data (actual and projected) for selected years from 1960 to 2050, the rate of change of U.S. population P can be modeled by

$$\frac{dP}{dt} = -0.0002187t^2 + 0.0276t + 1.98$$

million people per year, where $t = 0$ represents 1960.
(a) In what year does this rate of change reach its maximum?
(b) In 1960, the U.S. population was 181 million. Use this to find a model for $P(t)$.
(c) For 2025, the Census Bureau's predicted U.S. population is 348 million. What does the model predict?

56. *Consumer prices* The consumer price index (CPI) measures how prices have changed for consumers. With 1995 as a reference of 100, a year with CPI = 150 indicates that consumer costs in that year were 1.5 times the 1995 costs. With U.S. Department of Labor data for selected years from 1995 and projected to 2050, the rate of change of the CPI can be modeled by

$$\frac{dC}{dt} = 0.009t^2 - 0.096t + 4.85$$

dollars per year, where $t = 0$ represents 1990.
(a) Find the function that models $C(t)$, if the CPI was 175 in 2010.
(b) What does the model from part (a) predict for the consumer costs in 2030? How does this compare to 2010?

OBJECTIVE   12.2

- To evaluate integrals of the form $\int u^n \cdot u' \, dx = \int u^n \, du$ if $n \neq -1$

The Power Rule

▮ | APPLICATION PREVIEW |

In the previous section, we saw that total revenue could be found by integrating marginal revenue. That is,

$$R(x) = \int \overline{MR} \, dx$$

For example, if the marginal revenue for a product is given by

$$\overline{MR} = \frac{600}{\sqrt{3x + 1}} + 2$$

then

$$R(x) = \int \left[\frac{600}{\sqrt{3x + 1}} + 2 \right] dx$$

To evaluate this integral, however, we need a more general formula than the Powers of x Formula. (See Example 8.)

In this section, we will extend the Powers of x Formula to a rule for powers of a function of x.

Differentials   Our goal in this section is to extend the Powers of x Formula,

$$\int x^n \, dx = \frac{x^{n+1}}{n+1} + C, \quad n \neq -1$$

to powers of a function of x. In order to do this, we must understand the symbol dx.

Recall from Section 9.3, "Rates of Change and Derivatives," that the derivative of $y = f(x)$ with respect to x can be denoted by dy/dx. As we will see, there are advantages to using dy and dx as separate quantities whose ratio dy/dx equals $f'(x)$.

Differentials   If $y = f(x)$ is a differentiable function with derivative $dy/dx = f'(x)$, then the **differential of x** is dx, and the **differential of y** is dy, where

$$dy = f'(x) \, dx$$

Although differentials are useful in certain approximation problems, we are interested in the differential notation at this time.

EXAMPLE 1   Differentials

Find (a) dy if $y = x^3 - 4x^2 + 5$ and (b) du if $u = 5x^4 + 11$.

Solution

(a) $dy = f'(x) \, dx = (3x^2 - 8x) \, dx$

(b) If the dependent variable in a function is u, then $du = u'(x) \, dx$.

$$du = u'(x) \, dx = 20x^3 \, dx$$ ▮

The Power Rule   In terms of our goal of extending the Powers of x Formula, we would suspect that if x is replaced by a function of x, then dx should be replaced by the differential of that function. Let's see whether this is true.

Recall that if $y = [u(x)]^n$, the derivative of y is

$$\frac{dy}{dx} = n[u(x)]^{n-1} \cdot u'(x)$$

Using this formula for derivatives, we can see that

$$\int n[u(x)]^{n-1} \cdot u'(x)\, dx = [u(x)]^n + C$$

It is easy to see that this formula is equivalent to the following formula, which is called the **Power Rule for Integration.**

Power Rule for Integration	$$\int [u(x)]^n \cdot u'(x)\, dx = \frac{[u(x)]^{n+1}}{n+1} + C, \quad n \neq -1$$

Using the fact that

$$du = u'(x)\, dx \quad \text{or} \quad du = u'dx$$

we can write the Power Rule in the following alternative form.

Power Rule (Alternative Form)	If $u = u(x)$, then $$\int u^n\, du = \frac{u^{n+1}}{n+1} + C, \quad n \neq -1$$

Note that this formula has the same form as the formula

$$\int x^n\, dx = \frac{x^{n+1}}{n+1} + C, \quad n \neq -1$$

with the *function u substituted for x and du substituted for dx.*

EXAMPLE 2   Power Rule

Evaluate $\int (3x^2 + 4)^5 \cdot 6x\, dx$.

Solution
To use the Power Rule, we must be sure that we have the function $u(x)$, its derivative $u'(x)$, and n.

$$u = 3x^2 + 4, \quad n = 5$$
$$u' = 6x$$

All required parts are present, so the integral is of the form

$$\int (3x^2 + 4)^5\, 6x\, dx = \int u^5 \cdot u'dx = \int u^5\, du$$
$$= \frac{u^6}{6} + C = \frac{(3x^2 + 4)^6}{6} + C$$

We can check the integration by noting that the derivative of

$$\frac{(3x^2 + 4)^6}{6} + C \text{ is } (3x^2 + 4)^5 \cdot 6x$$

EXAMPLE 3   Power Rule

Evaluate $\int \sqrt{2x + 3} \cdot 2 \, dx$.

Solution

If we let $u = 2x + 3$, then $u' = 2$, and so we have

$$\int \sqrt{2x + 3} \cdot 2 \, dx = \int \sqrt{u} \, u' \, dx = \int \sqrt{u} \, du$$

$$= \int u^{1/2} \, du = \frac{u^{3/2}}{3/2} + C$$

Because $u = 2x + 3$, we have

$$\int \sqrt{2x + 3} \cdot 2 \, dx = \frac{2}{3}(2x + 3)^{3/2} + C$$

Check: The derivative of $\frac{2}{3}(2x + 3)^{3/2} + C$ is $(2x + 3)^{1/2} \cdot 2$. ✔

Some members of the family of functions given by

$$\int \sqrt{2x + 3} \cdot 2 \, dx = \frac{2}{3}(2x + 3)^{3/2} + C$$

are shown in Figure 12.4. Note from the graphs that the domain of each function is $x \geq -3/2$. This is because $2x + 3$ must be nonnegative so that $(2x + 3)^{3/2} = (\sqrt{2x + 3})^3$ is a real number.

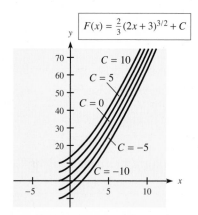

Figure 12.4

EXAMPLE 4   Power Rule

Evaluate $\int x^3(5x^4 + 11)^9 \, dx$.

Solution

If we let $u = 5x^4 + 11$, then $u' = 20x^3$. Thus we do not have an integral of the form $\int u^n \cdot u' \, dx$, as we had in Example 2 and Example 3; the factor 20 is not in the integrand. To get the integrand in the correct form, we can multiply by 20 and divide it out as follows:

$$\int x^3(5x^4 + 11)^9 \, dx = \int (5x^4 + 11)^9 \cdot x^3 \, dx = \int (5x^4 + 11)^9 \left(\frac{1}{20}\right)(20x^3) \, dx$$

Because $\frac{1}{20}$ is a constant factor, we can factor it outside the integral sign, getting

$$\int (5x^4 + 11)^9 \cdot x^3 \, dx = \frac{1}{20} \int (5x^4 + 11)^9 (20x^3) \, dx$$

$$= \frac{1}{20} \int u^9 \cdot u' \, dx = \frac{1}{20} \cdot \frac{u^{10}}{10} + C$$

$$= \frac{1}{200}(5x^4 + 11)^{10} + C$$

EXAMPLE 5 Power Rule

Evaluate $\int 5x^2\sqrt{x^3-4}\,dx$.

Solution

If we let $u = x^3 - 4$, then $u' = 3x^2$. Thus we need the factor 3, rather than 5, in the integrand. If we first reorder the factors and then multiply by the constant factor 3 (and divide it out), we have

$$\int \sqrt{x^3-4}\cdot 5x^2\,dx = \int \sqrt{x^3-4}\cdot\frac{5}{3}(3x^2)\,dx$$

$$= \frac{5}{3}\int (x^3-4)^{1/2}\cdot 3x^2\,dx$$

This integral is of the form $\frac{5}{3}\int u^{1/2}\cdot u'\,dx$, resulting in

$$\frac{5}{3}\cdot\frac{u^{3/2}}{3/2} + C = \frac{5}{3}\cdot\frac{(x^3-4)^{3/2}}{3/2} + C = \frac{10}{9}(x^3-4)^{3/2} + C$$

Note that we can factor a constant outside the integral sign to obtain the integrand in the form we seek. But if a variable must be factored outside the integral to obtain the form $u^n\cdot u'\,dx$, we *cannot* use this form and must try something else.

EXAMPLE 6 Power Rule Fails

Evaluate $\int (x^2+4)^2\,dx$.

Solution

If we let $u = x^2 + 4$, then $u' = 2x$. Because we would have to introduce a variable to get u' in the integral, we cannot solve this problem by using the Power Rule. We must find another method. We can evaluate this integral by squaring and then integrating term by term.

$$\int (x^2+4)^2\,dx = \int (x^4+8x^2+16)\,dx$$

$$= \frac{x^5}{5} + \frac{8x^3}{3} + 16x + C$$

Note that if we tried to introduce the factor $2x$ into the integral in Example 6, we would get

$$\int (x^2+4)^2\,dx = \int (x^2+4)^2\cdot\frac{1}{2x}(2x)\,dx$$

Although it is tempting to factor $1/2x$ outside the integral and use the Power Rule, this leads to an "answer" that does not check. That is, the derivative of the "answer" is not the integrand. (Try it and see.) To emphasize again, we can introduce only *a constant factor* to get an integral in the proper form.

EXAMPLE 7 Power Rule

Evaluate:

(a) $\int (2x^2-4x)^2(x-1)\,dx$ (b) $\int \frac{x^2-1}{(x^3-3x)^3}\,dx$

Solution

(a) If we want to treat this as an integral of the form $\int u^n u'\,dx$, we will have to let $u = 2x^2 - 4x$. Then $u' = 4x - 4$. Multiplying and dividing by 4 will give us this form.

$$\int (2x^2 - 4x)^2 (x - 1)\, dx = \int (2x^2 - 4x)^2 \cdot \frac{1}{4} \cdot 4(x - 1)\, dx$$

$$= \frac{1}{4} \int (2x^2 - 4x)^2 (4x - 4)\, dx$$

$$= \frac{1}{4} \int u^2 u'\, dx = \frac{1}{4} \cdot \frac{u^3}{3} + C$$

$$= \frac{1}{4} \frac{(2x^2 - 4x)^3}{3} + C$$

$$= \frac{1}{12}(2x^2 - 4x)^3 + C$$

(b) This integral can be treated as $\int u^{-3} u'\, dx$ if we let $u = x^3 - 3x$.

$$\int \frac{x^2 - 1}{(x^3 - 3x)^3}\, dx = \int (x^3 - 3x)^{-3}(x^2 - 1)\, dx$$

Then $u' = 3x^2 - 3$ and we can multiply and divide by 3 to get the form we need.

$$= \int (x^3 - 3x)^{-3} \cdot \frac{1}{3} \cdot 3(x^2 - 1)\, dx$$

$$= \frac{1}{3} \int (x^3 - 3x)^{-3}(3x^2 - 3)\, dx$$

$$= \frac{1}{3} \left[\frac{(x^3 - 3x)^{-2}}{-2} \right] + C$$

$$= \frac{-1}{6(x^3 - 3x)^2} + C$$

✓ **CHECKPOINT**

1. Which of the following can be evaluated with the Power Rule?
    (a) $\int (4x^2 + 1)^{10}(8x\, dx)$
    (b) $\int (4x^2 + 1)^{10}(x\, dx)$
    (c) $\int (4x^2 + 1)^{10}(8\, dx)$
    (d) $\int (4x^2 + 1)^{10}\, dx$

2. Which of the following is equal to $\int (2x^3 + 5)^{-2}(6x^2\, dx)$?
    (a) $\dfrac{[(2x^4)/4 + 5x]^{-1}}{-1} \cdot \dfrac{6x^3}{3} + C$
    (b) $\dfrac{(2x^3 + 5)^{-1}}{-1} \cdot \dfrac{6x^3}{3} + C$
    (c) $\dfrac{(2x^3 + 5)^{-1}}{-1} + C$

3. True or false: Constants can be factored outside the integral sign.
4. Evaluate the following.
    (a) $\int (x^3 + 9)^5(3x^2\, dx)$
    (b) $\int (x^3 + 9)^{15}(x^2\, dx)$
    (c) $\int (x^3 + 9)^2(x\, dx)$

EXAMPLE 8   **Revenue** | APPLICATION PREVIEW |

Suppose that the marginal revenue for a product is given by

$$\overline{MR} = \frac{600}{\sqrt{3x + 1}} + 2$$

Find the total revenue function.

Solution

$$R(x) = \int \overline{MR}\, dx = \int \left[\frac{600}{(3x + 1)^{1/2}} + 2 \right] dx$$

$$= \int 600(3x + 1)^{-1/2}\, dx + \int 2\, dx$$

$$= 600 \left(\frac{1}{3} \right) \int (3x + 1)^{-1/2}(3\, dx) + 2 \int dx$$

$$= 200 \frac{(3x + 1)^{1/2}}{1/2} + 2x + K$$

$$= 400 \sqrt{3x + 1} + 2x + K$$

We know that $R(0) = 0$, so we have

$$0 = 400 \sqrt{1} + 0 + K \quad \text{or} \quad K = -400$$

Thus the total revenue function is

$$R(x) = 400 \sqrt{3x + 1} + 2x - 400$$

Note in Example 8 that even though $R(0) = 0$, the constant of integration K was *not* 0. This is because $x = 0$ does not necessarily mean that $u(x)$ will also be 0.

The formulas we have stated and used in this and the previous section are all the result of "reversing" derivative formulas. We summarize in the following box.

Integration Formula		Derivative Formula
$\int dx = x + C$	because	$\dfrac{d}{dx}(x + C) = 1$
$\int x^n\, dx = \dfrac{x^{n+1}}{n + 1} + C\ (n \neq -1)$	because	$\dfrac{d}{dx}\left(\dfrac{x^{n+1}}{n + 1} + C \right) = x^n$
$\int [u(x) + v(x)]\, dx = \int u(x)\, dx + \int v(x)\, dx$	because	$\dfrac{d}{dx}[u(x) + v(x)] = \dfrac{du}{dx} + \dfrac{dv}{dx}$
$\int [u(x) - v(x)]\, dx = \int u(x)\, dx - \int v(x)\, dx$	because	$\dfrac{d}{dx}[u(x) - v(x)] = \dfrac{du}{dx} - \dfrac{dv}{dx}$
$\int cf(x)\, dx = c \int f(x)\, dx$	because	$\dfrac{d}{dx}[cf(x)] = c\left(\dfrac{df}{dx} \right)$
$\int u^n u'\, dx = \dfrac{u^{n+1}}{n + 1} + C\ (n \neq -1)$	because	$\dfrac{d}{dx}\left(\dfrac{u^{n+1}}{n + 1} + C \right) = u^n \cdot u'$

Note that there are no integration formulas that correspond to "reversing" the derivative formulas for a product or for a quotient. This means that functions that may be easy to differentiate can be quite difficult (or even impossible) to integrate. Hence, in general, integration is a more difficult process than differentiation. In fact, some functions whose derivatives can be readily found cannot be integrated, such as

$$f(x) = \sqrt{x^3 + 1} \ \text{ and } \ g(x) = \frac{2x^2}{x^4 + 1}$$

In the next section, we will add to this list of basic integration formulas by "reversing" the derivative formulas for exponential and logarithmic functions.

1. Expressions (a) and (b) can be evaluated with the Power Rule. Expressions (c) and (d) do not fit the format of the Power Rule.

2. (c) $\int (2x^3 + 5)^{-2}(6x^2)\, dx = -(2x^3 + 5)^{-1} + C$

3. True

4. (a) $\int (x^3 + 9)^5(3x^2\, dx) = \dfrac{(x^3 + 9)^6}{6} + C$

   (b) $\int (x^3 + 9)^{15}(x^2\, dx) = \dfrac{(x^3 + 9)^{16}}{48} + C$

   (c) $\int (x^3 + 9)^2(x\, dx) = \int (x^6 + 18x^3 + 81)x\, dx = \dfrac{x^8}{8} + \dfrac{18x^5}{5} + \dfrac{81x^2}{2} + C$

| EXERCISES | 12.2

In Problems 1 and 2, find du.

1. $u = 2x^5 + 9$

2. $u = 3x^4 - 4x^3$

In each of Problems 3 and 4, one of parts (a) and (b) can be integrated with the Power Rule and the other cannot. Integrate the part that can be done with the Power Rule, and explain why the Power Rule cannot be used to evaluate the other.

3. (a) $\int (3x^4 - 7)^{12}\,(12x\, dx)$ (b) $\int (5x^3 + 11)^7\,(15x^2\, dx)$

4. (a) $\int 5(6 + 5x)^{10}\, dx$ (b) $\int 14x^2(2x^7 + 9)^6\, dx$

Evaluate the integrals in Problems 5–34. Check your results by differentiation.

5. $\int (x^2 + 3)^3\, 2x\, dx$

6. $\int (3x^3 + 1)^4\, 9x^2\, dx$

7. $\int (5x^3 + 11)^4\, 15x^2\, dx$

8. $\int (8x^4 + 5)^3\,(32x^3)\, dx$

9. $\int (3x - x^3)^2\,(3 - 3x^2)\, dx$

10. $\int (4x^2 - 3x)^4\,(8x - 3)\, dx$

11. $\int 4x^3(7x^4 + 12)^3\, dx$

12. $\int 9x^5(3x^6 - 4)^6\, dx$

13. $\int 7(4x - 1)^6\, dx$

14. $\int 3(5 - x)^{-3}\, dx$

15. $\int 8x^5(4x^6 + 15)^{-3}\, dx$

16. $\int 5x^3(3x^4 + 7)^{-4}\, dx$

17. $\int (x - 1)(x^2 - 2x + 5)^4\, dx$

18. $\int (2x^3 - x)(x^4 - x^2)^6\, dx$

19. $\int 2(x^3 - 1)(x^4 - 4x + 3)^{-5}\, dx$

20. $\int 3(x^5 - 2x)(x^6 - 6x^2 + 7)^{-2}\, dx$

21. $\int 7x^3\sqrt{x^4 + 6}\, dx$

22. $\int 3x\sqrt{5 - x^2}\, dx$

23. $\int (x^3 + 1)^2\,(3x\, dx)$

24. $\int (x^2 - 5)^2\,(2x^2\, dx)$

25. $\int (3x^4 - 1)^2\, 12x\, dx$

26. $\int (2x^4 + 3)^2\,(8x\, dx)$

27. $\int \sqrt{x^3 - 3x}\,(x^2 - 1)\, dx$

28. $\int \sqrt[3]{x^2 + 2x}\,(x + 1)\, dx$

29. $\displaystyle\int \frac{3x^4\, dx}{(2x^5 - 5)^4}$

30. $\displaystyle\int \frac{5x^3\, dx}{(x^4 - 8)^3}$

31. $\displaystyle\int \frac{x^3 - 1}{(x^4 - 4x)^3}\, dx$

32. $\displaystyle\int \frac{3x^5 - 2x^3}{(x^6 - x^4)^5}\, dx$

33. $\displaystyle\int \frac{x^2 - 4x}{\sqrt{x^3 - 6x^2 + 2}}\, dx$

34. $\displaystyle\int \frac{x^2 + 1}{\sqrt{x^3 + 3x + 10}}\, dx$

35. If $\int f(x)\, dx = (7x - 13)^{10} + C$, find $f(x)$.

36. If $\int g(x)\, dx = (5x^2 + 2)^6 + C$, find $g(x)$.

In Problems 37 and 38, (a) evaluate each integral and (b) graph the members of the solution family for $C = -5$, $C = 0$, and $C = 5$.

37. $\int x(x^2 - 1)^3\, dx$

38. $\int (3x - 11)^{1/3}\, dx$

Each of Problems 39 and 40 has the form $\int f(x)\, dx$.
(a) Evaluate each integral to obtain a family of functions.
(b) Find and graph the family member that passes through the point $(0, 2)$. Call that function $F(x)$.
(c) Find any x-values where $f(x)$ is not defined but $F(x)$ is.
(d) At the x-values found in part (c), what kind of tangent line does $F(x)$ have?

39. $\displaystyle\int \frac{3\, dx}{(2x - 1)^{3/5}}$

40. $\displaystyle\int \frac{x^2\, dx}{(x^3 - 1)^{1/3}}$

In each of Problems 41 and 42, a family of functions is given, together with the graphs of some functions in the family. Write the indefinite integral that gives the family.

41. $F(x) = (x^2 - 1)^{4/3} + C$

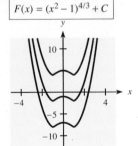

$F(x) = (x^2 - 1)^{4/3} + C$

42. $F(x) = 54(4x^2 + 9)^{-1} + C$

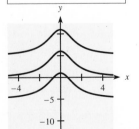

$F(x) = 54(4x^2 + 9)^{-1} + C$

In parts (a)–(c) of Problems 43 and 44, three integrals are given. Integrate those that can be done by the methods studied so far. Additionally, as part (d), give your own example of an integral that looks as though it might use the Power Rule but that cannot be integrated by using methods studied so far.

43. (a) $\displaystyle\int \frac{7x^3\,dx}{(x^3+4)^2}$ (b) $\displaystyle\int \frac{7x^2\,dx}{(x^3+4)^2}$

   (c) $\displaystyle\int \frac{\sqrt{x^2+4}}{x}\,dx$

44. (a) $\displaystyle\int \frac{(2x^5+1)^{7/2}}{3x^4}\,dx$ (b) $\displaystyle\int 10x(2x^5+1)^{7/2}\,dx$

   (c) $\displaystyle\int \frac{5x^3}{(2x^4+1)^{7/2}}\,dx$

APPLICATIONS

45. *Revenue* Suppose that the marginal revenue for a product is given by

$$\overline{MR} = \frac{-30}{(2x+1)^2} + 30$$

   where x is the number of units and revenue is in dollars. Find the total revenue.

46. *Revenue* The marginal revenue for a new calculator is given by

$$\overline{MR} = 60,000 - \frac{40,000}{(10+x)^2}$$

   where x represents hundreds of calculators and revenue is in dollars. Find the total revenue function for these calculators.

47. *Physical productivity* The total physical output of a number of machines or workers is called *physical productivity* and is a function of the number of machines or workers. If $P = f(x)$ is the productivity, dP/dx is the marginal physical productivity. If the marginal physical productivity for bricklayers is $dP/dx = 90(x+1)^2$, where P is the number of bricks laid per day and x is the number of bricklayers, find the physical productivity of 4 bricklayers. (*Note:* $P = 0$ when $x = 0$.)

48. *Production* The rate of production of a new line of products is given by

$$\frac{dx}{dt} = 200\left[1 + \frac{400}{(t+40)^2}\right]$$

   where x is the number of items and t is the number of weeks the product has been in production.
   (a) Assuming that $x = 0$ when $t = 0$, find the total number of items produced as a function of time t.
   (b) How many items were produced in the fifth week?

49. *Data entry speed* The rate of change in data entry speed of the average student is $ds/dx = 5(x+1)^{-1/2}$, where x is the number of lessons the student has had and s is in entries per minute.

(a) Find the data entry speed as a function of the number of lessons if the average student can complete 10 entries per minute with no lessons ($x = 0$).
(b) How many entries per minute can the average student complete after 24 lessons?

50. *Productivity* Because a new employee must learn an assigned task, production will increase with time. Suppose that for the average new employee, the rate of performance is given by

$$\frac{dN}{dt} = \frac{1}{2\sqrt{t+1}}$$

   where N is the number of units completed t hours after beginning a new task. If 2 units are completed after 3 hours, how many units are completed after 8 hours?

51. *Film attendance* An excellent film with a very small advertising budget must depend largely on word-of-mouth advertising. In this case, the rate at which weekly attendance might grow can be given by

$$\frac{dA}{dt} = \frac{-100}{(t+10)^2} + \frac{2000}{(t+10)^3}$$

   where t is the time in weeks since release and A is attendance in millions.
   (a) Find the function that describes weekly attendance at this film.
   (b) Find the attendance at this film in the tenth week.

52. *Product quality and advertising* An inferior product with a large advertising budget does well when it is introduced, but sales decline as people discontinue use of the product. Suppose that the rate of weekly sales revenue is given by

$$S'(t) = \frac{400}{(t+1)^3} - \frac{200}{(t+1)^2}$$

   where S is sales in thousands of dollars and t is time in weeks.
   (a) Find the function that describes the weekly sales.
   (b) Find the sales for the first week and the ninth week.

53. *Demographics* Because of job outsourcing, a Pennsylvania town predicts that its public school population will decrease at the rate

$$\frac{dN}{dx} = \frac{-300}{\sqrt{x+9}}$$

   where x is the number of years and N is the total school population. If the present population ($x = 0$) is 8000, what population size is expected in 7 years?

54. *Franchise growth* A new fast-food firm predicts that the number of franchises for its products will grow at the rate

$$\frac{dn}{dt} = 9\sqrt{t+1}$$

   where t is the number of years, $0 \le t \le 10$. If there is one franchise ($n = 1$) at present ($t = 0$), how many franchises are predicted for 8 years from now?

55. **Obesity** The rate of change of the percent of obese Americans who are severely obese can be modeled by the function

$$f'(x) = \frac{375}{(0.743x + 6.97)^2}$$

percentage points per year, where x is the number of years after 1980 (*Source: American Journal of Preventive Medicine* 42, June 2012).

(a) If 8.3% of obese Americans were severely obese in 1995, find the function $f(x)$ that gives the percent of obese Americans who are severely obese. Use three significant digits.

(b) Use the function from part (a) to predict the percent of obese American adults who will be severely obese in 2025.

56. **Social Security beneficiaries** Suppose the rate of change of the number of Social Security beneficiaries (in millions per year) can be modeled by

$$\frac{dB}{dt} = 0.07149(0.1t + 1)^2 - 0.67114(0.1t + 1) + 2.2016$$

where t is the number of years past 1950.

Year	Number of Beneficiaries (millions)	Year	Number of Beneficiaries (millions)
1950	2.9	2000	44.8
1960	14.3	2010	53.3
1970	25.2	2020	68.8
1980	35.1	2030	82.7
1990	39.5		

Source: Social Security Administration

(a) Use integration and the data point for 2000 to find the function $B(t)$ that models the millions of Social Security beneficiaries.

(b) The data in the table give the millions of Social Security beneficiaries for selected years from 1950 and projected to 2030. Graph $B(t)$ from part (a) with the data in the table; let $t = 0$ represent 1950.

(c) How well does the model fit the data?

57. **Females in the workforce** Suppose the rate of change of the percent p of total U.S. workers who are female can be modeled by

$$\frac{dp}{dt} = \frac{2154.18}{(1.38t + 64.1)^2}$$

percentage points per year, where t is the number of years past 1950.

Year	% Female	Year	% Female
1950	29.6	2010	47.9
1960	33.4	2015	48.3
1970	38.1	2020	48.1
1980	42.5	2030	48.0
1990	45.2	2040	47.9
2000	46.6	2050	47.9

Source: U.S. Census Bureau

(a) Use integration and the data point for 2040 to find the function $p(t)$ that models the percent of the workforce that is female.

(b) For selected years from 1950 and projected to 2050, the data in the table show the percent of total U.S. workers who are female. With t as the number of years past 1950, graph these data with the model found in part (a).

(c) Comment on the model's fit to the data.

58. **Energy use** Energy use per dollar of GDP in the United States indexed to 1980 means that energy use for any year is viewed as a percent of the use per dollar of GDP in 1980. The table shows the energy use per dollar of GDP, as a percent, for selected years from 1985 and projected to 2035. Suppose the rate of change of energy use per dollar of GDP can be modeled by

$$\frac{dE}{dt} = 0.00468(0.4t + 2)^2 - 0.104(0.4t + 2) - 0.56$$

percentage points per year, where t is the number of years past 1980.

Energy Use per Dollar of GDP

Year	Percent	Year	Percent
1985	83	2015	51
1990	79	2020	45
1995	75	2025	41
2000	67	2030	37
2005	60	2035	34
2010	56		

Source: U.S. Department of Energy

(a) Use integration and the data point for 1990 to find the function $E(t)$ that models the energy use per dollar of GDP. Use two significant digits.

(b) Let $t = 0$ represent 1980 and graph the model from part (a) with the data in the table.

(c) Find the model's predicted energy use per dollar of GDP in 2025.

OBJECTIVES

12.3

- To evaluate integrals of the form $\int e^u u' \, dx$ or, equivalently, $\int e^u \, du$
- To evaluate integrals of the form $\int \dfrac{u'}{u} \, dx$ or, equivalently, $\int \dfrac{1}{u} \, du$

Integrals Involving Exponential and Logarithmic Functions

■ | **APPLICATION PREVIEW** | ■

As the real estate market emerges from the crisis that began in 2008, the rate of growth of the market value of a home is expected to exceed the inflation rate. Suppose, for example, that home prices in a selected area are projected to increase at an average annual rate of 8%. Then the rate of change of the value of a house in this area that cost $200,000 can be modeled by

$$\frac{dV}{dt} = 15.4e^{0.077t}$$

where V is the value of the home in thousands of dollars and t is the time in years since the home was purchased. To find the market value of such a home 10 years after it was purchased, we would first have to integrate dV/dt. That is, we must be able to integrate an exponential. (See Example 3.)

In this section, we consider integration formulas that result in natural logarithms and formulas for integrating exponentials.

Integrals Involving Exponential Functions

We know that

$$\frac{d}{dx}(e^x) = e^x \quad \text{and} \quad \frac{d}{dx}(e^u) = e^u \cdot u'$$

The corresponding integrals are given by the following.

Exponential Formula

If u is a function of x,

$$\int e^u \cdot u' \, dx = \int e^u \, du = e^u + C$$

In particular, $\int e^x \, dx = e^x + C.$

EXAMPLE   1   **Integral of an Exponential**

Evaluate $\int 5e^x \, dx.$

Solution
$\int 5e^x \, dx = 5\int e^x \, dx = 5e^x + C$

EXAMPLE   2   **Integral of $e^u \, du$**

Evaluate: (a) $\int 2xe^{x^2} \, dx$ (b) $\int \dfrac{x^2 \, dx}{e^{x^3}}$

Solution
(a)   Letting $u = x^2$ implies that $u' = 2x$, and the integral is of the form $\int e^u \cdot u' dx$. Thus

$$\int 2xe^{x^2} \, dx = \int e^{x^2}(2x) \, dx = \int e^u \cdot u' dx = e^u + C = e^{x^2} + C$$

(b) In order to use $\int e^u \cdot u' \, dx$, we write the exponential in the numerator. Thus

$$\int \frac{x^2 \, dx}{e^{x^3}} = \int e^{-x^3} (x^2 \, dx)$$

This is *almost* of the form $\int e^u \cdot u' \, dx$. Letting $u = -x^3$ gives $u' = -3x^2$. Thus

$$\int e^{-x^3}(x^2 \, dx) = -\frac{1}{3}\int e^{-x^3}(-3x^2 \, dx) = -\frac{1}{3} e^{-x^3} + C = \frac{-1}{3e^{x^3}} + C$$

✓ **CHECKPOINT**

1. True or false:

   (a) $\int e^{x^2}(2x \, dx) = e^{x^2} \cdot x^2 + C$ (b) $\int e^{-3x} \, dx = -\frac{1}{3} e^{-3x} + C$

   (c) $\int \frac{dx}{e^{3x}} = \frac{1}{3}\left(\frac{1}{e^{3x}}\right) + C$ (d) $\int e^{3x+1}(3 \, dx) = \frac{e^{3x+2}}{3x+2} + C$

EXAMPLE 3 **Real Estate Inflation | APPLICATION PREVIEW |**

Suppose the rate of change of the value of a house that cost $200,000 in 2015 can be modeled by

$$\frac{dV}{dt} = 15.4e^{0.077t}$$

where V is the market value of the home in thousands of dollars and t is the time in years since 2015.

(a) Find the function that expresses the value V in terms of t.
(b) Find the predicted value in 2025 (after 10 years).

Solution

(a) $V = \int \frac{dV}{dt} \, dt = \int 15.4e^{0.077t} \, dt$

$$V = 15.4 \int e^{0.077t}\left(\frac{1}{0.077}\right)(0.077 \, dt)$$

$$V = 15.4\left(\frac{1}{0.077}\right)\int e^{0.077t}(0.077 \, dt)$$

$$V = 200e^{0.077t} + C$$

Using $V = 200$ (thousand) when $t = 0$, we have

$$200 = 200 + C$$
$$0 = C$$

Thus we have the value as a function of time given by

$$V = 200e^{0.077t}$$

(b) The value after 10 years is found by using $t = 10$.

$$V = 200e^{0.077(10)} = 200e^{0.77} \approx 431.95$$

Thus, in 2025, the predicted value of the home is $431,950.

EXAMPLE 4 **Graphs of Functions and Integrals**

Figure 12.5 shows the graphs of $g(x) = 5e^{-x^2}$ and $h(x) = -10xe^{-x^2}$. One of these functions is $f(x)$ and the other is $\int f(x) \, dx$ with $C = 0$.

(a) Decide which of $g(x)$ and $h(x)$ is $f(x)$ and which is $\int f(x) \, dx$.
(b) How can the graph of $f(x)$ be used to locate and classify the extrema of $\int f(x) \, dx$?

(c) What feature of the graph of $f(x)$ occurs at the same x-values as the inflection points of the graph of $\int f(x)\,dx$?

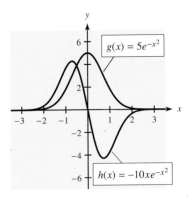

Figure 12.5

Solution

(a) The graph of $h(x)$ looks like the graph of $g'(x)$ because $h(x) > 0$ where $g(x)$ is increasing, $h(x) < 0$ where $g(x)$ is decreasing, and $h(x) = 0$ where $g(x)$ has its maximum. However, if $h(x) = g'(x)$, then, equivalently,

$$\int h(x)\,dx = \int g'(x)\,dx$$
$$= g(x) + C$$

so $h(x) = f(x)$ and $g(x) = \int f(x)\,dx$. We can verify this by noting that

$$\int -10xe^{-x^2}\,dx = 5\int e^{-x^2}(-2x\,dx)$$
$$= 5e^{-x^2} + C$$

(b) We know that $f(x)$ is the derivative of $\int f(x)\,dx$, so, as we saw in part (a), the x-intercepts of $f(x)$ locate the critical values and extrema of $\int f(x)\,dx$.

(c) The first derivative of $\int f(x)\,dx$ is $f(x)$, and its second derivative is $f'(x)$. Hence the inflection points of $\int f(x)\,dx$ occur where $f'(x) = 0$. But $f(x)$ has its extrema where $f'(x) = 0$. Thus the extrema of $f(x)$ occur at the same x-values as the inflection points of $\int f(x)\,dx$. ■

Integrals Involving Logarithmic Functions

Recall that the Power Rule for integrals applies only if $n \neq -1$. That is,

$$\int u^n u'\,dx = \frac{u^{n+1}}{n+1} + C \quad \text{if } n \neq -1$$

The following formula applies when $n = -1$.

Logarithmic Formula

If u is a function of x, then

$$\int u^{-1} u'\,dx = \int \frac{u'}{u}\,dx = \int \frac{1}{u}\,du = \ln|u| + C$$

In particular, $\int \dfrac{1}{x}\,dx = \ln|x| + C.$

We use the absolute value of u in the integral because the logarithm is defined only when the quantity is positive. This logarithmic formula is a direct result of the fact that

$$\frac{d}{dx}(\ln|u|) = \frac{1}{u} \cdot u'$$

We can see this result by considering the following.

$$\text{For } u > 0: \quad \frac{d}{dx}(\ln|u|) = \frac{d}{dx}(\ln u) = \frac{1}{u} \cdot u'$$

$$\text{For } u < 0: \quad \frac{d}{dx}(\ln|u|) = \frac{d}{dx}[\ln(-u)] = \frac{1}{(-u)} \cdot (-u') = \frac{1}{u} \cdot u'$$

In addition to this verification, we can graphically illustrate the need for the absolute value sign. Figure 12.6(a) shows that $f(x) = 1/x$ is defined for $x \neq 0$, and from Figures 12.6(b) and 12.6(c), we see that $F(x) = \int 1/x\, dx = \ln|x|$ is also defined for $x \neq 0$, but $y = \ln x$ is defined only for $x > 0$.

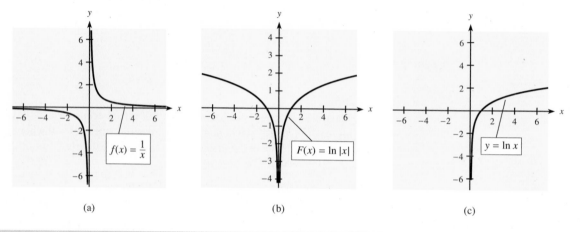

Figure 12.6   (a)   (b)   (c)

EXAMPLE 5   Integral Resulting in a Logarithmic Function

Evaluate $\displaystyle\int \frac{4}{4x + 8}\, dx$.

Solution
This integral is of the form

$$\int \frac{u'}{u}\, dx = \ln|u| + C$$

with $u = 4x + 8$ and $u' = 4$. Thus

$$\int \frac{4}{4x + 8}\, dx = \ln|4x + 8| + C$$ ∎

Figure 12.7 shows several members of the family

$$F(x) = \int \frac{4\, dx}{4x + 8} = \ln|4x + 8| + C$$

We can choose different values for C and use a graphing utility to graph families of curves such as those in Figure 12.7.

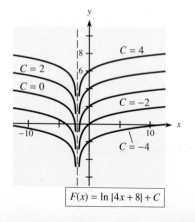

Figure 12.7

$F(x) = \ln|4x + 8| + C$

EXAMPLE 6   Integral of du/u

Evaluate $\int \dfrac{x-3}{x^2-6x+1}\,dx$.

Solution

This integral is of the form $\int (u'/u)\,dx$, *almost*. If we let $u = x^2 - 6x + 1$, then $u' = 2x - 6$. If we multiply (and divide) the numerator by 2, we get

$$\int \frac{x-3}{x^2-6x+1}\,dx = \frac{1}{2}\int \frac{2(x-3)}{x^2-6x+1}\,dx$$

$$= \frac{1}{2}\int \frac{2x-6}{x^2-6x+1}\,dx$$

$$= \frac{1}{2}\int \frac{u'}{u}\,dx = \frac{1}{2}\ln|u| + C$$

$$= \frac{1}{2}\ln|x^2-6x+1| + C$$
■

EXAMPLE 7   Population Growth

Because the world contains only about 10 billion acres of arable land, world population is limited. Suppose that world population is limited to 40 billion people and that the rate of population growth per year is given by

$$\frac{dP}{dt} = k(40 - P)$$

where P is the population in billions at time t and k is a positive constant. Then the relationship between the year and the population during that year is given by the integral

$$t = \frac{1}{k}\int \frac{1}{40-P}\,dP$$

where $40 - P > 0$ because 40 billion is the population's upper limit.

(a) Evaluate this integral to find the relationship.
(b) Use properties of logarithms and exponential functions to write P as a function of t.

Solution

(a) $t = \dfrac{1}{k}\int \dfrac{1}{40-P}\,dP = -\dfrac{1}{k}\int \dfrac{-dP}{40-P} = -\dfrac{1}{k}\ln|40-P| + C_1$

(b) $t = -\dfrac{1}{k}\ln(40-P) + C_1$ because $40 - P > 0$ means $|40 - P| = 40 - P$.

Solving this equation for P requires converting to exponential form.

$$-k(t - C_1) = \ln(40 - P)$$
$$e^{C_1 k - kt} = 40 - P$$
$$e^{C_1 k} \cdot e^{-kt} = 40 - P$$

Because $e^{C_1 k}$ is an unknown constant, we replace it with C and solve for P.

$$P = 40 - Ce^{-kt}$$
■

If an integral contains a fraction in which the degree of the numerator is equal to or greater than that of the denominator, we should divide the denominator into the numerator as a first step.

EXAMPLE 8 | **Integral Requiring Division**

Evaluate $\int \dfrac{x^4 - 2x^3 + 4x^2 - 7x - 1}{x^2 - 2x}\, dx$.

Solution

Because the numerator is of higher degree than the denominator, we begin by dividing $x^2 - 2x$ into the numerator.

$$
\begin{array}{r}
x^2 \qquad\quad +4 \\
x^2 - 2x\overline{\smash{\big)}\,x^4 - 2x^3 + 4x^2 - 7x - 1} \\
\underline{x^4 - 2x^3} \qquad\qquad\qquad \\
4x^2 - 7x - 1 \\
\underline{4x^2 - 8x} \qquad\quad \\
x - 1
\end{array}
$$

Thus

$$
\int \frac{x^4 - 2x^3 + 4x^2 - 7x - 1}{x^2 - 2x}\, dx = \int\left(x^2 + 4 + \frac{x - 1}{x^2 - 2x}\right) dx
$$

$$
= \int(x^2 + 4)\, dx + \frac{1}{2}\int \frac{2(x - 1)\, dx}{x^2 - 2x}
$$

$$
= \frac{x^3}{3} + 4x + \frac{1}{2}\ln|x^2 - 2x| + C
$$

✓ CHECKPOINT

2. True or false:

(a) $\displaystyle\int \frac{3x^2\, dx}{x^3 + 4} = \ln|x^3 + 4| + C$

(b) $\displaystyle\int \frac{2x\, dx}{\sqrt{x^2 + 1}} = \ln|\sqrt{x^2 + 1}| + C$

(c) $\displaystyle\int \frac{2}{x}\, dx = 2\ln|x| + C$

(d) $\displaystyle\int \frac{x}{x + 1}\, dx = x\int \frac{1}{x + 1}\, dx = x\ln|x + 1| + C$

(e) To evaluate $\displaystyle\int \frac{4x}{4x + 1}\, dx$, our first step is to divide $4x + 1$ into $4x$.

3. (a) Divide $4x + 1$ into $4x$.

(b) Evaluate $\displaystyle\int \frac{4x}{4x + 1}\, dx$.

✓ CHECKPOINT
ANSWERS

1. (a) False. The correct solution is $e^{x^2} + C$ (see Example 2a).

(b) True

(c) False; $\displaystyle\int \frac{dx}{e^{3x}} = \frac{-1}{3e^{3x}} + C$

(d) False; $\displaystyle\int e^{3x+1}(3\, dx) = e^{3x+1} + C$

2. (a) True

(b) False; $\displaystyle\int \frac{2x\, dx}{\sqrt{x^2 + 1}} = 2(x^2 + 1)^{1/2} + C$

(c) True

(d) False. We cannot factor the variable x outside the integral sign.

(e) True

3. (a) $\dfrac{4x}{4x + 1} = 1 - \dfrac{1}{4x + 1}$

(b) $\displaystyle\int \frac{4x\, dx}{4x + 1} = \int\left(1 - \frac{1}{4x + 1}\right) dx = x - \frac{1}{4}\ln|4x + 1| + C$

| EXERCISES | 12.3

Evaluate the integrals in Problems 1–32.

1. $\int 3e^{3x}\,dx$

2. $\int 4e^{4x}\,dx$

3. $\int e^{-x}\,dx$

4. $\int e^{2x}\,dx$

5. $\int 1000e^{0.1x}\,dx$

6. $\int 1600e^{0.4x}\,dx$

7. $\int 840e^{-0.7x}\,dx$

8. $\int 250e^{-0.5x}\,dx$

9. $\int x^3 e^{3x^4}\,dx$

10. $\int xe^{2x^2}\,dx$

11. $\int \dfrac{3}{e^{2x}}\,dx$

12. $\int \dfrac{4}{e^{1-2x}}\,dx$

13. $\int \dfrac{x^5}{e^{2-3x^6}}\,dx$

14. $\int \dfrac{x^3}{e^{4x^4}}\,dx$

15. $\int \left(e^{4x} - \dfrac{3}{e^{x/2}}\right)dx$

16. $\int \left(xe^{3x^2} - \dfrac{5}{e^{x/3}}\right)dx$

17. $\int \dfrac{3x^2}{x^3 + 4}\,dx$

18. $\int \dfrac{8x^7}{x^8 - 1}\,dx$

19. $\int \dfrac{dz}{4z + 1}$

20. $\int \dfrac{y}{y^2 + 1}\,dy$

21. $\int \dfrac{6x^3}{2x^4 + 1}\,dx$

22. $\int \dfrac{7x^2}{4x^3 - 9}\,dx$

23. $\int \dfrac{4x}{5x^2 - 4}\,dx$

24. $\int \dfrac{5x^2}{3x^3 - 8}\,dx$

25. $\int \dfrac{3x^2 - 2}{x^3 - 2x}\,dx$

26. $\int \dfrac{4x^3 + 2x}{x^4 + x^2}\,dx$

27. $\int \dfrac{z^2 + 1}{z^3 + 3z + 17}\,dz$

28. $\int \dfrac{(x + 2)\,dx}{x^2 + 4x - 9}$

29. $\int \dfrac{x^3 - x^2 + 1}{x - 1}\,dx.$

30. $\int \dfrac{2x^3 + x^2 + 2x + 3}{2x + 1}\,dx$

31. $\int \dfrac{x^2 + x + 3}{x^2 + 3}\,dx$

32. $\int \dfrac{x^4 - 2x^2 + x}{x^2 - 2}\,dx$

In Problems 33 and 34, graphs of two functions labeled $g(x)$ and $h(x)$ are given. Decide which is the graph of $f(x)$ and which is one member of the family $\int f(x)\,dx$. Check your conclusions by evaluating the integral.

33.
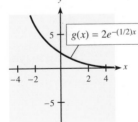
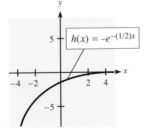

34.
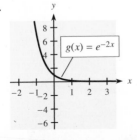
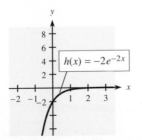

In Problems 35 and 36, a function $f(x)$ and its graph are given. Find the family $F(x) = \int f(x)\,dx$ and graph the member that satisfies $F(0) = 0$.

35.
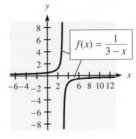

36.
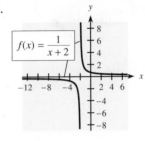

In Problems 37–40, a family of functions is given and graphs of some members are shown. Find the function $f(x)$ such that the family is given by $\int f(x)\,dx$.

37. $F(x) = x + \ln|x| + C$

38. $F(x) = -\ln(x^2 + 4) + C$

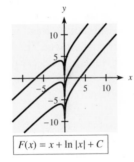
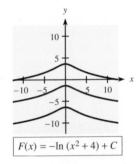

39. $F(x) = 5xe^{-x} + C$

40. $F(x) = e^{0.4x} + e^{-0.4x} + C$

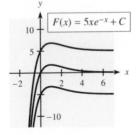
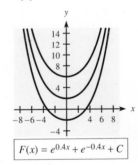

In parts (a)–(d) of Problems 41 and 42, integrate those that can be done by the methods studied so far.

41. (a) $\int xe^{x^3}\,dx$

(b) $\int \dfrac{x + 2}{x^2 + 2x + 7}\,dx$

(c) $\int \dfrac{x^2 + 2x}{x^3 + 3x^2 + 7}\,dx$

(d) $\int 5x^3 e^{2x^4}\,dx$

42. (a) $\int \dfrac{3x - 1}{x^3 - x + 2}\,dx$

(b) $\int \dfrac{3x - 1}{6x^2 - 4x + 9}\,dx$

(c) $\int 5\sqrt{x}\,e^{\sqrt{x}}\,dx$

(d) $\int 6xe^{-x^2/8}\,dx$

APPLICATIONS

43. *Revenue* Suppose that the marginal revenue from the sale of x units of a product is $\overline{MR} = R'(x) = 6e^{0.01x}$. What is the revenue in dollars from the sale of 100 units of the product?

44. *Concentration of a drug* Suppose that the rate at which the concentration of a drug in the blood changes with respect to time t is given by

$$C'(t) = \frac{c}{b-a}(be^{-bt} - ae^{-at}), \quad t \geq 0$$

where a, b, and c are constants depending on the drug administered, with $b > a$. Assuming that $C(t) = 0$ when $t = 0$, find the formula for the concentration of the drug in the blood at any time t.

45. *Radioactive decay* The rate of disintegration of a radioactive substance can be described by

$$\frac{dn}{dt} = n_0(-K)e^{-Kt}$$

where n_0 is the number of radioactive atoms present when time t is 0, and K is a positive constant that depends on the substance involved. Using the fact that the constant of integration is 0, integrate dn/dt to find the number of atoms n that are still radioactive after time t.

46. *Radioactive decay* Radioactive substances decay at a rate that is proportional to the amount present. Thus, if k is a constant and the amount present is x, the decay rate is

$$\frac{dx}{dt} = kx \quad (t \text{ in hours})$$

This means that the relationship between the time and the amount of substance present can be found by evaluating the integral

$$t = \int \frac{dx}{kx}$$

(a) Evaluate the integral to find the relationship.
(b) Use properties of logarithms and exponential functions to write x as a function of t.

47. *Memorization* The rate of vocabulary memorization of the average student in a foreign language course is given by

$$\frac{dv}{dt} = \frac{40}{t+1}$$

where t is the number of continuous hours of study, $0 < t \leq 4$, and v is the number of words. How many words would the average student memorize in 3 hours?

48. *Population growth* The rate of growth of world population can be modeled by

$$\frac{dN}{dt} = N_0 re^{rt}, \quad r < 1$$

where t is the time in years from the present and N_0 and r are constants. What function describes world population if the present population is N_0?

49. *Compound interest* If $\$P$ is invested for n years at 10% compounded continuously, the rate at which the future value is growing is

$$\frac{dS}{dn} = 0.1Pe^{0.1n}$$

(a) What function describes the future value at the end of n years?
(b) In how many years will the future value double?

50. *Temperature changes* When an object is moved from one environment to another, its temperature T changes at a rate given by

$$\frac{dT}{dt} = kCe^{kt}$$

where t is the time in the new environment (in hours), C is the temperature difference (old – new) between the two environments, and k is a constant. If the temperature of the object (and the old environment) is 70°F, and $C = -10°F$, what function describes the temperature T of the object t hours after it is moved?

51. *Blood pressure in the aorta* The rate at which blood pressure decreases in the aorta of a normal adult after a heartbeat is

$$\frac{dp}{dt} = -46.645e^{-0.491t}$$

where t is time in seconds.
(a) What function describes the blood pressure in the aorta if $p = 95$ when $t = 0$?
(b) What is the blood pressure 0.1 second after a heartbeat?

52. *Sales and advertising* A store finds that its sales decline after the end of an advertising campaign, with its daily sales for the period declining at the rate $S'(t) = -1477.8e^{-0.2t}$, $0 \leq t \leq 35$, where t is the number of days since the end of the campaign. Suppose that $S = 7389$ units when $t = 0$.
(a) Find the function that describes the number of daily sales t days after the end of the campaign.
(b) Find the total number of sales 10 days after the end of the advertising campaign.

53. *Life expectancy* Suppose the rate of change of the expected life span l at birth of people born in the United States can be modeled by

$$\frac{dl}{dt} = \frac{14.304}{t+20}$$

where t is the number of years past 1920.

Year	Life Span (years)	Year	Life Span (years)
1920	54.1	1994	75.7
1930	59.7	1996	76.1
1940	62.9	1998	76.7
1950	68.2	2000	76.9
1960	69.7	2001	77.2
1970	70.8	2003	77.5
1975	72.6	2005	77.9
1980	73.7	2010	78.1
1985	74.7	2015	78.9
1990	75.4	2020	79.5
1992	75.5		

Source: National Center for Health Statistics

(a) Use integration and the data point for 2000 to find the function that models the life span.
(b) The data in the table give the expected life spans for people born in various years. Graph the function from part (a) with the data, with $t = 0$ representing 1920.
(c) How well does the model fit the data?

54. **U.S. households with cable/satellite TV** Suppose the rate of change of the percent P of U.S. households with cable/satellite TV can be modeled by

$$\frac{dP}{dt} = \frac{46.3}{t + 5}$$

where t is the number of years past 1975.
(a) Use integration and the data point for 2010 to find the function $P(t)$ that models the percent of U.S. households with cable/satellite TV.
(b) How well does the model from part (a) fit the data in the table?
(c) If the model remains valid, use it to predict the percent of U.S. households with cable/satellite TV in 2018.

Year	Percent	Year	Percent
1980	22.6	2000	67.8
1985	46.2	2005	85.7
1990	59.0	2010	90.6
1995	65.7	2013	90.5

Source: Nielsen Media Research

55. **Consumer price index** The Social Security Administration makes projections about the consumer price index (CPI) in order to understand the effects of inflation on Social Security benefits and to plan for cost-of-living increases. Suppose the rate of change of the CPI can be modeled with the function

$$\frac{dC}{dt} = 3.087e^{0.0384t}$$

dollars per year, where C is the consumer price index and t is the number of years past 1990.
(a) Does the model for the rate reflect the fact that the Social Security Administration's data (actual and projected for selected years from 1995 to 2070) in the table show that the CPI is increasing? Explain.

(b) Use integration and the table's data point for 2005 to find the function that models the Social Security Administration's CPI figures.
(c) Find and interpret $C(35)$ and $C'(35)$.

Year	CPI	Year	CPI
1995	100.00	2035	465.98
2000	118.21	2040	566.94
2005	143.67	2045	689.77
2010	174.80	2050	839.21
2015	212.67	2055	1021.02
2020	258.74	2060	1242.23
2025	314.80	2065	1511.36
2030	383.00	2070	1838.81

Source: Social Security Administration

56. **Average annual wage** The following table shows the U.S. average annual wage in thousands of dollars for selected years from 2012 and projected to 2050. Suppose the rate of change of the U.S. average annual wage can be modeled by

$$\frac{dW}{dt} = 1.66e^{0.0395t}$$

thousand dollars per year, where t is the number of years past 2010.
(a) Use the data to find the average rates of change of the U.S. average annual wage from 2014 to 2016 and from 2016 to 2018. Which of these average rates better approximates the instantaneous rate in 2016?
(b) Use the data point from 2016 to find the function that models $W(t)$. Use three significant digits.

Year	Average Annual Wage ($ thousands)	Year	Average Annual Wage ($ thousands)
2012	44.6	2030	93.2
2014	48.6	2035	113.2
2016	53.3	2040	137.6
2018	58.7	2045	167.1
2020	63.7	2050	202.5
2025	76.8		

Source: Social Security Administration

OBJECTIVES   **12.4**

- To use integration to find total cost functions from information involving marginal cost
- To optimize profit, given information regarding marginal cost and marginal revenue
- To use integration to find national consumption functions from information about marginal propensity to consume and marginal propensity to save

Applications of the Indefinite Integral in Business and Economics

▌| APPLICATION PREVIEW |▐

If we know that the consumption of a nation is $9 trillion when income is $0 and the marginal propensity to save is 0.25, we can easily find the marginal propensity to consume and use integration to find the national consumption function. (See Example 6.)

In this section, we also use integration to derive total cost and profit functions from the marginal cost and marginal revenue functions. One of the reasons for the marginal approach in economics is that firms can observe marginal changes in real life. If they know the marginal cost and the total cost when a given quantity is sold, they can develop their total cost function.

Total Cost and Profit   We know that the marginal cost for a commodity is the derivative of the total cost function— that is, $\overline{MC} = C'(x)$, where $C(x)$ is the total cost function. Thus if we have the marginal cost function, we can integrate (or "reverse" the process of differentiation) to find the total cost. That is, $C(x) = \int \overline{MC}\, dx$.

If, for example, the marginal cost is $\overline{MC} = 4x + 3$, the total cost is given by

$$C(x) = \int \overline{MC}\, dx$$

$$= \int (4x + 3)\, dx$$

$$= 2x^2 + 3x + K$$

where K represents the constant of integration. We know that the total revenue is 0 if no items are produced, but the total cost may not be 0 if nothing is produced. The fixed costs accrue whether goods are produced or not. Thus the value of the constant of integration depends on the fixed costs FC of production.

Thus we cannot determine the total cost function from the marginal cost unless additional information is available to help us determine the fixed costs.

EXAMPLE   1   Total Cost

Suppose the marginal cost function for a month for a certain product is $\overline{MC} = 3x + 50$, where x is the number of units and cost is in dollars. If the fixed costs related to the product amount to $10,000 per month, find the total cost function for the month.

Solution
The total cost function is

$$C(x) = \int (3x + 50)\, dx$$

$$= \frac{3x^2}{2} + 50x + K$$

The constant of integration K is found by using the fact that $C(0) = FC = 10,000$. Thus

$$3(0)^2 + 50(0) + K = 10,000, \text{ so } K = 10,000$$

and the total cost for the month is given by

$$C(x) = \frac{3x^2}{2} + 50x + 10,000$$

EXAMPLE 2   Cost

Suppose monthly records show that the rate of change of the cost (that is, the marginal cost) for a product is $\overline{MC} = 3(2x + 25)^{1/2}$, where x is the number of units and cost is in dollars. If the fixed costs for the month are \$11,125, what would be the total cost of producing 300 items per month?

Solution

We can integrate the marginal cost to find the total cost function.

$$C(x) = \int \overline{MC}\, dx = \int 3(2x + 25)^{1/2}\, dx$$

$$= 3 \cdot \left(\frac{1}{2}\right) \int (2x + 25)^{1/2}(2\, dx)$$

$$= \left(\frac{3}{2}\right) \frac{(2x + 25)^{3/2}}{3/2} + K$$

$$= (2x + 25)^{3/2} + K$$

We can find K by using the fact that fixed costs are \$11,125.

$$C(0) = 11{,}125 = (25)^{3/2} + K$$
$$11{,}125 = 125 + K, \text{ or } K = 11{,}000$$

Thus the total cost function is

$$C(x) = (2x + 25)^{3/2} + 11{,}000$$

and the cost of producing 300 items per month is

$$C(300) = (625)^{3/2} + 11{,}000$$
$$= 26{,}625 \text{ (dollars)}$$

It can be shown that the profit is usually maximized when $\overline{MR} = \overline{MC}$. To see that this does not always give us a maximum *positive* profit, consider the following facts concerning the manufacture of one particular product over the period of a month.

1. The marginal revenue is $\overline{MR} = 400 - 30x$.
2. The marginal cost is $\overline{MC} = 20x + 50$.
3. When 5 units are produced and sold, the total cost is \$1750. The profit *should* be maximized when $\overline{MR} = \overline{MC}$, or when $400 - 30x = 20x + 50$. Solving for x gives $x = 7$. To see whether our profit is maximized when 7 units are produced and sold, let us examine the profit function.

The profit function is given by $P(x) = R(x) - C(x)$, where

$$R(x) = \int \overline{MR}\, dx \text{ and } C(x) = \int \overline{MC}\, dx$$

Integrating, we get

$$R(x) = \int (400 - 30x)\, dx = 400x - 15x^2 + K$$

but $R(0) = 0$ gives $K = 0$ for this total revenue function, so

$$R(x) = 400x - 15x^2$$

The total cost function is

$$C(x) = \int (20x + 50)\, dx = 10x^2 + 50x + K$$

The value of fixed cost can be determined by using the fact that 5 units cost \$1750. This tells us that $C(5) = 1750 = 250 + 250 + K$, so $K = 1250$.

Thus the total cost is $C(x) = 10x^2 + 50x + 1250$. Thus, the profit is

$$P(x) = R(x) - C(x) = (400x - 15x^2) - (10x^2 + 50x + 1250)$$

Simplifying gives

$$P(x) = 350x - 25x^2 - 1250$$

We have found that $\overline{MR} = \overline{MC}$ if $x = 7$, and the graph of $P(x)$ is a parabola that opens downward, so profit is maximized at $x = 7$. But if $x = 7$, profit is

$$P(7) = 2450 - 1225 - 1250 = -25$$

That is, the production and sale of 7 items result in a loss of $25.

The preceding discussion indicates that even though setting $\overline{MR} = \overline{MC}$ may optimize profit, it does not indicate the level of profit or loss, as forming the profit function does.

If this firm is in a competitive market and its optimal level of production results in a loss, it has two options. It can continue to produce at the optimal level in the short run until it can lower or eliminate its fixed costs, even though it is losing money; or it can take a larger loss (its fixed cost) by stopping production. Producing 7 units causes a loss of $25 per month, and ceasing production results in a loss of $1250 (the fixed cost) per month. If this firm and many others like it cease production, the supply will be reduced, causing an eventual increase in price. The firm can resume production when the price increase indicates that it can make a profit.

EXAMPLE 3   Maximum Profit

Suppose that each week a company has a product with $\overline{MR} = 200 - 4x$, $\overline{MC} = 50 + 2x$, and the total cost of producing 10 units is $700. At what level should this company hold its weekly production in order to maximize profits?

Solution
Setting $\overline{MR} = \overline{MC}$, we can solve for the production level that maximizes profit.

$$200 - 4x = 50 + 2x$$
$$150 = 6x$$
$$25 = x$$

The level of production that should optimize profit is 25 units. To see whether 25 units maximizes profits or minimizes the losses (in the short run), we must find the total revenue and total cost functions.

$$R(x) = \int (200 - 4x)\,dx = 200x - 2x^2 + K$$
$$= 200x - 2x^2, \text{ because } K = 0$$
$$C(x) = \int (50 + 2x)\,dx = 50x + x^2 + K$$

We find K by noting that $C(x) = 700$ when $x = 10$.

$$700 = 50(10) + (10)^2 + K$$

so $K = 100$. Thus the cost is given by $C = C(x) = 50x + x^2 + 100$.

Thus the profit function is $P(x) = R(x) - C(x) = -3x^2 + 150x - 100$, whose graph is concave down, and the profit function is maximized at $x = 25$. The maximum weekly profit is $P(25) = \$1775$. ◼

Calculator Note   If it is difficult to solve $\overline{MC} = \overline{MR}$ analytically, we can use a graphing calculator to solve this equation by finding the point of intersection of the graphs of $\overline{MC}$ and $\overline{MR}$. We may also be able to integrate $\overline{MC}$ and $\overline{MR}$ to find the functions $C(x)$ and $R(x)$ and then use a graphing calculator to graph them. From the graphs of $C(x)$ and $R(x)$ we can learn about these functions—and hence about profit. ◼

EXAMPLE 4   **Cost, Revenue, and Profit**

Suppose that $\overline{MC} = 1.01(x + 190)^{0.01}$ and $\overline{MR} = (1/\sqrt{2x + 1}) + 2$, where x is the number of thousands of units and both revenue and cost are in thousands of dollars. Suppose further that fixed costs are $100,236 and that production is limited to at most 180 thousand units.

(a) Determine $C(x)$ and $R(x)$ and graph them to determine whether a profit can be made.

(b) Estimate the level of production that yields maximum profit, and find the maximum profit.

Solution

(a) $C(x) = \displaystyle\int \overline{MC}\ dx = \int 1.01(x + 190)^{0.01}\ dx$

$$= 1.01\frac{(x + 190)^{1.01}}{1.01} + K$$

When we say that fixed costs equal $100,236, we mean $C(0) = 100.236$.

$$100.236 = C(0) = (190)^{1.01} + K$$
$$100.236 = 200.236 + K$$
$$-100 = K$$

Thus $C(x) = (x + 190)^{1.01} - 100$.

$$R(x) = \int \overline{MR}\ dx = \int [(2x + 1)^{-1/2} + 2]dx$$

$$= \frac{1}{2}\int (2x + 1)^{-1/2}(2\ dx) + \int 2\ dx$$

$$= \frac{1}{2}\left[\frac{(2x + 1)^{1/2}}{1/2}\right] + 2x + K$$

$R(0) = 0$ means

$$0 = R(0) = (1)^{1/2} + 0 + K, \quad \text{or} \quad K = -1$$

Thus $R(x) = (2x + 1)^{1/2} + 2x - 1$.

The graphs of $C(x)$ and $R(x)$ are shown in Figure 12.8. (The x-range is chosen to include the production range from 0 to 180 (thousand) units. The y-range is chosen to extend beyond fixed costs of about 100 thousand dollars.)

From the figure we see that a profit can be made as long as the number of units sold exceeds about 95 (thousand). We could locate this break even value more precisely by using INTERSECT.

(b) From the graph we also see that $R(x) - C(x) = P(x)$ is at its maximum at the right edge of the graph. Because production is limited to at most 180 thousand units, profit will be maximized when $x = 180$ and the maximum profit is

$$P(180) = R(180) - C(180)$$
$$= [(361)^{1/2} + 360 - 1] - [(370)^{1.01} - 100]$$
$$\approx 85.46\ \text{(thousand dollars)}$$

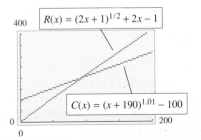

400

$R(x) = (2x + 1)^{1/2} + 2x - 1$

$C(x) = (x + 190)^{1.01} - 100$

0 200

Figure 12.8 0

1. True or false:
   (a) If $C(x) = \int \overline{MC}\, dx$, then the constant of integration equals the fixed costs.
   (b) If $R(x) = \int \overline{MR}\, dx$, then the constant of integration equals 0.

2. Find $C(x)$ if $\overline{MC} = \dfrac{100}{\sqrt{x+1}}$ and fixed costs are \$8000.

National Consumption and Savings

The consumption function is one of the basic ingredients in a larger discussion of how an economy can have persistent high unemployment or persistent high inflation. This study is often called **Keynesian analysis,** after its founder John Maynard Keynes (pronounced "canes").

If C represents national consumption (in trillions of dollars), then a **national consumption function** has the form $C = f(y)$, where y is disposable national income (also in trillions of dollars). The **marginal propensity to consume** is the derivative of the national consumption function with respect to y, or $dC/dy = f'(y)$. For example, suppose that

$$C = f(y) = 0.8y + 6$$

is a national consumption function; then the marginal propensity to consume is $f'(y) = 0.8$.

If we know the marginal propensity to consume, we can integrate with respect to y to find national consumption:

$$C = \int f'(y)\, dy = f(y) + K$$

We can find the unique national consumption function if we have additional information to help us determine the value of K, the constant of integration.

EXAMPLE 5   National Consumption

If consumption is \$6 trillion when disposable income is \$0, and if the marginal propensity to consume is $dC/dy = 0.3 + 0.4/\sqrt{y}$, find the national consumption function.

Solution
If

$$\frac{dC}{dy} = 0.3 + \frac{0.4}{\sqrt{y}}$$

then

$$C = \int \left(0.3 + \frac{0.4}{\sqrt{y}}\right) dy$$

$$= \int (0.3 + 0.4y^{-1/2})\, dy$$

$$= 0.3y + 0.4\,\frac{y^{1/2}}{1/2} + K = 0.3y + 0.8y^{1/2} + K$$

Now, if $C = 6$ when $y = 0$, then $6 = 0.3(0) + 0.8\sqrt{0} + K$. Thus the constant of integration is $K = 6$, and the consumption function is

$$C = 0.3y + 0.8\sqrt{y} + 6 \quad \text{(trillions of dollars)}$$

If S represents national savings, we can assume that the disposable national income is given by $y = C + S$, or $S = y - C$. Then the **marginal propensity to save** is $dS/dy = 1 - dC/dy$.

| EXAMPLE 6 | **Consumption and Savings** | APPLICATION PREVIEW |

If the consumption is $9 trillion when income is $0, and if the marginal propensity to save is 0.25, find the consumption function.

Solution
If $dS/dy = 0.25$, then $0.25 = 1 - dC/dy$, or $dC/dy = 0.75$. Thus

$$C = \int 0.75 \, dy = 0.75y + K$$

If $C = 9$ when $y = 0$, then $9 = 0.75(0) + K$, or $K = 9$. Then the consumption function is $C = 0.75y + 9$ (trillions of dollars).   ■

✓ CHECKPOINT

3. If the marginal propensity to save is

$$\frac{dS}{dy} = 0.7 - \frac{0.4}{\sqrt{y}}$$

find the marginal propensity to consume.
4. Find the national consumption function if the marginal propensity to consume is

$$\frac{dC}{dy} = \frac{1}{\sqrt{y+4}} + 0.2$$

and national consumption is $6.8 trillion when disposable income is $0.

**✓ CHECKPOINT
ANSWERS**

1. (a) False. $C(0)$ equals the fixed costs. It may or may not be the constant of integration.
    (b) False. We use $R(0) = 0$ to determine the constant of integration, but it may be nonzero.
2. $C(x) = 200\sqrt{x+1} + 7800$
3. $\dfrac{dC}{dy} = 1 - \dfrac{dS}{dy} = 0.3 + \dfrac{0.4}{\sqrt{y}}$
4. $C(y) = 2\sqrt{y+4} + 0.2y + 2.8$.

| EXERCISES | 12.4

TOTAL COST AND PROFIT

In Problems 1–12, cost, revenue, and profit are in dollars and x is the number of units.

1. If the daily marginal cost for a product is $\overline{MC} = 2x + 100$, with fixed costs amounting to $200, find the total cost function for each day.
2. If the monthly marginal cost for a product is $\overline{MC} = x + 30$ and the related fixed costs are $5000, find the total cost function for the month.
3. If the marginal cost for a product is $\overline{MC} = 4x + 2$ and the production of 10 units results in a total cost of $300, find the total cost function.
4. If the marginal cost for a product is $\overline{MC} = 3x + 50$ and the total cost of producing 20 units is $2000, find the total cost function.
5. If the marginal cost for a product is $\overline{MC} = 4x + 40$ and the total cost of producing 25 units is $3000, find the cost of producing 30 units.

6. If the marginal cost for producing a product is $\overline{MC} = 5x + 10$, with a fixed cost of $800, find the cost of producing 20 units.
7. A firm knows that its marginal cost for a product is $\overline{MC} = 3x + 20$, that its marginal revenue is $\overline{MR} = 44 - 5x$, and that the cost of production of 80 units is $11,400.
    (a) Find the optimal level of production.
    (b) Find the profit function.
    (c) Find the profit or loss at the optimal level.
8. A certain firm's marginal cost for a product is $\overline{MC} = 6x + 60$, its marginal revenue is $\overline{MR} = 180 - 2x$, and its total cost of production of 10 items is $1000.
    (a) Find the optimal level of production.
    (b) Find the profit function.
    (c) Find the profit or loss at the optimal level of production.
    (d) Should production be continued for the short run?
    (e) Should production be continued for the long run?

9. Suppose that the marginal revenue for a product is $\overline{MR} = 900$ and the marginal cost is $\overline{MC} = 30\sqrt{x + 4}$, with a fixed cost of $1000.
   (a) Find the profit or loss from the production and sale of 5 units.
   (b) How many units will result in a maximum profit?

10. Suppose that the marginal cost for a product is $\overline{MC} = 60\sqrt{x + 1}$ and its fixed cost is $340.00. If the marginal revenue for the product is $\overline{MR} = 80x$, find the profit or loss from production and sale of
    (a) 3 units.   (b) 8 units.

11. The average cost of a product changes at the rate

    $$\overline{C}'(x) = -6x^{-2} + 1/6$$

    and the average cost of 6 units is $10.00.
    (a) Find the average cost function.
    (b) Find the average cost of 12 units.

12. The average cost of a product changes at the rate

    $$\overline{C}'(x) = \frac{-10}{x^2} + \frac{1}{10}$$

    and the average cost of 10 units is $20.00.
    (a) Find the average cost function.
    (b) Find the average cost of 20 units.

13. Suppose that marginal cost for a certain product is given by $\overline{MC} = 1.05(x + 180)^{0.05}$ and marginal revenue is given by $\overline{MR} = (1/\sqrt{0.5x + 4}) + 2.8$, where x is in thousands of units and both revenue and cost are in thousands of dollars. Fixed costs are $200,000 and production is limited to at most 200 thousand units.
    (a) Find $C(x)$ and $R(x)$.
    (b) Graph $C(x)$ and $R(x)$ to determine whether a profit can be made.
    (c) Determine the level of production that yields maximum profit, and find the maximum profit (or minimum loss).

14. Suppose that the marginal cost for a certain product is given by $\overline{MC} = 1.02(x + 200)^{0.02}$ and marginal revenue is given by $\overline{MR} = (2/\sqrt{4x + 1}) + 1.75$, where x is in thousands of units and revenue and cost are in thousands of dollars. Suppose further that fixed costs are $150,000 and production is limited to at most 200 thousand units.
    (a) Find $C(x)$ and $R(x)$.
    (b) Graph $C(x)$ and $R(x)$ to determine whether a profit can be made.
    (c) Determine what level of production yields maximum profit, and find the maximum profit (or minimum loss).

NATIONAL CONSUMPTION AND SAVINGS

15. If consumption is $7 trillion when disposable income is $0 and if the marginal propensity to consume is 0.80, find the national consumption function (in trillions of dollars).

16. If national consumption is $9 trillion when income is $0 and if the marginal propensity to consume is 0.30, what is consumption when disposable income is $20 trillion?

17. If consumption is $8 trillion when income is $0 and if the marginal propensity to consume is

    $$\frac{dC}{dy} = 0.3 + \frac{0.2}{\sqrt{y}}$$

    find the national consumption function.

18. If consumption is $5 trillion when disposable income is $0 and if the marginal propensity to consume is

    $$\frac{dC}{dy} = 0.4 + \frac{0.3}{\sqrt{y}}$$

    find the national consumption function.

19. If consumption is $6 trillion when disposable income is $0 and if the marginal propensity to consume is

    $$\frac{dC}{dy} = \frac{1}{\sqrt{y + 1}} + 0.4$$

    find the national consumption function.

20. If consumption is $5.8 trillion when disposable income is $0 and if the marginal propensity to consume is

    $$\frac{dC}{dy} = \frac{1}{\sqrt{2y + 9}} + 0.8$$

    find the national consumption function.

21. Suppose that the marginal propensity to consume is

    $$\frac{dC}{dy} = 0.7 - e^{-2y}$$

    and that consumption is $5.65 trillion when disposable income is $0. Find the national consumption function.

22. Suppose that the marginal propensity to consume is

    $$\frac{dC}{dy} = 0.04 + \frac{\ln(y + 1)}{y + 1}$$

    and that consumption is $6.04 trillion when disposable income is $0. Find the national consumption function.

23. Suppose that the marginal propensity to save is

    $$\frac{dS}{dy} = 0.15$$

    and that consumption is $5.15 trillion when disposable income is $0. Find the national consumption function.

24. Suppose that the marginal propensity to save is

    $$\frac{dS}{dy} = 0.22$$

    and that consumption is $8.6 trillion when disposable income is $0. Find the national consumption function.

25. Suppose that the marginal propensity to save is

    $$\frac{dS}{dy} = 0.2 - \frac{1}{\sqrt{3y + 7}}$$

and that consumption is $6 trillion when disposable income is $0. Find the national consumption function.

26. If consumption is $3 trillion when disposable income is $0 and if the marginal propensity to save is

$$\frac{dS}{dy} = 0.2 + e^{-1.5y}$$

find the national consumption function.

- To show that a function is the solution to a differential equation
- To use integration to find the general solution to a differential equation
- To find particular solutions to differential equations using given conditions
- To solve separable differential equations
- To solve applied problems involving separable differential equations

Differential Equations

■ | APPLICATION PREVIEW |■

Carbon-14 dating, used to determine the age of fossils, is based on three facts. First, the half-life of carbon-14 is 5730 years. Second, the amount of carbon-14 in any living organism is essentially constant. Third, when an organism dies, the rate of change of carbon-14 in the organism is proportional to the amount present. If y represents the amount of carbon-14 present in the organism, then we can express the rate of change of carbon-14 by the differential equation

$$\frac{dy}{dt} = ky$$

where k is a constant and t is time in years. In this section, we study methods that allow us to find a function y that satisfies this differential equation, and then we use that function to date a fossil. (See Example 6.)

Recall that we introduced the derivative as an instantaneous rate of change and denoted the instantaneous rate of change of y with respect to time as dy/dt. For many growth or decay processes, such as carbon-14 decay, the rate of change of the amount of a substance with respect to time is proportional to the amount present. As we noted above, this can be represented by the equation

$$\frac{dy}{dt} = ky \quad (k = \text{constant})$$

An equation of this type, where y is an unknown function of x or t, is called a **differential equation** because it contains derivatives (or differentials). In this section, we restrict ourselves to differential equations where the highest derivative present in the equation is the first derivative. These differential equations are called **first-order differential equations.** Examples are

$$f'(x) = \frac{1}{x+1}, \quad \frac{dy}{dt} = 2t, \text{ and } x\,dy = (y+1)\,dx$$

Solution of Differential Equations The solution to a differential equation is a function [say $y = f(x)$] that, when used in the differential equation, results in an identity.

EXAMPLE 1 **Differential Equation**

Show that $y = 4e^{-5t}$ is a solution to $dy/dt + 5y = 0$.

Solution

We must show that substituting $y = 4e^{-5t}$ into the equation $dy/dt + 5y = 0$ results in an identity:

$$\frac{d}{dt}(4e^{-5t}) + 5(4e^{-5t}) = 0$$
$$-20e^{-5t} + 20e^{-5t} = 0$$
$$0 = 0$$

Thus $y = 4e^{-5t}$ is a solution.

Now that we know what it means for a function to be a solution to a differential equation, let us consider how to find solutions.

The most elementary differential equations are of the form

$$\frac{dy}{dx} = f(x)$$

where $f(x)$ is a continuous function. These equations are elementary to solve because the solutions are found by integration:

$$y = \int f(x)\, dx$$

EXAMPLE 2   Solving a Differential Equation

Find the solution of

$$f'(x) = \frac{1}{x+1}$$

Solution
The solution is

$$f(x) = \int f'(x)\, dx = \int \frac{1}{x+1}\, dx = \ln|x+1| + C \qquad \blacksquare$$

The solution in Example 2, $f(x) = \ln|x+1| + C$, is called the **general solution** because every solution to the equation has this form, and different values of C give different **particular solutions.** Figure 12.9 shows the graphs of several members of the family of solutions to this differential equation. (We cannot, of course, show all of them.)

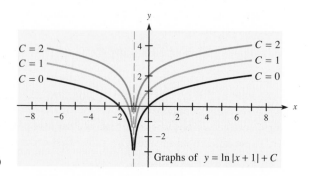

Figure 12.9

Graphs of $y = \ln|x+1| + C$

We can find a particular solution to a differential equation when we know that the solution must satisfy additional conditions, such as **initial conditions** or **boundary conditions.** For instance, to find the particular solution to

$$f'(x) = \frac{1}{x+1} \quad \text{with the condition that} \quad f(-2) = 2$$

we use $f(-2) = 2$ in the general solution, $f(x) = \ln|x+1| + C$.

$$2 = f(-2) = \ln|-2 + 1| + C$$
$$2 = \ln|-1| + C \text{ so } C = 2$$

Thus the particular solution is

$$f(x) = \ln|x+1| + 2$$

and is shown in Figure 12.9 with $C = 2$.

We frequently denote the value of the solution function $y = f(t)$ at the initial time $t = 0$ as $y(0)$ instead of $f(0)$.

✓ CHECKPOINT

1. Given $f'(x) = 2x - [1/(x + 1)]$, $f(0) = 4$,
   (a) find the general solution to the differential equation.
   (b) find the particular solution that satisfies $f(0) = 4$.

Just as we can find the differential of both sides of an equation, we can find the solution to a differential equation of the form

$$g(y)\, dy = f(x)\, dx$$

by integrating both sides.

EXAMPLE 3   Differential Equation with a Boundary Condition

Solve $3y^2\, dy = 2x\, dx$, if $y(1) = 2$.

Solution
We find the general solution by integrating both sides.

$$\int 3y^2\, dy = \int 2x\, dx$$
$$y^3 + C_1 = x^2 + C_2$$
$$y^3 = x^2 + C, \qquad \text{where } C = C_2 - C_1$$

By using $y(1) = 2$, we can find C.

$$2^3 = 1^2 + C$$
$$7 = C$$

Thus the particular solution is given implicitly by

$$y^3 = x^2 + 7$$

Calculator Note

We can use a graphing calculator to help solve a differential equation of the form $\dfrac{dy}{dx} = f(x)$, with boundary conditions. After integrating to get $y = F(x) + C$, we can use SOLVER with the known x-value and corresponding y-value to find C. The general solution to $\dfrac{dy}{dx} = 2x - 4$ is $y = x^2 - 4x + C$. The value of C that satisfies the boundary condition $y = 8$ when $x = -1$ is found to be 3, as shown in Figure 12.10; thus the particular solution is $y = x^2 - 4x + 3$. Details can be found in Appendix A, Section 12.5.

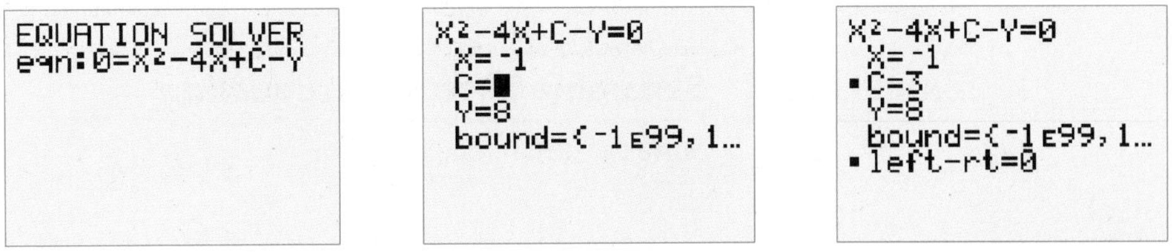

```
EQUATION SOLVER
eqn:0=X²-4X+C-Y
```

```
X²-4X+C-Y=0
 X=-1
 C=■
 Y=8
 bound={-1E99,1…
```

```
X²-4X+C-Y=0
 X=-1
▪C=3
 Y=8
 bound={-1E99,1…
▪left-rt=0
```

Figure 12.10

Separable Differential Equations

It is frequently necessary to change the form of a differential equation before it can be solved by integrating both sides.

For example, the equation

$$\frac{dy}{dx} = y^2$$

cannot be solved by simply integrating both sides of the equation with respect to x because we cannot evaluate $\int y^2\, dx$.

However, we can multiply both sides of $dy/dx = y^2$ by dx/y^2 to obtain an equation that has all terms containing y on one side of the equation and all terms containing x on the other side. That is, we obtain

$$\frac{dy}{y^2} = dx$$

Separable Differential Equations

When a differential equation can be equivalently expressed in the form

$$g(y)\, dy = f(x)\, dx$$

we say that the equation is **separable.**

The solution of a separable differential equation is obtained by integrating both sides of the equation after the variables have been separated.

EXAMPLE 4   Separable Differential Equation

Solve the differential equation

$$(x^2 y + x^2)\, dy = x^3\, dx$$

Solution

To write the equation in separable form, we first factor x^2 from the left side and divide both sides by it.

$$x^2(y + 1)\, dy = x^3\, dx$$

$$(y + 1)\, dy = \frac{x^3}{x^2}\, dx$$

The equation is now separated, so we integrate both sides.

$$\int (y + 1)\, dy = \int x\, dx$$

$$\frac{y^2}{2} + y + C_1 = \frac{x^2}{2} + C_2$$

This equation, as well as the equation

$$y^2 + 2y - x^2 = C, \quad \text{where} \quad C = 2(C_2 - C_1)$$

gives the solution implicitly.

Note that we need not write both C_1 and C_2 when we integrate because it is always possible to combine the two constants into one.

EXAMPLE 5   Separable Differential Equation

Solve the differential equation

$$\frac{dy}{dt} = ky \quad (k = \text{constant})$$

Solution

To solve the equation, we write it in separated form and integrate both sides:

$$\frac{dy}{y} = k\, dt \implies \int \frac{dy}{y} = \int k\, dt \implies \ln|y| = kt + C_1$$

Assuming that $y > 0$ and writing this equation in exponential form gives

$$y = e^{kt + C_1}$$

$$y = e^{kt} \cdot e^{C_1} = Ce^{kt}, \quad \text{where } C = e^{C_1}$$

This solution,

$$y = Ce^{kt}$$

is the general solution to the differential equation $dy/dt = ky$ because all solutions have this form, with different values of C giving different particular solutions. The case of $y < 0$ is covered by values of $C < 0$.   ■

✓ CHECKPOINT

2. True or false:
    (a) The general solution to $dy = (x/y)\, dx$ can be found from

    $$\int y\, dy = \int x\, dx$$

    (b) The first step in solving $dy/dx = -2xy^2$ is to separate it.
    (c) The equation $dy/dx = -2xy^2$ separates as $y^2\, dy = -2x\, dx$.
3. Suppose that $(xy + x)(dy/dx) = x^2 y + y$.
    (a) Separate this equation.   (b) Find the general solution.

In many applied problems that can be modeled with differential equations, we know conditions that allow us to obtain a particular solution.

Applications of Differential Equations   We now consider two applications that can be modeled by differential equations. These are radioactive decay and one-container mixture problems (as a model for drugs in an organ).

EXAMPLE 6   **Carbon-14 Dating | APPLICATION PREVIEW |**

When an organism dies, the rate of change of the amount of carbon-14 present is proportional to the amount present and is represented by the differential equation

$$\frac{dy}{dt} = ky$$

where y is the amount present, k is a constant, and t is time in years. If we denote the initial amount of carbon-14 in an organism as y_0, then $y = y_0$ represents the amount present at time $t = 0$ (when the organism dies). Suppose that anthropologists discover a fossil that contains 1% of the initial amount of carbon-14. Find the age of the fossil. (Recall that the half-life of carbon-14 is 5730 years.)

Solution
We must find a particular solution to

$$\frac{dy}{dt} = ky$$

subject to the fact that when $t = 0$, $y = y_0$, and we must determine the value of k on the basis of the half-life of carbon-14 ($t = 5730$ years, $y = \frac{1}{2}y_0$ units.) From Example 5, we know that the general solution to the differential equation $dy/dt = ky$ is $y = Ce^{kt}$. Using $y = y_0$ when $t = 0$, we obtain $y_0 = C$, so the equation becomes $y = y_0 e^{kt}$. Then using $t = 5730$ and $y = \frac{1}{2}y_0$ in this equation gives

$$\frac{1}{2}y_0 = y_0 e^{5730k} \quad \text{or} \quad 0.5 = e^{5730k}$$

Rewriting this equation in logarithmic form and then solving for k, we get

$$\ln(0.5) = 5730k$$
$$-0.69315 \approx 5730k$$
$$-0.00012097 \approx k$$

Thus the equation we seek is

$$y = y_0 e^{-0.00012097t}$$

Using the fact that $y = 0.01y_0$ when the fossil was discovered, we can find its age t by solving

$$0.01y_0 = y_0 e^{-0.00012097t} \quad \text{or} \quad 0.01 = e^{-0.00012097t}$$

Rewriting this in logarithmic form and then solving give

$$\ln{(0.01)} = -0.00012097t$$
$$-4.6051702 = -0.00012097t$$
$$38{,}069 \approx t$$

Thus the fossil is approximately 38,069 years old.   ∎

The differential equation $dy/dt = ky$, which describes the decay of radioactive substances in Example 6, also models the rate of growth of an investment that is compounded continuously and the rate of decay of purchasing power due to inflation.

Another application of differential equations comes from a group of applications called *one-container mixture problems*. In problems of this type, there is a substance whose amount in a container is changing with time, and the goal is to determine the amount of the substance at any time t. The differential equations that model these problems are of the following form:

$$\begin{bmatrix} \text{Rate of change} \\ \text{of the amount} \\ \text{of the substance} \end{bmatrix} = \begin{bmatrix} \text{Rate at which} \\ \text{the substance} \\ \text{enters the container} \end{bmatrix} - \begin{bmatrix} \text{Rate at which} \\ \text{the substance} \\ \text{leaves the container} \end{bmatrix}$$

We consider this application as it applies to the amount of a drug in an organ.

EXAMPLE 7   Drug in an Organ

A liquid carries a drug into an organ of volume 300 cubic centimeters at a rate of 5 cubic centimeters per second, where the liquid becomes well-mixed and then leaves the organ at the same rate. If the concentration of the drug in the entering liquid is 0.1 grams per cubic centimeter, and if x represents the amount of drug in the organ at any time t, then using the fact that the rate of change of the amount of the drug in the organ, dx/dt, equals the rate at which the drug enters minus the rate at which it leaves, we have

$$\frac{dx}{dt} = \left(\frac{5 \text{ cc}}{\text{s}}\right)\left(\frac{0.1 \text{ g}}{\text{cc}}\right) - \left(\frac{5 \text{ cc}}{\text{s}}\right)\left(\frac{x \text{ g}}{300 \text{ cc}}\right)$$

or

$$\frac{dx}{dt} = 0.5 - \frac{x}{60} = \frac{30}{60} - \frac{x}{60} = \frac{30 - x}{60}, \text{ in grams per second}$$

Find the amount of the drug in the organ as a function of time t.

Solution

Multiplying both sides of the equation $\dfrac{dx}{dt} = \dfrac{30 - x}{60}$ by $\dfrac{dt}{(30 - x)}$ gives

$$\frac{dx}{30 - x} = \frac{1}{60} dt$$

The equation is now separated, so we can integrate both sides.

$$\int \frac{dx}{30 - x} = \int \frac{1}{60} dt$$

$$-\ln{(30 - x)} = \frac{1}{60}t + C_1 \quad (30 - x > 0)$$

$$\ln{(30 - x)} = -\frac{1}{60}t - C_1$$

Rewriting this in exponential form gives

$$30 - x = e^{-t/60 - C_1} = e^{-t/60} \cdot e^{-C_1}$$

Letting $C = e^{-C_1}$ yields

$$30 - x = Ce^{-t/60}$$

so

$$x = 30 - Ce^{-t/60}$$

and we have the desired function.   ■

✓ CHECKPOINT ANSWERS

1. (a) $f(x) = \int \left(2x - \dfrac{1}{x+1}\right) dx = x^2 - \ln|x+1| + C$

   (b) $f(x) = x^2 - \ln|x+1| + 4$

2. (a) True

   (b) True

   (c) False. It separates as $dy/y^2 = -2x\, dx$.

3. (a) $\dfrac{y+1}{y}\, dy = \dfrac{x^2+1}{x}\, dx$ (b) $y + \ln|y| = \dfrac{x^2}{2} + \ln|x| + C$

| EXERCISES | 12.5

In Problems 1–4, show that the given function is a solution to the differential equation.

1. $y = x^2$; $4y - 2xy' = 0$
2. $y = x^3$; $3y - xy' = 0$
3. $y = 3x^2 + 1$; $2y\, dx - x\, dy = 2\, dx$
4. $y = 4x^3 + 2$; $3y\, dx - x\, dy = 6\, dx$

In Problems 5–10, use integration to find the general solution to each differential equation.

5. $dy = xe^{x^2+1}\, dx$
6. $dy = x^2 e^{x^3-1}\, dx$
7. $2y\, dy = 4x\, dx$
8. $4y\, dy = 4x^3\, dx$
9. $3y^2\, dy = (2x - 1)\, dx$
10. $4y^3\, dy = (3x^2 + 2x)\, dx$

In Problems 11–14, find the particular solution.

11. $y' = e^{x-3}$; $y(0) = 2$
12. $y' = e^{2x+1}$; $y(0) = e$
13. $dy = \left(\dfrac{1}{x} - x\right) dx$; $y(1) = 0$
14. $dy = \left(x^2 - \dfrac{1}{x+1}\right) dx$; $y(0) = \dfrac{1}{3}$

In Problems 15–28, find the general solution to the given differential equation.

15. $\dfrac{dy}{dx} = \dfrac{x^2}{y}$
16. $y^3\, dx = \dfrac{dy}{x^3}$
17. $dx = x^3 y\, dy$
18. $dy = x^2 y^3\, dx$
19. $dx = (x^2 y^2 + x^2)\, dy$
20. $dy = (x^2 y^3 + xy^3)\, dx$
21. $y^2\, dx = x\, dy$
22. $y\, dx = x\, dy$
23. $\dfrac{dy}{dx} = \dfrac{x}{y}$
24. $\dfrac{dy}{dx} = \dfrac{x^2 + x}{y+1}$
25. $(x + 1)\dfrac{dy}{dx} = y$
26. $x^2 y\dfrac{dy}{dx} = y^2 + 1$

27. $(x^2 + 2)e^{y^2} dx = xy\, dy$
28. $e^{4x}(y + 1)\, dx + e^{2x} y\, dy = 0$

In Problems 29–36, find the particular solution to each differential equation.

29. $\dfrac{dy}{dx} = \dfrac{x^2}{y^3}$ when $x = 1, y = 1$
30. $\dfrac{dy}{dx} = \dfrac{x+1}{xy}$ when $x = 1, y = 3$
31. $2y^2\, dx = 3x^2\, dy$ when $x = 2, y = -1$
32. $(x + 1)\, dy = y^2\, dx$ when $x = 0, y = 2$
33. $x^2 e^{2y}\, dy = (x^3 + 1)\, dx$ when $x = 1, y = 0$
34. $y' = \dfrac{1}{xy}$ when $x = 1, y = 3$
35. $2xy\dfrac{dy}{dx} = y^2 + 1$ when $x = 1, y = 2$
36. $xe^y\, dx = (x + 1)\, dy$ when $x = 0, y = 0$

APPLICATIONS

37. *Allometric growth* If x and y are measurements of certain parts of an organism, then the rate of change of y with respect to x is proportional to the ratio of y to x. That is, if k is a constant, then these measurements satisfy

$$\frac{dy}{dx} = k\frac{y}{x}$$

which is referred to as an allometric law of growth. Solve this differential equation.

38. *Bimolecular chemical reactions* A bimolecular chemical reaction is one in which two chemicals react to form another substance. Suppose that one molecule of each of the two chemicals reacts to form two molecules of a new substance. If x represents the number

of molecules of the new substance at time t, then the rate of change of x is proportional to the product of the numbers of molecules of the original chemicals available to be converted. That is, if each of the chemicals initially contained A molecules, then

$$\frac{dx}{dt} = k(A - x)^2$$

where k is a constant. If 40% of the initial amount A is converted after 1 hour, how long will it be before 90% is converted?

Compound interest **In Problems 39 and 40, use the following information.**

When interest is compounded continuously, the rate of change of the amount x of the investment is proportional to the amount present. In this case, the proportionality constant is the annual interest rate r (as a decimal); that is,

$$\frac{dx}{dt} = rx$$

39. (a) If $10,000 is invested at 6%, compounded continuously, find an equation for the future value of the investment as a function of time t in years.
   (b) What is the future value of the investment after 1 year? After 5 years?
   (c) How long will it take for the investment to double?

40. (a) If $2000 is invested at 8%, compounded continuously, find an equation for the future value of the investment as a function of time t, in years.
   (b) How long will it take for the investment to double?
   (c) What will be the future value of this investment after 35 years?

41. *Investing* When the interest on an investment is compounded continuously, the investment grows at a rate that is proportional to the amount in the account, so that if the amount present is P, then

$$\frac{dP}{dt} = kP$$

where P is in dollars, t is in years, and k is a constant. If $100,000 is invested (when $t = 0$) and the amount in the account after 15 years is $211,700, find the function that gives the value of the investment as a function of t. What is the interest rate on this investment?

42. *Investing* When the interest on an investment is compounded continuously, the investment grows at a rate that is proportional to the amount in the account. If $20,000 is invested (when $t = 0$) and the amount in the account after 22 years is $280,264, find the function that gives the value of the investment as a function of t. What is the interest rate on this investment?

43. *Bacterial growth* Suppose that the growth of a certain population of bacteria satisfies

$$\frac{dy}{dt} = ky$$

where y is the number of organisms, t is the number of hours, and k is a constant. If initially there are 10,000 organisms and the number triples after 2 hours, how long will it be before the population reaches 100 times the original population?

44. *Bacterial growth* Suppose that, for a certain population of bacteria, growth occurs according to

$$\frac{dy}{dt} = ky \quad (t \text{ in hours, } k \text{ constant})$$

If the doubling rate depends on temperature, find how long it takes for the number of bacteria to reach 50 times the original number at each given temperature in parts (a) and (b).
   (a) At 90°F, the number doubles after 30 minutes ($\frac{1}{2}$ hour).
   (b) At 40°F, the number doubles after 3 hours.

45. *Sales and pricing* Suppose that in a certain company, the relationship between the price per unit p of its product and the weekly sales volume y, in thousands of dollars, is given by

$$\frac{dy}{dp} = -\frac{2}{5}\left(\frac{y}{p + 8}\right)$$

Solve this differential equation if $y = 8$ when $p = 24$.

46. *Sales and pricing* Suppose that a chain of auto service stations, Quick-Oil, Inc., has found that the relationship between its price p for an oil change and its monthly sales volume y, in thousands of dollars, is

$$\frac{dy}{dp} = -\frac{1}{2}\left(\frac{y}{p + 5}\right)$$

Solve this differential equation if $y = 18$ when $p = 20$.

47. *Half-life* A breeder reactor converts uranium-238 into an isotope of plutonium-239 at a rate proportional to the amount present at any time. After 10 years, 0.03% of the radioactivity has dissipated (that is, 0.9997 of the initial amount remains). Suppose that initially there is 100 pounds of this substance. Find the half-life.

48. *Radioactive decay* A certain radioactive substance has a half-life of 50 hours. Find how long it will take for 90% of the radioactivity to be dissipated if the amount of material x satisfies

$$\frac{dx}{dt} = kx \quad (t \text{ in hours, } k \text{ constant})$$

49. *Drug in an organ* Suppose that a liquid carries a drug into a 100-cc organ at a rate of 5 cc/s and leaves the organ at the same rate. Suppose that the concentration of the drug entering is 0.06 g/cc. If initially there is no drug in the organ, find the amount of drug in the organ as a function of time t.

50. *Drug in an organ* Suppose that a liquid carries a drug into a 250-cc organ at a rate of 10 cc/s and leaves the organ at the same rate. Suppose that the concentration

of the drug entering is 0.15 g/cc. Find the amount of drug in the organ as a function of time t if initially there is none in the organ.

51. **Drug in an organ** Suppose that a liquid carries a drug with concentration 0.1 g/cc into a 200-cc organ at a rate of 5 cc/s and leaves the organ at the same rate. If initially there is 10 g of the drug in the organ, find the amount of drug in the organ as a function of time t.

52. **Drug in an organ** Suppose that a liquid carries a drug with concentration 0.05 g/cc into a 150-cc organ at a rate of 6 cc/s and leaves at the same rate. If initially there is 1.5 g of drug in the organ, find the amount of drug in the organ as a function of time t.

53. **Tumor volume** Let V denote the volume of a tumor, and suppose that the growth rate of the tumor satisfies

$$\frac{dV}{dt} = 0.2Ve^{-0.1t}$$

If the initial volume of the tumor is 1.86 units, find an equation for V as a function of t.

54. **Gompertz curves** The differential equation

$$\frac{dx}{dt} = x(a - b \ln x)$$

where x represents the number of objects at time t, and a and b are constants, is the model for Gompertz curves. Recall from Section 5.3, "Solutions of Exponential Equations," that Gompertz curves can be used to study growth or decline of populations, organizations, and revenue from sales of a product, as well as forecast equipment maintenance costs. Solve the differential equation to obtain the Gompertz curve formula

$$x = e^{a/b}e^{-ce^{-bt}}$$

55. **Cell growth** If V is the volume of a spherical cell, then in certain cell growth and for some fetal growth models, the rate of change of V is given by

$$\frac{dV}{dt} = kV^{2/3}$$

where k is a constant depending on the organism. If $V = 0$ when $t = 0$, find V as a function of t.

56. **Atmospheric pressure** The rate of change of atmospheric pressure P with respect to the altitude above sea level h is proportional to the pressure. That is,

$$\frac{dP}{dh} = kP \quad (k \text{ constant})$$

Suppose that the pressure at sea level is denoted by P_0, and at 18,000 ft the pressure is half what it is at sea level. Find the pressure, as a percent of P_0, at 25,000 ft.

57. **Newton's law of cooling** Newton's law of cooling (and warming) states that the rate of change of temperature $u = u(t)$ of an object is proportional to the temperature difference between the object and its surroundings, where T is the constant temperature of the surroundings. That is,

$$\frac{du}{dt} = k(u - T) \quad (k \text{ constant})$$

Suppose an object at 0°C is placed in a room where the temperature is 20°C. If the temperature of the object is 8°C after 1 hour, how long will it take for the object to reach 18°C?

58. **Newton's law of cooling** Newton's law of cooling can be used to estimate time of death. (Actually the estimate may be quite rough because cooling does not begin until metabolic processes have ceased.) Suppose a corpse is discovered at noon in a 70°F room and at that time the body temperature is 96.1°F. If at 1:00 P.M. the body temperature is 94.6°F, use Newton's law of cooling to estimate the time of death.

59. **Fossil-fuel emissions** Carbon in the atmosphere is due to carbon dioxide (CO_2) emissions from fossil-fuel burning and is considered to be a primary contributor to climate change. Using data from the Organization for Economic Cooperation and Development (OECD) for selected years from 1980 and projected to 2050, global CO_2 emissions can be modeled by the differential equation

$$\frac{dE}{dt} = 0.0164E$$

where t is the number of years past 1980 and E is global CO_2 emissions (in gigatons).
(a) Solve this differential equation and find a particular solution that satisfies $E(0) = 18.5$.
(b) Graph your solution and compare it with the following graph that shows OECD's data and projections.

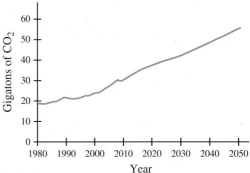

Global CO_2 Emissions: 1980–2050

Source: OECD Environmental Outlook Baseline

60. **Gross domestic product** With Social Security Administration data from 1995 and projected to 2070, the billions of dollars G of gross domestic product (GDP) can be modeled by

$$\frac{dG}{dt} = 0.05317G, \quad G(15) = 12{,}145$$

where t is the number of years past 1990 (and thus $G(15)$ is the GDP in 2005).

(a) Find the particular solution to this differential equation.

(b) The Social Security Administration's forecast for the 2020 GDP is $27,683 billion. What does the model predict for the 2020 GDP?

 61. *Impact of inflation* The impact of a 5% inflation rate on an $80,000-per-year pension can be severe. If P represents the purchasing power (in dollars) of an $80,000 pension, then the effect of a 5% inflation rate can be modeled by the differential equation

$$\frac{dP}{dt} = -0.05P, \quad P(0) = 80,000$$

where t is in years.

(a) Find the particular solution to this differential equation.

(b) Find the purchasing power after 15 years.

62. *Alzheimer's disease* The rate of change of the number of Americans over age 65 with Alzheimer's disease (in millions per year) for the years 2000 through 2050 can be modeled by the differential equation

$$\frac{dA}{dt} = 0.0228A$$

where t is the number of years after 2000 (*Source:* Alzheimer's Association).

(a) Given that in 2010, Alzheimer's disease affected 5.15 million Americans over age 65, find the particular solution to the differential equation. This solution models the number of Americans over age 65 with Alzheimer's disease.

(b) Use the model found in part (a) to project the number of Americans over age 65 with Alzheimer's disease in 2035.

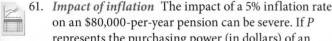

Chapter 12 Summary & Review

KEY TERMS AND FORMULAS

Section 12.1

General antiderivative of $f'(x)$ (p. 739)
$$f(x) + C$$
Indefinite integral (p. 739)
$$\int f(x)\, dx$$
Powers of x Formula (p. 740)
$$\int x^n\, dx = \frac{x^{n+1}}{n+1} + C \; (n \neq -1)$$

Integration Formulas (p. 741)
$$\int dx = x + C$$
$$\int cu(x)\, dx = c\int u(x)\, dx \quad (c = \text{a constant})$$
$$\int [u(x) \pm v(x)]\, dx = \int u(x)\, dx \pm \int v(x)\, dx$$

Section 12.2

Differentials (p. 746)
Power Rule (p. 747)

$$\int [u(x)]^n\, u'(x)\, dx = \frac{[u(x)]^{n+1}}{n+1} + C \; (n \neq -1)$$

Section 12.3

Exponential Formula (p. 755)
$$\int e^u u'\, dx = \int e^u\, du = e^u + C$$

Logarithmic Formula (p. 757)
$$\int \frac{u'}{u}\, dx = \int \frac{1}{u}\, du = \ln |u| + C$$

Section 12.4

Total cost (p. 764)
$$C(x) = \int \overline{MC}\, dx$$
Total revenue (p. 765)
$$R(x) = \int \overline{MR}\, dx$$
Profit (p. 766)
National consumption (p. 768)
$$C = \int f'(y)\, dy = \int \frac{dC}{dy}\, dy$$

Marginal propensity to consume (p. 768)
$$\frac{dC}{dy}$$
Marginal propensity to save (p. 768)
$$\frac{dS}{dy} = 1 - \frac{dC}{dy}$$

Section 12.5

Differential equations (p. 771)
General solution
Particular solution

First order (p. 771)
$$\frac{dy}{dx} = f(x) \Rightarrow y = \int f(x)\, dx$$

Separable (p. 774)

$$g(y)\, dy = f(x)\, dx \implies \int g(y)\, dy = \int f(x)\, dx$$

Radioactive decay (p. 775)

$$\frac{dy}{dt} = ky$$

Drug in an organ (p. 776)

Rate = (rate in) − (rate out)

REVIEW EXERCISES

Sections 12.1–12.3

Evaluate the integrals in Problems 1–26.

1. $\int x^6\, dx$

2. $\int x^{1/2}\, dx$

3. $\int (12x^3 - 3x^2 + 4x + 5)\, dx$

4. $\int 7(x^2 - 1)^2\, dx$

5. $\int 7x(x^2 - 1)^2\, dx$

6. $\int (x^3 - 3x^2)^5(x^2 - 2x)\, dx$

7. $\int (x^3 + 4)^2 3x\, dx$

8. $\int 5x^2(3x^3 + 7)^6\, dx$

9. $\int \frac{x^2}{x^3 + 1}\, dx$

10. $\int \frac{x^2}{(x^3 + 1)^2}\, dx$

11. $\int \frac{x^2\, dx}{\sqrt[3]{x^3 - 4}}$

12. $\int \frac{x^2\, dx}{x^3 - 4}$

13. $\int \frac{x^3 + 1}{x^2}\, dx$

14. $\int \frac{x^3 - 3x + 1}{x - 1}\, dx$

15. $\int y^2 e^{y^3}\, dy$

16. $\int (3x - 1)^{12}\, dx$

17. $\int \frac{3x^2}{2x^3 - 7}\, dx$

18. $\int \frac{5\, dx}{e^{4x}}$

19. $\int (x^3 - e^{3x})\, dx$

20. $\int x e^{1 + x^2}\, dx$

21. $\int \frac{6x^7}{(5x^8 + 7)^3}\, dx$

22. $\int \frac{7x^3}{\sqrt{1 - x^4}}\, dx$

23. $\int \left(\frac{e^{2x}}{2} + \frac{2}{e^{2x}} \right) dx$

24. $\int \left[x - \frac{1}{(x + 1)^2} \right] dx$

25. (a) $\int (x^2 - 1)^4 x\, dx$
    (b) $\int (x^2 - 1)^{10} x\, dx$
    (c) $\int (x^2 - 1)^7 3x\, dx$
    (d) $\int (x^2 - 1)^{-2/3} x\, dx$

26. (a) $\int \frac{2x\, dx}{x^2 - 1}$
    (b) $\int \frac{2x\, dx}{(x^2 - 1)^2}$
    (c) $\int \frac{3x\, dx}{\sqrt{x^2 - 1}}$
    (d) $\int \frac{3x\, dx}{x^2 - 1}$

Section 12.5

In Problems 27–32, find the general solution to each differential equation.

27. $\dfrac{dy}{dt} = 4.6e^{-0.05t}$

28. $dy = (64 + 76x - 36x^2)\, dx$

29. $\dfrac{dy}{dx} = \dfrac{4x}{y - 3}$

30. $t\, dy = \dfrac{dt}{y + 1}$

31. $\dfrac{dy}{dx} = \dfrac{x}{e^y}$

32. $\dfrac{dy}{dt} = \dfrac{4y}{t}$

In Problems 33 and 34, find the particular solution to each differential equation.

33. $y' = \dfrac{x^2}{y + 1}, \quad y(0) = 4$

34. $y' = \dfrac{2x}{1 + 2y}, \quad y(2) = 0$

APPLICATIONS

Section 12.1

35. *Revenue* If the marginal revenue for a month for a product is $\overline{MR} = 0.06x + 12$ dollars per unit, find the total revenue from the sale of $x = 800$ units of the product.

36. *Productivity* Suppose that the rate of change of production of the average worker at a factory is given by

$$\frac{dp}{dt} = 27 + 24t - 3t^2, \quad 0 \le t \le 8$$

where p is the number of units the worker produces in t hours. How many units will the average worker produce in an 8-hour shift? (Assume that $p = 0$ when $t = 0$.)

Section 12.2

37. *Oxygen levels in water* The rate of change of the oxygen level (in mmol/l) per month in a body of water after an oil spill is given by

$$P'(t) = 400 \left[\frac{5}{(t + 5)^2} - \frac{50}{(t + 5)^3} \right]$$

where t is the number of months after the spill. What function gives the oxygen level P at any time t if $P = 400$ mmol/l when $t = 0$?

38. *Bacterial growth* A population of bacteria grows at the rate

$$\frac{dp}{dt} = \frac{100{,}000}{(t + 100)^2}$$

where p is the population and t is time. If the population is 1000 when $t = 1$, write the equation that gives the size of the population at any time t.

Section 12.3

39. *Market share* The rate of change of the market share (as a percent) a firm expects for a new product is

$$\frac{dy}{dt} = 2.4e^{-0.04t}$$

where t is the number of months after the product is introduced.
   (a) Write the equation that gives the expected market share y at any time t. (Note that $y = 0$ when $t = 0$.)
   (b) What market share does the firm expect after 1 year?

40. *Revenue* If the marginal revenue for a product is

$$\overline{MR} = \frac{800}{x + 2}, \text{ find the total revenue function.}$$

Section 12.4

41. *Cost* The marginal cost for a product is $\overline{MC} = 6x + 4$ dollars per unit, and the cost of producing 100 items is $31,400.
   (a) Find the fixed costs.
   (b) Find the total cost function.

42. *Profit* Suppose a product has a daily marginal revenue $\overline{MR} = 46$ and a daily marginal cost $\overline{MC} = 30 + \frac{1}{5}x$, both in dollars per unit. If the daily fixed cost is $200, how many units will give maximum profit and what is the maximum profit?

43. *National consumption* If consumption is $8.5 trillion when disposable income is $0, and if the marginal propensity to consume is

$$\frac{dC}{dy} = \frac{1}{\sqrt{2y + 16}} + 0.6$$

find the national consumption function.

44. *National consumption* Suppose that the marginal propensity to save is

$$\frac{dS}{dy} = 0.2 - 0.1e^{-2y}$$

and consumption is $7.8 trillion when disposable income is $0. Find the national consumption function.

Section 12.5

45. *Allometric growth* For many species of fish, the length L and weight W of a fish are related by

$$\frac{dW}{dL} = \frac{3W}{L}$$

The general solution to this differential equation expresses the allometric relationship between the length and weight of a fish. Find the general solution.

46. *Investment* When the interest on an investment is compounded continuously, the investment grows at a rate that is proportional to the amount in the account, so that if the amount present is P, then

$$\frac{dP}{dt} = kP \quad (k \text{ constant})$$

where P is in dollars, t is in years, and k is a constant.
   (a) Solve this differential equation to find the relationship.
   (b) Use properties of logarithms and exponential functions to write P as a function of t.
   (c) If $50,000 is invested (when $t = 0$) and the amount in the account after 10 years is $135,914, find the function that gives the value of the investment as a function of t.
   (d) In part (c), what does the value of k represent?

47. *Fossil dating* Radioactive beryllium is sometimes used to date fossils found in deep-sea sediment. The amount of radioactive material x satisfies

$$\frac{dx}{dt} = kx$$

Suppose that 10 units of beryllium are present in a living organism and that the half-life of beryllium is 4.6 million years. Find the age of a fossil if 20% of the original radioactivity is present when the fossil is discovered.

48. *Drug in an organ* Suppose that a liquid carries a drug into a 120-cc organ at a rate of 4 cc/s and leaves the organ at the same rate. If initially there is no drug in the organ and if the concentration of drug in the liquid is 3 g/cc, find the amount of drug in the organ as a function of time.

49. *Chemical mixture* A 300-gal tank initially contains a solution with 100 lb of a chemical. A mixture containing 2 lb/gal of the chemical enters the tank at 3 gal/min, and the well-stirred mixture leaves at the same rate. Find an equation that gives the amount of the chemical in the tank as a function of time. How long will it be before there is 500 lb of chemical in the tank?

Chapter 12   TEST

Evaluate the integrals in Problems 1–11.

1. $\int (6x^2 + 8x - 7)\, dx$

2. $\int (11 - 2x^3)\, dx$

3. $\int 5(x^2 - 1)\, dx$

4. $\int \left(4 + \sqrt{x} - \dfrac{1}{x^2}\right) dx$

5. $\int 6x^2(7 + 2x^3)^9\, dx$

6. $\int 5x^2(4x^3 - 7)^9\, dx$

7. $\int (3x^2 - 6x + 1)^{-3}(2x - 2)\, dx$

8. $\int \left(e^x + \dfrac{5}{x} - 1\right) dx$

9. $\int \dfrac{s^3}{2s^4 - 5}\, ds$

10. $\int 100e^{-0.01x}\, dx$

11. $\int 5y^3 e^{2y^4 - 1}\, dy$

12. Evaluate $\int \dfrac{4x^2}{2x + 1}\, dx$. Use long division.

13. If $\int f(x)\, dx = 2x^3 - x + 5e^x + C$, find $f(x)$.

14. Find the general solution to $f'(x) = \dfrac{x^2}{3} - \dfrac{5}{8}$.

In Problems 15 and 16, find the particular solution to each differential equation.

15. $y' = 4x^3 + 3x^2$, if $y(0) = 4$

16. $\dfrac{dy}{dx} = e^{4x}$, if $y(0) = 2$

17. Find the general solution of the separable differential equation $\dfrac{dy}{dx} = x^3 y^2$.

18. Suppose the rate of growth of the population of a city is predicted to be

$$\frac{dp}{dt} = 2000t^{1.04}$$

where p is the population and t is the number of years past 2015. If the population in the year 2015 is 50,000, what is the predicted population in the year 2025?

19. Suppose that the marginal cost for x units of a product is $\overline{MC} = 4x + 50$, the marginal revenue is $\overline{MR} = 500$, and the cost of the production and sale of 10 units is $1000. What is the profit function for this product?

20. Suppose the marginal propensity to save is given by

$$\frac{dS}{dy} = 0.22 - \frac{0.25}{\sqrt{0.5y + 1}}$$

and national consumption is $6.6 trillion when disposable income is $0. Find the national consumption function.

21. A certain radioactive material has a half-life of 100 days. If the amount of material present, x, satisfies

$$\frac{dx}{dt} = kx$$

where t is in days and k is constant, how long will it take for 90% of the radioactivity to dissipate?

22. Suppose that a liquid carries a drug with concentration 0.1 g/cc into a 160-cc organ at the rate of 4 cc/sec and leaves at the same rate. Find the amount of drug in the organ as a function of time t, if initially there is none in the organ.

I. Employee Production Rate

The manager of a plant has been instructed to hire and train additional employees to manufacture a new product. She must hire a sufficient number of new employees so that within 30 days they will be producing 2500 units of the product each day.

Because a new employee must learn an assigned task, production will increase with training. Suppose that research on similar projects indicates that production increases with training according to the learning curve, so that for the average employee, the rate of production per day is given by

$$\frac{dN}{dt} = be^{-at}$$

where N is the number of units produced per day after t days of training and a and b are constants that depend on the project. Because of experience with a similar project, the manager expects the rate for this project to be

$$\frac{dN}{dt} = 2.5e^{-0.05t}$$

The manager tested her training program with 5 employees and learned that the average employee could produce 11 units per day after 5 days of training. On the basis of this information, she must decide how many employees to hire and begin to train so that a month from now they will be producing 2500 units of the product per day. She estimates that it will take her 10 days to hire the employees, and thus she will have 15 days remaining to train them. She also expects a 10% attrition rate during this period.

How many employees would you advise the plant manager to hire? Check your advice by answering the following questions.

1. Use the expected rate of production and the results of the manager's test to find the function relating N and t—that is, $N = N(t)$.
2. Find the number of units the average employee can produce after 15 days of training. How many such employees would be needed to maintain a production rate of 2500 units per day?
3. Explain how you would revise this last result to account for the expected 10% attrition rate. How many new employees should the manager hire?

II. Supply and Demand

If p is the price in dollars of a given commodity at time t, then we can think of price as a function of time. Similarly, the number of units demanded by consumers q_d at any time, and the number of units supplied by producers q_s at any time, may also be considered as functions of time as well as functions of price.

Both the quantity demanded and the quantity supplied depend not only on the price at the time, but also on the direction and rate of change that consumers and producers ascribe to prices. For example, even when prices are high, if consumers feel that prices are rising, the demand may rise. Similarly, if prices are low but producers feel they may go lower, the supply may rise.

If we assume that prices are determined in the marketplace by supply and demand, then the equilibrium price is the one we seek.

Suppose the supply and demand functions for a certain commodity in a competitive market are given, in hundreds of units, by

$$q_s = 30 + p + 5\frac{dp}{dt}$$

$$q_d = 51 - 2p + 4\frac{dp}{dt}$$

where dp/dt denotes the rate of change of the price with respect to time. If, at $t = 0$, the market equilibrium price is \$12, we can express the market equilibrium price as a function of time.

Our goals are as follows.

A. To express the market equilibrium price as a function of time.
B. To determine whether there is price stability in the marketplace for this item (that is, to determine whether the equilibrium price approaches a constant over time).

To achieve these goals, do the following.

1. Set the expressions for q_s and q_d equal to each other.
2. Solve this equation for $\dfrac{dp}{dt}$.
3. Write this equation in the form $f(p)\,dp = g(t)\,dt$.
4. Integrate both sides of this separated differential equation.
5. Solve the resulting equation for p in terms of t.
6. Use the fact that $p = 12$ when $t = 0$ to find C, the constant of integration, and write the market equilibrium price p as a function of time t.
7. Find $\lim\limits_{t \to \infty} p$, which gives the price we can expect this product to approach. If this limit is finite, then for this item there is price stability in the marketplace. If $p \to \infty$ as $t \to \infty$, then price will continue to increase until economic conditions change.

13

Definite Integrals: Techniques of Integration

We saw some applications of the indefinite integral in Chapter 12. In this chapter we define the definite integral and discuss a theorem and techniques that are useful in evaluating or approximating it. We will also see how it can be used in many interesting applications, such as consumer's and producer's surplus and total value, present value, and future value of continuous income streams. Improper integrals can be used to find the capital value of a continuous income stream.

The topics and some representative applications discussed in this chapter include the following.

SECTIONS		APPLICATIONS
13.1	Area Under a Curve	Ore production, health care expenses
13.2	The Definite Integral: The Fundamental Theorem of Calculus	Income streams, product life
13.3	Area Between Two Curves	Gini coefficient of income, average cost
13.4	Applications of Definite Integrals in Business and Economics	Continuous income streams, consumer's surplus, producer's surplus
13.5	Using Tables of Integrals	Hepatitis C treatments, cost
13.6	Integration by Parts	Equipment present value, demographics
13.7	Improper Integrals and Their Applications	Capital value, probability density functions
13.8	Numerical Integration Methods: The Trapezoidal Rule and Simpson's Rule	Pharmaceutical testing, income distributions

Chapter **Warm-Up**

Prerequisite Problem Type	For Section	Answer	Section for Review
Simplify: $\dfrac{1}{n^3}\left[\dfrac{n(n+1)(2n+1)}{6} - \dfrac{2n(n+1)}{2} + n\right]$	**13.1**	$\dfrac{2n^2 - 3n + 1}{6n^2}$	0.7 Fractions
(a) If $F(x) = \dfrac{x^4}{4} + 4x + C$, what is $\quad F(4) - F(2)$? (b) If $F(x) = -\dfrac{1}{9}\ln\left(\dfrac{9 + \sqrt{81 - 9x^2}}{3x}\right)$ $\quad$ what is $F(3) - F(2)$?	**13.2** **13.5**	(a) 68 (b) $\dfrac{1}{9}\ln\left(\dfrac{3 + \sqrt{5}}{2}\right)$	1.2 Function notation
Find the limit: (a) $\lim\limits_{n\to\infty} \dfrac{n^2 + n}{2n^2}$ (b) $\lim\limits_{n\to\infty} \dfrac{2n^2 - 3n + 1}{6n^2}$ (c) $\lim\limits_{b\to\infty}\left(1 - \dfrac{1}{b}\right)$ (d) $\lim\limits_{b\to\infty}\left(\dfrac{-100{,}000}{e^{0.10b}} + 100{,}000\right)$	**13.1** **13.7**	(a) $\frac{1}{2}$ (b) $\frac{1}{3}$ (c) 1 (d) 100,000	9.2 Limits at infinity
Find the derivative of $y = \ln x$.	**13.6**	$\dfrac{1}{x}$	11.1 Derivatives of logarithmic functions
Integrate: (a) $\int (x^3 + 4)dx$ (b) $\int x\sqrt{x^2 - 9}\, dx$ (c) $\int e^{2x}\, dx$	**13.2–13.7**	(a) $\dfrac{x^4}{4} + 4x + C$ (b) $\frac{1}{3}(x^2 - 9)^{3/2} + C$ (c) $\frac{1}{2}e^{2x} + C$	12.1, 12.2, 12.3 Integration

13.1

- To use the sum of areas of rectangles to approximate the area under a curve
- To use Σ notation to denote sums
- To find the exact area under a curve

Area Under a Curve

■ | APPLICATION PREVIEW |■

One way to find the accumulated production (such as the production of ore from a mine) over a period of time is to graph the rate of production as a function of time and find the area under the resulting curve over a specified time interval. For example, if a coal mine produces at a rate of 30 tons per day, the production over 10 days ($30 \cdot 10 = 300$) could be represented by the area under the line $y = 30$ between $x = 0$ and $x = 10$ (see Figure 13.1).

Using area to determine the accumulated production is very useful when the rate-of-production function varies at different points in time. For example, if the rate of production (in tons per day) is represented by

$$y = 100e^{-0.1x}$$

where x represents the number of days, then the area under the curve (and above the x-axis) from $x = 0$ to $x = 10$ represents the total production over the 10-day period (see Figure 13.2(a) and Example 1). In order to determine the accumulated production and to solve other types of problems, we need a method for finding areas under curves. That is the goal of this section.

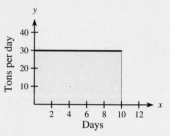

Figure 13.1

Area Under a Curve   To estimate the accumulated production for the example in the Application Preview, we approximate the area under the graph of the production rate function. We can find a rough approximation of the area under this curve by fitting two rectangles to the curve as shown in Figure 13.2(b). The area of the first rectangle is $5 \cdot 100 = 500$ square units, and the area of the second rectangle is $(10 - 5)[100e^{-0.1(5)}] \approx 303.27$ square units, so this rough approximation is 803.27 square units or 803.27 tons of ore. This approximation is clearly larger than the exact area under the curve. Why?

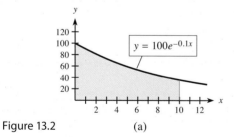

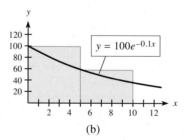

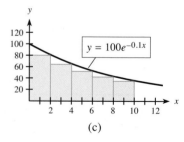

Figure 13.2   (a)   (b)   (c)

EXAMPLE 1   Ore Production | APPLICATION PREVIEW |

Find another, more accurate approximation of the tons of ore produced by approximating the area under the curve in Figure 13.2(a). Fit five rectangles with equal bases inside the area under the curve $y = 100e^{-0.1x}$, and use them to approximate the area under the curve from $x = 0$ to $x = 10$ (see Figure 13.2(c)).

Solution

Each of the five rectangles has base 2, and the height of each rectangle is the value of the function at the right-hand endpoint of the interval forming its base. Thus the areas of the rectangles are as follows.

Rectangle	Base	Height	Area = Base × Height
1	2	$100e^{-0.1(2)} \approx 81.87$	$2(81.87) = 163.74$
2	2	$100e^{-0.1(4)} \approx 67.03$	$2(67.03) = 134.06$
3	2	$100e^{-0.1(6)} \approx 54.88$	$2(54.88) = 109.76$
4	2	$100e^{-0.1(8)} \approx 44.93$	$2(44.93) = 89.86$
5	2	$100e^{-0.1(10)} \approx 36.79$	$2(36.79) = 73.58$

The area under the curve is approximately equal to

$$163.74 + 134.06 + 109.76 + 89.86 + 73.58 = 571$$

so approximately 571 tons of ore are produced in the 10-day period. The area is actually 632.12, to 2 decimal places (or 632.12 tons of ore), so the approximation 571 is smaller than the actual area but is much better than the one we obtained with just two rectangles. In general, if we use bases of equal width, the approximation of the area under a curve improves when more rectangles are used.   ■

Suppose that we wish to find the area between the curve $y = 2x$ and the x-axis from $x = 0$ to $x = 1$ (see Figure 13.3). As we saw in Example 1, one way to approximate this area is to use the areas of rectangles whose bases are on the x-axis and whose heights are the vertical distances from points on their bases to the curve. We can divide the interval $[0, 1]$ into n equal subintervals and use them as the bases of n rectangles whose heights are determined by the curve (see Figure 13.4). The width of each of these rectangles is $1/n$. Using the function value at the right-hand endpoint of each subinterval as the height of the rectangle, we get n rectangles as shown in Figure 13.4. Because part of each rectangle lies above the curve, the sum of the areas of the rectangles will overestimate the area.

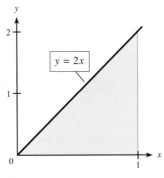

Figure 13.3

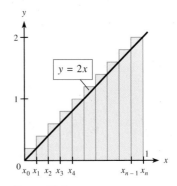

Figure 13.4

Then, with $y = f(x) = 2x$ and subinterval width $1/n$, the areas of the rectangles are as shown in the following table.

Rectangle	Base	Right Endpoint	Height	Area = Base × Height
1	$\dfrac{1}{n}$	$x_1 = \dfrac{1}{n}$	$f(x_1) = 2\left(\dfrac{1}{n}\right)$	$\dfrac{1}{n} \cdot \dfrac{2}{n} = \dfrac{2}{n^2}$
2	$\dfrac{1}{n}$	$x_2 = \dfrac{2}{n}$	$f(x_2) = 2\left(\dfrac{2}{n}\right)$	$\dfrac{1}{n} \cdot \dfrac{4}{n} = \dfrac{4}{n^2}$
3	$\dfrac{1}{n}$	$x_3 = \dfrac{3}{n}$	$f(x_3) = 2\left(\dfrac{3}{n}\right)$	$\dfrac{1}{n} \cdot \dfrac{6}{n} = \dfrac{6}{n^2}$
$\vdots$				
i	$\dfrac{1}{n}$	$x_i = \dfrac{i}{n}$	$f(x_i) = 2\left(\dfrac{i}{n}\right)$	$\dfrac{1}{n} \cdot \dfrac{2i}{n} = \dfrac{2i}{n^2}$
$\vdots$				
n	$\dfrac{1}{n}$	$x_n = \dfrac{n}{n}$	$f(x_n) = 2\left(\dfrac{n}{n}\right)$	$\dfrac{1}{n} \cdot \dfrac{2n}{n} = \dfrac{2n}{n^2}$

Note that $2i/n^2$ gives the area of the ith rectangle for *any* value of i. Thus for any value of n, this area can be approximated by the sum

$$A \approx \frac{2}{n^2} + \frac{4}{n^2} + \frac{6}{n^2} + \cdots + \frac{2i}{n^2} + \cdots + \frac{2n}{n^2}$$

In particular, we have the following approximations of this area for specific values of n (the number of rectangles).

$$n = 5: \quad A \approx \frac{2}{25} + \frac{4}{25} + \frac{6}{25} + \frac{8}{25} + \frac{10}{25} = \frac{30}{25} = 1.20$$

$$n = 10: \quad A \approx \frac{2}{100} + \frac{4}{100} + \frac{6}{100} + \cdots + \frac{20}{100} = \frac{110}{100} = 1.10$$

$$n = 100: \quad A \approx \frac{2}{10,000} + \frac{4}{10,000} + \frac{6}{10,000} + \cdots + \frac{200}{10,000}$$
$$= \frac{10,100}{10,000} = 1.01$$

Figure 13.5 shows the rectangles associated with each of these approximations ($n = 5$, 10, and 100) to the area under $y = 2x$ from $x = 0$ to $x = 1$. For larger n, the rectangles closely approximate the area under the curve.

Figure 13.5

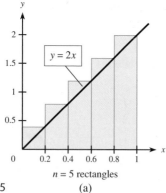

$n = 5$ rectangles
(a)

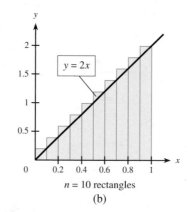

$n = 10$ rectangles
(b)

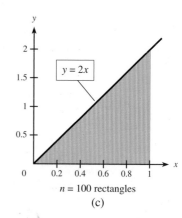

$n = 100$ rectangles
(c)

We can find this sum for any n more easily if we observe that the common denominator is n^2 and that the numerator is twice the sum of the first n terms of an arithmetic sequence with first term 1 and last term n. As you may recall from Section 6.1, "Simple Interest; Sequences," the first n terms of this arithmetic sequence add to $n(n + 1)/2$. Thus the area is approximated by

$$A \approx \frac{2(1 + 2 + 3 + \cdots + n)}{n^2} = \frac{2[n(n+1)/2]}{n^2} = \frac{n+1}{n}$$

Using this formula, we see the following.

$$n = 5: \quad A \approx \frac{5+1}{5} = \frac{6}{5} = 1.20$$

$$n = 10: \quad A \approx \frac{10+1}{10} = \frac{11}{10} = 1.10$$

$$n = 100: \quad A \approx \frac{100+1}{100} = \frac{101}{100} = 1.01$$

As Figure 13.5 indicated, as n gets larger, the number of rectangles increases, the area of each rectangle decreases, and the approximation becomes more accurate. If we let n increase without bound, the approximation approaches the exact area.

$$A = \lim_{n \to \infty} \frac{n+1}{n} = \lim_{n \to \infty} \left(1 + \frac{1}{n} \right) = 1$$

We can see that this area is correct, for we are computing the area of a triangle with base 1 and height 2. The formula for the area of a triangle gives

$$A = \frac{1}{2}bh = \frac{1}{2} \cdot 1 \cdot 2 = 1$$

Summation Notation

A special notation exists that uses the Greek letter Σ (capital sigma) to express the sum of numbers or expressions. (We used sigma notation informally in Chapter 8, "Further Topics in Probability; Data Description.") We may indicate the sum of the n numbers $a_1, a_2, a_3, a_4, \ldots, a_n$ by

$$\sum_{i=1}^{n} a_i = a_1 + a_2 + a_3 + \cdots + a_n$$

This may be read as "The sum of a_i as i goes from 1 to n." The subscript i in a_i is replaced first by 1, then by 2, then by 3, ... , until it reaches the value above the sigma. The i is called the **index of summation,** and it starts with the lower limit, 1, and ends with the upper limit, n. For example, if $x_1 = 2$, $x_2 = 3$, $x_3 = -1$, and $x_4 = -2$, then

$$\sum_{i=1}^{4} x_i = x_1 + x_2 + x_3 + x_4 = 2 + 3 + (-1) + (-2) = 2$$

The area of the triangle under $y = 2x$ that we discussed above was approximated by

$$A \approx \frac{2}{n^2} + \frac{4}{n^2} + \frac{6}{n^2} + \cdots + \frac{2i}{n^2} + \cdots + \frac{2n}{n^2}$$

Using **sigma notation,** we can write this sum as

$$A \approx \sum_{i=1}^{n} \left(\frac{2i}{n^2} \right)$$

Sigma notation allows us to represent the sums of the areas of the rectangles in an abbreviated fashion. Some formulas that simplify computations involving sums follow.

Sum Formulas

I. $\displaystyle\sum_{i=1}^{n} 1 = n$

II. $\displaystyle\sum_{i=1}^{n} cx_i = c \sum_{i=1}^{n} x_i \quad (c = \text{constant})$

III. $\displaystyle\sum_{i=1}^{n} (x_i + y_i) = \sum_{i=1}^{n} x_i + \sum_{i=1}^{n} y_i$

IV. $\displaystyle\sum_{i=1}^{n} i = \frac{n(n+1)}{2}$

V. $\displaystyle\sum_{i=1}^{n} i^2 = \frac{n(n+1)(2n+1)}{6}$

We have found that the area of the triangle discussed above was approximated by

$$A \approx \sum_{i=1}^{n} \frac{2i}{n^2}$$

We can use formulas II and IV to simplify this sum as follows.

$$\sum_{i=1}^{n} \frac{2i}{n^2} = \frac{2}{n^2} \sum_{i=1}^{n} i = \frac{2}{n^2} \left[\frac{n(n+1)}{2} \right] = \frac{n+1}{n}$$

Note that this is the same formula we obtained previously using other methods.

We can also use these sum formulas to evaluate a particular sum. For example,

$$\sum_{i=1}^{100}(2i^2 - 3) = \sum_{i=1}^{100} 2i^2 - \sum_{i=1}^{100} 3(1) \qquad \text{Formula III}$$

$$= 2\sum_{i=1}^{100} i^2 - 3\sum_{i=1}^{100} 1 \qquad \text{Formula II}$$

$$= 2\left[\frac{100(101)(201)}{6}\right] - 3(100) \qquad \text{Formulas I and V with } n = 100$$

$$= 676{,}400$$

Areas and Summation Notation

The following example shows that we can find the area by evaluating the function at the left-hand endpoints of the subintervals.

EXAMPLE 2   Area Under a Curve

Use rectangles to find the area under $y = x^2$ (and above the x-axis) from $x = 0$ to $x = 1$.

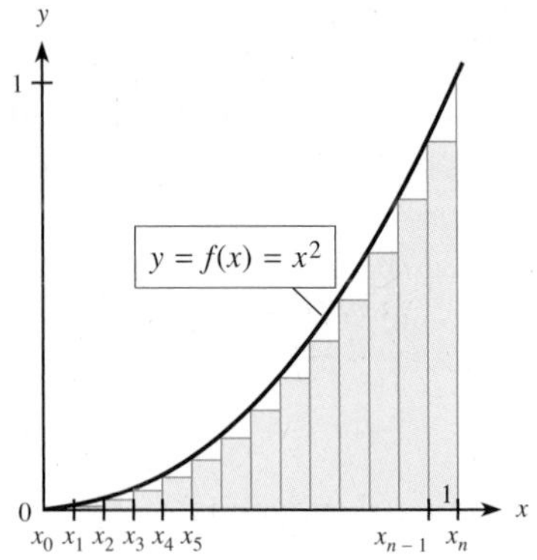

Solution
We again divide the interval $[0, 1]$ into n equal subintervals of length $1/n$. If we evaluate the function at the left-hand endpoints of these subintervals to determine the heights of the rectangles, the sum of the areas of the rectangles will underestimate the area (see Figure 13.6). Thus we have the information shown in the following table.

Figure 13.6

Rectangle	Base	Left Endpoint	Height	Area = Base $\times$ Height
1	$\dfrac{1}{n}$	$x_0 = 0$	$f(x_0) = 0$	$\dfrac{1}{n}\cdot 0 = 0$
2	$\dfrac{1}{n}$	$x_1 = \dfrac{1}{n}$	$f(x_1) = \dfrac{1}{n^2}$	$\dfrac{1}{n}\cdot\dfrac{1}{n^2} = \dfrac{1}{n^3}$
3	$\dfrac{1}{n}$	$x_2 = \dfrac{2}{n}$	$f(x_2) = \dfrac{4}{n^2}$	$\dfrac{1}{n}\cdot\dfrac{4}{n^2} = \dfrac{4}{n^3}$
4	$\dfrac{1}{n}$	$x_3 = \dfrac{3}{n}$	$f(x_3) = \dfrac{9}{n^2}$	$\dfrac{1}{n}\cdot\dfrac{9}{n^2} = \dfrac{9}{n^3}$
$\vdots$				
i	$\dfrac{1}{n}$	$x_{i-1} = \dfrac{i-1}{n}$	$\dfrac{(i-1)^2}{n^2}$	$\dfrac{(i-1)^2}{n^3}$
$\vdots$				
n	$\dfrac{1}{n}$	$x_{n-1} = \dfrac{n-1}{n}$	$\dfrac{(n-1)^2}{n^2}$	$\dfrac{(n-1)^2}{n^3}$

Thus: Area $= A \approx 0 + \dfrac{1}{n^3} + \dfrac{4}{n^3} + \dfrac{9}{n^3} + \cdots + \dfrac{(i-1)^2}{n^3} + \cdots + \dfrac{(n-1)^2}{n^3}$.

Note that $(i-1)^2/n^3 = (i^2 - 2i + 1)/n^3$ gives the area of the ith rectangle for *any* value of i. The sum of these areas may be written as

$$S = \sum_{i=1}^{n}\frac{i^2 - 2i + 1}{n^3} = \frac{1}{n^3}\left(\sum_{i=1}^{n} i^2 - 2\sum_{i=1}^{n} i + \sum_{i=1}^{n} 1\right) \qquad \text{Formulas II and III}$$

$$= \frac{1}{n^3}\left[\frac{n(n+1)(2n+1)}{6} - \frac{2n(n+1)}{2} + n\right] \qquad \text{Formulas V, IV, and I}$$

$$= \frac{2n^3 + 3n^2 + n}{6n^3} - \frac{n^2 + n}{n^3} + \frac{n}{n^3} = \frac{2n^2 - 3n + 1}{6n^2}$$

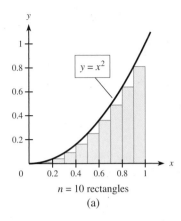

$y = x^2$

$n = 10$ rectangles
(a)

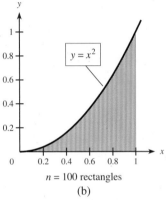

$y = x^2$

$n = 100$ rectangles
(b)

Figure 13.7

We can use this formula to find the approximate area (value of S) for different values of n.

$$\text{If } n = 10: \quad \text{Area} \approx S(10) = \frac{200 - 30 + 1}{600} = 0.285$$

$$\text{If } n = 100: \quad \text{Area} \approx S(100) = \frac{20{,}000 - 300 + 1}{60{,}000} = 0.328$$

Figure 13.7 shows the rectangles associated with each of these approximations.

As Figure 13.7 shows, the larger the value of n, the better the value of the sum approximates the exact area under the curve. If we let n increase without bound, we find the exact area.

$$A = \lim_{n \to \infty} \left(\frac{2n^2 - 3n + 1}{6n^2} \right) = \lim_{n \to \infty} \left(\frac{2 - \frac{3}{n} + \frac{1}{n^2}}{6} \right) = \frac{1}{3}$$

Note that the approximations with $n = 10$ and $n = 100$ are less than $\frac{1}{3}$. This is because all the rectangles are *under* the curve (see Figure 13.7). ■

Thus we see that we can determine the area under a curve $y = f(x)$ from $x = a$ to $x = b$ by dividing the interval $[a, b]$ into n equal subintervals of width $(b - a)/n$ and evaluating

$$A = \lim_{n \to \infty} S_R = \lim_{n \to \infty} \sum_{i=1}^{n} f(x_i) \left(\frac{b - a}{n} \right) \quad \text{(using right-hand endpoints)}$$

or

$$A = \lim_{n \to \infty} S_L = \lim_{n \to \infty} \sum_{i=1}^{n} f(x_{i-1}) \left(\frac{b - a}{n} \right) \quad \text{(using left-hand endpoints)}$$

✓ **CHECKPOINT**

$y = 3x - x^2$

1. For the interval $[0, 2]$, determine whether the following statements are true or false.
   (a) For 4 subintervals, each subinterval has width $\frac{1}{2}$.
   (b) For 200 subintervals, each subinterval has width $\frac{1}{100}$.
   (c) For n subintervals, each subinterval has width $\frac{2}{n}$.
   (d) For n subintervals, $x_0 = 0$, $x_1 = \frac{2}{n}$, $x_2 = 2\left(\frac{2}{n}\right)$, $\ldots$, $x_i = i\left(\frac{2}{n}\right)$, $\ldots$, $x_n = 2$.

2. Find the area under $y = f(x) = 3x - x^2$ from $x = 0$ to $x = 2$ using right-hand endpoints (see the figure). To accomplish this, use $\frac{b - a}{n} = \frac{2}{n}$, $x_i = \frac{2i}{n}$, and $f(x) = 3x - x^2$; find and simplify the following.
   (a) $f(x_i)$
   (b) $f(x_i)\left(\frac{b-a}{n}\right)$
   (c) $S_R = \sum_{i=1}^{n} f(x_i)\left(\frac{b-a}{n}\right)$
   (d) $A = \lim_{n \to \infty} S_R = \lim_{n \to \infty} \sum_{i=1}^{n} f(x_i)\left(\frac{b-a}{n}\right)$

Technology Note

Graphing calculators and Excel can be used to approximate the area under a curve. These technologies are especially useful when summations approximating the area do not simplify easily. Example 3 shows steps for using both a graphing calculator and Excel to approximate an area. See Appendices A and B, Section 13.1, for additional details. ■

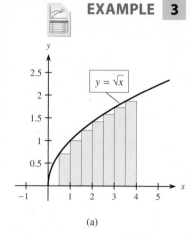

(a)

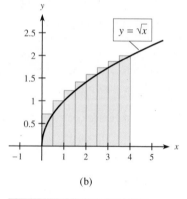

(b)

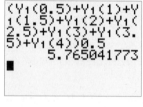

(c)

Figure 13.8

EXAMPLE 3 Estimating an Area with Technology

Approximate the area under the graph of $f(x) = \sqrt{x}$ on the interval $[0, 4]$ by using $n = 8$ rectangles with:

(a) a graphing calculator and both left-hand and right-hand endpoints.
(b) Excel and left-hand endpoints.

Solution

Figure 13.8(a) shows the graph of $y = \sqrt{x}$ with 8 rectangles whose heights are determined by evaluating the function at the left-hand endpoint of each interval (the first of these rectangles has height 0). Figure 13.8(b) shows the same graph with 8 rectangles whose heights are determined at the right-hand endpoints.

(a) To approximate the area with a graphing calculator, first enter the function as Y1. With $n = 8$, each rectangle has base width $(4 - 0)/8 = 0.5$. The subdivision values are $x_0 = 0$, $x_1 = 0.5$, $x_2 = 1, \ldots x_7 = 3.5$, $x_8 = 4$. Because $Y1(x_i)$ on a graphing calculator gives $\sqrt{x_i}$, the height of a rectangle at x_i, the left approximation of the area is

$$S_L = \sum_{i=1}^{8} f(x_{i-1})\Delta x = [Y1(x_0) + Y1(x_1) + \cdots + Y1(x_{8-1})]\Delta x$$

$$= [Y1(0) + Y1(0.5) + Y1(1) + Y1(1.5) + Y1(2) + Y1(2.5) + Y1(3) + Y1(3.5)](0.5) \approx 4.765$$

Figure 13.8(c) shows the right approximation of the area to be $S_R \approx 5.765$.

(b) To find the approximate area with Excel using the left-hand endpoints of rectangles with base width 0.5, we proceed as follows.
1. Enter "x" and the eight x-values of the left-hand endpoints of each rectangle in Column B, starting with 0.
2. Enter the formula for the function, =SQRT(B2), in C2 and fill down column C to get the function values that are the heights of the rectangles.
3. Enter the rectangle width, 0.5, in each cell of Column D.
4. Enter the formula for the area of each rectangle, =C2*D2, in E2 and fill down to get the area of each rectangle.
5. Enter "Total" in A10 and the formula =SUM(E2:E9) in E10 to get the approximate area, $S_L \approx 4.765$.

The following Excel output shows this S_L calculation. Note that S_R can be found by removing Row 2, adding a new row with $x = 4$, and proceeding as above.

	A	B	C	D	E
1	Rectangle	x	y	Width	Area
2	1	0	0	0.5	0
3	2	0.5	0.707107	0.5	0.353553
4	3	1	1	0.5	0.5
5	4	1.5	1.224745	0.5	0.612372
6	5	2	1.414214	0.5	0.707107
7	6	2.5	1.581139	0.5	0.790569
8	7	3	1.732051	0.5	0.866025
9	8	3.5	1.870829	0.5	0.935414
10	Total				4.765042

1. All parts are true.

2. (a) $f(x_i) = \dfrac{6i}{n} - \dfrac{4i^2}{n^2}$

(b) $f(x_i) \dfrac{b-a}{n} = \dfrac{12i}{n^2} - \dfrac{8i^2}{n^3}$

(c) $S_R = \dfrac{12}{n^2}\left[\dfrac{n(n+1)}{2}\right] - \dfrac{8}{n^3}\left[\dfrac{n(n+1)(2n+1)}{6}\right]$

$= \dfrac{6(n+1)}{n} - \dfrac{4(n+1)(2n+1)}{3n^2}$

(d) $A = \lim\limits_{n\to\infty} \sum\limits_{i=1}^{n} f(x_i)\dfrac{b-a}{n} = 6 - \dfrac{8}{3} = \dfrac{10}{3}$

| EXERCISES | 13.1

In Problems 1–4, approximate the area under each curve over the specified interval by using the indicated number of subintervals (or rectangles) and evaluating the function at the *right-hand* endpoints of the subintervals. (See Example 1.)

1. $f(x) = 4x - x^2$ from $x = 0$ to $x = 2$; 2 subintervals
2. $f(x) = x^3$ from $x = 0$ to $x = 3$; 3 subintervals
3. $f(x) = 9 - x^2$ from $x = 1$ to $x = 3$; 4 subintervals
4. $f(x) = x^2 + x + 1$ from $x = -1$ to $x = 1$; 4 subintervals

In Problems 5–8, approximate the area under each curve by evaluating the function at the *left-hand* endpoints of the subintervals.

5. $f(x) = 4x - x^2$ from $x = 0$ to $x = 2$; 2 subintervals
6. $f(x) = x^3$ from $x = 0$ to $x = 3$; 3 subintervals
7. $f(x) = 9 - x^2$ from $x = 1$ to $x = 3$; 4 subintervals
8. $f(x) = x^2 + x + 1$ from $x = -1$ to $x = 1$; 4 subintervals

When the area under $f(x) = x^2 + x$ from $x = 0$ to $x = 2$ is approximated, the formulas for the sum of n rectangles using *left-hand* endpoints and *right-hand* endpoints are

Left-hand endpoints: $S_L = \dfrac{14}{3} - \dfrac{6}{n} + \dfrac{4}{3n^2}$

Right-hand endpoints: $S_R = \dfrac{14n^2 + 18n + 4}{3n^2}$

Use these formulas to answer Problems 9–13.

9. Find $S_L(10)$ and $S_R(10)$.
10. Find $S_L(100)$ and $S_R(100)$.
11. Find $\lim\limits_{n\to\infty} S_L$ and $\lim\limits_{n\to\infty} S_R$.
12. Compare the right-hand and left-hand values by finding $S_R - S_L$ for $n = 10$, for $n = 100$, and as $n \to \infty$. (Use Problems 9–11.)
13. Because $f(x) = x^2 + x$ is increasing over the interval from $x = 0$ to $x = 2$, function values at the right-hand endpoints are maximum values for each subinterval, and function values at the left-hand endpoints are minimum values for each subinterval. How would

the approximate area using $n = 10$ and *any* other point within each subinterval compare with $S_L(10)$ and $S_R(10)$? What would happen to the area result as $n \to \infty$ if any other point in each subinterval were used?

In Problems 14–19, find the value of each sum.

14. $\sum\limits_{k=1}^{3} x_k$, if $x_1 = 1, x_2 = 3, x_3 = -1, x_4 = 5$

15. $\sum\limits_{i=1}^{4} x_i$, if $x_1 = 3, x_2 = -1, x_3 = 3, x_4 = -2$

16. $\sum\limits_{i=3}^{5}(i^2 + 1)$ 17. $\sum\limits_{j=2}^{5}(j^2 - 3)$

18. $\sum\limits_{i=4}^{7}\left(\dfrac{i-3}{i^2}\right)$ 19. $\sum\limits_{j=0}^{4}(j^2 - 4j + 1)$

In Problems 20–25, use the sum formulas I–V to express each of the following without the summation symbol. In Problems 20–23, find the numerical value.

20. $\sum\limits_{k=1}^{50} 1$ 21. $\sum\limits_{j=1}^{60} 3$

22. $\sum\limits_{k=1}^{50}(6k^2 + 5)$ 23. $\sum\limits_{k=1}^{30}(k^2 + 4k)$

24. $\sum\limits_{i=1}^{n}\left(1 - \dfrac{i^2}{n^2}\right)\left(\dfrac{2}{n}\right)$ 25. $\sum\limits_{i=1}^{n}\left(1 - \dfrac{2i}{n} + \dfrac{i^2}{n^2}\right)\left(\dfrac{3}{n}\right)$

Use the function $y = 2x$ from $x = 0$ to $x = 1$ and n equal subintervals with the function evaluated at the *left-hand* endpoint of each subinterval for Problems 26 and 27.

26. What is the area of the
    (a) first rectangle?   (b) second rectangle?
    (c) ith rectangle?

27. (a) Find a formula for the sum of the areas of the n rectangles (call this S). Then find
    (b) $S(10)$.   (c) $S(100)$.
    (d) $S(1000)$.   (e) $\lim\limits_{n\to\infty} S$.

28. How do your answers to Problems 27(a)–(e) compare with the corresponding calculations in the discussion (after Example 1) of the area under $y = 2x$ using *right-hand* endpoints?

29. For parts (a)–(e), use the function $y = x^2$ from $x = 0$ to $x = 1$ with n equal subintervals and the function evaluated at the *right-hand* endpoints.
    (a) Find a formula for the sum of the areas of the n rectangles (call this S). Then find
    (b) $S(10)$.   (c) $S(100)$.
    (d) $S(1000)$.   (e) $\lim_{n \to \infty} S$.

30. How do your answers to Problems 29(a)–(e) compare with the corresponding calculations in Example 2?

31. Use rectangles to find the area between $y = x^2 - 6x + 8$ and the x-axis from $x = 0$ to $x = 2$. Divide the interval $[0, 2]$ into n equal subintervals so that each subinterval has length $2/n$.

32. Use rectangles to find the area between $y = 4x - x^2$ and the x-axis from $x = 0$ to $x = 4$. Divide the interval $[0, 4]$ into n equal subintervals so that each subinterval has length $4/n$.

APPLICATIONS

33. **Per capita health care expenses** The annual per capita out-of-pocket expenses (to the nearest dollar) for U.S. health care for selected years from 2013 and projected to 2021 are shown in the table and figure.
    (a) Use $n = 4$ equal subdivisions and left-hand endpoints to estimate the area under the graph from 2013 to 2021.
    (b) What does this area represent in terms of per capita out-of-pocket expenses for U.S. health care?

Year	Dollars
2013	1020
2015	1022
2017	1097
2019	1209
2021	1325

Source: U.S. Centers for Medicare and Medicaid Services

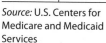
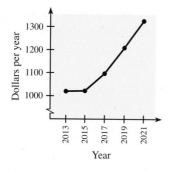

34. **Oil imports** Crude oil and petroleum products are imported continuously by the United States. The following table and figure show the net expenditures for U.S. oil imports for selected years (in billions of dollars per year) (*Source:* Energy Information Administration).
    (a) Use $n = 5$ equal subdivisions and left-hand endpoints to estimate the area under the graph from 2014 to 2024.
    (b) What does this area represent in terms of U.S. oil imports?

Year	Billions of Dollars
2014	219.5
2016	192.0
2018	191.0
2020	198.9
2022	214.5
2024	228.4

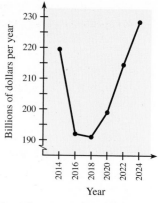

35. **Speed trials** The figure gives the times that it takes a Porsche 911 to reach speeds from 0 mph to 100 mph, in increments of 10 mph, with a curve connecting them. The area under this curve from $t = 0$ seconds to $t = 14$ seconds represents the total amount of distance traveled over the 14-second period. Count the squares under the curve to estimate this distance. This car will travel 1/4 mile in 14 seconds, to a speed of 100.2 mph. Is your estimate close to this result? (Be careful with time units.)

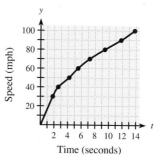

Source: Motor Trend

36. **Speed trials** The figure gives the times that it takes a Mitsubishi Eclipse GSX to reach speeds from 0 mph to 100 mph, in increments of 10 mph, with a curve connecting them. The area under this curve from $t = 0$ seconds to $t = 21.1$ seconds represents the total amount of distance traveled over the 21.1-second period. Count the squares under the curve to estimate this distance. This car will travel 1/4 mile in 15.4 seconds, to a speed of 89.0 mph, so your estimate should be more than 1/4 mile. Is it? (Be careful with time units.)

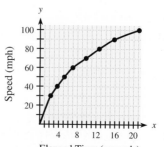

Source: Road & Track

37. *Pollution monitoring* Suppose the presence of phosphates in certain waste products dumped into a lake promotes the growth of algae. Rampant growth of algae affects the oxygen supply in the water, so an environmental group wishes to estimate the area of algae growth. The group measures the length across the algae growth (see the figure) and obtains the data (in feet) in the table.

x	Length	x	Length
0	0	50	27
10	15	60	24
20	18	70	23
30	18	80	0
40	30		

Use 8 rectangles with bases of 10 feet and lengths measured at the left-hand endpoints to approximate the area of the algae growth.

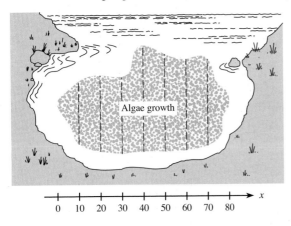

38. *Drug levels in the blood* The manufacturer of a medicine wants to test how a new 300-milligram capsule is released into the bloodstream. After a volunteer is given a capsule, blood samples are drawn every half-hour, and the number of milligrams of the drug in the bloodstream is calculated. The results obtained are shown in the table.

Time t (hr)	N(t) (mg)	Time t (hr)	N(t) (mg)
0	0	2.0	178.3
0.5	247.3	2.5	113.9
1.0	270.0	3.0	56.2
1.5	236.4	3.5	19.3

Use 7 rectangles, each with height $N(t)$ at the left endpoint and with width 0.5 hr, to estimate the area under the graph representing these data. Divide this area by 3.5 hr to estimate the average drug level over this time period.

39. *Emissions* With U.S. Department of Energy data for selected years from 2000 and projected to 2030, sulphur dioxide emissions from electricity generation (in millions of short tons per year) can be modeled by

$$E(x) = 0.0112x^2 + 0.612x + 11.9$$

where x is the number of years past 2000. Use $n = 10$ equal subdivisions and right-hand endpoints to approximate the area under the graph of $E(x)$ between $x = 10$ and $x = 15$. What does this area represent?

40. *Per capita income* The per capita personal income (in dollars per year) in the United States for selected years from 1960 and projected to 2018 can be modeled by

$$I(t) = 13.93t^2 + 136.8t + 1971$$

where t is the number of years past 1960 (*Source*: U.S. Bureau of Labor Statistics). Use $n = 10$ equal subdivisions with right-hand endpoints to approximate the area under the graph of $I(t)$ between $t = 50$ and $t = 55$. What does this area represent?

OBJECTIVES

13.2

- To evaluate definite integrals using the Fundamental Theorem of Calculus
- To use definite integrals to find the area under a curve

The Definite Integral: The Fundamental Theorem of Calculus

■ | APPLICATION PREVIEW | ■

Suppose that money flows continuously into a slot machine at a casino and grows at a rate given by $A'(t) = 100e^{0.1t}$, where t is the time in hours and $0 \leq t \leq 10$. Then the definite integral

$$\int_0^{10} 100\,e^{0.1t}\,dt$$

gives the total amount of money that accumulates over the 10-hour period, if no money is paid out. (See Example 6.)

(*continued*)

In the previous section, we used the sums of areas of rectangles to approximate the areas under curves. In this section, we will see how such sums are related to the definite integral and how to evaluate definite integrals. In addition, we will see how definite integrals can be used to solve several types of applied problems.

Riemann Sums and the Definite Integral

In the previous section, we saw that we could determine the area under a curve $y = f(x)$ over a closed interval $[a, b]$ by using equal subintervals and the function values at either the left-hand endpoints or the right-hand endpoints of the subintervals. In fact, we can use subintervals that are not of equal length, and we can use any point within each subinterval, denoted by x_i^*, to determine the height of each rectangle. For the ith rectangle (for any i), if we denote the width as Δx_i then the height is $f(x_i^*)$ and the area is $f(x_i^*)\Delta x_i$. Then, if $[a, b]$ is divided into n subintervals, the sum of the areas of the n rectangles is

$$S = \sum_{i=1}^{n} f(x_i^*)\, \Delta x_i$$

Such a sum is called a **Riemann sum** of f on $[a, b]$. Increasing the number of subintervals and making sure that every interval becomes smaller will in the long run improve the estimation. Thus for any subdivision of $[a, b]$ and any x_i^*, the exact area is given by

$$A = \lim_{\substack{\max \Delta x_i \to 0 \\ (n \to \infty)}} \sum_{i=1}^{n} f(x_i^*)\, \Delta x_i \qquad \text{provided that this limit exists}$$

In addition to giving the exact area, this limit of the Riemann sum has other important applications and is called the **definite integral** of $f(x)$ over interval $[a, b]$.

Definite Integral

If f is a function on the interval $[a, b]$, then, for any subdivision of $[a, b]$ and any choice of x_i^* in the ith subinterval, the **definite integral** of f from a to b is

$$\int_a^b f(x)dx = \lim_{\substack{\max \Delta x_i \to 0 \\ (n \to \infty)}} \sum_{i=1}^{n} f(x_i^*)\, \Delta x_i$$

If f is a continuous function, and $\Delta x_i \to 0$ as $n \to \infty$, then the limit exists and we say that f is integrable on $[a, b]$.

Note that for some intervals, values of f may be negative. In this case, the product $f(x_i^*)\Delta x_i$ will be negative and can be thought of geometrically as a "signed area." (Remember that area is a positive number.) Thus a definite integral can be thought of geometrically as the sum of signed areas, just as a derivative can be thought of geometrically as the slope of a tangent line. In the case where $f(x)$ is positive for all x from a to b, the definite integral equals the area between the graph of $y = f(x)$ and the x-axis.

Fundamental Theorem of Calculus

The obvious question is how this definite integral is related to the indefinite integral (the antiderivative) discussed in Chapter 12. The connection between these two concepts is the most important result in calculus, because it connects derivatives, indefinite integrals, and definite integrals.

To help see the connection, consider the marginal revenue function

$$R'(x) = 300 - 0.2x$$

and the revenue function

$$R(x) = \int (300 - 0.2x)\, dx = 300x - 0.1x^2$$

that is the indefinite integral of the marginal revenue function. (See Figure 13.9(a).)

Using this revenue function, we can find the revenue from the sale of 1000 units to be

$$R(1000) = 300(1000) - 0.1(1000)^2 = 200,000 \text{ dollars}$$

and the revenue from the sale of 500 units to be

$$R(500) = 300(500) - 0.1(500)^2 = 125,000 \text{ dollars}$$

Thus the additional revenue received from the sale of 500 units to 1000 units is

$$200,000 - 125,000 = 75,000 \text{ dollars}$$

If we used the definition of the definite integral (or geometry) to find the area under the graph of the marginal revenue function from $x = 500$ to $x = 1000$, we would find that the area is 75,000 (see Figure 13.9(b)). Note in Figure 13.9(b) that the shaded area is given by

Area = Area (Δ with base from 500 to 1500) − Area (Δ with base from 1000 to 1500)

Recall that the area of a triangle is $\frac{1}{2}$(base)(height) so we have

$$\text{Area} = \tfrac{1}{2}(1000)(200) - \tfrac{1}{2}(500)(100) = 75,000$$

Because the area under $R'(x)$ is also given by the definite integral, we can find this additional revenue when sales are increased from 500 to 1000 by evaluating

$$\int_{500}^{1000} (300 - 0.2x)\, dx$$

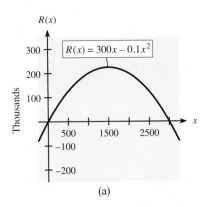

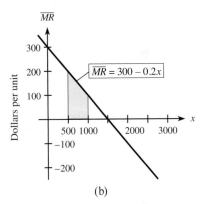

Figure 13.9   (a)   (b)

In general, the definite integral

$$\int_a^b f(x)\, dx$$

can be used to find the change in the function $F(x)$ when x changes from a to b, where $f(x)$ is the derivative of $F(x)$. This result is the **Fundamental Theorem of Calculus.**

Fundamental Theorem of Calculus	Let f be a continuous function on the closed interval $[a, b]$; then the definite integral of f exists on this interval, and $$\int_a^b f(x)\, dx = F(b) - F(a)$$ where F is any function such that $F'(x) = f(x)$ for all x in $[a, b]$.

Stated differently, this theorem says that if the function F is an indefinite integral of a function f that is continuous on the interval $[a, b]$, then

$$\int_a^b f(x)\, dx = F(b) - F(a)$$

Thus, we apply the Fundamental Theorem of Calculus by using the following two steps.

1. Integration of $f(x)$: $\displaystyle\int_a^b f(x)\,dx = F(x)\Big|_a^b$

2. Evaluation of $F(x)$: $F(x)\Big|_a^b = F(b) - F(a)$

EXAMPLE 1   Definite Integral

Evaluate $\displaystyle\int_2^4 (x^3 + 4)\,dx$.

Solution

1. $\displaystyle\int_2^4 (x^3 + 4)\,dx = \frac{x^4}{4} + 4x + C\,\Big|_2^4$

2. $\qquad = \left[\frac{(4)^4}{4} + 4(4) + C\right] - \left[\frac{(2)^4}{4} + 4(2) + C\right]$

$\qquad = (64 + 16 + C) - (4 + 8 + C)$

$\qquad = 68$ (Note that the C's subtract out.)

Note that the Fundamental Theorem states that F can be *any* indefinite integral of f, so we need not add the constant of integration to the integral.

EXAMPLE 2   Fundamental Theorem

Evaluate $\displaystyle\int_1^3 (3x^2 + 6x)\,dx$.

Solution

$$\int_1^3 (3x^2 + 6x)\,dx = x^3 + 3x^2\,\Big|_1^3$$

$$= (3^3 + 3\cdot 3^2) - (1^3 + 3\cdot 1^2)$$

$$= 54 - 4 = 50$$

Properties   The properties of definite integrals given next follow from properties of summations.

1. $\displaystyle\int_a^b [f(x) \pm g(x)]\,dx = \int_a^b f(x)\,dx \pm \int_a^b g(x)\,dx$

2. $\displaystyle\int_a^b kf(x)\,dx = k\int_a^b f(x)\,dx,$ where k is a constant

The following example uses both of these properties.

EXAMPLE 3   Definite Integral

Evaluate $\displaystyle\int_3^5 (\sqrt{x^2 - 9} + 2)x\,dx$.

Solution

$$\int_3^5 (\sqrt{x^2 - 9} + 2)x\,dx = \int_3^5 \sqrt{x^2 - 9}(x\,dx) + \int_3^5 2x\,dx$$

$$= \frac{1}{2}\int_3^5 (x^2 - 9)^{1/2}(2x\,dx) + \int_3^5 2x\,dx$$

$$= \frac{1}{2}\left[\frac{2}{3}(x^2 - 9)^{3/2}\right]_3^5 + x^2\,\Big|_3^5$$

$$= \frac{1}{3}[(16)^{3/2} - (0)^{3/2}] + (25 - 9)$$

$$= \frac{64}{3} + 16 = \frac{64}{3} + \frac{48}{3} = \frac{112}{3} \qquad \blacksquare$$

In the integral $\int_a^b f(x)\, dx$, we call a the *lower limit* and b the *upper limit* of integration. Although we developed the definite integral with the assumption that the lower limit was less than the upper limit, the following properties permit us to evaluate the definite integral even when that is not the case.

3. $\displaystyle\int_a^a f(x)\, dx = 0$

4. If f is integrable on $[a, b]$, then

$$\int_b^a f(x)\, dx = -\int_a^b f(x)\, dx$$

The following examples illustrate these properties.

EXAMPLE 4   Properties of Definite Integrals

(a) Evaluate $\displaystyle\int_4^4 x^2\, dx.$ (b) Compare $\displaystyle\int_2^4 3x^2\, dx$ and $\displaystyle\int_4^2 3x^2\, dx.$

Solution

(a) $\displaystyle\int_4^4 x^2\, dx = \frac{x^3}{3}\Big|_4^4 = \frac{4^3}{3} - \frac{4^3}{3} = 0$

This illustrates Property 3.

(b) $\displaystyle\int_2^4 3x^2\, dx = x^3\Big|_2^4 = 4^3 - 2^3 = 56$ and $\displaystyle\int_4^2 3x^2\, dx = x^3\Big|_4^2 = 2^3 - 4^3 = -56$

This illustrates Property 4. $\blacksquare$

Another property of definite integrals is called the additive property.

5. If f is continuous on some interval containing a, b, and c,* then

$$\int_a^c f(x)\, dx + \int_c^b f(x)\, dx = \int_a^b f(x)\, dx$$

EXAMPLE 5   Properties of Definite Integrals

Show that $\displaystyle\int_2^3 4x\, dx + \int_3^5 4x\, dx = \int_2^5 4x\, dx.$

Solution

$$\int_2^3 4x\, dx = 2x^2\Big|_2^3 = 18 - 8 = 10 \quad \text{and} \quad \int_3^5 4x\, dx = 2x^2\Big|_3^5 = 50 - 18 = 32$$

Thus $\displaystyle\int_2^5 4x\, dx = 2x^2\Big|_2^5 = 50 - 8 = 42 = \int_2^3 4x\, dx + \int_3^5 4x\, dx$ $\blacksquare$

The Definite Integral and Areas

Let us now return to area problems, to see the relationship between the definite integral and the area under a curve. By the formula for the area of a triangle or by using rectangles and the limit definition of area, the area under the curve (line) $y = x$ from $x = 0$

*Note that c need not be between a and b.

to $x = 1$ can be shown to be $\frac{1}{2}$ (see Figure 13.10(a)). Using the definite integral to find the area gives

$$A = \int_0^1 x\, dx = \left.\frac{x^2}{2}\right|_0^1 = \frac{1}{2} - 0 = \frac{1}{2}$$

In the previous section, we used rectangles to find that the area under $y = x^2$ from $x = 0$ to $x = 1$ was $\frac{1}{3}$ (see Figure 13.10(b)). Using the definite integral, we get

$$A = \int_0^1 x^2\, dx = \left.\frac{x^3}{3}\right|_0^1 = \frac{1}{3} - 0 = \frac{1}{3}$$

which agrees with the answer obtained previously.

However, not every definite integral represents the area between the curve and the x-axis over an interval. For example,

$$\int_0^2 (x - 2)\, dx = \left.\frac{x^2}{2} - 2x\right|_0^2 = (2 - 4) - (0) = -2$$

This would indicate that the area between the curve and the x-axis is negative, but area must be positive. A look at the graph of $y = x - 2$ (see Figure 13.10(c)) shows us what is happening.

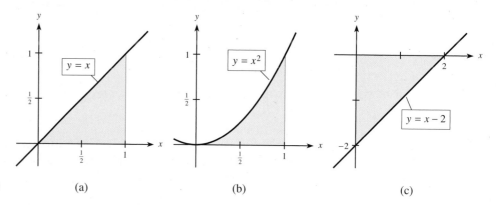

Figure 13.10   (a)   (b)   (c)

The region bounded by $y = x - 2$ and the x-axis between $x = 0$ and $x = 2$ is a triangle whose base is 2 and whose height is 2, so its area is $\frac{1}{2}bh = \frac{1}{2}(2)(2) = 2$. The integral has value -2 because $y = x - 2$ lies below the x-axis from $x = 0$ to $x = 2$, and the function values over the interval $[0, 2]$ are negative. Thus the value of the definite integral over *this* interval does not represent the area between the curve and the x-axis but, rather, represents the "signed" area as mentioned previously.

In general, the definite integral will give the **area under the curve** and above the x-axis only when $f(x) \geq 0$ for all x in $[a, b]$.

Area Under a Curve

If f is a continuous function on $[a, b]$ and $f(x) \geq 0$ on $[a, b]$, then the exact area between $y = f(x)$ and the x-axis from $x = a$ to $x = b$ is given by

$$\underset{\text{(shaded)}}{\text{Area}} = \int_a^b f(x)\, dx$$

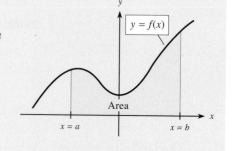

Note also that if $f(x) \leq 0$ for all x in $[a, b]$, then

$$\int_a^b f(x)\, dx = -\text{Area (between } f(x) \text{ and the } x\text{-axis)}$$

Calculator Note The approximate area under the graph of $y = f(x)$ and above the x-axis can be found with a calculator. After we enter the function in the $Y =$ menu and graph it, we choose $\int f(x)dx$ under 2nd CALC, press ENTER, and select the lower and upper x-values for the interval on which we seek the area. Figure 13.11 shows the steps in finding the area under the graph of $f(x) = x^2 + 1$ from $x = 0$ to $x = 3$ after the function is graphed. See Appendix A, Section 13.2, for details.

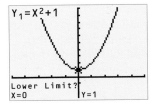
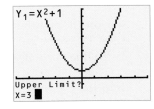
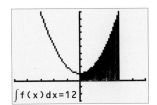

Figure 13.11

✓ CHECKPOINT

1. True or false:
    (a) For any integral, we can omit the constant of integration (the $+C$).
    (b) $-\int_{-1}^{3} f(x)\, dx = \int_{3}^{-1} f(x)\, dx$, if f is integrable on $[-1, 3]$.

    (c) The area between $f(x)$ and the x-axis on the interval $[a, b]$ is given by $\int_{a}^{b} f(x)\, dx$.

2. Evaluate:

    (a) $\displaystyle\int_{0}^{3} (x^2 + 1)\, dx$ (b) $\displaystyle\int_{0}^{3} (x^2 + 1)^4 x\, dx$

If the rate of growth of some function with respect to time t is $f'(t)$, then the total growth of the function during the period from $t = 0$ to $t = k$ can be found by evaluating the definite integral

$$\int_{0}^{k} f'(t)\, dt = f(t)\Big|_{0}^{k} = f(k) - f(0)$$

For nonnegative rates of growth, this definite integral (and thus growth) is the same as the area under the graph of $f'(t)$ from $t = 0$ to $t = k$.

EXAMPLE 6   **Income Stream** | **APPLICATION PREVIEW** |

Suppose that money flows continuously into a slot machine at a casino and grows at a rate given by

$$A'(t) = 100e^{0.1t}$$

where t is time in hours and $0 \le t \le 10$. Find the total amount that accumulates in the machine during the 10-hour period, if no money is paid out.

Solution
The total dollar amount is given by

$$A = \int_{0}^{10} 100e^{0.1t}\, dt = \frac{100}{0.1}\int_{0}^{10} e^{0.1t}(0.1)\, dt = 1000e^{0.1t}\Big|_{0}^{10} = 1000e - 1000 \approx 1718.28$$

Probability Density Functions   In Section 8.4, "Normal Probability Distribution," we stated that the total area under the normal curve is 1 and that the area under the curve from value x_1 to value x_2 represents the probability that a score chosen at random will lie between x_1 and x_2.

The normal distribution is an example of a **continuous probability distribution** because the values of the random variable are considered over intervals rather than at discrete values. The statements above relating probability and area under the graph apply to other continuous probability distributions determined by **probability density functions.** In fact, if x is a continuous random variable with probability density function $f(x)$, then the probability that x is between a and b is

$$\Pr(a \le x \le b) = \int_a^b f(x)\, dx$$

EXAMPLE 7   Product Life

Suppose the probability density function for the life of a computer component is $f(x) = 0.10e^{-0.10x}$, where $x \ge 0$ is the number of years the component is in use. Find the probability that the component will last between 3 and 5 years.

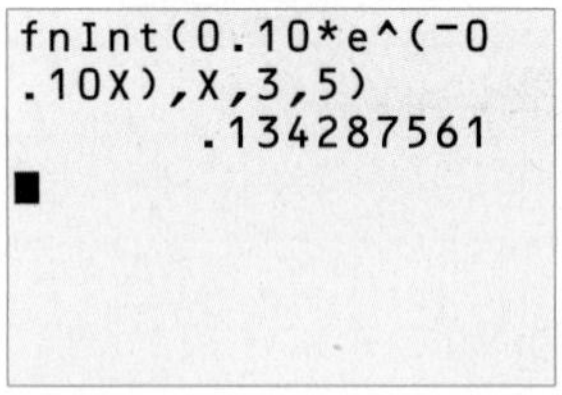

```
fnInt(0.10*e^(-0
.10X),X,3,5)
          .134287561
■
```

Solution

The probability that the component will last between 3 and 5 years is the area under the graph of the function between $x = 3$ and $x = 5$. The probability is given by

$$\Pr(3 \le x \le 5) = \int_3^5 0.10e^{-0.10x}\, dx = -e^{-0.10x}\Big|_3^5 = -e^{-0.5} + e^{-0.3}$$

$$\approx -0.6065 + 0.7408 = 0.1343$$

Figure 13.12

Calculator Note Using fnInt($f(x)$, x, a, b) under MATH on a graphing calculator gives the definite integral of the function $y = f(x)$ from $x = a$ to $x = b$. This feature can be used to evaluate definite integrals directly or to check those done with the Fundamental Theorem of Calculus. Figure 13.12 shows the numerical integration feature applied to the integral in Example 7. Note that when this answer is rounded to 4 decimal places, the results agree. See Appendix A, Section 13.2 for details. ■

✓ CHECKPOINT ANSWERS

1. (a) False
   (b) True   (c) False; this is true only if $f(x) \ge 0$ on $[a, b]$

2. (a) $\displaystyle\int_0^3 (x^2 + 1)\, dx = \frac{x^3}{3} + x\Big|_0^3 = 12$

   (b) $\displaystyle\int_0^3 (x^2 + 1)^4 x\, dx = \frac{1}{2} \cdot \frac{(x^2 + 1)^5}{5}\Big|_0^3 = \frac{1}{10}(10^5 - 1) = 9999.9$

| EXERCISES | 13.2

Evaluate the definite integrals in Problems 1–30.

1. $\displaystyle\int_0^3 4x\, dx$

2. $\displaystyle\int_0^1 8x\, dx$

3. $\displaystyle\int_2^4 dx$

4. $\displaystyle\int_1^5 2\, dy$

5. $\displaystyle\int_2^4 x^3\, dx$

6. $\displaystyle\int_0^5 x^2\, dx$

7. $\displaystyle\int_0^5 4\sqrt[3]{x^2}\, dx$

8. $\displaystyle\int_2^4 3\sqrt{x}\, dx$

9. $\displaystyle\int_1^4 (10 - 4x)\, dx$

10. $\displaystyle\int_{-1}^4 (6x - 9)\, dx$

11. $\displaystyle\int_2^4 (4x^3 - 6x^2 - 5x)\, dx$

12. $\displaystyle\int_0^2 (x^4 - 5x^3 + 2x)\, dx$

13. $\displaystyle\int_3^4 (x - 4)^9\, dx$

14. $\displaystyle\int_{-1}^0 (x + 2)^{13}\, dx$

15. $\displaystyle\int_2^4 (x^2 + 2)^3 x\, dx$

16. $\displaystyle\int_0^1 5x^2\,(4x^3 - 2)^4\, dx$

17. $\displaystyle\int_{-1}^2 (x^3 - 3x^2)^3 (x^2 - 2x)\, dx$

18. $\displaystyle\int_0^3 (2x - x^2)^4 (1 - x)\, dx$

19. $\displaystyle\int_{-2}^2 15x^3\,(x^4 - 6)^6\, dx$

20. $\displaystyle\int_0^4 (3x^2 - 2)^4 x\, dx$

21. $\displaystyle\int_0^4 \sqrt{4x + 9}\, dx$

22. $\displaystyle\int_0^2 \sqrt[3]{2x^3 - 8}\, x^2\, dx$

23. $\displaystyle\int_1^3 \frac{3}{y^2}\, dy$

24. $\displaystyle\int_1^2 \frac{5}{z^3}\, dz$

25. $\displaystyle\int_0^1 e^{3x}\, dx$

26. $\displaystyle\int_0^2 e^{4x-3}\, dx$

27. $\displaystyle\int_1^e \frac{4}{z}\, dz$

28. $\displaystyle\int_1^{5e} 3y^{-1}\, dy$

29. $\int_0^2 8x^2 e^{-x^3}\, dx$

30. $\int_0^1 \dfrac{3x^3\, dx}{4x^4 + 9}$

In Problems 31–34, evaluate each integral (a) with the Fundamental Theorem of Calculus and (b) with a graphing calculator (as a check).

31. $\int_3^6 \dfrac{x}{3x^2 + 4}\, dx$

32. $\int_0^2 \dfrac{x}{x^2 + 4}\, dx$

33. $\int_1^2 \dfrac{x^2 + 3}{x}\, dx$

34. $\int_1^4 \dfrac{4\sqrt{x} + 5}{\sqrt{x}}\, dx$

35. In the figures, which of the shaded regions (A, B, C, or D) has the area given by

(a) $\int_a^b f(x)\, dx$?

(b) $-\int_a^b f(x)\, dx$?

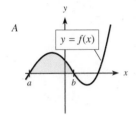

A

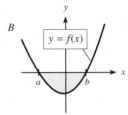

B

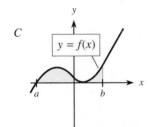

C

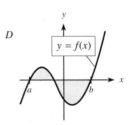

D

36. For which of the following functions $f(x)$ does

$$\int_0^2 f(x)\, dx$$

give the area between the graph of $f(x)$ and the x-axis from $x = 0$ to $x = 2$?

(a) $f(x) = x^2 + 1$ (b) $f(x) = -x^2$

(c) $f(x) = x - 1$

In Problems 37–40, (a) write the integral that describes the area of the shaded region and (b) find the area.

37.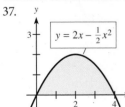

$y = 2x - \frac{1}{2}x^2$

38.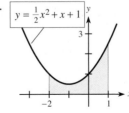

$y = \frac{1}{2}x^2 + x + 1$

39.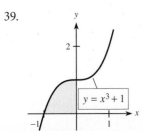

$y = x^3 + 1$

40.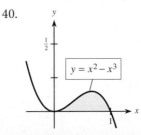

$y = x^2 - x^3$

41. Find the area between the curve $y = -x^2 + 3x - 2$ and the x-axis from $x = 1$ to $x = 2$.

42. Find the area between the curve $y = x^2 + 3x + 2$ and the x-axis from $x = -1$ to $x = 3$.

43. Find the area between the curve $y = xe^{x^2}$ and the x-axis from $x = 1$ to $x = 3$.

44. Find the area between the curve $y = e^{-x}$ and the x-axis from $x = -1$ to $x = 1$.

In Problems 45 and 46, use the figures to decide which of $\int_0^a f(x)\, dx$ or $\int_0^a g(x)\, dx$ is larger or if they are equal. Explain your choices.

45.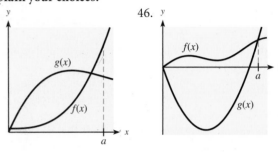

46.

In Problems 47–52, use properties of definite integrals.

47. How does $\int_{-1}^{-3} x\sqrt{x^2 + 1}\, dx$ compare with

$$\int_{-3}^{-1} x\sqrt{x^2 + 1}\, dx?$$

48. If $\int_{-1}^0 x^3\, dx = -\frac{1}{4}$ and $\int_0^1 x^3\, dx = \frac{1}{4}$, what does $\int_{-1}^1 x^3\, dx$ equal?

49. If $\int_1^2 (2x - x^2)\, dx = \frac{2}{3}$ and $\int_2^4 (2x - x^2)\, dx = -\frac{20}{3}$, what does $\int_1^4 (x^2 - 2x)\, dx$ equal?

50. If $\int_1^2 (2x - x^2)\, dx = \frac{2}{3}$, what does $\int_1^2 6(2x - x^2)\, dx$ equal?

51. Evaluate $\int_4^4 \sqrt{x^2 - 2}\, dx$.

52. Evaluate $\int_2^2 (x^3 + 4x)^{-6}\, dx$.

APPLICATIONS

53. *Depreciation* The rate of depreciation of a building is given by $D'(t) = 3000(20 - t)$ dollars per year, $0 \le t \le 20$; see the figure on the next page.

(a) Use the graph to find the total depreciation of the building over the first 10 years ($t = 0$ to $t = 10$).

(b) Use a definite integral to find the total depreciation over the first 10 years.

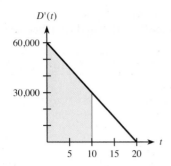

54. **Depreciation** The rate of depreciation of a building is given by $D'(t) = 3000(20 - t)$ dollars per year, $0 \le t \le 20$; see the figure in Problem 53.
    (a) Use the graph to find the total depreciation of the building over the first 20 years.
    (b) Use a definite integral to find the total depreciation over the first 20 years.
    (c) Use the graph to find the total depreciation between 10 years and 20 years and check it with a definite integral.

55. **Sales and advertising** A store finds that its sales revenue changes at a rate given by
$$S'(t) = -30t^2 + 360t \quad \text{dollars per day}$$
where t is the number of days after an advertising campaign ends and $0 \le t \le 30$.
    (a) Find the total sales for the first week after the campaign ends ($t = 0$ to $t = 7$).
    (b) Find the total sales for the second week after the campaign ends ($t = 7$ to $t = 14$).

56. **Health care costs** The total annual health care costs in the United States (actual and projected, in billions of dollars) for selected years are given in the table. The equation
$$y = 4.447x^2 - 9.108x + 1055.4$$
models the annual health care costs, y (in billions of dollars per year), as a function of the number of years past 1990, x. Use a definite integral and this model to find the total cost of health care over the period 2005–2015.

Year	Cost	Year	Cost
2000	1353	2012	2931
2003	1735	2015	3541
2006	2113	2018	4353
2009	2509		

Source: U.S. Department of Health and Human Services

57. **Total income** The income from an oil change service chain can be considered as flowing continuously at an annual rate given by
$$f(t) = 10,000e^{0.02t} \quad \text{(dollars/year)}$$
Find the total income for this chain over the first 2 years (from $t = 0$ to $t = 2$).

58. **Total income** Suppose that a vending machine service company models its income by assuming that money flows continuously into the machines, with the annual rate of flow given by
$$f(t) = 120e^{0.01t}$$
in thousands of dollars per year. Find the total income from the machines over the first 3 years.

59. CO_2 **Emissions** Using U.S. Energy Information Administration data from 2010 and projected to 2030, the carbon dioxide emissions from biomass energy combustion (in millions of metric tons per year) can be modeled by
$$C(t) = 19.12t + 319.0$$
where t is the number of years past 2010. Evaluate $\int_0^{10} C(t)\, dt$ and tell what it represents.

60. **Health services and supplies expenditures** The per capita expenditures for U.S. health services and supplies (in dollars per year) for selected years from 2000 and projected to 2018 can be modeled by
$$H(t) = 4676e^{0.053t}$$
where t is the number of years past 2000 (*Source:* U.S. Centers for Medicare and Medicaid Services). Assuming the model remains valid, evaluate $\int_{10}^{20} H(t)\, dt$ and tell what it represents.

Velocity of blood **In Problems 61 and 62, the velocity of blood through a vessel is given by $v = K(R^2 - r^2)$, where K is a constant, R is the (constant) radius of the vessel, and r is the distance of the particular corpuscle from the center of the vessel. The rate of flow can be found by measuring the volume of blood that flows past a point in a given time period. This volume, V, is given by**
$$V = \int_0^R v\,(2\pi r\, dr)$$

61. If $R = 0.30$ cm and $v = (0.30 - 3.33r^2)$ cm/s, find the volume.

62. Develop a general formula for V by evaluating
$$V = \int_0^R v(2\pi r\, dr)$$
using $v = K(R^2 - r^2)$.

Production **In Problems 63 and 64, the rate of production of a new line of products is given by**
$$\frac{dx}{dt} = 200\left[1 + \frac{400}{(t + 40)^2}\right]$$
where x is the number of items produced and t is the number of weeks the products have been in production.

63. How many units were produced in the first 5 weeks?
64. How many units were produced in the sixth week?
65. **Testing** The time t (in minutes) needed to read an article appearing on a foreign-language placement test is given by the probability density function

$$f(t) = 0.012t^2 - 0.0012t^3, \qquad 0 \le t \le 10$$

For a test taker chosen at random, find the probability that this person takes 8 minutes or more to read the article.

66. *Response time* In a small city the response time t (in minutes) of the fire company is given by the probability density function

$$f(t) = \frac{60t^2 - 4t^3}{16{,}875}, \qquad 0 \le t \le 15$$

For a fire chosen at random, find the probability that the response time is 10 minutes or less.

67. *Customer service* The duration t (in minutes) of customer service calls received by a certain company is given by the probability density function

$$f(t) = 0.3e^{-0.3t}, \qquad t \ge 0$$

Find the probability that a call selected at random lasts
(a) 3 minutes or less.   (b) between 5 and 10 minutes.

68. *Product life* The useful life of a car battery t (in years) is given by the probability density function

$$f(t) = 0.2e^{-0.2t}, \qquad t \ge 0$$

Find the probability that a battery chosen at random lasts
(a) 2 years or less.   (b) between 4 and 6 years.

69. **Modeling** *Gasoline usage* In the United States, gasoline for motor vehicles is used continuously, and that usage, in billions of gallons per year, is shown in the table for selected years from 2014 and projected to 2024.
(a) Find a quadratic function that models these data, with t as the number of years past 2014. Report the model as $G(t)$ with three significant digits.

(b) Evaluate $\int_0^{10} G(t)\,dt$ and tell what it represents.

Year	2014	2016	2018	2020	2022	2024
Gasoline Usage (billions of gallons)	132.8	133.5	130.4	126.1	121.2	116.3

Source: U.S. Department of Energy

70. **Modeling** *SRT Viper acceleration* Table 13.1(a) shows the time in seconds that an SRT Viper requires to reach various speeds up to 100 mph. Table 13.1(b) shows the same data, but with speeds in miles per second.
(a) Fit a power model to the data in Table 13.1(b).
(b) Use a definite integral from 0 to 9.2 of the function you found in part (a) to find the distance traveled by the Viper as it went from 0 mph to 100 mph in 9.2 seconds.

| TABLE 13.1 |

(a)		(b)	
Time (seconds)	Speed (mph)	Time (seconds)	Speed (mi/s)
1.7	30	1.7	0.00833
2.4	40	2.4	0.01111
3.2	50	3.2	0.01389
4.1	60	4.1	0.01667
5.8	70	5.8	0.01944
6.2	80	6.2	0.02222
7.8	90	7.8	0.02500
9.2	100	9.2	0.02778

Source: Motor Trend

- To find the area between two curves
- To find the average value of a function

Area Between Two Curves

| APPLICATION PREVIEW |

In economics, the **Lorenz curve** is used to represent the inequality of income distribution among different groups in the population of a country. The curve is constructed by plotting the cumulative percent of families at or below a given income level and the cumulative percent of total personal income received by these families. For example, the table on the next page shows the coordinates of some points on the Lorenz curve $y = L(x)$ that divide the income (for the United States in 2012) into 5 equal income levels (quintiles). The point $(0.40, 0.115)$ is on the Lorenz curve because the families with incomes in the bottom 40% of the country received 11.5% of the total income in 2012. The graph of the Lorenz curve $y = L(x)$ is shown in Figure 13.13. *(continued)*

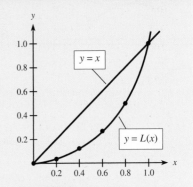

Figure 13.13

U.S. Income Distribution for 2012 (Points on the Lorenz Curve)	
x, Cumulative Proportion of Families Below Income Level	**y = L(x), Cumulative Proportion of Total Income**
0	0
0.20	0.032
0.40	0.115
0.60	0.259
0.80	0.489
1	1

Source: U.S. Bureau of the Census

Equality of income would result if each family received an equal proportion of the total income, so that the bottom 20% would receive 20% of the total income, the bottom 40% would receive 40%, and so on. The Lorenz curve representing this would have the equation $y = x$.

The inequality of income distribution is measured by the Gini coefficient of income, which measures how far the Lorenz curve falls below $y = x$. It is defined as

$$\frac{\text{Area between } y = x \text{ and } y = L(x)}{\text{Area below } y = x}$$

Because the area of the triangle below $y = x$ and above the x-axis from $x = 0$ to $x = 1$ is $1/2$, the Gini coefficient of income is

$$\frac{\text{Area between } y = x \text{ and } y = L(x)}{1/2} = 2 \cdot [\text{area between } y = x \text{ and } y = L(x)]$$

In this section we will use the definite integral to find the area between two curves. We will use the area between two curves to find the Gini coefficient of income (see Example 4) and to find average cost, average revenue, average profit, and average inventory.

Area Between Two Curves

We have used the definite integral to find the area of the region between a curve and the x-axis over an interval where the curve lies above the x-axis. We can easily extend this technique to finding the area between two curves over an interval.

Suppose that the graphs of both $y = f(x)$ and $y = g(x)$ lie above the x-axis and that the graph of $y = f(x)$ lies above $y = g(x)$ throughout the interval from $x = a$ to $x = b$; that is, $f(x) \geq g(x)$ on $[a, b]$. (See Figure 13.14.)

Then Figures 13.15(a) and 13.15(b) show the areas under $y = f(x)$ and $y = g(x)$. Figure 13.15(c) shows how the difference of these two areas can be used to find the area of the region between the graphs of $y = f(x)$ and $y = g(x)$. That is,

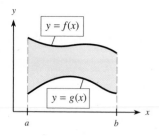

Figure 13.14

$$\text{Area between the curves} = \int_a^b f(x)\,dx - \int_a^b g(x)\,dx$$

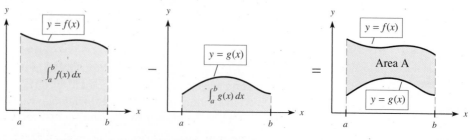

Figure 13.15   (a)   (b)   (c)

Although Figure 13.15(c) shows the graphs of both $y = f(x)$ and $y = g(x)$ lying above the x-axis, this difference of their integrals will always give the area between their graphs if both functions are continuous and if $f(x) \geq g(x)$ on the interval $[a, b]$. Using the fact that

$$\int_a^b f(x)\, dx - \int_a^b g(x)\, dx = \int_a^b [f(x) - g(x)]\, dx$$

we have the following result for the **area between two curves.**

Area Between Two Curves	If f and g are continuous functions on $[a, b]$ and if $f(x) \geq g(x)$ on $[a, b]$, then the area of the region bounded by $y = f(x)$, $y = g(x)$, $x = a$, and $x = b$ is $$A = \int_a^b [f(x) - g(x)]\, dx$$

EXAMPLE 1 Area Between Two Curves

Find the area of the region bounded by $y = x^2 + 4$, $y = x$, $x = 0$, and $x = 3$.

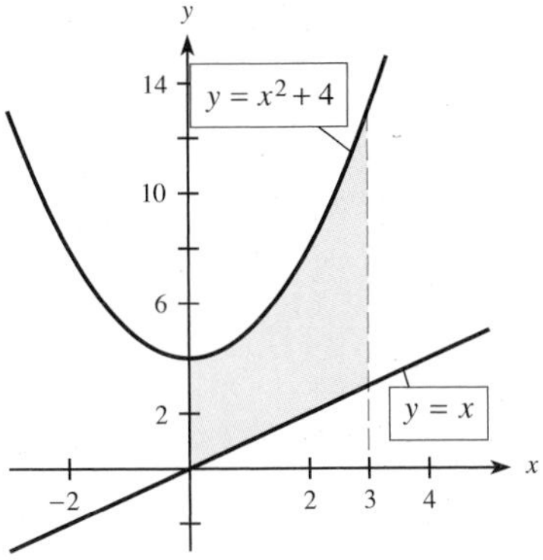

Solution

We first sketch the graphs of the functions on the same set of axes. The graph of the region is shown in Figure 13.16. Because $y = x^2 + 4$ lies above $y = x$ in the interval from $x = 0$ to $x = 3$, the area is

$$A = \int (\text{top curve} - \text{bottom curve})\, dx$$

$$A = \int_0^3 [(x^2 + 4) - x]\, dx = \frac{x^3}{3} + 4x - \frac{x^2}{2}\Big|_0^3$$

$$= \left(9 + 12 - \frac{9}{2}\right) - (0 + 0 - 0) = 16\tfrac{1}{2}\ \text{square units} \qquad \blacksquare$$

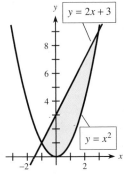

Figure 13.16

We are sometimes asked to find the area enclosed by two curves. In this case, we find the points of intersection of the curves to determine a and b.

EXAMPLE 2 Area Enclosed by Two Curves

Find the area enclosed by $y = x^2$ and $y = 2x + 3$.

Solution

We first find a and b by finding the x-coordinates of the points of intersection of the graphs. Setting the y-values equal gives

$$x^2 = 2x + 3$$
$$x^2 - 2x - 3 = 0$$
$$(x - 3)(x + 1) = 0$$
$$x = 3, \quad x = -1$$

Thus with $a = -1$ and $b = 3$, we sketch the graphs of these functions on the same set of axes. Because the graphs do not intersect on the interval $(-1, 3)$, we can determine which function is larger on this interval by evaluating $2x + 3$ and x^2 at any value c where $-1 < c < 3$. Figure 13.17 shows the region between the graphs, with $2x + 3 \geq x^2$ from $x = -1$ to $x = 3$. The area of the enclosed region is

$$A = \int_{-1}^3 [(2x + 3) - x^2]\, dx = x^2 + 3x - \frac{x^3}{3}\Big|_{-1}^3$$

$$= (9 + 9 - 9) - \left(1 - 3 + \frac{1}{3}\right) = 10\tfrac{2}{3}\ \text{square units} \qquad \blacksquare$$

Figure 13.17

Some graphs enclose two or more regions because they have more than two points of intersection.

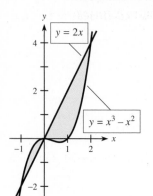

Figure 13.18

EXAMPLE 3   A Region with Two Sections

Find the area of the region enclosed by the graphs of

$$y = f(x) = x^3 - x^2 \qquad \text{and} \qquad y = g(x) = 2x.$$

Solution

To find the points of intersection of the graphs, we solve $f(x) = g(x)$, or $x^3 - x^2 = 2x$.

$$x^3 - x^2 - 2x = 0$$
$$x(x - 2)(x + 1) = 0$$
$$x = 0, x = 2, x = -1$$

Graphing these functions between $x = -1$ and $x = 2$, we see that for any x-value in the interval $(-1, 0)$, $f(x) \geq g(x)$, so $f(x) \geq g(x)$ for the region enclosed by the curves from $x = -1$ to $x = 0$. But evaluating the functions for any x-value in the interval $(0, 2)$ shows that $f(x) \leq g(x)$ for the region enclosed by the curves from $x = 0$ to $x = 2$. See Figure 13.18.

Thus we need one integral to find the area of the region from $x = -1$ to $x = 0$ and a second integral to find the area from $x = 0$ to $x = 2$. The area is found by summing these two integrals.

$$A = \int_{-1}^{0} [(x^3 - x^2) - (2x)] \, dx + \int_{0}^{2} [(2x) - (x^3 - x^2)] \, dx$$

$$= \int_{-1}^{0} (x^3 - x^2 - 2x) \, dx + \int_{0}^{2} (2x - x^3 + x^2) \, dx$$

$$= \left(\frac{x^4}{4} - \frac{x^3}{3} - x^2 \right) \Bigg|_{-1}^{0} + \left(x^2 - \frac{x^4}{4} + \frac{x^3}{3} \right) \Bigg|_{0}^{2}$$

$$= \left[(0) - \left(\frac{1}{4} - \frac{-1}{3} - 1 \right) \right] + \left[\left(4 - \frac{16}{4} + \frac{8}{3} \right) - (0) \right] = \frac{37}{12}$$

The area between the curves is $\frac{37}{12}$ square units.

Calculator Note   We've seen how to perform numerical integration with a graphing calculator. Appendix A, Section 13.3, shows two methods of approximating the area between two curves with a graphing calculator.

✓ CHECKPOINT

1. True or false:
   (a) Over the interval $[a, b]$, the area between the continuous functions $f(x)$ and $g(x)$ is

   $$\int_{a}^{b} [f(x) - g(x)] \, dx$$

   (b) If $f(x) \geq g(x)$ and the area between $f(x)$ and $g(x)$ is given by

   $$\int_{a}^{b} [f(x) - g(x)] \, dx$$

   then $x = a$ and $x = b$ represent the left and right boundaries, respectively, of the region.

   (c) To find points of intersection of $f(x)$ and $g(x)$, solve $f(x) = g(x)$.

2. Consider the functions $f(x) = x^2 + 3x - 9$ and $g(x) = \frac{1}{4}x^2$.

   (a) Find the points of intersection of $f(x)$ and $g(x)$.
   (b) Determine which function is greater than the other between the points found in part (a).
   (c) Set up the integral used to find the area between the curves in the interval between the points found in part (a).
   (d) Find the area.

EXAMPLE   4   Income Distribution | APPLICATION PREVIEW |

The inequality of income distribution is measured by the **Gini coefficient** of income, which is defined as

$$\frac{\text{Area between } y = x \text{ and } y = L(x)}{\text{Area below } y = x} = \frac{\int_0^1 [x - L(x)] \, dx}{1/2}$$

$$= 2\int_0^1 [x - L(x)] \, dx$$

The function $y = L(x) = 2.409x^4 - 3.400x^3 + 2.164x^2 - 0.1746x$ models the 2012 income distribution data in the Application Preview.

(a)   Use this $L(x)$ to find the Gini coefficient of income for 2012.
(b)   If the Census Bureau Gini coefficient of income for 1991 is 0.428, during which year is the distribution of income more nearly equal?

Solution

(a)   The Gini coefficient of income for 2012 is

$$2\int_0^1 [x - L(x)] \, dx = 2\int_0^1 [x - (2.409x^4 - 3.400x^3 + 2.164x^2 - 0.1746x)] \, dx$$

$$= 2[-0.4818x^5 + 0.8500x^4 - 0.7213x^3 + 0.5873x^2]_0^1 \approx 0.468$$

(b)   Absolute equality of income would occur if the Gini coefficient of income were 0; and smaller coefficients indicate more nearly equal incomes. Thus the distribution of income was more nearly equal in 1991 than in 2012.   ■

Average Value   If the graph of $y = f(x)$ lies on or above the x-axis from $x = a$ to $x = b$, then the area between the graph and the x-axis is

$$A = \int_a^b f(x) \, dx \quad \text{(See Figure 13.19(a))}$$

The area A is also the area of a rectangle with base equal to $b - a$ and height equal to the **average value** (or average height) of the function $y = f(x)$ (see Figure 13.19(b)).

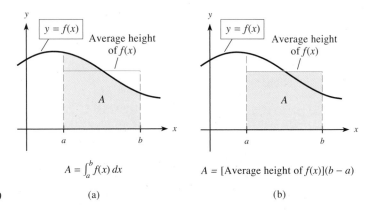

$$A = \int_a^b f(x) \, dx \qquad\qquad A = [\text{Average height of } f(x)](b - a)$$

Figure 13.19   (a)   (b)

Thus the average value of the function in Figure 13.19 is

$$\frac{A}{b - a} = \frac{1}{b - a}\int_a^b f(x) \, dx$$

Even if $f(x) \leq 0$ on all or part of the interval $[a, b]$, we can find the average value by using the integral. Thus we have the following.

Average Value	The **average value** of a continuous function $y = f(x)$ over the interval $[a, b]$ is

$$\text{Average value} = \frac{1}{b - a}\int_a^b f(x)\, dx$$

EXAMPLE 5   Average Cost

Suppose that the cost in dollars for x table lamps is given by $C(x) = 400 + x + 0.3x^2$.

(a) What is the average value of $C(x)$ for 10 to 20 lamps?
(b) Find the average cost per unit if 40 lamps are produced.

Solution

(a) The average value of $C(x)$ from $x = 10$ to $x = 20$ is

$$\frac{1}{20 - 10}\int_{10}^{20}(400 + x + 0.3x^2)\, dx = \frac{1}{10}\left(400x + \frac{x^2}{2} + 0.1x^3\right)\bigg|_{10}^{20}$$

$$= \frac{1}{10}[(8000 + 200 + 800) - (4000 + 50 + 100)]$$

$$= 485 \quad \text{(dollars)}$$

Thus for any number of lamps between 10 and 20 the *average total cost* for that number of lamps is $485.

(b) The average cost per unit if 40 units are produced is the average cost function evaluated at $x = 40$. The average cost function is

$$\overline{C}(x) = \frac{C(x)}{x} = \frac{400}{x} + 1 + 0.3x$$

Thus the *average cost per unit* if 40 units are produced is

$$\overline{C}(40) = \frac{400}{40} + 1 + 0.3(40) = 23 \quad \text{[dollars per unit (i.e., lamp)]}$$ ∎

✓ **CHECKPOINT**   3. Find the average value of $f(x) = x^2 - 4$ over $[-1, 3]$.

EXAMPLE 6   Average Value of a Function

Consider the functions $f(x) = x^2 - 4$ and $g(x) = x^3 - 4x$. For each function, do the following.

(a) Graph the function on the interval $[-3, 3]$.
(b) On the graph, "eyeball" the average value (height) of each function on $[-2, 2]$.
(c) Compute the average value of the function over the interval $[-2, 2]$.

Solution
For $f(x) = x^2 - 4$:

(a) The graph of $f(x) = x^2 - 4$ is shown in Figure 13.20(a).
(b) The average height of $f(x)$ over $[-2, 2]$ appears to be near -2.
(c) The average value of $f(x)$ over the interval is given by

$$\frac{1}{2 - (-2)}\int_{-2}^{2}(x^2 - 4)\, dx = \frac{1}{4}\left(\frac{x^3}{3} - 4x\right)\bigg|_{-2}^{2}$$

$$= \left(\frac{8}{12} - 2\right) - \left(-\frac{8}{12} + 2\right) = \frac{4}{3} - 4 = -\frac{8}{3} = -2\tfrac{2}{3}$$

For $g(x) = x^3 - 4x$:

(a) The graph of $g(x) = x^3 - 4x$ is shown in Figure 13.20(b).

(b) The average height of $g(x)$ over $[-2, 2]$ appears to be approximately 0.

(c) The average value of $g(x)$ is given by

$$\frac{1}{2 - (-2)}\int_{-2}^{2} (x^3 - 4x)\, dx = \frac{1}{4}\left(\frac{x^4}{4} - \frac{4x^2}{2}\right)\Bigg|_{-2}^{2}$$

$$= \left(\frac{16}{16} - \frac{16}{8}\right) - \left(\frac{16}{16} - \frac{16}{8}\right) = 0$$

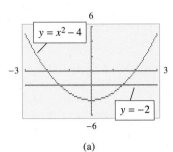

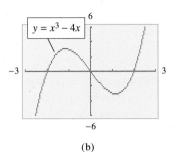

Figure 13.20   (a)   (b)

✓ **CHECKPOINT ANSWERS**

1. (a) False. This is true only if $f(x) \geq g(x)$ over $[a, b]$.
   (b) True   (c) True

2. (a) Solve $f(x) = g(x)$, or $x^2 + 3x - 9 = \frac{1}{4}x^2$. The solutions are $x = -6$ and $x = 2$.

   The points of intersection are $(-6, 9)$ and $(2, 1)$.
   (b) $g(x) \geq f(x)$ on the interval $[-6, 2]$.

   (c) $A = \int_{-6}^{2}\left[\frac{1}{4}x^2 - (x^2 + 3x - 9)\right] dx$

   (d) $A = \frac{x^3}{12} - \frac{x^3}{3} - \frac{3x^2}{2} + 9x\Bigg|_{-6}^{2} = 64$ square units

3. $\frac{1}{3 - (-1)}\int_{-1}^{3}(x^2 - 4)\, dx = -\frac{5}{3}$

| EXERCISES | 13.3

For each shaded region in Problems 1–6, (a) form the integral that represents the area of the shaded region and (b) find the area of the region.

1.
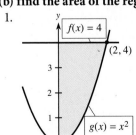

2.
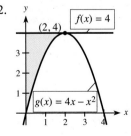

3.
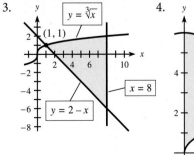

4.
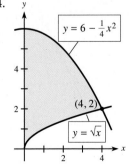

5.

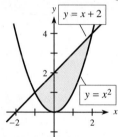

$y = 4 - x^2$

$(2, 0)$

$y = \frac{1}{4}x^3 - 2$

6.

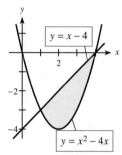

$y = x^3 + 1$

$x = 1$

$y = -x - 1$

For each shaded region in Problems 7–12, (a) find the points of intersection of the curves, (b) form the integral that represents the area of the shaded region, and (c) find the area of the shaded region.

7.

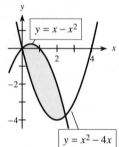

$y = x + 2$

$y = x^2$

8.

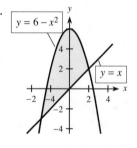

$y = x - 4$

$y = x^2 - 4x$

9.

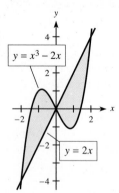

$y = x - x^2$

$y = x^2 - 4x$

10.

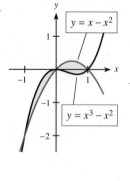

$y = 6 - x^2$

$y = x$

11.

$y = x^3 - 2x$

$y = 2x$

12.

$y = x - x^2$

$y = x^3 - x^2$

In Problems 13–26, equations are given whose graphs enclose a region. In each problem, find the area of the region.

13. $f(x) = x^2 + 2$; $g(x) = -x^2$; $x = 0$; $x = 2$
14. $f(x) = x^2$; $g(x) = -\frac{1}{10}(10 + x)$; $x = 0$; $x = 3$

15. $y = x^3 - 1$; $y = x - 1$; to the right of the y-axis
16. $y = x^2 - 2x + 1$; $y = x^2 - 5x + 4$; $x = 2$
17. $y = \frac{1}{2}x^2$; $y = x^2 - 2x$
18. $y = x^2$; $y = 4x - x^2$
19. $h(x) = x^2$; $k(x) = \sqrt{x}$
20. $g(x) = 1 - x^2$; $h(x) = x^2 + x$
21. $f(x) = x^3$; $g(x) = x^2 + 2x$
22. $f(x) = x^3$; $g(x) = 2x - x^2$
23. $f(x) = \frac{3}{x}$; $g(x) = 4 - x$
24. $f(x) = \frac{6}{x}$; $g(x) = -x - 5$
25. $y = \sqrt{x + 3}$; $x = -3$; $y = 2$
26. $y = \sqrt{4 - x}$; $x = 4$; $y = 3$

In Problems 27–32, find the average value of each function over the given interval.

27. $f(x) = 9 - x^2$ over $[0, 3]$
28. $f(x) = 2x - x^2$ over $[0, 2]$
29. $f(x) = x^3 - x$ over $[-1, 1]$
30. $f(x) = \frac{1}{2}x^3 + 1$ over $[-2, 0]$
31. $f(x) = \sqrt{x} - 2$ over $[1, 4]$
32. $f(x) = \sqrt[3]{x}$ over $[-8, -1]$

33. Use a graphing calculator or computer to find the area between the curves $y = f(x) = x^3 - 4x$ and $y = g(x) = x^2 - 4$.

34. Use a graphing calculator or computer to find the area between the curves $f(x) = \sqrt[3]{x}$ and $g(x) = x^3 - x$.

APPLICATIONS

35. *Average profit* For the product whose total cost and total revenue are shown in the figure, represent total revenue by $R(x)$ and total cost by $C(x)$ and write an integral that gives the average profit for the product over the interval from x_0 to x_1.

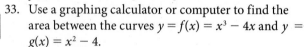
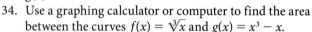

36. *Sales and advertising* The figure shows the sales growth rates under different levels of distribution and advertising from a to b. Set up an integral to determine the extra sales growth if $4 million is used in advertising rather than $2 million.

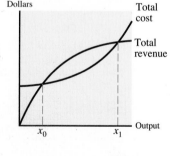

37. *Cost* The cost of producing x smart phones is $C(x) = x^2 + 400x + 2000$.

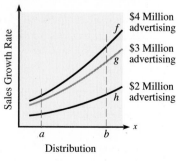

(a) Use $C(x)$ to find the average cost of producing 1000 smart phones.

(b) Find the average value of the cost function $C(x)$ over the interval from 0 to 1000.

38. **Inventory management** The figure shows how an inventory of a product is depleted each quarter of a given year. What is the average inventory per month for the first 3 months for this product? (Assume that the graph is a line joining (0, 1300) and (3, 100).)

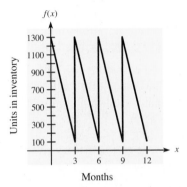

Months

39. **Sales and advertising** The number of daily sales of a product was found to be given by

$$S = 100xe^{-x^2} + 100$$

x days after the start of an advertising campaign for this product.

(a) Find the average daily sales during the first 20 days of the campaign—that is, from $x = 0$ to $x = 20$.

(b) If no new advertising campaign is begun, what is the average number of sales per day for the next 10 days (from $x = 20$ to $x = 30$)?

40. **Demand** The demand function for a certain product is given by

$$p = 500 + \frac{1000}{q + 1}$$

where p is the price and q is the number of units demanded. Find the average price as demand ranges from 49 to 99 units.

41. **Social Security beneficiaries** With data from the Social Security Trustees Report for selected years from 1950 and projected to 2030, the number of Social Security beneficiaries (in millions) can be modeled by

$$B(t) = 0.00024t^3 - 0.026t^2 + 1.6t + 2.2$$

where t is the number of years past 1950. Use the model to find the average number of Social Security beneficiaries per year (actual and predicted) between

(a) 1980 and 2000.   (b) 2010 and 2030.

42. **Total income** Suppose that the income from a slot machine in a casino flows continuously at a rate of

$$f(t) = 100e^{0.1t}$$

where t is the time in hours since the casino opened. Then the total income during the first 10 hours is given by

$$\int_0^{10} 100e^{0.1t}\, dt$$

Find the average income over the first 10 hours.

43. **Drug levels in the blood** A drug manufacturer has developed a time-release capsule with the number of milligrams of the drug in the bloodstream given by

$$S = 30x^{18/7} - 240x^{11/7} + 480x^{4/7}$$

where x is in hours and $0 \le x \le 4$. Find the average number of milligrams of the drug in the bloodstream for the first 4 hours after a capsule is taken.

44. **Income distribution** With data from the U.S. Census Bureau, the Lorenz curves for the income distribution for 2012 in the United States for White non-Hispanic households and for Asian households are given below. Find the Gini coefficient of income for both groups, and compare the distribution of income for these groups.

White non-Hispanic:

$$y = 2.344x^4 - 3.340x^3 + 2.154x^2 - 0.1588x + 0.0005952$$

Asian:

$$y = 2.240x^4 - 3.222x^3 + 2.169x^2 - 0.1871x + 0.0004762$$

45. **Income distribution** With U.S. Census Bureau data, the Lorenz curves for the 2012 income distribution in the United States for Black households and for Hispanic households (of any race) are given below. Find the Gini coefficient of income for both groups, and compare the distribution of income for these groups.

Black:

$$y = 2.292x^4 - 3.127x^3 + 2.006x^2 - 0.1710x + 0.0005556$$

Hispanic:

$$y = 1.830x^4 - 2.564x^3 + 1.806x^2 - 0.1137x + 0.0003512$$

46. **Income distribution** In an effort to make the distribution of income more nearly equal, the government of a country passes a tax law that changes the Lorenz curve from $y = 0.99x^{2.1}$ for one year to $y = 0.32x^2 + 0.68x$ for the next year. Find the Gini coefficient of income for both years, and compare the distributions of income before and after the tax law was passed. Interpret the result.

47. **Income distribution** The Lorenz curves for the income distribution in the United States for all races for 2012 and for 2004 are given below (*Source:* U.S. Census Bureau). Find the Gini coefficient of income for both groups, and compare the distribution of income for these groups.

2012: $y = x^{2.760}$ 2004: $y = x^{2.671}$

48. **Income distribution** Suppose the Gini coefficient of income for a certain country is $2/5$. Find the value of p if the Lorenz curve for this country is

$$L(x) = \frac{1}{3}x + \frac{2}{3}x^p.$$

49. *Gini coefficient* If the Lorenz curve for the income distribution for a given year is $L(x) = x^p$, use integration to find a simple formula for the corresponding Gini coefficient.

50. *Lorenz curve* If the Lorenz curve for the income distribution for a given year has the form $L(x) = x^p$ and the Gini coefficient is G, find a formula for p.

OBJECTIVES

- To use definite integrals to find total income, present value, and future value of continuous income streams
- To use definite integrals to find the consumer's surplus
- To use definite integrals to find the producer's surplus

13.4

Applications of Definite Integrals in Business and Economics

| APPLICATION PREVIEW |

Suppose the oil pumped from a well is considered as a continuous income stream with annual rate of flow (in thousands of dollars per year) at time t years given by

$$f(t) = 600e^{-0.2(t+5)}$$

The company can use a definite integral involving $f(t)$ to estimate the well's present value over the next 10 years (see Example 2).

The definite integral can be used in a number of applications in business and economics. In addition to the present value, the definite integral can be used to find the total income over a fixed number of years from a continuous income stream.

Definite integrals also can be used to determine the savings realized in the marketplace by some consumers (called consumer's surplus) and some producers (called producer's surplus).

Continuous Income Streams

An oil company's profits depend on the amount of oil that can be pumped from a well. Thus we can consider a pump at an oil field as producing a **continuous stream of income** for the owner. Because both the pump and the oil field "wear out" with time, the continuous stream of income is a function of time. Suppose $f(t)$ is the (annual) *rate* of flow of income from this pump; then we can find the total income from the rate of income by using integration. In particular, the total income for k years is given by

$$\text{Total income} = \int_0^k f(t)\, dt$$

EXAMPLE 1   Oil Revenue

Now ——————————→ 10 Years From Now

$2,594,000 = Total Income

A small oil company considers the continuous pumping of oil from a well as a continuous income stream with its annual rate of flow at time t given by

$$f(t) = 600e^{-0.2t}$$

in thousands of dollars per year. Find an estimate of the total income from this well over the next 10 years.

Solution

$$\text{Total income} = \int_0^{10} f(t)\, dt = \int_0^{10} 600e^{-0.2t}\, dt$$

$$= \left. \frac{600}{-0.2} e^{-0.2t} \right|_0^{10} \approx 2594 \quad \text{(to the nearest integer)}$$

Thus the total income is approximately \$2,594,000.

In addition to the total income from a continuous income stream, the **present value** of the stream is also important. The present value is the value today of a continuous income stream that will be providing income in the future. The present value is useful in deciding when to replace machinery or what new equipment to select.

To find the present value of a continuous stream of income with rate of flow $f(t)$, we first graph the function $f(t)$ and divide the time interval from 0 to k into n subintervals of width Δt_i, $i = 1$ to n.

The total amount of income is the area under this curve between $t = 0$ and $t = k$. We can approximate the amount of income in each subinterval by finding the area of the rectangle in that subinterval. (See Figure 13.21.)

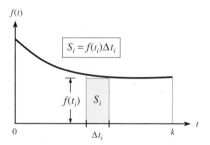

Figure 13.21

We have shown that the future value S that accrues if $\$P$ is invested for t years at an annual rate r, compounded continuously, is $S = Pe^{rt}$. Thus the present value of the investment that yields the single payment of $\$S$ after t years is

$$P = \frac{S}{e^{rt}} = Se^{-rt}$$

The contribution to S in the ith subinterval is $S_i = f(t_i)\,\Delta t_i$, and the present value of this amount is $P_i = f(t_i)\,\Delta t_i e^{-rt_i}$. Thus the total present value of S can be approximated by

$$\sum_{i=1}^{n} f(t_i)\,\Delta t_i e^{-rt_i}$$

This approximation improves as $\Delta t_i \to 0$ with the present value given by

$$\lim_{\Delta t_i \to 0} \sum_{i=1}^{n} f(t_i)\,\Delta t_i e^{-rt_i}$$

This limit gives the **present value** as a definite integral.

Present Value of a Continuous Income Stream	If $f(t)$ is the rate of continuous income flow earning interest at rate r, compounded continuously, then the **present value of the continuous income stream** is $$\text{Present value} = \int_{0}^{k} f(t)e^{-rt}\, dt$$ where $t = 0$ to $t = k$ is the time interval.

EXAMPLE 2   Present Value | APPLICATION PREVIEW |

Suppose that the oil company in Example 1 is planning to sell one of its wells because of its remote location. Suppose further that the company wants to use the present value of this well over the next 10 years to help establish its selling price. If the company determines that the annual rate of flow is

$$f(t) = 600e^{-0.2(t+5)}$$

in thousands of dollars per year, and if money is worth 10%, compounded continuously, find this present value.

Solution

$$\text{Present value} = \int_0^{10} f(t)e^{-rt}\, dt$$

$$= \int_0^{10} 600e^{-0.2(t+5)}e^{-0.1t}\, dt = \int_0^{10} 600e^{-0.3t-1}\, dt$$

If $u = -0.3t - 1$, then $u' = -0.3$ and we get

$$\frac{1}{-0.3}\int 600e^{-0.3t-1}(-0.3\, dt) = \frac{600}{-0.3}e^{-0.3t-1}\bigg|_0^{10}$$

$$= -2000\,(e^{-4} - e^{-1}) \approx 699 \quad \text{(to the nearest integer)}$$

Thus the present value is \$699,000. ∎

Recall that the future value of a continuously compounded investment at rate r after k years is Pe^{rk}, where P is the amount invested (or the present value). Thus, for a continuous income stream, the **future value** is found as follows.

Future Value of a Continuous Income Stream	If $f(t)$ is the rate of continuous income flow for k years earning interest at rate r, compounded continuously, then the **future value of the continuous income stream** is
	$$FV = e^{rk}\int_0^k f(t)e^{-rt}\, dt$$

EXAMPLE 3   Future Value

If the rate of flow of income from an asset is $1000e^{0.02t}$, in millions of dollars per year, and if the income is invested at 6% compounded continuously, find the future value of the asset 4 years from now.

Solution
The future value is given by

$$FV = e^{rk}\int_0^k f(t)e^{-rt}\, dt$$

$$= e^{(0.06)4}\int_0^4 1000e^{0.02t}e^{-0.06t}\, dt = e^{0.24}\int_0^4 1000e^{-0.04t}\, dt$$

$$= e^{0.24}(-25,000e^{-0.04t})\big|_0^4 = -25,000e^{0.24}(e^{-0.16} - 1)$$

$$\approx 4699.05 \quad \text{(millions of dollars)} \quad ∎$$

✓ CHECKPOINT

1. Suppose that a continuous income stream has an annual rate of flow given by $f(t) = 5000e^{-0.01t}$, and suppose that money is worth 7% compounded continuously. Create the integral used to find
   (a) the total income for the next 5 years.
   (b) the present value for the next 5 years.
   (c) the future value 5 years from now.

Consumer's Surplus   Suppose that the demand for a product is given by $p = f(x)$ and that the supply of the product is described by $p = g(x)$. The price p_1 where the graphs of these functions intersect is the **equilibrium price** (see Figure 13.22(a)). As the demand curve shows, some consumers (but not all) would be willing to pay more than $\$p_1$ for the product.

For example, some consumers would be willing to buy x_3 units if the price were $\$p_3$. Those consumers willing to pay more than $\$p_1$ are benefiting from the lower price. The total gain for all those consumers willing to pay more than $\$p_1$ is called the **consumer's surplus,** and under proper assumptions the area of the shaded region in Figure 13.22(a) represents this consumer's surplus.

Looking at Figure 13.22(b), we see that if the demand curve has equation $p = f(x)$, the consumer's surplus is given by the area between $f(x)$ and the x-axis from 0 to x_1, *minus* the area of the rectangle denoted *TR*:

$$CS = \int_0^{x_1} f(x)\,dx - p_1 x_1$$

Note that with equilibrium price p_1 and equilibrium quantity x_1, the product $p_1 x_1$ is the area of the rectangle that represents the total dollars spent by consumers and received as revenue by producers (see Figure 13.22(b)).

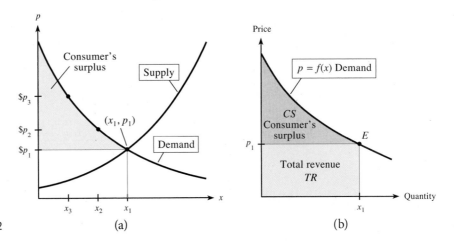

Figure 13.22

(a)   (b)

EXAMPLE 4   Consumer's Surplus

The demand function for x units of a product is $p = f(x) = 1020/(x + 1)$ dollars. If the equilibrium price is $20, what is the consumer's surplus?

Solution
We must first find the quantity that will be purchased at this price. Letting $p = 20$ and solving for x, we get

$$20 = \frac{1020}{x + 1} \quad \text{so} \quad 20(x + 1) = 1020$$

Thus $20x + 20 = 1020$ or $20x = 1000$ so $x = 50$

Thus the equilibrium point is $(x_1, p_1) = (50, 20)$. The consumer's surplus is given by

$$CS = \int_0^{x_1} f(x)\,dx - p_1 x_1 = \int_0^{50} \frac{1020}{x + 1}\,dx - 20 \cdot 50$$

$$= 1020 \ln |x + 1| \Big|_0^{50} - 1000$$

$$= 1020(\ln 51 - \ln 1) - 1000$$

$$\approx 4010.46 - 1000 = 3010.46$$

The consumer's surplus is $3010.46.

EXAMPLE 5   Consumer's Surplus

A product's demand function is $p = f(x) = \sqrt{49 - 6x}$ and its supply function is $p = g(x) = x + 1$, where p is the price per unit in dollars and x is the number of units. Find the equilibrium point and the consumer's surplus there.

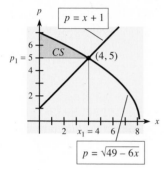

Figure 13.23

Solution

We can determine the equilibrium point by solving the two equations simultaneously.

$$\sqrt{49 - 6x} = x + 1$$
$$49 - 6x = (x + 1)^2$$
$$0 = x^2 + 8x - 48$$
$$0 = (x + 12)(x - 4)$$
$$x = 4 \quad \text{or} \quad x = -12$$

Thus the equilibrium quantity is $x_1 = 4$ and the equilibrium price is $p_1 = \$5$ ($x = -12$ is not a solution). The graphs of the supply and demand functions are shown in Figure 13.23.

The consumer's surplus is given by

$$CS = \int_0^4 f(x)\,dx - p_1 x_1 = \int_0^4 \sqrt{49 - 6x}\,dx - 5 \cdot 4$$

$$= -\frac{1}{6}\int_0^4 \sqrt{49 - 6x}\,(-6\,dx) - 20 = -\frac{1}{9}(49 - 6x)^{3/2}\Big|_0^4 - 20$$

$$= -\frac{1}{9}[(25)^{3/2} - (49)^{3/2}] - 20 = -\frac{1}{9}(125 - 343) - 20 \approx 4.22$$

The consumer's surplus is $4.22.

In a monopoly market there is no market equilibrium point, and we find consumer's surplus at the point where the monopoly has maximum profit.

EXAMPLE 6   Monopoly Market

Suppose a monopoly has its total cost (in dollars) for a product given by $C(x) = 6000 + 0.2x^3$. Suppose also that demand is given by $p = f(x) = 300 - 0.1x$, where p is in dollars and x is the number of units. Find the consumer's surplus at the point where the monopoly has maximum profit.

Solution

We must first find the point where the profit function is maximized. Because the demand for x units is $p = 300 - 0.1x$, the total revenue is

$$R(x) = (300 - 0.1x)x = 300x - 0.1x^2$$

Thus the profit function is

$$P(x) = R(x) - C(x)$$
$$P(x) = 300x - 0.1x^2 - (6000 + 0.2x^2)$$
$$P(x) = 300x - 6000 - 0.3x^2$$

Then $P'(x) = 300 - 0.6x$. So $0 = 300 - 0.6x$ has the solution $x = 500$.

Because $P''(500) = -0.6 < 0$, the profit for the monopoly is maximized when $x = 500$ units are sold at price $p = 300 - 0.1(500) = 250$ dollars per unit.

The consumer's surplus at $x_1 = 500$, $p_1 = 250$ is given by

$$CS = \int_0^{500} f(x)\,dx - 500 \cdot 250 = \int_0^{500} (300 - 0.1x)\,dx - 125{,}000$$

$$= 300x - \frac{0.1x^2}{2}\Big|_0^{500} - 125{,}000 = (150{,}000 - 12{,}500) - 125{,}000 = 12{,}500$$

The consumer's surplus is $12,500.

MONEY ON THE TABLE

Another way to understand consumer surplus is through the popular concept of "money on the table." Imagine you have $500 to spend on an iPad. When you get to the Apple store, the model you want costs only $450. You're delighted to spend only $450, but you **would** have spent all your money if the iPad had been priced at $500. In this example, Apple left money—$50 to be exact—on the table (or in your pocket). That's a consumer's surplus.

GRAVY

Now think of the iPad example in another way. Suppose Apple meets all its profit targets at a price of $400 per iPad. The fact that you pay $450 means that there is a producer surplus. Apple would sell that iPad at the $400 price point, but because you were willing to part with $450, the company receives an extra $50—or gravy—for its bottom line. That is, a producer's surplus.

Producer's Surplus

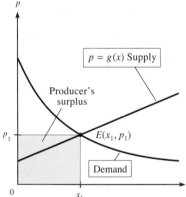

Figure 13.24

When a product is sold at the equilibrium price, p_1, some producers will also benefit, for they would have sold the product at a lower price. The area between the line $p = p_1$ and the supply curve (from $x = 0$ to $x = x_1$) gives the producer's surplus (see Figure 13.24).

If the supply function is $p = g(x)$, the **producer's surplus** is given by the area between the graph of $p = g(x)$ and the x-axis from 0 to x_1 *subtracted from* $p_1 x_1$, the area of the rectangle shown in Figure 13.24.

$$PS = p_1 x_1 - \int_0^{x_1} g(x)\, dx$$

Note that $p_1 x_1$ represents the total revenue at the equilibrium point.

EXAMPLE   7   Producer's Surplus

Suppose that the supply function for x million units of a product is $p = x^2 + x$ dollars per unit. If the equilibrium price is $20 per unit, what is the producer's surplus?

Solution

Because $p = 20$, we can find the equilibrium quantity x as follows:

$$20 = x^2 + x$$
$$0 = x^2 + x - 20$$
$$0 = (x + 5)(x - 4)$$
$$x = -5, \quad x = 4$$

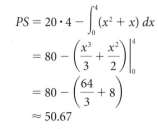

Figure 13.25

The equilibrium point is $x_1 = 4$ million units, $p_1 = \$20$. The producer's surplus is given by

$$PS = 20 \cdot 4 - \int_0^4 (x^2 + x)\, dx$$

$$= 80 - \left(\frac{x^3}{3} + \frac{x^2}{2} \right) \Bigg|_0^4$$

$$= 80 - \left(\frac{64}{3} + 8 \right)$$

$$\approx 50.67$$

The producer's surplus is $50.67 million. See Figure 13.25.

EXAMPLE 8   Producer's Surplus

The demand function for a product is $p = \sqrt{49 - 6x}$ and the supply function is $p = x + 1$. Find the producer's surplus.

Solution

We found the equilibrium point for these functions to be $(4, 5)$ in Example 5 (see Figure 13.23 earlier in this section). The producer's surplus is

$$PS = 5 \cdot 4 - \int_0^4 (x + 1)\, dx = 20 - \left(\frac{x^2}{2} + x\right)\Big|_0^4$$
$$= 20 - (8 + 4) = 8$$

The producer's surplus is $8.

✓ CHECKPOINT

2. Suppose that for a certain product, the supply function is $p = f(x)$, the demand function is $p = g(x)$, and the equilibrium point is (x_1, p_1). Decide whether the following are true or false.

    (a) $CS = \int_0^{x_1} f(x)\, dx - p_1 x_1$ (b) $PS = \int_0^{x_1} f(x)\, dx - p_1 x_1$

3. If demand is $p = \dfrac{100}{x + 1}$, supply is $p = x + 1$, and the market equilibrium is $(9, 10)$, create the integral used to find the
    (a) consumer's surplus.   (b) producer's surplus.

EXAMPLE 9   Consumer's Surplus and Producer's Surplus

Suppose that for x units of a certain product, the demand function is $p = 200e^{-0.01x}$ dollars and the supply function is $p = \sqrt{200x + 49}$ dollars.

(a) Use a graphing calculator to find the market equilibrium point.
(b) Find the consumer's surplus.
(c) Find the producer's surplus.

Solution

(a) Solving $200e^{-0.01x} = \sqrt{200x + 49}$ is very difficult using algebraic techniques. Using SOLVER or INTERSECT on a graphing calculator gives $x = 60$, to the nearest unit, with a price of $109.76. (See Figure 13.26(a).)

(b) The consumer's surplus, shown in Figure 13.26(b), is

$$\int_0^{60} 200e^{-0.01x}\, dx - 109.76(60) = (-20{,}000e^{-0.01x})\Big|_0^{60} - 6585.60$$
$$= -20{,}000e^{-0.6} + 20{,}000 - 6585.60$$
$$\approx 2438.17 \text{ dollars}$$

(c) The producer's surplus, shown in Figure 13.26(b), is

$$60(109.76) - \int_0^{60} \sqrt{200x + 49}\, dx = 6585.60 - \frac{1}{200}\left[\frac{(200x + 49)^{3/2}}{3/2}\right]_0^{60}$$
$$= 6585.60 - \frac{1}{300}[(12{,}049^{3/2} - 49^{3/2})] \approx 2178.10 \text{ dollars}$$

Note that we also could have evaluated these definite integrals with the numerical integration feature of a graphing calculator, and we would have obtained the same results.

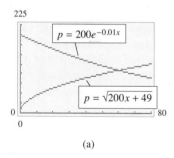

225

$p = 200e^{-0.01x}$

$p = \sqrt{200x + 49}$

0   80
0

(a)

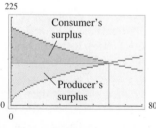

225

Consumer's surplus

Producer's surplus

0   80
0

(b)

Figure 13.26

✓ CHECKPOINT
ANSWERS

1. (a) $\int_0^5 5000e^{-0.01t}\,dt$

   (b) $\int_0^5 (5000e^{-0.01t})(e^{-0.07t})\,dt = \int_0^5 5000e^{-0.08t}\,dt$

   (c) $e^{(0.07)(5)}\int_0^5 (5000e^{-0.01t})(e^{-0.07t})\,dt = e^{0.35}\int_0^5 5000e^{-0.08t}\,dt$

2. (a) False. Consumer's surplus uses the demand function, so

   $$CS = \int_0^{x_1} g(x)\,dx - p_1 x_1$$

   (b) False. Producer's surplus uses the supply function, but the formula is

   $$PS = p_1 x_1 - \int_0^{x_1} f(x)\,dx$$

3. (a) $CS = \int_0^9 \dfrac{100}{x+1}\,dx - 90$ (b) $PS = 90 - \int_0^9 (x+1)\,dx$

| EXERCISES | 13.4

CONTINUOUS INCOME STREAMS

1. Find the total income over the next 10 years from a continuous income stream that has an annual rate of flow at time t given by $f(t) = 12{,}000e^{0.01t}$ (dollars per year).

2. Find the total income over the next 8 years from a continuous income stream with an annual rate of flow at time t given by $f(t) = 8500e^{-0.2t}$ (dollars per year).

3. Suppose that a steel company views the production of its continuous caster as a continuous income stream with a monthly rate of flow at time t given by

   $f(t) = 24{,}000e^{0.03t}$ (dollars per month)

   Find the total income from this caster in the first year.

4. Suppose that the Quick-Fix Car Service franchise finds that the income generated by its stores can be modeled by assuming that the income is a continuous stream with a monthly rate of flow at time t given by

   $f(t) = 10{,}000e^{0.02t}$ (dollars per month)

   Find the total income from a Quick-Fix store for the first 2 years of operation.

5. A small brewery considers the output of its bottling machine as a continuous income stream with an annual rate of flow at time t given by

   $f(t) = 80e^{-0.1t}$

   in thousands of dollars per year. Find the income from this stream for the next 10 years.

6. A company that services a number of vending machines considers its income as a continuous stream with an annual rate of flow at time t given by

   $f(t) = 120e^{-0.4t}$

   in thousands of dollars per year. Find the income from this stream over the next 5 years.

7. A franchise models the profit from its store as a continuous income stream with a monthly rate of flow at time t given by

   $f(t) = 3000e^{0.004t}$ (dollars per month)

   When a new store opens, its manager is judged against the model, with special emphasis on the second half of the first year. Find the total profit for the second 6-month period ($t = 6$ to $t = 12$).

8. The Medi Spa franchise has a continuous income stream with a monthly rate of flow modeled by $f(t) = 20{,}000e^{0.03t}$ (dollars per month). Find the total income for years 2 through 5.

9. A continuous income stream has an annual rate of flow at time t given by

   $f(t) = 12{,}000e^{0.04t}$ (dollars per year)

   If money is worth 8% compounded continuously, find the present value of this stream for the next 8 years.

10. A continuous income stream has an annual rate of flow at time t given by

    $f(t) = 9000e^{0.12t}$ (dollars per year)

    Find the present value of this income stream for the next 10 years, if money is worth 6% compounded continuously.

11. The income from an established chain of laundromats is a continuous stream with its annual rate of flow at time t given by $f(t) = 630{,}000$ (dollars per year). If money is worth 7% compounded continuously, find the present value and future value of this chain over the next 5 years.

12. The profit from an insurance agency can be considered as a continuous income stream with an annual rate of flow at time t given by $f(t) = 840{,}000$ (dollars per year). Find the present value and future value of this agency

over the next 12 years, if money is worth 8% compounded continuously.

13. Suppose that a printing firm considers its production as a continuous income stream. If the annual rate of flow at time t is given by

$$f(t) = 97.5e^{-0.2(t+3)}$$

in thousands of dollars per year, and if money is worth 6% compounded continuously, find the present value and future value of the presses over the next 10 years.

14. Suppose that a vending machine company is considering selling some of its machines. Suppose further that the income from these particular machines is a continuous stream with an annual rate of flow at time t given by

$$f(t) = 12e^{-0.4(t+3)}$$

in thousands of dollars per year. Find the present value and future value of the machines over the next 5 years if money is worth 10% compounded continuously.

15. A 58-year-old couple are considering opening a business of their own. They will either purchase an established Gift and Card Shoppe or open a new Wine Boutique. The Gift Shoppe has a continuous income stream with an annual rate of flow at time t given by

$$G(t) = 30,000 \qquad \text{(dollars per year)}$$

and the Wine Boutique has a continuous income stream with a projected annual rate of flow at time t given by

$$W(t) = 21,600e^{0.08t} \qquad \text{(dollars per year)}$$

The initial investment is the same for both businesses, and money is worth 10% compounded continuously. Find the present value of each business over the next 7 years (until the couple reach age 65) to see which is the better buy.

16. If the couple in Problem 15 plan to keep the business until age 70 (for the next 12 years), find each present value to see which business is the better buy in this case.

CONSUMER'S SURPLUS

In Problems 17–26, p and C are in dollars and x is the number of units.

17. The demand function for a product is $p = 34 - x^2$. If the equilibrium price is $9 per unit, what is the consumer's surplus?

18. The demand function for a product is $p = 100 - 4x$. If the equilibrium price is $40 per unit, what is the consumer's surplus?

19. The demand function for a product is $p = 200/(x + 2)$. If the equilibrium quantity is 8 units, what is the consumer's surplus?

20. The demand function for a certain product is $p = 100/(1 + 2x)$. If the equilibrium quantity is 12 units, what is the consumer's surplus?

21. The demand function for a certain product is $p = 81 - x^2$ and the supply function is $p = x^2 + 4x + 11$. Find the equilibrium point and the consumer's surplus there.

22. The demand function for a product is $p = 49 - x^2$ and the supply function is $p = 4x + 4$. Find the equilibrium point and the consumer's surplus there.

23. If the demand function for a product is $p = 12/(x + 1)$ and the supply function is $p = 1 + 0.2x$, find the consumer's surplus under pure competition.

24. If the demand function for a good is $p = 110 - x^2$ and the supply function for it is $p = 2 - \frac{6}{5}x + \frac{1}{5}x^2$, find the consumer's surplus under pure competition.

25. A monopoly has a total cost function $C = 1000 + 120x + 6x^2$ for its product, which has demand function $p = 360 - 3x - 2x^2$. Find the consumer's surplus at the point where the monopoly has maximum profit.

26. A monopoly has a total cost function $C = 500 + 2x^2 + 10x$ for its product, which has demand function $p = -\frac{1}{3}x^2 - 2x + 30$. Find the consumer's surplus at the point where the monopoly has maximum profit.

PRODUCER'S SURPLUS

In Problems 27–36, p is in dollars and x is the number of units.

27. Suppose that the supply function for a good is $p = 4x^2 + 2x + 2$. If the equilibrium price is $422 per unit, what is the producer's surplus there?

28. Suppose that the supply function for a good is $p = 0.1x^2 + 3x + 20$. If the equilibrium price is $36 per unit, what is the producer's surplus there?

29. If the supply function for a commodity is $p = 10e^{x/3}$, what is the producer's surplus when 15 units are sold?

30. If the supply function for a commodity is $p = 40 + 100(x + 1)^2$, what is the producer's surplus at $x = 20$?

31. Find the producer's surplus at market equilibrium for a product if its demand function is $p = 81 - x^2$ and its supply function is $p = x^2 + 4x + 11$.

32. Find the producer's surplus at market equilibrium for a product if its demand function is $p = 49 - x^2$ and its supply function is $p = 4x + 4$.

33. Find the producer's surplus at market equilibrium for a product with demand function $p = 12/(x + 1)$ and supply function $p = 1 + 0.2x$.

34. Find the producer's surplus at market equilibrium for a product with demand function $p = 110 - x^2$ and supply function $p = 2 - \frac{6}{5}x + \frac{1}{5}x^2$.

35. The demand function for a certain product is $p = 144 - 2x^2$ and the supply function is $p = x^2 + 33x + 48$. Find the producer's surplus at the equilibrium point.

36. The demand function for a product is $p = 280 - 4x - x^2$ and the supply function for it is $p = 160 + 4x + x^2$. Find the producer's surplus at the equilibrium point.

OBJECTIVE

13.5

- To use tables of integrals to evaluate certain integrals

Using Tables of Integrals

■ **| APPLICATION PREVIEW |** ■

The rate of change of worldwide sales (in billions of dollars per year) of Hepatitis C treatments, from 2012 and projected to 2016, is given by

$$\frac{dS}{dt} = 0.575(1.234^t)$$

where t equals the number of years after 2010 and S is projected to be $9.7 billion in 2016 (*Source:* Evaluate Pharma). Finding the function that models the worldwide sales of Hepatitis C treatments requires evaluating the integral

$$S(t) = \int 0.575(1.234^t)dt$$

Evaluating this integral is made easier by using a formula such as those given in Table 13.2. (See Example 4.) The examples in this section illustrate how some of these formulas are used.

The formulas in Table 13.2, and others listed in other resources, extend the number of integrals that can be evaluated. Using the formulas is not quite as easy as it may sound because finding the correct formula and using it properly may present problems.

■ **| TABLE 13.2** ■

INTEGRATION FORMULAS

1. $\int u^n\, du = \dfrac{u^{n+1}}{n+1} + C,\ \text{for } n \neq -1$

2. $\int \dfrac{du}{u} = \int u^{-1}\, du = \ln|u| + C$

3. $\int a^u\, du = a^u \log_a e + C = \dfrac{a^u}{\ln a} + C$

4. $\int e^u\, du = e^u + C$

5. $\int \dfrac{du}{a^2 - u^2} = \dfrac{1}{2a} \ln\left|\dfrac{a+u}{a-u}\right| + C$

6. $\int \sqrt{u^2 + a^2}\, du = \dfrac{1}{2}(u\sqrt{u^2+a^2} + a^2 \ln|u + \sqrt{u^2+a^2}|) + C$

7. $\int \sqrt{u^2 - a^2}\, du = \dfrac{1}{2}(u\sqrt{u^2-a^2} - a^2 \ln|u + \sqrt{u^2-a^2}|) + C$

8. $\int \dfrac{du}{\sqrt{u^2+a^2}} = \ln|u + \sqrt{u^2+a^2}| + C$

9. $\int \dfrac{du}{u\sqrt{a^2-u^2}} = -\dfrac{1}{a} \ln\left|\dfrac{a+\sqrt{a^2-u^2}}{u}\right| + C$

10. $\int \dfrac{du}{\sqrt{u^2-a^2}} = \ln|u + \sqrt{u^2-a^2}| + C$

11. $\int \dfrac{du}{u\sqrt{a^2+u^2}} = -\dfrac{1}{a} \ln\left|\dfrac{a+\sqrt{a^2+u^2}}{u}\right| + C$

12. $\int \dfrac{u\,du}{au+b} = \dfrac{u}{a} - \dfrac{b}{a^2} \ln|au+b| + C$

(*continued*)

███ | **TABLE 13.2** ███████████████████████

INTEGRATION FORMULAS (*continued*)

13. $\displaystyle\int \frac{du}{u(au + b)} = \frac{1}{b} \ln \left| \frac{u}{au + b} \right| + C$

14. $\displaystyle\int \ln u \, du = u(\ln u - 1) + C$

15. $\displaystyle\int \frac{u \, du}{(au + b)^2} = \frac{1}{a^2} \left(\ln|au + b| + \frac{b}{au + b} \right) + C$

16. $\displaystyle\int u\sqrt{au + b} \, du = \frac{2(3au - 2b)(au + b)^{3/2}}{15a^2} + C$

EXAMPLE 1   **Using an Integration Formula**

Evaluate $\displaystyle\int \frac{dx}{\sqrt{x^2 + 4}}$.

Solution
We must find a formula in Table 13.2 that is of the same form as this integral. We see that formula 8 has the desired form, *if* we let $u = x$ and $a = 2$. Thus

$$\int \frac{dx}{\sqrt{x^2 + 4}} = \ln\left|x + \sqrt{x^2 + 4}\right| + C$$

■

EXAMPLE 2   **Fitting Integration Formulas**

Evaluate (a) $\displaystyle\int_1^2 \frac{dx}{x^2 + 2x}$ and   (b) $\displaystyle\int \ln (2x + 1) \, dx$.

Solution
(a) There does not appear to be any formula that has exactly the same form as this integral. But if we rewrite the integral as

$$\int_1^2 \frac{dx}{x(x + 2)}$$

we see that formula 13 will work. Letting $u = x$, $a = 1$, and $b = 2$, we get

$$\int_1^2 \frac{dx}{x(x + 2)} = \frac{1}{2} \ln \left| \frac{x}{x + 2} \right| \Big|_1^2 = \frac{1}{2} \ln \left| \frac{2}{4} \right| - \frac{1}{2} \ln \left| \frac{1}{3} \right|$$

$$= \frac{1}{2} \left(\ln \frac{1}{2} - \ln \frac{1}{3} \right)$$

$$= \frac{1}{2} \ln \frac{3}{2} \approx 0.2027$$

(b) This integral has the form of formula 14, with $u = 2x + 1$. But if $u = 2x + 1$, du must be represented by the differential of $2x + 1$ (that is, $2 \, dx$). Thus

$$\int \ln (2x + 1) \, dx = \frac{1}{2} \int \ln (2x + 1)(2 \, dx)$$

$$= \frac{1}{2} \int \ln (u) \, du = \frac{1}{2} u[\ln (u) - 1] + C$$

$$= \frac{1}{2} (2x + 1)[\ln (2x + 1) - 1] + C$$

■

✓ **CHECKPOINT**

1. Can both $\int \dfrac{dx}{\sqrt{x^2 - 4}}$ and $-\int \dfrac{dx}{\sqrt{4 - x^2}}$ be evaluated with formula 10 in Table 13.2?

2. Determine the formula used to evaluate $\int \dfrac{3x}{4x - 5}\,dx$, and show how the formula would be applied.

3. True or false: In order for us to use a formula, the given integral must correspond exactly to the formula, including du.

4. True or false: $\int \dfrac{dx}{x^2(3x^2 - 7)}$ can be evaluated with formula 13.

5. True or false: $\int \dfrac{dx}{(6x + 1)^2}$ can be evaluated with either formula 1 or formula 15.

6. True or false: $\int \sqrt{x^2 + 4}\,dx$ can be evaluated with formula 1, formula 6, or formula 16.

EXAMPLE 3   **Fitting an Integration Formula**

Evaluate $\displaystyle\int_1^2 \dfrac{dx}{x\sqrt{81 - 9x^2}}$.

Solution

This integral is similar to that of formula 9 in Table 13.2. Letting $a = 9$, letting $u = 3x$, and multiplying the numerator and denominator by 3 give the proper form.

$$\int_1^2 \dfrac{dx}{x\sqrt{81 - 9x^2}} = \int_1^2 \dfrac{3\,dx}{3x\sqrt{81 - 9x^2}}$$

$$= -\dfrac{1}{9} \ln \left| \dfrac{9 + \sqrt{81 - 9x^2}}{3x} \right| \Bigg|_1^2$$

$$= \left[-\dfrac{1}{9} \ln \left(\dfrac{9 + \sqrt{45}}{6} \right) \right] - \left[-\dfrac{1}{9} \ln \left(\dfrac{9 + \sqrt{72}}{3} \right) \right]$$

$$\approx 0.088924836$$

Remember that the formulas given in Table 13.2 represent only a very small sample of all possible integration formulas. Additional formulas may be found in books of mathematical tables or online.

EXAMPLE 4   **Hepatitis C | APPLICATION PREVIEW |**

The rate of change of worldwide sales (in billions of dollars per year) of Hepatitis C treatments, from 2012 and projected to 2016, is given by

$$\dfrac{dS}{dt} = 0.575(1.234^t)$$

where t equals the number of years after 2010 and S is projected to be \$9.7 billion in 2016 (*Source:* Evaluate Pharma).

(a) Find the function that models the worldwide sales of Hepatitis C treatments.
(b) Find the predicted worldwide sales in 2020.

Solution

(a) To find $S(t)$, we integrate $\dfrac{dS}{dt}$ by using formula 3, with $a = 1.234$ and $u = t$.

$$S(t) = \int 0.575(1.234^t)\,dt = 0.575 \int (1.234^t)\,dt = 0.575\dfrac{1.234^t}{\ln 1.234} + C \approx 2.735(1.234^t) + C$$

We use the projected sales of $9.7 billion for 2016 (when $t = 6$) to find the value of C.

$$9.7 = 2.735(1.234^6) + C \Rightarrow 9.7 \approx 9.7 + C \Rightarrow C \approx 0$$

Thus $S(t) = 2.735(1.234^t)$.

(b) The projected sales in 2020 (when $t = 10$) are $S(10) = 2.735(1.234^{10}) \approx \$22.4\,$billion. ∎

Calculator Note   Numerical integration with a graphing calculator is especially useful in evaluating definite integrals when the formulas for the integrals are difficult to use or are not available. For example, evaluating the definite integral in Example 3 with the numerical integration feature of a graphing calculator gives 0.08892484. The decimal approximation of the answer found in Example 3 is 0.088924836, so the numerical approximation of the answer agrees for the first 8 decimal places. ∎

✓ CHECKPOINT
ANSWERS

1. No; $\sqrt{4 - x^2}$ cannot be written in the form $\sqrt{u^2 - a^2}$.

2. Use formula 12: $\dfrac{3x}{4} + \dfrac{15}{16} \ln |4x - 5| + C$

3. True.

4. False. With $u = x^2$, we must have $du = 2x\, dx$.

5. False. The integral can be evaluated with formula 1, but not with formula 15.

6. False. The integral can be evaluated only with formula 6.

| EXERCISES | 13.5

Evaluate the integrals in Problems 1–32. Identify the formula used.

1. $\displaystyle \int \frac{dx}{16 - x^2}$

2. $\displaystyle \int \frac{dx}{x(3x + 5)}$

3. $\displaystyle \int_1^4 \frac{dx}{x\sqrt{9 + x^2}}$

4. $\displaystyle \int \frac{dx}{x\sqrt{9 - x^2}}$

5. $\displaystyle \int \ln w\, dw$

6. $\displaystyle \int 4(3^x)\, dx$

7. $\displaystyle \int_0^2 \frac{q\, dq}{6q + 9}$

8. $\displaystyle \int_1^5 \frac{dq}{q\sqrt{25 + q^2}}$

9. $\displaystyle \int \frac{dv}{v(3v + 8)}$

10. $\displaystyle \int_0^3 \sqrt{x^2 + 16}\, dx$

11. $\displaystyle \int_5^7 \sqrt{x^2 - 25}\, dx$

12. $\displaystyle \int \frac{x\, dx}{(3x + 2)^2}$

13. $\displaystyle \int w\sqrt{4w + 5}\, dw$

14. $\displaystyle \int \frac{dy}{\sqrt{9 + y^2}}$

15. $\displaystyle \int x\, 5^{x^2}\, dx$

16. $\displaystyle \int \sqrt{9x^2 + 4}\, dx$

17. $\displaystyle \int_0^3 x\sqrt{x^2 + 4}\, dx$

18. $\displaystyle \int x\sqrt{x^4 - 36}\, dx$

19. $\displaystyle \int \frac{5\, dx}{x\sqrt{4 - 9x^2}}$

20. $\displaystyle \int x\, e^{x^2}\, dx$

21. $\displaystyle \int \frac{dx}{\sqrt{9x^2 - 4}}$

22. $\displaystyle \int \frac{dx}{25 - 4x^2}$

23. $\displaystyle \int \frac{3x\, dx}{(2x - 5)^2}$

24. $\displaystyle \int_0^1 \frac{x\, dx}{6 - 5x}$

25. $\displaystyle \int \frac{dx}{\sqrt{(3x + 1)^2 + 1}}$

26. $\displaystyle \int \frac{dx}{9 - (2x + 3)^2}$

27. $\displaystyle \int_0^3 x\sqrt{(x^2 + 1)^2 + 9}\, dx$

28. $\displaystyle \int_1^e x \ln x^2\, dx$

29. $\displaystyle \int \frac{x\, dx}{7 - 3x^2}$

30. $\displaystyle \int_0^1 \frac{e^x}{1 + e^x}\, dx$

31. $\displaystyle \int \frac{dx}{\sqrt{4x^2 + 7}}$

32. $\displaystyle \int e^{2x}\sqrt{3e^x + 1}\, dx$

 Use formulas or numerical integration with a graphing calculator or computer to evaluate the definite integrals in Problems 33–36.

33. $\displaystyle \int_2^3 \frac{e^{\sqrt{x-1}}}{\sqrt{x - 1}}\, dx$

34. $\displaystyle \int_2^4 \frac{3x}{\sqrt{x^4 - 9}}\, dx$

35. $\displaystyle \int_0^1 \frac{x^3\, dx}{(4x^2 + 5)^2}$

36. $\displaystyle \int_0^1 (e^x + 1)^3 e^x\, dx$

APPLICATIONS

37. *Producer's surplus* If the supply function for x units of a commodity is $p = 40 + 100 \ln (x + 1)^2$ dollars, what is the producer's surplus at $x = 20$?

38. *Consumer's surplus* If the demand function for wheat is $p = \dfrac{1500}{\sqrt{x^2 + 1}} + 4$ dollars, where x is the number of hundreds of bushels of wheat, what is the consumer's surplus at $x = 7$, $p = 216.13$?

39. *Cost* (a) If the marginal cost for x units of a good is $\overline{MC} = \sqrt{x^2 + 9}$ (dollars per unit) and if the fixed cost is $300, what is the total cost function of the good?
(b) What is the total cost of producing 4 units of this good?

40. *Consumer's surplus* Suppose that the demand function for an appliance is

$$p = \frac{400q + 400}{(q + 2)^2}$$

where q is the number of units and p is in dollars. What is the consumer's surplus if the equilibrium price is $19 and the equilibrium quantity is 18?

41. *Income streams* Suppose that when a new oil well is opened, its production is viewed as a continuous income stream with monthly rate of flow

$$f(t) = 10 \ln (t + 1) - 0.1t$$

where t is time in months and $f(t)$ is in thousands of dollars per month. Find the total income over the next 10 years (120 months).

42. *Spread of disease* An isolated community of 1000 people susceptible to a certain disease is exposed when one member returns carrying the disease. If x represents the number infected with the disease at time t (in days), then the rate of change of x is proportional to the product of the number infected, x, and the number still susceptible, $1000 - x$. That is,

$$\frac{dx}{dt} = kx(1000 - x) \quad \text{or} \quad \frac{dx}{x(1000 - x)} = k\, dt$$

(a) If $k = 0.001$, integrate both sides to solve this differential equation.
(b) Find how long it will be before half the population of the community is affected.
(c) Find the rate of new cases, dx/dt, every other day for the first 13 days.

OBJECTIVE

13.6

- To evaluate integrals using the method of integration by parts

Integration by Parts

| APPLICATION PREVIEW |

If the value of oil produced by a piece of oil extraction equipment is considered a continuous income stream with an annual rate of flow (in dollars per year) at time t in years given by

$$f(t) = 300{,}000 - 2500t, \quad 0 \le t \le 10$$

and if money can be invested at 8%, compounded continuously, then the present value of the piece of equipment is

$$\int_0^{10} (300{,}000 - 2500t)e^{-0.08t}\, dt$$

$$= 300{,}000 \int_0^{10} e^{-0.08t}\, dt - 2500 \int_0^{10} te^{-0.08t}\, dt$$

The first integral can be evaluated with the formula for the integral of $e^u\, du$. Evaluating the second integral can be done by using integration by parts, which is a special technique that involves rewriting an integral in a form that can be evaluated. (See Example 6.)

Integration by parts is an integration technique that uses a formula that follows from the Product Rule for derivatives (actually differentials) as follows:

$$\frac{d}{dx}(uv) = u\frac{dv}{dx} + v\frac{du}{dx} \quad \text{so} \quad d(uv) = u\, dv + v\, du$$

Rearranging the differential form and integrating both sides give the following.

$$u\, dv = d(uv) - v\, du$$
$$\int u\, dv = \int d(uv) - \int v\, du$$
$$\int u\, dv = uv - \int v\, du$$

Integration by Parts Formula

$$\int u\, dv = uv - \int v\, du$$

Integration by parts is very useful if the integral we seek to evaluate can be treated as the product of one function, u, and the differential dv of a second function, so that the two integrals $\int dv$ and $\int v\,du$ can be found. Let us consider an example using this method.

EXAMPLE 1   **Integration by Parts**

Evaluate $\int xe^x\,dx$.

Solution

We cannot evaluate this integral using methods we have learned. But we can "split" the integrand into two parts, setting one part equal to u and the other part equal to dv. This "split" must be done in such a way that $\int dv$ and $\int v\,du$ can be evaluated. Letting $u = x$ and letting $dv = e^x dx$ are possible choices. If we make these choices, we have

$$u = x \quad dv = e^x\,dx$$
$$du = 1\,dx \quad v = \int e^x\,dx = e^x$$

Then

$$\int xe^x\,dx = uv - \int v\,du$$
$$= xe^x - \int e^x\,dx = xe^x - e^x + C$$

We see that choosing $u = x$ and $dv = e^x dx$ worked in evaluating $\int xe^x dx$ in Example 1. If we had chosen $u = e^x$ and $dv = x\,dx$, the results would not have been so successful.

How can we select u and dv to make integration by parts work? As a general guideline, we do the following.

First identify the types of functions occurring in the problem in the order

Logarithm,   Polynomial (or Power of x),   Radical,   Exponential*

Thus, in Example 1, we had x and e^x, a polynomial and an exponential.

Second, choose u to equal the function whose type occurs first on the list; hence in Example 1 we chose $u = x$. Then dv equals the rest of the integrand (and always includes dx) so that $u\,dv$ equals the original integrand. A helpful way to remember the order of the function types that help us choose u is the sentence

"Lazy People Rarely Excel."

in which the first letters, LPRE, coordinate with the order and types of functions. Consider the following examples.

EXAMPLE 2   **Integration by Parts**

Evaluate $\int x \ln x\,dx$.

Solution

The integral contains a logarithm ($\ln x$) and a polynomial (x). Thus, choose $u = \ln x$ and $dv = x\,dx$. Then

$$du = \frac{1}{x}\,dx \quad \text{and} \quad v = \frac{x^2}{2}$$

so $$\int x \ln x\,dx = u \cdot v - \int v\,du = (\ln x)\,\frac{x^2}{2} - \int \frac{x^2}{2} \cdot \frac{1}{x}\,dx$$

$$= \frac{x^2}{2}\ln x - \int \frac{x}{2}\,dx = \frac{x^2}{2}\ln x - \frac{x^2}{4} + C$$

---

*This order is related to the ease with which the function types can be integrated.

Note that letting $dv = \ln x\, dx$ is contrary to our guidelines and would lead to great difficulty in evaluating $\int dv$ and $\int v\, du$.

EXAMPLE 3 Integration by Parts

Evaluate $\int \ln x^2\, dx$.

Solution

The only function in this problem is a logarithm, the first function type on our list for choosing u. Thus

$$\text{Choose:} \qquad u = \ln x^2 \qquad\qquad dv = dx$$

$$\text{Calculate:} \qquad du = \frac{2x}{x^2}\, dx = \frac{2}{x}\, dx \qquad v = x$$

Then

$$\int \ln x^2\, dx = x \ln x^2 - \int x \cdot \frac{2}{x}\, dx = x \ln x^2 - 2x + C$$

Note that if we write $\ln x^2$ as $2 \ln x$, we can also evaluate this integral using formula 14 in Table 13.2 in the previous section, so integration by parts would not be needed.

✓ CHECKPOINT

1. True or false: In evaluating $\int u\, dv$ by parts,
   (a) the parts u and dv are selected and the parts du and v are calculated.
   (b) the differential (often dx) is always chosen as part of dv.
   (c) the parts du and v are found from u and dv as follows:

$$du = u'dx \quad \text{and} \quad v = \int dv$$

   (d) For $\int \dfrac{3x}{e^{2x}}\, dx$, we could choose $u = 3x$ and $dv = e^{2x} dx$.

2. For $\int \dfrac{\ln x}{x^4}\, dx$,
   (a) identify u and dv.
   (b) find du and v.
   (c) complete the evaluation of the integral.

Sometimes it is necessary to repeat integration by parts to complete the evaluation. When this occurs, at each use of integration by parts it is important to choose u and dv consistently according to our guidelines.

EXAMPLE 4 Repeated Integration by Parts

Evaluate $\int x^2 e^{2x}\, dx$.

Solution

Choose $u = x^2$ and $dv = e^{2x} dx$. Then we calculate $du = 2x\, dx$ and $v = \frac{1}{2} e^{2x}$. Thus

$$\int x^2 e^{2x}\, dx = \frac{1}{2} x^2 e^{2x} - \int x e^{2x}\, dx$$

We cannot evaluate $\int x e^{2x}\, dx$ directly, but this new integral is simpler than the original, and a second integration by parts will be successful. Choosing $u = x$ and $dv = e^{2x} dx$, we calculate $du = dx$ and $v = \frac{1}{2} e^{2x}$. Thus

$$\int x^2 e^{2x}\, dx = \frac{1}{2} x^2 e^{2x} - \left(\frac{1}{2} x e^{2x} - \int \frac{1}{2} e^{2x}\, dx \right)$$

$$= \frac{1}{2} x^2 e^{2x} - \frac{1}{2} x e^{2x} + \frac{1}{4} e^{2x} + C$$

$$= \frac{1}{4} e^{2x} (2x^2 - 2x + 1) + C$$

The most obvious choices for u and dv are not always the correct ones, as the following example shows. Integration by parts may still involve some trial and error.

EXAMPLE 5   A Tricky Integration by Parts

Evaluate $\int x^3\sqrt{x^2+1}\, dx$.

Solution

Here we have a polynomial (x^3) and a radical. However, if we choose $u = x^3$, then the resulting $dv = \sqrt{x^2+1}\, dx$ cannot be integrated easily. But we can use $\sqrt{x^2+1}$ as part of dv, and we can evaluate $\int dv$ if we let $dv = x\sqrt{x^2+1}\, dx$. Then

$$u = x^2 \qquad dv = (x^2+1)^{1/2}x\, dx$$

$$du = 2x\, dx \qquad v = \int (x^2+1)^{1/2}(x\, dx) = \frac{1}{2}\int (x^2+1)^{1/2}(2x\, dx)$$

$$v = \frac{1}{2}\frac{(x^2+1)^{3/2}}{3/2} = \frac{1}{3}(x^2+1)^{3/2}$$

Then $\int x^3\sqrt{x^2+1}\, dx = \frac{x^2}{3}(x^2+1)^{3/2} - \int \frac{1}{3}(x^2+1)^{3/2}(2x\, dx)$

$$= \frac{x^2}{3}(x^2+1)^{3/2} - \frac{1}{3}\frac{(x^2+1)^{5/2}}{5/2} + C$$

$$= \frac{x^2}{3}(x^2+1)^{3/2} - \frac{2}{15}(x^2+1)^{5/2} + C$$

EXAMPLE 6   Income Stream | APPLICATION PREVIEW |

Suppose that the value of oil produced by a piece of oil extraction equipment is considered a continuous income stream with an annual rate of flow (in dollars per year) at time t in years given by

$$f(t) = 300{,}000 - 2500t, \quad 0 \le t \le 10$$

and that money is worth 8%, compounded continuously. Find the present value of the piece of equipment.

Solution

The present value of the piece of equipment is given by

$$\int_0^{10} (300{,}000 - 2500t)e^{-0.08t}\, dt = 300{,}000\int_0^{10} e^{-0.08t}\, dt - 2500\int_0^{10} te^{-0.08t}\, dt$$

$$= \frac{300{,}000}{-0.08}e^{-0.08t}\Big|_0^{10} - 2500\int_0^{10} te^{-0.08t}\, dt$$

The value of the first integral is

$$\frac{300{,}000}{-0.08}e^{-0.08t}\Big|_0^{10} = \frac{300{,}000}{-0.08}e^{-0.8} - \frac{300{,}000}{-0.08}$$

$$\approx -1{,}684{,}983.615 + 3{,}750{,}000 = 2{,}065{,}016.385$$

The second of these integrals can be evaluated by using integration by parts, with $u = t$ and $dv = e^{-0.08t}\, dt$. Then $du = 1\, dt$ and $v = \dfrac{e^{-0.08t}}{-0.08}$, and this integral is

$$-2500\int_0^{10} te^{-0.08t}\, dt = -2500\frac{te^{-0.08t}}{-0.08}\Big|_0^{10} + 2500\int_0^{10} \frac{e^{-0.08t}}{-0.08}\, dt$$

$$= \frac{2500}{0.08}te^{-0.08t}\Big|_0^{10} + \frac{2500}{0.0064}e^{-0.08t}\Big|_0^{10}$$

$$= \frac{2500}{0.08}10e^{-0.8} + \frac{2500}{0.0064}e^{-0.8} - \frac{2500}{0.0064} \approx -74{,}690.572$$

Thus the sum of the integrals is

$$2{,}065{,}016.385 + (-74{,}690.572) = 1{,}990{,}325.813$$

so the present value of this piece of equipment is $1,990,325.81.

One further note about integration by parts. It can be very useful on certain types of problems, but not on all types. Don't attempt to use integration by parts when easier methods are available.

Calculator Note   Using the numerical integration feature of a graphing calculator to evaluate the integral in Example 6 gives the present value of $1,990,325.81, so this answer is the same as that found in Example 6.

✓ CHECKPOINT ANSWERS

1. (a) True   (b) True   (c) True
   (d) False. The *product* of u and dv must equal the original integrand.
2. (a) $u = \ln x$ and $dv = x^{-4}\, dx$
   (b) $du = \dfrac{1}{x}\, dx$ and $v = \displaystyle\int x^{-4}\, dx = \dfrac{x^{-3}}{-3}$
   (c) $-\dfrac{\ln x}{3x^3} - \dfrac{1}{9x^3} + C$

| EXERCISES | 13.6

In Problems 1–16, use integration by parts to evaluate the integral.

1. $\displaystyle\int xe^{2x}\, dx$

2. $\displaystyle\int xe^{-x}\, dx$

3. $\displaystyle\int x^2 \ln x\, dx$

4. $\displaystyle\int x^3 \ln x\, dx$

5. $\displaystyle\int_4^6 q\sqrt{q-4}\, dq$

6. $\displaystyle\int_0^1 y(1-y)^{3/2}\, dy$

7. $\displaystyle\int \frac{\ln x}{x^2}\, dx$

8. $\displaystyle\int \frac{\ln(x-1)}{\sqrt{x-1}}\, dx$

9. $\displaystyle\int_1^e \ln x\, dx$

10. $\displaystyle\int \frac{x}{\sqrt{x-3}}\, dx$

11. $\displaystyle\int x \ln(2x-3)\, dx$

12. $\displaystyle\int x \ln(4x)\, dx$

13. $\displaystyle\int q^3 \sqrt{q^2-3}\, dq$

14. $\displaystyle\int \frac{x^3}{\sqrt{9-x^2}}\, dx$

15. $\displaystyle\int_0^4 x^3 \sqrt{x^2+9}\, dx$

16. $\displaystyle\int \sqrt{x} \ln x\, dx$

In Problems 17–24, use integration by parts to evaluate the integral. Note that evaluation may require integration by parts more than once.

17. $\displaystyle\int x^2 e^{-x}\, dx$

18. $\displaystyle\int_0^1 4x^2 e^x\, dx$

19. $\displaystyle\int_0^2 3x^3 e^{x^2}\, dx$

20. $\displaystyle\int x^3 e^x\, dx$

21. $\displaystyle\int x^3 \ln^2 x\, dx$

22. $\displaystyle\int \frac{x^2}{\sqrt{x-3}}\, dx$

23. $\displaystyle\int e^{2x} \sqrt{e^x+1}\, dx$

24. $\displaystyle\int_1^2 (\ln x)^2\, dx$

In Problems 25–30, match each of the integrals with the formula or method (I–IV) that should be used to evaluate it. Then evaluate the integral.

I. Integration by parts

II. $\displaystyle\int e^u\, du$

III. $\displaystyle\int \frac{du}{u}$

IV. $\displaystyle\int u^n\, du$

25. $\displaystyle\int xe^{x^2}\, dx$

26. $\displaystyle\int \frac{x}{9-4x^2}\, dx$

27. $\displaystyle\int e^x \sqrt{e^x+1}\, dx$

28. $\displaystyle\int 4x^2 e^{x^3}\, dx$

29. $\displaystyle\int_0^4 \frac{t}{e^t}\, dt$

30. $\displaystyle\int x^2 \sqrt{x-1}\, dx$

APPLICATIONS

31. **Producer's surplus** If the supply function for x units of a commodity is $p = 30 + 50 \ln(2x+1)^2$ dollars, what is the producer's surplus at $x = 30$?

32. **Cost** If the marginal cost function for x units of a product is $\overline{MC} = 1 + 3 \ln(x+1)$ dollars per unit and if the fixed cost is $100, find the total cost function.

33. **Present value** Suppose that a machine's production can be considered as a continuous income stream with annual rate of flow at time t given by

$$f(t) = 10{,}000 - 500t \quad \text{(dollars per year)}$$

If money is worth 10%, compounded continuously, find the present value of the machine over the next 5 years.

34. *Present value* Suppose that the production of a machine used to mine coal is considered as a continuous income stream with annual rate of flow at time t given by

$$f(t) = 280{,}000 - 14{,}000t \quad \text{(dollars per year)}$$

If money is worth 7%, compounded continuously, find the present value of this machine over the next 8 years.

35. *Income distribution* Suppose the Lorenz curve for the distribution of income of a certain country is given by

$$y = xe^{x-1}$$

Find the Gini coefficient of income.

36. *Income streams* Suppose the income from an Internet access business is a continuous income stream with annual rate of flow given by

$$f(t) = 100te^{-0.1t}$$

in thousands of dollars per year. Find the total income over the next 10 years.

37. *Demographics* The number of millions of White non-Hispanic individuals in the U.S. civilian non-institutional population age 16 and older for selected years from 1980 and projected to 2050 can be modeled by

$$P(x) = 96.1 + 17.4 \ln x$$

with x equal to the number of years past 1970 (*Source:* U.S. Census Bureau). Find the projected average population over the years 2020 through 2030.

38. *Natural gas* The percent of natural gas in the United States extracted from shale rock from 2012 and projected to 2040 is given by the function $y = -0.843 + 13.8 \ln x$, where x is the is the number of years after 2010 (*Source:* Energy Information Administration). What does this model project the average percent to be for the years 2015 through 2030?

OBJECTIVES

13.7

- To evaluate improper integrals
- To apply improper integrals to continuous income streams and to probability density functions

Improper Integrals and Their Applications

■ | APPLICATION PREVIEW | ■

We saw in Section 13.4, "Applications of Definite Integrals in Business and Economics," that the present value of a continuous income stream over a fixed number of years can be found by using a definite integral. When this notion is extended to an infinite time interval, the result is called the capital value of the income stream and is given by

$$\text{Capital value} = \int_{0}^{\infty} f(t)e^{-rt}\, dt$$

where $f(t)$ is the annual rate of flow at time t, and r is the annual interest rate, compounded continuously. This is called an improper integral. For example, the capital value of a trust that provides $f(t) = \$10{,}000$ per year indefinitely (when interest is 10% compounded continuously) is found by evaluating an improper integral. (See Example 2.)

Improper Integral   Some applications of calculus to statistics or business (such as capital value) involve definite integrals over intervals of infinite length (**improper integrals**). The area of a region that extends infinitely to the left or right along the x-axis (see Figure 13.27) could be described by an improper integral.

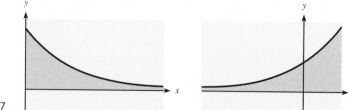
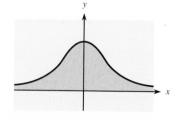

Figure 13.27

To see how to find such an area and hence evaluate an improper integral, let us consider how to find the area between the curve $y = 1/x^2$ and the x-axis to the right of $x = 1$.

To find the area under this curve from $x = 1$ to $x = b$, where b is any number greater than 1 (see Figure 13.28), we evaluate

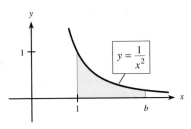

Figure 13.28

$$A = \int_1^b \frac{1}{x^2}\, dx = \frac{-1}{x}\Bigg|_1^b = \frac{-1}{b} - \left(\frac{-1}{1}\right) = 1 - \frac{1}{b}$$

Note that the larger b is, the closer the area is to 1. If $b = 100$, $A = 0.99$; if $b = 1000$, $A = 0.999$; and if $b = 1{,}000{,}000$, $A = 0.999999$.

We can represent the area of the region under $1/x^2$ to the right of 1 using the notation

$$\lim_{b \to \infty} \int_1^b \frac{1}{x^2}\, dx = \lim_{b \to \infty}\left(1 - \frac{1}{b}\right)$$

where $\lim\limits_{b \to \infty}$ represents the limit as b gets larger without bound. Note that $\dfrac{1}{b} \to 0$ as $b \to \infty$, so

$$\lim_{b \to \infty}\left(1 - \frac{1}{b}\right) = 1 - 0 = 1$$

Thus the area under the curve $y = 1/x^2$ to the right of $x = 1$ is 1.

In general, we define the area under a curve $y = f(x)$ to the right of $x = a$, with $f(x) \geq 0$, to be

$$\text{Area} = \lim_{b \to \infty}(\text{area from } a \text{ to } b) = \lim_{b \to \infty}\int_a^b f(x)\, dx$$

This motivates the definition that follows.

Improper Integral

$$\int_a^\infty f(x)\, dx = \lim_{b \to \infty}\int_a^b f(x)\, dx$$

If the limit defining the **improper integral** is a unique finite number, we say that the integral *converges;* otherwise, we say that the integral *diverges.*

EXAMPLE 1   Improper Integrals

Evaluate the following improper integrals, if they converge.

(a) $\displaystyle\int_1^\infty \frac{1}{x^3}\, dx$ (b) $\displaystyle\int_1^\infty \frac{1}{x}\, dx$

Solution

(a) $\displaystyle\int_1^\infty \frac{1}{x^3}\, dx = \lim_{b \to \infty}\int_1^b x^{-3}\, dx = \lim_{b \to \infty}\left(\frac{x^{-2}}{-2}\right)\Bigg|_1^b$

$$= \lim_{b \to \infty}\left[\frac{-1}{2b^2} - \left(\frac{-1}{2(1)^2}\right)\right] = \lim_{b \to \infty}\left(\frac{-1}{2b^2} + \frac{1}{2}\right)$$

Notice that $1/(2b^2) \to 0$ as $b \to \infty$, so the limit converges to $0 + \frac{1}{2} = \frac{1}{2}$. That is,

$$\int_1^\infty \frac{1}{x^3}\, dx = \frac{1}{2}$$

(b) $\displaystyle\int_1^\infty \frac{1}{x}\, dx = \lim_{b \to \infty}\int_1^b \frac{1}{x}\, dx = \lim_{b \to \infty}\left(\ln|x|\right)\Bigg|_1^b = \lim_{b \to \infty}(\ln b - \ln 1)$

Here, $\ln b$ increases without bound as $b \to \infty$, so the limit diverges and we write

$$\int_1^\infty \frac{1}{x}\, dx = \infty$$

From Example 1 we can conclude that the area under the curve $y = 1/x^3$ to the right of $x = 1$ is $\frac{1}{2}$ whereas the corresponding area under the curve $y = 1/x$ is infinite. (We have already seen that the corresponding area under $y = 1/x^2$ is 1.)

As Figure 13.29 shows, the graphs of $y = 1/x^2$ and $y = 1/x$ look similar, but the graph of $y = 1/x^2$ gets "close" to the x-axis much more rapidly than the graph of $y = 1/x$. The area under $y = 1/x$ does not converge to a finite number because as $x \to \infty$ the graph of $1/x$ does not approach the x-axis rapidly enough.

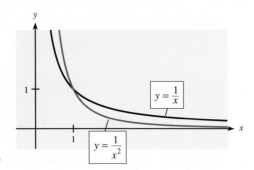

Figure 13.29

EXAMPLE 2   **Capital Value | APPLICATION PREVIEW |**

Suppose that an organization wants to establish a trust fund that will provide a continuous income stream with an annual rate of flow at time t given by $f(t) = 10,000$ dollars per year. If the interest rate remains at 10% compounded continuously, find the capital value of the fund.

Solution
The capital value of the fund uses the formula

$$\text{Capital value} = \int_0^\infty f(t)\, e^{-rt}\, dt$$

where $f(t)$ is the annual rate of flow at time t, and r is the annual interest rate, compounded continuously.

$$\int_0^\infty 10,000 e^{-0.10t}\, dt = \lim_{b \to \infty} \int_0^b 10,000 e^{-0.10t}\, dt = \lim_{b \to \infty} \left(-100,000 e^{-0.10t} \right)\Big|_0^b$$

$$= \lim_{b \to \infty} \left(\frac{-100,000}{e^{0.10b}} + 100,000 \right) = 100,000$$

Thus the capital value of the fund is $100,000.

Another term for a fund such as the one in Example 2 is a **perpetuity.** Usually the rate of flow of a perpetuity is a constant. If the rate of flow is a constant A, it can be shown that the capital value is given by A/r (see Problem 37 in the 13.7 Exercises).

✓ CHECKPOINT

1. True or false:

    (a) $\displaystyle\lim_{b \to \infty} \frac{1}{b^p} = 0$ if $p > 0$ $\qquad$ (b) If $p > 0$, $b^p \to \infty$ as $b \to \infty$

    (c) $\displaystyle\lim_{b \to \infty} e^{-pb} = 0$ if $p > 0$

2. Evaluate the following (if they exist).

    (a) $\displaystyle\int_1^\infty \frac{1}{x^{4/3}}\, dx = \lim_{b \to \infty} \int_1^b x^{-4/3}\, dx$ $\quad$ (b) $\displaystyle\int_0^\infty \frac{dx}{\sqrt{x+1}}$

Two **additional improper integrals** involving infinite limits are defined as follows:

Additional Improper Integrals	$$\int_{-\infty}^{b} f(x)\,dx = \lim_{a\to\infty}\int_{-a}^{b} f(x)\,dx$$

The integral converges if the limit is finite. Otherwise it diverges.

$$\int_{-\infty}^{\infty} f(x)\,dx = \lim_{a\to\infty}\int_{-a}^{c} f(x)\,dx + \lim_{b\to\infty}\int_{c}^{b} f(x)\,dx$$

for any finite constant c. (Often, c is chosen to be 0.) If both limits are finite, the improper integral converges; otherwise, it diverges.

EXAMPLE 3   Improper Integrals

Evaluate the following integrals.

(a) $\displaystyle\int_{-\infty}^{4} e^{3x}\,dx$

(b) $\displaystyle\int_{-\infty}^{\infty} \frac{x^3}{(x^4+3)^2}\,dx$

Solution

(a) $\displaystyle\int_{-\infty}^{4} e^{3x}\,dx = \lim_{a\to\infty}\int_{-a}^{4} e^{3x}\,dx = \lim_{a\to\infty}\left[\left(\frac{1}{3}\right)e^{3x}\right]_{-a}^{4}$

$\displaystyle= \lim_{a\to\infty}\left[\left(\frac{1}{3}\right)e^{12} - \left(\frac{1}{3}\right)e^{-3a}\right] = \lim_{a\to\infty}\left[\left(\frac{1}{3}\right)e^{12} - \left(\frac{1}{3}\right)\left(\frac{1}{e^{3a}}\right)\right]$

$\displaystyle= \frac{1}{3}e^{12} \quad \text{(because } 1/e^{3a} \to 0 \text{ as } a\to\infty\text{)}$

(b) $\displaystyle\int_{-\infty}^{\infty} \frac{x^3}{(x^4+3)^2}\,dx = \lim_{a\to\infty}\int_{-a}^{0} \frac{x^3}{(x^4+3)^2}\,dx + \lim_{b\to\infty}\int_{0}^{b} \frac{x^3}{(x^4+3)^2}\,dx$

$\displaystyle= \lim_{a\to\infty}\left[\frac{1}{4}\frac{(x^4+3)^{-1}}{-1}\right]_{-a}^{0} + \lim_{b\to\infty}\left[\frac{1}{4}\frac{(x^4+3)^{-1}}{-1}\right]_{0}^{b}$

$\displaystyle= \lim_{a\to\infty}\left[-\frac{1}{4}\left(\frac{1}{3} - \frac{1}{a^4+3}\right)\right] + \lim_{b\to\infty}\left[-\frac{1}{4}\left(\frac{1}{b^4+3} - \frac{1}{3}\right)\right]$

$\displaystyle= -\frac{1}{12} + 0 + 0 + \frac{1}{12} = 0 \left(\text{since } \lim_{a\to\infty}\frac{1}{a^4+3} = 0 \text{ and } \lim_{b\to\infty}\frac{1}{b^4+3} = 0\right)$ ∎

Probability   We noted in Chapter 8, "Further Topics in Probability; Data Description," that the sum of the probabilities for a probability distribution (a **probability density function**) equals 1. In particular, we stated that the area under the normal probability curve is 1. The normal distribution is an example of a continuous probability distribution because the values of the random variable are considered over intervals rather than at discrete values. There are many important continuous probability distributions besides the normal distribution, but all such distributions satisfy the following definition.

Probability Density Function	If $f(x) \geq 0$ for all x, then f is a **probability density function** for a continuous random variable x if and only if

$$\int_{-\infty}^{\infty} f(x)\,dx = 1$$

We have noted previously that when $f(x)$ is a continuous probability density function, then

$$Pr(a \leq x \leq b) = \int_{a}^{b} f(x)\,dx$$

EXAMPLE 4 **Product Life**

Suppose the probability density function for the life span of a computer component is given by

$$f(x) = \begin{cases} 0.10e^{-0.10x} & \text{if } x \geq 0 \\ 0 & \text{if } x < 0 \end{cases}$$

(a) Verify that $f(x)$ is a probability density function.
(b) Find the probability that such a component lasts more than 3 years.

Solution

(a) To verify that $f(x)$ is a probability density function, we show $\int_{-\infty}^{\infty} f(x)\,dx = 1$:

$$\int_{-\infty}^{\infty} f(x)\,dx = \int_{-\infty}^{0} f(x)\,dx + \int_{0}^{\infty} f(x)\,dx = \int_{-\infty}^{0} 0\,dx + \int_{0}^{\infty} 0.10e^{-0.10x}\,dx$$

$$= 0 + \lim_{b \to \infty} \int_{0}^{b} 0.10e^{-0.10x}\,dx$$

$$= \lim_{b \to \infty}(-e^{-0.10x})\Big|_{0}^{b} = \lim_{b \to \infty}(-e^{-0.10b} + 1) = 1$$

(b) The probability that the component lasts more than 3 years is given by

$$\Pr(x \geq 3) = \int_{3}^{\infty} 0.10e^{-0.10x}\,dx$$

$$= \lim_{b \to \infty} \int_{3}^{b} 0.10e^{-0.10x}\,dx = \lim_{b \to \infty}(-e^{-0.10x})\Big|_{3}^{b}$$

$$= \lim_{b \to \infty}(-e^{-0.10b} + e^{-0.3}) = e^{-0.3} \approx 0.7408 \quad \blacksquare$$

In Section 8.3, "Discrete Probability Distributions; The Binomial Distribution," we found the expected value (mean) of a discrete probability distribution using the formula

$$E(x) = \sum x\Pr(x)$$

For continuous probability distributions, such as the normal probability distribution, the **expected value**, or **mean**, can be found by evaluating the improper integral

$$\int_{-\infty}^{\infty} xf(x)\,dx$$

Mean (Expected Value)	If x is a continuous random variable with probability density function f, then the **mean (expected value)** of the probability distribution is $$\mu = \int_{-\infty}^{\infty} xf(x)\,dx$$

The normal distribution density function, in standard form, is

$$f(x) = \frac{1}{\sqrt{2\pi}} e^{-x^2/2}$$

so the mean of the normal probability distribution is given by

$$\mu = \int_{-\infty}^{\infty} x\left(\frac{1}{\sqrt{2\pi}} e^{-x^2/2}\right) dx$$

$$= \lim_{a \to \infty} \int_{-a}^{0} \frac{1}{\sqrt{2\pi}} xe^{-x^2/2}\,dx + \lim_{b \to \infty} \int_{0}^{b} \frac{1}{\sqrt{2\pi}} xe^{-x^2/2}\,dx$$

$$= \lim_{a \to \infty} \frac{1}{\sqrt{2\pi}} \left(-e^{-x^2/2} \right) \Big|_{-a}^{0} + \lim_{b \to \infty} \frac{1}{\sqrt{2\pi}} \left(-e^{-x^2/2} \right) \Big|_{0}^{b}$$

$$= \frac{1}{\sqrt{2\pi}} (-1 + 0) + \frac{1}{\sqrt{2\pi}} (0 + 1) = 0$$

This verifies the statement in Chapter 8 that the mean of the standard normal distribution is 0.

✓ **CHECKPOINT ANSWERS**

1. (a) True   (b) True   (c) True

2. (a) $\int_1^\infty \frac{1}{x^{4/3}} \, dx = 3$

   (b) $\lim_{b \to \infty} \int_0^b (x+1)^{-1/2} \, dx = \lim_{b \to \infty} (2\sqrt{b+1} - 2)$; diverges

| EXERCISES | 13.7

In Problems 1–20, evaluate the improper integrals that converge.

1. $\int_1^\infty \frac{dx}{x^6}$

2. $\int_1^\infty \frac{1}{x^4} \, dx$

3. $\int_1^\infty \frac{dt}{t^{3/2}}$

4. $\int_5^\infty \frac{dx}{(x-1)^3}$

5. $\int_1^\infty e^{-x} \, dx$

6. $\int_0^\infty x^2 e^{-x^3} \, dx$

7. $\int_1^\infty \frac{dt}{t^{1/3}}$

8. $\int_1^\infty \frac{1}{\sqrt{x}} \, dx$

9. $\int_0^\infty e^{3x} \, dx$

10. $\int_1^\infty x e^{x^2} \, dx$

11. $\int_{-\infty}^{-1} \frac{10}{x^2} \, dx$

12. $\int_{-\infty}^{-2} \frac{x}{\sqrt{(x^2-1)^3}} \, dx$

13. $\int_{-\infty}^0 x^2 e^{-x^3} \, dx$

14. $\int_{-\infty}^0 \frac{x}{(x^2+1)^2} \, dx$

15. $\int_{-\infty}^{-1} \frac{6}{x} \, dx$

16. $\int_{-\infty}^{-2} \frac{5}{3x+5} \, dx$

17. $\int_{-\infty}^\infty \frac{2x}{(x^2+1)^2} \, dx$

18. $\int_{-\infty}^\infty \frac{9x^5}{(3x^6+7)^2} \, dx$

19. $\int_{-\infty}^\infty x^3 e^{-x^4} \, dx$

20. $\int_{-\infty}^\infty x^4 e^{-x^5} \, dx$

21. For what value of c does $\int_0^\infty \frac{c}{e^{0.5t}} \, dt = 1$?

22. For what value of c does $\int_{10}^\infty \frac{c}{x^3} \, dx = 1$?

In Problems 23–26, find the area, if it exists, of the region under the graph of $y = f(x)$ and to the right of $x = 1$.

23. $f(x) = \dfrac{x}{e^{x^2}}$

24. $f(x) = \dfrac{1}{\sqrt[5]{x^3}}$

25. $f(x) = \dfrac{1}{\sqrt[3]{x^5}}$

26. $f(x) = \dfrac{1}{x\sqrt{x}}$

27. Show that the function

$$f(x) = \begin{cases} \dfrac{200}{x^3} & \text{if } x \geq 10 \\ 0 & \text{otherwise} \end{cases}$$

is a probability density function.

28. Show that

$$f(t) = \begin{cases} 3e^{-3t} & \text{if } t \geq 0 \\ 0 & \text{if } t < 0 \end{cases}$$

is a probability density function.

29. For what value of c is the function

$$f(x) = \begin{cases} c/x^2 & \text{if } x \geq 1 \\ 0 & \text{otherwise} \end{cases}$$

a probability density function?

30. For what value of c is the function

$$f(x) = \begin{cases} c/x^3 & \text{if } x \geq 100 \\ 0 & \text{otherwise} \end{cases}$$

a probability density function?

31. Find the value of c so that

$$f(x) = \begin{cases} ce^{-x/4} & x \geq 0 \\ 0 & x < 0 \end{cases}$$

is a probability density function.

32. Find the value of c (in terms of k) so that

$$f(x) = \begin{cases} ce^{-kx} & \text{if } x \geq 0 \\ 0 & \text{if } x < 0 \end{cases}$$

is a probability density function.

33. Find the mean of the probability distribution if the probability density function is

$$f(x) = \begin{cases} \dfrac{200}{x^3} & \text{if } x \geq 10 \\ 0 & \text{otherwise} \end{cases}$$

34. Find the mean of the probability distribution if the probability density function is

$$f(x) = \begin{cases} c/x^4 & \text{if } x \ge 10 \\ 0 & \text{otherwise} \end{cases}$$

35. Find the area below the graph of $y = f(x)$ and above the x-axis for $f(x) = 24xe^{-3x}$. Use the graph of $y = f(x)$ to find the interval for which $f(x) \ge 0$ and the graph of the integral of $f(x)$ over this interval to find the area.

36. Find the area below the graph of $y = f(x)$ and above the x-axis for $f(x) = x^2e^{-x}$ and $x \ge 0$. Use the graph of the integral of $f(x)$ over this interval to find the area.

APPLICATIONS

37. **Capital value** Suppose that a continuous income stream has an annual rate of flow at time t given by $f(t) = A$, where A is a constant. If the interest rate is r (as a decimal, $r > 0$), compounded continuously, show that the capital value of the stream is A/r.

38. **Capital value** Suppose that a donor wishes to provide a cash gift to a hospital that will generate a continuous income stream with an annual rate of flow at time t given by $f(t) = \$20,000$ per year. If the annual interest rate is 12% compounded continuously, find the capital value of this perpetuity.

39. **Capital value** Suppose that a business provides a continuous income stream with an annual rate of flow at time t given by $f(t) = 120e^{0.04t}$ in thousands of dollars per year. If the interest rate is 9% compounded continuously, find the capital value of the business.

40. **Capital value** Suppose that the output of the machinery in a factory can be considered as a continuous income stream with annual rate of flow at time t given by $f(t) = 450e^{-0.09t}$ (in thousands of dollars per year). If the annual interest rate is 6% compounded continuously, find the capital value of the machinery.

41. **Capital value** A business has a continuous income stream with an annual rate of flow at time t given by $f(t) = 56,000e^{0.02t}$ (dollars per year). If the interest rate is 10% compounded continuously, find the capital value of the business.

42. **Capital value** Suppose that a business provides a continuous income stream with an annual rate of flow at time t given by $f(t) = 10,800e^{0.06t}$ (dollars per year).

If money is worth 12% compounded continuously, find the capital value of the business.

43. **Repair time** In a manufacturing process involving several machines, the average down time t (in hours) for a machine that needs repair has the probability density function

$$f(t) = 0.5e^{-0.5t} \quad t \ge 0$$

Find the probability that a failed machine's down time is
(a) 2 hours or more.   (b) 8 hours or more.

44. **Customer service** The duration t (in minutes) of customer service calls received by a certain company is given by the probability density function

$$f(t) = 0.4e^{-0.4t} \quad t \ge 0$$

Find the probability that a call selected at random lasts
(a) 4 minutes or more.
(b) 10 minutes or more.

45. **Quality control** The probability density function for the life span of an electronics part is $f(t) = 0.08e^{-0.08t}$, where t is the number of months in service. Find the probability that any given part of this type lasts longer than 24 months.

46. **Warranties** A transmission repair firm that wants to offer a lifetime warranty on its repairs has determined that the probability density function for transmission failure after repair is $f(t) = 0.3e^{-0.3t}$, where t is the number of months after repair. What is the probability that a transmission chosen at random will last
(a) 3 months or less?
(b) more than 3 months?

47. **Radioactive waste** Suppose that the rate at which a nuclear power plant produces radioactive waste is proportional to the number of years it has been operating, according to $f(t) = 500t$ (in pounds per year). Suppose also that the waste decays exponentially at a rate of 3% per year. Then the amount of radioactive waste that will accumulate in b years is given by

$$\int_0^b 500te^{-0.03(b-t)}\, dt$$

(a) Evaluate this integral.
(b) How much waste will accumulate in the long run? Take the limit as $b \to \infty$ in part (a).

Numerical Integration Methods: The Trapezoidal Rule and Simpson's Rule

▌| APPLICATION PREVIEW |▌

A pharmaceutical company tests the body's assimilation of a new drug by administering a 200-milligram dose and collecting the following data from blood samples (t is time in hours, and $R(t)$ gives the assimilation of the drug in milligrams per hour).

t	0	0.5	1.0	1.5	2.0	2.5	3.0
$R(t)$	0.0	15.3	32.3	51.0	74.8	102.0	130.9

The company would like to find the amount of drug assimilated in 3 hours, which is given by

$$\int_0^3 R(t)\, dt$$

In this section we develop straightforward and quite accurate methods for evaluating integrals such as this, even when, as with $R(t)$, only function data, rather than a formula, are given. (See Example 4.)

We have studied several techniques for integration and have even used tables to evaluate some integrals. Yet some functions that arise in practical problems cannot be integrated by using any formula. For any function $f(x) \geq 0$ on an interval $[a, b]$, however, we have seen that a definite integral can be viewed as an area and that we can usually approximate the area and hence the integral (see Figure 13.30). One such approximation method uses rectangles, as we saw when we defined the definite integral. In this section, we consider two other **numerical integration methods** to approximate a definite integral: the **Trapezoidal Rule** and **Simpson's Rule.**

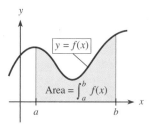

Figure 13.30

Trapezoidal Rule
To develop the Trapezoidal Rule formula, we assume that $f(x) \geq 0$ on $[a, b]$ and subdivide the interval $[a, b]$ into n equal pieces, each of length $(b - a)/n = h$. Then, within each subdivision, we can approximate the area by using a trapezoid. As shown in Figure 13.31 on the next page, we can use the formula for the area of a trapezoid to approximate the area of the first subdivision. Continuing in this way for each trapezoid, we have

$$\int_a^b f(x)\, dx \approx A_1 + A_2 + A_3 + \cdots + A_{n-1} + A_n$$

$$= \left[\frac{f(x_0) + f(x_1)}{2}\right] h + \left[\frac{f(x_1) + f(x_2)}{2}\right] h +$$

$$\left[\frac{f(x_2) + f(x_3)}{2}\right] h + \cdots + \left[\frac{f(x_{n-1}) + f(x_n)}{2}\right] h$$

$$= \frac{h}{2}\left[f(x_0) + f(x_1) + f(x_1) + f(x_2) + f(x_2) + \cdots + f(x_{n-1}) + f(x_{n-1}) + f(x_n)\right]$$

This can be simplified to obtain the **Trapezoidal Rule**.

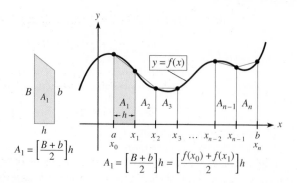

$$A_1 = \left[\frac{B+b}{2}\right]h$$

$$A_1 = \left[\frac{B+b}{2}\right]h = \left[\frac{f(x_0)+f(x_1)}{2}\right]h$$

Figure 13.31

Trapezoidal Rule

If f is continuous on $[a, b]$, then

$$\int_a^b f(x)\,dx \approx \frac{h}{2}[f(x_0) + 2f(x_1) + 2f(x_2) + \cdots + 2f(x_{n-1}) + f(x_n)]$$

where $h = \dfrac{b-a}{n}$ and n is the number of equal subdivisions of $[a, b]$.

Despite the fact that we used areas to develop the Trapezoidal Rule, we can use this rule to evaluate definite integrals even if $f(x) < 0$ on all or part of $[a, b]$.

EXAMPLE   1   **Trapezoidal Rule**

Use the Trapezoidal Rule to approximate $\displaystyle\int_1^3 \frac{1}{x}\,dx$ with

(a) $n = 4$.
(b) $n = 8$.

Solution

First, we note that this integral can be evaluated directly:

$$\int_1^3 \frac{1}{x}\,dx = \ln|x| \ \bigg|_1^3 = \ln 3 - \ln 1 = \ln 3 \approx 1.099$$

(a)   The interval $[1, 3]$ must be divided into 4 equal subintervals of width $h = \frac{3-1}{4} = \frac{2}{4} = \frac{1}{2}$ as follows:

$$
\begin{array}{ccccccccc}
1 & & 1.5 & & 2 & & 2.5 & & 3 \\
\vert & & \vert & & \vert & & \vert & & \vert \\
x_0 & & x_1 & & x_2 & & x_3 & & x_4
\end{array}
$$

Thus, from the Trapezoidal Rule, we have

$$\int_1^3 \frac{1}{x}\,dx \approx \frac{h}{2}[f(x_0) + 2f(x_1) + 2f(x_2) + 2f(x_3) + f(x_4)]$$

$$= \frac{1/2}{2}[f(1) + 2f(1.5) + 2f(2) + 2f(2.5) + f(3)]$$

$$= \frac{1}{4}\left[1 + 2\left(\frac{1}{1.5}\right) + 2\left(\frac{1}{2}\right) + 2\left(\frac{1}{2.5}\right) + \frac{1}{3}\right] \approx 1.117$$

(b)   In this case, the interval $[1, 3]$ is divided into 8 equal subintervals of width $h = \frac{3-1}{8} = \frac{2}{8} = \frac{1}{4}$ as follows.

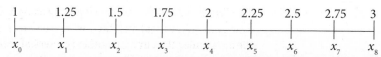

Thus, from the Trapezoidal Rule, we have

$$\int_1^3 \frac{1}{x}\, dx \approx \frac{h}{2}[f(x_0) + 2f(x_1) + 2f(x_2) + 2f(x_3) + 2f(x_4) + 2f(x_5)$$
$$+ 2f(x_6) + 2f(x_7) + f(x_8)]$$
$$= \frac{1/4}{2}\left[\frac{1}{1} + 2\left(\frac{1}{1.25}\right) + 2\left(\frac{1}{1.5}\right) + 2\left(\frac{1}{1.75}\right) + 2\left(\frac{1}{2}\right) + 2\left(\frac{1}{2.25}\right)\right.$$
$$\left. + 2\left(\frac{1}{2.5}\right) + 2\left(\frac{1}{2.75}\right) + \frac{1}{3}\right]$$
$$\approx 1.103$$

In Example 1, because we know that the value of the integral is $\ln 3 \approx 1.099$, we can measure the accuracy of each approximation. We can see that a larger value of n (namely, $n = 8$) produced a more accurate approximation to $\ln 3$. In general, larger values of n produce more accurate approximations, but they also make computations more difficult.

Technology Note

The TABLE feature of a graphing calculator can be used to find the values of $f(x_0), f(x_1), f(x_2)$, etc. needed for the Trapezoidal Rule. Figure 13.32(a) shows the TABLE set up for Example 1(b), and Figures 13.32(b) and 13.32(c) show the TABLE values.

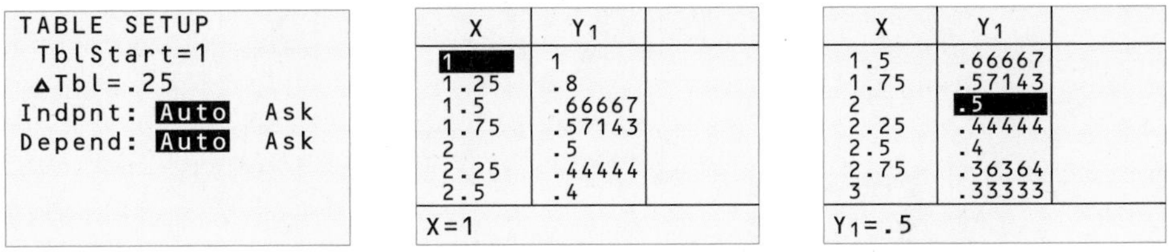

```
TABLE SETUP
TblStart=1
ΔTbl=.25
Indpnt: Auto  Ask
Depend: Auto  Ask
```

X	Y₁
1	1
1.25	.8
1.5	.66667
1.75	.57143
2	.5
2.25	.44444
2.5	.4

X=1

X	Y₁
1.5	.66667
1.75	.57143
2	.5
2.25	.44444
2.5	.4
2.75	.36364
3	.33333

Y₁=.5

Figure 13.32   (a)   (b)   (c)

Thus for the Trapezoidal Rule, we get

$$\int_1^3 \frac{1}{x}\, dx \approx \frac{0.25}{2}\left[\begin{array}{c}1(1) + 2(0.8) + 2(0.66667) + 2(0.57143) + 2(0.5) \\ + 2(0.44444) + 2(0.4) + 2(0.36364) + 1(0.33333)\end{array}\right] \approx 1.103$$

just as we did in Example 1.

An Excel spreadsheet also can be very useful in finding the numerical integral with the Trapezoidal Rule. A discussion of the use of Excel in finding numerical integrals is given in the Online Excel Guide that accompanies this text.

Because the exact value of an integral is rarely available when an approximation is used, it is important to have some way to judge the accuracy of an answer. The following formula, which we state without proof, can be used to bound the error that results from using the Trapezoidal Rule.

Trapezoidal Rule Error

The error E in using the Trapezoidal Rule to approximate $\int_a^b f(x)\, dx$ satisfies

$$|E| \le \frac{(b-a)^3}{12n^2}\left[\max_{a \le x \le b}|f''(x)|\right]$$

where n is the number of equal subdivisions of $[a, b]$.

For a numerical method to be worthwhile, there must be some way of assessing its accuracy. Hence, this formula is important. We leave its application, however, to more advanced courses.

Simpson's Rule   The Trapezoidal Rule was developed by using a line segment to approximate the function over each subinterval and then using the areas under the line segments to approximate the area under the curve. Another numerical method, **Simpson's Rule,** uses a parabola to approximate the function over each pair of subintervals (see Figure 13.33) and then uses the areas under the parabolas to approximate the area under the curve. Because Simpson's Rule is based on pairs of subintervals, n must be even.

Simpson's Rule (n Even)	If $f(x)$ is continuous on $[a, b]$, and if $[a, b]$ is divided into an *even* number n of equal subdivisions, then $$\int_a^b f(x)\, dx \approx \frac{h}{3}[f(x_0) + 4f(x_1) + 2f(x_2) + 4f(x_3) + \cdots + 2f(x_{n-2}) + 4f(x_{n-1}) + f(x_n)]$$ where $h = \dfrac{b - a}{n}$.

We leave the derivation of Simpson's Rule to more advanced courses.

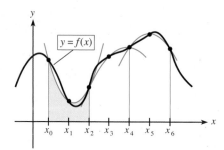

Figure 13.33

EXAMPLE 2   Simpson's Rule

Use Simpson's Rule with $n = 4$ to approximate $\displaystyle\int_1^3 \frac{1}{x}\, dx$.

Solution
Because $n = 4$ is even, Simpson's Rule can be used, and the interval is divided into four subintervals of length $h = \frac{3-1}{4} = \frac{1}{2}$ as follows:

$$
\begin{array}{ccccc}
1 & 1.5 & 2 & 2.5 & 3 \\
x_0 & x_1 & x_2 & x_3 & x_4
\end{array}
$$

$$\int_1^3 \frac{1}{x}\, dx \approx \frac{1/2}{3}[f(1) + 4f(1.5) + 2f(2) + 4f(2.5) + f(3)]$$

$$= \frac{1}{6}\left[\frac{1}{1} + 4\left(\frac{1}{1.5}\right) + 2\left(\frac{1}{2}\right) + 4\left(\frac{1}{2.5}\right) + \frac{1}{3}\right]$$

$$= 1.100$$

Note that the result of Example 2 is better than both the $n = 4$ and the $n = 8$ Trapezoidal Rule approximations done in Example 1. In general, Simpson's Rule is more accurate than the Trapezoidal Rule for a given number of subdivisions. We can determine the accuracy of Simpson's Rule approximations by using the following formula.

Simpson's Rule Error Formula	The error E in using Simpson's Rule to approximate $\displaystyle\int_a^b f(x)\, dx$ satisfies $$|E| \le \frac{(b - a)^5}{180n^4}\left[\max_{a \le x \le b}\left|f^{(4)}(x)\right|\right]$$ where n is the number of equal subdivisions of $[a, b]$ and $f^{(4)}(x)$ is the fourth derivative of $f(x)$.

The presence of the factor $180n^4$ in the denominator indicates that the error will often be quite small for even a modest value of n. Although Simpson's Rule often leads to more accurate results than the Trapezoidal Rule for a fixed choice of n, the Trapezoidal Rule is sometimes used because its error is more easily determined than that of Simpson's Rule or, more important, because the number of subdivisions is odd.

✓ CHECKPOINT

1. Suppose [1, 4] is divided into 6 equal subintervals.
   (a) Find the width h of each subinterval.
   (b) Find the subdivision points.
2. True or false:
   (a) When the Trapezoidal Rule is used, the number of subdivisions, n, must be even.
   (b) When Simpson's Rule is used, the number of subdivisions, n, must be even.
3. We can show that $\int_0^2 4x^3 \, dx = 16$. Use $n = 4$ subdivisions to find
   (a) the Trapezoidal Rule approximation of this integral.
   (b) the Simpson's Rule approximation of this integral.

Technology Note As mentioned previously for the Trapezoidal Rule, the TABLE feature of a graphing calculator can be used in exactly the same way with Simpson's Rule. An Excel spreadsheet also can be very useful with Simpson's Rule. ∎

 EXAMPLE 3 **Simpson's Rule**

Use Simpson's Rule with $n = 10$ subdivisions to approximate

$$\int_1^6 [\ln (x)]^2 \, dx$$

Solution

With $y_1 = [\ln (x)]^2$ and $h = \dfrac{b - a}{n} = \dfrac{6 - 1}{10} = 0.5$, we can use a graphing calculator with ΔTbl $= 0.5$ to find the following table of values.

x	y_1		x	y_1
1	0		4	1.9218
1.5	0.1644		4.5	2.2622
2	0.48045		5	2.5903
2.5	0.83959		5.5	2.9062
3	1.2069		6	3.2104
3.5	1.5694			

Thus Simpson's Rule yields

$$\int_1^6 [\ln (x)]^2 \, dx$$

$$\approx \frac{0.5}{3} \left[\begin{array}{l} 1(0) + 4(0.1644) + 2(0.48045) + 4(0.83959) + 2(1.2069) + 4(1.5694) \\ \quad + 2(1.9218) + 4(2.2622) + 2(2.5903) + 4(2.9062) + 1(3.2104) \end{array} \right]$$

$$\approx 7.7627$$ ∎

One advantage that both the Trapezoidal Rule and Simpson's Rule offer is that they may be used when only function values at the subdivision points are known and the function formula itself is not known. This can be especially useful in applied problems.

EXAMPLE 4   Pharmaceutical Testing | APPLICATION PREVIEW |

A pharmaceutical company tests the body's ability to assimilate a drug. The test is done by administering a 200-milligram dose and then, every half-hour, monitoring the rate of assimilation. The following table gives the data; t is time in hours, and $R(t)$ is the rate of assimilation in milligrams per hour.

t	0	0.5	1.0	1.5	2.0	2.5	3.0
$R(t)$	0.0	15.3	32.3	51.0	74.8	102.0	130.9

To find the total amount of the drug (in milligrams) that is assimilated in the first 3 hours, the company must find

$$\int_0^3 R(t)\, dt$$

Use Simpson's Rule to approximate this definite integral.

Solution

The values of t correspond to the endpoints of the subintervals, and the values of $R(t)$ correspond to function values at those endpoints. From the table we see that $h = \frac{1}{2}$ and $n = 6$ (even); thus Simpson's Rule is applied as follows:

$$\int_0^3 R(t)\, dt \approx \frac{h}{3}[R(t_0) + 4R(t_1) + 2R(t_2) + 4R(t_3) + 2R(t_4) + 4R(t_5) + R(t_6)]$$

$$= \frac{1}{6}[0 + 4(15.3) + 2(32.3) + 4(51.0) + 2(74.8) + 4(102.0) + 130.9]$$

$$\approx 169.7 \text{ mg}$$

Thus at the end of 3 hours, the body has assimilated approximately 169.7 milligrams of the 200-milligram dose. ∎

Note in Example 4 that the Trapezoidal Rule could also be used, because it relies only on the values of $R(t)$ at the subdivision endpoints.

In practice, these kinds of approximations are usually done with a computer, where the computations can be done quickly, even for large values of n. In addition, numerical methods (and hence computer programs) exist that approximate the errors, even in cases such as Example 4 where the function is not known. We leave any further discussion of error approximation formulas and additional numerical techniques for a more advanced course.

✓ CHECKPOINT ANSWERS

1. (a) $h = \dfrac{1}{2} = 0.5$

   (b) $x_0 = 1, x_1 = 1.5, x_2 = 2, x_3 = 2.5, x_4 = 3, x_5 = 3.5, x_6 = 4$
2. (a) False. With the Trapezoidal Rule, n can be even or odd.
   (b) True
3. (a) 17
   (b) 16

| EXERCISES | 13.8

For each interval $[a, b]$ and value of n given in Problems 1–6, find h and the values of $x_0, x_1, \ldots x_n$.

1. $[0, 2]\ n = 4$
2. $[0, 4]\ n = 8$
3. $[1, 4]\ n = 6$
4. $[2, 5]\ n = 9$
5. $[-1, 4]\ n = 5$
6. $[-1, 2]\ n = 6$

For each integral in Problems 7–12, do the following.
(a) Approximate its value by using the Trapezoidal Rule.
(b) Approximate its value by using Simpson's Rule.
(c) Find its exact value by integration.

(d) State which approximation is more accurate. (Round each result to 2 decimal places.)

7. $\int_0^3 x^2\,dx$; $n = 6$

8. $\int_0^1 x^3\,dx$; $n = 4$

9. $\int_1^2 \frac{1}{x^2}\,dx$; $n = 4$

10. $\int_1^4 \frac{1}{x}\,dx$; $n = 6$

11. $\int_0^4 x^{1/2}\,dx$; $n = 8$

12. $\int_0^2 x^{3/2}\,dx$; $n = 8$

In Problems 13–18, approximate each integral by
(a) the Trapezoidal Rule.
(b) Simpson's Rule.
Use $n = 4$ and round answers to 3 decimal places.

13. $\int_0^2 \sqrt{x^3 + 1}\,dx$

14. $\int_0^2 \frac{dx}{\sqrt{4x^3 + 1}}$

15. $\int_0^1 e^{-x^2}\,dx$

16. $\int_0^1 e^{x^2}\,dx$

17. $\int_1^5 \ln(x^2 - x + 1)\,dx$

18. $\int_1^5 \ln(x^2 + x + 2)\,dx$

Use the table of values given in each of Problems 19–22 to approximate $\int_a^b f(x)\,dx$. Use Simpson's Rule whenever n is even; otherwise, use the Trapezoidal Rule. Round answers to 1 decimal place.

19. Find $\int_1^4 f(x)\,dx$.

20. Find $\int_1^2 f(x)\,dx$.

x	f(x)
1	1
1.6	2.2
2.2	1.8
2.8	2.9
3.4	4.6
4.0	2.1

x	f(x)
1	1
1.2	0.5
1.4	0.3
1.6	0.1
1.8	0.8
2.0	0.1

21. Find $\int_{1.2}^{3.6} f(x)\,dx$.

22. Find $\int_0^{1.8} f(x)\,dx$.

x	f(x)
1.2	6.1
1.6	4.8
2.0	3.1
2.4	2.0
2.8	2.8
3.2	5.6
3.6	9.7

x	f(x)
0	8.8
0.3	4.6
0.6	1.5
0.9	0
1.2	0.7
1.5	2.8
1.8	7.6

APPLICATIONS

In Problems 23–30, round all calculations to 2 decimal places.

23. *Total income* Suppose that the production from an assembly line can be considered as a continuous income stream with annual rate of flow given by

$$f(t) = 100\frac{e^{0.1t}}{t + 1}\quad\text{(in thousands of dollars per year)}$$

Use Simpson's Rule with $n = 4$ to approximate the total income over the first 2 years, given by

$$\text{Total income} = \int_0^2 \frac{100e^{0.1t}}{t + 1}\,dt$$

24. *Present value* Suppose that the rate of flow of a continuous income stream is given by $f(t) = 500t$ (in thousands of dollars per year). If money is worth 7% compounded continuously, then the present value of this stream over the next 5 years is given by

$$\text{Present value} = \int_0^5 500t\, e^{-0.07t}\,dt$$

Use the Trapezoidal Rule with $n = 5$ to approximate this present value.

25. *Cost* Suppose that a company's total cost (in dollars) of producing x items is given by $C(x) = (x^2 + 1)^{3/2} + 1000$. Use the Trapezoidal Rule with $n = 3$ to approximate the average total cost for the production of $x = 30$ to $x = 33$ items.

26. *Demand* Suppose that the demand for q units of a certain product at $\$p$ per unit is given by

$$p = 850 + \frac{100}{q^2 + 1}$$

Use Simpson's Rule with $n = 6$ to approximate the average price as demand ranges from 3 to 9 items.

Supply and demand **Use the supply and demand schedules in Problems 27 and 28, with p in dollars and x as the number of units.**

Supply Schedule		Demand Schedule	
x	p	x	p
0	120	0	2400
10	260	10	1500
20	380	20	1200
30	450	30	950
40	540	40	800
50	630	50	730
60	680	60	680
70	720	70	640

27. Use Simpson's Rule to approximate the producer's surplus at market equilibrium. Note that market equilibrium can be found from the tables.

28. Use Simpson's Rule to approximate the consumer's surplus at market equilibrium.

29. *Production* Suppose that the rate of production of a product (in units per week) is measured at the end of each of the first 5 weeks after start-up, and the data in the table are obtained.

Weeks t	Rate $R(t)$	Weeks t	Rate $R(t)$
0	250.0	3	243.3
1	247.6	4	241.3
2	245.4	5	239.5

Approximate the total number of units produced in the first 5 weeks.

30. **Drug levels in the blood** The manufacturer of a medicine wants to test how a new 300-milligram capsule is released into the bloodstream. After a volunteer is given a capsule, blood samples are drawn every half-hour, and the number of milligrams of the drug in the bloodstream is calculated. The results obtained are as follows.

Time t (hr)	$N(t)$ (mg)	Time t (hr)	$N(t)$ (mg)
0	0	2.0	178.3
0.5	247.3	2.5	113.9
1.0	270	3.0	56.2
1.5	236.4	3.5	19.3

Approximate the *average* number of milligrams in the bloodstream during the first $3\frac{1}{2}$ hours.

Income distribution If the Lorenz curves for years a and b are given by $L_a(x)$ and $L_b(x)$, respectively, then from year a to year b, the change in the Gini coefficient $(G_b - G_a)$ is given by

$$2\int_0^1 [L_a(x) - L_b(x)]\, dx$$

In Problems 31 and 32, complete the following.
(a) Use the data in the table and make a new table for x and the corresponding values of $[L_a(x) - L_b(x)]$ for the given years, where year b is 2012.
(b) Use the table from part (a) in the Trapezoidal Rule to approximate

$$2\int_0^1 [L_a(x) - L_b(x)]\, dx$$

which gives $G_b - G_a$, the difference of the Gini coefficients for the two years.
(c) Is the value of the integral positive or negative? In which year was the income more equally distributed?

x	0.0	0.2	0.4	0.6	0.8	1.0
			$L(x)$ for Blacks			
2012	0.0	0.028	0.106	0.247	0.482	1.0
1990	0.0	0.031	0.111	0.260	0.511	1.0

x	0.0	0.2	0.4	0.6	0.8	1.0
			$L(x)$ for Asians			
2012	0.0	0.030	0.119	0.269	0.506	1.0
2009	0.0	0.027	0.109	0.253	0.483	1.0

Source: U.S. Bureau of the Census

31. Complete parts (a)–(c) above for Blacks in 1990 and 2012.
32. Complete parts (a)–(c) above for Asians in 2009 and 2012.
33. **Pollution monitoring** Suppose that the presence of phosphates in certain waste products dumped into a lake promotes the growth of algae. Rampant growth of algae affects the oxygen supply in the water, so an environmental group wishes to estimate the area of algae growth. Group members measure across the algae growth (see Figure 13.34) and obtain the data (in feet) in the table.

x	Width w	x	Width w
0	0	50	27
10	15	60	24
20	18	70	23
30	18	80	0
40	30		

(a) Can either the Trapezoidal Rule or Simpson's Rule be used to calculate the area of the algae growth?
(b) When either the Trapezoidal Rule or Simpson's Rule can be used, which is usually more accurate?
(c) Use Simpson's Rule to approximate the area of the algae growth.

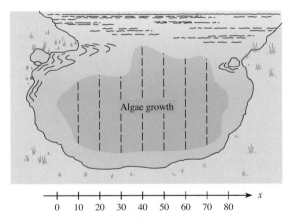

Figure 13.34

34. **Development costs** A land developer is planning to dig a small lake and build a group of homes around it. To estimate the cost of the project, the area of the lake

must be calculated from the proposed measurements (in feet) given in Figure 13.35 and in the data in the table. Use Simpson's Rule to approximate the area of the lake.

x	Width $w(x)$
0	0
100	300
200	200
300	400
400	0

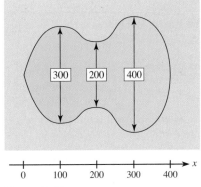

Figure 13.35

Chapter 13 Summary & Review

KEY TERMS AND FORMULAS

Section 13.1

Sigma (summation) notation (p. 791)

$$\sum_{i=1}^{n} a_i = a_1 + a_2 + \cdots + a_n$$

Formulas (p. 791)

$$\sum_{i=1}^{n} 1 = n; \quad \sum_{i=1}^{n} i = \frac{n(n+1)}{2}$$

$$\sum_{i=1}^{n} i^2 = \frac{n(n+1)(2n+1)}{6}$$

Area (p. 792)

Right-hand endpoints

$$\lim_{n \to \infty} \sum_{i=1}^{n} f(x_i) \frac{b-a}{n}; \quad x_i = a + i\left(\frac{b-a}{n}\right)$$

Left-hand endpoints

$$\lim_{n \to \infty} \sum_{i=1}^{n} f(x_{i-1}) \frac{b-a}{n}$$

Section 13.2

Riemann sum (p. 798)

$$\sum_{i=1}^{n} f(x_i^*) \Delta x_i$$

Definite integral (p. 798)

$$\int_a^b f(x)\,dx = \lim_{\substack{\max \Delta x_i \to 0 \\ (n \to \infty)}} \sum_{i=1}^{n} f(x_i^*) \Delta x_i$$

Fundamental Theorem of Calculus (p. 799)

$$\int_a^b f(x)\,dx = F(b) - F(a), \text{ where } F'(x) = f(x)$$

Definite integral properties (pp. 800–801)

$$\int_a^a f(x)\,dx = 0$$

$$\int_b^a f(x)\,dx = -\int_a^b f(x)\,dx$$

$$\int_a^b [f(x) \pm g(x)]\,dx = \int_a^b f(x)\,dx \pm \int_a^b g(x)\,dx$$

$$\int_a^b kf(x)\,dx = k\int_a^b f(x)\,dx$$

$$\int_a^c f(x)\,dx + \int_c^b f(x)\,dx = \int_a^b f(x)\,dx$$

Area under $f(x)$, where $f(x) \geq 0$ (p. 802)

$$A = \int_a^b f(x)\,dx$$

$f(x)$ is a probability density function (p. 803)

$$\Pr(a \leq x \leq b) = \int_a^b f(x)\,dx$$

Section 13.3

Lorenz curve, $L(x)$ (p. 807)
Area between $f(x)$ and $g(x)$, where $f(x) \geq g(x)$ (p. 808)

$$A = \int_a^b [f(x) - g(x)]\,dx$$

Gini coefficient (p. 811)

$$2\int_0^1 [x - L(x)]\,dx$$

Average value of f over $[a, b]$ (p. 811)

$$\frac{1}{b-a} \int_a^b f(x)\,dx$$

Section 13.4

Continuous income streams (p. 816)

Total income

$$\int_0^k f(t)\, dt \quad \text{(for } k \text{ years)}$$

Present value

$$\int_0^k f(t)e^{-rt}\, dt, \quad \text{where } r \text{ is the interest rate}$$

Future value

$$e^{rk}\int_0^k f(t)e^{-rt}\, dt$$

Consumer's surplus [demand is $f(x)$] (p. 818)

$$CS = \int_0^{x_1} f(x)\, dx - p_1 x_1$$

Producer's surplus [supply is $g(x)$] (p. 821)

$$PS = p_1 x_1 - \int_0^{x_1} g(x)\, dx$$

Section 13.5

Integration by formulas (p. 825)

See Table 13.2.

Section 13.6

Integration by parts (p. 829)

$$\int u\, dv = uv - \int v\, du$$

Section 13.7

Improper integrals (p. 834)

$$\int_a^\infty f(x)\, dx = \lim_{b\to\infty} \int_a^b f(x)\, dx$$

$$\int_{-\infty}^b f(x)\, dx = \lim_{a\to\infty} \int_{-a}^b f(x)\, dx$$

$$\int_{-\infty}^\infty f(x)\, dx = \int_{-\infty}^c f(x)\, dx + \int_c^\infty f(x)\, dx$$

Capital value of a continuous income stream (p. 836)

$$\int_0^\infty f(t)e^{-rt}\, dt$$

Probability density function, $f(x)$ (p. 837)

$$f(x) \geq 0 \text{ and } \int_{-\infty}^\infty f(x)\, dx = 1$$

Mean

$$u = \int_{-\infty}^\infty xf(x)\, dx$$

Section 13.8

Trapezoidal Rule (p. 841)

$$\int_a^b f(x)\, dx \approx \frac{h}{2} [f(x_0) + 2f(x_1) + \cdots + 2f(x_{n-1}) + f(x_n)]$$

where $h = \dfrac{b-a}{n}$

Error formula

$$|E| \leq \frac{(b-a)^3}{12n^2}\left[\max_{a\leq x\leq b}|f''(x)|\right]$$

Simpson's Rule (p. 843)

$$\int_a^b f(x)\, dx \approx \frac{h}{3} [f(x_0) + 4f(x_1) + 2f(x_2) + 4f(x_3)$$
$$+ \cdots + 2f(x_{n-2}) + 4f(x_{n-1}) + f(x_n)],$$

where n is even and $h = \dfrac{b-a}{n}$

Error formula

$$|E| \leq \frac{(b-a)^5}{180n^4}\left[\max_{a\leq x\leq b}|f^{(4)}(x)|\right]$$

REVIEW EXERCISES

Section 13.1

1. Calculate $\displaystyle\sum_{k=1}^{8} (k^2 + 1)$.

2. Use formulas to simplify

$$\sum_{i=1}^n \frac{3i}{n^3}$$

3. Use 6 subintervals of the same size to approximate the area under the graph of $y = 3x^2$ from $x = 0$ to $x = 1$.

Use the right-hand endpoints of the subintervals to find the heights of the rectangles.

4. Use rectangles to find the exact area under the graph of $y = 3x^2$ from $x = 0$ to $x = 1$. Use n equal subintervals.

Section 13.2

5. Use a definite integral to find the area under the graph of $y = 3x^2$ from $x = 0$ to $x = 1$.

6. Find the area between the graph of $y = x^3 - 4x + 5$ and the x-axis from $x = 1$ to $x = 3$.

Evaluate the integrals in Problems 7–18.

7. $\int_1^4 4\sqrt{x^3}\, dx$

8. $\int_{-3}^2 (x^3 - 3x^2 + 4x + 2)\, dx$

9. $\int_0^5 (x^3 + 4x)\, dx$

10. $\int_{-1}^3 (3x + 4)^{-2}\, dx$

11. $\int_{-3}^{-1} (x + 1)\, dx$

12. $\int_2^3 \frac{x^2}{2x^3 - 7}\, dx$

13. $\int_{-1}^2 (x^2 + x)\, dx$

14. $\int_1^4 \left(\frac{1}{x} + \sqrt{x}\right) dx$

15. $\int_0^2 5x^2(6x^3 + 1)^{1/2}\, dx$

16. $\int_0^1 \frac{x}{x^2 + 1}\, dx$

17. $\int_0^1 e^{-2x}\, dx$

18. $\int_0^1 xe^{x^2}\, dx$

Section 13.3

Find the area between the curves in Problems 19–22.

19. $y = x^2 - 3x + 2$ and $y = x^2 + 4$ from $x = 0$ to $x = 5$
20. $y = x^2$ and $y = 4x + 5$
21. $y = x^3$ and $y = x$ from $x = -1$ to $x = 0$
22. $y = x^3 - 1$ and $y = x - 1$

Section 13.5

Evaluate the integrals in Problems 23–26, using the formulas in Table 13.2.

23. $\int \sqrt{x^2 - 4}\, dx$

24. $\int_0^1 3^x\, dx$

25. $\int x \ln x^2\, dx$

26. $\int \frac{dx}{x(3x + 2)}$

Section 13.6

In Problems 27–30, use integration by parts to evaluate.

27. $\int x^5 \ln x\, dx$

28. $\int x^2 e^{-2x}\, dx$

29. $\int \frac{x\, dx}{\sqrt{x + 5}}$

30. $\int_1^e \ln x\, dx$

Section 13.7

Evaluate the improper integrals in Problems 31–34.

31. $\int_1^\infty \frac{1}{x + 1}\, dx$

32. $\int_{-\infty}^{-1} \frac{200}{x^3}\, dx$

33. $\int_0^\infty 5e^{-3x}\, dx$

34. $\int_{-\infty}^0 \frac{x}{(x^2 + 1)^2}\, dx$

Section 13.8

35. Evaluate $\int_1^3 \frac{2}{x^3}\, dx$

   (a) exactly.

(b) by using the Trapezoidal Rule with $n = 4$ (to 3 decimal places).
(c) by using Simpson's Rule with $n = 4$ (to 3 decimal places).

36. Use the Trapezoidal Rule with $n = 5$ to approximate

$$\int_0^1 \frac{4\, dx}{x^2 + 1}$$

Round your answer to 3 decimal places.

37. Use the table that follows to approximate

$$\int_1^{2.2} f(x)\, dx$$

by using Simpson's Rule. Round your answer to 1 decimal place.

x	$f(x)$
1	0
1.3	2.8
1.6	5.1
1.9	4.2
2.2	0.6

38. Suppose that a definite integral is to be approximated and it is found that to achieve a specified accuracy, n must satisfy $n \geq 4.8$. What is the smallest n that can be used, if
   (a) the Trapezoidal Rule is used?
   (b) Simpson's Rule is used?

APPLICATIONS

Section 13.2

39. *Maintenance* Maintenance costs for buildings increase as the buildings age. If the rate of increase in maintenance costs for a building is

$$M'(t) = \frac{14{,}000}{\sqrt{t + 16}}$$

where M is in dollars and t is time in years, $0 \leq t \leq 15$, find the total maintenance cost for the first 9 years ($t = 0$ to $t = 9$).

40. *Quality control* Suppose the probability density function for the life expectancy of a battery is

$$f(x) = \begin{cases} 1.4e^{-1.4x} & x \geq 0 \\ 0 & x < 0 \end{cases}$$

Find the probability that the battery lasts 2 years or less.

Section 13.3

41. *Savings* The future value of $1000 invested in a savings account at 10%, compounded continuously, is $S = 1000e^{0.1t}$, where t is in years. Find the average amount in the savings account during the first 5 years.

42. *Income streams* Suppose the total income in dollars from a machine is given by

$$I = 50e^{0.2t}, \quad 0 \le t \le 4, t \text{ in hours}$$

Find the average income over this 4-hour period.

43. *Income distribution* In 1969, after the "Great Society" initiatives of the Johnson administration, the Lorenz curve for the U.S. income distribution was $L(x) = x^{2.1936}$. In 2000, after the stock market's historic 10-year growth, the Lorenz curve for the U.S. income distribution was $L(x) = x^{2.4870}$. Find the Gini coefficient of income for both years, and determine in which year income was more equally distributed.

Section 13.4

44. *Consumer's surplus* The demand function for a product under pure competition is $p = \sqrt{64 - 4x}$, and the supply function is $p = x - 1$, where x is the number of units and p is in dollars.
    (a) Find the market equilibrium.
    (b) Find the consumer's surplus at market equilibrium.
45. *Producer's surplus* Find the producer's surplus at market equilibrium for Problem 44.
46. *Income streams* Find the total income over the next 10 years from a continuous income stream that has an annual flow rate at time t given by $f(t) = 125e^{0.05t}$ in thousands of dollars per year.
47. *Income streams* Suppose that a machine's production is considered a continuous income stream with an annual rate of flow at time t given by $f(t) = 150e^{-0.2t}$ in thousands of dollars per year. Money is worth 8%, compounded continuously.
    (a) Find the present value of the machine's production over the next 5 years.
    (b) Find the future value of the production 5 years from now.

Section 13.5

48. *Average cost* Suppose the cost function for x units of a product is given by $C(x) = \sqrt{40{,}000 + x^2}$ dollars. Find the average cost over the first 150 units.
49. *Producer's surplus* Suppose the supply function for x units of a certain lamp is given by

$$p = 0.02x + 50.01 - \frac{10}{\sqrt{x^2 + 1}}$$

where p is in dollars. Find the producer's surplus if the equilibrium price is $70 and the equilibrium quantity is 1000.

Section 13.6

50. *Income streams* Suppose the present value of a continuous income stream over the next 5 years is given by

$$P = 9000 \int_0^5 te^{-0.08t} \, dt, P \text{ in dollars, } t \text{ in years}$$

Find the present value.

51. *Cost* If the marginal cost for x units of a product is $\overline{MC} = 3 + 60(x + 1) \ln(x + 1)$ dollars per unit and if the fixed cost is $2000, find the total cost function.

Section 13.7

52. *Quality control* Find the probability that a phone lasts more than 1 year if the probability density function for its life expectancy is given by

$$f(x) = \begin{cases} 1.4e^{-1.4x} & x \ge 0 \\ 0 & x < 0 \end{cases}$$

53. *Capital value* Find the capital value of a business if its income is considered a continuous income stream with annual rate of flow given by

$$f(t) = 120e^{0.03t}$$

in thousands of dollars per year, and the current interest rate is 6% compounded continuously.

Section 13.8

54. *Total income* Suppose that a continuous income stream has an annual rate of flow $f(t) = 100e^{-0.01t^2}$ (in thousands of dollars per year). Use Simpson's Rule with $n = 4$ to approximate the total income from this stream over the next 2 years.
55. *Revenue* A company has the data shown in the table from the sale of its product.

x	$\overline{MR}$
0	0
2	480
4	720
6	720
8	480
10	0

If x represents hundreds of units and revenue R is in hundreds of dollars, approximate the total revenue from the sale of 1000 units by approximating

$$\int_0^{10} \overline{MR} \, dx$$

with the Trapezoidal Rule.

Chapter 13 TEST

1. Use left-hand endpoints and $n = 4$ subdivisions to approximate the area under $f(x) = \sqrt{4 - x^2}$ on the interval $[0, 2]$.

2. Consider $f(x) = 5 - 2x$ from $x = 0$ to $x = 1$ with n equal subdivisions.
   (a) If $f(x)$ is evaluated at right-hand endpoints, find a formula for the sum, S, of the areas of the n rectangles.
   (b) Find $\lim\limits_{n \to \infty} S$.

3. Express the area in Quadrant I under $y = 12 + 4x - x^2$ (shaded in the figure) as an integral. Then evaluate the integral to find the area.

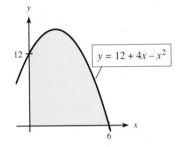

4. Evaluate the following definite integrals. (Do not approximate with technology.)
   (a) $\displaystyle\int_0^4 (9 - 4x)\,dx$ (b) $\displaystyle\int_0^3 x(8x^2 + 9)^{-1/2}\,dx$

   (c) $\displaystyle\int_1^4 \frac{5}{4x - 1}\,dx$ (d) $\displaystyle\int_1^\infty \frac{7}{x^2}\,dx$

   (e) $\displaystyle\int_4^4 \sqrt{x^3 + 10}\,dx$ (f) $\displaystyle\int_0^1 5x^2 e^{2x^3}\,dx$

5. Use integration by parts to evaluate the following.
   (a) $\displaystyle\int 3xe^x\,dx$ (b) $\displaystyle\int x \ln(2x)\,dx$

6. If $\displaystyle\int_1^4 f(x)\,dx = 3$ and $\displaystyle\int_3^4 f(x)\,dx = 7$, find $\displaystyle\int_1^3 2f(x)\,dx$.

7. Use Table 13.2 in Section 13.5 to evaluate each of the following.
   (a) $\displaystyle\int \ln(2x)\,dx$ (b) $\displaystyle\int x\sqrt{3x - 7}\,dx$

8. Use the numerical integration feature of a graphing calculator to approximate $\displaystyle\int_1^4 \sqrt{x^3 + 10}\,dx$.

9. Suppose the supply function for a product is $p = 40 + 0.001x^2$ and the demand function is $p = 120 - 0.2x$, where x is the number of units and p is the price in dollars. If the market equilibrium price is $80, find (a) the consumer's surplus and (b) the producer's surplus.

10. Suppose a continuous income stream has an annual rate of flow $f(t) = 85e^{-0.01t}$, in thousands of dollars per year, and the current interest rate is 7% compounded continuously.
    (a) Find the total income over the next 12 years.
    (b) Find the present value over the next 12 years.
    (c) Find the capital value of the stream.

11. Find the area between $y = 2x + 4$ and $y = x^2 - x$.

12. The figure shows typical supply and demand curves. On the figure, sketch and shade the region whose area represents the consumer's surplus.

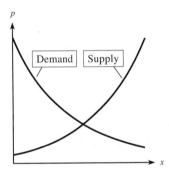

13. In an effort to make the distribution of income more nearly equal, the government of a country passes a tax law that changes the Lorenz curve from $y = 0.998x^{2.6}$ for one year to $y = 0.57x^2 + 0.43x$ for the next year. Find the Gini coefficient of income for each year and determine whether the distribution of income is more or less equitable after the tax law is passed. Interpret the result.

14. With data from the U.S. Department of Energy, the number of billions of barrels of U.S. crude oil produced per year from 2010 and projected to 2030 can be modeled by the function
    $$f(t) = -0.001t^2 + 0.037t + 1.94$$
    where t is the number of years past 2010.
    (a) Find the total number of barrels of oil produced from 2010 to 2020.
    (b) Find the average number of barrels produced per year from 2010 to 2018.

15. Use the Trapezoidal Rule to approximate

$$\int_1^3 (x \ln x)\, dx$$

with $n = 4$ subintervals.

16. The environmental effects of a chemical spill in the Clarion River can be estimated from the river's flow volume. To measure this flow volume, it is necessary to find the cross-sectional area of the river at a point downriver from the spill. The distance across the river is 240 feet, and the table gives depth measurements, $D(x)$, of the river at various distances x from one bank of the river. Use these measurements with Simpson's Rule to estimate the cross-sectional area.

x	0	40	80	120	160	200	240
$D(x)$	0	22	35	48	32	24	0

I. Purchasing Electrical Power (Modeling)

In order to plan its purchases of electrical power from suppliers over the next 5 years, the PAC Electric Company needs to model its load data (demand for power by its customers) and use this model to predict future loads. The company pays for the electrical power each month on the basis of the peak load (demand) at any point during the month. The table gives, for the years 1998–2016, the load in megawatts (that is, in millions of watts) for the month when the maximum load occurred and the load in megawatts for the month when the minimum load occurred. The maximum loads occurred in summer, and the minimum loads occurred in spring or fall.

Year	Maximum Monthly Load	Minimum Monthly Load
1998	40.9367	19.4689
1999	45.7127	22.1504
2000	48.0460	25.3670
2001	56.1712	28.7254
2002	55.5793	31.0460
2003	62.4285	31.3838
2004	76.6536	34.8426
2005	73.8214	38.4544
2006	74.8844	40.6080
2007	83.0590	47.3621
2008	88.3914	45.8393
2009	88.7704	48.7956
2010	94.2620	48.3313
2011	105.1596	52.7710
2012	95.8301	54.4757
2013	97.8854	55.2210
2014	102.8912	55.1360
2015	109.5541	57.2162
2016	111.2516	58.3216

The company wishes to predict the average monthly load over the next 5 years so that it can plan its future monthly purchases. To assist the company, proceed as follows.

1. (a) Using the years and the maximum monthly load given for each year, graph the data, with x representing the number of years past 1998 and y representing the maximum load in megawatts.
   (b) Find the equation that best fits the data, using both a quadratic model and a cubic model.
   (c) Graph the data and both of these models from 1998 to 2018 (that is, from $x = 0$ to $x = 20$).
2. Do the two models appear to fit the data equally well in the interval 1998–2018? Which model appears to be a better predictor for the next decade?
3. Use the quadratic model to predict the maximum monthly load in the year 2021. How can this value be used by the company? Should this number be used to plan monthly power purchases for each month in 2021?

4. To create a "typical" monthly load function:
   (a) Create a table with the year as the independent variable and the average of the maximum and minimum monthly loads as the dependent variable.
   (b) Find the quadratic model that best fits these data points, using $x = 0$ to represent 1998.
5. Use a definite integral with the typical monthly load function to predict the average monthly load over the years 2018–2023.
6. What factors in addition to the average monthly load should be considered when the company plans future purchases of power?

II. Retirement Planning

A 52-year-old client asks an accountant how to plan for his future retirement at age 62. He expects income from Social Security in the amount of $21,600 per year and a retirement pension of $40,500 per year from his employer. He wants to make monthly contributions to an investment plan that pays 8%, compounded monthly, for 10 years so that he will have a total income of $83,700 per year for 30 years. What will the size of the monthly contributions have to be to accomplish this goal, if it is assumed that money will be worth 8%, compounded continuously, throughout the period after he is 62?

To help you answer this question, complete the following.

1. How much money must the client withdraw annually from his investment plan during his retirement so that his total income goal is met?
2. How much money S must the client's account contain when he is 62 so that it will generate this annual amount for 30 years? (*Hint:* S can be considered the present value over 30 years of a continuous income stream with the amount you found in Question 1 as its annual rate of flow.)
3. The monthly contribution R that would, after 10 years, amount to the present value S found in Question 2 can be obtained from the formula

$$R = S\left[\frac{i}{(1 + i)^n - 1}\right]$$

where i represents the monthly interest rate and n the number of months. Find the client's monthly contribution, R.

14

Functions of Two or More Variables

In this chapter we will extend our study of functions to functions of two or more variables. We will use these concepts to solve problems in the management, life, and social sciences. In particular, we will discuss joint cost functions, utility functions that describe the customer satisfaction derived from the consumption of two products, Cobb-Douglas production functions, and wind chill temperatures as a function of air temperature and wind speed.

We will use derivatives with respect to one of two variables (called partial derivatives) to find marginal cost, marginal productivity, marginal utility, marginal demand, and other rates of change. We will use partial derivatives to find maxima and minima of functions of two variables, and we will use Lagrange multipliers to optimize functions of two variables subject to a condition that constrains these variables. These skills are used to maximize profit, production, and utility and to minimize cost subject to constraints.

The topics and some representative applications discussed in this chapter include the following.

SECTIONS

APPLICATIONS

14.1 **Functions of Two or More Variables** — Utility, Cobb-Douglas production functions

14.2 **Partial Differentiation** — Marginal cost, marginal sales

14.3 **Applications of Functions of Two Variables in Business and Economics**
Joint cost and marginal cost
Production functions
Demand functions
— Joint cost, marginal cost, crop harvesting, marginal productivity, marginal demand

14.4 **Maxima and Minima**
Linear regression
— Maximum profit, inventory

14.5 **Maxima and Minima of Functions Subject to Constraints: Lagrange Multipliers**
— Maximum utility, maximum production, marginal productivity of money

Chapter   Warm-Up

Prerequisite Problem Type	For Section	Answer	Section for Review
If $y = f(x)$, x is the independent variable and y is the _____ variable.	14.1	Dependent	1.2 Functions
What is the domain of $f(x) = \dfrac{3x}{x-1}$?	14.1	All reals except $x = 1$	1.2 Domains
If $C(x) = 5 + 5x$, what is $C(0.20)$?	14.1	6	1.2 Function notation
(a) Solve for x and y: $\begin{cases} 0 = 50 - 2x - 2y \\ 0 = 60 - 2x - 4y \end{cases}$ (b) Solve for x and y: $\begin{cases} x = 2y \\ x + y - 9 = 0 \end{cases}$	14.4 14.5	(a) $x = 20,\ y = 5$ (b) $x = 6,\ y = 3$	1.5 Systems of equations
If $z = 4x^2 + 5x^3 - 7$, what is $\dfrac{dz}{dx}$?	14.2–14.5	$\dfrac{dz}{dx} = 8x + 15x^2$	9.4 Derivatives
If $f(x) = (x^2 - 1)^2$, what is $f'(x)$?	14.2	$f'(x) = 4x(x^2 - 1)$	9.6 Derivatives
If $z = 10y - \ln y$, what is $\dfrac{dz}{dy}$?	14.2	$\dfrac{dz}{dy} = 10 - \dfrac{1}{y}$	11.1 Derivatives of logarithmic functions
If $z = 5x^2 + e^x$, what is $\dfrac{dz}{dx}$?	14.2	$\dfrac{dz}{dx} = 10x + e^x$	11.2 Derivatives of exponential functions
Find the slope of the tangent to $y = 4x^3 - 4e^x$ at $(0, -4)$.	14.2	-4	9.3, 9.4, 11.2 Derivatives

OBJECTIVES	**14.1**

- To find the domain of a function of two or more variables
- To evaluate a function of two or more variables given values for the independent variables

Functions of Two or More Variables

▮ | APPLICATION PREVIEW | ▮

The relations we have studied up to this point have been limited to two variables, with one of the variables assumed to be a function of the other. But there are many instances where one variable may depend on two or more other variables. For example, a company's quantity of output or production Q (measured in units or dollars) can be modeled according to the equation

$$Q = AK^\alpha L^{1-\alpha}$$

where A is a constant, K is the company's capital investment, L is the size of the labor force (in work-hours), and α is a constant with $0 < \alpha < 1$. Functions of this type are called **Cobb-Douglas production functions,** and they are frequently used in economics. For example, suppose the Cobb-Douglas production function for a company is given by

$$Q = 4K^{0.4}L^{0.6}$$

where Q is thousands of dollars of production value, K is hundreds of dollars of capital investment, and L is hours of labor. We could use this function to determine the production value for a given amount of capital investment and available work-hours of labor. We could also find how production is affected by changes in capital investment or available work-hours. (See Example 5.)

In addition, the demand function for a commodity frequently depends on the price of the commodity, available income, and prices of competing goods. Other examples from economics will be presented later in this chapter.

We write $z = f(x, y)$ to state that z is a function of both x and y. The variables x and y are called the **independent variables** and z is called the **dependent variable.** Thus the function f associates with each pair of possible values for the independent variables (x and y) exactly one value of the dependent variable (z).

The equation $z = x^2 - xy$ defines z as a function of x and y. We can denote this by writing $z = f(x, y) = x^2 - xy$. The domain of the function is the set of all ordered pairs (of real numbers), and the range is the set of all real numbers.

EXAMPLE 1 Domain

Give the domain of the function

$$g(x, y) = \frac{x^2 - 3y}{x - y}$$

Solution
The domain of the function is the set of ordered pairs that do not give a 0 denominator. That is, the domain is the set of all ordered pairs where the first and second elements are not equal (that is, where $x \neq y$). ▪

✓ **CHECKPOINT**

1. Find the domain of the function

$$f(x, y) = \frac{2}{\sqrt{x^2 - y^2}}$$

We graph the function $z = f(x, y)$ by using three dimensions. We can construct a three-dimensional coordinate space by drawing three mutually perpendicular axes, as in

Figure 14.1. By setting up scales of measurement along the three axes from the origin of each axis, we can determine the three coordinates (x, y, z) for any point P. The point shown in Figure 14.1 is $+2$ units in the x-direction, $+3$ units in the y-direction, and $+4$ units in the z-direction, so the coordinates of the point are $(2, 3, 4)$.

The pairs of axes determine the three **coordinate planes;** the xy-plane, the yz-plane, and the xz-plane. The planes divide the space into eight **octants.** The point $P(2, 3, 4)$ is in the first octant.

If we are given a function $z = f(x, y)$, we can find the z-value corresponding to $x = a$ and $y = b$ by evaluating $f(a, b)$.

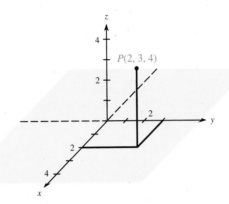

Figure 14.1

EXAMPLE 2   Function Values

If $z = f(x, y) = x^2 - 4xy + xy^3$, find the following.

(a) $f(1, 2)$ (b) $f(2, 5)$ (c) $f(-1, 3)$

Solution

(a) $f(1, 2) = 1^2 - 4(1)(2) + (1)(2)^3 = 1$
(b) $f(2, 5) = 2^2 - 4(2)(5) + (2)(5)^3 = 214$
(c) $f(-1, 3) = (-1)^2 - 4(-1)(3) + (-1)(3)^3 = -14$

✓ CHECKPOINT   2. If $f(x, y, z) = x^2 + 2y - z$, find $f(2, 3, 4)$.

EXAMPLE 3   Cost

A small furniture company's cost (in dollars) to manufacture 1 unit of several different all-wood items is given by

$$C(x, y) = 5 + 5x + 22y$$

where x represents the number of board-feet of material used and y represents the number of work-hours of labor required for assembly and finishing. A certain bookcase uses 20 board-feet of material and requires 2.5 work-hours for assembly and finishing. Find the cost of manufacturing this bookcase.

Solution
The cost is

$$C(20, 2.5) = 5 + 5(20) + 22(2.5) = 160 \quad \text{dollars}$$

For a given function $z = f(x, y)$, we can construct a table of values by assigning values to x and y and finding the corresponding values of z. To each pair of values for x and y there corresponds a unique value of z, and thus a unique point in a three-dimensional coordinate system. From a table of values such as this, a finite number of points can be

| TABLE 14.1 |

x	y	z
−2	0	0
0	−2	0
−1	0	3
0	−1	3
−1	−1	2
0	0	4
1	0	3
0	1	3
1	1	2
2	0	0
0	2	0

plotted. All points that satisfy the equation form a "surface" in space. Because z is a function of x and y, lines parallel to the z-axis will intersect such a surface in at most one point. The graph of the equation $z = 4 - x^2 - y^2$ is the surface shown in Figure 14.2(a). The portion of the surface above the xy-plane resembles a bullet and is called a **paraboloid.** Some points on the surface are given in Table 14.1.

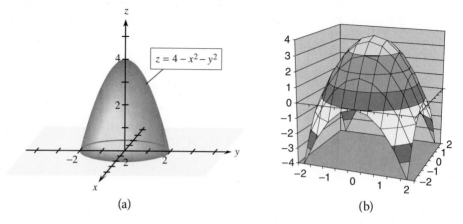

Figure 14.2   (a)   (b)

Spreadsheet Note

We can create the graphs of functions of two variables with Excel in a manner similar to that used to graph functions of one variable. Figure 14.2(b) shows the Excel graph of $z = 4 - x^2 - y^2$. Appendix B, Section 14.1, shows the steps used in graphing functions of two variables. See also the Online Excel Guide.

In practical applications of functions of two variables, we will have little need to construct the graphs of the surfaces. For this reason, we will not discuss methods of sketching the graphs. Although you will not be asked to sketch graphs of these surfaces, the fact that the graphs do *exist* will be used in studying relative maxima and minima of functions of two variables.

The properties of functions of one variable can be extended to functions of two variables. The precise definition of continuity for functions of two variables is technical and may be found in more advanced books. We will limit our study to functions that are continuous and have continuous derivatives in the domain of interest to us. We may think of continuous functions as functions whose graphs consist of surfaces without "holes" or "breaks" in them.

Utility Functions

Let the function $U = f(x, y)$ represent the **utility** (that is, satisfaction) derived by a consumer from the consumption of two goods, X and Y, where x and y represent the amounts of X and Y, respectively. Because we will assume that the utility function is continuous, a given level of utility can be derived from an infinite number of combinations of x and y. The graph of all points (x, y) that give the same utility is called an **indifference curve.** A set of indifference curves corresponding to different levels of utility is called an **indifference map** (see Figure 14.3).

EXAMPLE 4   **Utility**

Suppose that the utility function for two goods, X and Y, is $U = x^2y^2$ and a consumer purchases 10 units of X and 2 units of Y.

(a) What is the level of utility U for these two products?
(b) If the consumer purchases 5 units of X, how many units of Y must be purchased to retain the same level of utility?
(c) Graph the indifference curve for this level of utility.
(d) Graph the indifference curves for this utility function if $U = 100$ and if $U = 36$.

Figure 14.3

Solution

(a) If $x = 10$ and $y = 2$ satisfy the utility function, then the level of utility is
$$U = 10^2 \cdot 2^2 = 400.$$

(b) If x is 5, y must satisfy $400 = 5^2y^2$, so $y = \pm 4$, and 4 units of Y must be purchased.

(c) The indifference curve for $U = 400$ is $400 = x^2y^2$. The graph for positive x and y is shown in Figure 14.3.

(d) The indifference map in Figure 14.3 contains these indifference curves. ■

Technology Note A graphing calculator or Excel can be used to graph each indifference curve in the indifference map shown in Figure 14.3. To graph the indifference curve for a given value of U, we must recognize that y will be positive and solve for y to express it as a function of x. Try it for $U = 100$. Does your graph agree with the one shown in Figure 14.3? ■

Sometimes functions of two variables are studied by fixing a value for one variable and graphing the resulting function of a single variable. We'll do this in Section 14.3 with production functions.

EXAMPLE 5 **Production** | **APPLICATION PREVIEW** |

Suppose a company has the Cobb-Douglas production function

$$Q = 4K^{0.4}L^{0.6}$$

where Q is thousands of dollars of production value, K is hundreds of dollars of capital investment per week, and L is work-hours of labor per week.

(a) If current capital investment is \$72,900 per week and work-hours are 3072 per week, find the current weekly production value.

(b) If weekly capital investment is increased to \$97,200 and new employees are hired so that there are 4096 total weekly work-hours, find the percent increase in the production value.

Solution

(a) Capital investment of \$72,900 means that $K = 729$. We use this value and $L = 3072$ in the production function.

$$Q = 4(729)^{0.4}(3072)^{0.6} = 6912$$

Thus the weekly production value is \$6,912,000.

(b) In this case we use $K = 972$ and $L = 4096$.

$$Q = 4(972)^{0.4}(4096)^{0.6} = 9216$$

This is an increase in production value of $9216 - 6912 = 2304$, which is equivalent to a weekly increase of

$$\frac{2304}{6912} = 0.33\tfrac{1}{3} = 33\tfrac{1}{3}\%$$

■

✓ CHECKPOINT ANSWERS

1. $x^2 > y^2$; that is, where $|x| > |y|$.
2. $f(2, 3, 4) = 6$

| **EXERCISES** | **14.1**

Give the domain of each function in Problems 1–8.

1. $z = x^2 + y^2$
2. $z = 4x - 3y$
3. $z = \dfrac{4x - 3}{y}$
4. $z = \dfrac{x + y^2}{\sqrt{x}}$
5. $z = \dfrac{4x^3y - x}{2x - y}$
6. $z = \sqrt{x - y}$
7. $q = \sqrt{p_1} + 3p_2$
8. $q = 5p_1 - \sqrt{p_1 - p_2}$

In Problems 9–14, evaluate the functions at the given values of the independent variables.

9. $z = x^3 + 4xy + y^2$; $x = 1, y = -1$
10. $z = 4x^2 - 3xy^3$; $x = 2, y = 2$
11. $z = \dfrac{x - y}{x + y}$; $x = 4, y = -1$
12. $z = \dfrac{x^2 + xy}{x - y}$; $x = 3, y = 2$

13. $C(x_1, x_2) = 600 + 4x_1 + 6x_2$; $x_1 = 400, x_2 = 50$
14. $C(x_1, x_2) = 500 + 5x_1 + 7x_2$; $x_1 = 200, x_2 = 300$

In Problems 15–22, evaluate each function as indicated.

15. $q_1(p_1, p_2) = \dfrac{p_1 + 4p_2}{p_1 - p_2}$; find $q_1(40, 35)$.

16. $q_1(p_1, p_2) = \dfrac{5p_1 - p_2}{p_1 + 3p_2}$; find $q_1(50, 10)$.

17. $z(x, y) = xe^{x+y}$; find $z(3, -3)$.

18. $f(x, y) = ye^{2x} + y^2$; find $f(0, 7)$.

19. $f(x, y) = \dfrac{\ln(xy)}{x^2 + y^2}$; find $f(-3, -4)$.

20. $z(x, y) = x \ln y - y \ln x$; find $z(1, 1)$.

21. $w(x, y, z) = \dfrac{x^2 + 4yz}{xyz}$; find $w(1, 3, 1)$.

22. $f(w, x, y, z) = \dfrac{wx - yz^2}{xy - wz}$; find $f(2, 3, 1, -1)$.

APPLICATIONS

23. *Investment* The future value S of an investment earning 6% compounded continuously is a function of the principal P and the length of time t that the principal has been invested. It is given by

$$S = f(P, t) = Pe^{0.06t}$$

Find $f(2000, 20)$, and interpret your answer.

24. *Amortization* If $100,000 is borrowed to purchase a home, then the monthly payment R is a function of the interest rate i (expressed as a percent) and the number of years n before the mortgage is paid. It is given by

$$R = f(i, n) = 100{,}000\left[\dfrac{0.01(i/12)}{1 - (1 + 0.01(i/12))^{-12n}}\right]$$

Find $f(7.25, 30)$ and interpret your answer.

25. *Wilson's lot size formula* In economics, the most economical quantity Q of goods (TVs, dresses, gallons of paint, etc.) for a store to order is given by Wilson's lot size formula

$$Q = f(K, M, h) = \sqrt{2KM/h}$$

where K is the cost of placing the order, M is the number of items sold per week, and h is the weekly holding cost for each item (the cost of storage space, utilities, taxes, security, etc.). Find $f(200, 625, 1)$ and interpret your answer.

26. *Gas law* Suppose that a gas satisfies the universal gas law, $V = nRT/P$, with n equal to 10 moles of the gas and R, the universal gas constant, equal to 0.082054. What is V if $T = 10$ K (kelvins, the units in which temperature is measured on the Kelvin scale) and $P = 1$ atmosphere?

Temperature-humidity models **There are different models for measuring the effects of high temperature**

and humidity. Two of these are the Summer Simmer Index (S) and the Apparent Temperature (A),* and they are given by

$$S = 1.98T - 1.09(1 - H)(T - 58) - 56.9$$
$$A = 2.70 + 0.885T - 78.7H + 1.20TH$$

where T is the air temperature (in degrees Fahrenheit) and H is the relative humidity (expressed as a decimal). Use these models in Problems 27 and 28.

27. At the Dallas–Fort Worth Airport, the average daily temperatures and humidities for July are

   Maximum: 97.8°F with 44% humidity
   Minimum: 74.7°F with 80% humidity**

   Calculate the Summer Simmer Index S and the Apparent Temperature A for both the average daily maximum and the average daily minimum temperature.

28. In Orlando, Florida, the following represent the average daily temperatures and humidities for August.

   Maximum: 91.6°F with 60% humidity
   Minimum: 73.4°F with 92% humidity**

   Calculate the Summer Simmer Index S and the Apparent Temperature A for both the average daily maximum and the average daily minimum temperature.

29. *Mortgage* The tables show that a monthly mortgage payment, R, is a function of the amount financed, A, in thousands of dollars; the duration of the loan, n, in years; and the annual interest rate, r, as a percent. If $R = f(A, n, r)$, use the tables to find the following, and then write a sentence of explanation for each.
   (a) $f(90, 20, 8)$ (b) $f(160, 15, 9)$

8% Annual Percent Rate
Monthly Payments (Principal and Interest)

Amount Financed	10 Years	15 Years	20 Years	25 Years	30 Years
$50,000	$606.64	$477.83	$418.22	$385.91	$366.88
60,000	727.97	573.39	501.86	463.09	440.26
70,000	849.29	668.96	585.51	540.27	513.64
80,000	970.62	764.52	669.15	617.45	587.01
90,000	1091.95	860.09	752.80	694.63	660.39
100,000	1213.28	955.65	836.44	771.82	733.76
120,000	1455.94	1146.78	1003.72	926.18	880.52
140,000	1698.58	1337.92	1171.02	1080.54	1027.28
160,000	1941.24	1529.04	1338.30	1234.90	1174.02
180,000	2183.90	1720.18	1505.60	1389.26	1320.78
200,000	2426.56	1911.30	1672.88	1543.64	1467.52

---

*W. Bosch and L. G. Cobb, "Temperature-Humidity Indices," UMAP Unit 691, *The UMAP Journal*, 10, no. 3 (Fall 1989): 237–256.
**James Ruffner and Frank Bair (eds.), *Weather of U.S. Cities*, Gale Research Co., Detroit, MI, 1987.

9% Annual Percent Rate
Monthly Payments (Principal and Interest)

Amount Financed	10 Years	15 Years	20 Years	25 Years	30 Years
$50,000	$633.38	$507.13	$449.86	$419.60	$402.31
60,000	760.05	608.56	539.84	503.52	482.77
70,000	886.73	709.99	629.81	587.44	563.24
80,000	1013.41	811.41	719.78	671.36	643.70
90,000	1140.08	912.84	809.75	755.28	724.16
100,000	1266.76	1014.27	899.73	839.20	804.62
120,000	1520.10	1217.12	1079.68	1007.04	965.54
140,000	1773.46	1419.96	1259.62	1174.88	1126.48
160,000	2026.82	1622.82	1439.56	1342.72	1287.40
180,000	2280.16	1825.68	1619.50	1510.56	1448.32
200,000	2533.52	2028.54	1799.46	1678.40	1609.24

Source: The Mortgage Money Guide, Federal Trade Commission

30. **Wind chill** Wind and cold temperatures combine to make the air temperature feel colder than it actually is. This combination is reported as *wind chill*. The table shows the latest wind chill calculations from the National Weather Service and indicates that wind chill temperatures, *WC*, are a function of wind speed, *s*, and air temperature, *t*. If $WC = f(s, t)$, use the table to find and interpret each of the following.
  (a) $f(25, 5)$ (b) $f(15, -15)$

	Air Temperature (°F)						
Wind Speed (mph)	35	25	15	5	−5	−15	−25
5	31	19	7	−5	−16	−28	−40
15	25	13	0	−13	−26	−39	−51
25	23	9	−4	−17	−31	−44	−58
35	21	7	−7	−21	−34	−48	−62
45	19	5	−9	−23	−37	−51	−65

Source: National Weather Service

31. **Utility** Suppose that the utility function for two goods *X* and *Y* is given by $U = xy^2$, and a consumer purchases 9 units of *X* and 6 units of *Y*.
  (a) If the consumer purchases 9 units of *Y*, how many units of *X* must be purchased to retain the same level of utility?
  (b) If the consumer purchases 81 units of *X*, how many units of *Y* must be purchased to retain the same level of utility?
  (c) Graph the indifference curve for the utility level found in parts (a) and (b). Use the graph to confirm your answers to parts (a) and (b).

32. **Utility** Suppose that an indifference curve for two goods, *X* and *Y*, has the equation $xy = 400$.
  (a) If 25 units of *X* are purchased, how many units of *Y* must be purchased to remain on this indifference curve?
  (b) Graph this indifference curve and confirm your results in part (a).

33. **Production** Suppose that a company's production for *Q* units of its product is given by the Cobb-Douglas production function
$$Q = 30K^{1/4}L^{3/4}$$
where *K* is dollars of capital investment and *L* is labor hours.
  (a) Find *Q* if $K = \$10,000$ and $L = 625$ hours.
  (b) Show that if *both K* and *L* are doubled, then the output is doubled.
  (c) If capital investment is held at $10,000, graph *Q* as a function of *L*.

34. **Production** Suppose that a company's production for *Q* units of its product is given by the Cobb-Douglas production function
$$Q = 70K^{2/3}L^{1/3}$$
where *K* is dollars of capital investment and *L* is labor hours.
  (a) Find *Q* if $K = \$64,000$ and $L = 512$ hours.
  (b) Show that if both *K* and *L* are halved, then *Q* is also halved.
  (c) If capital investment is held at $64,000, graph *Q* as a function of *L*.

35. **Production** Suppose that the number of units of a good produced, *z*, is given by $z = 20xy$, where *x* is the number of machines working properly and *y* is the average number of work-hours per machine. Find the production for a week in which
  (a) 12 machines are working properly and the average number of work-hours per machine is 30.
  (b) 10 machines are working properly and the average number of work-hours per machine is 25.

36. **Profit** The Kirk Kelly Kandy Company makes two kinds of candy, Kisses and Kreams. The profit, in dollars, for the company is given by
$$P(x, y) = 10x + 6.4y - 0.001x^2 - 0.025y^2$$
where *x* is the number of pounds of Kisses sold per week and *y* is the number of pounds of Kreams. What is the company's profit if it sells
  (a) 20 pounds of Kisses and 10 pounds of Kreams?
  (b) 100 pounds of Kisses and 16 pounds of Kreams?
  (c) 10,000 pounds of Kisses and 256 pounds of Kreams?

37. **Epidemic** The cost per day to society of an epidemic is
$$C(x, y) = 20x + 200y$$
where *C* is in dollars, *x* is the number of people infected on a given day, and *y* is the number of people who die on a given day. If 14,000 people are infected and 20 people die on a given day, what is the cost to society?

38. **Pesticide** An area of land is to be sprayed with two brands of pesticide: *x* liters of brand 1 and *y* liters of brand 2. If the number of thousands of insects killed is given by
$$f(x, y) = 10,000 - 6500e^{-0.01x} - 3500e^{-0.02y}$$
how many insects would be killed if 80 liters of brand 1 and 120 liters of brand 2 were used?

OBJECTIVES

- To find partial derivatives of functions of two or more variables
- To evaluate partial derivatives of functions of two or more variables at given points
- To use partial derivatives to find slopes of tangents to surfaces
- To find and evaluate second- and higher-order partial derivatives of functions of two variables

14.2

Partial Differentiation

| APPLICATION PREVIEW |

Suppose that a company's sales are related to its television advertising by

$$s = 20{,}000 + 10nt + 20n^2$$

where n is the number of commercials per day and t is the length of the commercials in seconds. To find the instantaneous rate of change of sales with respect to the number of commercials per day if the company is running ten 30-second commercials, we find the partial derivative of s with respect to n. (See Example 7.)

In this section we find partial derivatives of functions of two or more variables and use these derivatives to find rates of change and slopes of tangents to surfaces. We will also find second- and higher-order partial derivatives of functions of two variables.

First-Order Partial Derivatives

If $z = f(x, y)$, we find the **partial derivative** of z with respect to x (denoted $\partial z/\partial x$) by treating the variable y as a constant and taking the derivative of $z = f(x, y)$ with respect to x. We can also take the partial derivative of z with respect to y by holding the variable x constant and taking the derivative of $z = f(x, y)$ with respect to y. We denote this derivative as $\partial z/\partial y$. Note that dz/dx represents the derivative of a function of one variable, x, and that $\partial z/\partial x$ represents the partial derivative of a function of two or more variables. Notations used to represent the partial derivative of $z = f(x, y)$ with respect to x are

$$\frac{\partial z}{\partial x}, \quad \frac{\partial f}{\partial x}, \quad \frac{\partial}{\partial x} f(x, y), \quad f_x(x, y), \quad f_x, \quad \text{and} \quad z_x$$

and notations used to represent the partial derivative of $z = f(x, y)$ with respect to y are

$$\frac{\partial z}{\partial y}, \quad \frac{\partial f}{\partial y}, \quad \frac{\partial}{\partial y} f(x, y), \quad f_y(x, y), \quad f_y, \quad \text{and} \quad z_y$$

EXAMPLE 1 Partial Derivatives

If $z = 4x^2 + 5x^2y^2 + 6y^3 - 7$, find $\partial z/\partial x$ and $\partial z/\partial y$.

Solution

To find $\partial z/\partial x$, we hold y constant so that the term $6y^3$ is constant, and its derivative is 0; the term $5x^2y^2$ is viewed as the constant $(5y^2)$ times (x^2), so its derivative is $(5y^2)(2x) = 10xy^2$. Thus

$$\frac{\partial z}{\partial x} = 8x + 10xy^2$$

Similarly for $\partial z/\partial y$, the term $4x^2$ is constant and the partial derivative of $5x^2y^2$ is the constant $5x^2$ times the derivative of y^2, so its derivative is $(5x^2)(2y) = 10x^2y$. Thus

$$\frac{\partial z}{\partial y} = 10x^2y + 18y^2$$

EXAMPLE 2 Marginal Cost

Suppose that the total cost of manufacturing a product is given by

$$C(x, y) = 5 + 5x + 2y$$

where x represents the cost of 1 ounce of materials and y represents the labor cost in dollars per hour.

(a) Find the rate at which the total cost changes with respect to the material cost x. This is the marginal cost of the product with respect to the material cost.
(b) Find the rate at which the total cost changes with respect to the labor cost y. This is the marginal cost of the product with respect to the labor cost.

Solution
(a) The rate at which the total cost changes with respect to the material cost x is the partial derivative found by treating the y-variable as a constant and taking the derivative of both sides of $C(x, y) = 5 + 5x + 2y$ with respect to x.

$$\frac{\partial C}{\partial x} = 0 + 5 + 0, \text{ or } \frac{\partial C}{\partial x} = 5$$

The partial derivative $\partial C/\partial x = 5$ tells us that a change of \$1 in the cost of materials will cause an increase of \$5 in total costs, *if* labor costs remain constant.
(b) The rate at which the total cost changes with respect to the labor cost y is the partial derivative found by treating the x-variable as a constant and taking the derivative of both sides of $C(x, y) = 5 + 5x + 2y$ with respect to y.

$$\frac{\partial C}{\partial y} = 0 + 0 + 2, \text{ or } \frac{\partial C}{\partial y} = 2$$

The partial derivative $\partial C/\partial y = 2$ tells us that a change of \$1 in the cost of labor will cause an increase of \$2 in total costs, *if* material costs remain constant. ■

EXAMPLE 3   Partial Derivatives

If $z = x^2y + e^x - \ln y$, find z_x and z_y.

Solution

$$z_x = \frac{\partial z}{\partial x} = 2xy + e^x$$

$$z_y = \frac{\partial z}{\partial y} = x^2 - \frac{1}{y}$$

■

EXAMPLE 4   Power Rule

If $f(x, y) = (x^2 - y^2)^2$, find the following.

(a) f_x
(b) f_y

Solution
(a) $f_x = 2(x^2 - y^2)2x = 4x^3 - 4xy^2$
(b) $f_y = 2(x^2 - y^2)(-2y) = -4x^2y + 4y^3$

■

✓ CHECKPOINT

1. If $z = 100x + 10xy - y^2$, find the following.
   (a) z_x
   (b) $\frac{\partial z}{\partial y}$

EXAMPLE 5   **Quotient Rule**

If $q = \dfrac{p_1 p_2 + 2p_1}{p_1 p_2 - 2p_2}$, find $\partial q / \partial p_1$.

Solution

$$
\begin{aligned}
\frac{\partial q}{\partial p_1} &= \frac{(p_1 p_2 - 2p_2)(p_2 + 2) - (p_1 p_2 + 2p_1)p_2}{(p_1 p_2 - 2p_2)^2} \\
&= \frac{p_1 p_2^2 + 2p_1 p_2 - 2p_2^2 - 4p_2 - p_1 p_2^2 - 2p_1 p_2}{(p_1 p_2 - 2p_2)^2} \\
&= \frac{-2p_2^2 - 4p_2}{p_2^2(p_1 - 2)^2} \\
&= \frac{-2p_2(p_2 + 2)}{p_2^2(p_1 - 2)^2} \\
&= \frac{-2(p_2 + 2)}{p_2(p_1 - 2)^2}
\end{aligned}
$$

We may evaluate partial derivatives by substituting values for x and y in the same way we did with derivatives of functions of one variable. For example, if $\partial z / \partial x = 2x - xy$, the value of the partial derivative with respect to x at $x = 2, y = 3$ is

$$
\frac{\partial z}{\partial x}\bigg|_{(2,\,3)} = 2(2) - 2(3) = -2
$$

Other notations used to denote evaluation of partial derivatives with respect to x at (a, b) are

$$
\frac{\partial}{\partial x} f(a, b) \text{ and } f_x(a, b)
$$

We denote the evaluation of partial derivatives with respect to y at (a, b) by

$$
\frac{\partial z}{\partial y}\bigg|_{(a,\,b)}, \ \frac{\partial}{\partial y} f(a, b), \text{ or } f_y(a, b)
$$

EXAMPLE 6   **Partial Derivative at a Point**

Find the partial derivative of $f(x, y) = x^2 + 3xy + 4$ with respect to x at the point $(1, 2, 11)$.

Solution

$$
\begin{aligned}
f_x(x, y) &= 2x + 3y \\
f_x(1, 2) &= 2(1) + 3(2) = 8
\end{aligned}
$$

✓ **CHECKPOINT**   2. If $g(x, y) = 4x^2 - 3xy + 10y^2$, find the following.

(a) $\dfrac{\partial g}{\partial x}(1, 3)$ (b) $g_y(4, 2)$

EXAMPLE 7   **Marginal Sales | APPLICATION PREVIEW |**

Suppose that a company's sales are related to its television advertising by

$$
s = 20{,}000 + 10nt + 20n^2
$$

where n is the number of commercials per day and t is the length of the commercials in seconds. Find the partial derivative of s with respect to n, and use the result to find the instantaneous rate of change of sales with respect to the number of commercials per day, if the company is currently running ten 30-second commercials per day.

Solution
The partial derivative of s with respect to n is $\partial s/\partial n = 10t + 40n$. At $n = 10$ and $t = 30$, the rate of change in sales is

$$\left.\frac{\partial s}{\partial n}\right|_{\substack{n=10 \\ t=30}} = 10(30) + 40(10) = 700$$

Thus, increasing the number of 30-second commercials by 1 would result in approximately 700 additional sales. This is the marginal sales with respect to the number of commercials per day at $n = 10$, $t = 30$. ◾

We have seen that the partial derivative $\partial z/\partial x$ is found by holding y constant and taking the derivative of z with respect to x and that the partial derivative $\partial z/\partial y$ is found by holding x constant and taking the derivative of z with respect to y. We now give formal definitions of these **partial derivatives.**

Partial Derivatives

The **partial derivative** of $z = f(x, y)$ with respect to x at the point (x, y) is

$$\frac{\partial z}{\partial x} = \frac{\partial}{\partial x}f(x, y) = \lim_{h \to 0}\frac{f(x + h, y) - f(x, y)}{h}$$

provided this limit exists.
The **partial derivative** of $z = f(x, y)$ with respect to y at the point (x, y) is

$$\frac{\partial z}{\partial y} = \frac{\partial}{\partial y}f(x, y) = \lim_{h \to 0}\frac{f(x, y + h) - f(x, y)}{h}$$

provided this limit exists.

We have already stated that the graph of $z = f(x, y)$ is a surface in three dimensions. The partial derivative with respect to x of such a function may be thought of as the slope of the tangent to the surface at a point (x, y, z) on the surface in the *positive direction of the x-axis.* That is, if a plane parallel to the xz-plane cuts the surface, passing through the point (x_0, y_0, z_0), the line in the plane that is tangent to the surface will have a slope equal to $\partial z/\partial x$ evaluated at the point. Thus

$$\left.\frac{\partial z}{\partial x}\right|_{(x_0, y_0)}$$

represents the slope of the tangent to the surface at (x_0, y_0, z_0), in the positive direction of the x-axis (see Figure 14.4 on the next page).

Similarly,

$$\left.\frac{\partial z}{\partial y}\right|_{(x_0, y_0)} = \frac{\partial}{\partial y}f(x_0, y_0)$$

represents the slope of the tangent to the surface at (x_0, y_0, z_0) in the positive direction of the y-axis (see Figure 14.5 on the next page).

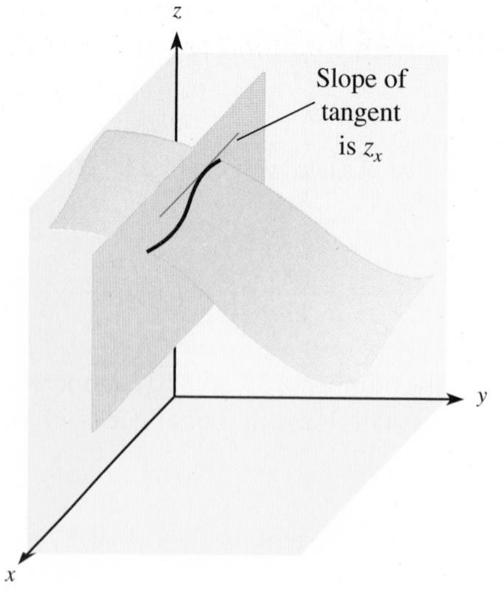

Figure 14.4

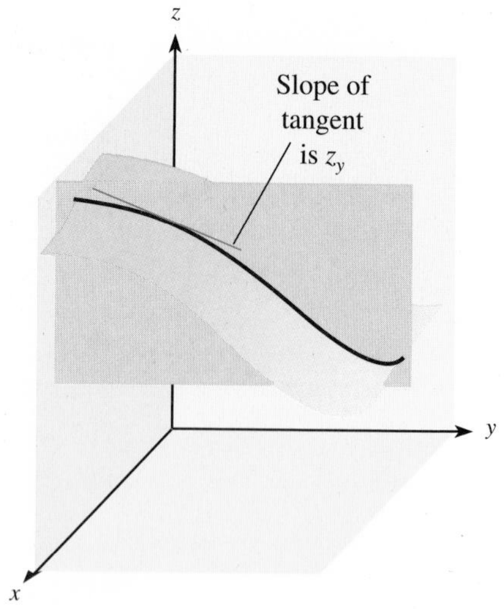

Figure 14.5

EXAMPLE 8 Slopes of Tangents

Let $z = 4x^3 - 4e^x + 4y^2$ and let P be the point $(0, 2, 12)$. Find the slope of the tangent to z at the point P in the positive direction of (a) the x-axis and (b) the y-axis.

Solution

(a) The slope of z at P in the positive x-direction is given by $\dfrac{\partial z}{\partial x}$, evaluated at P.

$$\frac{\partial z}{\partial x} = 12x^2 - 4e^x \text{ and } \frac{\partial z}{\partial x}\bigg|_{(0,\,2)} = 12(0)^2 - 4e^0 = -4$$

This tells us that z *decreases* approximately 4 units for an increase of 1 unit in x at this point.

(b) The slope of z at P in the positive y-direction is given by $\dfrac{\partial z}{\partial y}$, evaluated at P.

$$\frac{\partial z}{\partial y} = 8y \text{ and } \frac{\partial z}{\partial y}\bigg|_{(0,\,2)} = 8(2) = 16$$

Thus, at the point P $(0, 2, 12)$, the function *increases* approximately 16 units in the z-value for a unit increase in y. ▪

Up to this point, we have considered derivatives of functions of two variables. We can easily extend the concept to functions of three or more variables. We can find the partial derivative with respect to any one independent variable by taking the derivative of the function with respect to that variable while holding all other independent variables constant.

EXAMPLE 9 Functions of Four Variables

If $u = f(w, x, y, z) = 3x^2y + w^3 - 4xyz$, find the following.

(a) $\dfrac{\partial u}{\partial w}$ (b) $\dfrac{\partial u}{\partial x}$ (c) $\dfrac{\partial u}{\partial y}$ (d) $\dfrac{\partial u}{\partial z}$

Solution

(a) $\dfrac{\partial u}{\partial w} = 3w^2$ (b) $\dfrac{\partial u}{\partial x} = 6xy - 4yz$

(c) $\dfrac{\partial u}{\partial y} = 3x^2 - 4xz$ (d) $\dfrac{\partial u}{\partial z} = -4xy$ ▪

EXAMPLE 10   **Functions of Three Variables**

If $C = 4x_1 + 2x_1^2 + 3x_2 - x_1 x_2 + x_3^2$, find the following.

(a) $\dfrac{\partial C}{\partial x_1}$ (b) $\dfrac{\partial C}{\partial x_2}$ (c) $\dfrac{\partial C}{\partial x_3}$

Solution

(a) $\dfrac{\partial C}{\partial x_1} = 4 + 4x_1 - x_2$ (b) $\dfrac{\partial C}{\partial x_2} = 3 - x_1$ (c) $\dfrac{\partial C}{\partial x_3} = 2x_3$

✓ **CHECKPOINT**   3. If $f(w, x, y, z) = 8xy^2 + 4yz - xw^2$, find

(a) $\dfrac{\partial f}{\partial x}$ (b) $\dfrac{\partial f}{\partial w}$ (c) $\dfrac{\partial f}{\partial y}(1, 2, 1, 3)$ (d) $\dfrac{\partial f}{\partial z}(0, 2, 1, 3)$

High-Order Partial Derivatives

Just as we have taken derivatives of derivatives to obtain higher-order derivatives of functions of one variable, we may also take partial derivatives of partial derivatives to obtain higher-order partial derivatives of a function of more than one variable. If $z = f(x, y)$, then the partial derivative functions z_x and z_y are called *first partials*. Partial derivatives of z_x and z_y are called *second partials*, so $z = f(x, y)$ has *four* **second partial derivatives.** The notations for these second partial derivatives follow.

Second Partial Derivatives

$z_{xx} = \dfrac{\partial^2 z}{\partial x^2} = \dfrac{\partial}{\partial x}\left(\dfrac{\partial z}{\partial x}\right)$:   both derivatives taken with respect to x

$z_{yy} = \dfrac{\partial^2 z}{\partial y^2} = \dfrac{\partial}{\partial y}\left(\dfrac{\partial z}{\partial y}\right)$:   both derivatives taken with respect to y

$z_{xy} = \dfrac{\partial^2 z}{\partial y\, \partial x} = \dfrac{\partial}{\partial y}\left(\dfrac{\partial z}{\partial x}\right)$:   first derivative taken with respect to x, second with respect to y

$z_{yx} = \dfrac{\partial^2 z}{\partial x\, \partial y} = \dfrac{\partial}{\partial x}\left(\dfrac{\partial z}{\partial y}\right)$:   first derivative taken with respect to y, second with respect to x

EXAMPLE 11   **Second Partial Derivatives**

If $z = x^3 y - 3xy^2 + 4$, find each of the second partial derivatives of the function.

Solution

Because $z_x = 3x^2 y - 3y^2$ and $z_y = x^3 - 6xy$,

$$z_{xx} = \frac{\partial}{\partial x}(3x^2 y - 3y^2) = 6xy$$

$$z_{xy} = \frac{\partial}{\partial y}(3x^2 y - 3y^2) = 3x^2 - 6y$$

$$z_{yy} = \frac{\partial}{\partial y}(x^3 - 6xy) = -6x$$

$$z_{yx} = \frac{\partial}{\partial x}(x^3 - 6xy) = 3x^2 - 6y$$

Note that z_{xy} and z_{yx} are equal for the function in Example 11. This will always occur if the derivatives of the function are continuous.

| $z_{xy} = z_{yx}$ | If the second partial derivatives z_{xy} and z_{yx} of a function $z = f(x, y)$ are continuous at a point, they are equal there. |

EXAMPLE 12 Second Partial Derivatives

Find each of the second partial derivatives of $z = x^2y + e^{xy}$.

Solution

Because $z_x = 2xy + e^{xy} \cdot y = 2xy + ye^{xy}$,

$$z_{xx} = 2y + e^{xy} \cdot y^2 = 2y + y^2e^{xy}$$
$$z_{xy} = 2x + (e^{xy} \cdot 1 + ye^{xy} \cdot x)$$
$$= 2x + e^{xy} + xye^{xy}$$

Because $z_y = x^2 + e^{xy} \cdot x = x^2 + xe^{xy}$,

$$z_{yx} = 2x + (e^{xy} \cdot 1 + xe^{xy} \cdot y)$$
$$= 2x + e^{xy} + xye^{xy}$$
$$z_{yy} = 0 + xe^{xy} \cdot x = x^2e^{xy}$$

✓ **CHECKPOINT**

4. If $z = 4x^3y^4 + 4xy$, find the following.
   (a) z_{xx} (b) z_{yy} (c) z_{xy} (d) z_{yx}

5. If $z = x^2 + 4e^{xy}$, find z_{xy}.

We can find partial derivatives of orders higher than the second. For example, we can find the third-order partial derivatives z_{xyx} and z_{xyy} for the function in Example 11 from the second derivative $z_{xy} = 3x^2 - 6y$.

$$z_{xyx} = 6x$$
$$z_{xyy} = -6$$

EXAMPLE 13 Third Partial Derivatives

If $y = 4y \ln x + e^{xy}$, find z_{xyy}.

Solution

$$z_x = 4y \cdot \frac{1}{x} + e^{xy} \cdot y$$

$$z_{xy} = 4 \cdot \frac{1}{x} \cdot 1 + e^{xy} \cdot 1 + y \cdot e^{xy} \cdot x = \frac{4}{x} + e^{xy} + xye^{xy}$$

$$z_{xyy} = 0 + e^{xy} \cdot x + xy \cdot e^{xy} \cdot x + e^{xy} \cdot x = x^2ye^{xy} + 2xe^{xy}$$

✓ **CHECKPOINT ANSWERS**

1. (a) $z_x = 100 + 10y$
   (b) $\frac{\partial z}{\partial y} = 10x - 2y$

2. (a) $\frac{\partial g}{\partial x} = 8x - 3y$ and $\frac{\partial g}{\partial x}(1, 3) = 8(1) - 3(3) = -1$
   (b) $g_y = -3x + 20y$ and $g_y(4, 2) = -3(4) + 20(2) = 28$

3. (a) $\frac{\partial f}{\partial x} = 8y^2 - w^2$ (b) $\frac{\partial f}{\partial w} = -2xw$
   (c) $\frac{\partial f}{\partial y} = 16xy + 4z$ and $\frac{\partial f}{\partial y}(1, 2, 1, 3) = 44$
   (d) $\frac{\partial f}{\partial z} = 4y$ and $\frac{\partial f}{\partial z}(0, 2, 1, 3) = 4$

4. $z_x = 12x^2y^4 + 4y$ and $z_y = 16x^3y^3 + 4x$
   (a) $z_{xx} = 24xy^4$ (b) $z_{yy} = 48x^3y^2$
   (c) $z_{xy} = 48x^2y^3 + 4$ (d) $z_{yx} = 48x^2y^3 + 4$
5. $z_x = 2x + 4e^{xy}(y) = 2x + 4ye^{xy}$
   $z_{xy} = 4xye^{xy} + 4e^{xy}$

| EXERCISES | 14.2

1. If $z = x^4 - 5x^2 + 6x + 3y^3 - 5y + 7$, find $\dfrac{\partial z}{\partial x}$ and $\dfrac{\partial z}{\partial y}$.

2. If $z = x^5 - 6x + 4y^4 - y^2$, find $\dfrac{\partial z}{\partial x}$ and $\dfrac{\partial z}{\partial y}$.

3. If $z = x^3 + 4x^2y + 6y^2$, find z_x and z_y.

4. If $z = 3xy + y^2$, find z_x and z_y.

5. If $f(x, y) = (x^3 + 2y^2)^3$, find $\dfrac{\partial f}{\partial x}$ and $\dfrac{\partial f}{\partial y}$.

6. If $f(x, y) = (xy^3 + y)^2$, find $\dfrac{\partial f}{\partial x}$ and $\dfrac{\partial f}{\partial y}$.

7. If $f(x, y) = \sqrt{2x^2 - 5y^2}$, find f_x and f_y.

8. If $g(x, y) = \sqrt{xy - x}$, find g_x and g_y.

9. If $C(x, y) = 600 - 4xy + 10x^2y$, find $\dfrac{\partial C}{\partial x}$ and $\dfrac{\partial C}{\partial y}$.

10. If $C(x, y) = 1000 - 4x + xy^2$, find $\dfrac{\partial C}{\partial x}$ and $\dfrac{\partial C}{\partial y}$.

11. If $Q(s, t) = \dfrac{2s - 3t}{s^2 + t^2}$, find $\dfrac{\partial Q}{\partial s}$ and $\dfrac{\partial Q}{\partial t}$.

12. If $q = \dfrac{5p_1 + 4p_2}{p_1 + p_2}$, find $\dfrac{\partial q}{\partial p_1}$ and $\dfrac{\partial q}{\partial p_2}$.

13. If $z = e^{2x} + y \ln x$, find z_x and z_y.

14. If $z = \ln(1 + x^2y) - ye^{-x}$, find z_x and z_y.

15. If $f(x, y) = \ln(xy + 1)$, find $\dfrac{\partial f}{\partial x}$ and $\dfrac{\partial f}{\partial y}$.

16. If $f(x, y) = 100e^{xy}$, find $\dfrac{\partial f}{\partial x}$ and $\dfrac{\partial f}{\partial y}$.

17. Find the partial derivative of
$$f(x, y) = 4x^3 - 5xy + y^2$$
with respect to x at the point $(1, 2, -2)$.

18. Find the partial derivative of
$$f(x, y) = 3x^2 + 4x + 6xy$$
with respect to y at $x = 2$, $y = -1$.

19. Find the slope of the tangent in the positive x-direction to the surface $z = 5x^3 - 4xy$ at the point $(1, 2, -3)$.

20. Find the slope of the tangent in the positive y-direction to the surface $z = x^3 - 5xy$ at $(2, 1, -2)$.

21. Find the slope of the tangent in the positive y-direction to the surface $z = 3x + 2y - 7e^{xy}$ at $(3, 0, 2)$.

22. Find the slope of the tangent in the positive x-direction to the surface $z = 2xy + \ln(4x + 3y)$ at $(1, -1, -2)$.

23. If $u = f(w, x, y, z) = y^2 - x^2z + 4x$, find the following.
   (a) $\dfrac{\partial u}{\partial w}$ (b) $\dfrac{\partial u}{\partial x}$ (c) $\dfrac{\partial u}{\partial y}$ (d) $\dfrac{\partial u}{\partial z}$

24. If $u = x^2 + 3xy + xz$, find the following.
   (a) u_x (b) u_y (c) u_z

25. If $C(x_1, x_2, x_3) = 4x_1^2 + 5x_1x_2 + 6x_2^2 + x_3$, find the following.
   (a) $\dfrac{\partial C}{\partial x_1}$ (b) $\dfrac{\partial C}{\partial x_2}$ (c) $\dfrac{\partial C}{\partial x_3}$

26. If $f(x, y, z) = 2x\sqrt{yz - 1} + x^2z^3$, find the following.
   (a) $\dfrac{\partial f}{\partial x}$ (b) $\dfrac{\partial f}{\partial y}$ (c) $\dfrac{\partial f}{\partial z}$

27. If $z = x^2 + 4x - 5y^3$, find the following.
   (a) z_{xx} (b) z_{xy} (c) z_{yx} (d) z_{yy}

28. If $z = x^3 - 5y^2 + 4y + 1$, find the following.
   (a) z_{xx} (b) z_{xy} (c) z_{yx} (d) z_{yy}

29. If $z = x^2y - 4xy^2$, find the following.
   (a) z_{xx} (b) z_{xy} (c) z_{yx} (d) z_{yy}

30. If $z = xy^2 + 4xy - 5$, find the following.
   (a) z_{xx} (b) z_{xy} (c) z_{yx} (d) z_{yy}

31. If $f(x, y) = x^2 + e^{xy}$, find the following.
   (a) $\dfrac{\partial^2 f}{\partial x^2}$ (b) $\dfrac{\partial^2 f}{\partial y\, \partial x}$ (c) $\dfrac{\partial^2 f}{\partial x\, \partial y}$ (d) $\dfrac{\partial^2 f}{\partial y^2}$

32. If $z = xe^{xy}$, find the following.
   (a) z_{xx} (b) z_{yy} (c) z_{xy} (d) z_{yx}

33. If $f(x, y) = y^2 - \ln xy$, find the following.
   (a) $\dfrac{\partial^2 f}{\partial x^2}$ (b) $\dfrac{\partial^2 f}{\partial y\, \partial x}$ (c) $\dfrac{\partial^2 f}{\partial x\, \partial y}$ (d) $\dfrac{\partial^2 f}{\partial y^2}$

34. If $f(x, y) = x^3 + \ln(xy - 1)$, find the following.
   (a) $\dfrac{\partial^2 f}{\partial x^2}$ (b) $\dfrac{\partial^2 f}{\partial y^2}$ (c) $\dfrac{\partial^2 f}{\partial x\, \partial y}$ (d) $\dfrac{\partial^2 f}{\partial y\, \partial x}$

35. If $f(x, y) = x^3y + 4xy^4$, find $\left. \dfrac{\partial^2}{\partial x^2} f(x, y) \right|_{(1, -1)}$.

36. If $f(x, y) = x^4y^2 + 4xy$, find $\left. \dfrac{\partial^2}{\partial y^2} f(x, y) \right|_{(1, 2)}$.

37. If $f(x, y) = \dfrac{2x}{x^2 + y^2}$, find the following.
   (a) $\left. \dfrac{\partial^2 f}{\partial x^2} \right|_{(-1, 4)}$ (b) $\left. \dfrac{\partial^2 f}{\partial y^2} \right|_{(-1, 4)}$

38. If $f(x, y) = \dfrac{2y^2}{3xy + 4}$, find the following.
   (a) $\left. \dfrac{\partial^2 f}{\partial x^2} \right|_{(1, -2)}$ (b) $\left. \dfrac{\partial^2 f}{\partial y^2} \right|_{(1, -2)}$

39. If $z = x^2y + ye^{x^2}$, find $z_{yx}|_{(1, 2)}$.

40. If $z = xy^3 + x \ln y^2$, find $z_{xy}|_{(1, 2)}$.

41. If $z = x^2 - xy + 4y^3$, find z_{xyx}.

42. If $z = x^3 - 4x^2y + 5y^3$, find z_{yyx}.

43. If $w = 4x^3y + y^2z + z^3$, find the following.
   (a) w_{xxy} (b) w_{xyx} (c) w_{xyz}

44. If $w = 4xyz + x^3y^2z + x^3$, find the following.
   (a) w_{xyz} (b) w_{xzz} (c) w_{yyz}

APPLICATIONS

45. **Mortgage** When a homeowner has a 25-year variable-rate mortgage loan, the monthly payment R is a function of the amount of the loan A and the current interest rate i (as a percent); that is, $R = f(A, i)$. Interpret each of the following.
   (a) $f(100{,}000, 8) = 771.82$

   (b) $\dfrac{\partial f}{\partial i}(100{,}000, 8) = 66.25$

46. **Mass transportation ridership** Suppose that in a certain city, the number of people N using the mass transportation system is a function of the fare f and the daily cost of downtown parking p, so that $N = N(f, p)$. Interpret each of the following.
   (a) $N(5, 20) = 65{,}000$ (b) $\dfrac{\partial N}{\partial f}(5, 20) = -4000$

   (c) $\dfrac{\partial N}{\partial p}(5, 20) = 2500$

47. **Wilson's lot size formula** In economics, the most economical quantity Q of goods (TVs, dresses, gallons of paint, etc.) for a store to order is given by Wilson's lot size formula

   $$Q = \sqrt{2KM/h}$$

   where K is the cost of placing the order, M is the number of items sold per week, and h is the weekly holding costs for each item (the cost of storage space, utilities, taxes, security, etc.).
   (a) Explain why $\dfrac{\partial Q}{\partial M}$ will be positive.

   (b) Explain why $\dfrac{\partial Q}{\partial h}$ will be negative.

48. **Cost** Suppose that the total cost (in dollars) of producing a product is $C(x, y) = 25 + 2x^2 + 3y^2$, where x is the cost per pound for material and y is the cost per hour for labor.
   (a) If material costs are held constant, at what rate will the total cost increase for each $1-per-hour increase in labor?
   (b) If the labor costs are held constant, at what rate will the total cost increase for each increase of $1 in material cost?

49. **Pesticide** Suppose that the number of thousands of insects killed by two brands of pesticide is given by

   $$f(x, y) = 10{,}000 - 6500e^{-0.01x} - 3500e^{-0.02y}$$

where x is the number of liters of brand 1 and y is the number of liters of brand 2. What is the rate of change of insect deaths with respect to the number of liters of brand 1 if 100 liters of each brand are currently being used? What does this mean?

50. **Profit** Suppose that the profit (in dollars) from the sale of Kisses and Kreams is given by

   $$P(x, y) = 10x + 6.4y - 0.001x^2 - 0.025y^2$$

   where x is the number of pounds of Kisses and y is the number of pounds of Kreams. Find $\partial P/\partial y$, and give the approximate rate of change of profit with respect to the number of pounds of Kreams that are sold if 100 pounds of Kisses and 16 pounds of Kreams are currently being sold. What does this mean?

51. **Utility** If $U = f(x, y)$ is the utility function for goods X and Y, the *marginal utility* of X is $\partial U/\partial x$ and the *marginal utility* of Y is $\partial U/\partial y$. If $U = x^2y^2$, find the marginal utility of
   (a) X.   (b) Y.

52. **Utility** If the utility function for goods X and Y is $U = xy + y^2$, find the marginal utility of
   (a) X.   (b) Y.

53. **Production** Suppose that the output Q (in units) of a certain company is $Q = 75K^{1/3}L^{2/3}$, where K is the capital expenditures in thousands of dollars and L is the number of labor hours. Find $\partial Q/\partial K$ and $\partial Q/\partial L$ when capital expenditures are $729,000 and the labor hours total 5832. Interpret each answer.

54. **Production** Suppose that the production Q (in gallons of paint) of a paint manufacturer can be modeled by $Q = 140K^{1/2}L^{1/2}$, where K is the company's capital expenditures in thousands of dollars and L is the size of the labor force (in hours worked). Find $\partial Q/\partial K$ and $\partial Q/\partial L$ when capital expenditures are $250,000 and the labor hours are 1225. Interpret each answer.

Wind chill Dr. Paul Siple conducted studies testing the effect of wind on the formation of ice at various temperatures and developed the concept of *wind chill*, which we hear reported during winter weather reports. In 2001, the National Weather Service introduced the new wind chill index given in the table. This new index is more accurate at measuring how cold it feels when it is windy. For example, the table shows that an air temperature of 15°F together with a wind speed of 35 mph feels the same as an air temperature of −7°F when there is no wind.

	Air Temperature (°F)						
Wind Speed (mph)	35	25	15	5	−5	−15	−25
5	31	19	7	−5	−16	−28	−40
15	25	13	0	−13	−26	−39	−51
25	23	9	−4	−17	−31	−44	−58
35	21	7	−7	−21	−34	−48	−62
45	19	5	−9	−23	−37	−51	−65

Source: National Weather Service

The wind chill WC has been modeled by

$$WC = 35.74 + 0.6215t - 35.75s^{0.16} + 0.4275ts^{0.16}$$

where s is the wind speed and t is the actual air temperature. Use this model to answer Problems 55 and 56.

55. (a) To see how the wind chill temperature changes with wind speed, find $\partial WC/\partial s$.

(b) Find $\partial WC/\partial s$ when the temperature is 10°F and the wind speed is 25 mph. What does this mean?

56. (a) To see how wind chill temperature changes with temperature, find $\partial WC/\partial t$.

(b) Find $\partial WC/\partial t$ when the temperature is 10°F and the wind speed is 25 mph. What does this mean?

OBJECTIVES

- To evaluate cost functions at given levels of production
- To find marginal costs from total cost and joint cost functions
- To find marginal productivity for given production functions
- To find marginal demand functions from demand functions for a pair of related products

14.3

Applications of Functions of Two Variables in Business and Economics

| APPLICATION PREVIEW |

Suppose that the joint cost function for two commodities is

$$C = 50 + x^2 + 8xy + y^3$$

where x and y represent the quantities of each commodity and C is the total cost for the two commodities. We can take the partial derivative of C with respect to x to find the marginal cost of the first product and we can take the partial derivative of C with respect to y to find the marginal cost of the second product. (See Example 1.) This is one of three types of applications that we will consider in this section. In the second case, we consider production functions and revisit Cobb-Douglas production functions. Marginal productivity is introduced and its meaning is explained. Finally, we consider demand functions for two products in a competitive market. Partial derivatives are used to define marginal demands, and these marginals are used to classify the products as competitive or complementary.

Joint Cost and Marginal Cost

Suppose that a firm produces two products using the same inputs in different proportions. In such a case the **joint cost function** is of the form $C = Q(x, y)$, where x represents the quantity of product X and y represents the quantity of product Y. Then $\partial C/\partial x$ is the **marginal cost** of the first product and $\partial C/\partial y$ is the marginal cost of the second product.

EXAMPLE 1   Joint Cost | APPLICATION PREVIEW |

If the joint cost function for two products is

$$C = Q(x, y) = 50 + x^2 + 8xy + y^3$$

where x represents the quantity of product X and y represents the quantity of product Y, find the marginal cost with respect to the number of units of:

(a) Product X (b) Product Y (c) Product X at (5, 3)   (d) Product Y at (5, 3)

Solution

(a) The marginal cost with respect to the number of units of product X is $\dfrac{\partial C}{\partial x} = 2x + 8y$.

(b) The marginal cost with respect to the number of units of product Y is $\dfrac{\partial C}{\partial y} = 8x + 3y^2$.

(c) $\dfrac{\partial C}{\partial x}\bigg|_{(5, 3)} = 2(5) + 8(3) = 34$

Thus if 5 units of product X and 3 units of product Y are produced, the total cost will increase approximately \$34 for a unit increase in product X if y is held constant.

(d) $\left.\dfrac{\partial C}{\partial y}\right|_{(5,\,3)} = 8(5) + 3(3)^2 = 67$

Thus if 5 units of product X and 3 units of product Y are produced, the total cost will increase approximately \$67 for a unit increase in product Y if x is held constant. ∎

✓ **CHECKPOINT**

1. If the joint cost in dollars for two products is given by

$$C = 100 + 3x + 10xy + y^2$$

find the marginal cost with respect to (a) the number of units of Product X and (b) the number of units of Product Y at $(7, 3)$.

Production Functions An important problem in economics concerns how the factors necessary for production determine the output of a product. For example, the output of a product depends on available labor, land, capital, material, and machines. If the amount of output z of a product depends on the amounts of two inputs x and y, then the quantity z is given by the **production function** $z = f(x, y)$.

EXAMPLE 2   Crop Harvesting

Suppose that it is known that z bushels of a crop can be harvested according to the function

$$z = \frac{21(6xy + 4x^2 - 3y)}{2x + 0.01y}$$

when $100x$ work-hours of labor are employed on y acres of land. What would be the output (in bushels) if 200 work-hours were used on 300 acres?

Solution
Because $z = f(x, y)$,

$$f(2, 300) = \frac{(21)[6(2)(300) + 4(2)^2 - 3(300)]}{2(2) + 3}$$

$$= \frac{(21)[3600 + 16 - 900]}{7} = 8148 \quad \text{(bushels)} \quad ∎$$

If $z = f(x, y)$ is a production function, $\partial z/\partial x$ represents the rate of change in the output z with respect to input x while input y remains constant. This partial derivative is called the **marginal productivity of x**. The partial derivative $\partial z/\partial y$ is the **marginal productivity of y** and measures the rate of change of z with respect to input y.

Marginal productivity (for either input) will be positive over a wide range of inputs, but it increases at a decreasing rate, and it may eventually reach a point where it no longer increases and begins to decrease.

EXAMPLE 3   Production

If a production function is given by $z = 5x^{1/2}y^{1/4}$, find the marginal productivity of

(a) x.   (b) y.

Solution

(a) $\dfrac{\partial z}{\partial x} = \dfrac{5}{2}x^{-1/2}y^{1/4}$ (b) $\dfrac{\partial z}{\partial y} = \dfrac{5}{4}x^{1/2}y^{-3/4}$

Note that the marginal productivity of x is positive for all values of x but that it decreases as x gets larger (because of the negative exponent). The same is true for the marginal productivity of y. ∎

✓ CHECKPOINT

2. If the production function for a product is

$$P = 10x^{1/4}y^{1/2}$$

find the marginal productivity of x.

Calculator Note

If we have a production function and fix a value for one variable, then we can use a graphing calculator to analyze the marginal productivity with respect to the other variable. Example 4 illustrates the use of a graphing calculator to analyze the marginal productivity of x and of y. ■

EXAMPLE 4   Production

Suppose the Cobb-Douglas production function for a company is given by

$$z = 100x^{1/4}y^{3/4}$$

where x is the company's capital investment (in dollars) and y is the size of the labor force (in work-hours).

(a) Find the marginal productivity of x.
(b) If the current labor force is 625 work-hours, substitute $y = 625$ in your answer to part (a) and graph the result.
(c) From the graph in part (b), what can be said about the effect on production of additional capital investment when the work-hours remain at 625?
(d) Find the marginal productivity of y.
(e) If current capital investment is $10,000, substitute $x = 10,000$ in your answer to part (d) and graph the result.
(f) From the graph in part (e), what can be said about the effect on production of additional work-hours when capital investment remains at $10,000?

Solution

(a) $z_x = 25x^{-3/4}y^{3/4}$
(b) If $y = 625$, then z_x becomes

$$z_x = 25x^{-3/4}(625)^{3/4} = 25\left(\frac{1}{x^{3/4}}\right)(125) = \frac{3125}{x^{3/4}}$$

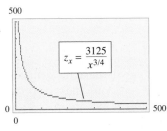

Figure 14.6

The graph of z_x can be limited to Quadrant I because the capital investment is $x > 0$, and hence $z_x > 0$. Knowledge of asymptotes can help us determine range values for x and z_x that give an accurate graph. See Figure 14.6.

(c) Figure 14.6 shows that $z_x > 0$ for $x > 0$. This means that any increases in capital investment will result in increases in productivity. However, Figure 14.6 also shows that z_x is decreasing for $x > 0$, which means that as capital investment increases, productivity increases, but at a slower rate.

(d) $z_y = 75x^{1/4}y^{-1/4}$
(e) If $x = 10,000$, then z_y becomes

$$z_y = 75(10,000)^{1/4}\left(\frac{1}{y^{1/4}}\right) = \frac{750}{y^{1/4}}$$

The graph is shown in Figure 14.7(a) on the next page.

(f) Figure 14.7(a) also shows that $z_y > 0$ when $y > 0$, so increasing work-hours increases productivity. Note that z_y is decreasing for $y > 0$, but Figure 14.7(b) shows that it does so more slowly than z_x. This indicates that increases in work-hours have a diminishing impact on productivity, but still a more significant one than increases in capital expenditures. ■

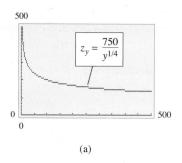

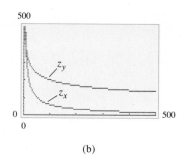

Figure 14.7   (a)   (b)

Demand Functions   Suppose that two products are sold at prices p_1 and p_2 (both in dollars) in a competitive market consisting of a fixed number of consumers with given tastes and incomes. Then the amount of each *one* of the products demanded by the consumers is dependent on the prices of *both* products on the market. If q_1 represents the demand for the number of units of the first product, then $q_1 = f(p_1, p_2)$ is the **demand function** for that product. The graph of such a function is called a **demand surface.** An example of a demand function in two variables is $q_1 = 400 - 2p_1 - 4p_2$. Here q_1 is a function of two variables p_1 and p_2. If $p_1 = \$10$ and $p_2 = \$20$, the demand would equal $400 - 2(10) - 4(20) = 300$ units.

EXAMPLE 5   Demand

The demand functions for two products are

$$q_1 = 50 - 5p_1 - 2p_2$$
$$q_2 = 100 - 3p_1 - 8p_2$$

where q_1 and q_2 are the numbers of units and p_1 and p_2 are in dollars.

(a)   What is the demand for each of the products if the price of the first is $p_1 = \$5$ and the price of the second is $p_2 = \$8$?

(b)   Find a pair of prices p_1 and p_2 such that the demands for product 1 and product 2 are equal.

Solution

(a)
$$q_1 = 50 - 5(5) - 2(8) = 9$$
$$q_2 = 100 - 3(5) - 8(8) = 21$$

Thus if these are the prices, the demand for product 2 is higher than the demand for product 1.

(b)   We want q_1 to equal q_2. Setting $q_1 = q_2$, we see that

$$50 - 5p_1 - 2p_2 = 100 - 3p_1 - 8p_2$$
$$6p_2 - 50 = 2p_1$$
$$p_1 = 3p_2 - 25$$

Now, any pair of positive values that satisfies this equation will make the demands equal. Letting $p_2 = 10$, we see that $p_1 = 5$ will satisfy the equation. Thus the prices $p_1 = 5$ and $p_2 = 10$ will make the demands equal. The prices $p_1 = 2$ and $p_2 = 9$ will also make the demands equal. Many pairs of values (that is, all those satisfying $p_1 = 3p_2 - 25$) will equalize the demands.   ■

If the demand functions for a pair of related products, product 1 and product 2, are $q_1 = f(p_1, p_2)$ and $q_2 = g(p_1, p_2)$, respectively, then the partial derivatives of q_1 and q_2 are called **marginal demand functions.**

$\dfrac{\partial q_1}{\partial p_1}$ is the marginal demand of q_1 with respect to p_1.

$\dfrac{\partial q_1}{\partial p_2}$ is the marginal demand of q_1 with respect to p_2.

$\dfrac{\partial q_2}{\partial p_1}$ is the marginal demand of q_2 with respect to p_1.

$\dfrac{\partial q_2}{\partial p_2}$ is the marginal demand of q_2 with respect to p_2.

For typical demand functions, if the price of product 2 is fixed, the demand for product 1 will decrease as its price p_1 increases. In this case the marginal demand of q_1 with respect to p_1 will be negative; that is, $\partial q_1/\partial p_1 < 0$. Similarly, $\partial q_2/\partial p_2 < 0$.

But what about $\partial q_2/\partial p_1$ and $\partial q_1/\partial p_2$? If $\partial q_2/\partial p_1$ and $\partial q_1/\partial p_2$ are both positive, the two products are **competitive** because an increase in price p_1 will result in an increase in demand for product 2 (q_2) if the price p_2 is held constant and an increase in price p_2 will increase the demand for product 1 (q_1) if p_1 is held constant. Stated more simply, an increase in the price of one of the two products will result in an increased demand for the other, so the products are in competition. For example, an increase in the price of a Japanese automobile will result in an increase in demand for an American automobile if the price of the American automobile is held constant.

If $\partial q_2/\partial p_1$ and $\partial q_1/\partial p_2$ are both negative, the products are **complementary** because an increase in the price of one product will cause a decrease in demand for the other product if the price of the second product doesn't change. Under these conditions, a *decrease* in the price of product 1 will result in an *increase* in the demand for product 2, and a decrease in the price of product 2 will result in an increase in the demand for product 1. For example, a decrease in the price of gasoline will result in an increase in the demand for large automobiles.

If the signs of $\partial q_2/\partial p_1$ and $\partial q_1/\partial p_2$ are different, the products are neither competitive nor complementary. This situation rarely occurs but is possible.

EXAMPLE 6 Demand

The demand functions for two related products, product 1 and product 2, are given by

$$q_1 = 400 - 5p_1 + 6p_2 \quad \text{and} \quad q_2 = 250 + 4p_1 - 5p_2$$

(a) Determine the four marginal demands.
(b) Are product 1 and product 2 complementary or competitive?

Solution

(a) $\dfrac{\partial q_1}{\partial p_1} = -5; \quad \dfrac{\partial q_2}{\partial p_2} = -5; \quad \dfrac{\partial q_1}{\partial p_2} = 6; \quad \dfrac{\partial q_2}{\partial p_1} = 4$

(b) Because $\partial q_1/\partial p_2$ and $\partial q_2/\partial p_1$ are positive, products 1 and 2 are competitive.

✓ CHECKPOINT   3. If the demand functions for two products are

$$q_1 = 200 - 3p_1 - 4p_2 \quad \text{and} \quad q_2 = 50 - 6p_1 - 5p_2$$

find the marginal demand of
(a) q_1 with respect to p_1.   (b) q_2 with respect to p_2.

✓ CHECKPOINT
ANSWERS

1. (a) $\dfrac{\partial C}{\partial x} = 3 + 10y$

(b) $\dfrac{\partial C}{\partial y} = 10x + 2y \qquad \dfrac{\partial C}{\partial y}(7, 3) = 76$ (dollars per unit)

2. $\dfrac{\partial P}{\partial x} = \dfrac{2.5y^{1/2}}{x^{3/4}}$

3. (a) $\dfrac{\partial q_1}{\partial p_1} = -3$ (b) $\dfrac{\partial q_2}{\partial p_2} = -5$

| EXERCISES | 14.3

JOINT COST AND MARGINAL COST

1. The cost (in dollars) of manufacturing one item is given by

$$C(x, y) = 30 + 3x + 5y$$

where x is the cost of 1 hour of labor and y is the cost of 1 pound of material.
  (a) If the hourly cost of labor is $20, and the material costs $3 per pound, what is the cost of manufacturing one of these items?
  (b) Find and interpret the partial derivative of C with respect to x.

2. The manufacture of 1 unit of a product has a cost (in dollars) given by

$$C(x, y, z) = 10 + 8x + 3y + z$$

where x is the cost of 1 pound of one raw material, y is the cost of 1 pound of a second raw material, and z is the cost of 1 work-hour of labor.
  (a) If the cost of the first raw material is $16 per pound, the cost of the second raw material is $8 per pound, and labor costs $18 per work-hour, what will it cost to produce 1 unit of the product?
  (b) Find and interpret the partial derivative of C with respect to x.

3. The total cost of producing 1 unit of a product is

$$C(x, y) = 30 + 2x + 4y + \frac{xy}{50} \quad \text{dollars}$$

where x is the cost per pound of raw materials and y is the cost per hour of labor.
  (a) If labor costs are held constant, at what rate will the total cost increase for each increase of $1 per pound in material cost?
  (b) If material costs are held constant, at what rate will the total cost increase for each $1 per hour increase in labor costs?

4. The total cost of producing an item is

$$C(x, y) = 40 + 4x + 6y + \frac{x^2 y}{100} \quad \text{dollars}$$

where x is the cost per pound of raw materials and y is the cost per hour for labor. How will an increase of
  (a) $1 per pound of raw materials affect the total cost?
  (b) $1 per hour in labor costs affect the total cost?

5. The total cost of producing 1 unit of a product is given by

$$C(x, y) = 20x + 70y + \frac{x^2}{1000} + \frac{xy^2}{100} \quad \text{dollars}$$

where x represents the cost per pound of raw materials and y represents the hourly rate for labor. The present cost for raw materials is $10 per pound and the present hourly rate for labor is $24. How will an increase of
  (a) $1 per pound for raw materials affect the total cost?
  (b) $1 per hour in labor costs affect the total cost?

6. The total cost of producing 1 unit of a product is given by

$$C(x, y) = 30 + 0.5x^2 + 30y - xy \quad \text{dollars}$$

where x is the hourly labor rate and y is the cost per pound of raw materials. The current hourly rate is $25, and the raw materials cost $6 per pound. How will an increase of
  (a) $1 per pound for the raw materials affect the total cost?
  (b) $1 in the hourly labor rate affect the total cost?

7. The joint cost (in dollars) for two products is given by

$$C(x, y) = 30 + x^2 + 3y + 2xy$$

where x represents the quantity of product X produced and y represents the quantity of product Y produced.
  (a) Find and interpret the marginal cost with respect to x if 8 units of product X and 10 units of product Y are produced.
  (b) Find and interpret the marginal cost with respect to y if 8 units of product X and 10 units of product Y are produced.

8. The joint cost (in dollars) for products X and Y is given by

$$C(x, y) = 40 + 3x^2 + y^2 + xy$$

where x represents the quantity of X and y represents the quantity of Y.
  (a) Find and interpret the marginal cost with respect to x if 20 units of product X and 15 units of product Y are produced.
  (b) Find and interpret the marginal cost with respect to y if 20 units of X and 15 units of Y are produced.

9. If the joint cost function for two products is

$$C(x, y) = x\sqrt{y^2 + 1} \quad \text{dollars}$$

  (a) find the marginal cost (function) with respect to x.
  (b) find the marginal cost with respect to y.

10. Suppose the joint cost function for x units of product X and y units of product Y is given by

$$C(x, y) = 2500\sqrt{xy + 1} \quad \text{dollars}$$

Find the marginal cost with respect to
  (a) x. (b) y.

11. Suppose that the joint cost function for two products is

$$C(x, y) = 1200 \ln (xy + 1) + 10{,}000 \quad \text{dollars}$$

Find the marginal cost with respect to
  (a) x.   (b) y.

12. Suppose that the joint cost function for two products is

$$C(x, y) = y \ln (x + 1) \quad \text{dollars}$$

Find the marginal cost with respect to
  (a) x.   (b) y.

PRODUCTION FUNCTIONS

13. Suppose that the production function for a product is $z = \sqrt{4xy}$, where x represents the number of work-hours per month and y is the number of available machines. Determine the marginal productivity of
    (a) x.   (b) y.

14. Suppose the production function for a product is

    $$z = 60x^{2/5}y^{3/5}$$

    where x is the capital expenditures and y is the number of work-hours. Find the marginal productivity of
    (a) x.   (b) y.

15. Suppose that the production function for a product is $z = \sqrt{x}\ln(y + 1)$, where x represents the number of work-hours and y represents the available capital (per week). Find the marginal productivity of
    (a) x.   (b) y.

16. Suppose that a company's production function for a certain product is

    $$z = (x + 1)^{1/2}\ln(y^2 + 1)$$

    where x is the number of work-hours of unskilled labor and y is the number of work-hours of skilled labor. Find the marginal productivity of
    (a) x.   (b) y.

For Problems 17–19, suppose that the number of crates of an agricultural product is given by

$$z = \frac{11xy - 0.0002x^2 - 5y}{0.03x + 3y}$$

where x is the number of hours of labor and y is the number of acres of the crop.

17. Find the output when $x = 300$ and $y = 500$.
18. Find and interpret the marginal productivity of the number of acres of the crop (y) when $x = 300$ and $y = 500$.
19. Find and interpret the marginal productivity of the number of hours of labor (x) when $x = 300$ and $y = 500$.
20. If a production function is given by $z = 12x^{3/4}y^{1/3}$, find the marginal productivity of
    (a) x.   (b) y.

21. Suppose the Cobb-Douglas production function for a company is given by

    $$z = 400x^{3/5}y^{2/5}$$

    where x is the company's capital investment and y is the size of the labor force (in work-hours).
    (a) Find the marginal productivity of x.
    (b) If the current labor force is 1024 work-hours, substitute $y = 1024$ in your answer to part (a) and graph the result.
    (c) Find the marginal productivity of y.
    (d) If the current capital investment is $59,049, substitute $x = 59,049$ in your answer to part (c) and graph the result.

(e) Interpret the graphs in parts (b) and (d) with regard to what they say about the effects on productivity of an increased capital investment (part (b)) and of an increased labor force (part (d)).

22. Suppose the Cobb-Douglas production function for a company is given by

    $$z = 300x^{2/3}y^{1/3}$$

    where x is the company's capital investment and y is the size of the labor force (in work-hours).
    (a) Find the marginal productivity of x.
    (b) If the current labor force is 729 work-hours, substitute $y = 729$ in your answer to part (a) and graph the result.
    (c) Find the marginal productivity of y.
    (d) If the current capital investment is $27,000, substitute $x = 27,000$ in your answer to part (c) and graph the result.
    (e) Interpret the graphs in parts (b) and (d) with regard to what they say about the effects on productivity of an increased capital investment (part (b)) and of an increased labor force (part (d)).

DEMAND FUNCTIONS

In Problems 23–26, prices p_1 and p_2 are in dollars and q_1 and q_2 are numbers of units.

23. The demand functions for two products are given by

    $$q_1 = 300 - 8p_1 - 4p_2$$
    $$q_2 = 400 - 5p_1 - 10p_2$$

    Find the demand for each of the products if the price of the first is $p_1 = \$10$ and the price of the second is $p_2 = \$8$.

24. The demand functions for two products are given by

    $$q_1 = 900 - 9p_1 + 2p_2$$
    $$q_2 = 1200 + 6p_1 - 10p_2$$

    Find the demands q_1 and q_2 if $p_1 = \$10$ and $p_2 = \$12$.

25. Find a pair of prices p_1 and p_2 such that the demands for the two products in Problem 23 will be equal.

26. Find a pair of prices p_1 and p_2 such that the demands for the two products in Problem 24 will be equal.

In Problems 27–30, the demand functions for q_A and q_B units of two related products, A and B, are given. Complete parts (a)–(e) for each problem. Assume p_A and p_B are in dollars.
(a) Find the marginal demand of q_A with respect to p_A.
(b) Find the marginal demand of q_A with respect to p_B.
(c) Find the marginal demand of q_B with respect to p_B.
(d) Find the marginal demand of q_B with respect to p_A.
(e) Are the two goods competitive or complementary?

27. $\begin{cases} q_A = 400 - 3p_A - 2p_B \\ q_B = 250 - 5p_A - 6p_B \end{cases}$

28. $\begin{cases} q_A = 600 - 4p_A + 6p_B \\ q_B = 1200 + 8p_A - 4p_B \end{cases}$

29.
$$\begin{cases} q_A = 5000 - 50p_A - \dfrac{600}{p_B + 1} \\[2mm] q_B = 10{,}000 - \dfrac{400}{p_A + 4} + \dfrac{400}{p_B + 4} \end{cases}$$

30.
$$\begin{cases} q_A = 2500 + \dfrac{600}{p_A + 2} - 40p_B \\[2mm] q_B = 3000 - 100p_A + \dfrac{400}{p_B + 5} \end{cases}$$

31. The markets for new cars and for used cars are related. As new car sales increase, the available supply of used cars (trade-ins) increases. This tends to decrease the price of used cars. As the prices of used cars decline, typically the demand for new cars also declines.

(a) Does this analysis suggest that new cars and used cars are complementary or competitive products? Explain.

(b) Suppose a large automobile dealership has the following monthly market demand and supply functions for its new and used cars.

Demand for New Cars:

$$p_{NEW} = 78{,}000 - 30q_{NEW} + 2p_{USED}$$

Supply of New Cars:

$$p_{NEW} = 20{,}000 + 10q_{NEW}$$

Demand for Used Cars:

$$p_{USED} = 3000 - 4q_{USED} + 0.05p_{NEW}$$

Supply of Used Cars:

$$p_{USED} = 2000 + q_{USED}$$

(i) Solve each demand equation for the quantity in terms of the prices for new cars and used cars.

(ii) Calculate the necessary partial derivatives to confirm or refute your answer in part (a).

- To find relative maxima, minima, and saddle points of functions of two variables
- To apply linear regression formulas

Maxima and Minima

▮ | APPLICATION PREVIEW |

Adele Lighting manufactures 20-inch lamps and 31-inch lamps. Suppose that x is the number of thousands of 20-inch lamps and that the demand for these is given by $p_1 = 50 - x$, where p_1 is in dollars. Similarly, suppose that y is the number of thousands of 31-inch lamps and that the demand for these is given by $p_2 = 60 - 2y$, where p_2 is also in dollars. Adele Lighting's joint cost function for these lamps is $C = 2xy$ (in thousands of dollars). Therefore, Adele Lighting's profit (in thousands of dollars) is a function of the two variables x and y. In order to determine Adele's maximum profit, we need to develop methods for finding maximum values for a function of two variables. (See Example 4.)

In this section we will find relative maxima and minima of functions of two variables, and use them to solve applied optimization problems. We will also use minimization of functions of two variables to develop the linear regression formulas. Recall that we first used linear regression in Chapter 2.

Maxima and Minima   In our study of differentiable functions of one variable, we saw that for a relative maximum or minimum to occur at a point, the tangent line to the curve had to be horizontal at that point. The function $z = f(x, y)$ describes a surface in three dimensions. If all partial derivatives of $f(x, y)$ exist, then there must be a horizontal plane tangent to the surface at a point in order for the surface to have a relative maximum at that point (see Figure 14.8(a)) or a minimum at that point (see Figure 14.8(b)). But if the plane tangent to the surface at the point is horizontal, then all the tangent lines to the surface at that point must also be horizontal, for they lie in the tangent plane. In particular, the tangent line in the direction of the x-axis will be horizontal, so $\partial z/\partial x = 0$ at the point; and the tangent line in the direction of the y-axis will be horizontal, so $\partial z/\partial y = 0$ at the point. Thus those points where *both* $\partial z/\partial x = 0$ and $\partial z/\partial y = 0$ are called **critical points** for the surface.

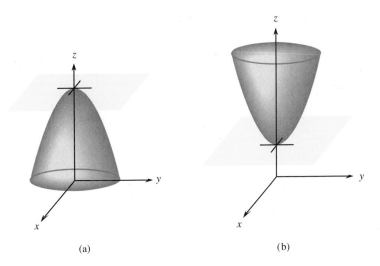

Figure 14.8   (a)   (b)

How can we determine whether a critical point is a relative maximum, a relative minimum, or neither of these? Finding that $\partial^2 z/\partial x^2 < 0$ and $\partial^2 z/\partial y^2 < 0$ is not enough to tell us that we have a relative maximum. The "second-derivative" test we must use involves the values of all of the second partial derivatives at the critical point (a, b).

We state, without proof, the result that determines whether there is a relative maximum, a relative minimum, or neither at the critical point (a, b).

Test for Maxima and Minima

Let $z = f(x, y)$ be a function for which both

$$\frac{\partial z}{\partial x} = 0 \text{ and } \frac{\partial z}{\partial y} = 0 \text{ at a point } (a, b)$$

and suppose that all second partial derivatives are continuous there. Define

$$D = \frac{\partial^2 z}{\partial x^2} \cdot \frac{\partial^2 z}{\partial y^2} - \left(\frac{\partial^2 z}{\partial x \partial y}\right)^2 = (z_{xx})(z_{yy}) - (z_{xy})^2$$

Evaluate D at the critical point (a, b), and conclude the following:

1. If $D(a, b) > 0$:
   (a) $z_{xx}(a, b) > 0 \Rightarrow$ a relative minimum occurs at (a, b). [Note $z_{yy}(a, b) > 0$ also.]
   (b) $z_{xx}(a, b) < 0 \Rightarrow$ a relative maximum occurs at (a, b). [Note $z_{yy}(a, b) < 0$ also.]
2. $D(a, b) < 0$, then neither a relative maximum nor a relative minimum occurs at (a, b).
3. $D(a, b) = 0$, then the test fails; investigate the function near the point.

We can test for relative maxima and minima by using the following procedure.

Maxima and Minima of $z = f(x, y)$

Procedure	Example
To find relative maxima and minima of $z = f(x, y)$:	Test $z = 4 - 4x^2 - y^2$ for relative maxima and minima.
1. Find $\partial z/\partial x$ and $\partial z/\partial y$.	1. $\dfrac{\partial z}{\partial x} = -8x;\ \dfrac{\partial z}{\partial y} = -2y$

(continued)

Procedure	Example
2. Find the point(s) that satisfy *both* $\partial z/\partial x = 0$ and $\partial z/\partial y = 0$. These are the critical points.	2. $\dfrac{\partial z}{\partial x} = 0$ if $x = 0$. $\quad \dfrac{\partial z}{\partial y} = 0$ if $y = 0$. The critical point is $(0, 0, 4)$.
3. Find all second partial derivatives.	3. $\dfrac{\partial^2 z}{\partial x^2} = -8; \quad \dfrac{\partial^2 z}{\partial y^2} = -2; \quad \dfrac{\partial^2 z}{\partial x\,\partial y} = \dfrac{\partial^2 z}{\partial y\,\partial x} = 0$
4. Evaluate D at each critical point.	4. At $(0, 0)$, $D = (-8)(-2) - 0^2 = 16$.
5. Use the test for maxima and minima to determine whether relative maxima or minima occur.	5. $D > 0$, $\partial^2 z/\partial x^2 < 0$, and $\partial^2 z/\partial y^2 < 0$. A relative maximum occurs at $(0, 0)$. See Figure 14.9.

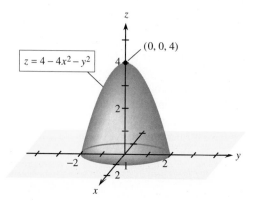

$z = 4 - 4x^2 - y^2$

$(0, 0, 4)$

Figure 14.9

EXAMPLE 1   Relative Maxima and Minima

Let $z = x^2 + y^2 - 2x + 1$. Find the relative maxima and minima of z, if they exist.

Solution

1. $\dfrac{\partial z}{\partial x} = 2x - 2; \quad \dfrac{\partial z}{\partial y} = 2y$

2. $\dfrac{\partial z}{\partial x} = 0$ if $x = 1$. $\quad \dfrac{\partial z}{\partial y} = 0$ if $y = 0$.
   Both are 0 if $x = 1$ and $y = 0$, so the critical point is $(1, 0, 0)$.

3. $\dfrac{\partial^2 z}{\partial x^2} = 2; \quad \dfrac{\partial^2 z}{\partial y^2} = 2; \quad \dfrac{\partial^2 z}{\partial x \partial y} = \dfrac{\partial^2 z}{\partial y\, \partial x} = 0$

4. At $(1, 0)$, $D = 2 \cdot 2 - 0^2 = 4$.

5. $D > 0$, $\partial^2 z/\partial x^2 > 0$, and $\partial^2 z/\partial y^2 > 0$. A relative minimum occurs at $(1, 0)$. (See Figure 14.10.)

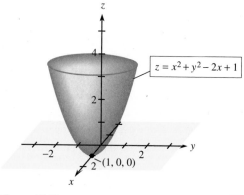

$z = x^2 + y^2 - 2x + 1$

$(1, 0, 0)$

Figure 14.10

EXAMPLE 2   Saddle Points

Let $z = y^2 - x^2$. Find the relative maxima and minima of z, if they exist.

Solution

1. $\dfrac{\partial z}{\partial x} = -2x; \quad \dfrac{\partial z}{\partial y} = 2y$

2. $\dfrac{\partial z}{\partial x} = 0$ if $x = 0$; $\quad \dfrac{\partial z}{\partial y} = 0$ if $y = 0$.
   Thus both equal 0 if $x = 0$, $y = 0$. The critical point is $(0, 0, 0)$.

3. $\dfrac{\partial^2 z}{\partial x^2} = -2;\quad \dfrac{\partial^2 z}{\partial y^2} = 2;\quad \dfrac{\partial^2 z}{\partial x\,\partial y} = \dfrac{\partial^2 z}{\partial y\,\partial x} = 0$

4. $D = (-2)(2) - 0 = -4$

5. $D < 0$, so the critical point is neither a relative maximum nor a relative minimum. As Figure 14.11 shows, the surface formed has the shape of a saddle. For this reason, critical points that are neither relative maxima nor relative minima are called **saddle points**.

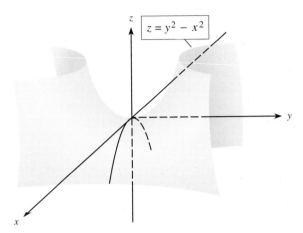

$z = y^2 - x^2$

Figure 14.11

The following example involves a surface with two critical points.

EXAMPLE 3   Maxima and Minima

Let $z = x^3 + y^2 + 6xy + 24x$. Find the relative maxima, relative minima, and saddle points for z, if they exist.

Solution

1. $\dfrac{\partial z}{\partial x} = 3x^2 + 6y + 24\quad$ and $\quad\dfrac{\partial z}{\partial y} = 2y + 6x$

2. $\dfrac{\partial z}{\partial x} = 0$ if $0 = 3x^2 + 6y + 24,\quad$ and $\quad\dfrac{\partial z}{\partial y} = 0$ if $0 = 2y + 6x$.

   But $0 = 2y + 6x$ means $y = -3x$. Because *both* partial derivatives must equal zero, we substitute $(-3x)$ for y in $0 = 3x^2 + 6y + 24$ to obtain

   $$0 = 3x^2 + 6(-3x) + 24 = 3x^2 - 18x + 24$$
   $$0 = 3(x^2 - 6x + 8) = 3(x - 2)(x - 4)$$

   Thus, the solutions are $x = 2$ and $x = 4$.
   When $x = 2$, $y = -3x = -6$, and $z = 2^3 + (-6)^2 + 6(2)(-6) + 24(2) = 20$, so one critical point is $(2, -6, 20)$.
   When $x = 4$, $y = -3x = -12$, and $z = 4^3 + (-12)^2 + 6(4)(-12) + 24(4) = 16$, so another critical point is $(4, -12, 16)$.

3. $\dfrac{\partial^2 z}{\partial x^2} = 6x;\quad \dfrac{\partial^2 z}{\partial y^2} = 2;\quad \dfrac{\partial^2 z}{\partial x\,\partial y} = \dfrac{\partial^2 z}{\partial y\,\partial x} = 6$

4. $D = (6x)(2) - 6^2$

5. At $(2, -6)$, $\dfrac{\partial^2 z}{\partial x^2} = 6 \cdot 2 = 12 > 0$, $\dfrac{\partial^2 z}{\partial y^2} = 2 > 0$, and $D = (6 \cdot 2)(2) - 6^2 = -12 < 0$, so a saddle point occurs at $(2, -6, 20)$.

   At $(4, -12)$, $\dfrac{\partial^2 z}{\partial x^2} = 6 \cdot 4 = 24 > 0$, $\dfrac{\partial^2 z}{\partial y^2} = 2 > 0$, and $D = (6 \cdot 4)(2) - 6^2 = 12 > 0$, so a relative minimum occurs at $(4, -12, 16)$.

✓ CHECKPOINT   Suppose that $z = 4 - x^2 - y^2 + 2x - 4y$.
1. Find z_x and z_y.
2. Solve $z_x = 0$ and $z_y = 0$ simultaneously to find the critical point(s) for the graph of this function.
3. Test the point(s) for relative maxima and minima.

EXAMPLE 4   **Maximum Profit | APPLICATION PREVIEW |**

Adele Lighting manufactures 20-inch lamps and 31-inch lamps. Suppose that x is the number of thousands of 20-inch lamps and that the demand for these is given by $p_1 = 50 - x$, where p_1 is in dollars. Similarly, suppose that y is the number of thousands of 31-inch lamps and that the demand for these is given by $p_2 = 60 - 2y$, where p_2 is also in dollars. Adele Lighting's joint cost function for these lamps is $C = 2xy$ (in thousands of dollars). Therefore, Adele Lighting's profit (in thousands of dollars) is a function of the two variables x and y. Determine Adele's maximum profit.

Solution

The profit function is $P(x, y) = p_1x + p_2y - C(x, y)$. Thus,

$$P(x, y) = (50 - x)x + (60 - 2y)y - 2xy$$
$$= 50x - x^2 + 60y - 2y^2 - 2xy$$

gives the profit in thousands of dollars. To maximize the profit, we proceed as follows.

$$P_x = 50 - 2x - 2y \text{ and } P_y = 60 - 4y - 2x$$

Solving $P_x = 0$ and $P_y = 0$ simultaneously, we have

$$\begin{cases} 0 = 50 - 2x - 2y \\ 0 = 60 - 2x - 4y \end{cases}$$

Subtraction gives $-10 + 2y = 0$, so $y = 5$. With $y = 5$, the equation $0 = 50 - 2x - 2y$ becomes $0 = 40 - 2x$, so $x = 20$. Now

$$P_{xx} = -2, P_{yy} = -4, \text{ and } P_{xy} = -2, \text{ and}$$
$$D = (P_{xx})(P_{yy}) - (P_{xy})^2 = (-2)(-4) - (-2)^2 = 4$$

Because $P_{xx} < 0$, $P_{yy} < 0$, and $D > 0$, the values $x = 20$ and $y = 5$ yield maximum profit. Therefore, when $x = 20$ and $y = 5$, $p_1 = 30$, $p_2 = 50$, and the maximum profit is

$$P(20, 5) = 600 + 250 - 200 = 650$$

That is, Adele Lighting's maximum profit is $650,000 when the company sells 20,000 of the 20-inch lamps at $30 each and 5000 of the 31-inch lamps at $50 each.   ■

Linear Regression   We have used different types of functions to model cost, revenue, profit, demand, supply, and other real-world relationships. Sometimes we have used calculus to study the behavior of these functions, finding, for example, marginal cost, marginal revenue, producer's surplus, and so on. We now have the mathematical tools to understand and develop the formulas that graphing calculators and other technology use to find the equations for linear models.

The formulas used to find the equation of the straight line that is the best fit for a set of data are developed using max-min techniques for functions of two variables. This line is called the **regression line.** In Figure 14.12, we define line ℓ to be the best fit for the data points (that is, the regression line) if the sum of the squares of the differences between the actual y-values of the data points and the y-values of the points on the line is a minimum.

In general, to find the equation of the regression line, we assume that the relationship between x and y is approximately linear and that we can find a straight line with equation

$$\hat{y} = a + bx$$

where the values of $\hat{y}$ will approximate the y-values of the points we know. That is, for each given value of x, the point $(x, \hat{y})$ will be on the line. For any given x-value, x_i, we are interested in the deviation between the y-value of the data point (x_i, y_i) and the $\hat{y}$-value from the equation, $\hat{y}_i$, that results when x_i is substituted for x. These deviations are of the form

$$d_i = \hat{y}_i - y_i, \quad \text{for } i = 1, 2, \ldots, n$$

See Figure 14.12 for a general case with the deviations exaggerated.

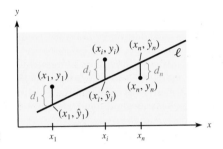

Figure 14.12

To measure the deviations in a way that accounts for the fact that some of the y-values will be above the line and some below, we will say that the line that is the best fit for the data is the one for which the sum of the squares of the deviations is a minimum. That is, we seek the a and b in the equation

$$\hat{y} = a + bx$$

such that the sum of the squares of the deviations,

$$S = \sum_{i=1}^{n}(\hat{y}_i - y_i)^2 = \sum_{i=1}^{n}[(bx_i + a) - y_i]^2$$
$$= (bx_1 + a - y_1)^2 + (bx_2 + a - y_2)^2 + \cdots + (bx_n + a - y_n)^2$$

is a minimum. The procedure for determining a and b is called the **method of least squares.**
We seek the values of b and a that make S a minimum, so we find the values that make

$$\frac{\partial S}{\partial b} = 0 \text{ and } \frac{\partial S}{\partial a} = 0$$

where

$$\frac{\partial S}{\partial b} = 2(bx_1 + a - y_1)x_1 + 2(bx_2 + a - y_2)x_2 + \cdots + 2(bx_n + a - y_n)x_n$$

$$\frac{\partial S}{\partial a} = 2(bx_1 + a - y_1) + 2(bx_2 + a - y_2) + \cdots + 2(bx_n + a - y_n)$$

Setting each equation equal to 0, dividing by 2, and using sigma notation give

$$0 = b\sum_{i=1}^{n}x_i^2 + a\sum_{i=1}^{n}x_i - \sum_{i=1}^{n}x_i y_i \tag{1}$$

$$0 = b\sum_{i=1}^{n}x_i + a\sum_{i=1}^{n}1 - \sum_{i=1}^{n}y_i \tag{2}$$

We can write Equations (1) and (2) as follows:

$$\sum_{i=1}^{n}x_i y_i = a\sum_{i=1}^{n}x_i + b\sum_{i=1}^{n}x_i^2 \tag{3}$$

$$\sum_{i=1}^{n}y_i = an + b\sum_{i=1}^{n}x_i \tag{4}$$

Multiplying Equation (3) by n and Equation (4) by $\sum_{i=1}^{n}x_i$ permits us to begin to solve for b.

$$n\sum_{i=1}^{n}x_i y_i = na\sum_{i=1}^{n}x_i + nb\sum_{i=1}^{n}x_i^2 \tag{5}$$

$$\sum_{i=1}^{n}x_i\sum_{i=1}^{n}y_i = na\sum_{i=1}^{n}x_i + b\left(\sum_{i=1}^{n}x_i\right)^2 \tag{6}$$

Subtracting Equation (5) from Equation (6) gives

$$\sum_{i=1}^{n} x_i \sum_{i=1}^{n} y_i - n \sum_{i=1}^{n} x_i y_i = b \left(\sum_{i=1}^{n} x_i \right)^2 - nb \sum_{i=1}^{n} x_i^2$$

$$= b \left[\left(\sum_{i=1}^{n} x_i \right)^2 - n \sum_{i=1}^{n} x_i^2 \right]$$

Thus

$$b = \frac{\sum_{i=1}^{n} x_i \sum_{i=1}^{n} y_i - n \sum_{i=1}^{n} x_i y_i}{\left(\sum_{i=1}^{n} x_i \right)^2 - n \sum_{i=1}^{n} x_i^2}$$

and, from Equation (4),

$$a = \frac{\sum_{i=1}^{n} y_i - b \sum_{i=1}^{n} x_i}{n}$$

It can be shown that these values for b and a give a minimum value for S, so we have the following.

Linear Regression Equation

Given a set of data points (x_1, y_1), (x_2, y_2), ..., (x_n, y_n), the equation of the line that is the best fit for these data is

$$\hat{y} = a + bx$$

where

$$b = \frac{\sum_{i=1}^{n} x_i \sum_{i=1}^{n} y_i - n \sum_{i=1}^{n} x_i y_i}{\left(\sum_{i=1}^{n} x_i \right)^2 - n \sum_{i=1}^{n} x_i^2}, \quad a = \frac{\sum_{i=1}^{n} y_i - b \sum_{i=1}^{n} x_i}{n}$$

EXAMPLE 5   Inventory

The data in the table show the relation between the diameter of a partial roll of blue denim material at MacGregor Mills and the actual number of yards remaining on the roll. Use linear regression to find the linear equation that gives the number of yards as a function of the diameter.

Diameter (inches)	Yards/Roll	Diameter (inches)	Yards/Roll
14.0	120	22.5	325
15.0	145	24.0	360
16.5	170	24.5	380
17.75	200	25.25	405
18.5	220	26.0	435
19.8	255	26.75	460
20.5	270	27.0	470
22.0	305	28.0	500

Solution

Let x be the diameter of the partial roll and y be the number of yards on the roll. Before finding the values for a and b, we evaluate some parts of the formulas.

$$n = 16$$

$$\sum_{i=1}^{n} x_i = 348.05$$

$$\sum_{i=1}^{n} x_i^2 = 7871.48$$

$$\sum_{i=1}^{n} y_i = 5020$$

$$\sum_{i=1}^{n} x_i y_i = 117{,}367.75$$

$$b = \frac{\displaystyle\sum_{i=1}^{n} x_i \sum_{i=1}^{n} y_i - n \sum_{i=1}^{n} x_i y_i}{\left(\displaystyle\sum_{i=1}^{n} x_i\right)^2 - n \sum_{i=1}^{n} x_i^2}$$

$$= \frac{(348.05)(5020) - 16(117{,}367.75)}{(348.05)^2 - 16(7871.48)} \approx 27.1959$$

$$a = \frac{\displaystyle\sum_{i=1}^{n} y_i - b \sum_{i=1}^{n} x_i}{n}$$

$$= \frac{(5020) - (27.1959)(348.05)}{16} \approx -277.8458$$

Thus the linear equation that can be used to estimate the number of yards of denim remaining on a roll is

$$\hat{y} = -277.85 + 27.196x$$

Note that if we use the linear regression capability of a graphing utility, we obtain exactly the same equation.

✓ CHECKPOINT

4. Use linear regression to write the equation of the line that is the best fit for the following points.

x	50	25	10	5
y	2	4	10	20

Finally, we note that formulas for models other than linear ones, such as power models ($y = ax^b$), exponential models ($y = ab^x$), and logarithmic models [$y = a + b \ln(x)$], can also be developed with the least-squares method. That is, we apply max-min techniques for functions of two variables to minimize the sum of the squares of the deviations.

✓ CHECKPOINT
ANSWERS

1. $z_x = -2x + 2$, $z_y = -2y - 4$
2. $(1, -2, 9)$.
3. $(1, -2, 9)$ is a relative maximum.
4. $\hat{y} = -0.33x + 16.5$

| EXERCISES | 14.4

In Problems 1–16, find each function's relative maxima, relative minima, and saddle points, if they exist.

1. $z = 9 - x^2 - y^2$
2. $z = 16 - 4x^2 - 9y^2$
3. $z = x^2 + y^2 + 4$
4. $z = x^2 + y^2 - 4$
5. $z = x^2 - y^2 + 4x - 6y + 11$
6. $z = 4y^2 - x^2 + 4y + 10x + 12$
7. $z = x^2 + y^2 - 2x + 4y + 5$
8. $z = 4x^2 + y^2 + 4x + 1$
9. $z = x^2 + 6xy + y^2 + 16x$
10. $z = x^2 - 4xy + y^2 - 6y$
11. $z = 24 - x^2 + xy - y^2 + 36y$
12. $z = 46 - x^2 + 2xy - 4y^2$

13. $z = x^2 + xy + y^2 - 4y + 10x$
14. $z = x^2 + 5xy + 10y^2 + 8x - 40y$
15. $z = x^3 + y^3 - 6xy$
16. $z = 6xy - x^3 - y^2$

In Problems 17 and 18, use the points given in the tables to write the equation of the line that is the best fit for the points.

17.

x	3	4	5	6
y	15	22	28	32

18.

x	10	20	30	40
y	2	6	5	6

APPLICATIONS

19. *Profit* Suppose that the quarterly profit from the sale of Kisses and Kreams is given by

$$P(x, y) = 10x + 6.4y - 0.001x^2 - 0.025y^2 \quad \text{dollars}$$

where x is the number of pounds of Kisses and y is the number of pounds of Kreams. Selling how many pounds of Kisses and Kreams will maximize profit? What is the maximum profit?

20. *Profit* The profit from the sales of two products is given by

$$P(x, y) = 20x + 70y - x^2 - y^2 \quad \text{dollars}$$

where x is the number of units of product 1 sold and y is the number of units of product 2. Selling how much of each product will maximize profit? What is the maximum profit?

21. *Nutrition* A new food is designed to add weight to mature beef cattle. The weight in pounds is given by $W = 13xy(20 - x - 2y)$, where x is the number of units of the first ingredient and y is the number of units of the second ingredient. How many units of each ingredient will maximize the weight? What is the maximum weight?

22. *Profit* The profit for a grain crop is related to fertilizer and labor. The profit per acre is

$$P = 100x + 40y - 5x^2 - 2y^2 \quad \text{dollars}$$

where x is the number of units of fertilizer and y is the number of work-hours. What values of x and y will maximize the profit? What is the maximum profit?

23. *Production* Suppose that

$$P = 3.78x^2 + 1.5y^2 - 0.09x^3 - 0.01y^3 \quad \text{tons}$$

is the production function for a product with x units of one input and y units of a second input. Find the values of x and y that will maximize production. What is the maximum production?

24. *Production* Suppose that x units of one input and y units of a second input result in

$$P = 40x + 50y - x^2 - y^2 - xy$$

units of a product. Determine the inputs x and y that will maximize P. What is the maximum production?

25. *Profit* Suppose that a manufacturer produces two brands of a product, brand 1 and brand 2. Suppose the demand for brand 1 is $x = 70 - p_1$ thousand units and the demand for brand 2 is $y = 80 - p_2$ thousand units, where p_1 and p_2 are prices in dollars. If the joint cost function is $C = xy$, in thousands of dollars, how many of each brand should be produced to maximize profit? What is the maximum profit?

26. *Profit* Suppose that a firm produces two products, A and B, that sell for $a and $b, respectively, with the total cost of producing x units of A and y units of B equal to $C(x, y)$. Show that when the profit from these products is maximized,

$$\frac{\partial C}{\partial x}(x, y) = a \text{ and } \frac{\partial C}{\partial y}(x, y) = b$$

27. *Manufacturing* Find the values for each of the dimensions of an open-top box of length x, width y, and height $500{,}000/(xy)$ (in inches) such that the box requires the least amount of material to make.

28. *Manufacturing* Find the values for each of the dimensions of a closed-top box of length x, width y, and height z (in inches) if the volume equals 27,000 cubic inches and the box requires the least amount of material to make. (*Hint:* First write the height in terms of x and y, as in Problem 27.)

29. *Profit* A company manufactures two products, A and B. If x is the number of thousands of units of A and y is the number of thousands of units of B, then the cost and revenue in thousands of dollars are

$$C(x, y) = 2x^2 - 2xy + y^2 - 7x - 10y + 11$$
$$R(x, y) = 5x + 8y$$

Find the number of each type of product that should be manufactured to maximize profit. What is the maximum profit?

30. *Production* Let x be the number of work-hours required and let y be the amount of capital required to produce z units of a product. Show that the average production per work-hour, z/x, is maximized when

$$\frac{\partial z}{\partial x} = \frac{z}{x}$$

Use $z = f(x, y)$ and assume that a maximum exists.

The manager of the Sea Islands Chicken Shack is interested in finding new ways to improve sales and profitability. Currently Sea Islands offers both "eat-in" and "take-out" chicken dinners at $3.25 each, sells 6250 dinners per week, and has short-run weekly costs (in dollars) given by

$$C = 500 + 1.2x$$

where x is the total number of eat-in and take-out chicken dinners.

The Sea Islands manager recently commissioned a local consulting firm to study the eat-in and take-out market demand per week. The study results provided the following weekly demand estimates for each of the two market segments:

$$\text{Eat-in:} \quad x = 6000 - 1000p_x$$
$$\text{Take-out:} \quad y = 10{,}000 - 2000p_y$$

where x is the number of eat-in dinners, with p_x as the price of each, and y is the number of take-out dinners, with p_y as the price of each. Use this information in Problems 31 and 32.

31. (a) If Sea Islands Chicken Shack prices chicken dinners differently for eat-in and take-out customers, how many dinners per week would it expect to sell to each type of customer in order to maximize weekly profit?
   (b) What prices should Sea Islands charge each market segment to maximize the total weekly profit, and what is that profit?
   (c) Would it be more profitable for Sea Islands to continue charging $3.25 per dinner (and sell 6250 dinners per week) or to change the policy and to price differently for each type of customer? Explain.

32. When eat-in and take-out dinners are considered separately, Sea Islands Chicken Shack's short-run weekly cost function becomes

$$C = 500 + 0.60x + 1.60y$$

   (a) Use this revised cost function to find the number of eat-in and take-out chicken dinners that would give maximum profit.
   (b) What price should be charged for each type of dinner now, and what is the maximum weekly profit?
   (c) Based on the results of parts (a) and (b), which is the best pricing strategy? Explain.

33. **Earnings and gender** The data in the table show the average earnings of year-round full-time workers by gender for several different levels of educational attainment.

Average Annual Earnings	
Males	Females
$21,659	$17,023
26,277	19,162
35,725	26,029
41,875	30,816
44,404	33,481
57,220	41,681
71,530	51,316
82,401	68,875

Source: U.S. Bureau of the Census

(a) Use linear regression to find the linear equation that is the best for these data, with x representing the earnings for males and y representing the earnings for females.
(b) Find and interpret the slope of the linear regression line.
(c) If you made a similar linear model with data from 1965, how do you think its slope would compare with the slope found in part (b)? Explain.

34. **Retirement benefits** The table gives the approximate benefits for PepsiCo executives who earned an average of $250,000 per year during the last 5 years of service, based on the number of years of service, from 15 years to 45 years.
   (a) Use linear regression to find the linear equation that is the best fit for the data.
   (b) Use the equation to find the expected annual retirement benefits after 38 years of service.
   (c) Write a sentence that interprets the slope of the linear regression line.

Years of Service	Annual Retirement
25	$109,280
30	121,130
35	132,990
40	145,490
45	160,790

Source: TRICON Salaried Employees Retirement Plan

35. **World population** The table gives the actual or projected world population in billions for selected years from 2000 to 2050.
   (a) Use linear regression to find the linear equation that is the best fit for the data, with x equal to the number of years past 2000.
   (b) Use the equation to predict the population in 2018.
   (c) Write a sentence that interprets the slope of the linear regression line.

Years	Population (billions)
2000	6.08
2010	6.82
2020	7.54
2030	8.18
2040	8.72
2050	9.19

Source: U.S. Bureau of the Census, International Data Base

36. **Student loans** The following table shows the balance of federal direct student loans (in billions of dollars) for selected years from 2011 and projected to 2023.

(a) Find the linear regression equation for the federal direct student loan balance as a function of the years past 2010. Report the model with three significant digit coefficients.

(b) What does the reported model predict for this balance in 2025?

(c) Write a sentence that interprets the slope of your linear regression equation.

Years	Amount ($ billion)
2011	702
2013	940
2015	1220
2017	1500
2019	1775
2021	2000
2023	2250

Source: U.S. Office of Management and Budget

OBJECTIVE

- To find the maximum or minimum value of a function of two or more variables subject to a condition that constrains the variables

14.5

Maxima and Minima of Functions Subject to Constraints: Lagrange Multipliers

| APPLICATION PREVIEW |

Many practical problems require that a function of two or more variables be maximized or minimized subject to certain conditions, or constraints, that limit the variables involved. For example, a firm will want to maximize its profits within the limits (constraints) imposed by its production capacity. Similarly, a city planner may want to locate a new building to maximize access to public transportation yet may be constrained by the availability and cost of building sites.

Specifically, suppose that the utility function for commodities X and Y is given by $U = x^2y^2$, where x and y are the numbers of units (in thousands) of X and Y, respectively. If p_1 and p_2 represent the prices in dollars of X and Y, respectively, and I represents the consumer's income available to purchase these two commodities, the equation $p_1x + p_2y = I$ is called the *budget constraint*. If the price of X is \$2, the price of Y is \$4, and the income available is \$40 thousand, then the budget constraint is $2x + 4y = 40$. Thus we seek to maximize the consumer's utility $U = x^2y^2$ subject to the budget constraint $2x + 4y = 40$. (See Example 4.) In this section we develop methods to solve this type of constrained maximum or minimum.

We can obtain maxima and minima for a function $z = f(x, y)$ subject to the constraint $g(x, y) = 0$ by using the method of **Lagrange multipliers,** named for the famous eighteenth-century mathematician Joseph Louis Lagrange. Lagrange multipliers can be used with functions of two or more variables when the constraints are given by an equation.

In order to find the critical values of a function $f(x, y)$ subject to the constraint $g(x, y) = 0$, we will use the new variable λ to form the **objective function**

$$F(x, y, \lambda) = f(x, y) + \lambda g(x, y)$$

It can be shown that the critical values of $F(x, y, \lambda)$ will satisfy the constraint $g(x, y)$ and will also be critical points of $f(x, y)$. Thus we need only find the critical points of $F(x, y, \lambda)$ to find the required critical points.

To find the critical points of $F(x, y, \lambda)$, we must find the points that make all the partial derivatives equal to 0. That is, the points must satisfy

$$\partial F/\partial x = 0, \quad \partial F/\partial y = 0, \quad \text{and} \quad \partial F/\partial \lambda = 0$$

Because $F(x, y, \lambda) = f(x, y) + \lambda g(x, y)$, these equations may be written as

$$\frac{\partial f}{\partial x} + \lambda \frac{\partial g}{\partial x} = 0$$

$$\frac{\partial f}{\partial y} + \lambda \frac{\partial g}{\partial y} = 0$$

$$g(x, y) = 0$$

Finding the values of x and y that satisfy these three equations simultaneously gives the critical values.

This method will not tell us whether the critical points correspond to maxima or minima, but this can be determined either from the physical setting for the problem or by testing according to a procedure similar to that used for unconstrained maxima and minima. The following examples illustrate the use of Lagrange multipliers.

EXAMPLE 1 Maxima Subject to Constraints

Find the maximum value of $z = x^2y$ subject to $x + y = 9$, $x \geq 0$, $y \geq 0$.

Solution
The function to be maximized is $f(x, y) = x^2y$. The constraint is $g(x, y) = 0$, where $g(x, y) = x + y - 9$. The objective function is

$$F(x, y, \lambda) = f(x, y) + \lambda g(x, y) = x^2y + \lambda(x + y - 9)$$

Thus

$$\frac{\partial F}{\partial x} = 2xy + \lambda(1) = 0, \ \text{ or } \ 2xy + \lambda = 0$$

$$\frac{\partial F}{\partial y} = x^2 + \lambda(1) = 0, \ \text{ or } \ x^2 + \lambda = 0$$

$$\frac{\partial F}{\partial \lambda} = 0 + 1(x + y - 9) = 0, \ \text{ or } \ x + y - 9 = 0$$

Solving the first two equations for λ and substituting give

$$\lambda = -2xy$$
$$\lambda = -x^2$$
$$2xy = x^2$$
$$2xy - x^2 = 0$$
$$x(2y - x) = 0$$
$$x = 0 \ \text{ or } \ x = 2y$$

Because $x = 0$ could not make $z = x^2y$ a maximum, we substitute $x = 2y$ into $x + y - 9 = 0$.

$$2y + y = 9 \qquad \Longrightarrow \qquad y = 3$$
$$\text{and} \qquad x = 2y \qquad \Longrightarrow \qquad x = 6$$

Thus the function $z = x^2y$ is maximized at 108 when $x = 6$, $y = 3$, if the constraint is $x + y = 9$. Testing values near $x = 6$, $y = 3$, and satisfying the constraint shows that the function is maximized there. (Try $x = 5.5$, $y = 3.5$; $x = 7$, $y = 2$; and so on.) ∎

EXAMPLE 2 Minima Subject to Constraints

Find the minimum value of the function $z = x^3 + y^3 + xy$ subject to the constraint $x + y - 4 = 0$.

Solution
The function to be minimized is $f(x, y) = x^3 + y^3 + xy$. The constraint function is $g(x, y) = x + y - 4$. The objective function is

$$F(x, y, \lambda) = f(x, y) + \lambda g(x, y) = x^3 + y^3 + xy + \lambda(x + y - 4)$$

Then

$$\frac{\partial F}{\partial x} = 3x^2 + y + \lambda = 0$$

$$\frac{\partial F}{\partial y} = 3y^2 + x + \lambda = 0$$

$$\frac{\partial F}{\partial \lambda} = x + y - 4 = 0$$

Solving the first two equations for λ and substituting, we get

$$\lambda = -(3x^2 + y)$$
$$\lambda = -(3y^2 + x)$$
$$3x^2 + y = 3y^2 + x$$

Solving $x + y - 4 = 0$ for y gives $y = 4 - x$. Substituting for y in the equation above, we get

$$3x^2 + (4 - x) = 3(4 - x)^2 + x$$
$$3x^2 + 4 - x = 48 - 24x + 3x^2 + x$$
$$22x = 44 \text{ or } x = 2$$

Thus when $x + y - 4 = 0$, $x = 2$ and $y = 2$ give the minimum value $z = 20$ because other values that satisfy the constraint give larger z-values. ∎

✓ CHECKPOINT

Find the minimum value of $f(x, y) = x^2 + y^2 - 4xy$, subject to the constraint $x + y = 10$, by

1. forming the objective function $F(x, y, \lambda)$,
2. finding $\dfrac{\partial F}{\partial x}, \dfrac{\partial F}{\partial y}$, and $\dfrac{\partial F}{\partial \lambda}$,
3. setting the three partial derivatives (from Question 2) equal to 0, and solving the equations simultaneously for x and y,
4. finding the value of $f(x, y)$ at the critical values of x and y.

We can also use Lagrange multipliers to find the maxima and minima of functions of three (or more) variables, subject to two (or more) constraints. The method involves using two multipliers, one for each constraint, to form an objective function $F = f + \lambda g_1 + \mu g_2$. We leave further discussion for more advanced courses.

We can easily extend the method to functions of three or more variables, as the following example shows.

EXAMPLE 3 Minima of Function of Three Variables

Find the minimum value of the function $w = x + y^2 + z^2$, subject to the constraint $x + y + z = 1$.

Solution

The function to be minimized is $f(x, y, z) = x + y^2 + z^2$. The constraint is $g(x, y, z) = 0$, where $g(x, y, z) = x + y + z - 1$. The objective function is

$$F(x, y, z, \lambda) = f(x, y, z) + \lambda g(x, y, z)$$

or

$$F(x, y, z, \lambda) = x + y^2 + z^2 + \lambda(x + y + z - 1)$$

Then

$$\frac{\partial F}{\partial x} = 1 + \lambda = 0$$

$$\frac{\partial F}{\partial y} = 2y + \lambda = 0$$

$$\frac{\partial F}{\partial z} = 2z + \lambda = 0$$

$$\frac{\partial F}{\partial \lambda} = x + y + z - 1 = 0$$

Solving the first three equations simultaneously gives

$$\lambda = -1, \quad y = \frac{1}{2}, \quad z = \frac{1}{2}$$

Substituting these values in the fourth equation (which is the constraint), we get $x + \frac{1}{2} + \frac{1}{2} - 1 = 0$, so $x = 0, y = \frac{1}{2}, z = \frac{1}{2}$. Thus $w = \frac{1}{2}$ is the minimum value because other values of x, y, and z that satisfy $x + y + z = 1$ give larger values of w.   ∎

EXAMPLE 4 Utility | APPLICATION PREVIEW |

Find x and y that maximize the utility function $U = x^2y^2$ subject to the budget constraint $2x + 4y = 40$.

Solution
First we rewrite the constraint as $2x + 4y - 40 = 0$. Then the objective function is

$$F(x, y, \lambda) = x^2y^2 + \lambda(2x + 4y - 40)$$

$$\frac{\partial F}{\partial x} = 2xy^2 + 2\lambda, \qquad \frac{\partial F}{\partial y} = 2x^2y + 4\lambda, \qquad \frac{\partial F}{\partial \lambda} = 2x + 4y - 40$$

Setting these partial derivatives equal to 0 and solving give

$$-\lambda = xy^2 = x^2y/2, \ \text{ or } \ xy^2 - x^2y/2 = 0$$

so

$$xy(y - x/2) = 0$$

yields $x = 0$, $y = 0$, or $x = 2y$. Neither $x = 0$ nor $y = 0$ maximizes utility. If $x = 2y$, then $0 = 2x + 4y - 40$ becomes

$$0 = 4y + 4y - 40 \quad \text{or} \quad 40 = 8y$$

Thus $y = 5$ and $x = 10$.
    Testing values near $x = 10$, $y = 5$ shows that these values maximize utility at $U = 2500$.   ∎

Figure 14.13 shows the budget constraint $2x + 4y = 40$ from Example 4 graphed with the indifference curves for $U = x^2y^2$ that correspond to $U = 500$, $U = 2500$, and $U = 5000$.
    Whenever an indifference curve intersects the budget constraint, that utility level is attainable within the budget. Note that the highest attainable utility (such as $U = 2500$, found in Example 4) corresponds to the indifference curve that touches the budget constraint at exactly one point—that is, the curve that has the budget constraint as a tangent line. Note also that utility levels greater than $U = 2500$ are not attainable within the budget because the indifference curve "misses" the budget constraint line (as for $U = 5000$).

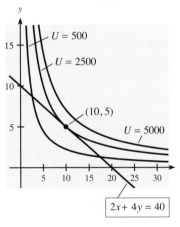

Figure 14.13

EXAMPLE 5 Production

Suppose that the Cobb-Douglas production function for a certain manufacturer gives the number of units of production z according to

$$z = f(x, y) = 100x^{4/5}y^{1/5}$$

where x is the number of units of labor and y is the number of units of capital. Suppose further that labor costs \$160 per unit, capital costs \$200 per unit, and the total cost for capital and labor is limited to \$100,000, so that production is constrained by

$$160x + 200y = 100{,}000$$

Find the number of units of labor and the number of units of capital that maximize production.

Solution
The objective function is

$$F(x, y, \lambda) = 100x^{4/5}y^{1/5} + \lambda(160x + 200y - 100{,}000)$$

$$\frac{\partial F}{\partial x} = 80x^{-1/5}y^{1/5} + 160\lambda, \qquad \frac{\partial F}{\partial y} = 20x^{4/5}y^{-4/5} + 200\lambda$$

$$\frac{\partial F}{\partial \lambda} = 160x + 200y - 100{,}000$$

Setting these partial derivatives equal to 0 and solving give

$$\lambda = \frac{-80x^{-1/5}y^{1/5}}{160} = \frac{-20x^{4/5}y^{-4/5}}{200} \quad \text{or} \quad \frac{y^{1/5}}{2x^{1/5}} = \frac{x^{4/5}}{10y^{4/5}}$$

This means $5y = x$. Using this in $\dfrac{\partial F}{\partial \lambda} = 0$ gives

$$160(5y) + 200y - 100{,}000 = 0$$
$$1000y = 100{,}000$$
$$y = 100$$
$$x = 5y = 500$$

Thus production is maximized at $z = 100(500)^{4/5}(100)^{1/5} \approx 36{,}239$ when $x = 500$ (units of labor) and $y = 100$ (units of capital). See Figure 14.14.

In problems of this type, economists call the value of $-\lambda$ the **marginal productivity of money.** In this case,

$$-\lambda = \frac{y^{1/5}}{2x^{1/5}} = \frac{(100)^{0.2}}{2(500)^{0.2}} \approx 0.362$$

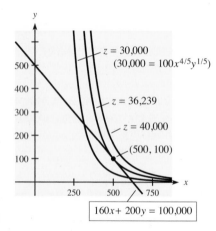

Figure 14.14

This means that each additional dollar spent on production results in approximately 0.362 additional unit produced.

Finally, Figure 14.14 shows the graph of the constraint, together with some production function curves that correspond to different production levels.

Spreadsheet Note The Excel tool called Solver can be used to find maxima and minima of functions subject to constraints. For details of this solution method, see Appendix B, Section 14.5, and the Online Excel Guide. In Example 5, the numbers of units of labor and capital that maximize production if the production is given by the Cobb-Douglas function

$$P(x, y) = 100x^{4/5}y^{1/5}$$

and if capital and labor are constrained by

$$160x + 200y = 100{,}000$$

were found to be 500 and 100, respectively. The following Excel screen shows the entries for beginning the solution of this problem.

	A	B	C
1	**Variables**		
2			
3	units labor (x)	0	
4	units capital (y)	0	
5			
6	**Objective**		
7			
8	Maximize production	= 100*B3^(4/5)*B4^(1/5)	
9			
10	**Constraint**		
11		Amount used	Available
12	Cost	= 160*B3 + 200*B4	100000
13			

The solution, found with Solver, is shown below.

	A	B	C
1	**Variables**		
2			
3	units labor (x)	500	
4	units capital (y)	100	
5		/	
6	**Objective**		
7			
8	Maximize production	36238.98	
9			
10	**Constraint**		
11		Amount used	Available
12	Cost	100000	100000
13			

✓ **CHECKPOINT SOLUTIONS**

1. $F(x, y, \lambda) = x^2 + y^2 - 4xy + \lambda(10 - x - y)$
2. $\dfrac{\partial F}{\partial x} = 2x - 4y - \lambda, \quad \dfrac{\partial F}{\partial y} = 2y - 4x - \lambda, \quad \dfrac{\partial F}{\partial \lambda} = 10 - x - y$
3. $x = 5$ and $y = 5$.
4. $f(5, 5) = -50$ is the minimum

| EXERCISES | 14.5

1. Find the minimum value of $z = x^2 + y^2$ subject to the condition $x + y = 6$.
2. Find the minimum value of $z = 4x^2 + y^2$ subject to the constraint $x + y = 5$.
3. Find the minimum value of $z = 3x^2 + 5y^2 - 2xy$ subject to the constraint $x + y = 5$.
4. Find the maximum value of $z = 2xy - 3x^2 - 5y^2$ subject to the constraint $x + y = 5$.
5. Find the maximum value of $z = x^2y$ subject to $x + y = 6$, $x \geq 0$, $y \geq 0$.
6. Find the maximum value of the function $z = x^3y^2$ subject to $x + y = 10$, $x \geq 0$, $y \geq 0$.
7. Find the maximum value of the function $z = 2xy - 2x^2 - 4y^2$ subject to the condition $x + 2y = 8$.
8. Find the minimum value of $z = 2x^2 + y^2 - xy$ subject to the constraint $2x + y = 8$.
9. Find the maximum value of $z = xy$ subject to $9x^2 + 25y^2 = 450$; $x \geq 0$, $y \geq 0$.
10. Find the maximum value of $z = xy^2$ subject to $2x^2 + y^2 = 600$; $x \geq 0$, $y \geq 0$.
11. Find the minimum value of $w = x^2 + y^2 + z^2$ subject to the constraint $x + y + z = 3$.
12. Find the minimum value of $w = x^2 + y^2 + z^2$ subject to the condition $2x - 4y + z = 21$.
13. Find the maximum value of $w = xz + y$ subject to the constraint $x^2 + y^2 + z^2 = 1$.
14. Find the maximum value of $w = x^2yz$ subject to the constraint $4x + y + z = 4$, $x \geq 0$, $y \geq 0$, and $z \geq 0$.

APPLICATIONS

15. *Utility* Suppose that the utility function for two commodities is given by $U = xy^2$ and that the budget constraint is $3x + 6y = 18$. What values of x and y will maximize utility?

16. *Utility* Suppose that the budget constraint in Problem 15 is $5x + 20y = 90$. What values of x and y will maximize $U = xy^2$?

17. *Utility* Suppose that the utility function for two products is given by $U = x^2y$, and the budget constraint is $2x + 3y = 120$. Find the values of x and y that maximize utility. Check by graphing the budget constraint with the indifference curve for maximum utility and with two other indifference curves.

18. *Utility* Suppose that the utility function for two commodities is given by $U = x^2y^3$, and the budget constraint is $10x + 15y = 250$. Find the values of x and y that maximize utility. Check by graphing the budget constraint with the indifference curve for maximum utility and with two other indifference curves.

19. *Production* A company has the Cobb-Douglas production function

$$z = 400x^{0.6}y^{0.4}$$

where x is the number of units of labor, y is the number of units of capital, and z is the units of production. Suppose labor costs \$150 per unit, capital costs \$100 per unit, and the total cost of labor and capital is limited to \$100,000.
   (a) Find the number of units of labor and the number of units of capital that maximize production.
   (b) Find the marginal productivity of money and interpret it.
   (c) Graph the constraint with the optimal value for production and with two other z-values (one smaller than the optimal value and one larger).

20. *Production* Suppose a company has the Cobb-Douglas production function

$$z = 100^{0.75}y^{0.25}$$

where x is the number of units of labor, y is the number of units of capital, and z is the units of production. Suppose further that labor costs \$90 per unit, capital costs \$150 per unit, and the total costs of labor and capital are limited to \$90,000.
   (a) Find the number of units of labor and the number of units of capital that maximize production.
   (b) Find the marginal productivity of money and interpret it.
   (c) Graph the constraint with the optimal value for production and with two other z-values (one smaller than the optimal value and one larger).

21. *Cost* A firm has two plants, X and Y. Suppose that the cost of producing x units at plant X is $x^2 + 1200$ dollars and the cost of producing y units of the same product at plant Y is given by $3y^2 + 800$ dollars. If the firm has an order for 1200 units, how many should it produce at each plant to fill this order and minimize the cost of production?

22. *Cost* Suppose that the cost of producing x units at plant X is $(3x + 4)x$ dollars and that the cost of producing y units of the same product at plant Y is $(2y + 8)y$ dollars. If the firm that owns the plants has an order for 149 units, how many should it produce at each plant to fill this order and minimize its cost of production?

23. *Revenue* On the basis of past experience a company has determined that its monthly sales revenue (in dollars) is related to its advertising according to the formula $s = 20x + y^2 + 4xy$, where x is the amount spent on print advertising and y is the amount spent on cable advertising. If the company plans to spend \$30,000 per month on these two means of advertising, how much should it spend on each method to maximize its monthly sales revenue?

24. *Manufacturing* Find the dimensions x, y, and z (in inches) of the rectangular box with the largest volume that satisfies

$$3x + 4y + 12z = 12$$

25. *Manufacturing* Find the dimensions (in centimeters) of the box with square base, open top, and volume 500,000 cubic centimeters that requires the least materials.

26. *Manufacturing* Show that a box with a square base, an open top, and a fixed volume requires the least material to build if it has a height equal to one-half the length of one side of the base.

Chapter 14 **Summary & Review**

KEY TERMS AND FORMULAS

Section 14.1

Cobb-Douglas production function (p. 860)
$$Q = AK^\alpha L^{1-\alpha}$$
Function of two variables (p. 860)
$$z = f(x, y)$$
Independent variables: x and y
Dependent variable: z
Domain

Coordinate planes (p. 861)
Utility (p. 862)
   Indifference curve
   Indifference map

Section 14.2

First-order partial derivative (p. 866)
  With respect to x
$$z_x = \frac{\partial z}{\partial x}$$
  With respect to y
$$z_y = \frac{\partial z}{\partial y}$$

Higher-order partial derivatives (p. 871)
  Second partial derivatives
   z_{xx}, z_{yy}, z_{xy}, and z_{yx}

Section 14.3

Joint cost function (p. 875)
$$C = Q(x, y)$$
Marginal cost (p. 875)
Marginal productivity (p. 876)

Demand function (p. 878)
Marginal demand functions (p. 878)
Competitive products (p. 879)
Complementary products (p. 879)

Section 14.4

Critical values for maxima and minima (p. 882)
$$\text{Solve simultaneously} \begin{cases} z_x = 0 \\ z_y = 0 \end{cases}.$$
Test for critical values (p. 883)
  Use $D(x, y) = (z_{xx})(z_{yy}) - (z_{xy})^2$

Linear regression (p. 886)
$$\hat{y} = a + bx$$
$$b = \frac{\Sigma x \cdot \Sigma y - n\Sigma xy}{(\Sigma x)^2 - n\Sigma x^2}$$
$$a = \frac{\Sigma y - b\Sigma x}{n}$$

Section 14.5

Lagrange multipliers (p. 892)
Objective function (p. 892)

$$F(x, y, \lambda) = f(x, y) + \lambda g(x, y)$$
Maxima and minima subject to constraints (p. 893)

REVIEW EXERCISES

Section 14.1

1. What is the domain of $z = \dfrac{3}{2x - y}$?

2. What is the domain of $z = \dfrac{3x + 2\sqrt{y}}{x^2 + y^2}$?

3. If $w(x, y, z) = x^2 - 3yz$, find $w(2, 3, 1)$.

4. If $Q(K, L) = 70K^{2/3}L^{1/3}$, find $Q(64,000, 512)$.

Section 14.2

5. Find $\dfrac{\partial z}{\partial x}$ if $z = 5x^3 + 6xy + y^2$.

6. Find $\dfrac{\partial z}{\partial y}$ if $z = 12x^5 - 14x^3y^3 + 6y^4 - 1$.

In Problems 7–12, find z_x and z_y.

7. $z = 4x^2y^3 + \dfrac{x}{y}$

8. $z = \sqrt{x^2 + 2y^2}$

9. $z = (xy + 1)^{-2}$

10. $z = e^{x^2y^3}$

11. $z = e^{xy} + y \ln x$

12. $z = e^{\ln xy}$

13. Find the partial derivative of $f(x, y) = 4x^3 - 5xy^2 + y^3$ with respect to x at the point $(1, 2, -8)$.

14. Find the slope of the tangent in the x-direction to the surface $z = 5x^4 - 3xy^2 + y^2$ at $(1, 2, -3)$.

In Problems 15–18, find the second partials.

(a) z_{xx} (b) z_{yy} (c) z_{xy} (d) z_{yx}

15. $z = x^2y - 3xy$

16. $z = 3x^3y^4 - \dfrac{x^2}{y^2}$

17. $z = x^2e^{y^2}$

18. $z = \ln(xy + 1)$

Section 14.4

19. Test $z = 16 - x^2 - xy - y^2 + 24y$ for maxima and minima.

20. Test $z = x^3 + y^3 - 12x - 27y$ for maxima and minima.

Section 14.5

21. Find the minimum value of $z = 4x^2 + y^2$ subject to the constraint $x + y = 10$.

22. Find the maximum value of $z = x^4y^2$ subject to the constraint $x + y = 9$, $x \geq 0$, $y \geq 0$.

APPLICATIONS

Sections 14.1–14.2

23. *Utility* Suppose that the utility function for two goods X and Y is given by $U = x^2y$.
    (a) Write the equation of the indifference curve for a consumer who purchases 6 units of X and 15 units of Y.
    (b) If the consumer purchases 60 units of Y, how many units of X must be purchased to retain the same level of utility?

24. *Savings plans* The accumulated value A of a monthly savings plan over a 20-year period is a function of the monthly contribution R and the interest rate $r\%$, compounded monthly, according to

$$A = f(R, r) = \dfrac{1200R\left[\left(1 + \dfrac{r}{1200}\right)^{240} - 1\right]}{r}$$

(a) Find the accumulated value of a plan that contributes $100 per month with interest rate 6%.
(b) Interpret $f(250, 7.8) \approx 143{,}648$.
(c) Interpret $\dfrac{\partial A}{\partial r}(250, 7.8) \approx 17{,}770$.
(d) Find $\dfrac{\partial A}{\partial R}(250, 7.8)$ and interpret the result.

25. *Retirement benefits* The monthly benefit B (in thousands of dollars) from a retirement account that is invested at 9% compounded monthly is a function of the account value V (also in thousands of dollars) and the number of years t that benefits are paid, and it can be approximated by

$$B = f(V, t) = \dfrac{3V}{400 - 400e^{-0.0897t}}$$

(a) Find the benefit if the account value is $1,000,000 and the monthly benefits last for 20 years.
(b) Find and interpret $\dfrac{\partial B}{\partial V}(1000, 20)$.
(c) Find and interpret $\dfrac{\partial B}{\partial t}(1000, 20)$.

26. *Advertising and sales* The number of units of sales of a product, S, is a function of the dollars spent for advertising, A, and the product's price, p. Suppose $S = f(A, p)$ is the function relating these quantities.
    (a) Explain why $\dfrac{\partial S}{\partial A} > 0$.
    (b) Do you think $\dfrac{\partial S}{\partial p}$ is positive or negative? Explain.

Section 14.3

27. *Cost* The joint cost, in dollars, for two products is given by $C(x, y) = x^2\sqrt{y^2 + 13}$. Find the marginal cost with respect to
    (a) x if 20 units of x and 6 units of y are produced.
    (b) y if 20 units of x and 6 units of y are produced.

28. *Production* Suppose that the production function for a company is given by

$$Q = 80K^{1/4}L^{3/4}$$

where Q is the output (in hundreds of units), K is the capital expenditures (in thousands of dollars), and L is the work-hours. Find $\partial Q/\partial K$ and $\partial Q/\partial L$ when expenditures are $625,000 and total work-hours are 4096. Interpret the results.

29. *Marginal demand* The demand functions for two related products, product A and product B, are given by

$$q_A = 400 - 2p_A - 3p_B$$
$$q_B = 300 - 5p_A - 6p_B$$

where p_A and p_B are the respective prices in dollars.
(a) Find the marginal demand of q_A with respect to p_A.
(b) Find the marginal demand of q_B with respect to p_B.
(c) Are the products complementary or competitive?

30. *Marginal demand* Suppose that the demand functions for two related products, A and B, are given by

$$q_A = 800 - 40p_A - \frac{2}{p_B + 1}$$

$$q_B = 1000 - \frac{10}{p_A + 4} - 30p_B$$

where p_A and p_B are the respective prices in dollars. Determine whether the products are competitive or complementary.

Section 14.4

31. *Profit* The weekly profit (in dollars) from the sale of two products is given by

$$P(x, y) = 40x + 80y - x^2 - y^2$$

where x is the number of units of product 1 and y is the number of units of product 2. Selling how much of each product will maximize profit? Find the maximum weekly profit.

32. *Cost* Suppose a company has two separate plants that manufacture the same item. Suppose x is the amount produced at plant I and y is the amount at plant II. If the total cost function for the two plants is

$$C(x, y) = 22{,}500 - 12x - 30y + 0.03x^2 + 0.01y^2$$

find the production allocation that minimizes the company's total cost.

Section 14.5

33. *Utility* If the utility function for two commodities is $U = x^2y$, and the budget constraint is $4x + 5y = 60$, find the values of x and y that maximize utility.

34. *Production* Suppose a company has the Cobb-Douglas production function

$$z = 300x^{2/3}y^{1/3}$$

where x is the number of units of labor, y is the number of units of capital, and z is the units of production. Suppose labor costs are $50 per unit, capital costs are $50 per unit, and total costs are limited to $75,000.
   (a) Find the number of units of labor and the number of units of capital that maximize production.
   (b) Find the marginal productivity of money and interpret your result.

(c) Graph the constraint with the production function when $z = 180{,}000$, $z = 300{,}000$, and when the z-value is optimal.

Section 14.4

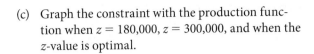

35. *U.S. average wage* The table gives the average annual U.S. wage from 2011 and projected through 2021.
   (a) Use linear regression to find the best linear model for the wage as a function of the number of years after 2010. Report the model with four significant digits.
   (b) Use the reported model to predict the average annual wage in 2025.

Year	Wage ($)	Year	Wage ($)
2011	43,009	2017	55,989
2012	44,644	2018	58,698
2013	46,496	2019	61,179
2014	48,595	2020	63,676
2015	50,893	2021	66,161
2016	53,317		

Source: Social Security Administration

36. The table gives the number of women and the number of men (both in millions) in the U.S. civilian labor force for selected years from 1950 and projected to 2050.
   (a) Let x represent the millions of women and y represent the millions of men and find the linear regression model for $y = f(x)$. Report the model to three significant digits.
   (b) Find and interpret the slope of the reported model.

U.S. Civilian Workforce (in millions): 1950–2050

Women	Men	Women	Men
18.4	43.8	75.5	82.2
23.2	46.4	78.6	84.2
31.5	51.2	79.3	85.4
45.5	61.5	81.6	88.5
56.8	69.0	86.5	94.0
65.7	75.2	91.5	100.3

Source: U.S. Bureau of Labor Statistics

Chapter 14　TEST

1. Consider the function $f(x, y) = \dfrac{2x + 3y}{\sqrt{x^2 - y}}$.

   (a) Find the domain of $f(x, y)$.
   (b) Evaluate $f(-4, 12)$.

2. Find all first and second partial derivatives of
$$z = f(x, y) = 5x - 9y^2 + 2(xy + 1)^5$$

3. Let $z = 6x^2 + x^2y + y^2 - 4y + 9$. Find the pairs (x, y) that are critical points for z, and then classify each as a relative maximum, a relative minimum, or a saddle point.

4. Suppose a company's monthly production value Q, in thousands of dollars, is given by the Cobb-Douglas production function
$$Q = 10K^{0.45}L^{0.55}$$
   where K is thousands of dollars of capital investment per month and L is the total hours of labor per month. Capital investment is currently \$10,000 per month and monthly work-hours of labor total 1590.
   (a) Find the monthly production value (to the nearest thousand dollars).
   (b) Find the marginal productivity with respect to capital investment, and interpret your result.
   (c) Find the marginal productivity with respect to total hours of labor, and interpret your result.

5. The monthly payment R on a loan is a function of the amount borrowed, A, in thousands of dollars; the length of the loan, n, in years; and the annual interest rate, r, as a percent. Thus $R = f(A, n, r)$. In parts (a) and (b), write a sentence that explains the practical meaning of each mathematical statement.
   (a) $f(94.5, 25, 7) = \$667.91$
   (b) $\dfrac{\partial f}{\partial r}(94.5, 25, 7) = \$60.28$
   (c) Would $\dfrac{\partial f}{\partial n}(94.5, 25, 7)$ be positive, negative, or zero? Explain.

6. Let $f(x, y) = 2e^{x^2y^2}$. Find $\dfrac{\partial^2 f}{\partial x \, \partial y}$.

7. Suppose the demand functions for two products are
$$q_1 = 300 - 2p_1 - 5p_2 \text{ and } q_2 = 150 - 4p_1 - 7p_2$$
   where q_1 and q_2 represent quantities demanded and p_1 and p_2 represent prices. What calculations enable us to decide whether the products are competitive or complementary? Are these products competitive or complementary?

8. Suppose a store sells two brands of disposable razors and the profit for these is a function of their two selling prices. The type 1 razor sells for \$x, the type 2 sells for \$y, and profit is given by
$$P = 915x - 30x^2 - 45xy + 975y - 30y^2 - 3500$$
   Find the selling prices that maximize profit. Find the maximum profit.

9. Find x and y that maximize the utility function $U = x^3y$ subject to the budget constraint $30x + 20y = 8000$.

10. In the last half of the twentieth century the U.S. population grew more diverse both racially and ethnically, with persons of Hispanic origin representing one of the fastest growing segments. The table gives the percent of the U.S. civilian non-institutional population of Hispanic origin for selected years from 1980 and projected to 2050.
    (a) Find the linear regression line for these data. Use $x = 0$ to represent 1980.
    (b) How well does the regression line fit the data?
    (c) Using the linear regression equation, predict the percent of the U.S. civilian non-institutional population of Hispanic origin in 2025.

Year	Percent
1980	5.7
1990	8.4
2000	10.7
2010	13.0
2015	14.1
2020	15.3
2030	17.8
2040	20.3
2050	22.8

Source: U.S. Bureau of the Census

Extended Applications & Group Projects

I. Advertising

To model sales of its tires, the manufacturer of GRIPPER tires used the quadratic equation $S = a_0 + a_1 x + a_2 x^2 + b_1 y$, where S is regional sales in millions of dollars, x is TV advertising expenditures in millions of dollars, and y is other promotional expenditures in millions of dollars. (See the Extended Application/Group Project "Marginal Return to Sales," in Chapter 9.)

Although this model represents the relationship between advertising and sales dollars for small changes in advertising expenditures, it is clear to the vice president of advertising that it does not apply to large expenditures for TV advertising on a national level. He knows from experience that increased expenditures for TV advertising result in more sales, but at a decreasing rate of return for the product.

The vice president is aware that some advertising agencies model the relationship between advertising and sales by the function

$$S = b_0 + b_1(1 - e^{-ax}) + c_1 y$$

where $a > 0$, S is sales in millions of dollars, x is TV advertising expenditures in millions of dollars, and y is other promotional expenditures in millions of dollars.* The equation

$$S_n = 24.58 + 325.18(1 - e^{-x/14}) + b_1 y$$

has the form mentioned previously as being used by some advertising agencies. For TV advertising expenditures up to $20 million, this equation closely approximates

$$S_1 = 30 + 20x - 0.4x^2 + b_1 y$$

which, in the Extended Application/Group Project in Chapter 9, was used with fixed promotional expenses to describe advertising and sales in Region 1.

To help the vice president decide whether this is a better model for large expenditures, answer the following questions.

1. What is $\partial S_1/\partial x$? Does this indicate that sales might actually decline after some amount is spent on TV advertising? If so, what is this amount?
2. Does the quadratic model $S_1(x, y)$ indicate that sales will become negative after some amount is spent on TV advertising? Does this model cease to be useful in predicting sales after a certain point?
3. What is $\partial S_n/\partial x$? Does this indicate that sales will continue to rise if additional money is devoted to TV advertising? Is S_n growing at a rate that is increasing or decreasing when promotional sales are held constant? Is S_n a better model for large expenditures?
4. If this model does describe the relationship between advertising and sales, and if promotional expenditures are held constant at y_0, is there an upper limit to the sales, even if an unlimited amount of money is spent on TV advertising? If so, what is it?

*Edwin Mansfield, *Managerial Economics* (New York: Norton, 1990).

II. Competitive Pricing

Retailers often sell different brands of competing products. Depending on the joint demand for the products, the retailer may be able to set prices that regulate demand and, therefore, influence profits.

Suppose HOME-ALL, Inc., a national chain of home improvement retailers, sells two competing brands of motion sensor outdoor light sets, Dark-B-Gone 100 and Croyle & James, which the chain purchases for $8 per set and $10 per set, respectively. HOME-ALL's research department has determined the following two monthly demand equations for these light sets:

$$D = 120 - 40d + 30c \quad \text{and} \quad C = 680 + 30d - 40c$$

where D is hundreds of Dark-B-Gone 100 light sets demanded at d per set and C is hundreds of Croyle & James light sets demanded at c per set. For what prices should HOME-ALL sell these light sets in order to maximize its monthly profit on these items?

To answer this question, complete the following.

1. Recall that revenue is a product's selling price per item times the number of items sold. With this in mind, formulate HOME-ALL's total revenue function for the two light set brands as a function of their prices.
2. Form HOME-ALL's profit function for the two light set brands (in terms of their selling prices).
3. Determine the price of each type of light set that will maximize HOME-ALL's profit.
4. Write a brief report to management that details your pricing recommendations and justifies them.

Graphing Calculator Guide

Operating the TI-83 and TI-84 Plus Calculators

Turning the Calculator On and Off

| ON | Turns the calculator on. |

| 2nd | ON | Turns the calculator off. |

Adjusting the Display Contrast

| 2nd | ▲ | Increases the contrast (darkens the screen). |

| 2nd | ▼ | Decreases the contrast (lightens the screen). |

Note: If the display begins to dim (especially during calculations) and you must adjust the contrast to 8 or 9 in order to see the screen, then batteries are low and you should replace them soon.

The TI-83 and TI-84 Plus keyboards are divided into four zones: graphing keys, editing keys, advanced function keys, and scientific calculator keys (Figure 1).

Courtesy of Texas Instruments

Home screen

Graphing keys

Editing keys (Allow you to edit expressions and variables)

Advanced function keys (Display menus that access advanced functions)

Scientific calculator keys

Figure 1

Chapter 1
Section 1.4 Graphing Equations

Entering Equations for Graphing

To graph an equation in the variables x and y, first solve the equation for y in terms of x. If the equation has variables other than x and y, solve for the dependent variable and replace the independent variable with x. Press the Y = key to access the function entry screen and enter the equation. To erase an equation, press CLEAR. To return to the home-screen, press 2nd MODE (QUIT).

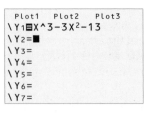

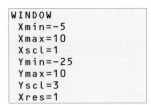

Setting Windows

The window defines the highest and lowest values of x and y on the graph of the function that will be shown on the screen. The values that define the viewing window can be set by using ZOOM keys. The standard window (ZOOM 6) is often appropriate. The standard window gives x- and y-values between -10 and 10.

To set the window manually, press the WINDOW key and enter the values that you want.

The Xscl (x scale) is the distance between tic marks on the x-axis, going in both directions from the origin, and the Yscl (y scale) is the distance between tic marks on the y-axis, going in both directions from the origin. The scales can be set individually, and are useful in visually determining x- and y-intercepts of graphs.

Graphing Equations

Determine an appropriate viewing window. The window should be set so that the important parts of the graph are shown and the unseen parts are suggested. Such a graph is called **complete**. Using the displayed coordinates from TRACE helps to determine an appropriate window. Pressing GRAPH or a ZOOM key will activate the graph.

Graph of $y = x^3 - 3x^2 - 13$

With standard window   Using $[-5, 10]$ by $[-25, 10]$

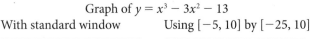

Section 1.4 Finding Function Values

Using TRACE on the Graph

Enter the function to be evaluated in Y_1. Choose a window so that it contains the x-value whose y-value you seek. Press TRACE and then enter the selected x-value followed by ENTER. The cursor will move to the selected value and give the resulting y-value if the selected x-value is in the window. If the selected x-value is not in the window, Err: INVALID occurs. If the x-value is in the window, the y-value will occur even if it is not visible in the window.

To evaluate $y = -x^2 + 8x + 9$ when $x = -5$ and when $x = 3$, graph the function using the window $[-10, 10]$ by $[-10, 30]$.

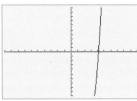

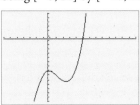

Using the TABLE ASK Feature

Enter the function with the Y = key. {Note: The = sign must be highlighted.) Press 2nd WINDOW (TBLSET), move the cursor to Ask opposite Indpnt:, and press ENTER. This allows you to input specific values for x. Then press 2nd TABLE and enter the specific values. Pressing DEL clears an entry. The table on the right evaluates $y = -x^2 + 8x + 9$ at -5 and at 3.

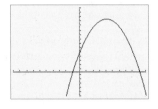

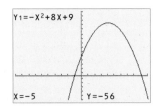

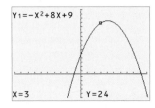

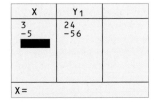

Making a Table of Values

If the Indpnt variable is on Auto, enter an initial x-value for the table in TblStart, and enter the desired change in the x-value as ΔTbl.

Enter 2nd TABLE to get a list of x-values and the corresponding y-values. The value of the function at the given value of x can be read from the table. Use the up or down arrows to find the x-values where the function is to be evaluated. The table on the right evaluates $y = -x^2 + 8x + 9$ for integer x-values from -3 to 3.

Section 1.4 Solving Linear Equations by the x-Intercept Method

To find the solution to $f(x) = 0$ (the x-value where the graph crosses the x-axis):

1. Set one side of the equation to 0 and enter the other side as Y_1 in the Y = menu.
2. Set the window so that the x-intercept to be located can be seen.
3. Press 2nd TRACE to access the CALC menu and select 2:zero.
4. Answer the question "*Left Bound?*" with ENTER after moving the cursor close to and to the left of an x-intercept.
5. Answer the question "*Right Bound?*" with ENTER after moving the cursor close to and to the right of this x-intercept.
6. To the question "*Guess?*" press ENTER. The coordinates of the x-intercept are displayed. The x-value is a solution.

The solution to $5x - 9 = 0$ is found to be $x = 1.8$.

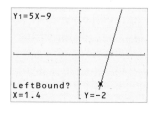
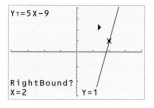
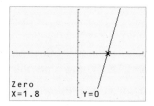

Section 1.5 Solving Systems of Equations in Two Variables

To solve a system of linear equations in two variables graphically:

1. Solve both equations for y.
2. Graph the first equation as Y_1 and the second as Y_2.
3. To find the point of intersection of the graphs:
   (a) Press 2nd TRACE to access the CALC menu and select 5:intersect.
   (b) Answer the question "*First curve?*" by pressing ENTER and "*Second curve?*" by pressing ENTER.
   (c) To the question "*Guess?*" press ENTER. The solution is shown on the right.

If the two lines intersect in one point, the coordinates give the x- and y-values of the solution.

The solution of the system at the right is $x = 2$, $y = 1$.

To solve $\begin{cases} 4x + 3y = 11 \\ 2x - 5y = -1 \end{cases}$ graphically, graph

$y_1 = -\dfrac{4}{3}x + \dfrac{11}{3}$ and $y_2 = \dfrac{2}{5}x + \dfrac{1}{5}$, then use Intersect.

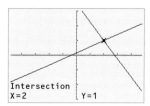

Chapter 2

Section 2.1 Solving Nonlinear Equations by the x-Intercept Method

To find the solutions to $f(x) = 0$ (the x-values where the graph crosses the x-axis):

1. Set one side of the equation to 0 and enter the other side as Y_1 in the Y = menu.
2. Set the window so that the x-intercepts to be located can be seen.
3. Press 2nd TRACE to access the CALC menu and select 2:zero.
4. Answer the question "*Left Bound?*" with ENTER after moving the cursor close to and to the left of an x-intercept.
5. Answer the question "*Right Bound?*" with ENTER after moving the cursor close to and to the right of this x-intercept.
6. To the question "*Guess?*" press ENTER. The coordinates of the x-intercept are displayed. The x-value is a solution.
7. Repeat to get all x-intercepts. The graph of a linear equation will cross the x-axis at most 1 time; the graph of quadratic equation will cross the x-axis at most 2 times, etc.

The solutions to $0 = -x^2 + 8x + 9$ are found to be 9 and -1.

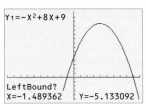
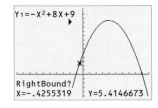

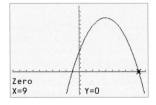
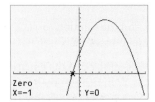

Section 2.1 Solving Nonlinear Equations by the Intersection Method

To solve nonlinear equations by the intersection method:

1. Graph the left side of the equation as Y_1 and the right side as Y_2.
2. Find a point of intersection of the graphs as shown in Section 1.5.
3. To find another point of intersection, repeat while keeping the cursor near the second point.

The solutions to $7x^2 = 16x - 4$ are found at the right.

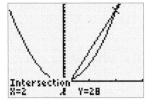
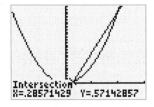

Section 2.2 Graphing Quadratic Functions

To graph a quadratic function:

1. Solve for y in terms of x and enter it in the Y = menu.
2. Find the coordinates of its vertex, with $x = -b/a$.
3. Set the window so the x-coordinate of its vertex is near its center and the y-coordinate is visible.
4. Press Graph.

To graph $P(x) = -0.1x^2 + 300x - 1200$, enter $y_1 = -0.1x^2 + 300x - 1200$ on a window with its center near $x = (-300)/[2(-0.1)] = 1500$ and with $y = P(1500) = 223{,}800$ visible, and graph.

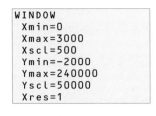
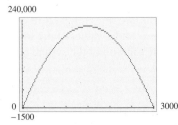

Section 2.4 Graphing Polynomial Functions

To graph a polynomial function:

1. See Table 2.1 in the text to determine possible shapes for the graph.
   (The graph of the function $y = x^3 - 16x$ has one of four shapes in the table.)
2. Graph the function in a window large enough to see the shape of the complete graph. This graph is like the graph of Degree 3(b) in the table in Section 2.4.
3. If necessary, adjust the window for a better view of the graph.

$[-40, 40]$ by $[-100, 100]$ $[-6, 6]$ by $[-30, 30]$

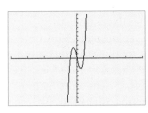   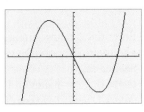

$y = x^3 - 16x$ $y = x^3 - 16x$
                   (Better view)

Section 2.4 Graphing Rational Functions

To graph a rational function:

1. Determine the vertical and horizontal asymptotes.
2. Set the window so that the x range is centered near the x-value of the vertical asymptote.
3. Set the window so that the horizontal asymptote is near the center of the y range.
4. Graph the function in a window large enough to see the shape of the complete graph.
5. If necessary, adjust the window for a better view of the graph.

To graph $y = \dfrac{12x + 8}{3x - 9}$, set the center of the window near the vertical asymptote $x = 3$ and near the horizontal asymptote $y = 4$.

   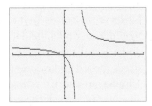

Section 2.4 Graphing Piecewise Defined Functions

A piecewise defined function is defined differently over two or more intervals.

To graph a piecewise defined function $y = \begin{cases} f(x) \text{ if } x \le a \\ g(x) \text{ if } x > a \end{cases}$

1. Go to the Y = key and enter

   $Y_1 = f(x)/(x \le a)$ and $Y_2 = g(x)/(x > a)$

   (The inequality symbols are found under the TEST menu.)
2. Graph the function using an appropriate window.
3. Evaluating a piecewise defined function at a given value of x requires that the correct equation ("piece") be selected.

To graph $y = \begin{cases} x + 7 \text{ if } x \le -5 \\ -x + 2 \text{ if } x > -5 \end{cases}$

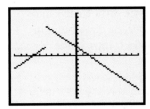

Section 2.5 Modeling

A. To Create a Scatter Plot

1. Press STAT and under EDIT press 1:Edit. This brings you to the screen where you enter data into lists.
2. Enter the *x*-values (input) in the column headed L1 and the corresponding *y*-values (output) in the column headed L2.
3. Go to the Y = menu and turn off or clear any functions entered there. To turn off a function, move the cursor over the = sign and press ENTER.
4. Press 2nd STAT PLOT, 1:Plot 1. Highlight ON, and then highlight the first graph type (Scatter Plot), Enter Xlist:L1, Ylist:L2, and pick the point plot mark you want.
5. Choose an appropriate WINDOW for the graph and press GRAPH, or press ZOOM, 9:ZoomStat to plot the data points.

B. To Find an Equation That Models a Set of Data Points

1. Observe the scatter plot to determine what type function would best model the data. Press STAT, move to CALC, and select the function type to be used to model the data.
2. Press the VARS key, move to Y-VARS, and select 1:Function and 1:Y_1. Press ENTER.
   The coefficients of the equation will appear on the screen and the regression equation will appear as Y_1 on the Y = screen.

Pressing ZOOM 9 shows how well the model fits the data.

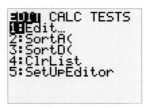

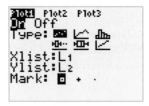

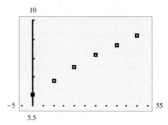

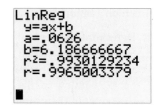

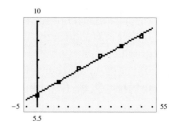

The Model is $y = 0.0626x + 6.19$.

Chapter 3
Section 3.1 Entering Data into Matrices

To enter data into matrices, press the MATRIX key. Move the cursor to EDIT. Enter the number of the matrix into which the data is to be entered. Enter the dimensions of the matrix, and enter the value for each entry of the matrix. Press ENTER after each entry.
For example, we enter the matrix below as [A].

$$\begin{bmatrix} 1 & 2 & 3 \\ 2 & -2 & 1 \\ 3 & 1 & -2 \end{bmatrix}$$

1. Enter 3's to set the dimension, and enter the numbers.
2. To perform operations with the matrix or leave the editor, first press 2nd QUIT.
3. To view the matrix, press MATRIX, the number of the matrix, and ENTER.

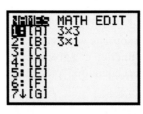

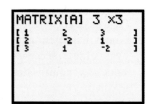

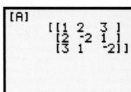

Section 3.1 Operations with Matrices

To find the sum of two matrices, [A] and [D], enter
[A] + [D], and press ENTER. For example, the sum

$$\begin{bmatrix} 1 & 2 & 3 \\ 2 & -2 & 1 \\ 3 & 1 & -2 \end{bmatrix} + \begin{bmatrix} 7 & -3 & 2 \\ 4 & -5 & 3 \\ 0 & 2 & 1 \end{bmatrix} \text{ is shown at right.}$$

To find the difference, enter [A] − [D] and press ENTER.
We can multiply a matrix [D] by a real number (scalar) k by
entering k [D].

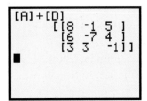

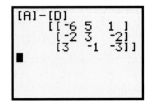

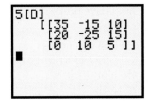

Section 3.2 Multiplying Two Matrices

To find the product of two matrices, [C] times [A], enter
[C][A] and press ENTER. For example, we compute the
product

$$\begin{bmatrix} 1 & 2 & 4 \\ -3 & 2 & -1 \end{bmatrix} \begin{bmatrix} 1 & 2 & 3 \\ 2 & -2 & 1 \\ 3 & 1 & -2 \end{bmatrix} \text{ at right.}$$

Note that entering [A][C] gives an error message. [A][C]
cannot be computed because the dimensions do not match
in this order.

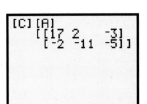

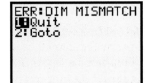

Section 3.3 Solution of Systems—
Reduced Echelon Form

To solve a 3 × 3 system:
1. Enter the coefficients and the constants in an
   augmented matrix.
2. Under the MATRIX menu, choose MATH and B:rref,
   then enter the matrix to be reduced followed by ")", and
   press ENTER.
3. (a) If each row in the coefficient matrix (first 3 columns)
        contains a 1 with the other elements 0's, the solution
        is unique and the number in column 4 of a row is the
        value of the variable corresponding to a 1 in that row.

For example, the system $\begin{cases} 2x - y + z = 6 \\ x + 2y - 3z = 9 \\ 3x \quad\quad - 3z = 15 \end{cases}$ is solved
at right.

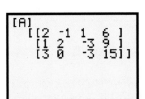

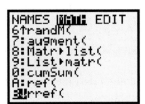

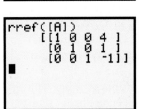

The solution to the system above is unique: $x = 4$, $y = 1$,
and $z = -1$.
   (b) If the reduced matrix has all zeros in the third row,
        the solution is nonunique.
   (c) If the reduced matrix has 3 zeros in a row and a
        nonzero element in the fourth column in that row,
        there is no solution.

Section 3.4 Finding the Inverse of a Matrix

To find the inverse of a matrix:
1. Enter the elements of the matrix using MATRIX and EDIT. Press 2nd QUIT.
2. Press MATRIX, the number of the matrix, and ENTER, then press the x^{-1} key and ENTER.

   For example, the inverse of $E = \begin{bmatrix} 2 & 0 & 2 \\ -1 & 0 & 1 \\ 4 & 2 & 0 \end{bmatrix}$

   is shown at right.
3. To see the entries as fractions, press MATH, press 1:Frac, and press ENTER.

Not all matrices have inverses. Matrices that do not have inverses are called singular matrices. See matrix F at the right.

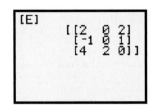

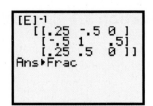

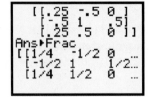

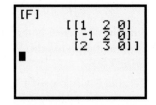

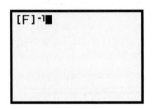

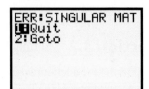

Section 3.4 Solving Systems of Linear Equations with Matrix Inverses

The matrix equation $AX = B$ can be solved by computing $X = A^{-1}B$ if a unique solution exists.

The solution to $\begin{cases} 25x + 20y + 50z = 15{,}000 \\ 25x + 50y + 100z = 27{,}500 \\ 250x + 50y + 500z = 92{,}250 \end{cases}$

is found at the right to be $x = 254$, $y = 385$, $z = 19$.

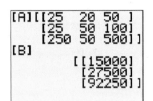

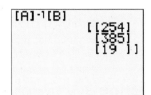

Chapter 4
Section 4.1 Graphing Solution Regions of Linear Inequalities

To graph the solution of a linear inequality in two variables, first solve the inequality for the dependent variable and enter the other side of the inequality in Y_1, so that $Y_1 = f(x)$. If the inequality has the form $y \leq f(x)$, shade the region below the graphed line and if the inequality has the form $y \geq f(x)$, shade the region above the line.

In this example, we shade the region above the line with SHADE under the DRAW menu and enter Shade (Y_1, 10) on the home screen. See Section 4.1 on the next page for an alternate method of shading.

To solve $4x - 2y \leq 6$, convert it to $y \geq 2x - 3$ and graph $Y_1 = 2x - 3$.

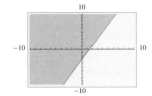

Section 4.1 Graphing Solutions of Systems of Linear Inequalities

To graph the solution region for a system of linear inequalities in two variables, write the inequalities as equations solved for y, and graph the equations.

For example, to find the region defined by the inequalities

$$\begin{cases} 5x + 2y \le 54 \\ 2x + 4y \le 60 \\ x \ge 0, y \ge 0 \end{cases}$$

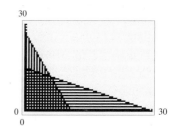

1. Choose a window with x min $= 0$ and y min $= 0$ because the inequalities $x \ge 0, y \ge 0$ limit the graph to Quadrant I.
2. Solve the equations for y getting $y = 27 - 5x/2$ and $y = 15 - x/2$. Enter these equations as Y_1 and Y_2.
3. Because both inequalities have "$y \le$," we move our cursor to the left of Y_1 and press ENTER until a triangle shaded like that in the figure appears and then repeat this with Y_2. Pressing GRAPH then shows the solution to the system.
4. Using TRACE or INTERSECT with the pair of equations and finding the intercepts give the corners of the solution region, where the borders intersect. These corners of the region are (0, 0) (0, 15), (6, 12), and (10.8, 0).

Section 4.2 Linear Programming

To solve a linear programming problem involving constraints in two variables:

1. Graph the constraint inequalities as equations, solved for y.
2. Test points to determine the region and use TRACE or INTERSECT to find each of the corners of the region, where the borders intersect.
3. Then evaluate the objective function at each of the corners.

For example, to maximize $f = 5x + 11y$ subject to the constraints

$$\begin{cases} 5x + 2y \le 54 \\ 2x + 4y \le 60 \\ x \ge 0, y \ge 0 \end{cases}$$

we graphically find the constraint region (as shown above), and evaluate the objective function at the coordinates of each of the corners of the region. Evaluating $f = 5x + 11y$ at each of the corners determines where this objective function is maximized or minimized.

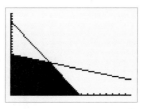

The corners of the region determined by the inequalities are (0, 0) (0, 15), (6, 12), and (10.8, 0).

At (0, 0), $f = 0$ At (0, 15), $f = 165$
At (6,12), $f = 162$ At (10.8, 0), $f = 54$

The maximum value of f is 165 at $x = 0, y = 15$.

Chapter 5

Section 5.1 Graphing Exponential Functions

1. Enter the function as Y_1 in the Y = menu.
2. Set the x-range centered at $x = 0$.
3. Set the y-range to reflect the function's range of $y > 0$.

Note that some graphs (such as the graph of $y = 4^x$ shown here) appear to eventually merge with the negative x-axis. Adjusting the window can show that these graphs never touch the x-axis. For more complicated exponential functions, it may be helpful to use TABLE to find a useful window.

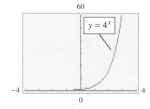

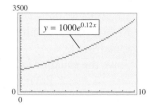

Section 5.1 Modeling with Exponential Functions

1. Create a scatter plot for the data.
2. Choose STAT, then CALC. Scroll down to 0:ExpReg and press ENTER, then VARS, Y-VARS, FUNCTION, Y_1, and ENTER.

   (Recall that this both calculates the requested exponential model and enters its equation as Y_1 in the Y = menu.)

The last screen shows how well the model fits the data.

Find the exponential model for the following data.

x	1	2	3	4	5	6	7	8	9	10
y	43	38	33	29	25	22	19	15	14	12

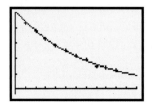

Section 5.2 Graphing Base e and Base 10 Logarithmic Functions

Enter the function as Y_1 in the Y = menu.

1. For $y = \ln(x)$ use the LN key.
2. For $y = \log(x)$ use the LOG key.
3. Set the window x-range to reflect that the function's domain is $x > 0$.
4. Center the window y-range at $y = 0$.

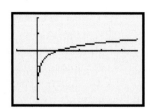

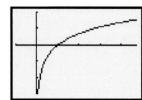

Section 5.2 Modeling with Logarithmic Functions

1. Create a scatter plot for the data.
2. Choose STAT, then CALC. Scroll down to 9:LnReg and press ENTER, then VARS, Y-VARS, FUNCTION, Y1, and ENTER.

The last figure on the right shows how well the model fits the data.

Find the logarithmic model for the following data.

x	10	20	30	38
y	2.21	3.79	4.92	5.77

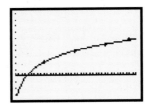

Section 5.2 Graphing Logarithmic Functions with Other Bases

1. Use a change of base formula to rewrite the logarithmic function with base 10 or base e.

$$\log_b x = \frac{\log x}{\log b} \quad \text{or} \quad \log_b x = \frac{\ln x}{\ln b}$$

2. Proceed as described above for graphing base e and base 10 logarithms.

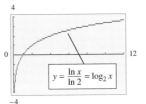

Chapter 6
Section 6.2 Future Value of a Lump Sum

To find the future value of a lump-sum investment:

1. Press the APPS key and select Finance, press ENTER.
2. Select TVM Solver, press ENTER.
3. Set N = the total number of periods, set I% = the annual percentage rate.
4. Set the PV = the lump sum preceded by a "−" to indicate the lump sum is leaving your possession.
5. Set PMT = 0 and set both P/Y and C/Y = the number of compounding periods per year.
6. Put the cursor on FV and press ALPHA ENTER to get the future value.

The future value of $10,000 invested at 9.8% compounded quarterly for 17 years is shown at the right.

 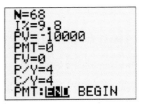

Section 6.3 Future Value of an Annuity

To find the future value of an ordinary annuity:

1. Press the APPS key and select Finance, press ENTER.
2. Select TVM Solver, press ENTER.
3. Set N = the total number of periods, set I% = the annual percentage rate.
4. Set the PV = 0 and set both P/Y and C/Y = the number of compounding periods per year. END should be highlighted.
5. Set PMT = the periodic payment preceded by a "−" to indicate the lump sum is leaving your possession.
6. Put the cursor on FV and press ALPHA ENTER to get the future value.
   The future value of an ordinary annuity of $200 deposited at the end of each quarter for $2\frac{1}{4}$ years, with interest at 4% compounded quarterly, is shown.

For annuities due, all steps are the same except that BEGIN is highlighted.

Ordinary annuity   Annuity due

Section 6.4 Present Value of an Annuity

To find the present value of an ordinary annuity:

1. Press the APPS key and select Finance, press ENTER.
2. Select TVM Solver, press ENTER.
3. Set N = the total number of periods, set I% = the annual percentage rate.
4. Set the FV = 0 and set both P/Y and C/Y = the number of compounding periods per year. END should be highlighted.
5. Set PMT = the periodic payment.
6. Put the cursor on PV and press ALPHA ENTER to get the present value.

The lump sum that needs to be deposited to receive $1000 at the end of each month for 16 years if the annuity pays 9%, compounded monthly is shown at the right.
For annuities due, BEGIN is highlighted.

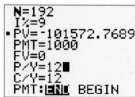

Ordinary annuity   Annuity due

Section 6.5 Finding Payments to Amortize a Loan

To find the size of periodic payments to amortize a loan:

1. Press the APPS key and select Finance, press ENTER.
2. Select TVM Solver, press ENTER.
3. Set N = the total number of periods, set I% = the APR.
4. Set the PV = loan value and set both P/Y and C/Y = the number of periods per year. END should be highlighted.
5. Set FV = 0.
6. Put the cursor on PMT and press ALPHA ENTER to get the payment.

To repay a loan of $10,000 in 5 annual payments with annual interest at 10%, each payment must be $2637.97.

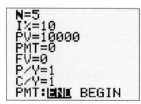

Section 6.5 Finding the Number of Payments to Amortize a Loan

To find the number of payments needed to amortize a loan:

1. Press the APPS key and select Finance, press ENTER.
2. Select TVM Solver, press ENTER.
3. Set the PV = loan value, set both P/Y and C/Y = the number of periods per year, set I% = the APR.
4. Set PMT = required payment and set FV = 0.
5. Put the cursor on N and press ALPHA ENTER to get the number of payments.

The number of monthly payments to pay a $2500 credit card loan with $55 payments and 18% interest is 76.9 months, or 6 years, 5 months.

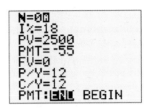 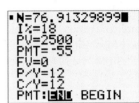

Chapter 7
Section 7.5 Evaluating Factorials

To evaluate factorials:

1. Enter the number whose factorial is to be calculated.
2. Choose MATH, then PRB. Scroll to 4: ! and press ENTER.

Press ENTER again to find the factorial.
7! is shown on the right.

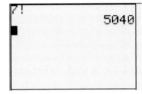

Section 7.5 Evaluating Permutations

To evaluate permutations:

1. For a "permutation of n objects taken r at a time" (such as $_{20}P_4$), first enter the value of n (such as $n = 20$).
2. Choose MATH, then PRB. Scroll to 2: nPr and press ENTER.
3. Enter the value of r (such as $r = 4$), and press ENTER to find the value of nPr.
   $_{20}P_4$ is shown on the right.

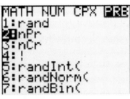 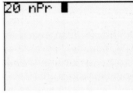

Section 7.5 Evaluating Combinations

To evaluate combinations:

1. For a "combination of n objects taken r at a time" (such as $_{20}C_4$), first enter the value of n (such as $n = 20$).
2. Choose MATH, then PRB. Scroll to 3:nCr and press ENTER.
3. Enter the value of r (such as $r = 4$), and press ENTER to find the value of nCr.

Note that $nCr = nC(n - r)$, as the figure on the right shows for $_{20}C_4$ and $_{20}C_{16}$.

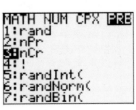 

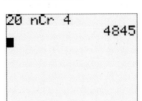 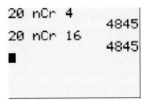

Section 7.6 Finding Probabilities Using Permutations and Combinations

To solve a probability problem that involves permutations or combinations:

1. Determine if permutations or combinations should be used.
2. Enter the ratios of permutations or combinations to find the probability.
3. If desired, use MATH, then 1:Frac to get the probability as a fraction.

If there are 5 defective computer chips in a box of 10, the probability that 2 chips drawn together from the box will both be defective is

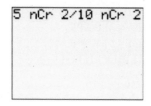
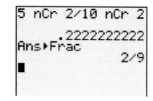

Section 7.7 Evaluating Markov Chains

To evaluate a Markov chain:

1. Enter the initial probability vector as matrix A and the transition matrix as matrix B.
2. To find the probabilities for the nth state, calculate $[A][B]^n$. The 3rd state, $[A][B]^3$, is shown on the right.

```
[A]
        [[.4 .4 .2]]
[B]
        [[.5 .4 .1]
         [.4 .5 .1]
         [.3 .3 .4]]
■
```

```
[A][B]^3
[[.4278 .4278 ...
```

Section 7.7 Finding Steady-State Vectors for Markov Chains

If the transition matrix contains only positive entries, the probabilities will approach a steady-state vector, which is found as follows.

1. Calculate and store $[C] = [B] - [I]$, where $[B]$ is the regular transition matrix and $[I]$ is the appropriately sized identity matrix.
2. Solve $[C]^T = [0]$ as follows:
   (a) Find $[C]^T$ with MATRIX, then MATH, 2:^{T}.
   (b) Find rref$[C]^T$.
3. Choose the solutions that add to 1 because they are probabilities.

```
identity(3)→[I]
        [[1 0 0]
         [0 1 0]
         [0 0 1]]
■
```

```
[B]-[I]
        [[-.5 .4  .1 ]
         [.4  -.5 .1 ]
         [.3  .3  -.6]]
Ans→[C]■
```

```
rref([C]ᵀ)
        [[1 0 -3]
         [0 1 -3]
         [0 0  0]]
```

Thus $x = 3z$ and $y = 3z$, and $3z + 3z + z = 1$ gives $z = 1/7$, so the probabilities are given in the steady-state vector $\left[\dfrac{3}{7} \ \dfrac{3}{7} \ \dfrac{1}{7}\right]$.

Chapter 8
Section 8.1 Binomial Probabilities

To find binomial probabilities:

1. 2nd DISTR A:binompdf(n,p,x) computes the probability of x success in n trials of a binomial experiment with probability of success p. Using MATH 1:Frac gives the probabilities as fractions.
   The probability of 3 heads in 6 tosses of a fair coin is found using 2nd DISTR, binompdf(6,.5,3).
2. The probabilities can be computed for more than one number in one command, using 2nd DISTR, binompdf($n,p,\{x_1,x_2,\ldots\}$).
   The probabilities of 4, 5, or 6 heads in 6 tosses of a fair coin are found using 2nd DISTR, binompdf(6,.5,{4,5,6}).
3. 2nd DISTR, binomcdf(n,p,x) computes the probability that the number of successes is less than or equal to x for the binomial distribution with n trials and probability of success p.

   The probability of 4 or fewer heads in 6 tosses of a fair coin is found using 2nd DISTR, binomcdf(6,.5,4).

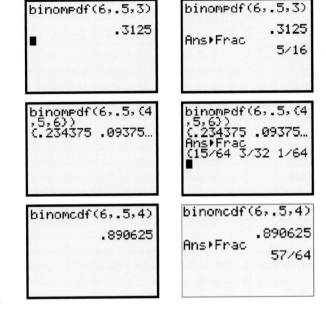

Section 8.2 Histograms

To find a frequency histogram for a set of data:

1. Press STAT, EDIT, 1:edit to enter each number in a column headed by L1 and the corresponding frequency of each number in L2.
2. Press 2nd STAT PLOT, 1:Plot 1. Highlight ON, and then press ENTER on the histogram icon. Enter L1 in xlist and L2 in Feq.
3. Press ZOOM, 9:ZoomStat or press Graph with an appropriate window.
4. If the data is given in interval form, a histogram can be created using the steps above with the class marks used to represent the intervals.

The frequency histogram for the scores 38, 37, 36, 40, 35, 40, 38, 37, 36, 37, 39, 38 is shown.

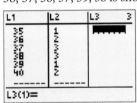

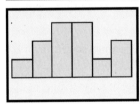

Section 8.2 Finding the Mean and Standard Deviation of Raw Data

To find descriptive statistics for a set of data:

1. Enter the data in list L1.
2. To find the mean and standard deviation of the data in L1, press STAT, move to CALC, and press 1:1-Var Stats, and ENTER.

The mean and sample standard deviation of the data 1, 2, 3, 3, 4, 4, 4, 4, 5, 5, 6, 7 are $\bar{x} = 4$ and $s \approx 1.65$.

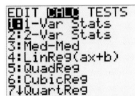

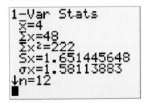

Section 8.2 Finding the Mean and Standard Deviation of Grouped Data

To find descriptive statistics for a set of data:

1. Enter the data in list L1 and the frequencies in L2.
2. To find the mean and standard deviation of the data in L1, press STAT, move to CALC, and press 1:1-Var Stats L1, L2, and ENTER.

The mean and standard deviation for the data in the table are shown on the right.

Salary	Number	Salary	Number
$59,000	1	$31,000	1
30,000	2	75,000	1
26,000	7	35,000	1
34,000	2		

The mean is $34,000 and the sample standard deviation is $14,132.84.

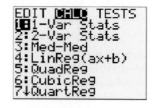

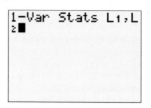

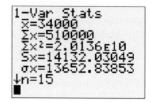

Section 8.4 Calculating Normal Probabilites

To calculate normal probabilities:

The command 2nd DISTR, 2:normalcdf(lowerbound, upperbound, μ, σ) gives the probability that x lies between the lower bound and the upper bound when the mean is μ and the standard deviation is σ.

The probability that a score lies between 33 and 37 when the mean is 35 and the standard deviation is 2 is found below.

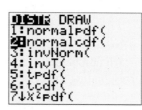

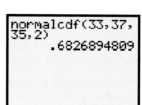

Section 8.4 Graphing Normal Distributions

To graph the normal distribution, press Y = and enter 2nd DIST 1:normalpdf(x, μ, σ) into Y_1.

Then set the window values xmin and xmax so the mean μ falls between them and press ZOOM, 0:Zoomfit.

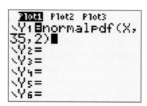

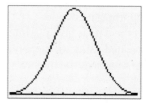

Chapter 9
Section 9.1 Evaluating Limits

To evaluate $\lim\limits_{x \to c} f(x)$

1. Enter the function as Y_1 in the Y = menu.
2. Set the x-range so it contains $x = c$.
3. Evaluate $f(x)$ for several x-values near $x = c$ and on each side by c by using one of the following methods.
   (a) Graphical Evaluation
   TRACE and ZOOM near $x = c$. If the values of y approach the same number L as x approaches c from the left and the right, there is evidence that the limit is L.
   (b) Numerical Evaluation
   Use TBLSET with Indpnt set to *Ask*. Enter values very close to and on both sides of c. The y-values will approach the same limit as above.

Evaluate $\lim\limits_{x \to 3} \dfrac{x^2 - 9}{x - 3}$.

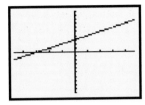
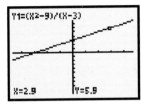

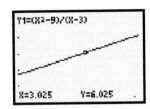
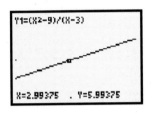

The y-values seem to approach 6.

The limit as x approaches 3 of $f(x)$ appears to be 6.

X	Y1
2.998	5.998
2.999	5.999
3	ERROR
3.001	6.001
3.002	6.002
3.003	6.003
3.004	6.004

Y1=ERROR

Section 9.1 Limits of Piecewise Functions

Enter the function, then use one of the methods for evaluating limits discussed above.

Find $\lim\limits_{x \to -5} f(x)$ where $f(x) = \begin{cases} x + 7 & \text{if } x \le -5 \\ -x + 2 & \text{if } x > -5 \end{cases}$

First enter $Y_1 = (x + 7)/(x \le -5)$ and $Y_2 = (-x + 2)/(x > -5)$.

Both methods indicate that the limit does not exist (DNE).

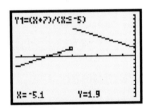
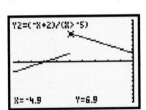

TABLE SETUP
TblStart=-5.003
△Tbl=1
Indpnt: Auto **Ask**
Depend: **Auto** Ask

X	Y1	Y2
-5.003	1.997	ERROR
-5.001	1.999	ERROR
-4.999	ERROR	6.999
-4.98	ERROR	6.98

X=

Section 9.2 Limits as $x \to \infty$

Enter the function as Y_1, then use large values of x with one of the methods for evaluating limits discussed above. Note: limits as $x \to -\infty$ are done similarly.

Evaluate $\lim\limits_{x \to \infty} \dfrac{3x - 2}{1 - 5x}$.

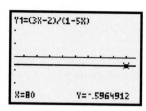

Both methods suggest that the limit is -0.6.

Sections 9.3–9.7 Approximating Derivatives

To find the numerical derivative (approximate derivative) of $f(x)$ at $x = c$, use Method 1 or Method 2.

Find the numerical derivative of $f(x) = x^3 - 2x^2$ at $x = 2$.

Method 1
1. Choose MATH, then 8:nDeriv(and press ENTER.
2. Enter the function, x, and the value of c, so the display shows nDeriv($f(x)$, x, c) then press ENTER. The approximate derivative at the specified value will be displayed.

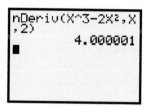

The numerical derivative is approximately 4.

Method 2
1. Enter the function as Y_1 in the Y = menu, and graph in a window that contains both c and $f(c)$.
2. Choose CALC by using 2nd TRACE, then 6:dy/dx, enter the x-value, c, and press ENTER. Then approximate derivative at the specified value will be displayed.

Warning: Both approximation methods above require that the derivative exists at $x = c$ and will give incorrect information when $f'(c)$ does not exist.

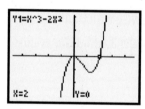
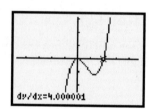

The value of the derivative is approximately 4.

Sections 9.3–9.7 Checking Derivatives

To check the correctness of the derivative function $f(x)$:

1. In the Y = menu, enter as Y_1 the derivative $f'(x)$ that you found, and graph it in a convenient window.
2. Enter the following as Y_2:nDeriv(f(x), x, x).
3. If the second graph lies on top of the first, the derivative is correct.

Verify the derivative of $f(x) = x^3 - 2x^2$ is $f'(x) = 3x^2 - 4x$.

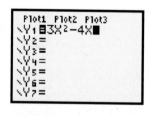
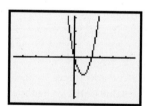

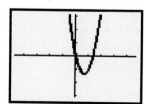

Section 9.8 Approximating the Second Derivative

To approximate $f''(c)$:

1. Enter $f(x)$ as Y_1 in the Y = menu.
2. Enter nDeriv(Y_1,x,x) as Y_2.
3. Estimate $f''(c)$ by using nDeriv(Y_2,x,c).

Find the second derivative of $f(x) = x^3 - 2x^2$ at $x = 2$.

 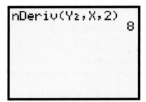

Thus $f''(2) = 8$.

Chapter 10
Section 10.1 Finding Critical Values

To find or approximate critical values of $f(x)$, that is
x-values that make the derivative equal to 0 or undefined:

I. Find the derivative of $f(x)$.

II. Use Method 1 or Method 2 to find the critical values.

Method 1

1. Enter the derivative in the Y = menu as Y_1 and graph it
   in a convenient window.
2. Find where $Y_1 = 0$ by one of the following:
   (a) Using TRACE to find the x-intercepts of Y_1.
   (b) Using 2nd CALC then 2:zero.
   (c) Using TBLSET then TABLE to find the values of x
       that give $Y_1 = 0$.
3. Use the graph of Y_1 (and TRACE or TABLE) to find the
   values of x that make the derivative undefined.

Method 2

1. Enter the derivative in the Y = menu as Y_1.
2. Press MATH and select Solver. Press the up arrow
   revealing EQUATION SOLVER equ:0 =, and enter Y_1
   (the derivative).
3. Press the down arrow or ENTER and the variable
   x appears with a value (not the solution). Move the
   cursor to the line containing the variable whose value
   is sought.
4. Press ALPHA SOLVE (ENTER). The value of the
   variable changes to the solution that is closest to that
   value originally shown.
5. To find additional solutions (if they exist), change
   the value of the variable and press ALPHA SOLVE
   (ENTER). The value of the variable changes to the
   solution of $Y_1 = 0$ that is closest to that value.
6. If appropriate, use 0:Solver to solve (Denominator of Y_1)
   = 0 to find the critical values for which the derivative is
   undefined.

Methods 1 and 2 show how to find the critical values of
$f(x) = \frac{1}{3}x^3 - 4x$. Note that the derivative is $f'(x) = x^2 - 4$.

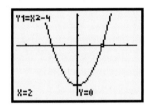 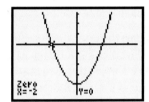

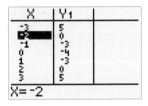

The only critical values are $x = 2$ and $x = -2$.

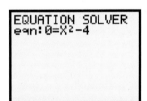 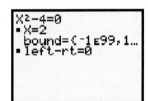

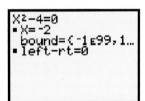

The only critical values are $x = 2$ and $x = -2$.

Section 10.1   Relative Maxima and Minima

To find relative maxima and minima:

1. In the $Y =$ menu enter the function as Y_1 and the derivative as Y_2.
2. Use TBLSET and TABLE to evaluate the derivative to the left and to the right of each critical value.
3. Use the signs of the values of the derivative to determine whether f is increasing or decreasing around the critical values, and thus to classify the critical values as relative maxima, relative minima, or horizontal points of inflection.
4. Graph the function to confirm your conclusions.

Find the relative maxima and minima of $f(x) = \dfrac{1}{3}x^3 - 4x$. Note that the derivative is $f'(x) = x^2 - 4$.

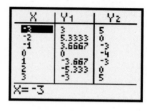 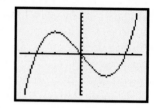

The relative max is at $(-2, 16/3)$, and the relative min is at $(2, -16/3)$.

Section 10.1   Critical Values and Viewing Windows

To use critical values to set a viewing window that shows a complete graph:

1. Once the critical values for a function have been found,
   (a) Enter the function as Y_1 and the derivative as Y_2. Use TABLE to determine where the function is increasing and where it is decreasing.
   (b) In WINDOW menu set x-min so that it is smaller than the smallest critical value and set x-max so that it is larger than the largest critical value.
2. Use TABLE to determine the y-coordinates of the critical values. Set y-min and y-max to contain the y-coordinates of the critical points.
3. Graph the function.

Let $f(x) = 0.0001x^3 + 0.003x^2 - 3.6x + 5$. Given that the critical values for $f(x)$ are $x = -120$ and $x = 100$, set a window that shows a complete graph and graph the function.

X	Y₁	Y₂
-150	275	2.25
-120	307.4	0
0	5	-3.6
100	-225	0
150	-130	4.05
X=0		

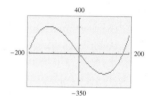

Section 10.2   Exploring f, f', and f'' Relationships

To explore relationships among the graphs of a function and its derivatives:

1. Find the functions for f' and f''.
2. Graph all three functions in the same window.
   • Notice that f increases when f' is above the x-axis $(+)$ and decreases when f' is below the x-axis $(-)$.
   • Notice that f is concave up when f'' is above the x-axis $(+)$ and is concave down when f'' is below the x-axis $(-)$.

Let $f(x) = x^3 - 9x^2 + 24x$. Graph f', f'', and f'' on the interval $[0, 5]$ to explore the relationships among these functions.

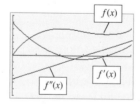

Sections 10.3–10.4 Finding Optimal Values

To find the optimal values of a function when the goal is not to produce a graph:

1. Enter the function as Y_1 in the Y = menu.
2. Select a window that includes the x-values of interest and graph the function.
3. While looking at the graph of the function, choose the CALC menu, scroll to 3:minimum or 4:maximum depending on which one is to be found, and press ENTER. This will result in a "*Left Bound?*" prompt.
   (a) Move the cursor to a point to the left of the point of interest. Press ENTER to select the left bound.
   (b) Move the cursor to the right of the point of interest. Press ENTER to select the right bound.
   (c) Press ENTER at the "*Guess?*" prompt. The resulting point is an approximation of the desired optimum value.

Let $f(x) = 100\sqrt{2304 + x^2} + 50(100 - x)$. Find the minimum value of $f(x)$ on the interval $[0, 100]$ for x. Note: the following screens use the window x: $[-10, 110]$ and y: $[-2500, 12500]$

The minimum value is $y \approx 9157$ and occurs when $x \approx 27.7$.

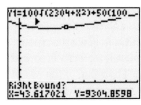 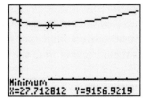

Note: Finding a maximum value works similarly.

Section 10.5 Asymptotes and Window Setting

To use asymptotes and critical values to set a viewing window that shows a complete graph:

1. Once the asymptotes and critical values for a function have been found,
   (a) Determine where the function is increasing and where it is decreasing (by using TABLE and with Y_1 as the function and Y_2 as the derivative).
   (b) Under WINDOW, set x-min so that it is smaller than the smallest x-value that is either a vertical asymptote or a critical value and set x-max so that it is larger than the largest of these important x-values.
2. Use TABLE or VALUE to determine the y-coordinates of the critical values. Set y-min and y-max so they contain the y-coordinates of any horizontal asymptotes and critical points.
3. Graph the function.

Let $f(x) = \dfrac{x^2}{x - 2}$. Given that $f(x)$ has the line $x = 2$ as a vertical asymptote, has no horizontal asymptote, and has critical values $x = 0$ and $x = 4$. Set the window and graph $y = f(x)$.

The critical points for $f(x)$ are $(0, 0)$ and $(4, 8)$. The window needs an x-range that contains 0, 2, and 4 and a y-range that contains 0 and 8.

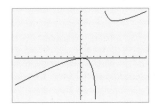

Chapter 11
Section 11.1 Derivatives of Logarithmic Functions

To check that the derivative of $y = \ln(2x^6 - 3x + 2)$

is $Y_1 = \dfrac{12x^5 - 3}{2x^6 - 3x + 2}$, we show that the graph of

$Y_2 = $ nDeriv$(\ln(2x^6 - 3x + 2), x, x)$ lies on the graph of

$y_1 = \dfrac{12x^5 - 3}{2x^6 - 3x + 2}$.

 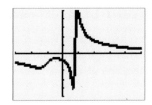

Section 11.1 Finding Critical Values

To find the critical values of a function $y = f(x)$:

1. Enter the function as Y_1 in the Y = menu and the derivative as Y_2.
2. Press MATH and select Solver. Press the up arrow revealing EQUATION SOLVER equ:0 = , and enter Y_2 (the derivative).
3. Press the down arrow or ENTER, and the variable x appears with a value (not the solution). Place the cursor on the line containing the variable whose value is sought.
4. Press ALPHA SOLVE (ENTER). The value of the variable changes to the solution of the equation that is closest to that value.
5. To find additional solutions (if they exist), change the value of the variable and press ALPHA SOLVE (ENTER). The value of the variable gives the solution of $Y_2 = 0$ that is closest to that value.

To find the critical values of $y = x^2 - 8 \ln x$, we solve
$$0 = 2x - \frac{8}{x}.$$

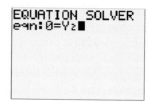

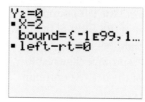
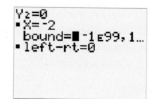

The two critical values are $x = 2$ and $x = -2$.

Section 11.1 Finding Optimal Values

To find the optimal values of a function $y = f(x)$:

1. Find the critical values of a function $y = f(x)$. (Use the steps for finding critical values discussed above.)
2. Graph $y = f(x)$ on a window containing the critical values.
3. The y-values at the critical values (if they exist) are the optimal values of the function.

To find the optimal values of $y = x^2 - 8 \ln x$, we solve
$$0 = 2x - \frac{8}{x}$$ and evaluate $y = x^2 - 8 \ln x$ at the solutions.

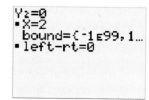

The two critical values are $x = 2$ and $x = -2$.

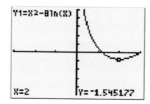
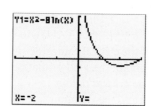

The minimum of $y = x^2 - 8 \ln x$ is -1.545 at $x = 2$. The function $y = x^2 - 8 \ln x$ is undefined for all negative values because $\ln x$ is undefined for all negative values. Thus, there is no optimum value of the function at $x = -2$.

Section 11.2 Derivatives of Exponential Functions

To check that the derivative of $y = 5^{x^2+x}$ is $y' = 5^{x^2+x}(2x + 1) \ln 5$, we show that the graph of $Y_2 = \text{nDeriv}(5^{x^2+x}, x, x)$ lies on the graph of $Y_1 = 5^{x^2+x}(2x + 1) \ln 5$.

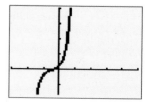

Section 11.2 Finding Critical Values

To find the critical values of a function $y = f(x)$:

1. Enter the function as Y_1 in the $Y =$ menu and the derivative as Y_2.
2. Press MATH and select Solver. Press the up arrow revealing EQUATION SOLVER equ:0 =, and enter Y_2 (the derivative).
3. Press the down arrow or ENTER and the variable appears with a value (not the solution). Place the cursor on the variable whose value is sought.
4. Press ALPHA SOLVE (ENTER). The value of the variable changes to the solution of the equation that is closest to that value.
5. To find additional solutions (if they exist), change the value of the variable and press ALPHA SOLVE (ENTER). The value of the variable gives the solution of $Y_2 = 0$ that is closest to that value.

To find the critical values of $y = e^x - 3x^2$, we solve $0 = e^x - 6x$.

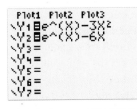 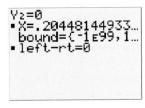

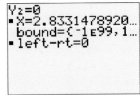 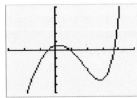

A relative maximum of $y = e^x - 3x^2$ occurs at $x \approx 0.204$, and a relative minimum occurs at $x \approx 2.833$.

Chapter 12
Section 12.1 Checking Powers of x Integrals

To check a computed indefinite integral with the command fnInt:

1. Enter the integral of $f(x)$ (without the $+ C$) as Y_1 in $Y =$ menu.
2. Move the cursor to Y_2, press MATH, 9:fnInt(and enter $f(x)$, x, 0, x) so the equation is $Y_2 = $ fnInt $(f(x), x, 0, x)$.
3. Pressing ENTER with the cursor to the left of Y_2 changes the thickness of the second graph, making it more evident that the second lies on top of the first.
4. Press GRAPH with an appropriate window. If the second graph lies on top of the first, the graphs agree and the computed integral checks.

Checking that the integral of $f(x) = x^2$ is $\int x^2\, dx = \dfrac{x^3}{3} + C$:

 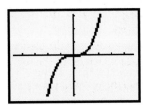

Sections 12.1–12.2 Families of Functions

To graph some functions in the family of indefinite integrals of $f(x)$:

1. Integrate $f(x)$.

2. Enter equations of the form $\int f(x)\, dx + C$ for different values of C.

3. Press GRAPH with an appropriate window. The graphs will be shifted up or down, depending on C.

The graphs of members of the family $y = \int (2x - 4)dx + C$ with $C = 0, 1, -2$, and 3.

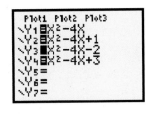 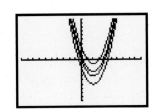

Section 12.5 Differential Equations

To solve initial value problems in differential equations:

1. Integrate $f(x)$, getting $y = F(x) + C$. If a value of x and a corresponding value of y in the integral $y = F(x) + C$ are known, this initial value can be used to find one function that satisfies the given conditions.
2. Press MATH and select Solver. Press the up arrow to see EQUATION SOLVER.
3. Set 0 equal the integral minus y, getting $0 = F(x) + C - y$, and press the down arrow.
4. Enter the given values of x and y, place the cursor on C, and press ALPHA, SOLVE. Replace C with this value to find the function satisfying the conditions.

To solve $\dfrac{dy}{dx} = 2x - 4$, we note that the integral of both sides is $y = x^2 - 4x + C$. If $y = 8$ when $x = -1$ in this integral, we can find C, and thus a unique solution, shown below.

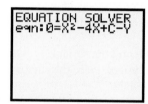
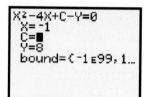

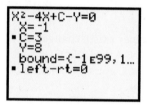

The unique solution is $y = x^2 - 4x + 3$.

Chapter 13
Section 13.1 Approximating Areas Under Curves Using Rectangles

To find the area under $y = f(x)$ and above the x-axis over a given interval using rectangles of equal width:

1. Enter $f(x)$ in Y_1 of the Y = menu and calculate the width of the base of each of the n rectangles.
2. Compute the area using left-hand endpoints.

$$S_L = \sum_{i=1}^{n} f(x_{i-1})\Delta x$$
$$= \left[Y_1(x_0) + Y_1(x_1) + \cdots + Y_1(x_{n-1})\right]\Delta x$$

3. Use a similar formula to compute the area using right-hand endpoints.

$$S_R = \sum_{i=1}^{n} f(x_i)\Delta x$$
$$= \left[Y_1(x_1) + Y_1(x_2) + \cdots + Y_1(x_n)\right]\Delta x$$

To find the left and right approximations for the area under $f(x) = \sqrt{x}$ from $x = 0$ to $x = 4$ using $n = 8$ equal subdivisions:

1. Each rectangle has base width $(4 - 0)/8 = 0.5$. The subdivision values are
   $$x_0 = 0, x_1 = 0.5, x_2 = 1, \cdots x_7 = 3.5, x_8 = 4.$$
2. Calculate the left approximation of the area as follows.
   $$S_L = 0.5[Y_1(0) + Y_1(0.5) + Y_1(1) + Y_1(1.5) + Y_1(2)$$
   $$+ Y_1(2.5) + Y_1(3) + Y_1(3.5)] \approx 4.765$$
   (See the screen below.)
3. Calculate the right approximation of the area as follows.
   $$S_R = [Y_1(0.5) + Y_1(1) + Y_1(1.5) + Y_1(2) + Y_1(2.5)$$
   $$+ Y_1(3) + Y_1(3.5) + Y_1(4)]0.5 \approx 5.765$$

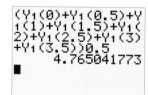
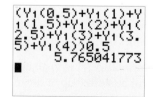

Section 13.2 Approximating Definite Integrals— Areas Under Curves

To approximate the area under the graph of $y = f(x)$ and above the x-axis:

1. Enter $f(x)$ under the Y = menu and graph the function with an appropriate window.

2. Press 2nd CALC and 7: $\int f(x)\, dx$.

3. Press ENTER. Move the cursor to, or enter, the lower limit (the left x-value).

4. Press ENTER. Move the cursor to, or enter, the upper limit (the right x-value).

5. Press ENTER. The area will be displayed.

The approximate area under the graph of $f(x) = x^2 + 1$ from $x = 0$ to $x = 3$ is found below.

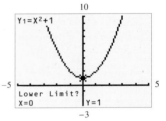

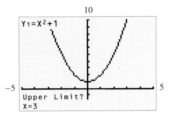
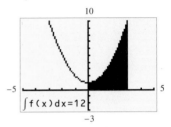

Section 13.2 Approximating Definite Integrals— Alternative Method

To approximate the definite integral of $f(x)$ from $x = a$ to $x = b$:

1. Press MATH, 9: fnInt(. Enter $f(x)$, x, a, b) so the display shows fnInt($f(x)$, x, a, b).

2. Press ENTER to find the approximation of the integral.

3. The approximation may be made closer than that in Step 2 by adding a fifth argument with a number (tolerance) smaller than 0.00001.

The approximation of $\int_{-1}^{3} (4x^2 - 2x)dx$ is found as follows.

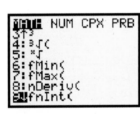
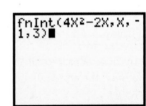

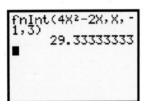

Section 13.3 Approximating the Area Between Two Curves

To approximate the area between the graphs of two functions:

1. Enter one function as Y_1 and the second as Y_2. Press GRAPH using a window that shows all points of intersection of the graphs.
2. Find the x-coordinates of the points of intersection of the graphs, using 2nd CALC: intersect.
3. Determine visually which graph is above the other over the interval between the points of intersection.
4. Press MATH, 9: fnInt(. Enter $f(x)$, x, a, b so the display shows fnInt($f(x)$, x, a, b) where $f(x)$ is $Y_2 - Y_1$ if the graph of Y_2 is above the graph of Y_1 between a and b, or $Y_1 - Y_2$ if Y_1 is above Y_2.

The area enclosed by the graphs of $y = 4x^2$ and $y = 8x$ is found as follows.

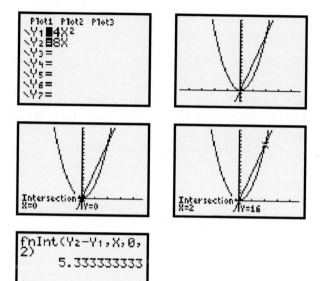

The area between the curves is 16/3.

Section 13.3 Approximating the Area Between Two Curves—Alternate Method

The area between the graphs can also be found by using 2nd CALC, $\int f(x)\, dx$.

1. Enter $Y_3 = Y_2 - Y_1$ where Y_2 is above Y_1.
2. Turn off the graphs of Y_1 and Y_2 and graph Y_3 with a window showing where $Y_3 > 0$.
3. Press 2nd CALC and $\int f(x)\, dx$.
4. Press ENTER. Move the cursor to, or enter, the lower limit (the left x-value).
5. Press ENTER. Move the cursor to, or enter, the upper limit (the right x-value).
6. Press ENTER. The area will be displayed.

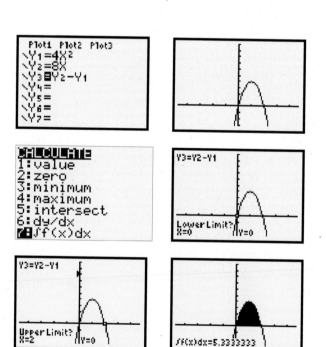

Excel Guide Part 1
Excel 2003

Excel Worksheet

When you start up Excel by using the instructions for your software and computer, the following screen will appear.

The components of the **spreadsheet** are shown, and the grid shown is called a **worksheet.** You can move to other worksheets by clicking on the tabs at the bottom.

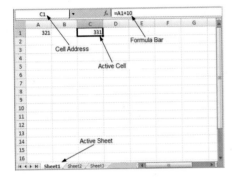

Addresses and Operations

Notice the letters at the top of the columns and the numbers identifying the rows. The cell addresses are given by the column and row; for example, the first cell has address A1. You can move from one cell to another with arrow keys, or you can select a cell by clicking on it. After you type an entry in a cell, press enter to accept it. To edit the contents of a cell, click on the cell and edit the contents in the formula bar at the top. To delete the contents, press the delete key.

The file operations such as "open a new file," "saving a file," and "printing a file" are similar to those in Word. For example, <CTRL>S saves a file. You can also format a cell entry by selecting it and using menus similar to those in Word.

Working with Cells

Cell entries, rows containing entries, and columns containing entries can be copied and pasted with the same commands as in Word. For example, a highlighted cell can be copied with <CTRL>C. Sometimes entries exceed the width of the cells containing them, especially if they are text. To widen the cells in a column, place the cursor at the right side of the column heading until you see the symbol ↔, then move the cursor to the right (moving it to the left makes the column more narrow). If entering a number results in #####, the number is too long for the cell, and the cell should be widened.

Chapter 1
Section 1.4 Entering Data and Evaluating Functions

The cells are identified by the column and the row. For example, the cell B3 is in the second column and the third row.

1. Put headings on the two columns.
2. Fill the inputs in Column A by hand or with a formula for them. The formula =A2+1 gives 2 in A3 when ENTER is pressed.
3. Moving the mouse to the lower right corner of A3 until there is a thin "+" sign and dragging the mouse down "fills down" all required entries in column A.
4. Enter the function formula for the function in B2. Use = 1000*(1.1)^A2 to represent $S = 1000(1.1)^t$. Pressing ENTER gives the value when $t = 1$.
5. Using Fill Down gives the output for all inputs.

	A	B
1	Year	Future Value
2	1	=1000*(1.1)^(A2)
3	=A2+1	

	A	B
1	Year	Future Value
2	1	1100
3	2	1210
4	3	1331

Section 1.4 Graphing a Function

1. Put headings on the two columns (x and $f(x)$, for example).
2. Fill the inputs (x-values) in Column A by hand or with a formula for them.
3. Enter the formula for the function in B2.
   Use = 6*A2 − 3 to represent $f(x) = 6x - 3$.
4. Select the cell containing the formula for the function (B2, for example).
5. Move the mouse to the lower right corner until there is a thin "+" sign.
6. Drag the mouse down to the last cell where formula is required, and press ENTER.
7. Highlight the two columns containing the values of x and $f(x)$.
8. Click the *Chart Wizard* icon and then select the *XY(Scatter)* chart with smooth curve option.
9. Click the Next button to get the *Chart Source Data* box. Then click Next to get the *Chart Options* box, and enter your chart title and labels for the x- and y-axes.
10. Click Next, select whether the graph should be within the current worksheet or on another, and click Finish.

	A	B
1	x	$f(x) = 6x - 3$
2	−5	−33
3	−2	−15
4	−1	−9
5	0	−3
6	1	3
7	3	15
8	5	27

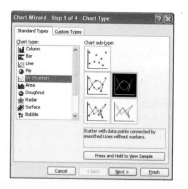

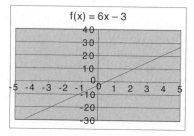

Section 1.5 Solving Systems of Two Equations in Two Variables

1. Write the two equations as linear functions in the form $y = mx + b$.
2. Enter the input variable x in cell A2 and the formula for each of the two equations in cells B2 and C2, respectively.
3. Enter $= B2 - C2$ in cell D2.
4. Use *Tools > Goal Seek*.
5. In the dialog box:
   a. Click the *Set Cell* box and click on the D2 cell.
   b. Enter 0 in the *To Value* box.
   c. Click the *By Changing Cell* box and click on the A2 cell.
6. Click OK in the *Goal Seek* dialog box, getting 0.
7. The x-value of the solution is in cell A2, and the y-value is in both B2 and C2.

The solution of the system

$$\begin{cases} 3x + 2y = 12 \\ 4x - 3y = -1 \end{cases}$$

is found to be $x = 2$, $y = 3$ using *Goal Seek* as follows.

	A	B	C	D
1	x	= 6 − 1.5x	= 1/3 + 4x/3	= y1 − y2
2	1	= 6 − 1.5*A2	= 1/3 + 4*A2/3	= B2 − C2

	A	B	C	D
1	x	= 6 − 1.5x	= 1/3 + 4x/3	= y1 − y2
2	2	3	3	0

Chapter 2
Section 2.1 Solving Quadratic Equations

To solve a quadratic equation of the form $f(x) = ax^2 + bx + c = 0$:

1. Enter x-values centered around the x-coordinate $x = \dfrac{-b}{2a}$ in column A and use the function formula to find the values of $f(x)$ in column B.

2. Graph the function, $f(x) = 2x^2 - 9x + 4$ in this case, and observe where the graph crosses the x-axis ($f(x)$ near 0).

3. Use *Tools > Goal Seek*, entering a cell address with a function value in column B at or near 0, enter the set cell to the value 0, and enter the changing cell.

4. Click OK to find the x-value of the solution in cell A2. The solution may be approximate. The spreadsheet shows $x = 0.5001$, which is an approximation of the exact solution $x = 0.5$.

5. After finding the first solution, repeat the process using a second function value at or near 0. The second solution is $x = 4$ in this case.

	A	B	C	D	E	F
1	x	f(x)=2x^2−9x+4				
2	−1	15				
3	0	4				
4	1	−3				
5	2	−6				
6	3	−5				
7	4	0				
8	5	9				
9						

	A	B	C	D	E	F
1	x	f(x)=2x^2−9x+4				
2	−1	15				
3	0	4				
4	0.500001	−7.287E−05				
5	2	−6				
6	3	−5				
7	4	0				
8	5	9				
9						

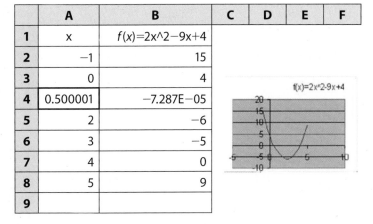

Sections 2.2 and 2.4 Graphing Polynomial Functions

To graph a polynomial function:

1. Use the function to create a table containing values for x and $f(x)$. (See Graphing a Function on page AP-28.)
2. Highlight the two columns containing the values of x and $f(x)$.
3. Click the *Chart Wizard* icon and then select the *XY(Scatter)* chart with smooth curve option.
4. Click the Next button to get the *Chart Source Data* box. Then click Next to get the *Chart Options* box, and enter your chart title and labels for the x- and y-axes.
5. Click Next, select whether the graph should be within the current worksheet or on another, and click Finish.

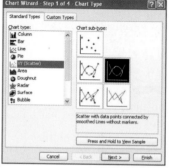

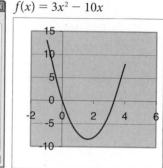

$f(x) = 3x^2 - 10x$

Section 2.4 Graphing Rational Functions

An Excel graph will connect all points corresponding to values in the table, so if the function you are graphing is undefined for some x-value a, enter x-values near this value and leave (or make) the corresponding $f(a)$ cell blank.

To graph $f(x) = \dfrac{1}{1-x}$, which is undefined at $x = 1$:

1. Generate a table for x-values from -1 to 3, with extra values near $x = 1$.
2. Generate the values of $f(x)$, and leave a blank cell for the $f(x)$ value for $x = 1$.
3. Select the table and plot the graph using *Chart Wizard*.

x	f(x)
−1	−0.5
−0.5	−0.6667
0	−1
0.5	−2
0.75	−4
0.9	−10
1	
1.1	10
1.25	4
1.5	2
2	1
2.5	0.666667
3	0.5

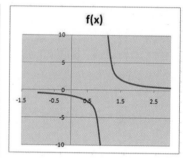

Section 2.5 Modeling

To create a scatter plot of data:

1. Enter the inputs (x-values) in Column A and the outputs (y-values) in Column B.
2. Highlight the two columns and use *Chart Wizard* to plot the points.
3. In Step 3 of the *Chart Wizard,* you can add the title and x- and y-axis labels.

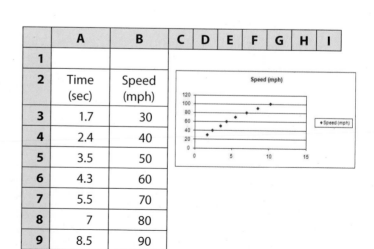

	A	B	C	D	E	F	G	H	I
1									
2	Time (sec)	Speed (mph)							
3	1.7	30							
4	2.4	40							
5	3.5	50							
6	4.3	60							
7	5.5	70							
8	7	80							
9	8.5	90							
10	10.2	100							
11									

To find the equation of a line or curve that best fits a given set of data points:

1. Place the scatter plot of the data in the worksheet.
2. Single-click on the scatter plot in the workbook.
3. From the *Chart* menu select *Add Trendline*.
4. Select the regression type that appears to be the best function fit for the scatter plot. [Note: If *Polynomial* is selected, choose the appropriate Order (degree).]
5. Click the *Options* tab and check the *Display equation on chart* box.
6. Click OK and the graph of the selected best-fit function will appear along with its equation.

The power function that models Corvette acceleration follows.

	A	B	C	D	E	F	G	H	I
1									
2	Time (sec)	Speed (mph)							
3	1.7	30							
4	2.4	40							
5	3.5	50							
6	4.3	60							
7	5.5	70							
8	7	80							
9	8.5	90							
10	10.2	100							
11									

Corvette Acceleration

$y = 21.875x^{0.6663}$

The speed of the Corvette is given by $y = 21.875x^{0.6663}$ where x is the time in seconds.

Chapter 3
Section 3.1 Operations with Matrices

Operations for 3×3 matrices can be used for other orders.

1. Type a name A in A1 to identify the first matrix.
2. Enter the matrix elements of matrix A in the cells B1:D3.
3. Type a name B in A5 to identify the second matrix.
4. Enter the matrix elements of matrix B in the cells B5:D7.
5. Type a name A + B in A9 to indicate the matrix sum.
6. Type the formula "=B1 + B5" in B9 and press ENTER.
7. Use Fill Across to copy this formula across the row to C9 and D9.
8. Use Fill Down to copy the row B9:D9 to B11:D11, which gives the sum.
9. To subtract the matrices, change the formula in B9 to "=B1 − B5" and proceed as with addition.

	A	B	C	D
1	A	1	2	3
2		4	5	6
3		7	8	9
4				
5	B	−2	−4	3
6		1	4	−5
7		3	6	−1
8				
9	A+B	−1	−2	6
10		5	9	1
11		10	14	8

Section 3.2 Multiplying Two Matrices

Steps for two 3×3 matrices:

1. Enter the names and elements of the matrices.
2. Enter the name AxB in A9 to indicate the matrix product.
3. Select a range of cells that is the correct size to contain the product (B9:D11 in this case).
4. Type "=mmult(" in the formula bar, and then select the cells containing the elements of matrix A (B1:D3).
5. Stay in the formula bar, type a comma and select the matrix B elements (B5:D7), and close the parentheses.
6. Hold the CTRL and SHIFT keys down and press ENTER, giving the product.

	A	B	C	D
1	A	1	2	3
2		4	5	6
3		7	8	9
4				
5	B	−2	−4	3
6		1	4	−5
7		3	6	−1
8				
9	AxB	=mmult(B1:D3		
10		MMULT(**array1,**		
11		array2)		

B9	fx	{(=MMULT(B1:D3,B5:D7)}			
	A	B	C	D	E
1	A	1	2	3	
2		4	5	6	
3		7	8	9	
4					
5	B	−2	−4	3	
6		1	4	−5	
7		3	6	−1	
8					
9	AxB	9	22	−10	
10		15	40	−19	
11		21	58	−28	

Section 3.4 Finding the Inverse of a Matrix

Steps for a 3 × 3 matrix:

1. Enter the name A in A1 and the elements of the matrix in B1:D3 as above.
2. Enter the name "Inverse(A)" in A5 and select a range of cells that is the correct size to contain the inverse [(B5:D7) in this case].
3. Enter "=minverse(", select matrix A (B1:D3), and close the parentheses.
4. Hold the CTRL and SHIFT keys down and press ENTER, getting the inverse.

SUM	*fx*	=minverse(B1:D3)		
	A	**B**	**C**	**D**
1	A	2	1	1
2		1	2	0
3		2	0	1
4				
5	inverse(A)	=minverse(B1:D3)		
6				
7				

Section 3.4 Solving Systems of Linear Equations with Matrix Inverses

A system of linear equations can be solved by multiplying the matrix containing the augment by the inverse of the coefficient matrix. The steps used to solve a 3 × 3 system follow.

1. Enter the coefficient matrix A in B1:D3.
2. Enter the name "inverse(A)" in A5 and compute the inverse of A in B5:D7.
3. Enter B in cell A9 and enter the augment matrix in B9:B11.
4. Enter X in A13 and select the cells B13:B15.
5. In the formula bar, type "=mmult(", then select matrix inverse(A) in B5:D7, type a comma, select matrix B in B9:B11, and close the parentheses.
6. Hold the CTRL and SHIFT keys down and press ENTER, getting the solution.
7. Matrix X gives the solution.

$$\text{The system } \begin{cases} 2x + y + z = 8 \\ x + 2y \quad\;\; = 6 \\ 2x + \quad\;\; z = 5 \end{cases} \text{ is solved as follows.}$$

B5	*fx*	{=MINVERSE(B1:D3)}		
	A	**B**	**C**	**D**
1	A	2	1	1
2		1	2	0
3		2	0	1
4				
5	inverse(A)	−2	1	2
6		1	0	−1
7		4	−2	−3

	SUM	*fx*	=mmult(B5:D7,B9:B11)		
	A	**B**	**C**	**D**	**E**
1	A	2	1	1	
2		1	2	0	
3		2	0	1	
4					
5	inverse(A)	−2	1	2	
6		1	0	−1	
7		4	−2	−3	
8					
9	B	8			
10		6			
11		5			
12					
13	X	=mmult(B5:D7,B9:B11)			
14					
15					

	B13	*fx*	{(=MMULT(B5:D7,B9:B11)}		
	A	**B**	**C**	**D**	**E**
1	A	2	1	1	
2		1	2	0	
3		2	0	1	
4					
5	inverse(A)	−2	1	2	
6		1	0	−1	
7		4	−2	−3	
8					
9	B	8			
10		6			
11		5			
12					
13	X	0			
14		3			
15		5			

The solution is $x = 0$, $y = 3$, $z = 5$.

Chapter 4
Sections 4.3–4.5 Linear Programming

Maximize $f = 12x + 26y + 40z$ subject to the constraints

$$\begin{cases} 5x + 7y + 10z \le 90{,}000 \\ x + 3y + 4z \le 30{,}000 \\ x + y + z \le 9000 \\ x \ge 0, y \ge 0, z \ge 0 \end{cases}$$

1. On a blank spreadsheet, type a heading in cell A1, followed by the variable descriptions in cells A3–A5 and the initial values (zeros) in cells B3–B5.
2. a. Enter the heading "Objective" in cell A7.
   b. Enter a description of the objective in cell A9 and the formula for the objective function in B9. The formula is =12*B3 + 26*B4 + 40*B5.
3. a. Type in the heading "Constraints" in A11 and descriptive labels in A13–A15.
   b. Enter the left side of the constraint inequalities in B13–B15 and the maximums from the right side in C13–C15.
4. Select *Solver* under the *Tools* menu. A dialog box will appear.*
5. a. Click the *Set Target Cell* box and B9 (containing the formula for the objective function).
   b. Check the button *Max* for maximization.
   c. Click the *By Changing Cells* box and select cells B3–B5.
6. a. Click the *Subject to Constraints* entry box. Press the Add button to add the first constraint.
   b. Click the left entry box and click cell B13 (containing the formula for the first constraint).
   c. Set the middle entry box to <=.
   d. Click the right entry box and C13 to enter the constraint.
   e. Click Add and repeat the Steps 6b–6d for the remaining constraints.
   f. Click in the left entry box for the constraints and select the variables in B3–B5. Set the middle entry to >=, and type 0 in the right entry box.
7. Click Solve in the *Solver* dialog box. A dialog box states that *Solver* found a solution. To see the *Solver* results, click *Keep Solver Solution* and also select Answer.
8. Go back to the spreadsheet. The new values in B3–B5 are the values of the variables that give the maximum, and the value in B9 is the maximum value of the objective function.

In *Solver*, minimization of an objective function is handled exactly the same as maximization, except min is checked and the inequality signs are $\ge$.

When mixed constraints are used, simply enter them with "mixed" inequalities.

*You may need to use an "Add-In" to access *Solver*.

	A	B	C
1	**Variables**		
2			
3	# small calculators (x)	0	
4	# medium calculators (y)	0	
5	# large calculators (z)	0	
6			
7	**Objective**		
8			
9	Maximize profit	=12*B3+26*B4+40*B5	
10			
11	**Constraints**		
12		Amount used	Maximum
13	Circuit components	=5*B3+7*B4+10*B5	90000
14	Labor	=B3+3*B4+4*B5	30000
15	Cases	=B3+B4+B5	9000

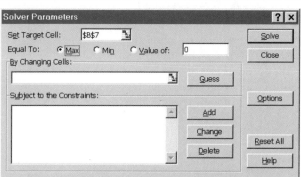

	A	B	C
1	Variables		
2			
3	# small calculators (x)	2000	
4	# medium calculators (y)	0	
5	# large calculators (z)	7000	
6			
7	Objective		
8			
9	Maximize profit	304000	
10			
11	Constraints		
12		Amount used	Maximum
13	Circuit components	80000	90000
14	Labor	30000	30000
15	Cases	9000	9000

Chapter 5

Section 5.1 Graphing Exponential Functions

Exponential functions are entered into Excel differently for base e than for bases other than e.

A. To graph $y = a^x$, use the formula =a^x.
B. To graph $y = e^x$, use the formula =exp(x).

To graph $f(x) = 1.5^x$ and $g(x) = e^x$ on the same axes:
1. Type x in cell A1 and numbers centered at 0 in Column A.
2. Type $f(x)$ in cell B1, enter the formula =1.5^A2 in cell B2 and fill down.
3. Type $g(x)$ in cell C1, enter the formula =exp(A2) in cell C2 and fill down.
4. Select the entire table and use *Chart Wizard* to graph as described in Section 1.4.

	A	B	C
	x	$f(x)$	$g(x)$
1			
2	−2	0.444444	0.135335
3	−1.5	0.544433	0.22313
4	−1	0.666667	0.367879
5	−0.5	0.816497	0.606531
6	0	1	1
7	0.5	1.224745	1.648721
8	1	1.5	2.718282
9	1.5	1.837117	4.481689
10	2	2.25	7.389056
11	2.5	2.755676	12.18249
12	3	3.375	20.08554

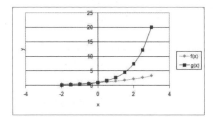

Section 5.1 Modeling with Exponential Functions

To model data with an exponential function:

1. Create a scatter plot for the data.
2. From the *Chart* menu, choose *Add Trendline*.
3. Check *exponential regression* type since that function type appears to be the best fit for the scatter plot.
4. Click the *Options* tab on this box and select *Display equation* on chart box.
5. Click Next and Finish to see the equation and graph.

The exponential model for the following data is shown.

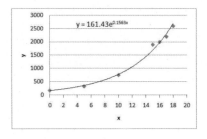

x	0	5	10	15	16	17	18
y	170	325	750	1900	2000	2200	2600

Section 5.2 Graphing Base *e* and Base 10
Logarithmic Functions

1. Create a table of values for *x*-values with $x > 0$ to reflect the function's domain.
2. For $y = \ln(x)$, use the formula $=\ln(x)$.
3. For $y = \log(x)$, use the formula $=\log 10(x)$.
4. Select the entire table and use *Chart Wizard* to graph.

	A	B
1	*x*	$f(x)=\ln(x)$
2	0.5	−0.6931472
3	0.9	−0.1053605
4	1	0
5	2	0.6931472
6	3	1.0986123
7	4	1.3862944
8	5	1.6094379
9	6	1.7917595
10	7	1.9459101
11	8	2.0894415
12	9	2.1972246
13	10	2.3025851
14	11	2.3978953

	A	B
1	*x*	$f(x)=\log(x)$
2	0.5	−0.30103
3	0.9	−0.0457575
4	1	0
5	2	0.30103
6	3	0.47712125
7	4	0.60205999
8	5	0.69897
9	6	0.77815125
10	7	0.84509804
11	8	0.90308999
12	9	0.95424251
13	10	1
14	11	1.04139269

f(x)=ln(x)

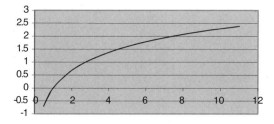

f(x)=log(x)

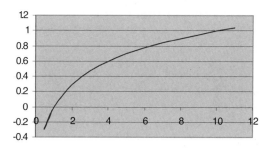

Section 5.2 Modeling Logarithmic Functions

1. Create a scatter plot for the data.
2. In the *Chart* menu, choose *Add Trendline* and click *logarithmic regression*.
3. Click the *options* tab, select *Display equation* and click Next and Finish.

x	*y*
10	2.207
20	3.972
30	5.004
40	5.737

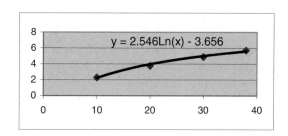

y = 2.546Ln(x) - 3.656

Section 5.2 Graphing Logarithmic Functions with Other Bases

1. The graph of a logarithm with any base b can be created by entering the formula =log(x, b).
2. Proceed as described above for graphing base e and base 10 logarithms.

	A	B
1	x	f(x)=log(x,7)
2	0.5	−0.356207187
3	0.9	−0.054144594
4	1	0
5	2	0.356207187
6	3	0.564575034
7	4	0.712414374
8	5	0.827087475
9	6	0.920782221
10	7	1
11	8	1.068621561
12	9	1.129150068
13	10	1.183294662
14	11	1.1232274406

The graph of $f(x) = \log_7 x$ is shown below.

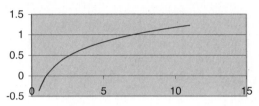

Chapter 6

Section 6.2 Finding the Future Value of a Lump Sum

To find the future value of a lump-sum investment:

1. Type the headings in Row 1, and enter their values (with the interest rate as a decimal) in Row 2. Enter the investment as a negative number.
2. Type the formula =D2/E2 in F2 to compute the rate per period.
3. Type the heading Future Value in A4.
4. In cell B4, type the formula =fv(F2,B2,C2,A2,0) to compute the future value.

This spreadsheet gives the future value of an investment of $10,000 for 17 years at 9.8%, compounded quarterly.

Principal	Number of Periods	Payment	Annual Rate	Periods per year	Periodic Rate
−10,000	68	0	0.098	4	0.0245
Future Value	$51,857.73				

Section 6.3 Finding the Future Value of an Annuity

To find the future value of an ordinary annuity:

1. Type the headings in Row 1, and enter their values (with the interest rate as a decimal) in Row 2. Enter the deposit as a negative payment.
2. Type the formula =D2/E2 in F2 to compute the rate per period.
3. Type the heading Future Value in A4.
4. In cell B4, type the formula =fv(F2,B2,C2,A2,0) to compute the future value.
   (The 0 indicates that the payments are made at the end of each period.)

For annuities due, use =fv(F2,B2,C2,A2,1).
The payments are made at the beginning of each period.

This spreadsheet gives the future value of an ordinary annuity of $200 deposited at the end of each quarter for $2\frac{1}{4}$ years, with interest at 4%, compounded quarterly.

Principal	Number of Periods	Payment	Annual Rate	Periods per year	Periodic Rate
0	9	−200	0.04	4	0.01
Ordinary annuity					
Future Value	$1,873.71				
Annuities due					
Future Value	$1,892.44				

Section 6.4 Finding the Present Value of an Annuity

To find the present value of an ordinary annuity:

1. Type the headings in Row 1, and enter their values (with the interest rate as a decimal) in Row 2.
2. Enter the formula =D2/E2 in F2 to compute the rate per period.
3. Type the heading Present Value in A4.
4. In cell B4, type the formula =pv(F2,B2,C2,A2,0) to compute the present value. (The 0 indicates that the payments are made at the end of each period.) The parentheses on the present value in cell B4 mean that it is negative (the money leaves the investor to start the annuity).

For annuities due, use =pv(F2,B2,C2,A2,1).
The payments are made at the beginning of each period.

This spreadsheet gives the lump sum deposit (present value) necessary to receive $1000 per month for 16 years, with interest at 9%, compounded monthly, for both an ordinary annuity and an annuity due.

Future Value	Number of Periods	Payment	Annual Rate	Periods per year	Periodic Rate
0	192	1000	0.09	12	0.0075
Ordinary annuity					
Present Value	($101,572.77)				
Annuities due					
Present value	($102,334.56)				

Section 6.5 Finding Payments to Amortize a Loan

To find the periodic payment to pay off a loan:

1. Type the headings in Row 1 and their values (with the interest rate as a decimal) in Row 2.
2. Enter the formula =D2/E2 in F2 to compute the rate per period.
3. Type the heading Payment in A4.
4. In cell B4, type the formula =Pmt(F2,B2,A2,C2,0) to compute the payment.

This spreadsheet gives the annual payment of a loan of $10,000 over 5 years when interest is 10% per year.

The parentheses indicates a payment out.

Loan Amount	Number of Periods	Future Value	Annual Rate	Periods per year	Periodic Rate
10000	5	0	0.1	1	0.1
Payment	($2,637.97)				

Chapter 8
Section 8.1 Binomial Probabilities

1. Type headings in cells A1:A3 and their respective values in cells B1:B3.
2. Use the function =binomdist(B1,B2,B3,cumulative) where
   - B1 is the number of successes.
   - B2 is the number of independent trials.
   - B3 is the probability of success in each trial.
   - True replaces cumulative if a cumulative probability is sought; it is replaced by false otherwise.
   - The probability of exactly 3 heads in 6 tosses is found by evaluating =binomdist(B1,B2,B3,false) in B4.
   - The probability of 3 or fewer heads in 6 tosses is found by evaluating =binomdist(B1,B2,B3,true) in B5.

This spreadsheet gives the probability of 3 heads in 6 tosses of a fair coin and the probability of 3 or fewer heads in 6 tosses.

	A	B
1	Number of successes	3
2	Number of trials	6
3	Probability of success	.5
4	Probability of 3 successes	.3125
5	Probability of 3 or fewer successes	.65625

Section 8.2 Bar Graphs

To construct a bar graph for the given table of test scores:

1. Copy the entries of the table to cells A2:B6.
2. Select the range A2:B6.
3. Click the *Chart Wizard* icon.
4. Select the graph option with the first sub-type.
5. Click Next.
6. Click Next through Steps 2–4. Note that you can add a title in Step 3.

	A	B
1	Grade Range	Frequency
2	90–100	2
3	80–89	5
4	70–79	3
5	60–69	3
6	0–59	2

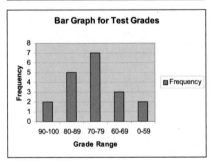

Section 8.2 Finding the Mean, Standard Deviation, and Median of Raw Data

To find the mean, standard deviation, and median of a raw data set:

1. Enter the data in Row 1 (cells A1:L1).
2. Type the heading Mean in cell A3.
3. Type the formula =average(A1:L1) in cell B4.
4. Type the heading Standard Deviation in cell A4.
5. In cell B4, type the formula =stdev(A1:L1).
6. In cell A5, type the heading Median.
7. In cell B5, type the formula =median(A1:L1).

The mean, standard deviation, and median for the data 1, 1, 1, 3, 3, 4, 4, 5, 6, 6, 7, 7 is shown below.

	A	B	C	D	E	F	G	H	I	J	K	L
1	1	1	1	3	3	4	4	5	6	6	7	7
2												
3	Mean	4										
4	Standard Deviation	2.2563										
5	Median	4										

Section 8.2 Finding the Mean and Standard Deviation of Grouped Data

To find the mean:

1. Enter the data and headings in the cells A1:C6.
2. In D1, type the heading Class mark*frequency.
3. In D2, type the formula =B2*C2.
4. Copy the formula in D2 to D3:D6.
5. In B7, type the heading Total.
6. In cell C7, type the formula for the total frequencies, =sum(C2:C6).
7. In cell D7, type the formula for the total, =sum(D2:D6).
8. In cell A8, type the heading Mean.
9. In cell A9, type in the formula =D7/C7.

To find the standard deviation:

10. In cell E1, type in the heading freq *(x − x_mean)^2.
11. In cell E2, type the formula =C2*(B2-A9)^2. (The A9 gives the value in A9; the reference doesn't change as we fill down.)
12. Copy the formula in E2 to E3:E6.
13. In cell E7, type the formula =sum(E2:E6).
14. In cell A10 type the heading Standard Deviation.
15. In cell A11, type the formula =sqrt(E7/(C7-1)).

Grade Range	Class Marks	Frequency
90–100	95	3
80–89	84.5	4
70–79	74.5	7
60–69	64.5	0
50–59	54.5	2

	A	B	C	D	E
1	Grade Range	Class Marks	Frequency	Class mark*frequency	Freq*(x-x_mean)^2
2	90-100	95	3	285	832.2919922
3	80-89	84.5	4	338	151.5976563
4	70-79	74.5	7	521.5	103.4208984
5	60-69	64.5	0	0	0
6	50-59	54.5	2	109	1137.048828
7		Total	16	1253.5	2224.359375
8	Mean				
9	78.344				
10	Standard Deviation				
11	12.177				

Section 8.4 Calculating Normal Probabilities

To calculate normal probabilities:

1. Type headings in A1:A4 and their respective values in cells B1:B4.
2. To find the probability that a score X is less than the x1 value in B3, enter the formula =normdist(B3,B1,B2,true) in cell B5.
3. To find the probability that X is less than the x2 value in B4, enter the formula =normdist(B4,B1,B2,true) in cell B6.
4. To find the probability that a score X is more than the value in B3 and less than the x2 value in B4, enter the formula =B6-B5 in cell B7.

Entries in B5, B6, and B7 give the probabilities of a score X being less than 100, less than 115, and between 100 and 115, respectively, when the mean is 100 and the standard deviation is 15.

	A	B
1	Mean	100
2	Standard Deviation	15
3	x1	100
4	x2	115
5	Pr(X<x1)	0.5
6	Pr(X<x2)	0.841345
7	Pr(x1<X<x2)	0.341345

Chapter 9
Sections 9.1–9.2 Evaluating Limits

To evaluate $\lim_{x \to c} f(x)$:

1. Make a table of values for $f(x)$ near $x = c$. Include values on both sides of $x = c$.
2. Use the table of values to predict the limit (or that the limit does not exist).

   Note: All limit evaluations with Excel use appropriate tables of values of $f(x)$. This is true when $f(x)$ is piecewise defined and for limits as $\to \infty$.

Find $\lim_{x \to 2} \dfrac{x^2 - 4}{x - 2}$.

	A	B	C	D
	x	**f(x)**	**x**	**f(x)**
1				
2	2.1	4.1	1.9	3.9
3	2.05	4.05	1.95	3.95
4	2.01	4.01	1.99	3.99
5	2.001	4.001	1.999	3.999

The tables suggest that $\lim_{x \to 2} \dfrac{x^2 - 4}{x - 2} = 4$.

Sections 9.3–9.7 Approximating Derivatives

To approximate $f'(c)$:

1. Numerically investigate the limit in the definition of derivative:
$$f'(x) = \lim_{h \to 0} \frac{f(x + h) - f(x)}{h}$$
2. Use the given $f(x)$ and $x = c$ to create a table of values for h near 0 (and on both sides of $h = 0$).

Note that rows 5 and 6 have the values of h closest to 0.

Investigate $f'(1)$ for $f(x) = x^3$.

	A	B	C	D	E
1	**h**	**1+h**	**f(1)**	**f(1+h)**	**(f(1+h)-f(1))/h**
2	0.1	1.1	1	1.331	3.31
3	0.01	1.01	1	1.030301	3.0301
4	0.001	1.001	1	1.003003	3.003001
5	0.0001	1.0001	1	1.0003	3.00030001
6	−0.0001	0.9999	1	0.9997	2.99970001
7	−0.001	0.999	1	0.997003	2.997001
8	−0.01	0.99	1	0.970299	2.9701
9	−0.1	0.9	1	0.729	2.71

The table suggests that $f'(1) = 3$, which is the actual value.
Note: Excel has no built-in derivative approximation tool.

Chapter 10
Section 10.1 Relative Maxima and Minima

1. Make a table with columns for x-values, the function, and the derivative.
2. Extend the table to include x-values to the left and to the right of all critical values.
3. Use the signs of the values of the derivative to determine whether f is increasing or decreasing around the critical values, and thus to classify the critical values as relative maxima, relative minima, or horizontal points of inflection. You may want to graph the function to confirm your conclusions.

The spreadsheet shows that the relative maxima or minima of $f(x) = x^2$ is 0 at $x = 0$. Note that the derivative is $f'(x) = 2x$.

	A	B	C
1	x	f(x)	f'(x)
2	−2	4	−4
3	−1.5	2.25	−3
4	−1	1	−2
5	−0.5	0.25	−1
6	0	0	0
7	0.5	0.25	1
8	1	1	2
9	1.5	2.25	3
10	2	4	4

Section 10.2 Exploring f, f', f'' Relationships

To explore relationships among the graphs of a function and its derivatives:

1. Find functions for f' and f''.
2. Graph all three functions on the same plot.
   - Notice that f increases when f' is above the x-axis ($+$) and decreases when f' is below the x-axis ($-$).
   - Notice that f is concave up when f'' is above the x-axis ($+$) and is concave down when f'' is below the x-axis ($-$).

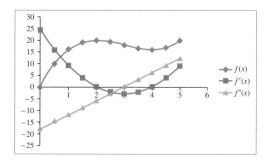

Let $f(x) = x^3 - 9x^2 + 24x$. Graph f, f', and f'' on the interval $[0, 5]$ to explore the relationships among these functions.

	A	**B**	**C**	**D**
1	x	$f(x)$	$f'(x)$	$f''(x)$
2	0	0	24	−18
3	0.5	9.875	15.75	−15
4	1	16	9	−12
5	1.5	19.125	3.75	−9
6	2	20	0	−6
7	2.5	19.375	−2.25	−3
8	3	18	−3	0
9	3.5	16.625	−2.25	3
10	4	16	0	6
11	4.5	16.875	3.75	9
12	5	20	9	12

Sections 10.3–10.4 Finding Optimal Values

To find the optimal value of a function when the goal is not to produce a graph:

1. Set up a spreadsheet that identifies the variable and the function whose optimal value is sought.
2. Choose *Tools* > *Solver*. Then, in the *Solver* dialog box,
   a. Set the *Target Cell* as that of the objective function.
   b. Check *Max* or *Min* according to your goal.
   c. Set the *Changing Cells* to reference the variable.
3. Click on the *Options* box. Make sure "Assume Linear Model" is not checked. Then click OK.
4. Click *Solve* in the *Solver* dialog box. You will get a dialog box stating that Solver found a solution. Save the solution if desired, then click OK.
5. The cells containing the variable and the function should now contain the optimal values.

Minimize area $A = x^2 + \dfrac{160}{x}$ for $x > 0$.

	A	**B**
1	**Variable**	
2		
3	x length of base	1
4		
5	**Objective**	
6		
7	Minimize Area	=B3^2+160/B3

	A	**B**
1	**Variable**	
2		
3	x length of base	4.3089
4		
5	**Objective**	
6		
7	Minimize Area	55.699

The function is minimized for $x = 4.3089$ and the minimum value is $A = 55.699$.

Chapter 13
Section 13.1 Approximating Definite Integrals Using Rectangles

To find the approximate area between $y = f(x)$ and the x-axis over an interval by using the right-hand endpoints of rectangles with equal base width:

1. Enter "x" and the x-values of the right-hand endpoints of each rectangle in Column B. Fill down can be used if the number of rectangles is large.
2. Enter the formula for the function, using cell addresses, in C2 and fill down column C to get the heights of the rectangles.
3. Enter the width of the rectangles in Column D.
4. Enter the formula for the area of each rectangle, =C2*D2, in E2 and fill down to get the area of each rectangle.
5. Enter "Total" in A10 and the formula =SUM(E2:E9) in E10 to get the approximate area.

To find the approximation for the area under $y = \sqrt{x}$ from $x = 0$ to $x = 4$ using $n = 8$ equal subdivisions (with base width 0.5) and right-hand endpoints:

1. Enter "x" and the eight x- values of the right-hand endpoints (starting with 0.5) in Column B.
2. Enter "y" in C1 and =SQRT(B2) in C2 and fill down column C.
3. Enter "width" in D1 and 0.5 below in Column D.
4. Enter "Area" in E1, =C2*D2 in E2, and fill down to get the area of each rectangle.
5. Enter "Total" in A10 and the formula =SUM(E2:E9) in E10 to get the approximate area, $S_R \approx 5.765$.

	A	B	C	D	E
1	Rectangle	x	y	Width	Area
2	1	0.5	0.707107	0.5	0.353553
3	2	1	1	0.5	0.5
4	3	1.5	1.224745	0.5	0.612372
5	4	2	1.414214	0.5	0.707107
6	5	2.5	1.581139	0.5	0.790569
7	6	3	1.732051	0.5	0.866025
8	7	3.5	1.870829	0.5	0.935414
9	8	4	2	0.5	1
10	Total				5.765042

Chapter 14
Section 14.1 Graphs of Functions of Two Variables

To create a surface plot for a function of two variables:

1. a. Generate appropriate *x*-values beginning in B1 and continuing *across*.
   b. Generate appropriate *y*-values beginning in A2 and continuing *down*.
2. Generate values for the function that correspond to the points (*x*, *y*) from Step 1 as follows:
   In cell B2, enter the function formula with B$1 used to represent *x* and $A2 to represent *y*. (See the online Excel Guide for additional information about the role and use of the $ in this step.) Then use fill down and fill across to complete the table.
3. Select the entire table of values. Click the *Chart Wizard* and choose *Surface in the Chart* menu.
4. Annotate the graph and click *Finish* to create the surface plot. You can move and view the surface from a different perspective by clicking into the resulting graph.

Let $f(x,y) = 10 - x^2 - y^2$. Plot the graph of this function for both *x* and *y* in the interval $[-2,2]$.

	A	B	C	D	E	F	G	H	I	J
1		−2	−1.5	−1	−0.5	0	0.5	1	1.5	2
2	−2									
3	−1.5									
4	−1									
5	−0.5									
6	0									
7	0.5									
8	1									
9	1.5									
10	2									

	A	B	C	D	E	F	G	H	I	J
1		−2	−1.5	−1	−0.5	0	0.5	1	1.5	2
2	−2	2	3.75	5	5.75	6	5.75	5	3.75	2
3	−1.5	3.75	5.5	6.75	7.5	7.75	7.5	6.75	5.5	3.75
4	−1	5	6.75	8	8.75	9	8.75	8	6.75	5
5	−0.5	5.75	7.5	8.75	9.5	9.75	9.5	8.75	7.5	5.75
6	0	6	7.75	9	9.75	10	9.75	9	7.75	6
7	0.5	5.75	7.5	8.75	9.5	9.75	9.5	8.75	7.5	5.75
8	1	5	6.75	8	8.75	9	8.75	8	6.75	5
9	1.5	3.75	5.5	6.75	7.5	7.75	7.5	6.75	5.5	3.75
10	2	2	3.75	5	5.75	6	5.75	5	3.75	2

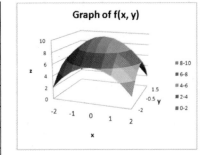

Graph of f(x, y)

Section 14.5 Constrained Optimization and Lagrange Multipliers

To solve a constrained optimization problem:

1. Set up the problem in Excel.
2. Choose *Tolls* > *Solver* and do the following:
   - Choose the objective function as the *Target Cell*.
   - Check *Max* or *Min* depending on the problem.
   - Choose the cells representing the variables for the *By Changing Cells* box.
   - Click on the *Constraints* box and press Add. Then enter the constraint equations.
3. Click on the *Options* box and make sure that "Assume Linear Model" is *not* checked. Then click OK.
4. Click *Solve* in the *Solver* dialog box. Then click OK to solve.

Maximize $P = 600l^{2/3}k^{1/3}$ subject to $40l + 100k = 3000$.

	A	B	C
1	**Variables**		
2			
3	units labor (l)	0	
4	units capital (k)	0	
5			
6			
7	**Objective**		
8			
9	Maximize production	=600*B3^(2/3)*B4^(1/3)	
10			
11	Constraint		
12		Amount used	Available
13	Cost	=40*B3+100*B4	3000

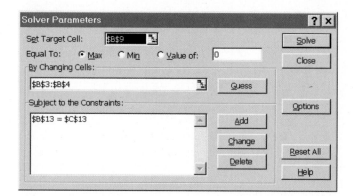

	A	B	C
1	**Variables**		
2			
3	units labor (l)	50.00000002	
4	units capital (k)	10	
5			
6			
7	**Objective**		
8			
9	Maximize product	17544.10644	
10			
11	**Constraint**		
12		Amount used	Available
13	Cost	3000.000001	3000

Excel Guide Part 2
Excel 2007 and 2010

Except where noted, the directions given will work in both Excel 2007 and Excel 2010.

Excel Worksheet

When you start up Excel by using the instructions for your software and computer, the following screen will appear.

The components of the **spreadsheet** are shown, and the grid shown is called a **worksheet.** You can move to other worksheets by clicking on the tabs at the bottom.

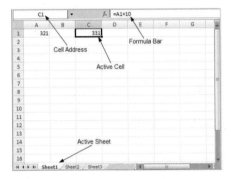

Addresses and Operations

Notice the letters at the top of the columns and the numbers identifying the rows. The cell addresses are given by the column and row; for example, the first cell has address A1. You can move from one cell to another with arrow keys, or you can select a cell by clicking on it. After you type an entry in a cell, press enter to accept it. To edit the contents of a cell, click on the cell and edit the contents in the formula bar at the top. To delete the contents, press the delete key.

The file operations such as "open a new file," "saving a file," and "printing a file" are similar to those in Word. For example, <CTRL>S saves a file. You can also format a cell entry by selecting it and using menus similar to those in Word.

Working with Cells

Cell entries, rows containing entries, and columns containing entries can be copied and pasted with the same commands as in Word. For example, a highlighted cell can be copied with <CTRL>C. Sometimes entries exceed the width of the cells containing them, especially if they are text. To widen the cells in a column, place the cursor at the right side of the column heading until you see the symbol ↔, then move the cursor to the right (moving it to the left makes the column more narrow). If entering a number results in #####, the number is too long for the cell, and the cell should be widened.

Chapter 1
Section 1.4 Entering Data and Evaluating Functions

The cells are identified by the column and the row. For example, the cell B3 is in the second column and the third row.

1. Put headings on the two columns.
2. Fill the inputs in Column A by hand or with a formula for them. The formula =A2+1 gives 2 in A3 when ENTER is pressed.
3. Moving the mouse to the lower right corner of A3 until there is a thin "+" sign and dragging the mouse down "fills down" all required entries in column A.
4. Enter the function formula for the function in B2. Use $= 1000*(1.1)\wedge A2$ to represent $S = 1000(1.1)^t$. Pressing ENTER gives the value when $t = 1$.
5. Using Fill Down gives the output for all inputs.

	A	B
1	Year	Future Value
2	1	=1000*(1.1)^(A2)
3	=A2+1	

	A	B
1	Year	Future Value
2	1	1100
3	2	1210
4	3	1331

Section 1.4 Graphing a Function

1. Put headings on the two columns (x and $f(x)$, for example).
2. Fill the inputs (x-values) in Column A by hand or with a formula for them.
3. Enter the formula for the function in B2. Use $= 6*A2 - 3$ to represent $f(x) = 6x - 3$.
4. Select the cell containing the formula for the function (B2, for example).
5. Move the mouse to the lower right corner until there is a thin "+" sign.
6. Drag the mouse down to the last cell where formula is required, and press ENTER.
7. Highlight the two columns containing the values of x and $f(x)$.
8. Click the Insert tab and then select the *Scatter* chart type with the smooth curve option.
9. Once the smooth curve option is selected, the chart will appear within the worksheet. The worksheet is now in design mode, where you can change the chart options.
10. Click on the Home tab to get back to the original view of the spreadsheet.

	A	B
1	x	$f(x) = 6x - 3$
2	−5	−33
3	−2	−15
4	−1	−9
5	0	−3
6	1	3
7	3	15
8	5	27

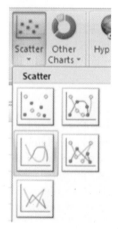

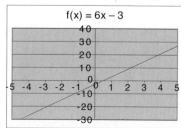

Section 1.5 Solving Systems of Two Equations in Two Variables

1. Write the two equations as linear functions in the form $y = mx + b$.
2. Enter the input variable x in cell A2 and the formula for each of the two equations in cells B2 and C2, respectively.
3. Enter $= B2 - C2$ in cell D2.
4. Start *Goal Seek* by clicking on the Data tab. Then select *What-If Analysis* in the Data Tools group, and then *Goal Seek*.
5. In the dialog box:
   a. Click the *Set Cell* box and click on the D2 cell.
   b. Enter 0 in the *To Value* box.
   c. Click the *By Changing Cell* box and click on the A2 cell.
6. Click OK in the *Goal Seek* dialog box, getting 0.
7. The x-value of the solution is in cell A2, and the y-value is in both B2 and C2.

The solution of the system

$$\begin{cases} 3x + 2y = 12 \\ 4x - 3y = -1 \end{cases}$$

is found to be $x = 2$, $y = 3$ using *Goal Seek* as follows.

	A	B	C	D
1	x	$= 6 - 1.5x$	$= 1/3 + 4x/3$	$= y1 - y2$
2	1	$= 6 - 1.5*A2$	$= 1/3 + 4*A2/3$	$= B2 - C2$

	A	B	C	D
1	x	$= 6 - 1.5x$	$= 1/3 + 4x/3$	$= y1 - y2$
2	2	3	3	0

Chapter 2
Section 2.1 Solving Quadratic Equations

To solve a quadratic equation of the form
$f(x) = ax^2 + bx + c = 0$

1. Enter x-values centered around the x-coordinate
   $x = \dfrac{-b}{2a}$ in column A and use the function formula
   to find the values of $f(x)$ in column B.
2. Graph the function, $f(x) = 2x^2 - 9x + 4$ in this case, and observe where the graph crosses the x-axis ($f(x)$ near 0). Insert the graph by selecting *Insert>Scatter>Smooth curve* option.
3. Use *Data>What-If>Goal Seek*, entering a cell address with a function value in column B at or near 0, enter the set cell to the value 0, and enter the changing cell.
4. Click OK to find the x-value of the solution in cell A2. The solution may be approximate. The spreadsheet shows $x = 0.5001$, which is an approximation of the exact solution $x = 0.5$.
5. After finding the first solution, repeat the process using a second function value at or near 0. The second solution is $x = 4$ in this case.

	A	B	C	D	E	F
1	x	$f(x)=2x^2-9x+4$				
2	−1	15				
3	0	4				
4	1	−3				
5	2	−6				
6	3	−5				
7	4	0				
8	5	9				
9						

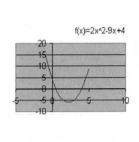

	A	B	C	D	E	F
1	x	$f(x)=2x^2-9x+4$				
2	−1	15				
3	0	4				
4	0.500001	−7.287E−05				
5	2	−6				
6	3	−5				
7	4	0				
8	5	9				
9						

Sections 2.2 and 2.4 Graphing Polynomial Functions

To graph a polynomial function:

1. Use the function to create a table containing values for x and $f(x)$. (See Graphing a Function on page AP-46.)
2. Highlight the two columns containing the values of x and $f(x)$.
3. Insert the graph by selecting *Insert > Scatter > Smooth curve* option.
4. Click on the Home tab to get back to the original view of the spreadsheet.

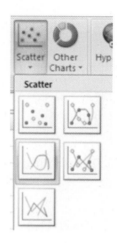

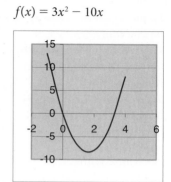

$f(x) = 3x^2 - 10x$

Section 2.4 Graphing Rational Functions

An Excel graph will connect all points corresponding to values in the table, so if the function you are graphing is undefined for some x-value a, enter x-values near this value and leave (or make) the corresponding $f(a)$ cell blank.

To graph $f(x) = \dfrac{1}{1-x}$, which is undefined at $x = 1$:

1. Generate a table for x-values from -1 to 3, with extra values near $x = 1$.
2. Generate the values of $f(x)$, and leave a blank cell for the $f(x)$ value for $x = 1$.
3. Select the table and plot the graph using *Insert > Scatter > Smooth curve* option.

x	f(x)
−1	−0.5
−0.5	−0.6666667
0	−1
0.5	−2
0.75	−4
0.9	−10
1	
1.1	10
1.25	4
1.5	2
2	1
2.5	0.66666667
3	0.5

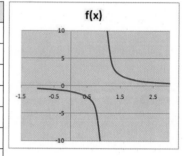

Section 2.5 Modeling

To create a scatter plot of data:

1. Enter the inputs (x-values) in Column A and the outputs (y-values) in Column B.
2. Highlight the two columns and use *Insert > Scatter* and choose the points only option to plot the points.
3. To add titles, click on the icons in the *Charts Layout* group.

	A	B	C	D	E	F	G	H	I
1									
2	Time (sec)	Speed (mph)							
3	1.7	30							
4	2.4	40							
5	3.5	50							
6	4.3	60							
7	5.5	70							
8	7	80							
9	8.5	90							
10	10.2	100							
11									

To find the equation of a line or curve that best fits a given set of data points:

1. Single-click on the scatter plot of the data in the worksheet.
2. Right-click on one of the data points.
3. Select *Add Trendline*.
4. Select the function type that appears to best fit and check the box that says *Display Equation on chart*. [Note: If *Polynomial* is selected, choose the appropriate Order (degree).]
5. Close the dialog box and you will see the graph of the selected function that is best fit along with its equation.

The power function that models Corvette acceleration follows.

	A	B	C	D	E	F	G	H	I
1									
2	Time (sec)	Speed (mph)							
3	1.7	30							
4	2.4	40							
5	3.5	50							
6	4.3	60							
7	5.5	70							
8	7	80							
9	8.5	90							
10	10.2	100							
11									

Corvette Acceleration

Speed (mph)

$y = 21.875x^{0.6663}$

The speed of the Corvette is given by $y = 21.875x^{0.6663}$ where x is the time in seconds.

Chapter 3
Section 3.1 Operations with Matrices

Operations for 3×3 matrices can be used for other orders.

1. Type a name A in A1 to identify the first matrix.
2. Enter the matrix elements of matrix A in the cells B1:D3.
3. Type a name B in A5 to identify the second matrix.
4. Enter the matrix elements of matrix B in the cells B5:D7.
5. Type a name A + B in A9 to indicate the matrix sum.
6. Type the formula "=B1 + B5" in B9 and press ENTER.
7. Use Fill Across to copy this formula across the row to C9 and D9.
8. Use Fill Down to copy the row B9:D9 to B11:D11, which gives the sum.
9. To subtract the matrices, change the formula in B9 to "=B1 − B5" and proceed as with addition.

	A	B	C	D
1	A	1	2	3
2		4	5	6
3		7	8	9
4				
5	B	−2	−4	3
6		1	4	−5
7		3	6	−1
8				
9	A+B	−1	−2	6
10		5	9	1
11		10	14	8

Section 3.2 Multiplying Two Matrices

Steps for two 3×3 matrices:

1. Enter the names and elements of the matrices.
2. Enter the name AxB in A9 to indicate the matrix product.
3. Select a range of cells that is the correct size to contain the product (B9:D11 in this case).
4. Type "=mmult(" in the formula bar, and then select the cells containing the elements of matrix A (B1:D3).
5. Stay in the formula bar, type a comma and select the matrix B elements (B5:D7), and close the parentheses.
6. Hold the CTRL and SHIFT keys down and press ENTER, giving the product.

	A	B	C	D
1	A	1	2	3
2		4	5	6
3		7	8	9
4				
5	B	−2	−4	3
6		1	4	−5
7		3	6	−1
8				
9	AxB	=mmult(B1:D3)		
10		MMULT(**array1**,		
11		array2)		

B9	fx	{(=MMULT(B1:D3,B5:D7)}			
	A	B	C	D	E
1	A	1	2	3	
2		4	5	6	
3		7	8	9	
4					
5	B	−2	−4	3	
6		1	4	−5	
7		3	6	−1	
8					
9	AxB	9	22	−10	
10		15	40	−19	
11		21	58	−28	

Section 3.4 Finding the Inverse of a Matrix

Steps for a 3×3 matrix:

1. Enter the name A in A1 and the elements of the matrix in B1:D3 as above.
2. Enter the name "Inverse(A)" in A5 and select a range of cells that is the correct size to contain the inverse [(B5:D7) in this case].
3. Enter "=minverse(", select matrix A(B1:D3), and close the parentheses.
4. Hold the CTRL and SHIFT keys down and press ENTER, getting the inverse.

SUM	*fx*	=minverse(B1:D3)		
	A	**B**	**C**	**D**
1	A	2	1	1
2		1	2	0
3		2	0	1
4				
5	inverse(A)	=minverse(B1:D3)		
6				
7				

B5	*fx*	{=MINVERSE(B1:D3)}		
	A	**B**	**C**	**D**
1	A	2	1	1
2		1	2	0
3		2	0	1
4				
5	inverse(A)	−2	1	2
6		1	0	−1
7		4	−2	−3

Section 3.4 Solving Systems of Linear Equations with Matrix Inverses

A system of linear equations can be solved by multiplying the matrix containing the augment by the inverse of the coefficient matrix. The steps used to solve a 3×3 system follow.

1. Enter the coefficient matrix A in B1:D3.
2. Enter the name "inverse(A)" in A5 and compute the inverse of A in B5:D7.
3. Enter B in cell A9 and enter the augment matrix in B9:B11.
4. Enter X in A13 and select the cells B13:B15.
5. In the formula bar, type "=mmult(", then select matrix inverse(A) in B5:D7, type a comma, select matrix B in B9:B11, and close the parentheses.
6. Hold the CTRL and SHIFT keys down and press ENTER, getting the solution.
7. Matrix X gives the solution.

The system $\begin{cases} 2x + y + z = 8 \\ x + 2y \quad = 6 \\ 2x + \quad z = 5 \end{cases}$ is solved as follows.

	SUM	*fx*	=mmult(B5:D7,B9:B11)		
	A	**B**	**C**	**D**	**E**
1	A	2	1	1	
2		1	2	0	
3		2	0	1	
4					
5	inverse(A)	−2	1	2	
6		1	0	−1	
7		4	−2	−3	
8					
9	B	8			
10		6			
11		5			
12					
13	X	=mmult(B5:D7,B9:B11)			
14					
15					

	B13	*fx*	{(=MMULT(B5:D7,B9:B11)}		
	A	**B**	**C**	**D**	**E**
1	A	2	1	1	
2		1	2	0	
3		2	0	1	
4					
5	inverse(A)	−2	1	2	
6		1	0	−1	
7		4	−2	−3	
8					
9	B	8			
10		6			
11		5			
12					
13	X	0			
14		3			
15		5			

The solution is $x = 0, y = 3, z = 5$.

Chapter 4
Sections 4.3–4.5 Linear Programming
(Excel 2007)

Maximize $f = 12x + 26y + 40z$ subject to the constraints

$$\begin{cases} 5x + 7y + 10z \le 90{,}000 \\ x + 3y + 4z \le 30{,}000 \\ x + y + z \le 9000 \\ x \ge 0,\ y \ge 0,\ z \ge 0 \end{cases}$$

1. On a blank spreadsheet, type a heading in cell A1, followed by the variable descriptions in cells A3–A5 and the initial values (zeros) in cells B3–B5.
2. a. Enter the heading "Objective" in cell A7.
   b. Enter a description of the objective in cell A9 and the formula for the objective function in B9. The formula is $= 12{*}B3 + 26{*}B4 + 40{*}B5$.
3. a. Type in the heading "Constraints" in A11 and descriptive labels in A13–A15.
   b. Enter the left side of the constraint inequalities in B13–B15 and the maximums from the right side in C13–C15.
4. Click on the *Data* tab and select *Solver* in the Analysis group. A dialog box will appear.*
5. a. Click the *Set Target Cell* box and B9 (containing the formula for the objective function).
   b. Check the button *Max* for maximization.
   c. Click the *By Changing Cells* box and select cells B3–B5.
6. a. Click the *Subject to Constraints* entry box. Press the Add button to add the first constraint.
   b. Click the left entry box and click cell B13 (containing the formula for the first constraint).
   c. Set the middle entry box to $<=$.
   d. Click the right entry box and C13 to enter the constraint.
   e. Click Add and repeat the Steps 6b–6d for the remaining constraints.
   f. Click in the left entry box for the constraints and select the variables in B3–B5. Set the middle entry to $>=$, and type 0 in the right entry box.
7. Select the Simplex LP option (this is the default) Click Solve in the *Solver* dialog box.
   A dialog box states that *Solver* found a solution. To see the *Solver* results, click *Keep Solver Solution* and also select Answer.
8. Go back to the spreadsheet. The new values in B3–B5 are the values of the variables that give the maximum, and the value in B9 is the maximum value of the objective function.

In *Solver*, *minimization* of an objective function is handled exactly the same as maximization, except *min* is checked and the inequality signs are $\ge$.

When *mixed constraints* are used, simply enter them with "mixed" inequalities.

*You may need to use an "Add-In" to access *Solver*.

	A	B	C
1	**Variables**		
2			
3	# small calculators (x)	0	
4	# medium calculators (y)	0	
5	# large calculators (z)	0	
6			
7	**Objective**		
8			
9	Maximize profit	=12*B3+26*B4+40*B5	
10			
11	**Constraints**		
12		Amount used	Maximum
13	Circuit components	=5*B3+7*B4+10*B5	90000
14	Labor	=B3+3*B4+4*B5	30000
15	Cases	=B3+B4+B5	9000

	A	B	C
1	**Variables**		
2			
3	# small calculators (x)	2000	
4	# medium calculators (y)	0	
5	# large calculators (z)	7000	
6			
7	**Objective**		
8			
9	Maximize profit	304000	
10			
11	**Constraints**		
12		Amount used	Maximum
13	Circuit components	80000	90000
14	Labor	30000	30000
15	Cases	9000	9000

Chapter 4
Sections 4.3–4.5 Linear Programming
(Excel 2010)

Maximize $f = 12x + 26y + 40z$ subject to the constraints

$$\begin{cases} 5x + 7y + 10z \leq 90{,}000 \\ x + 3y + 4z \leq 30{,}000 \\ x + y + z \leq 9000 \\ x \geq 0, y \geq 0, z \geq 0 \end{cases}$$

1. On a blank spreadsheet, type a heading in cell A1, followed by the variable descriptions in cells A3–A5 and the initial values (zeros) in cells B3–B5.
2. a. Enter the heading "Objective" in cell A7.
   b. Enter a description of the objective in cell A9 and the formula for the objective function in B9. The formula is =12*B3 + 26*B4 + 40*B5.
3. a. Type in the heading "Constraints" in A11 and descriptive labels in A13–A15.
   b. Enter the left side of the constraint inequalities in B13–B15 and the maximums from the right side in C13–C15.
4. Click on the *Data* tab and select *Solver* in the Analysis group. A dialog box will appear.*
5. a. Click the *Set Target Cell* box and B9 (containing the formula for the objective function).
   b. Check the button *Max* for maximization.
   c. Click the *By Changing Cells* box and select cells B3–B5.
6. a. Click the *Subject to Constraints* entry box. Press the Add button to add the first constraint.
   b. Click the left entry box and click cell B13 (containing the formula for the first constraint).
   c. Set the middle entry box to $\leq$.
   d. Click the right entry box and C13 to enter the constraint.
   e. Click Add and repeat the Steps 6b–6d for the remaining constraints.
   f. Click in the left entry box for the constraints and select the variables in B3–B5. Set the middle entry to $\geq$, and type 0 in the right entry box.
7. Select the Simplex LP option (this is the default). Click Solve in the *Solver* dialog box. A dialog box states that *Solver* found a solution. To see the *Solver* results, click *Keep Solver Solution* and also select Answer.
8. Go back to the spreadsheet. The new values in B3–B5 are the values of the variables that give the maximum, and the value in B9 is the maximum value of the objective function.

In *Solver*, *minimization* of an objective function is handled exactly the same as maximization, except *min* is checked and the inequality signs are $\geq$.

When *mixed constraints* are used, simply enter them with "mixed" inequalities.

*You may need to use an "Add-In" to access *Solver*.

	A	B	C
1	**Variables**		
2			
3	# small calculators (x)	0	
4	# medium calculators (y)	0	
5	# large calculators (z)	0	
6			
7	**Objective**		
8			
9	Maximize profit	=12*B3+26*B4+40*B5	
10			
11	**Constraints**		
12		Amount used	Maximum
13	Circuit components	=5*B3+7*B4+10*B5	90000
14	Labor	=B3+3*B4+4*B5	30000
15	Cases	=B3+B4+B5	9000

	A	B	C
1	**Variables**		
2			
3	# small calculators (x)	2000	
4	# medium calculators (y)	0	
5	# large calculators (z)	7000	
6			
7	**Objective**		
8			
9	Maximize profit	304000	
10			
11	**Constraints**		
12		Amount used	Maximum
13	Circuit components	80000	90000
14	Labor	30000	30000
15	Cases	9000	9000

Chapter 5

Section 5.1 Graphing Exponential Functions

Exponential functions are entered into Excel differently for base e than for bases other than e.

A. To graph $y = a^x$, use the formula =a^x.
B. To graph $y = e^x$, use the formula =exp(x).

To graph $f(x) = 1.5^x$ and $g(x) = e^x$ on the same axes:
1. Type x in cell A1 and numbers centered at 0 in Column A.
2. Type $f(x)$ in cell B1, enter the formula =1.5^A2 in cell B2 and fill down.
3. Type $g(x)$ in cell C1, enter the formula =exp(A2) in cell C2 and fill down.
4. Select the entire table and graph using *Insert>Scatter >Smooth curve with markers* option to graph the functions.

	A	B	C
	x	$f(x)$	$g(x)$
1	x	$f(x)$	$g(x)$
2	−2	0.444444	0.135335
3	−1.5	0.544433	0.22313
4	−1	0.666667	0.367879
5	−0.5	0.816497	0.606531
6	0	1	1
7	0.5	1.224745	1.648721
8	1	1.5	2.718282
9	1.5	1.837117	4.481689
10	2	2.25	7.389056
11	2.5	2.755676	12.18249
12	3	3.375	20.08554

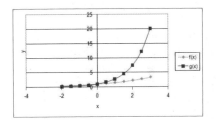

Section 5.1 Modeling with Exponential Functions

To model data with an exponential function:

1. Create a scatter plot for the data using *Insert>Scatter* and the points only option.
2. Single click on the chart and right click on one of the data points. Choose *Add Trendline.*
3. Check *Exponential* for the trendline type if that function type appears to be the best fit for the scatter plot, and check the box to Display the equation on the graph.
4. Close the dialog box to see the graph of the function and its equation.

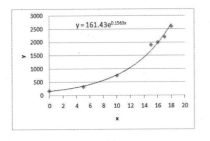

The exponential model for the following data is shown.

x	0	5	10	15	16	17	18
y	170	325	750	1900	2000	2200	2600

Section 5.2 Graphing Base *e* and Base 10 Logarithmic Functions

1. Create a table of values for *x*-values with $x > 0$ to reflect the function's domain.
2. For $y = \ln(x)$, use the formula =ln(x).
3. For $y = \log(x)$, use the formula =log10(x).
4. Select the entire table and graph using *Insert > Scatter > Smoothcurve* option.

	A	B
1	x	$f(x)=\ln(x)$
2	0.5	−0.6931472
3	0.9	−0.1053605
4	1	0
5	2	0.6931472
6	3	1.0986123
7	4	1.3862944
8	5	1.6094379
9	6	1.7917595
10	7	1.9459101
11	8	2.0894415
12	9	2.1972246
13	10	2.3025851
14	11	2.3978953

	A	B
1	x	$f(x)=\log(x)$
2	0.5	−0.30103
3	0.9	−0.0457575
4	1	0
5	2	0.30103
6	3	0.47712125
7	4	0.60205999
8	5	0.69897
9	6	0.77815125
10	7	0.84509804
11	8	0.90308999
12	9	0.95424251
13	10	1
14	11	1.04139269

$f(x)=\ln(x)$

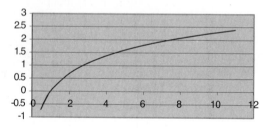

$f(x)=\log(x)$

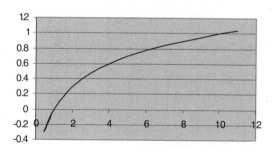

Section 5.2 Modeling Logarithmic Functions

1. Create a scatter plot for the data using *Insert > Scatter* and the points only option.
2. Single-click on the chart and right-click on one of the data points. Choose *Add Trendline* and check *Logarithmic* for *trendline* type.
3. Check the box to display the equation on the graph, and close the dialog box.

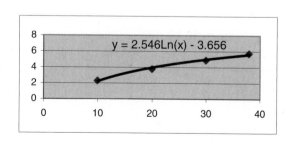

x	y
10	2.207
20	3.972
30	5.004
40	5.737

Section 5.2 Graphing Logarithmic Functions with Other Bases

1. The graph of a logarithm with any base b can be created by entering the formula $=\log(x, b)$.
2. Proceed as described above for graphing base e and base 10 logarithms.

	A	B
1	x	$f(x)=\log(x,7)$
2	0.5	−0.356207187
3	0.9	−0.054144594
4	1	0
5	2	0.356207187
6	3	0.564575034
7	4	0.712414374
8	5	0.827087475
9	6	0.920782221
10	7	1
11	8	1.068621561
12	9	1.129150068
13	10	1.183294662
14	11	1.1232274406

The graph of $f(x) =\log_7 x$ is shown below.

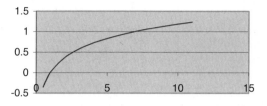

Chapter 6
Section 6.2 Finding the Future Value of a Lump Sum

To find the future value of a lump-sum investment:

1. Type the headings in Row 1, and enter their values (with the interest rate as a decimal) in Row 2. Enter the investment as a negative number.
2. Type the formula $=D2/E2$ in F2 to compute the rate per period.
3. Type the heading Future Value in A4.
4. In cell B4, type the formula $=fv(F2,B2,C2,A2,0)$ to compute the future value.

This spreadsheet gives the future value of an investment of $10,000 for 17 years at 9.8%, compounded quarterly.

Principal	Number of Periods	Payment	Annual Rate	Periods per year	Periodic Rate
−10,000	68	0	0.098	4	0.0245
Future Value	$51,857.73				

Section 6.3 Finding the Future Value of an Annuity

To find the future value of an ordinary annuity:

1. Type the headings in Row 1, and enter their values (with the interest rate as a decimal) in Row 2. Enter the deposit as a negative payment.
2. Type the formula $=D2/E2$ in F2 to compute the rate per period.
3. Type the heading Future Value in A4.
4. In cell B4, type the formula $=fv(F2,B2,C2,A2,0)$ to compute the future value.
(The 0 indicates that the payments are made at the end of each period.)

For annuities due, use $=fv(F2,B2,C2,A2,1)$.
The payments are made at the beginning of each period.

This spreadsheet gives the future value of an ordinary annuity of $200 deposited at the end of each quarter for $2\frac{1}{4}$ years, with interest at 4%, compounded quarterly.

Principal	Number of Periods	Payment	Annual Rate	Periods per year	Periodic Rate
0	9	−200	0.04	4	0.01
Ordinary annuity					
Future Value	$1,873.71				
Annuities due					
Future Value	$1,892.44				

Section 6.4 Finding the Present Value of an Annuity

To find the present value of an ordinary annuity:

1. Type the headings in Row 1, and enter their values (with the interest rate as a decimal) in Row 2.
2. Enter the formula =D2/E2 in F2 to compute the rate per period.
3. Type the heading Present Value in A4.
4. In cell B4, type the formula =pv(F2,B2,C2,A2,0) to compute the present value. (The 0 indicates that the payments are made at the end of each period.) The parentheses on the present value in cell B4 mean that it is negative (the money leaves the investor to start the annuity).

For annuities due, use =pv(F2,B2,C2,A2,1).
The payments are made at the beginning of each period.

This spreadsheet gives the lump sum deposit (present value) necessary to receive $1000 per month for 16 years, with interest at 9%, compounded monthly, for both an ordinary annuity and an annuity due.

Future Value	Number of Periods	Payment	Annual Rate	Periods per year	Periodic Rate
0	192	1000	0.09	12	0.0075
Ordinary annuity					
Present Value	($101,572.77)				
Annui-ties due					
Present value	($102,334.56)				

Section 6.5 Finding Payments to Amortize a Loan

To find the periodic payment to pay off a loan:

1. Type the headings in Row 1 and their values (with the interest rate as a decimal) in Row 2.
2. Enter the formula =D2/E2 in F2 to compute the rate per period.
3. Type the heading Payment in A4.
4. In cell B4, type the formula =Pmt(F2,B2,A2,C2,0) to compute the payment.

This spreadsheet gives the annual payment of a loan of $10,000 over 5 years when interest is 10% per year.

The parentheses indicates a payment out.

Loan Amount	Number of Periods	Future Value	Annual Rate	Periods per year	Periodic Rate
10000	5	0	0.1	1	0.1
Payment	($2,637.97)				

Chapter 8

Section 8.1 Binomial Probabilities

1. Type headings in cells A1:A3 and their respective values in cells B1:B3.
2. Use the function =binomdist(B1,B2,B3,cumulative) where
   - B1 is the number of successes.
   - B2 is the number of independent trials.
   - B3 is the probability of success in each trial.
   - True replaces cumulative if a cumulative probability is sought; it is replaced by false otherwise.
   - The probability of exactly 3 heads in 6 tosses is found by evaluating =binomdist(B1,B2,B3,false) in B4.
   - The probability of 3 or fewer heads in 6 tosses is found by evaluating =binomdist(B1,B2,B3,true) in B5.

This spreadsheet gives the probability of 3 heads in 6 tosses of a fair coin and the probability of 3 or fewer heads in 6 tosses.

	A	B
1	Number of successes	3
2	Number of trials	6
3	Probability of success	.5
4	Probability of 3 successes	.3125
5	Probability of 3 or fewer successes	.65625

Section 8.2 Bar Graphs

To construct a bar graph for the given table of test scores:

1. Copy the entries of the table to cells A2:B6.
2. Select the range A2:B6.
3. Click on the *Insert* tab. Then select *Column > Clustered Column* (first option), and the graph appears.
4. Single-click on the chart. You can add labels and change the title of the graph by clicking *Layout* under *Chart Tools*, and choosing the appropriate options.

	A	B
1	Grade Range	Frequency
2	90–100	2
3	80–89	5
4	70–79	3
5	60–69	3
6	0–59	2

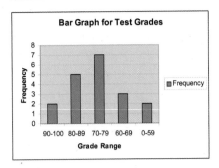

Section 8.2 Finding the Mean, Standard Deviation, and Median of Raw Data

To find the mean, standard deviation, and median of a raw data set:

1. Enter the data in Row 1 (cells A1:L1).
2. Type the heading Mean in cell A3.
3. Type the formula =average(A1:L1) in cell B4.
4. Type the heading Standard Deviation in cell A4.
5. In cell B4, type the formula =stdev(A1:L1).
6. In cell A5, type the heading Median.
7. In cell B5, type the formula =median(A1:L1).

The mean, standard deviation, and median for the data 1, 1, 1, 3, 3, 4, 4, 5, 6, 6, 7, 7 is shown below.

	A	B	C	D	E	F	G	H	I	J	K	L
1	1	1	1	3	3	4	4	5	6	6	7	7
2												
3	Mean	4										
4	Standard Deviation	2.2563										
5	Median	4										

Section 8.2 Finding the Mean and Standard Deviation of Grouped Data

To find the mean:

1. Enter the data and headings in the cells A1:C6.
2. In D1, type the heading Class mark*frequency.
3. In D2, type the formula =B2*C2.
4. Copy the formula in D2 to D3:D6.
5. In B7, type the heading Total.
6. In cell C7, type the formula for the total frequencies, =sum(C2:C6).
7. In cell D7, type the formula for the total, =sum(D2:D6).
8. In cell A8, type the heading Mean.
9. In cell A9, type in the formula =D7/C7.

To find the standard deviation:

10. In cell E1, type in the heading freq *(x − x_mean)^2.
11. In cell E2, type the formula =C2*(B2-A9)^2. (The A9 gives the value in A9; the reference doesn't change as we fill down.)
12. Copy the formula in E2 to E3:E6.
13. In cell E7, type the formula =sum(E2:E6).
14. In cell A10 type the heading Standard Deviation.
15. In cell A11, type the formula =sqrt(E7/(C7-1)).

Grade Range	Class Marks	Frequency
90–100	95	3
80–89	84.5	4
70–79	74.5	7
60–69	64.5	0
50–59	54.5	2

	A	B	C	D	E
1	Grade Range	Class Marks	Frequency	Class mark*frequency	Freq*(x-x_mean)^2
2	90-100	95	3	285	832.2919922
3	80-89	84.5	4	338	151.5976563
4	70-79	74.5	7	521.5	103.4208984
5	60-69	64.5	0	0	0
6	50-59	54.5	2	109	1137.048828
7			Total	16	1253.5	2224.359375
8	Mean				
9	78.344				
10	Standard Deviation				
11	12.177				

Section 8.4 Calculating Normal Probabilities

To calculate normal probabilities:

1. Type headings in A1:A4 and their respective values in cells B1:B4.
2. To find the probability that a score X is less than the x1 value in B3, enter the formula =normdist(B3,B1,B2,true) in cell B5.
3. To find the probability that X is less than the x2 value in B4, enter the formula =normdist(B4,B1,B2,true) in cell B6.
4. To find the probability that a score X is more than the value in B3 and less than the x2 value in B4, enter the formula =B6-B5 in cell B7.

Entries in B5, B6, and B7 give the probabilities of a score X being less than 100, less than 115, and between 100 and 115, respectively, when the mean is 100 and the standard deviation is 15.

	A	B
1	Mean	100
2	Standard Deviation	15
3	x1	100
4	x2	115
5	Pr(X<x1)	0.5
6	Pr(X<x2)	0.841345
7	Pr(x1<X<x2)	0.341345

Chapter 9
Sections 9.1–9.2 Evaluating Limits

To evaluate $\lim_{x \to c} f(x)$:

1. Make a table of values for $f(x)$ near $x = c$. Include values on both sides of $x = c$.
2. Use the table of values to predict the limit (or that the limit does not exist).

Note: All limit evaluations with Excel use appropriate tables of values of $f(x)$. This is true when $f(x)$ is piecewise defined and for limits as $\to \infty$.

Find $\lim_{x \to 2} \dfrac{x^2 - 4}{x - 2}$.

	A	B	C	D
	x	**f(x)**	**x**	**f(x)**
1				
2	2.1	4.1	1.9	3.9
3	2.05	4.05	1.95	3.95
4	2.01	4.01	1.99	3.99
5	2.001	4.001	1.999	3.999

The tables suggest that $\lim_{x \to 2} \dfrac{x^2 - 4}{x - 2} = 4$.

Sections 9.3–9.7 Approximating Derivatives

To approximate $f'(c)$:

1. Numerically investigate the limit in the definition of derivative:
$$f'(x) = \lim_{h \to 0} \frac{f(x + h) - f(x)}{h}$$

2. Use the given $f(x)$ and $x = c$ to create a table of values for h near 0 (and on both sides of $h = 0$).

Note that rows 5 and 6 have the values of h closest to 0.

Investigate $f'(1)$ for $f(x) = x^3$.

	A	B	C	D	E
1	**h**	**1+h**	**f(1)**	**f(1+h)**	**(f(1+h)-f(1))/h**
2	0.1	1.1	1	1.331	3.31
3	0.01	1.01	1	1.030301	3.0301
4	0.001	1.001	1	1.003003	3.003001
5	0.0001	1.0001	1	1.0003	3.00030001
6	−0.0001	0.9999	1	0.9997	2.99970001
7	−0.001	0.999	1	0.997003	2.997001
8	−0.01	0.99	1	0.970299	2.9701
9	−0.1	0.9	1	0.729	2.71

The table suggests that $f'(1) = 3$, which is the actual value.
Note: Excel has no built-in derivative approximation tool.

Chapter 10
Section 10.1 Relative Maxima and Minima

1. Make a table with columns for x-values, the function, and the derivative.
2. Extend the table to include x-values to the left and to the right of all critical values.
3. Use the signs of the values of the derivative to determine whether f is increasing or decreasing around the critical values, and thus to classify the critical values as relative maxima, relative minima, or horizontal points of inflection. You may want to graph the function to confirm your conclusions.

The spreadsheet shows that the relative minima of $f(x) = x^2$ is 0 at $x = 0$. Note that the derivative is $f'(x) = 2x$.

	A	B	C
1	x	f(x)	f'(x)
2	−2	4	−4
3	−1.5	2.25	−3
4	−1	1	−2
5	−0.5	0.25	−1
6	0	0	0
7	0.5	0.25	1
8	1	1	2
9	1.5	2.25	3
10	2	4	4

Section 10.2 Exploring f, f', f'' Relationships

To explore relationships among the graphs of a function and its derivatives:

1. Find functions for f' and f''.
2. Graph all three functions on the same plot.
   - Notice that f increases when f' is above the x-axis ($+$) and decreases when f' is below the x-axis ($-$).
   - Notice that f is concave up when f'' is above the x-axis ($+$) and is concave down when f'' is below the x-axis ($-$).

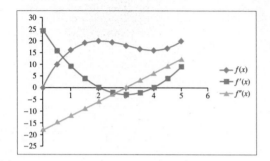

Let $f(x) = x^3 - 9x^2 + 24x$. Graph f, f', and f'' on the interval $[0, 5]$ to explore the relationships among these functions.

	A	B	C	D
	x	$f(x)$	$f'(x)$	$f''(x)$
1				
2	0	0	24	−18
3	0.5	9.875	15.75	−15
4	1	16	9	−12
5	1.5	19.125	3.75	−9
6	2	20	0	−6
7	2.5	19.375	−2.25	−3
8	3	18	−3	0
9	3.5	16.625	−2.25	−3
10	4	16	0	6
11	4.5	16.875	3.75	9
12	5	20	9	12

Sections 10.3–10.4 Finding Optimal Values

To find the optimal value of a function when the goal is not to produce a graph:

1. Set up a spreadsheet that identifies the variable and the function whose optimal value is sought.
2. Choose *Data > Analysis > Solver*. Then, in the Dialog Box
   a. Set the *Target Cell* as that of the objective function.
   b. Check *Max* or *Min* according to your goal.
   c. Set the *Changing Cells* to reference the variable.
3. Click on the *Options* box. Make sure "Assume Linear Model" is *not* checked. Then click OK.
4. Click *Solve* in the *Solver* dialog box. You will get a dialog box stating that *Solver* found a solution. Save the solution if desired, then click OK.
5. The cells containing the variable and the function should now contain the optimal values.

Minimize area $A = x^2 + \dfrac{160}{x}$ for $x > 0$.

	A	B
1	**Variable**	
2		
3	x length of base	1
4		
5	**Objective**	
6		
7	Minimize Area	=B3^2+160/B3

	A	B
1	**Variable**	
2		
3	x length of base	4.3089
4		
5	**Objective**	
6		
7	Minimize Area	55.699

The function is minimized for $x = 4.3089$ and the minimum value is $A = 55.699$.

Chapter 13

Section 13.1 Approximating Definite Integrals

Using Rectangles

To find the approximate area between $y = f(x)$ and the x-axis over an interval by using the right-hand endpoints of rectangles with equal base width:

1. Enter "x" and the x- values of the right-hand endpoints of each rectangle in Column B. Fill down can be used if the number of rectangles is large.
2. Enter the formula for the function, using cell addresses, in C2 and fill down column C to get the heights of the rectangles.
3. Enter the width of the rectangles in Column D.
4. Enter the formula for the area of each rectangle, =C2*D2, in E2 and fill down to get the area of each rectangle.
5. Enter "Total" in A10 and the formula =SUM(E2:E9) in E10 to get the approximate area.

To find the approximation for the area under $y = \sqrt{x}$ from $x = 0$ to $x = 4$ using $n = 8$ equal subdivisions (with base width 0.5) and right-hand endpoints:

1. Enter "x" and the eight x- values of the right-hand endpoints (starting with 0.5) in Column B.
2. Enter "y" in C1 and =SQRT(B2) in C2 and fill down column C.
3. Enter "width" in D1 and 0.5 below in Column D.
4. Enter "Area" in E1, =C2*D2 in E2, and fill down to get the area of each rectangle.
5. Enter "Total" in A10 and the formula =SUM(E2:E9) in E10 to get the approximate area, $S_R \approx 5.765$.

	A	B	C	D	E
1	Rectangle	x	y	Width	Area
2	1	0.5	0.707107	0.5	0.353553
3	2	1	1	0.5	0.5
4	3	1.5	1.224745	0.5	0.612372
5	4	2	1.414214	0.5	0.707107
6	5	2.5	1.581139	0.5	0.790569
7	6	3	1.732051	0.5	0.866025
8	7	3.5	1.870829	0.5	0.935414
9	8	4	2	0.5	1
10	Total				5.765042

Chapter 14

Section 14.1   Graphs of Functions of Two Variables

To create a surface plot for a function of two variables:

1. a. Generate appropriate *x*-values beginning in B1 and continuing *across*.
   b. Generate appropriate *y*-values beginning in A2 and continuing *down*.
2. Generate values for the function that correspond to the points (*x*, *y*) from Step 1 as follows:
   In cell B2, enter the function formula with B$1 used to represent *x* and $A2 to represent *y*. (See the online Excel Guide for additional information about the role and use of the $ in this step.) Then use fill down and fill across to complete the table.
3. Select the entire table of values. Click on *Insert > Other Charts > 3-D Surface* option, the first option in the Surface group.
4. To rotate the graph, click on the chart and click on *3-D rotation*. You can click on the various options present in the dialog box.

Let $f(x, y) = 10 - x^2 - y^2$. Plot the graph of this function for both *x* and *y* in the interval $[-2, 2]$.

	A	B	C	D	E	F	G	H	I	J
1		−2	−1.5	−1	−0.5	0	0.5	1	1.5	2
2	−2									
3	−1.5									
4	−1									
5	−0.5									
6	0									
7	0.5									
8	1									
9	1.5									
10	2									

	A	B	C	D	E	F	G	H	I	J
1		−2	−1.5	−1	−0.5	0	0.5	1	1.5	2
2	−2	2	3.75	5	5.75	6	5,75	5	3.75	2
3	−1.5	3.75	5.5	6.75	7.5	7.75	7.5	6.75	5.5	3.75
4	−1	5	6.75	8	8.75	9	8.75	8	6.75	5
5	−0.5	5.75	7.5	8.75	9.5	9.75	9.5	8.75	7.5	5.75
6	0	6	7.75	9	9.75	10	9.75	9	7.75	6
7	0.5	5.75	7.5	8.75	9.5	9.75	9.5	8.75	7.5	5.75
8	1	5	6.75	8	8.75	9	8.75	8	6.75	5
9	1.5	3.75	5.5	6.75	7.5	7.75	7.5	6.75	5.5	3.75
10	2	2	3.75	5	5.75	6	5.75	5	3.75	2

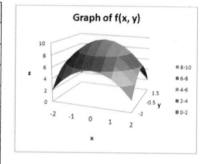

Graph of f(x, y)

Section 14.5 Constrained Optimization and Lagrange Multipliers (Excel 2007)

To solve a constrained optimization problem:

1. Set up the problem in Excel.
2. Choose *Data > Analysis > Solver* and do the following:
   - Choose the objective function as the *Target Cell*.
   - Check *Max* or *Min* depending on the problem.
   - Choose the cells representing the variables for the *By Changing Cells* box.
   - Click on the *Constraints* box and press Add. Then enter the constraint equations.
3. Click on the *Options* box and make sure that "Assume Linear Model" is *not* checked. Then click OK.
4. Click *Solve* in the *Solver* dialog box. Then click OK to solve.

Maximize $P = 600l^{2/3}k^{1/3}$ subject to $40l + 100k = 3000$.

	A	B	C
1	**Variables**		
2			
3	units labor (l)	0	
4	units capital (k)	0	
5			
6			
7	**Objective**		
8			
9	Maximize production	=600*B3^(2/3)*B4^(1/3)	
10			
11	**Constraint**		
12		Amount used	Available
13	Cost	=40*B3+100*B4	3000

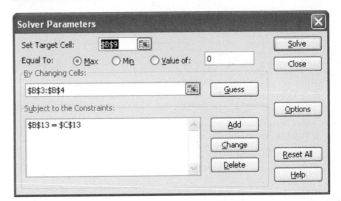

	A	B	C
1	**Variables**		
2			
3	units labor (l)	50.00000002	
4	units capital (k)	10	
5			
6			
7	**Objective**		
8			
9	Maximize product	17544.10644	
10			
11	**Constraint**		
12		Amount used	Available
13	Cost	3000.000001	3000

Section 14.5 Constrained Optimization and Lagrange Multipliers (Excel 2010)

To solve a constrained optimization problem:

1. Set up the problem in Excel.
2. Choose *Data > Analysis > Solver* and do the following:
   - Choose the objective function as the *Target Cell.*
   - Check *Max* or *Min* depending on the problem.
   - Choose the cells representing the variables for the *By Changing Cells* box.
   - Click on the *Constraints* box and press Add. Then enter the constraint equations.
3. Check the box making the variables nonnegative.
4. Select "GRG Nonlinear" as the solving method.
5. Click *Solve* in the *Solver* dialog box. Then click OK to solve.

Maximize $P = 600l^{2/3}k^{1/3}$ subject to $40l + 100k = 3000$.

	A	B	C
1	**Variables**		
2			
3	units labor (l)	0	
4	units capital (k)	0	
5			
6			
7	**Objective**		
8			
9	Maximize production	=600*B3^(2/3)*B4^(1/3)	
10			
11	**Constraint**		
12		Amount used	Available
13	Cost	=40*B3+100*B4	3000

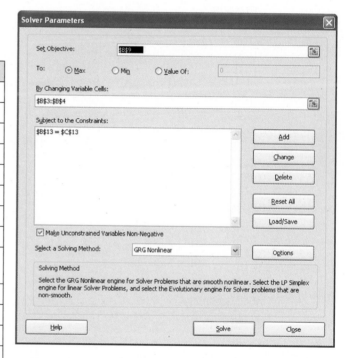

	A	B	C
1	**Variables**		
2			
3	units labor (l)	50.00000002	
4	units capital (k)	10	
5			
6			
7	**Objective**		
8			
9	Maximize product	17544.10644	
10			
11	**Constraint**		
12		Amount used	Available
13	Cost	3000.000001	3000

Areas Under the Standard Normal Curve

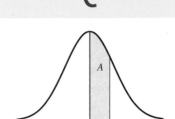

The value of A is the area under the standard normal curve between $z = 0$ and $z = z_0$, for $z_0 \geq 0$. Areas for negative values of z_0 are obtained by symmetry.

z_0	A	z_0	A	z_0	A	z_0	A
0.00	0.0000	0.43	0.1664	0.86	0.3051	1.29	0.4015
0.01	0.0040	0.44	0.1700	0.87	0.3078	1.30	0.4032
0.02	0.0080	0.45	0.1736	0.88	0.3106	1.31	0.4049
0.03	0.0120	0.46	0.1772	0.89	0.3133	1.32	0.4066
0.04	0.0160	0.47	0.1808	0.90	0.3159	1.33	0.4082
0.05	0.0199	0.48	0.1844	0.91	0.3186	1.34	0.4099
0.06	0.0239	0.49	0.1879	0.92	0.3212	1.35	0.4115
0.07	0.0279	0.50	0.1915	0.93	0.3238	1.36	0.4131
0.08	0.0319	0.51	0.1950	0.94	0.3264	1.37	0.4147
0.09	0.0359	0.52	0.1985	0.95	0.3289	1.38	0.4162
0.10	0.0398	0.53	0.2019	0.96	0.3315	1.39	0.4177
0.11	0.0438	0.54	0.2054	0.97	0.3340	1.40	0.4192
0.12	0.0478	0.55	0.2088	0.98	0.3365	1.41	0.4207
0.13	0.0517	0.56	0.2123	0.99	0.3389	1.42	0.4222
0.14	0.0557	0.57	0.2157	1.00	0.3413	1.43	0.4236
0.15	0.0596	0.58	0.2190	1.01	0.3438	1.44	0.4251
0.16	0.0636	0.59	0.2224	1.02	0.3461	1.45	0.4265
0.17	0.0675	0.60	0.2257	1.03	0.3485	1.46	0.4279
0.18	0.0714	0.61	0.2291	1.04	0.3508	1.47	0.4292
0.19	0.0754	0.62	0.2324	1.05	0.3531	1.48	0.4306
0.20	0.0793	0.63	0.2357	1.06	0.3554	1.49	0.4319
0.21	0.0832	0.64	0.2389	1.07	0.3577	1.50	0.4332
0.22	0.0871	0.65	0.2422	1.08	0.3599	1.51	0.4345
0.23	0.0910	0.66	0.2454	1.09	0.3621	1.52	0.4357
0.24	0.0948	0.67	0.2486	1.10	0.3643	1.53	0.4370
0.25	0.0987	0.68	0.2517	1.11	0.3665	1.54	0.4382
0.26	0.1026	0.69	0.2549	1.12	0.3686	1.55	0.4394
0.27	0.1064	0.70	0.2580	1.13	0.3708	1.56	0.4406
0.28	0.1103	0.71	0.2611	1.14	0.3729	1.57	0.4418
0.29	0.1141	0.72	0.2642	1.15	0.3749	1.58	0.4429
0.30	0.1179	0.73	0.2673	1.16	0.3770	1.59	0.4441
0.31	0.1217	0.74	0.2704	1.17	0.3790	1.60	0.4452
0.32	0.1255	0.75	0.2734	1.18	0.3810	1.61	0.4463
0.33	0.1293	0.76	0.2764	1.19	0.3830	1.62	0.4474
0.34	0.1331	0.77	0.2794	1.20	0.3849	1.63	0.4484
0.35	0.1368	0.78	0.2823	1.21	0.3869	1.64	0.4495
0.36	0.1406	0.79	0.2852	1.22	0.3888	1.65	0.4505
0.37	0.1443	0.80	0.2881	1.23	0.3907	1.66	0.4515
0.38	0.1480	0.81	0.2910	1.24	0.3925	1.67	0.4525
0.39	0.1517	0.82	0.2939	1.25	0.3944	1.68	0.4535
0.40	0.1554	0.83	0.2967	1.26	0.3962	1.69	0.4545
0.41	0.1591	0.84	0.2995	1.27	0.3980	1.70	0.4554
0.42	0.1628	0.85	0.3023	1.28	0.3997	1.71	0.4564

(continued)

z_0	A	z_0	A	z_0	A	z_0	A
1.72	0.4573	2.26	0.4881	2.80	0.4974	3.34	0.4996
1.73	0.4582	2.27	0.4884	2.81	0.4975	3.35	0.4996
1.74	0.4591	2.28	0.4887	2.82	0.4976	3.36	0.4996
1.75	0.4599	2.29	0.4890	2.83	0.4977	3.37	0.4996
1.76	0.4608	2.30	0.4893	2.84	0.4977	3.38	0.4996
1.77	0.4616	2.31	0.4896	2.85	0.4978	3.39	0.4997
1.78	0.4625	2.32	0.4898	2.86	0.4979	3.40	0.4997
1.79	0.4633	2.33	0.4901	2.87	0.4979	3.41	0.4997
1.80	0.4641	2.34	0.4904	2.88	0.4980	3.42	0.4997
1.81	0.4649	2.35	0.4906	2.89	0.4981	3.43	0.4997
1.82	0.4656	2.36	0.4909	2.90	0.4981	3.44	0.4997
1.83	0.4664	2.37	0.4911	2.91	0.4982	3.45	0.4997
1.84	0.4671	2.38	0.4913	2.92	0.4982	3.46	0.4997
1.85	0.4678	2.39	0.4916	2.93	0.4983	3.47	0.4997
1.86	0.4686	2.40	0.4918	2.94	0.4984	3.48	0.4997
1.87	0.4693	2.41	0.4920	2.95	0.4984	3.49	0.4998
1.88	0.4699	2.42	0.4922	2.96	0.4985	3.50	0.4998
1.89	0.4706	2.43	0.4925	2.97	0.4985	3.51	0.4998
1.90	0.4713	2.44	0.4927	2.98	0.4986	3.52	0.4998
1.91	0.4719	2.45	0.4929	2.99	0.4986	3.53	0.4998
1.92	0.4726	2.46	0.4931	3.00	0.4987	3.54	0.4998
1.93	0.4732	2.47	0.4932	3.01	0.4987	3.55	0.4998
1.94	0.4738	2.48	0.4934	3.02	0.4987	3.56	0.4998
1.95	0.4744	2.49	0.4936	3.03	0.4988	3.57	0.4998
1.96	0.4750	2.50	0.4938	3.04	0.4988	3.58	0.4998
1.97	0.4756	2.51	0.4940	3.05	0.4989	3.59	0.4998
1.98	0.4761	2.52	0.4941	3.06	0.4989	3.60	0.4998
1.99	0.4767	2.53	0.4943	3.07	0.4989	3.61	0.4998
2.00	0.4772	2.54	0.4945	3.08	0.4990	3.62	0.4999
2.01	0.4778	2.55	0.4946	3.09	0.4990	3.63	0.4999
2.02	0.4783	2.56	0.4948	3.10	0.4990	3.64	0.4999
2.03	0.4788	2.57	0.4949	3.11	0.4991	3.65	0.4999
2.04	0.4793	2.58	0.4951	3.12	0.4991	3.66	0.4999
2.05	0.4798	2.59	0.4952	3.13	0.4991	3.67	0.4999
2.06	0.4803	2.60	0.4953	3.14	0.4992	3.68	0.4999
2.07	0.4808	2.61	0.4955	3.15	0.4992	3.69	0.4999
2.08	0.4812	2.62	0.4956	3.16	0.4992	3.70	0.4999
2.09	0.4817	2.63	0.4957	3.17	0.4992	3.71	0.4999
2.10	0.4821	2.64	0.4959	3.18	0.4993	3.72	0.4999
2.11	0.4826	2.65	0.4960	3.19	0.4993	3.73	0.4999
2.12	0.4830	2.66	0.4961	3.20	0.4993	3.74	0.4999
2.13	0.4834	2.67	0.4962	3.21	0.4993	3.75	0.4999
2.14	0.4838	2.68	0.4963	3.22	0.4994	3.76	0.4999
2.15	0.4842	2.69	0.4964	3.23	0.4994	3.77	0.4999
2.16	0.4846	2.70	0.4965	3.24	0.4994	3.78	0.4999
2.17	0.4850	2.71	0.4966	3.25	0.4994	3.79	0.4999
2.18	0.4854	2.72	0.4967	3.26	0.4994	3.80	0.4999
2.19	0.4857	2.73	0.4968	3.27	0.4995	3.81	0.4999
2.20	0.4861	2.74	0.4969	3.28	0.4995	3.82	0.4999
2.21	0.4864	2.75	0.4970	3.29	0.4995	3.83	0.4999
2.22	0.4868	2.76	0.4971	3.30	0.4995	3.84	0.4999
2.23	0.4871	2.77	0.4972	3.31	0.4995	3.85	0.4999
2.24	0.4875	2.78	0.4973	3.32	0.4995	3.86	0.4999
2.25	0.4878	2.79	0.4974	3.33	0.4996		

Answers

Below are the answers to odd-numbered Section Exercises and all the Chapter Review and Chapter Test problems.

0.1 EXERCISES

1. $\in$ 3. $\notin$ 5. $\{1,2,3,4,5,6,7\}$

7. $\{x : x$ is a natural number greater than 2 and less than 8$\}$

9. $\varnothing, A, B$ 11. no   13. $D \subseteq C$

15. $A \subseteq B$ or $B \subseteq A$ 17. yes   19. no

21. A and B, B and D, C and D

23. $A \cap B = \{4, 6\}$ 25. $A \cap B = \varnothing$

27. $A \cup B = \{1, 2, 3, 4, 5\}$

29. $A \cup B = \{1, 2, 3, 4\} = B$

31. $A' = \{4, 6, 9, 10\}$

33. $A \cap B' = \{1, 2, 5, 7\}$

35. $(A \cup B)' = \{6, 9\}$

37. $A' \cup B' = \{1, 2, 4, 5, 6, 7, 9, 10\}$

39. $\{1, 2, 3, 5, 7, 9\}$ 41. $\{4, 6, 8, 10\}$

43. $A - B = \{1, 7\}$

45. $A - B = \varnothing$ or $\{\ \}$

47. (a) $L = \{00, 01, 04, 05, 06, 07, 10, 11, 12\}$
       $H = \{00, 01, 06, 07, 08, 10, 11, 12\}$
       $C = \{01, 02, 03, 08, 09\}$

     (b)   no

     (c) C' is the set of years when the percent change from low to high was 35% or less.

     (d) $\{00, 02, 03, 04, 05, 06, 07, 09, 10, 11, 12\}$ = the set of years when the high was 11,000 or less or the percent change was 35% or less.

     (e) $\{02, 03, 08, 09\}$ = the set of years when the low was 8000 or less and the percent change exceeded 35%.

49. (a)   130   (b)   840   (c)   520

51. (a)

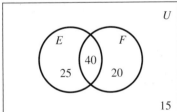

     (b)   40
     (c)   85
     (d)   25

53.

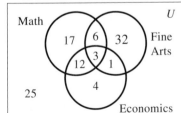

     (a)   25
     (b)   43
     (c)   53

55. (a) and (b)

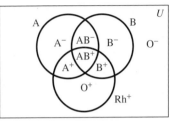

     (c) A^+: 34%; B^+: 9%; O^+: 38%; AB^+: 3%; O^-: 7%; A^-: 6%; B^-: 2%; AB^-: 1%

0.2 EXERCISES

1. (a) irrational   (b) rational, integer
     (c) rational, integer, natural   (d) meaningless

3. (a) Commutative   (b) Distributive
     (c) Associative   (d) Additive identity

5. $<$ 7. $<$ 9. $<$ 11. $>$

13. 11   15. 4   17. 2   19. $\frac{-4}{3}$ 21. 3   23. $\frac{17}{11}$

25. entire line

27. $(1, 3]$; half open

29. $(2, 10)$; open

31. $-3 \leq x < 5$

33. $x > 4$

35. $(-3, 4)$

37. $(4, \infty)$

39. $[-1, \infty)$

41. $(-\infty, 0) \cup (7, \infty)$

43. -0.000038585

45. 9122.387471   47. 3240.184509

49. (a) $1088.91 (b) $258.62   (c) $627.20

51. (a)   Eq. (1) = 2.69 billion
        Eq. (2) = 2.68 billion
        Eq. (1) is more accurate

     (b)   Eq. (1) = 3.73 billion
        Eq. (2) = 4.01 billion

53. (a) $82,401 \leq I \leq 171,850$; $171,851 \leq I \leq 373,650$; $I > 373,650$

     (b) $T = \$4681.25$ for $I = \$34,000$
        $T = \$16,781.25$ for $I = \$82,400$

     (c) $[4681.25, 16,781.25]$

0.3 EXERCISES

1. 256   3. -64 5. $\frac{1}{9}$ 7. $-\frac{9}{4}$

9. 2.0736   11. 0.1316872428   13. 6^8 15. $\frac{1}{10}$

17. $9^0 = 1$ 19. 3^9 21. $\left(\frac{3}{2}\right)^2 = \frac{9}{4}$

23. $-1/x^3$ 25. x/y^2 27. x^7

29. $x^{-2} = 1/x^2$ 31. x^4 33. y^{12} 35. x^{12}

37. x^2y^2 39. $16/x^{20}$ 41. $x^8/(16y^4)$

43. $-16a^2/b^2$ 45. $2/(xy^2)$ 47. $1/(x^9y^6)$

49. $(a^{18}c^{12})/b^6$

51. (a) $1/(2x^4)$ (b) $1/(16x^4)$ (c) $1/x^4$ (d) 8

53. x^{-1} 55. $8x^3$ 57. $\frac{1}{4}x^{-2}$ 59. $-\frac{1}{8}x^3$

61. $S = \$2114.81; I = \$914.81$

63. $S = \$9607.70; I = \$4607.70$

65. 7806.24

67. (a) 20, 40, 48

(b) \$1905, \$7373, \$12,669 billion

(c) \$24,922 billion

69. (a) 442, 976, 1072

(b) In 2020, 1090; 53 more

(c) Two possibilities might be more environmental
protections and the fact that there are only a limited number of species.

(d) There are only a limited number of species. Also,
below some threshold level the ecological balance
might be lost, perhaps resulting in an environmental catastrophe (which the equation could not
predict). Upper limit = 1095

71. (a) 10   (b) \$1385.5 billion

(c) \$2600.8 billion   (d) \$4304.3 billion

0.4 EXERCISES

1. (a) $\frac{16}{3} \approx 5.33$ (b) 1.2

3. (a) 8   (b) not real   5. $\frac{9}{4}$

7. (a) 16   (b) 1/16

9. $(6.12)^{4/9} \approx 2.237$

11. $m^{3/2}$ 13. $(m^2n^5)^{1/4}$ 15. $2\sqrt{x}$

17. $\sqrt[6]{x^7}$ 19. $-1/(4\sqrt[4]{x^5})$ 21. $y^{3/4}$

23. $z^{19/4}$ 25. $1/y^{5/2}$ 27. x

29. $1/y^{21/10}$ 31. $x^{1/2}$ 33. $1/x$

35. $8x^2$ 37. $8x^2y^2\sqrt{2y}$ 39. $2x^2y\sqrt[3]{5x^2y^2}$

41. $6x^2y\sqrt{x}$ 43. $42x^3y^2\sqrt{x}$ 45. $2xy^5/3$

47. $2b\sqrt[4]{b}/(3a^2)$ 49. $1/9$ 51. 7

53. $\sqrt[3]{6}/3$ 55. $\sqrt{mx}/x$ 57. $\sqrt[3]{mx^2}/x^2$

59. $-\frac{2}{3}x^{-2/3}$ 61. $3x^{3/2}$ 63. $(3\sqrt{x})/2$

65. $1/(2\sqrt{x})$

67. (a) $10^{8.5} = 10^{17/2} = \sqrt{10^{17}}$

(b) $10^{9.0} = 1,000,000,000$ (c) $10^{2.1} \approx 125.9$

69. (a) $S = 1000\sqrt{\left(1 + \dfrac{r}{100}\right)^5}$

(b) \$1173.26 (nearest cent)

71. (a) $P = 0.924\sqrt[100]{t^{13}}$

(b) 2005 to 2010; 0.1074 billion vs. 0.0209 billion.
By 2045 and 2050 the population will be much
larger than earlier in the 21st century, and there
is a limited number of people that any land can
support—in terms of both space and food.

73. 74 kg   75. 39,491

77. (a) 10   (b) 259

0.5 EXERCISES

1. (a) 2   (b) -1 (c) 10   (d) one

3. (a) 5   (b) -14 (c) 0   (d) several

5. (a) 5   (b) 0   (c) 2   (d) -5

7. -12 9. -296 11. $\frac{-7}{31}$

13. 87.4654 15. $21pq - 2p^2$ 17. $m^2 - 7n^2 - 3$

19. $3q + 12$ 21. $x^2 - 1$ 23. $35x^5$

25. $3rs$ 27. $2ax^4 + a^2x^3 + a^2bx^2$

29. $6y^2 - y - 12$ 31. $12 - 30x^2 + 12x^4$

33. $16x^2 + 24x + 9$ 35. $0.01 - 16x^2$

37. $36x^2 - 9$ 39. $x^4 - x^2 + \frac{1}{4}$

41. $0.1x^2 - 1.995x - 0.1$

43. $x^3 - 8$ 45. $x^8 + 3x^6 - 10x^4 + 5x^3 + 25x$

47. $3 + m + 2m^2n$ 49. $8x^3y^2/3 + 5/(3y) - 2x^2/(3y)$

51. $x^3 + 3x^2 + 3x + 1$

53. $8x^3 - 36x^2 + 54x - 27$

55. $x^2 - 2x + 5 - 11/(x + 2)$

57. $x^2 + 3x - 1 + (-4x + 2)/(x^2 + 1)$

59. (a) $9x^2 - 21x + 13$ (b) 5

61. $x + 2x^2$ 63. $x - x^{1/2} - 2$ 65. $x - 9$

67. $4x^2 + 4x$ 69. $55x$

71. (a) $49.95 + 0.49x$ (b) \$114.63

73. (a) $4000 - x$ (b) $0.10x$

(c) $0.08(4000 - x)$ (d) $0.10x + 0.08(4000 - x)$

75. $(15 - 2x)(10 - 2x)x$

0.6 EXERCISES

1. $3b(3a - 4a^2 + 6b)$ 3. $2x(2x + 4y^2 + y^3)$

5. $(7x^2 + 2)(x - 2)$ 7. $(6 + y)(x - m)$

9. $(x + 2)(x + 6)$ 11. $(x - 16)(x + 1)$

13. $(7x + 4)(x - 2)$ 15. $(x - 5)^2$

17. $(7a + 12b)(7a - 12b)$

19. (a) $(3x - 1)(3x + 8)$ (b) $(9x + 4)(x + 2)$

21. $x(4x - 1)$ 23. $(x^2 - 5)(x + 4)$

25. $(x - 3)(x + 2)$ 27. $2(x - 7)(x + 3)$

29. $2x(x - 2)^2$ 31. $(2x - 3)(x + 2)$

33. $3(x + 4)(x - 3)$ 35. $2x(x + 2)(x - 2)$

37. $(5x + 2)(2x + 3)$ 39. $(5x - 1)(2x - 9)$

41. $(y^2 + 4x^2)(y + 2x)(y - 2x)$

43. $(x + 2)^2(x - 2)^2$

45. $(2x + 1)(2x - 1)(x + 1)(x - 1)$

47. $x + 1$ 49. $1 + x$ 51. $(x + 1)^3$

53. $(x - 4)^3$ 55. $(x - 4)(x^2 + 4x + 16)$

57. $(3 + 2x)(9 - 6x + 4x^2)$

59. $P(1 + rt)$ 61. $m(c - m)$

63. (a) $p(10,000 - 100p); x = 10,000 - 100p$ (b) 6200

65. (a) $R = x(300 - x)$ (b) $300 - x$

0.7 EXERCISES

1. $2y^3/z$ 3. $\frac{1}{3}$ 5. $(x - 1)/(x - 3)$

7. $20x/y$ 9. $\frac{32}{3}$ 11. $3x + 9$

13. $\dfrac{-(x + 1)(x + 3)}{(x - 1)(x - 3)}$ 15. $15bc^2/2$ 17. $5y/(y - 3)$

19. $\dfrac{-x(x-3)(x+2)}{x+3}$

21. $\dfrac{1}{x+1}$

23. $\dfrac{4a-4}{a(a-2)}$

25. $\dfrac{-x^2+x+1}{x+1}$

27. $\dfrac{16a+15a^2}{12(x+2)}$

29. $\dfrac{79x+9}{30(x-2)}$

31. $\dfrac{9x+4}{(x-2)(x+2)(x+1)}$

33. $(7x-3x^3)/\sqrt{3-x^2}$

35. $\frac{1}{6}$

37. xy

39. $\dfrac{x+1}{x^2}$

41. $\dfrac{1}{\sqrt{a}}=\dfrac{\sqrt{a}}{a}$

43. (a) -12 (b) $\frac{25}{36}$

45. $2b^2-a$

47. $(1-2\sqrt{x}+x)/(1-x)$

49. $1/(\sqrt{x+h}+\sqrt{x})$

51. $(bc+ac+ab)/abc$

53. (a) $\dfrac{0.1x^2+55x+4000}{x}$

   (b) $0.1x^2+55x+4000$

55. $\dfrac{t^2+9t}{(t+3)^2}$

CHAPTER 0 REVIEW EXERCISES

1. yes   2. no   3. no
4. $\{1,2,3,4,9\}$ 5. $\{5,6,7,8,10\}$ 6. $\{1,2,3,4,9\}$
7. yes, $(A'\cup B')'=\{1,3\}=A\cap B$
8. (a) Commutative Property of Addition
   (b) Associative Property of Multiplication
   (c) Distributive Law
9. (a) irrational   (b) rational, integer
   (c) undefined
10. (a) $>$ (b) $<$ (c) $>$
11. 6   12. 142   13. 10
14. 5/4   15. 9   16. -29
17. 13/4   18. -10.62857888
19. (a) $[0,5]$, closed

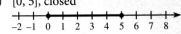

   (b) $[-3,7)$, half-open

   (c) $(-4,0)$, open

20. (a) $-1<x<16$ (b) $-12\le x\le 8$
   (c) $x<-1$
21. (a) 1   (b) $2^{-2}=1/4$ (c) 4^6 (d) 7
22. (a) $1/x^2$ (b) x^{10} (c) x^9 (d) $1/y^8$
   (e) y^6
23. $-x^2y^2/36$ 24. $9y^8/(4x^4)$ 25. $y^2/(4x^4)$
26. $-x^8z^4/y^4$ 27. $3x/(y^7z)$ 28. $x^5/(2y^3)$
29. (a) 4   (b) 2/7   (c) 1.1
30. (a) $x^{1/2}$ (b) $x^{2/3}$ (c) $x^{-1/4}$

31. (a) $\sqrt[7]{x^3}$ (b) $1/\sqrt{x}=\sqrt{x}/x$ (c) $-x\sqrt{x}$
32. (a) $5y\sqrt{2x}/2$ (b) $\sqrt[3]{x^2y}/x^2$
33. $x^{5/6}$ 34. y 35. $x^{17/4}$
36. $x^{11/3}$ 37. $x^{2/5}$ 38. x^2y^8
39. $2xy^2\sqrt{3xy}$ 40. $25x^3y^4\sqrt{2y}$
41. $6x^2y^4\sqrt[3]{5x^2y^2}$ 42. $8a^2b^4\sqrt{2a}$
43. $2xy$ 44. $4x\sqrt{3xy}/(3y^4)$
45. $-x-2$ 46. $-x^2-x$
47. $4x^3+xy+4y-4$ 48. $24x^5y^5$
49. $3x^2-7x+4$ 50. $3x^2+5x-2$
51. $4x^2-7x-2$ 52. $6x^2-11x-7$
53. $4x^2-12x+9$ 54. $16x^2-9$
55. $2x^4+2x^3-5x^2+x-3$ 56. $8x^3-12x^2+6x-1$
57. x^3-y^3 58. $(2/y)-(3xy/2)-3x^2$
59. $3x^2+2x-3+(-3x+7)/(x^2+1)$
60. $x^3-x^2+2x+7+21/(x-3)$
61. x^2-x 62. $2x-a$
63. $x^3(2x-1)$
64. $2(x^2+1)^2(1+x)(1-x)$
65. $(2x-1)^2$ 66. $(4+3x)(4-3x)$
67. $2x^2(x+2)(x-2)$ 68. $(x-7)(x+3)$
69. $(3x+2)(x-1)$ 70. $(x-3)(x-2)$
71. $(x-12)(x+2)$ 72. $(4x+3)(3x-8)$
73. $(2x+3)^2(2x-3)^2$ 74. $x^{2/3}+1$
75. (a) $\dfrac{x}{(x+2)}$ (b) $\dfrac{2xy(2-3xy)}{2x-3y}$

76. $\dfrac{x^2-4}{x(x+4)}$ 77. $\dfrac{(x+3)}{(x-3)}$

78. $\dfrac{x^2(3x-2)}{(x-1)(x+2)}$ 79. $(6x^2+9x-1)/(6x^2)$

80. $\dfrac{4x-x^2}{4(x-2)}$ 81. $-\dfrac{x^2+2x+2}{x(x-1)^2}$

82. $\dfrac{x(x-4)}{(x-2)(x+1)(x-3)}$ 83. $\dfrac{(x-1)^3}{x^2}$

84. $\dfrac{1-x}{1+x}$ 85. $3(\sqrt{x}+1)$

86. $2/(\sqrt{x}+\sqrt{x-4})$
87. (a)

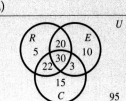

R: recognized
E: exercise
C: community involvement

   (b) 10   (c) 100
88. 52.55%   89. 16
90. (a) $4115.27   (b) $66,788.69
91. (a) 16   (b) $820.40   (c) $3281.60
92. (a) 1.1 inch, about quarter-sized
   (b) 104 mph
93. (a) $10,000\left[\dfrac{(0.0065)(1.0065)^n}{(1.0065)^n-1}\right]$ (b) $243.19 (for both)
94. (a) $S=k\sqrt[3]{A}$ (b) $\sqrt[3]{2.25}\approx 1.31$

95. $26x - 300 - 0.001x^2$

96. $1,450,000 - 3625x$

97. $(50 + x)(12 - 0.5x)$

98. (a) $\dfrac{12,000p}{100 - p}$

(b) $0. It costs nothing if no effort is made to remove pollution.

(c) $588,000

(d) Undefined. Removing 100% would be impossible, and the cost of getting close would be enormous.

99. $\dfrac{56x^2 + 1200x + 8000}{x}$

CHAPTER 0 TEST

1. (a) $\{3, 4, 6, 8\}$ (b) $\{3, 4\}; \{3, 6\};$ or $\{4, 6\}$
(c) $\{6\}$ or $\{8\}$

2. 21

3. (a) 8   (b) 1   (c) $\frac{1}{2}$ (d) -10
(e) 30   (f) $\frac{5}{6}$ (g) $\frac{2}{3}$ (h) -3

4. (a) $\sqrt[5]{x}$ (b) $1/\sqrt[4]{x^3}$

5. (a) $1/x^5$ (b) x^{21}/y^6

6. (a) $\dfrac{\sqrt{5x}}{5}$ (b) $2a^2b^2\sqrt{6ab}$ (c) $\dfrac{1 - 2\sqrt{x} + x}{1 - x}$

7. (a) 5   (b) -8 (c) -5

8. $(-2, 3]$

9. (a) $2x^2(4x - 1)$ (b) $(x - 4)(x - 6)$
(c) $(3x - 2)(2x - 3)$ (d) $2x^3(1 + 4x)(1 - 4x)$

10. (c); -2

11. $2x + 1 + \dfrac{2x - 6}{x^2 - 1}$

12. (a) $19y - 45$ (b) $-6t^6 + 9t^9$
(c) $4x^3 - 21x^2 + 13x - 2$ (d) $-18x^2 + 15x - 2$
(e) $4m^2 - 28m + 49$ (f) $\dfrac{x^4}{3x + 9}$
(g) $\dfrac{x^7}{81}$ (h) $\dfrac{6 - x}{x - 8}$
(i) $\dfrac{x^2 - 4x - 3}{x(x - 3)(x + 1)}$

13. $\dfrac{y - x}{y + xy^2}$

14. (a)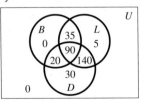

(b) 0
(c) 175

15. $4875.44 (nearest cent)

1.1 EXERCISES

1. $x = -9/4$ **3.** $x = 0$ **5.** $x = -32$
7. $x = -29/2$ **9.** $x = -5$ **11.** $x = 17/13$
13. $x = -1/3$ **15.** $x = 3$ **17.** $x = 5/4$

19. no solution   **21.** $x \approx -0.279$
23. $x \approx -1147.362$ **25.** $y = \frac{3}{4}x - \frac{15}{4}$
27. $y = -6x + \frac{22}{3}$ **29.** $t = \dfrac{S - P}{Pr}$

31. $x < 2$ **33.** $x < -4$ **35.** $x \le -1$
37. $x < -3$
39. $x < -6$
41. $x < 2$

43. 145 months
45. $3356.50   **47.** 440 packs, or 220,000 CDs
49. $29,600
51. (a) 82.3%   (b) $t \approx 25.6$, during 2016
53. 96
55. $90,000 at 9%; $30,000 at 13%
57. $2160/month; 8% increase
59. $x > 80$
61. $695 + 5.75x \le 900$; 35 or fewer
63. (a) $t = 38$ (b) $t \approx 47.6$ (c) in 2028
65. (a) $0.479 \le h \le 1$; $h = 1$ means 100% humidity
(b) $0 \le h \le 0.237$

1.2 EXERCISES

1. (a) To each x-value there corresponds exactly one y-value.
(b) domain: $\{-7, -1, 0, 3, 4.2, 9, 11, 14, 18, 22\}$
range: $\{0, 1, 5, 9, 11, 22, 35, 60\}$
(c) $f(0) = 1, f(11) = 35$

3. yes; to each x-value there corresponds exactly one y-value; domain $= \{1, 2, 3, 8, 9\}$, range $= \{-4, 5, 16\}$

5. The vertical-line test shows the graph represents a function of x.

7. The vertical line test shows the graph is not a function of x.

9. yes   **11.** no

13. (a) -10 (b) 6   (c) -34 (d) 2.8
15. (a) -3 (b) 1   (c) 13   (d) 6
17. (a) -251 (b) -128 (c) 22   (d) -4.25
19. (a) $63/8$ (b) 6   (c) -6
21. (a) no, $f(2 + 1) = f(3) = 13$ but $f(2) + f(1) = 10$
(b) $1 + x + h + x^2 + 2xh + h^2$
(c) no, $f(x) + f(h) = 2 + x + h + x^2 + h^2$
(d) no, $f(x) + h = 1 + x + x^2 + h$
(e) $1 + 2x + h$

23. (a) $-2x^2 - 4xh - 2h^2 + x + h$
(b) $-4x - 2h + 1$

25. (a) 10   (b) 6

27. (a) $(1, -3)$, yes   (b) $(3, -3)$, yes
(c) $b = a^2 - 4a$ (d) $x = 0, x = 4$, yes

29. domain: all reals; range: reals $y \ge 4$

31. domain: reals $x \ge -4$; range: reals $y \ge 0$

33. $x \ge 1, x \ne 2$ **35.** $-7 \le x \le 7$

37. (a) $3x + x^3$ (b) $3x - x^3$ (c) $3x^4$ (d) $\dfrac{3}{x^2}$

39. (a) $\sqrt{2x} + x^2$ (b) $\sqrt{2x} - x^2$

    (c) $x^2\sqrt{2x}$ (d) $\dfrac{\sqrt{2x}}{x^2}$

41. (a) $-8x^3$ (b) $1 - 2(x-1)^3$

    (c) $[(x-1)^3 - 1]^3$ (d) $(x-1)^6$

43. (a) $2\sqrt{x^4 + 5}$ (b) $16x^2 + 5$

    (c) $2\sqrt{2\sqrt{x}}$ (d) $4x$

45. (a) $f(20) = 103{,}000$ means that if $103{,}000 is borrowed, it can be repaid in 20 years (of $800-per-month payments).

    (b) no; $f(5 + 5) = f(10) = 69{,}000$, but $f(5) + f(5) = 40{,}000 + 40{,}000 = 80{,}000$

    (c) 15 years; $f(15) = 89{,}000$

47. (a) $866; $1080

    (b) $1160; starting benefits at age 68 provides $1160 per month.

    (c) $250; starting benefits at age 66 gives $250 more per month than starting at age 62.

49. (a) $W(100) \approx 155$ million; $O(140) \approx 120$ million

    (b) 166.3; in 2020 there are expected to be 166.3 million White, non-Hispanics in the civilian non-institutional labor force (CN-ILF).

    (c) 42.6; in 1990 there were 42.6 million non-Whites or Hispanics in the CN-ILF.

    (d) 79.5; in 2020 there are expected to be 79.5 million more White non-Hispanics in the CN-ILF than others.

    (e) 312.4; in 2050 the total size of the CN-ILF is expected to be 312.4 million.

    (f) $(W-O)(100)$ is greater; the graphs are further apart at $t = 100$ than at $t = 140$.

51. (a) Yes

    (b) All reals for F

    (c) Domain: $32 < F < 212$
       Range: $0 < C < 100$

    (d) $C = \left(\dfrac{40}{9}\right)^\circ$

53. (a) $C(10) = \$4210$

    (b) $C(100) = \$32{,}200$

    (c) The total cost of producing 100 items is $32,200.

55. (a) $0 \le p < 100$

    (b) $5972.73; to remove 45% of the particulate pollution would cost $5972.73.

    (c) $65,700; to remove 90% of the particulate pollution would cost $65,700.

    (d) $722,700; to remove 99% of the particulate pollution would cost $722,700.

    (e) $1,817,700; to remove 99.6% of the particulate pollution would cost $1,817,700.

57. (a) $(P \circ q)(t) = 180(1000 + 10t)$

$$-\dfrac{(1000 + 10t)^2}{100} - 200$$

    (b) $x = 1150$, $P = \$193{,}575$

59. (a) yes; the output of g (customers) is the input for f.

    (b) no; the output of f is revenue, and this is not the input for g.

(c) $f \circ g$: input (independent variable) is advertising dollars.
    output (dependent variable) is revenue dollars.
    $f \circ g$ shows how advertising dollars result in revenue dollars.

61. $L = 2x + 3200/x$

63. $R = (30 + x)(100 - 2x)$

1.3 EXERCISES

1. x-intercept 4   **3.** x-intercept 12
    y-intercept 3 y-intercept -7.5

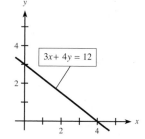 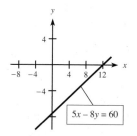

5. $m = 4$ **7.** $m = -1/2$ **9.** $m = 0$ **11.** 0

13. 0   **15.** (a) negative   (b) undefined

17. $m = 7/3$, $b = -1/4$ **19.** $m = 0$, $b = 3$

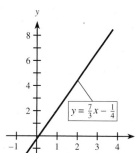 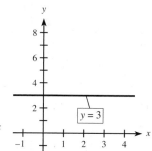

21. undefined slope,   **23.** $m = -2/3$, $b = 2$
    no y-intercept

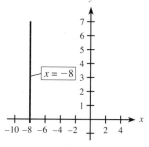 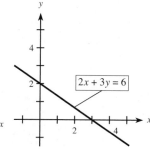

25.   **27.**

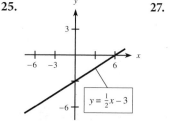 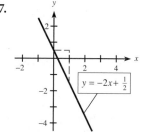

29.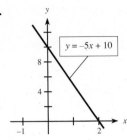

31.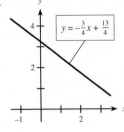

33.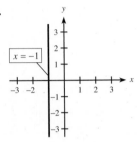

35. $y = 2x - 4$ **37.** $-x + 13y = 32$ **39.** $y = -3x - 12$
41. perpendicular   **43.** neither; same line
45. $y = -\frac{3}{5}x - \frac{41}{5}$ **47.** $y = -\frac{6}{5}x + \frac{23}{5}$
49. (a)

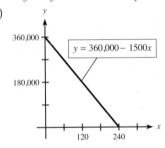

(b) 240 months
(c) After 60 months, the value of the building is $270,000.
51. (a) $m = 1.36; b = 55.98$
(b) The percent of the U.S. population with Internet service is changing at the rate of 1.36 percentage points per year.
(c)

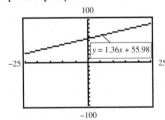

53. (a) $m = -0.065; b = 31.39$
(b) The F-intercept represents the percent of the world's land that was forest in 1990.
(c) -0.065 percentage points per year. This means that after 1990, the world forest area as a percent of land area changes by -0.065 percentage points per year.

55. (a) $m = 0.78$
(b) This means that the average annual earnings of females increases $0.78 for each $1 increase in the average annual earnings of males.
(c) $45,484
57. $y = 0.0838x + 16.37$
59. (a) $y = 1.296x - 2465$
(b) The U.S. civilian workforce changes at the rate of 1.296 million workers per years.
61. (a) $y = 5.183x - 10{,}349.10$
(b) The CPI is changing at the rate of about $5.18/year.
63. $p = 85{,}000 - 1700x$ **65.** $R = 3.2t - 0.2$
67. $y = 0.48x - 71$

1.4 EXERCISES

1.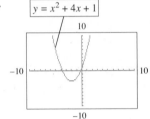

3.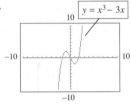

5.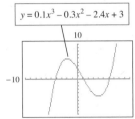

7.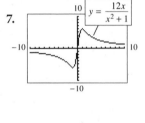

9.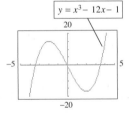

11. $y = (3x + 7)/(x^2 + 4)$
13. (a) $y = 0.01x^3 + 0.3x^2 - 72x + 150$

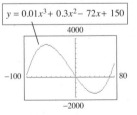

(b) standard window

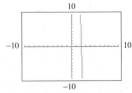

15. (a) $y = \dfrac{x + 15}{x^2 + 400}$

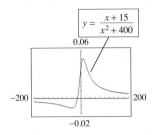

(b) standard window

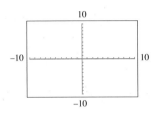

17. (a) x-intercept 30, y-intercept -0.03

(b)

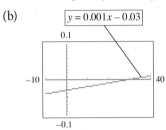

19.

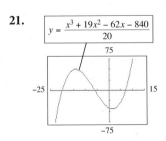

21.

23.

Linear

25.

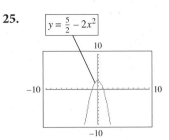

Not linear

27.

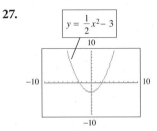

Not linear

29. $f(-2) = -18; f\left(\frac{3}{4}\right) = 0.734375$

31.

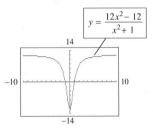

33.

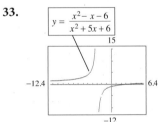

35. $x = 3.5$

37. either $x = 5$ or $x = -2$

39. (a) $-1.1098, 8.1098$ (b) $-1.1098, 8.1098$

41. (a)

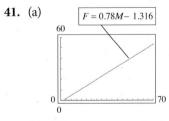

(b) When average annual earnings for males is $50,000, average annual earnings for females is $37,684.   (c) $47,434

43. (a)

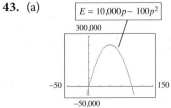

(b) $E \geq 0$ when $0 \leq p \leq 100$

45. (a)

$R = 28,000 - 0.08x$

(b) decreasing; as more people become aware of the product, there are fewer to learn about it, so the rate will decrease.

47. (a) x-min $= 0$, x-max $= 130$
(b) y-max $= 65$
(c)

$P = 0.000078x^3 - 0.0107x^2 - 0.182x + 64.68$

(d) The percent decreases from 57.4% in 1920 to 17.5% in 2000, and then it increases to 31.6% in 2030.

49. (a)

$C = \dfrac{285,000}{p} - 2850$

(b) Near $p = 0$, cost grows without bound.
(c) The coordinates of the point mean that obtaining stream water with 1% of the current pollution levels would cost $282,150.
(d) The p-intercept means that stream water with 100% of the current pollution levels would cost $0.

51. (a)

$T = 2.11x^3 - 93.01x^2 + 1373.9x + 394$

(b) increasing; the per capita tax burden is increasing

1.5 EXERCISES

1. one solution; $(-1, -2)$
3. infinitely many solutions (each point on the line)
5. $x = 2, y = 2$ **7.** no solution
9. $x = 10/3, y = 2$ **11.** $x = 14/11, y = 6/11$
13. $x = 4, y = -1$ **15.** $x = 3, y = -2$
17. $x = 2, y = 1$ **19.** $x = -52/7, y = -128/7$
21. $x = 1, y = 1$ **23.** dependent
25. $x = 4, y = 2$ **27.** $x = -1, y = 1$
29. $x = -17, y = 7, z = 5$
31. $x = 4, y = 12, z = -1$
33. $x = 44, y = -9, z = -1/2$

35. (a) $C = 10, F = 50$
(b) The tourist formula overestimates the F temperature—it adds 2°F to the estimate for each 1°C, but actual change is 1.8°C.
37. (a) $x + y = 1800$ (b) $20x$ (c) $30y$
(d) $20x + 30y = 42,000$
(e) 1200 tickets at $20; 600 tickets at $30
39. $68,000 at 18%; $77,600 at 10%
41. 10%: $27,000; 12%: $24,000
43. 4 oz of A, $6\frac{2}{3}$ oz of B
45. 4550 of species A, 1500 of species B
47. 7 cc of 20%; 3 cc of 5%
49. 10,000 at $40; 6000 at $60
51. 80 cc   **53.** 5 oz of A, 1 oz of B, 5 oz of C
55. 200 type A, 100 type B, 200 type C

1.6 EXERCISES

1. (a) $P(x) = 34x - 6800$ (b) $95,200
3. (a) $P(x) = 37x - 1850$
(b) $-$740, loss of $740   (c) 50
5. (a) $m = 5, b = 250$
(b) $\overline{MC} = 5$ means each additional unit produced costs $5.
(c) slope $=$ marginal cost; C-intercept $=$ fixed costs
(d) 5, 5
7. (a) 27
(b) $\overline{MR} = 27$ means each additional unit sold brings in $27.
(c) 27, 27
9. (a) $P(x) = 22x - 250$ (b) 22   (c) $\overline{MP} = 22$
(d) Each unit sold adds $22 to profits at all levels of production, so produce and sell as much as possible.
11. $P = 58x - 8500, \overline{MP} = 58$
13. (a) $C(x) = 35x + 6600$ (b) $R(x) = 60x$
(c) $P(x) = 25x - 6600$
(d) $C(200) = 13,600$ dollars is the cost of producing 200 helmets.
$R(200) = 12,000$ dollars is the revenue from sale of 200 helmets.
$P(200) = -1600$ dollars; will lose $1600 from production and sale of 200 helmets.
(e) $C(300) = 17,100$ dollars is the cost of producing 300 helmets.
$R(300) = 18,000$ dollars is the revenue from sale of 300 helmets.
$P(300) = 900$ dollars; will profit $900 from production and sale of 300 helmets.
(f) $\overline{MP} = 25$ dollars per unit; each additional unit produced and sold increases profit by $25.
15. (a) Revenue passes through the origin.
(b) $2000   (c) 400 units
(d) $\overline{MC} = 2.5$; $\overline{MR} = 7.5$
17. 33
19. (a) $R(x) = 12x$; $C(x) = 8x + 1600$
(b) 400 units

21. (a) $P(x) = 4x - 1600$
(b) $x = 400$ units to break even

23. (a) $C(x) = 4.5x + 1045$
(b) $R(x) = 10x$
(c) $P(x) = 5.5x - 1045$
(d) 190 surge protectors

25. (a) $R(x) = 54.90x$ (b) $C(x) = 14.90x + 20,200$
(c) 505

27. (a) R starts at origin and is the steeper line.
FC is a horizontal line.
VC starts at origin and is not as steep as R.
(See figure.)

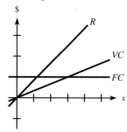

(b) C starts where FC meets the \$-axis and is parallel to VC. Where C meets R is the break-even point (BE). P starts on the \$-axis at the negative of FC and crosses the x-axis at BE. (See figure.)

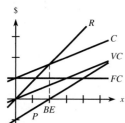

29. Demand decreases.
31. (a) 600   (b) 300   (c) shortage
33. 16 demanded, 25 supplied; surplus
35. $p = -2q/3 + 1060$
37. $p = 0.001q + 5$
39. (a) demand falls; supply rises   (b) (30, \$25)
41. (a) $q = 20$ (b) $q = 40$
(c) shortage, 20 units short
43. shortage   **45.** $q = 20, p = \$18$
47. $q = 10, p = \$180$ **49.** $q = 100, p = \$325$
51. (a) \$15   (b) $q = 100, p = \$100$
(c) $q = 50, p = \$110$ (d) yes
53. $q = 8; p = \$188$
55. $q = 500; p = \$40$
57. $q = 1200; p = \$15$

CHAPTER 1 REVIEW EXERCISES

1. $x = \frac{31}{3}$ **2.** $x = -13$ **3.** $x = -\frac{29}{8}$
4. $x = -\frac{1}{9}$ **5.** $x = 8$
6. no solution   **7.** $y = -\frac{2}{3}x - \frac{4}{3}$
8. $x \le 3$
9. $x \ge -20/3$

10. $x \ge -15/13$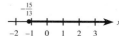
11. yes   **12.** no   **13.** yes
14. domain: reals $x \le 9$; range: reals $y \ge 0$
15. (a) 2   (b) 37   (c) 29/4
16. (a) 0   (b) 9/4   (c) 10.01
17. $9 - 2x - h$ **18.** yes   **19.** no
20. 4   **21.** $x = 0, x = 4$
22. (a) domain: $\{-2, -1, 0, 1, 3, 4\}$
range: $\{-3, 2, 4, 7, 8\}$
(b) 7   (c) $x = -1, x = 3$
(d)

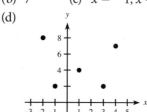

(e) no; for $y = 2$, there are two different x-values, -1 and 3.
23. (a) $x^2 + 3x + 5$ (b) $(3x + 5)/x^2$
(c) $3x^2 + 5$ (d) $9x + 20$
24. $x: 2, y: 5$ **25.** $x: 3/2, y: 9/5$

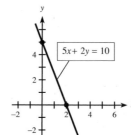 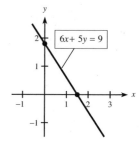

26. $x: -2, y:$ none

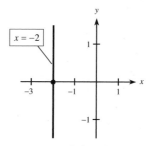

27. $m = 1$ **28.** undefined   **29.** $m = -\frac{2}{5}, b = 2$
30. $m = -\frac{4}{3}, b = 2$ **31.** $y = 4x + 2$ **32.** $y = -\frac{1}{2}x + 3$
33. $y = \frac{2}{5}x + \frac{9}{5}$ **34.** $y = -\frac{11}{8}x + \frac{17}{4}$ **35.** $x = -1$
36. $y = 4x + 2$ **37.** $y = \frac{4}{3}x + \frac{10}{3}$
38.

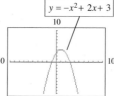

39.

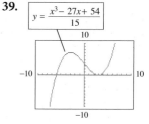

40. (a)

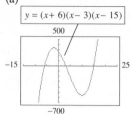

$y = (x+6)(x-3)(x-15)$

(b)

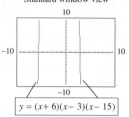

Standard window view

$y = (x+6)(x-3)(x-15)$

(c) The graph in (a) shows a complete graph. The graph in (b) shows a piece that rises toward the high point and a piece between the high and low points.

41. (a)

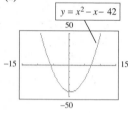

$y = x^2 - x - 42$

(b)

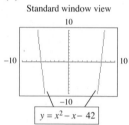

Standard window view

$y = x^2 - x - 42$

(c) The graph in (a) shows a complete graph. The one in (b) shows pieces that fall toward the minimum point and rise from it.

42. reals $x \geq -3$ with $x \neq 0$

43. $0.2857 \approx 2/7$

44. $x = 2, y = 1$

45. $x = 10, y = -1$

46. $x = 3, y = -2$

47. no solution

48. $x = 10, y = -71$

49. $x = 1, y = -1, z = 2$

50. $x = 11, y = 10, z = 9$

51. (a) 2020

(b) 21.5 years, to age 86.5

(c) $x = 80.5$; in 2031

52. 95%

53. 40,000 miles. He probably would drive more than 40,000 miles in 7 years, so he should buy diesel.

54. (a) yes   (b) no   (c) 4

55. (a) $565.44

(b) The monthly payment on a $70,000 loan is $494.75.

56. (a) $(P \circ q)(t) = 330(100 + 10t)$
$\quad - 0.05(100 + 10t)^2 - 5000$

(b) $x = 250, P = \$74{,}375$

57. $(W \circ L)(t) = 0.03[65 - 0.1(t - 25)^2]^3$

58. (a)

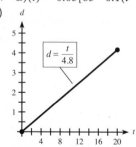

$d = \dfrac{t}{4.8}$

(b) When the time between seeing lightning and hearing thunder is 9.6 seconds, the storm is 2 miles away.

59.

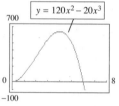

$H_c = 90 - T_a$

60. (a) $P = 58x - 8500$

(b) The profit increases by $58 for each unit sold.

61. (a) yes

(b) $m = 427, b = 4541$

(c) In 2000, average annual health care costs were $4541 per consumer.

(d) Average annual health care costs are changing at the rate of $427 per year.

62. $F = \frac{9}{5}C + 32$ or $C = \frac{5}{9}(F - 32)$

63. (a)

$y = 120x^2 - 20x^3$

(b) $0 \leq x \leq 6$

64. (a) $v^2 = 1960(h + 10)$

$\dfrac{v^2}{1960} = h + 10$

$h = \dfrac{1}{1960} v^2 - 10$

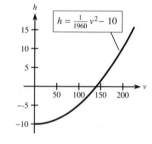

$h = \frac{1}{1960}v^2 - 10$

(b) 12.5 cm

65. $100,000 at 9.5%; $50,000 at 11%

66. 2.8 liters of 20%; 1.2 liters of 70%

67. (a) 12 supplied; 14 demanded   (b) shortfall

(c) increase

68.

p

Supply
$p = 6q + 60$

$(30, 240)$
Market Equilibrium

$p + 6q = 420$
Demand

q

69. (a) 38.80   (b) 61.30   (c) 22.50   (d) 200

70. (a) $C(x) = 22x + 1500$ (b) $R(x) = 52x$

(c) $P(x) = 30x - 1500$ (d) $\overline{MC} = 22$

(e) $\overline{MR} = 52$ (f) $\overline{MP} = 30$ (g) $x = 50$

71. (a) S: $p = 4q - 400$
$\quad$ D: $p = -4q + 1520$

(b) $q = 240; p = \$560$

72. $q = 700, p = \$80$

CHAPTER 1 TEST

1. $x = -6$ **2.** $x = 18/7$ **3.** $x = -3/7$
4. $x = -38$ **5.** $5 - 4x - 2h$
6. $x \geq -9$

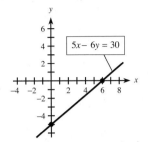

7. x: 6; y: -5 **8.** x: 3; y: 21/5

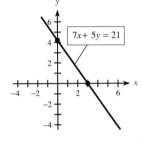

9. (a) domain: $x \geq -4$
          range: $f(x) \geq 0$
     (b) $2\sqrt{7}$ (c) 6
10. $y = -\frac{3}{2}x + \frac{1}{2}$ **11.** $m = -\frac{5}{4}$; $b = \frac{15}{4}$
12. (a) $x = -3$ (b) $y = -4x - 13$
13. (a) no; a vertical line intersects the curve twice.
     (b) yes; there is exactly one y-value for each x-value.
     (c) no; one value of x gives two y-values.
14. (a)

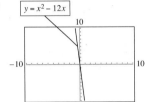

     (b)

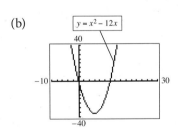

15. $x = -2$, $y = 2$
16. (a) $5x^3 + 2x^2 - 3x$ (b) $x + 2$
     (c) $5x^2 + 7x + 2$
17. (a)

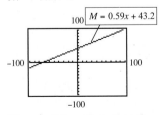

     (b) The model predicts that there will be 90.4 million
          men in the U.S. workforce in 2030.
     (c) 102.2 million
18. (a) 30   (b) $P = 8x - 1200$ (c) 150
     (d) $\overline{MP} = 8$; the sale of each additional unit gives
          $8 more profit.

19. (a) $R = 50x$
     (b) 19,000; it costs \$19,000 to produce 100 units.
     (c) 450
20. $q = 200$, $p = \$2500$
21. (a) 720,000; original value of the building
     (b) -2000; building depreciates \$2000 per month.
22. 400
23. 12,000 at 9%, 8000 at 6%

2.1 EXERCISES

1. $x^2 + 2x - 1 = 0$ **3.** $y^2 + 3y - 2 = 0$
5. $-2, 6$ **7.** $\frac{3}{2}, -\frac{3}{2}$ **9.** $0, 1$ **11.** $\frac{1}{2}$
13. (a) $2 + 2\sqrt{2}, 2 - 2\sqrt{2}$ (b) $4.83, -0.83$
15. no real solutions   **17.** $\sqrt{7}, -\sqrt{7}$ **19.** $4, -4$
21. $1, -9$ **23.** $-7, 3$ **25.** $8, -4$ **27.** $-\frac{7}{4}, \frac{3}{4}$
29. $-6, 2$ **31.** $(1 + \sqrt{31})/5, (1 - \sqrt{31})/5$
33. $-2, 5$ **35.** $-300, 100$ **37.** $0.69, -0.06$
39. $8, 1$ **41.** $1/2$ **43.** $-9, -10$
45. $x = 20$ or $x = 70$
47. (a) $x = 10$ or $x = 345\frac{5}{9}$
     (b) Yes; for any $x > 10$ and $x < 345\frac{5}{9}$
49. 6 seconds
51. (a) ± 50
     (b) $s = 50$; there is no particulate pollution in the air
          above the plant.
53. 97.0 mph   **55.** $t \approx 16.2$; in 2017   **57.** \$80
59. $x \approx 16.8$; in 2022
61. (a) 18   (b) ≈ 31
     (c) Speed triples, but K changes only by a factor of
          1.72.

2.2 EXERCISES

1. (a) $(-1, -\frac{1}{2})$ (b) minimum
     (c) -1 (d) $-\frac{1}{2}$
3. (a) $(1, 9)$ (b) maximum   (c) 1   (d) 9
5. (a) $(3, 9)$ (b) maximum   (c) 3   (d) 9
7. maximum, $(2, 1)$; zeros $(0, 0)$, $(4, 0)$; y-intercept $= 0$

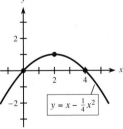

9. minimum, $(-2, 0)$; zero $(-2, 0)$; y-intercept $= 4$

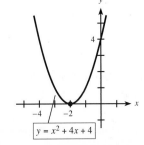

11. minimum, $(-1, -3\frac{1}{2})$; zeros $(-1 + \sqrt{7}, 0)$, $(-1 - \sqrt{7}, 0)$; y-intercept $= -3$

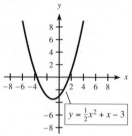

13. (a) 3 units to the right and 1 unit up
(b)

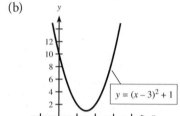

15. $y = (x + 2)^2 - 2$
(a) 2 units to the left and 2 units down
(b)

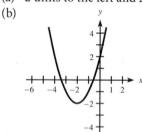

17. vertex $(1, -8)$; zeros $(-3, 0), (5, 0)$

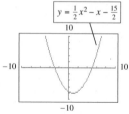

19. vertex $(-6, 3)$; no real zeros

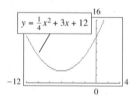

21. -5

23. vertex $(10, 64)$; zeros $(90, 0), (-70, 0)$

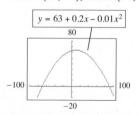

25. vertex $(0, -0.01)$; zeros $(10, 0), (-10, 0)$

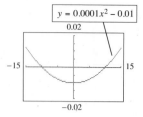

27. (a) $(1, -24)$ (b) $x \approx -0.732, 2.732$
29. (a) $x = 2$ (b) $x - 2$
(c) $(x - 2)(3x - 2)$ (d) $x = 2, \frac{2}{3}$
31. (a) 80 units (b) $540
33. 400 trees
35. dosage $= 500$ mg **37.** intensity $= 1.5$ lumens

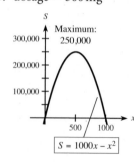
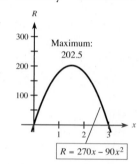

39. equation (a) $(384.62, 202.31)$ (b) $(54, 46)$
Projectile (a) goes higher.
41. (a) From b to c. The average rate is the same as the slope of the segment. Segment b to c is steeper.
(b) $d > b$ to make segment a to d be steeper (have greater slope).
43. (a)

Rent	Number Rented	Revenue
600	50	$30,000
620	49	$30,380
640	48	$30,720

(b) increase
(c) $R = (50 - x)(600 + 20x)$
(d) $800
45. (a) quadratic
(b) $a < 0$ because the graph opens downward.
(c) The vertex occurs after 2004 (when $x > 0$). Hence $-b/2a > 0$ and $a < 0$ means $b > 0$. The value $c = f(0)$, or the y-value in 2004. Hence $c > 0$.
47. 2010–2015: $399/year
2015–2020: $605/year
49.

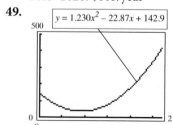

2.3 EXERCISES

1. $x = 40$ units, $x = 50$ units
3. $x = 50$ units, $x = 300$ units
5. $x = 15$ units; reject $x = 100$
7. $41,173.61
9. $87.50
11. $x = 55$, $P(55) = \$2025$
13. (a)

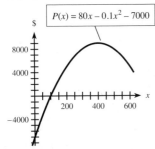

$P(x) = 80x - 0.1x^2 - 7000$

(b) $(400, 9000)$; maximum point   (c) positive
(d) negative   (e) closer to 0 as a gets closer to 400
15. (a) $P(x) = -x^2 + 350x - 15{,}000$; maximum profit is $15,625
   (b) no   (c) x-values agree
17. (a) $x = 28$ units, $x = 1000$ units
   (b) $651,041.67
   (c) $P(x) = -x^2 + 1028x - 28{,}000$; maximum profit is $236,196
   (d) $941.60
19. (a) $t \approx 5.1$, in 2012; $R \approx \$60.79$ billion
   (b) The data show a smaller revenue, $R = \$60.27$ billion, in 2011.
   (c)

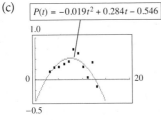

$R(t) = 0.271t^2 - 2.76t + 67.83$

   (d) The model fits the data quite well.
21. (a) $P(t) = -0.019t^2 + 0.284t - 0.546$
   (b) 2011
   (c)

$P(t) = -0.019t^2 + 0.284t - 0.546$

   (d) The model projects decreasing profits, and except for 2015, the data support this.
   (e) Management would be interested in increasing revenues or reducing costs (or both) to improve profits.

23. (a)

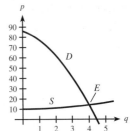

(b) See E on graph.   (c) $q = 4$, $p = \$14$
25. $q = 10$, $p = \$196$
27. $q = 216\frac{2}{3}$, $p = \$27.08$
29. $p = \$40$, $q = 30$
31. $q = 90$, $p = \$50$
33. $q = 70$, $p = \$62$

2.4 EXERCISES

1. b   **3.** f   **5.** j   **7.** k   **9.** a
11. c   **13.** (a) cubic   (b) quartic
15. e   **17.** b   **19.** d   **21.** g
23.

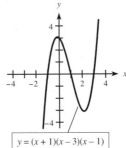

$y = (x + 1)(x - 3)(x - 1)$

25.

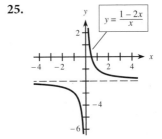

$y = \dfrac{1 - 2x}{x}$

27.

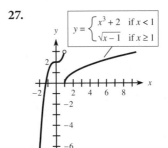

$y = \begin{cases} x^3 + 2 & \text{if } x < 1 \\ \sqrt{x - 1} & \text{if } x \geq 1 \end{cases}$

29. (a) $8/3$ (b) 9.9   (c) -999.999 (d) no
31. (a) 64   (b) 1   (c) 1000   (d) 0.027
33. (a) 2   (b) 4   (c) 0   (d) 2

35. (a)

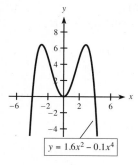

$y = 1.6x^2 - 0.1x^4$

(b) polynomial   (c) no asymptotes
(d) turning points at $x = 0$ and approximately
$x = -2.8$ and $x = 2.8$

37. (a)

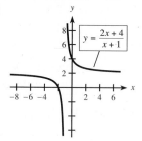

$y = \dfrac{2x + 4}{x + 1}$

(b) rational
(c) vertical: $x = -1$,
horizontal: $y = 2$
(d) no turning points

39. (a)

$f(x) = \begin{cases} -x & \text{if } x < 0 \\ 5x & \text{if } x \geq 0 \end{cases}$

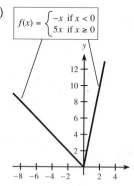

(b) piecewise defined
(c) no asymptotes
(d) turning point at
$x = 0$

41. (a) 6800; 11,200   (b) $0 < x < 27$
43. (a) upward   (b)

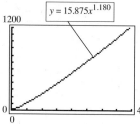

$y = 15.875x^{1.180}$

(c) $x \approx 37.7$; in 2018
45. (a) yes; $p = 100$
(b) $p \neq 100$
(c) $0 \leq p < 100$
(d) It increases without bound.
47. (a) $A(2) = 96$; $A(30) = 600$
(b) $0 < x < 50$
49. (a)

$y = \begin{cases} 5.59x^2 - 93.5x + 633 & \text{if } 0 \leq x \leq 55 \\ 6.56x^2 - 519x + 20{,}900 & \text{if } 55 < x \leq 90 \end{cases}$

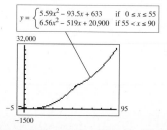

(b) $9933 billion ($9.933 trillion)
(c) $18,875 billion ($18.875 trillion)
51. (a) $P = \begin{cases} 49 & \text{if } 0 < x \leq 1 \\ 70 & \text{if } 1 < x \leq 2 \\ 91 & \text{if } 2 < x \leq 3 \\ 112 & \text{if } 3 < x \leq 4 \end{cases}$

(b) 70; it costs 70 cents to mail a 1.2 oz letter.
(c) Domain $0 < x \leq 4$; Range {49, 70, 91, 112}
(d) 70 cents and 91 cents
53. (a) $p = \dfrac{200}{2 + 0.1x}$

(b) yes, when $p = $100

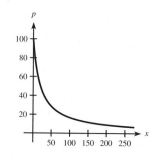

55. $C(x) = 30(x - 1) + \dfrac{3000}{x + 10}$

(a)

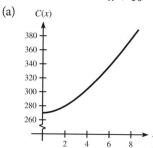

(b) Any turning point would indicate the minimum
or the maximum cost. In this case, $x = 0$ gives a
minimum.
(c) The y-intercept is the fixed cost of production.

2.5 EXERCISES

1. linear   **3.** quadratic   **5.** quartic
7. quadratic or power
9. $y = 2x - 3$ **11.** $y = 2x^2 - 1.5x - 4$

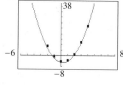

13. $y = x^3 - x^2 - 3x - 4$ **15.** $y = 2x^{0.5}$

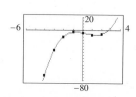

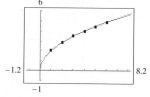

17. (a)

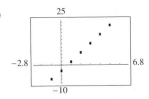

(b) linear
(c) $y = 5x - 3$

19. (a)

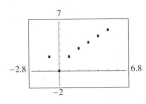

(b) quadratic
(c) $y = 0.09595x^2 + 0.4656x + 1.4758$

21. (a)

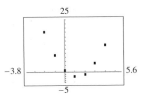

(b) quadratic
(c) $y = 2x^2 - 5x + 1$

23. (a)

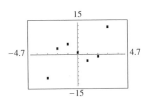

(b) cubic
(c) $y = x^3 - 5x + 1$

25. (a) $y = 154.0x + 35,860$ (b) 40,018,000   (c) 2070
27. (a) A linear function is best; $y = 327.6x + 9591$
(b) $15,160 billion
(c) $m = 327.6$ means the U.S. disposable income is increasing at the rate of about $327.6 billion per year.
29. (a) $y = 0.0052x^2 - 0.62x + 15$ (b) $x \approx 59.6$
(c) No, it is unreasonable to feel warmer for winds greater than 60 mph.
31. (a)

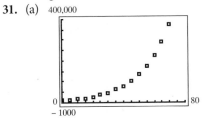

(b) $y = 106x^2 - 2870x + 28,500$
(c) $y = 1.70x^3 - 72.9x^2 + 1970x + 5270$
(d)

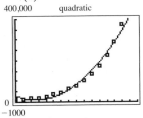
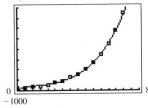

Cubic model fits better

33. (a) $y = 0.0157x + 2.01$
(b) $y = -0.00105x^2 + 0.0367x + 1.94$
(c)

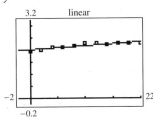

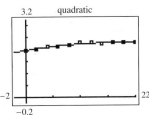

(d) The quadratic model is a slightly better fit.
35. (a) A cubic models looks best because of the two bends.
(b) $y = 0.864x^3 - 128x^2 + 6610x - 62,600$
(c)

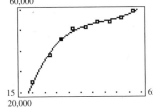

(d) 57
37. (a) $y = 0.0514x^{2.73}$
(b)

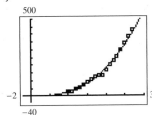

(c) $546 billion

CHAPTER 2 REVIEW EXERCISES

1. $x = 0, x = -\frac{5}{3}$ **2.** $x = 0, x = \frac{4}{3}$
3. $x = -2, x = -3$
4. $x = (-5 + \sqrt{47})/2, x = (-5 - \sqrt{47})/2$
5. no real solutions   **6.** $x = \sqrt{3}/2, x = -\sqrt{3}/2$
7. $\frac{5}{7}, -\frac{4}{5}$ **8.** $(-1 + \sqrt{2})/4, (-1 - \sqrt{2})/4$
9. $7/2, 100$ **10.** $13/5, 90$
11. no real solutions   **12.** $z = -9, z = 3$
13. $x = 8, x = -2$ **14.** $x = 3, x = -1$
15. $x = (-a \pm \sqrt{a^2 - 4b})/2$
16. $r = (2a \pm \sqrt{4a^2 + x^3c})/x$
17. $1.64, -7051.64$ **18.** $0.41, -2.38$

19. vertex $(-2, -2)$;
zeros $(0, 0)$, $(-4, 0)$

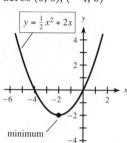

20. vertex $(0, 4)$;
zeros $(4, 0)$, $(-4, 0)$

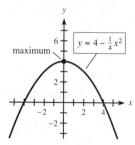

21. vertex $(\frac{1}{2}, \frac{25}{4})$;
zeros $(-2, 0)$, $(3, 0)$

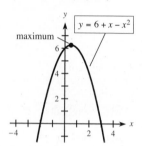

22. vertex $(2, 1)$;
no real zeros

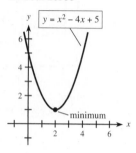

23. vertex $(-3, 0)$;
zero $(-3, 0)$

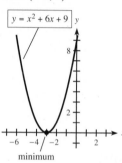

24. vertex $(\frac{3}{2}, 0)$;
zero $(\frac{3}{2}, 0)$

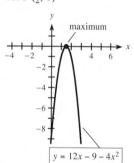

25. vertex $(0, -3)$;
zeros $(-3, 0)$, $(3, 0)$

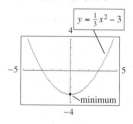

26. vertex $(0, 2)$;
no real zeros

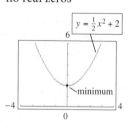

27. vertex $(-1, 4)$;
no real zeros

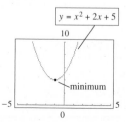

28. vertex $(\frac{7}{2}, \frac{9}{4})$;
zeros $(2, 0)$, $(5, 0)$

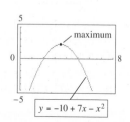

29. vertex $(100, 1000)$;
zeros $(0, 0)$, $(200, 0)$

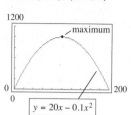

30. vertex $(75, -6.25)$;
zeros $(50, 0)$, $(100, 0)$

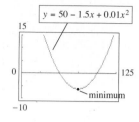

31. 20   **32.** 30
33. (a) $(1, -4\frac{1}{2})$ (b) $x = -2$, $x = 4$ (c) B
34. (a) $(0, 49)$ (b) $x = -7$, $x = 7$ (c) D
35. (a) $(7, 25)$ approximately, actual is $(7, 24\frac{1}{2})$
(b) $x = 0$, $x = 14$ (c) A
36. (a) $(-1, 9)$ (b) $x = -4$, $x = 2$ (c) C
37. (a)   (b)

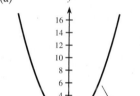
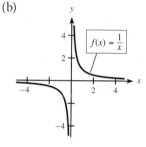

(c)

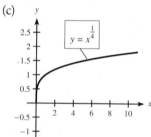

38. (a) 0   (b) 10,000   (c) -25 (d) 0.1
39. (a) -2 (b) 0   (c) 1   (d) 4
40.

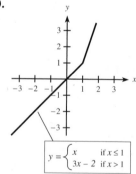

$$y = \begin{cases} x & \text{if } x \le 1 \\ 3x - 2 & \text{if } x > 1 \end{cases}$$

41. (a)   (b)

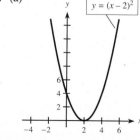
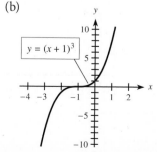

42. Turns: $(1, -5), (-3, 27)$

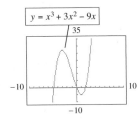

43. Turns: $(1.7, -10.4),$
$(-1.7, 10.4)$

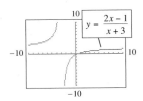

44. VA: $x = 2$;
HA: $y = 0$

45. VA: $x = -3$;
HA: $y = 2$

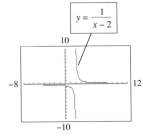

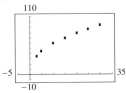

46. (a)

(b) $y = -2.1786x + 159.8571$
(c) $y = -0.0818x^2 - 0.2143x + 153.3095$

47. (a)

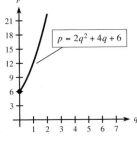

(b) $y = 2.1413x + 34.3913$
(c) $y = 22.2766x^{0.4259}$

48. (a) $t = -1.65, t = 3.65$ (b) Just $t = 3.65$
(c) at 3.65 seconds

49. $x = 20, x = 800$

50. (a) $t \approx 7.69$, in 2018; $E \approx 12.3$ million
(b) $t \approx 20.2$, in 2031

51. (a) $x = 200$
(b) $A = 30,000$ square feet

52.

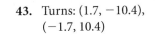

53.

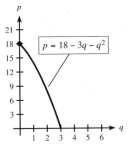

54. (a)

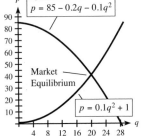

(b) $p = 41, q = 20$

55. $p = 400, q = 10$ **56.** $p = 10, q = 20$
57. $x = 46 + 2\sqrt{89} \approx 64.9, x = 46 - 2\sqrt{89} \approx 27.1$
58. $x = 15, x = 60$
59. max revenue $= \$2500$; max profit $= \$506.25$
60. max profit $= 12.25$; break-even at $x = 100, x = 30$

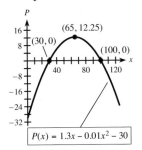

61. $x = 50, P(50) = 640$
62. (a) $C = 15,000 + 140x + 0.04x^2$;
$R = 300x - 0.06x^2$
(b) 100, 1500   (c) 2500
(d) $P = 160x - 15,000 - 0.1x^2$; max at 800
(e) at 2500: $P = -240,000$; at 800: $P = 49,000$
63. (a) power   (b) 21.8%
(c) 24.4; in 2025 about 24.4% of U.S. adults are
expected to have diabetes.
64. (a)

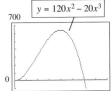

(b) $0 \le x \le 6$

65. (a) rational   (b) $0 \le p < 100$
(c) 0; it costs \$0 to remove no pollution.
(d) \$475,200
66. (a) $x = 12; C(12) = \$30.68$
(b) $x = 825; C(825) = \$1734.70$

67. (a) linear, quadratic, cubic, power
   (b) and (c)

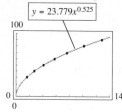

$y = 23.779x^{0.525}$

   (d) 55 mph   (e) 9.9 seconds

68. (a)

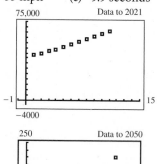

Data to 2021

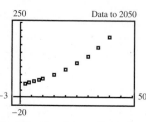

Data to 2050

   (b) Quadratic: $a(x) = 47.70x^2 + 1802x + 40{,}870$;
        $A(x) = 0.07294x^2 + 0.9815x + 44.45$
   (c) 2020 Data: \$63,676; $a(10) \approx \$63{,}664$–closer;
        $A(10) \approx \$61.564$ (\$61,564)
        2050 Data: 202.5 (\$202,500); $a(40) \approx \$189{,}292$;
        $A(40) \approx 200.413$ (\$200,413–closer)
   (d) $a = 150{,}000$ when $x \approx 32.5$, in 2043;
        $A = 150$ thousand when $x \approx 31.9$, in 2042

69. (a) $y = O(x) = 0.743x + 6.97$
   (b) $y = S(x) = 0.264x - 2.57$

   (c) $F(x) = \dfrac{0.264x - 2.57}{0.743x + 6.97}$. This is called a rational

        function and measures the fraction of obese
        adults who are severely obese.
   (d) Horizontal asymptote: $y \approx 0.355$ means that if
        this model remains valid far into the future, then
        the long-term projection is that about 0.355, or
        35.5%, of obese adults will be severely obese.

70. (a) Quadratic: $W(x) = -0.00903x^2 + 1.28x + 124$
        $O(x) = 0.00645x^2 + 1.02x + 20.0$
   (b) $x \approx 91.1$, in 2162

CHAPTER 2 TEST

1. (a)   (b)

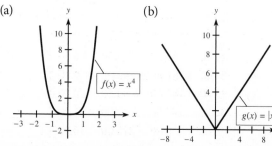

$f(x) = x^4$ $g(x) = |x|$

(c)

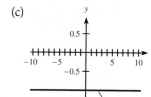

$h(x) = -1$

(d)

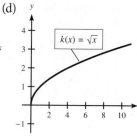

$k(x) = \sqrt{x}$

2. b; a

3.

4. (a)

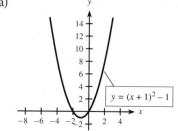

$y = (x + 1)^2 - 1$

(b)

$y = (x - 2)^3 + 1$

5. b; the function is cubic, $f(1) < 0$
6. (a) -10 (b) $-16\frac{1}{2}$ (c) -7
7.

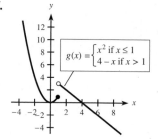

$g(x) = \begin{cases} x^2 & \text{if } x \le 1 \\ 4 - x & \text{if } x > 1 \end{cases}$

8. vertex $(-2, 25)$;
    zeros $-7, 3$

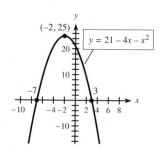

$(-2, 25)$
$y = 21 - 4x - x^2$

9. $x = 2, x = 1/3$

10. $x = \dfrac{-3 + 3\sqrt{3}}{2}, x = \dfrac{-3 - 3\sqrt{3}}{2}$

11. $x = 2/3$

12. c; $g(x)$ has a vertical asymptote at $x = -2$, as does graph c.

13. HA: $y = 0$; VA: $x = 5$

14. 42

15. (a) quartic   (b) cubic

16. (a)   model: $y = -0.3577x + 19.9227$
(b) $f(x) = 5.6$
(c) at $x = 55.7$

17. $q = 300, p = \$80$

18. (a) $P(x) = -x^2 + 250x - 15{,}000$
(b) 125 units, \$625   (c) 100 units, 150 units

19. (a) $f(15) = -19.5$ means that when the air temperature is 0°F and the wind speed is 15 mph, the air temperature feels like -19.5°F.
(b) -31.4°F

20. (a)

(b) Linear: $y = 26.8x + 695$;
Cubic: $y = 0.175x^3 - 5.27x^2 + 65.7x + 654$
(c) Linear: \$1258;   Cubic: \$1326. The cubic model is quite accurate, but both models are fairly close.
(d) The linear model increases steadily, but the cubic model rises rapidly for years past 2021.

3.1 EXERCISES

1. 3   **3.** A, F, Z **5.** $\begin{bmatrix} -1 & -2 & -3 \\ 1 & 0 & -1 \\ -2 & 3 & 4 \end{bmatrix}$

7. A, C, D, F, G, Z

9. 1   **11.** $\begin{bmatrix} 1 & 3 & 4 \\ 0 & 2 & 0 \\ -2 & 1 & 3 \end{bmatrix}$ **13.** $\begin{bmatrix} 0 & 0 & 0 \\ 0 & 0 & 0 \\ 0 & 0 & 0 \end{bmatrix}$

15. $\begin{bmatrix} 9 & 5 \\ 4 & 7 \end{bmatrix}$ **17.** $\begin{bmatrix} 0 & -2 & -5 \\ 4 & 2 & 0 \\ 2 & 3 & 7 \end{bmatrix}$ **19.** $\begin{bmatrix} 2 & 3 & 2 \\ 3 & 4 & 1 \\ 2 & 1 & 6 \end{bmatrix}$

21. impossible   **23.** $\begin{bmatrix} 3 & 3 & 9 & 0 \\ 12 & 6 & 3 & 3 \\ 9 & 6 & 0 & 3 \end{bmatrix}$

25. $\begin{bmatrix} 28 & 16 \\ 10 & 18 \end{bmatrix}$ **27.** impossible

29. $x = 3, y = 2, z = 3, w = 4$

31. $x = 4, y = 1, z = 3, w = 3$

33. $x = 2, y = 2, z = -3$

35. (a) $A = \begin{bmatrix} 70 & 78 & 14 & 15 & 82 \\ 14 & 13 & 20 & 9 & 65 \end{bmatrix}$

$B = \begin{bmatrix} 256 & 208 & 69 & 8 & 11 \\ 16 & 15 & 17 & 1 & 1 \end{bmatrix}$

(b) $A + B = \begin{bmatrix} 326 & 286 & 83 & 23 & 93 \\ 30 & 28 & 37 & 10 & 66 \end{bmatrix}$

(c) $\begin{bmatrix} 186 & 130 & 55 & -7 & -71 \\ 2 & 2 & -3 & -8 & -64 \end{bmatrix}$
more species in the United States

37. (a) $\begin{bmatrix} 11{,}041.7 & 8978.4 & 6461 \\ 8739.8 & 9159.6 & 6877.3 \\ 9798.1 & 9086.7 & 6448.4 \\ 9696.6 & 8926.7 & 6109.5 \end{bmatrix}$

(b) air pollution

39. (a) $\begin{bmatrix} 825 & 580 & 1560 \\ 810 & 650 & 350 \end{bmatrix}$ (b) $\begin{bmatrix} -75 & 20 & -140 \\ 10 & -50 & 50 \end{bmatrix}$

41. (a) $A = \begin{bmatrix} 74.7 & 79.9 \\ 75.7 & 80.6 \\ 76.5 & 81.3 \\ 77.1 & 81.8 \\ 77.7 & 82.4 \end{bmatrix}$ $B = \begin{bmatrix} 67.8 & 74.7 \\ 69.7 & 76.5 \\ 70.2 & 77.2 \\ 71.4 & 78.2 \\ 72.6 & 79.2 \end{bmatrix}$

(b) $A - B = \begin{bmatrix} 6.9 & 5.2 \\ 6 & 4.1 \\ 6.3 & 4.1 \\ 5.7 & 3.6 \\ 5.1 & 3.2 \end{bmatrix}$ (c) 2000

43. All answers are in quadrillion BTUs.

(a) $E - M = \begin{bmatrix} 0.41 & -0.18 & -0.33 \\ -1.49 & 0.09 & 1.56 \\ -13.29 & -12.05 & -12.56 \end{bmatrix}$
This represents the U.S. balance of trade for various energy types for selected years.

(b) $\frac{1}{12}C \approx \begin{bmatrix} 3.09 & 3.13 & 3.07 \\ 2.16 & 2.23 & 2.27 \\ 2.90 & 3.06 & 3.18 \end{bmatrix}$
This represents the average monthly U.S. consumption of various energy types for selected years.

(c) $C + E - M = \begin{bmatrix} 37.45 & 37.36 & 36.54 \\ 24.37 & 26.86 & 28.84 \\ 21.53 & 24.68 & 25.63 \end{bmatrix}$
This represents the total U.S production of various energy types for selected years.

45. (a) $\begin{bmatrix} 46.20 & 84.00 & 210.00 & 10.50 \\ 42.00 & 84.00 & 42.00 & 0.00 \\ 58.80 & 147.00 & 94.50 & 0.00 \\ 31.50 & 147.00 & 42.00 & 21.00 \\ 42.00 & 0.00 & 210.00 & 10.50 \end{bmatrix}$

(b) $\begin{bmatrix} 48.40 & 88.00 & 220.00 & 11.00 \\ 44.00 & 88.00 & 44.00 & 0.00 \\ 61.60 & 154.00 & 99.00 & 0.00 \\ 33.00 & 154.00 & 44.00 & 22.00 \\ 44.00 & 0.00 & 220.00 & 11.00 \end{bmatrix}$

47. (a) $A = \begin{bmatrix} 0 & 1 & 1 & 0 \\ 1 & 0 & 1 & 1 \\ 0 & 0 & 0 & 1 \\ 0 & 1 & 1 & 0 \end{bmatrix}$

(b) $B = \begin{bmatrix} 0 & 1 & 1 & 1 \\ 1 & 0 & 1 & 1 \\ 0 & 1 & 0 & 1 \\ 1 & 1 & 1 & 0 \end{bmatrix}$ (c) person 2

49. (a) $\begin{bmatrix} 80 & 75 \\ 58 & 106 \end{bmatrix}$ (b) $\begin{bmatrix} 176 & 127 \\ 139 & 143 \end{bmatrix}$

(c) $\begin{bmatrix} 10 & 4 \\ 7 & 2 \end{bmatrix}$ (d) $\begin{bmatrix} -10 & 19 \\ -7 & 20 \end{bmatrix}$ shortage, taken from inventory.

51. (a) 3, 4, 5, 6   (b) 1

53. Worker 1:   0.9625   Worker 2:   0.9375
Worker 3:   0.9125   Worker 4:   0.8875
Worker 5:   0.85   Worker 6:   0.875
Worker 7:   0.90   Worker 8:   0.925
Worker 9:   0.95
Worker 5 is least efficient; performs best at center 5

3.2 EXERCISES

1. (a) [32]   (b) [11   17]   3. $\begin{bmatrix} 29 & 25 \\ 10 & 12 \end{bmatrix}$

5. $\begin{bmatrix} 14 & 2 & 16 \\ 28 & 5 & 12 \end{bmatrix}$

7. $\begin{bmatrix} 7 & 5 & 3 & 2 \\ 14 & 9 & 11 & 3 \\ 13 & 10 & 12 & 3 \end{bmatrix}$ 9. impossible

11. $\begin{bmatrix} 13 & 9 & 3 & 4 \\ 9 & 7 & 16 & 1 \end{bmatrix}$ 13. $\begin{bmatrix} 9 & 7 & 16 \\ 5 & 17 & 20 \end{bmatrix}$

15. $\begin{bmatrix} 9 & 0 & 8 \\ 13 & 4 & 11 \\ 16 & 0 & 17 \end{bmatrix}$ 17. $\begin{bmatrix} 161 & 126 \\ 42 & 35 \end{bmatrix}$ 19. no

21. no   23. $\begin{bmatrix} -55 & 88 & 0 \\ -42 & 67 & 0 \\ 28 & -44 & 1 \end{bmatrix}$ 25. $\begin{bmatrix} 0 & 0 & 0 \\ 0 & 0 & 0 \\ 0 & 0 & 0 \end{bmatrix}$

27. $\begin{bmatrix} 0 & 0 & 0 \\ 0 & 0 & 0 \\ 0 & 0 & 0 \end{bmatrix}$ 29. A 31. Z

33. no (see Problem 25)

35. (a) $AB = \begin{bmatrix} 1 & 0 \\ 0 & 1 \end{bmatrix}$, $BA = \begin{bmatrix} 1 & 0 \\ 0 & 1 \end{bmatrix}$

(b) $ad - bc \neq 0$

37. $\begin{bmatrix} 2-2+2 \\ 6-4-4 \\ 4+0-2 \end{bmatrix} = \begin{bmatrix} 2 \\ -2 \\ 2 \end{bmatrix}$; solution

39. $\begin{bmatrix} 1+2+2 \\ 4+0+1 \\ 2+2+1 \end{bmatrix} = \begin{bmatrix} 5 \\ 5 \\ 5 \end{bmatrix}$; solution

41. $\begin{bmatrix} 1 & 0 & 0 & 0 & 0 \\ 0 & 1 & 0 & 0 & 0 \\ 0 & 0 & 1 & 0 & 0 \\ 0 & 0 & 0 & 1 & 0 \\ 0 & 0 & 0 & 0 & 1 \end{bmatrix}$ (Some entries may appear as decimal approximations of 0.)

43. $\begin{bmatrix} 31,680 & 36,960 \\ 42,500 & 47,600 \end{bmatrix}$
The entries represent the dealer's cost for each car.

45. (a) $A = \begin{bmatrix} 86.86 & 91.25 & 95.37 \\ 27.57 & 29.82 & 32.05 \\ 6.36 & 6.45 & 6.60 \end{bmatrix}$ and

$B = \begin{bmatrix} 91.1 & 0 & 0 \\ 0 & 86.0 & 0 \\ 0 & 0 & 82.5 \end{bmatrix}$

(b) $AB \approx \begin{bmatrix} 7913 & 7848 & 7868 \\ 2512 & 2565 & 2644 \\ 579.4 & 554.7 & 544.5 \end{bmatrix}$

(c) The 1–1 entry is the trillions of BTUs used in single-family households in 2015. The 2–3 entry is the trillions of BTUs predicted to be used in multi-family households in 2025.

(d) Using AB as stored in a calculator,
[1   1   1] $AB \approx$ [11,004.0   10,966.7   11,056.7]
Left to right, each entry gives the trillions of BTUs of delivered energy consumed by all households in 2015, 2020, and 2025.

47. (a) [0.55   0.45] After 5 years, M has 55% and S has 45% of the population.
(b) 10 years: $(PD)D = PD^2$; 15 years: PD^3
(c) 60% in M and 40% in S. Population proportions are stable.

49. 124, 149, 275, 334, 268, 327, 205, 249, 343, 417, 124, 149, 250, 304, 261, 316, 255, 310

51. (a)

$B = \begin{bmatrix} \frac{3}{4} & \frac{2}{5} & \frac{1}{4} \\ \frac{1}{4} & \frac{3}{5} & \frac{3}{4} \end{bmatrix}$; $D = \begin{bmatrix} 22 & 30 \\ 12 & 20 \\ 8 & 11 \end{bmatrix}$; $BD = \begin{bmatrix} 23.3 & 33.25 \\ 18.7 & 27.75 \end{bmatrix}$

Houston's need for black crude is 23,300 gal and for gold crude is 18,700 gal. Gulfport needs 33,250 gal of black and 27,750 gal of gold.

(b) $PBD = $ [135.945   197.5325]; Houston's cost = $135,945; Gulfport's cost = $197,532.50

53. (a) $B = \begin{bmatrix} 0.7 & 8.5 & 10.2 & 1.1 & 5.6 & 3.6 \\ 0.5 & 0.2 & 6.1 & 1.3 & 0.2 & 1.0 \\ 2.2 & 0.4 & 8.8 & 1.2 & 1.2 & 4.8 \\ 251.8 & 63.4 & 81.6 & 35.2 & 54.3 & 144.2 \\ 30.0 & 1.0 & 1.0 & 1.0 & 1.0 & 1.0 \\ 788.9 & 0 & 0 & 0 & 0 & 0 \end{bmatrix}$

(b) $A = \begin{bmatrix} 1.11 & 0 & 0 & 0 & 0 & 0 \\ 0 & 0.95 & 0 & 0 & 0 & 0 \\ 0 & 0 & 1.11 & 0 & 0 & 0 \\ 0 & 0 & 0 & 1.11 & 0 & 0 \\ 0 & 0 & 0 & 0 & 0.95 & 0 \\ 0 & 0 & 0 & 0 & 0 & 0.95 \end{bmatrix}$

3.3 EXERCISES

1. $\begin{bmatrix} 1 & -2 & -1 & -7 \\ 0 & 7 & 5 & 21 \\ 4 & 2 & 2 & 1 \end{bmatrix}$

3. $\begin{bmatrix} 1 & -3 & 4 & 2 \\ 2 & 0 & 2 & 1 \\ 1 & 2 & 1 & 1 \end{bmatrix}$ **5.** $x = 2, y = 1/2, z = -5$

7. $x = -5, y = 2, z = 1$

9. $x = 4, y = 1, z = -2$ **11.** $x = 15, y = -13, z = 2$

13. $x = 15, y = 0, z = 2$

15. $x = 1, y = 3, z = 1, w = 0$

17. no solution

19. (a) $x = (11 + 2z)/3, y = (-1 - z)/3$,
       $z =$ any real number
   (b) many possibilities, including $x = 11/3, y = -\frac{1}{3}$,
       $z = 0$ and $x = 13/3, y = -2/3, z = 1$

21. If a row of the matrix has all 0's in the coefficient matrix and a nonzero number in the augment, there is no solution.

23. $x = 0, y = -z, z =$ any real number

25. no solution

27. $x = 1 - z, y = \frac{1}{2}z, z =$ any real number

29. $x = 1, y = -1, z = 1$

31. $x = 2z - 2, y = 1 + z, z =$ any real number

33. $x = \frac{7}{2} - z, y = -\frac{1}{2}, z =$ any real number

35. $x = \frac{26}{5} - \frac{7}{5}z, y = \frac{4}{5} + \frac{2}{5}z, z =$ any real number

37. $x_1 = 20, x_2 = 60, x_3 = 40$

39. $x_1 = 1, x_2 = 0, x_3 = 1, x_4 = 0$

41. $x = 7/5, y = -3/5, z = w, w =$ any real number

43. no solution

45. $x_1 = 1 - 2x_4 - 3x_5, x_2 = 4 + 5x_4 + 7x_5$,
       $x_3 = -3 - 3x_4 - 5x_5, x_4 =$ any real number,
       $x_5 =$ any real number

47. $x = (b_2c_1 - b_1c_2)/(a_1b_2 - a_2b_1)$

49. beef: 2 cups; sirloin: 8 cups

51. (a) \$50,000 at 12%, \$85,000 at 10%, \$100,000 at 8%
   (b) \$6000 at 12%, \$8500 at 10%, \$8000 at 8%

53. AP = 1100, DT = 440, CA = 660

55. AF: 2 oz, FP: 2 oz, NMG: 1 oz

57. 2 of portfolio I, 2 of portfolio II

59. $\frac{3}{8}$ pound of red meat, 6 slices of bread, 4 glasses of milk

61. type I = 3(type IV), type II = 1000 − 2(type IV),
   type III = 500 − type IV, type IV = any integer
   satisfying $0 \le$ type IV ≤ 500

63. bacteria III = any amount satisfying
   $1800 \le$ bacteria III ≤ 2300
   bacteria I = 6900 − 3 (bacteria III)
   bacteria II = $\frac{1}{2}$(bacteria III) − 900

65. (a) $C = 2800 + 0.6R$
       $U = 7000 - R$
       $R =$ any integer satisfying $0 \le R \le 7000$
   (b) $R = 1000$: $C = 3400$
           $U = 6000$
           $R = 2000$: $C = 4000$
           $U = 5000$
   (c) Min $C = 2800$ when $R = 0$ and $U = 7000$
   (d) Max $C = 7000$ when $R = 7000$ and $U = 0$

67. There are three possibilities:
   (1) 4 of I and 2 of II
   (2) 5 of I, 1 of II, and 1 of III
   (3) 6 of I and 2 of III

3.4 EXERCISES

1. $\begin{bmatrix} 1 & 0 & 0 \\ 0 & 1 & 0 \\ 0 & 0 & 1 \end{bmatrix}$ **3.** Yes   **5.** $\begin{bmatrix} 2 & -7 \\ -1 & 4 \end{bmatrix}$

7. no inverse   **9.** $\begin{bmatrix} -\frac{1}{10} & \frac{7}{10} \\ \frac{1}{5} & -\frac{2}{5} \end{bmatrix}$

11. $\begin{bmatrix} \frac{1}{3} & 0 & 0 \\ 0 & \frac{1}{3} & -\frac{2}{3} \\ 0 & 0 & 1 \end{bmatrix}$ **13.** $\begin{bmatrix} -1 & 1 & 0 \\ 1 & 0 & 0 \\ -1 & 0 & 1 \end{bmatrix}$

15. $\begin{bmatrix} \frac{1}{3} & -\frac{1}{3} & \frac{1}{3} \\ -\frac{2}{3} & -\frac{1}{3} & \frac{7}{3} \\ \frac{1}{3} & \frac{2}{3} & -\frac{5}{3} \end{bmatrix}$

17. no inverse   **19.** no inverse

21. $\begin{bmatrix} 1 & 1 & 0 & 0 & -1 \\ -3 & 0 & -3 & 1 & 4 \\ 1 & -2 & 0 & 0 & 1 \\ -3 & 1 & -3 & 1 & 3 \\ -8 & 5 & -8 & 2 & 7 \end{bmatrix}$ **23.** $\begin{bmatrix} 13 \\ 5 \end{bmatrix}$ **25.** $\begin{bmatrix} 9 \\ 6 \\ 3 \end{bmatrix}$

27. $\begin{bmatrix} x \\ y \\ z \end{bmatrix} = \begin{bmatrix} 1 \\ 1 \\ 2 \end{bmatrix}$ **29.** $x = 2, y = 1$

31. $x = 1, y = 2$ **33.** $x = 1, y = 1, z = 1$

35. $x = 1, y = 3, z = 2$

37. $x_1 = 5.6, x_2 = 5.4, x_3 = 3.25, x_4 = 6.1, x_5 = 0.4$

39. (a) -2 **(b)** inverse exists

41. (a) 0 **(b)** no inverse

43. (a) -5 **(b)** inverse exists

45. (a) -19 **(b)** inverse exists

47. Hang on   **49.** Answers in back

51. $x_0 = 2400, y_0 = 1200$

53. (a) $A = 5.5$ mg and $B = 8.8$ mg for patient I
(b) $A = 10$ mg and $B = 16$ mg for patient II
55. \$68,000 at 18%, \$77,600 at 10%
57. (a) 2 Deluxe, 8 Premium, 32 Ultimate
(b) 22 Deluxe, 8 Premium, 22 Ultimate
[New] = [Old] + 8[Col. 1 of A^{-1}]
59. \$200,000 at 6%, \$300,000 at 8%, \$500,000 at 10%

61. (a) $\begin{bmatrix} 0 & 1 & 0 \\ 0 & 0 & 1 \\ 1 & 1 & 1 \end{bmatrix}$ (b) 108

63. (a) $M = \begin{bmatrix} 0 & 1 & 0 \\ 0 & 0 & 1 \\ 1 & 1 & 1 \end{bmatrix}$ (b) 30

3.5 EXERCISES

1. (a) 15   (b) 4   **3.** 8   **5.** 40
7. most: raw materials; least: fuels
9. raw materials, manufacturing, service
11. farm products = 200; machinery = 40
13. utilities = 200; manufacturing = 400
15. (a) agricultural products = 244; oil products = 732
(b) agricultural products = 0.4; oil products = 1.2
17. (a) mining = 106; manufacturing = 488
(b) mining = 1.4; manufacturing = 1.2
19. (a)

$\begin{matrix} & EC & C \\ A = & \begin{bmatrix} 0.3 & 0.6 \\ 0.2 & 0.2 \end{bmatrix} & \begin{matrix} \text{Electronic components} \\ \text{Computers} \end{matrix} \end{matrix}$

(b) electronic components = 1200; computers = 320
21. (a)

$\begin{matrix} & F & O \\ A = & \begin{bmatrix} 0.30 & 0.04 \\ 0.35 & 0.10 \end{bmatrix} & \begin{matrix} \text{Fishing} \\ \text{Oil} \end{matrix} \end{matrix}$

(b) fishing = 100; oil = 1250
23. development = \$21,000; promotional = \$12,000
25. engineering = \$15,000; computer = \$13,000
27. fishing = 400; agriculture = 500; mining = 400
29. electronics = 1240; steel = 1260; autos = 720
31. service = 90; manufacturing = 200; agriculture = 100
33. products = $\frac{7}{17}$ households;
machinery = $\frac{1}{17}$ households
35. government = $\frac{10}{19}$ households;
industry = $\frac{11}{19}$ households
37. (a)

$\begin{matrix} & M & U & H \\ A = & \begin{bmatrix} 0.5 & 0.4 & 0.3 \\ 0.4 & 0.5 & 0.3 \\ 0.1 & 0.1 & 0.4 \end{bmatrix} & & \begin{matrix} \text{Manufacturing} \\ \text{Utilities} \\ \text{Households} \end{matrix} \end{matrix}$

(b) manufacturing = 3 households;
utilities = 3 households

39. $\begin{bmatrix} 24 \\ 96 \\ 24 \\ 120 \\ 492 \\ 3456 \end{bmatrix}$ 3456 bolts, 492 braces, 120 panels

41. $\begin{bmatrix} 10 \\ 10 \\ 20 \\ 56 \\ 20 \\ 26 \\ 300 \end{bmatrix}$ 56 2 × 4s, 20 braces, 26 clamps, 300 nails

CHAPTER 3 REVIEW EXERCISES

1. 4   **2.** 0   **3.** A, B **4.** none   **5.** D, F, G, I

6. $\begin{bmatrix} -2 & 5 & 11 & -8 \\ -4 & 0 & 0 & -4 \\ 2 & 2 & -1 & -9 \end{bmatrix}$

7. zero matrix   **8.** order

9. $\begin{bmatrix} 6 & -1 & -9 & 3 \\ 10 & 3 & -1 & 4 \\ -2 & -2 & -2 & 14 \end{bmatrix}$ **10.** $\begin{bmatrix} 3 & -3 \\ 4 & -1 \\ 2 & -6 \\ 1 & -2 \end{bmatrix}$

11. $\begin{bmatrix} 2 & 1 \\ 5 & 1 \end{bmatrix}$ **12.** $\begin{bmatrix} 12 & -6 \\ 15 & 0 \\ 18 & 0 \\ 3 & 9 \end{bmatrix}$ **13.** $\begin{bmatrix} 4 & 0 \\ 0 & 4 \end{bmatrix}$

14. $\begin{bmatrix} 2 & -12 \\ -8 & -22 \end{bmatrix}$ **15.** $\begin{bmatrix} 9 & 20 \\ 4 & 5 \end{bmatrix}$ **16.** $\begin{bmatrix} 5 & 16 \\ 6 & 15 \end{bmatrix}$

17. $\begin{bmatrix} 2 & 37 & 61 & -55 \\ -2 & 9 & -3 & -20 \\ 10 & 10 & -14 & -30 \end{bmatrix}$ **18.** $\begin{bmatrix} 43 & -23 \\ 33 & -12 \\ -13 & 15 \end{bmatrix}$

19. $\begin{bmatrix} 10 & 16 \\ 15 & 25 \\ 18 & 30 \\ 6 & 11 \end{bmatrix}$ **20.** $\begin{bmatrix} 17 & 73 \\ 7 & 28 \end{bmatrix}$ **21.** $\begin{bmatrix} 3 & 7 \\ 23 & 42 \end{bmatrix}$

22. F   **23.** F   **24.** $\begin{bmatrix} -19 & 12 \\ -8 & 5 \end{bmatrix}$ **25.** F

26. (a) infinitely many solutions (bottom row of 0s)
(b) $x = 6 + 2z$
$y = 7 - 3z$
z = any real number
Two specific solutions:
If $z = 0$, then $x = 6$, $y = 7$.
If $z = 1$, then $x = 8$, $y = 4$.
27. (a) no solution (last row says $0 = 1$)
(b) no solution
28. (a) Unique (coefficient matrix is I_3)
(b) $x = 0$, $y = -10$, $z = 14$
29. $(1, 2, 1)$ **30.** $x = 22$, $y = 9$
31. $x = -3$, $y = 3$, $z = 4$
32. $x = -\frac{3}{2}$, $y = 7$, $z = -\frac{11}{2}$ **33.** no solution
34. $x = 2 - 2z$, $y = -1 - 2z$, z = any real number
35. $x = -2 + 8z$
$y = -2 + 3z$
z = any real number

36. $x_1 = 1, x_2 = 11, x_3 = -4, x_4 = -5$ **37.** yes

38. $\begin{bmatrix} \frac{1}{2} & \frac{1}{4} \\ \frac{5}{2} & \frac{7}{4} \end{bmatrix}$ **39.** $\begin{bmatrix} -1 & -2 & 8 \\ 1 & 2 & -7 \\ 1 & 1 & -4 \end{bmatrix}$

40. $\begin{bmatrix} 2 & 1 & -2 \\ 7 & 5 & -8 \\ -13 & -9 & 15 \end{bmatrix}$

41. $x = -33, y = 30, z = 19$

42. $x = 4, y = 5, z = -13$

43. $A^{-1} = \begin{bmatrix} -41 & 32 & 5 \\ 17 & -13 & -2 \\ -9 & 7 & 1 \end{bmatrix}; x = 4, y = -2, z = 2$

44. no   **45.** (a) 16   (b) yes, det $\neq 0$

46. (a) 0   (b) no, det $= 0$

47. $\begin{bmatrix} 250 & 140 \\ 480 & 700 \end{bmatrix}$ **48.** $\begin{bmatrix} 1030 & 800 \\ 700 & 1200 \end{bmatrix}$

49. (a) higher in June
    (b) higher in July

50. $\begin{array}{cc} \text{Men} & \text{Women} \end{array}$
$\begin{bmatrix} 865 & 885 \\ 210 & 270 \end{bmatrix} \begin{array}{l} \text{Robes} \\ \text{Hoods} \end{array}$

51. $\begin{bmatrix} 1750 \\ 480 \end{bmatrix} \begin{array}{l} \text{Robes} \\ \text{Hoods} \end{array}$

52. (a) $\begin{bmatrix} 13{,}500 & 12{,}400 \\ 10{,}500 & 10{,}600 \end{bmatrix}$

    (b) Department A should buy from Kink;
    Department B should buy from Ace.

53. (a) $[0.20 \quad 0.30 \quad 0.50]$

    (b) $\begin{bmatrix} 0.013469 \\ 0.013543 \\ 0.006504 \end{bmatrix}$

    (c) $[0.20\ 0.30\ 0.50]\begin{bmatrix} 0.013469 \\ 0.013543 \\ 0.006504 \end{bmatrix} = 0.20(0.013469) +$

    $0.30(0.013543) + 0.50(0.006504) = 0.0100087$

    (d) The historical return of the portfolio, 0.0100087, is the estimated expected monthly return of the portfolio. This is roughly 1% per month.

54. 400 fast food, 700 software, 200 pharmaceutical

55. (a) $A = 2C, B = 2000 - 4C,$
    $C = $ any integer satisfying $0 \leq C \leq 500$
    (b) yes; $A = 500, B = 1000, C = 250$
    (c) max $A = 1000$ when $B = 0, C = 500$

56. (a) 3 passenger, 4 transport, 4 jumbo
    (b) 1 passenger, 3 transport, 7 jumbo
    (c) column 2

57. (a) shipping $= 5680$; agriculture $= 1960$
    (b) shipping $= 0.4$; agriculture $= 1.8$

58. (a) $\begin{array}{cc} \quad S & C \end{array}$
    $A = \begin{bmatrix} 0.1 & 0.1 \\ 0.2 & 0.05 \end{bmatrix} \begin{array}{l} \text{Shoes} \\ \text{Cattle} \end{array}$

    (b) shoes $= 1000$; cattle $= 500$

59. mining $= 360$; manufacturing $= 320$; fuels $= 400$

60. government $= \frac{64}{93}$ households;
agriculture $= \frac{59}{93}$ households;
manufacturing $= \frac{40}{93}$ households

CHAPTER 3 TEST

1. $\begin{bmatrix} 3 & 1 & 5 \\ 1 & 3 & 6 \end{bmatrix}$ **2.** $\begin{bmatrix} -1 & 2 & 2 \\ 1 & -1 & 6 \end{bmatrix}$

3. $\begin{bmatrix} -12 & -16 & -155 \\ 5 & 12 & 87 \end{bmatrix}$ **4.** $\begin{bmatrix} 23 & 6 \\ 182 & 45 \\ 21 & 1 \end{bmatrix}$

5. $\begin{bmatrix} 0 & -7 \\ 26 & 1 \end{bmatrix}$ **6.** $\begin{bmatrix} -43 & -46 & -207 \\ 39 & 30 & -77 \\ 17 & 5 & -216 \end{bmatrix}$

7. $\begin{bmatrix} -2 & 3/2 \\ 1 & -1/2 \end{bmatrix}$ **8.** $\begin{bmatrix} -3 & 2 & 2 \\ 1 & 0 & -1 \\ 1/2 & -1/2 & 0 \end{bmatrix}$

9. $\begin{bmatrix} 5 \\ 14 \\ 15 \end{bmatrix}$ **10.** $x = -0.5, y = 0.5, z = 2.5$

11. $x = 4 - 1.8z, y = 0.2z, z = $ any real number

12. no solution

13. $x = 2, y = 2, z = 0, w = -2$

14. $x = 6w - 0.5, y = 0.5 - w, z = 2.5 - 3w, w = $ any real number

15. (a) $B = \$45{,}000, E = \$40{,}000$
    (b) $\$0 \leq H \leq \$25{,}000$ (so $B \geq 0$)
    (c) min $E = \$20{,}000$ when $H = \$0$ and $B = \$75{,}000$

16. (a) $\begin{bmatrix} 0.08 & 0.22 & 0.12 \\ 0.10 & 0.08 & 0.19 \\ 0.05 & 0.07 & 0.09 \\ 0.10 & 0.26 & 0.15 \\ 0.12 & 0.04 & 0.24 \end{bmatrix}$

    (b) 0.08, 0.22, 0.12 consumed by carnivores 1, 2, 3
    (c) plant 5 by 1, plant 4 by 2, plant 5 by 3

17. (a) $[1000 \quad 4000 \quad 2000 \quad 1000]$
    (b) $[45{,}000 \quad 55{,}000 \quad 90{,}000 \quad 70{,}000]$

    (c) $\begin{bmatrix} 5 \\ 3 \\ 4 \\ 4 \end{bmatrix}$ (d) $[\$1{,}030{,}000]$ (e) $\begin{array}{c} \\ A \\ B \\ C \\ D \end{array}\begin{bmatrix} \$ \\ 65 \\ 145 \\ 125 \\ 135 \end{bmatrix}$

18. (a) 121, 46, 247, 95, 261, 99, 287, 111, 179, 69, 169, 64
    (b) Frodo lives

19. growth, 2000; blue-chip, 400; utility, 400

20. (a) agriculture $= 245$; minerals $= 235$
    (b) agriculture $= 7$; minerals $= 1$
    (c) agriculture $= 0.5$; minerals $= 1.5$

21. profit $= $ households
nonprofit $= \frac{2}{3}$ households

$$\mathbf{22.}\quad\begin{matrix} & \text{Ag} & \text{M} & \text{F} & \text{S} & \\ \begin{bmatrix} 0.2 & 0.1 & 0.1 & 0.1 \\ 0.3 & 0.2 & 0.2 & 0.2 \\ 0.2 & 0.2 & 0.3 & 0.3 \\ 0.1 & 0.4 & 0.2 & 0.2 \end{bmatrix} & & & & & \begin{matrix} \text{Agriculture} \\ \text{Machinery} \\ \text{Fuel} \\ \text{Steel} \end{matrix} \end{matrix}$$

23. agriculture: 5000; machinery: 8000; fuel: 8000; steel: 7000

24. agriculture: $\frac{520}{699}$ households; steel: $\frac{236}{233}$ households; fuel: $\frac{159}{233}$ households

4.1 EXERCISES

1.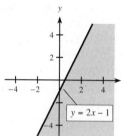

3.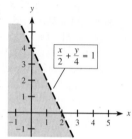

5.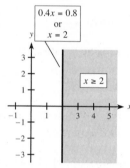

7. $(0, 0), (20, 10), (0, 15), (25, 0)$

9. $(5, 0), (15, 0), (6, 9), (2, 6)$

11. $(0, 5), (1, 2), (3, 1), (6, 0)$

13.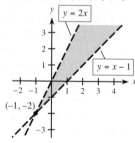

15.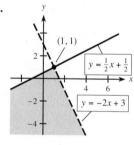

17.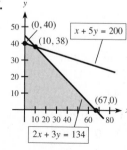

19.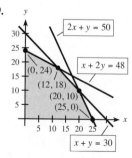

21.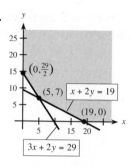

23.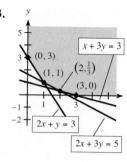

25.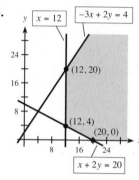

27. (a) Let $x =$ the number of deluxe models and $y =$ the number of economy models.
$$3x + 2y \le 24$$
$$\tfrac{1}{2}x + y \le 8$$
$$x \ge 0, y \ge 0$$

(b)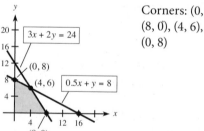

Corners: $(0, 0)$, $(8, 0), (4, 6)$, $(0, 8)$

29. (a) Let $x =$ the number of cord-type trimmers and $y =$ the number of cordless trimmers. Constraints are
$$x + y \le 300$$
$$2x + 4y \le 800$$
$$x \ge 0, y \ge 0$$

(b)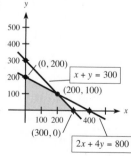

31. (a) Let x = the number of minutes on finance programs and y = the number of minutes on sports programs.
$$7x + 2y \geq 30$$
$$4x + 12y \geq 28$$
$$x \geq 0, y \geq 0$$

(b)
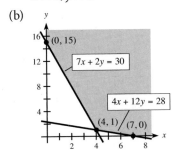

33. (a) Let x = the number of minutes of radio and y = the number of minutes of television. Constraints are
$$x + y \geq 80$$
$$0.006x + 0.09y \geq 2.16$$
$$x \geq 0, y \geq 0$$

(b)
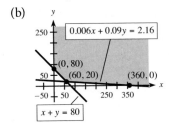

35. (a) Let x = the number of pounds of regular hot dogs and y = the number of pounds of all-beef hot dogs.
$$0.18x + 0.75y \leq 1020$$
$$0.2x + 0.2y \geq 500$$
$$0.3x \leq 600$$

(b)
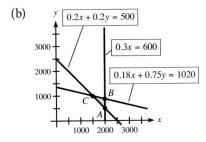

Solution region is triangle ABC, with
$A = (2000, 500)$
$B = (2000, 880)$
$C = (1500, 1000)$

4.2 EXERCISES

1. max $= 76$ at $(4, 4)$; min $= 0$ at $(0, 0)$
3. no max; min $= 11$ at $(1, 3)$
5. $(0, 0), (0, 20), (10, 18), (15, 10), (20, 0)$;
  max $= 66$ at $(10, 18)$; min $= 0$ at $(0, 0)$
7. $(0, 60), (10, 30), (20, 20), (70, 0)$; min $= 100$ at $(20, 20)$;
  no max

9. max $= 1260$ at $x = 12, y = 18$
11. min $= 66$ at $x = 0, y = 3$
13. max $= 22$ at $(2, 4)$
15. max $= 30$ on line between $(0, 5)$ and $(3, 4)$
17. min $= 32$ at $(2, 3)$
19. min $= 9$ at $(2, 3)$
21. max $= 10$ at $(2, 4)$
23. min $= 3100$ at $(40, 60)$
25. If x = the number of deluxe models and y = the number of economy models, then max $= \$792$ at $(4, 6)$.
27. If x = the number of cord-type trimmers and y = the number of cordless trimmers, then max $= \$9000$ at any point with integer coordinates on the segment joining $(0, 200)$ and $(200, 100)$, such as $(20, 190)$.
29. radio $= 60$, $TV = 20$, min $C = \$16,000$
31. inkjet $= 45$, laser $= 25$, max $P = \$3300$
33. 250 fish: 150 bass and 100 trout
35. (a) Max $P = \$1,260,120$ when corn $= 3749.5$ acres and soybeans $= 2251.5$ acres
  (b) Max $P = \$1,260,960$ when corn $= 3746$ acres and soybeans $= 2262$ acres
  (c) $\$120$/acre
37. 30 days for factory 1 and 20 days for factory 2; minimum cost $= \$700,000$
39. 60 days for location I and 70 days for location II; minimum cost $= \$86,000$
41. reg $= 2000$ lb; all-beef $= 880$ lb; maximum profit $= \$3320$
43. From Pittsburgh: 20 to Blairsville, 40 to Youngstown; From Erie: 15 to Blairsville, 0 to Youngstown; minimum cost $= \$1540$
45. (a) $R = \$366,000$ with 6 satellite and 17 full-service branches
  (b) Branches: used 23 of 25 possible; 2 not used (slack)
  New employees: hired 120 of 120 possible; 0 not hired (slack)
  Budget: used all $\$2.98$ million; $\$0$ not used (slack)
  (c) Additional new employees and additional budget. These items are completely used in the current optimal solution; more could change and improve the optimal solution.
  (d) Additional branches. The current optimal solution does not use all those allotted; more would just add to the extras.

4.3 EXERCISES

1. $3x + 5y + s_1 = 15$, $3x + 6y + s_2 = 20$

3.
$$\begin{bmatrix} 2 & 5 & 1 & 0 & 0 & | & 400 \\ 1 & 2 & 0 & 1 & 0 & | & 175 \\ -3 & -7 & 0 & 0 & 1 & | & 0 \end{bmatrix}$$

5.
$$\begin{bmatrix} 2 & 7 & 9 & 1 & 0 & 0 & 0 & | & 100 \\ 6 & 5 & 1 & 0 & 1 & 0 & 0 & | & 145 \\ 1 & 2 & 7 & 0 & 0 & 1 & 0 & | & 90 \\ -2 & -5 & -2 & 0 & 0 & 0 & 1 & | & 0 \end{bmatrix}$$

7. $x = 11, y = 9; f = 20$

9. $x = 0, y = 14, z = 11; f = 525$

11. (a) $x_1 = 0, x_2 = 45, s_1 = 14, s_2 = 0, f = 75$

(b) not complete

(c)
$$\left[\begin{array}{ccccc|c} \boxed{2} & 0 & 1 & -\frac{3}{4} & 0 & 14 \\ 3 & 1 & 0 & \frac{1}{3} & 0 & 45 \\ \hline -6 & 0 & 0 & 3 & 1 & 75 \end{array}\right]$$

$\frac{1}{2}R_1 \rightarrow R_1$, then $-3R_1 + R_2 \rightarrow R_2, 6R_1 + R_3 \rightarrow R_3$

13. (a) $x_1 = 0, x_2 = 0, s_1 = 200, s_2 = 400, s_3 = 350, f = 0$

(b) not complete

(c)
$$\left[\begin{array}{cccccc|c} \boxed{10} & 27 & 1 & 0 & 0 & 0 & 200 \\ 4 & 51 & 0 & 1 & 0 & 0 & 400 \\ 15 & 27 & 0 & 0 & 1 & 0 & 350 \\ \hline -8 & -7 & 0 & 0 & 0 & 1 & 0 \end{array}\right]$$

$\frac{1}{10}R_1 \rightarrow R_1$, then $-4R_1 + R_2 \rightarrow R_2$,
$-15R_1 + R_3 \rightarrow R_3, 8R_1 + R_4 \rightarrow R_4$

15. (a) $x_1 = 24, x_2 = 0, x_3 = 21,$
$s_1 = 16, s_2 = 0, s_3 = 0, f = 780$

(b) complete (no part (c))

17. (a) $x_1 = 0, x_2 = 0, x_3 = 12,$
$s_1 = 4, s_2 = 6, s_3 = 0, f = 150$

(b) not complete

(c)
$$\left[\begin{array}{ccccccc|c} 4 & 4 & 1 & 0 & 0 & 2 & 0 & 12 \\ \boxed{2} & \boxed{4} & 0 & 1 & 0 & 1 & 0 & 4 \\ -3 & -11 & 0 & 0 & 1 & -1 & 0 & 6 \\ \hline -3 & -3 & 0 & 0 & 0 & 4 & 1 & 150 \end{array}\right]$$

Either circled number may act as the next pivot entry, but only one of them. If 4 is used,
$\frac{1}{4}R_2 \rightarrow R_2$, then $-4R_2 + R_1 \rightarrow R_1$,
$11R_2 + R_3 \rightarrow R_3, 3R_2 + R_4 \rightarrow R_4$. If 2 is used,
$\frac{1}{2}R_2 \rightarrow R_2$, then $-4R_2 + R_1 \rightarrow R_1$,
$3R_2 + R_3 \rightarrow R_3, 3R_2 + R_4 \rightarrow R_4$.

19. $x = 0, y = 5; f = 50$

21. $x = 4, y = 3; f = 17$

23. $x = 4, y = 3; f = 11$

25. $x = 0, y = 2, z = 5; f = 40$

27. $x = 4, y = 2, z = 6; f = 16$

29. $x_1 = 36, x_2 = 24; x_3 = 0, x_4 = 8; f = 1728$

31. No solution; a new pivot cannot be found in column 2.

33. $x = 50, y = 10; f = 100$. Multiple solutions are possible; pivot with the 3-5 entry.

35. no solution

37. $x = 0, y = 50$ or $x = 40, y = 40; f = 600$

39. (a)
$$\left[\begin{array}{ccccc|c} 1 & 1 & 1 & 0 & 0 & 60 \\ 1 & 3 & 0 & 1 & 0 & 120 \\ \hline -40 & -60 & 0 & 0 & 1 & 0 \end{array}\right]$$

(b) Maximum profit is \$3000 with 30 inkjet and 30 laser printers.

41. 300 style-891, 450 style-917, maximum $P = \$15,525$

43. premium and light = 175 each; maximum $P = \$35,000$

45. Aries = 3, Belfair = 5, Wexford = 4; maximum $P = \$305,000$

47. (a) 26 one-bedroom; 40 two-bedroom; 48 three-bedroom

(b) \$100,200 per month

49. 21 newspapers, 13 radio; 230,000 exposures

51. \$1650 profit with 46 A, 20 B, 6 C

53. 20-in. LCDs = 40, 42-in. LCDs = 115, 42-in. plasma = 0, 50-in. plasma = 38; max $P = \$12,540$

4.4 EXERCISES

1. (a)
$$\left[\begin{array}{cc|c} 5 & 2 & 16 \\ 1 & 2 & 8 \\ 4 & 5 & g \end{array}\right] \text{ transpose } = \left[\begin{array}{cc|c} 5 & 1 & 4 \\ 2 & 2 & 5 \\ 16 & 8 & g \end{array}\right]$$

(b) maximize $f = 16x_1 + 8x_2$ subject to
$5x_1 + x_2 \leq 4, 2x_1 + 2x_2 \leq 5, x_1 \geq 0, x_2 \geq 0$.

3. (a)
$$\left[\begin{array}{cc|c} 1 & 2 & 30 \\ 1 & 4 & 50 \\ 7 & 3 & g \end{array}\right] \text{ transpose } = \left[\begin{array}{cc|c} 1 & 1 & 7 \\ 2 & 4 & 3 \\ 30 & 50 & g \end{array}\right]$$

(b) maximize $f = 30x_1 + 50x_2$ subject to
$x_1 + x_2 \leq 7, 2x_1 + 4x_2 \leq 3, x_1 \geq 0, x_2 \geq 0$

5. (a) $y_1 = 7, y_2 = 4, y_3 = 0$; min $g = 452$

(b) $x_1 = 15, x_2 = 0, x_3 = 29$; max $f = 452$

7. maximize $f = 11x_1 + 11x_2 + 16x_3$ subject to
$2x_1 + x_2 + x_3 \leq 2$
$x_1 + 3x_2 + 4x_3 \leq 10$
primal: $y_1 = 16, y_2 = 0; g = 32$ (min)
dual: $x_1 = 0, x_2 = 0, x_3 = 2; f = 32$ (max)

9. maximize $f = 11x_1 + 12x_2 + 6x_3$ subject to
$4x_1 + 3x_2 + 3x_3 \leq 3$
$x_1 + 2x_2 + x_3 \leq 1$
primal: $y_1 = 2, y_2 = 3; g = 9$ (min)
dual: $x_1 = 3/5, x_2 = 1/5, x_3 = 0; f = 9$ (max)

11. min = 28 at $x = 2, y = 0, z = 1$

13. $y_1 = 2/5, y_2 = 1/5, y_3 = 1/5; g = 16$ (min)

15. (a) minimize $g = 120y_1 + 50y_2$ subject to
$3y_1 + y_2 \geq 40$
$2y_1 + y_2 \geq 20$

(b) primal: $x_1 = 40, x_2 = 0, f = 1600$ (max)
dual: $y_1 = 40/3, y_2 = 0, g = 1600$ (min)

17. min = 480 at $y_1 = 0, y_2 = 0, y_3 = 16$

19. min = 90 at $y_1 = 0, y_2 = 3, y_3 = 1, y_4 = 0$

21. Atlanta = 150 hr, Fort Worth = 50 hr;
min C = \$210,000

23. line 1 for 4 hours, line 2 for 1 hour; \$1200

25. A = 12 weeks, B = 0 weeks, C = 0 weeks;
cost = \$12,000

27. factory 1: 50 days, factory 2: 0 days; min cost \$500,000

29. 105 minutes on radio, nothing on TV; min cost \$10,500

31. (a) Georgia package = 10
Union package = 20
Pacific package = 5

(b) \$4630

33. (a) min cost = \$16

(b) Many solutions are possible; two are: 16 oz of food I, 0 oz of food II, 0 oz of food III and 11 oz of food I, 1 oz of food II, 0 oz of food III.

35. Mon. = 8, Tues. = 0, Wed. = 5, Thurs. = 4,
Fri. = 5, Sat. = 0, Sun. = 3; min = 25

4.5 EXERCISES

1. $-3x + y \leq -5$

3. $-6x - y \leq -40$

5. (a) maximize $f = 2x + 3y$ subject to
$$7x + 4y \leq 28$$
$$3x - y \leq -2$$
$$x \geq 0, y \geq 0$$

(b) $\begin{bmatrix} 7 & 4 & 1 & 0 & 0 & | & 28 \\ 3 & -1 & 0 & 1 & 0 & | & -2 \\ -2 & -3 & 0 & 0 & 1 & | & 0 \end{bmatrix}$

7. (a) Maximize $-g = -3x - 8y$ subject to
$$4x - 5y \leq 50$$
$$x + y \leq 80$$
$$x - 2y \leq -4$$
$$x \geq 0, y \geq 0$$

(b) $\begin{bmatrix} 4 & -5 & 1 & 0 & 0 & 0 & | & 50 \\ 1 & 1 & 0 & 1 & 0 & 0 & | & 80 \\ 1 & -2 & 0 & 0 & 1 & 0 & | & -4 \\ 3 & 8 & 0 & 0 & 0 & 1 & | & 0 \end{bmatrix}$

9. $x = 6, y = 8, z = 12; f = 120$

11. $x = 10, y = 17; f = 57$

13. $x = 5, y = 7; f = 31$

15. $x = 5, y = 15; f = 45$

17. $x = 10, y = 20; f = 120$

19. $x = 20, y = 10, z = 0; f = 40$

21. $x = 5, y = 0, z = 3; f = 22$

23. $x = 70, y = 0, z = 40; f = 2100$

25. $x_1 = 20, x_2 = 10, x_3 = 20, x_4 = 80,$
$x_5 = 10, x_6 = 10; f = 3250$

27. regular = 2000 lb; beef = 880 lb; profit = $3320

29. Produce 200 of each at Monaca; produce 300
commercial components and 550 domestic furnaces
at Hamburg; profit = $355,250

31. 400 filters, 300 housing units; min cost = $5145

33. Produce 200 of each at Monaca; produce 300
commercial components and 550 domestic furnaces
at Hamburg; cost = $337,750

35. I = 3 million, II = 0, III = 3 million;
cost = $180,000

37. 2000 footballs, 0 soccer balls, 0 volleyballs; $60,000

CHAPTER 4 REVIEW EXERCISES

1.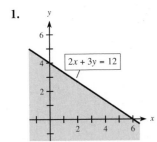

2.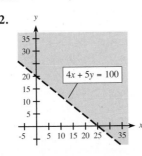

3.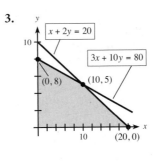

4.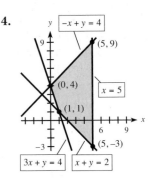

5. max = 25 at (5, 10); min = -12 at (12, 0)

6. max = 194 at (17, 23); min = 104 at (8, 14)

7. max = 140 at (20, 0); min = -52 at (20, 32)

8. min = 115 at (5, 7); no max exists, f can be made arbitrarily large

9. $f = 66$ at (6, 6)

10. $f = 43$ at (7, 9)

11. $f = 300$ at any point on the segment joining (0, 60) and
(50, 40)

12. $g = 24$ at (3, 3)

13. $g = 84$ at (64, 4)

14. $f = 76$ at (12, 8)

15. $f = 75$ at (15, 15)

16. $f = 168$ at (12, 7)

17. $f = 260$ at (60, 20)

18. $f = 360$ at (40, 30)

19. $f = 80$ at (20, 10)

20. $f = 640$ on the line between (160, 0) and (90, 70)

21. no solution

22. $g = 32$ at $y_1 = 2, y_2 = 3$

23. $g = 20$ at $y_1 = 4, y_2 = 2$

24. $g = 7$ at $y_1 = 1, y_2 = 5$

25. $g = 1180$ at $y_1 = 80, y_2 = 20$

26. $f = 165$ at $x = 20, y = 21$

27. $f = 54$ at $x = 6, y = 5$

28. $f = 270$ at (5, 3, 2)

29. $g = 140$ at $y_1 = 0, y_2 = 20, y_3 = 20$

30. $g = 1400$ at $y_1 = 0, y_2 = 100, y_3 = 100$

31. $f = 156$ at $x = 15, y = 2$

32. $f = 31$ at $x = 4, y = 5$

33. $f = 4380$ at (40, 10, 0, 0)

34. $f = 900$ at $x_1 = 25, x_2 = 50, x_3 = 25, x_4 = 0$

35. $g = 2020$ at $x_1 = 0, x_2 = 100, x_3 = 80, x_4 = 20$

36. $P = $29,500$ when 110 large and 75 small swing sets
are made

37. $C = $300,000$ when factory 1 operates 30 days, factory
2 operates 25 days

38. $P = $10,680$ at any pair of integer values on the seg-
ment joining (120, 0) and (90, 20)

39. $P = 960; I = 40, II = 20

40. $P = 1260; Jacob's ladders = 90,
locomotive engines = 30

41. (a) Let x_1 = the number of 27-in LCD sets,
x_2 = the number of 32-in LCD sets,
x_3 = the number of 42-in LCD sets,
x_4 = the number of 42-in plasma sets.

(b) Maximize $P = 80x_1 + 120x_2 + 160x_3 + 200x_4$
subject to
$$8x_1 + 10x_2 + 12x_3 + 15x_4 \le 1870$$
$$2x_1 + 4x_2 + 4x_3 + 4x_4 \le 530$$
$$x_1 + x_2 + x_3 + x_4 \le 200$$
$$x_3 + x_4 \le 100$$
$$x_2 \le 120$$

(c) $x_1 = 15, x_2 = 25, x_3 = 0, x_4 = 100$;
max profit = $24,200

42. food I = 0 oz, food II = 3 oz; C = $0.60 (min)

43. cost = $5.60; A = 40 lb, B = 0 lb

44. cost = $85,000; A = 20 days, B = 15 days, C = 0 days

45. pancake mix = 8000 lb; cake mix = 3000 lb;
profit = $3550

46. Texas: 55 desks, 65 computer tables; Louisiana:
75 desks, 65 computer tables; cost = $12,735

47. Midland: grade 1 = 486.5 tons, grade 2 = 0 tons;
Donora: grade 1 = 13.5 tons, grade 2 = 450 tons;
Cost = $271,905

CHAPTER 4 TEST

1. max = 120 at (0, 24)

2. (a)
$$C; \begin{bmatrix} 1 & 2 & 0 & 1 & 0 & -3/2 & 0 & 40 \\ 0 & \boxed{1} & 0 & -2 & 1 & 1/2 & 0 & 15 \\ 0 & 3 & 1 & -1 & 0 & 1/4 & 0 & 60 \\ \hline 0 & 0 & 0 & 4 & 0 & 6 & 1 & 220 \end{bmatrix}$$
$$-2R_2 + R_1 \to R_1 \quad -3R_2 + R_3 \to R_3$$

(b) A; pivot column is column 3, but new pivot is
undefined.

(c) B; $x_1 = 40, x_2 = 12, x_3 = 0, s_1 = 0$,
$s_2 = 20, s_3 = 0; f = 170$; This solution is not
optimal; the next pivot is the 3-6 entry.

3. (a)
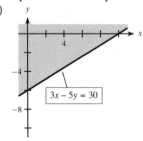

(b)
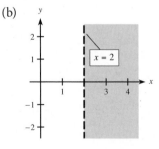

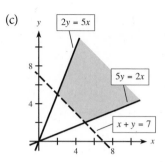

4. maximize $f = 100x_1 + 120x_2$ subject to
$$3x_1 + 4x_2 \le 2$$
$$5x_1 + 6x_2 \le 3$$
$$x_1 + 3x_2 \le 5$$
$$x_1 \ge 0, x_2 \ge 0$$

5. min = 21 at (1, 8); no max exists, f can be made
arbitrarily large

6. min = 136 at (28, 52)

7. maximize $-g = -7x - 3y$ subject to
$$x - 4y \le -4$$
$$x - y \le 5$$
$$2x + 3y \le 30$$

8. max: $x_1 = 17, x_2 = 15, x_3 = 0; f = 658$ (max)
min: $y_1 = 4, y_2 = 18, y_3 = 0; g = 658$ (min)

9. max: = 6300 at $x = 90, y = 0$

10. max $f = 6000$ at any point on the segment joining
(250, 150) and (350, 90)

11. max = 1188 at $x = 0, y = 16, z = 12$

12. If x = the number of barrels of lager and y = the
number of barrels of ale, then maximize $P = 35x + 30y$
subject to
$$3x + 2y \le 1200$$
$$2x + 2y \le 1000$$
P = $16,000 (max) at $x = 200, y = 300$

13. If x = the number of day calls and y = the number of
evening calls, then minimize $C = 3x + 4y$ subject to
$$0.3x + 0.3y \ge 150$$
$$0.1x + 0.3y \ge 120$$
$$x \le 0.5(x + y)$$
C = $1850 (min) at $x = 150, y = 350$

14. max profit = $15,000 when product 1 = 25 tons,
product 2 = 62.5 tons, product 3 = 0 tons,
product 4 = 12.5 tons

5.1 EXERCISES

1. (a) 3.162278   (b) 0.01296525

3. (a) 1.44225   (b) 7.3891

5.
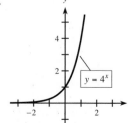

7.
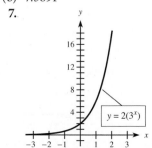

9.

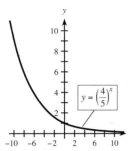

$$y = \left(\frac{4}{5}\right)^x$$

11.

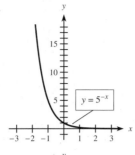

$$y = 5^{-x}$$

13.

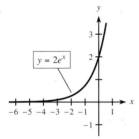

$$y = 2e^x$$

15.

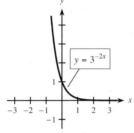

$$y = 3^{-2x}$$

17. (a) $y = 3(2.5)^{-x}$

(b) Decay. They have the form $y = C \cdot b^x$ for $0 < b < 1$ or $y = C \cdot a^{-x}$ for $a > 1$.

(c) The graphs are identical.

19. (a) and (b)

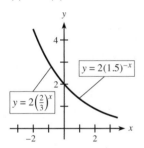

$$y = 2(1.5)^{-x}$$
$$y = 2\left(\frac{2}{3}\right)^x$$

(c) $(1.5)^{-x} = \left(\frac{3}{2}\right)^{-x} = \left(\frac{2}{3}\right)^x$

21. $y = \left(\frac{5}{4}\right)^{-x}$

23. All graphs have the same basic shape. For larger positive values of k, the graphs fall more sharply. For positive values of k nearer 0, the graphs fall more slowly.

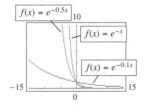

$f(x) = e^{-0.5x}$
$f(x) = e^{-x}$
$f(x) = e^{-0.1x}$

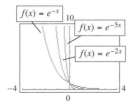

$f(x) = e^{-x}$
$f(x) = e^{-5x}$
$f(x) = e^{-2x}$

25. $y = f(x) + C$ is the same graph as $y = f(x)$ but shifted C units on the y-axis.

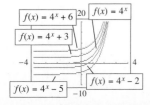

$f(x) = 4^x + 6$
$f(x) = 4^x$
$f(x) = 4^x + 3$
$f(x) = 4^x - 2$
$f(x) = 4^x - 5$

27. (a)

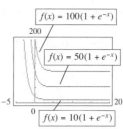

$f(x) = 100(1 + e^{-x})$
$f(x) = 50(1 + e^{-x})$
$f(x) = 10(1 + e^{-x})$

(b) As c changes, the y-intercept and the asymptote change.

29. \$1884.54

31.

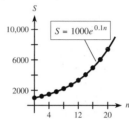

$S = 1000e^{0.1n}$

33. At time 0, the concentration is 0. The concentration rises rapidly for the first 4 minutes and then tends toward 100% as time nears 10 minutes.

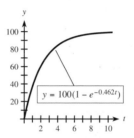

$y = 100(1 - e^{-0.462t})$

35.

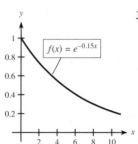

$f(x) = e^{-0.15x}$

As the TV sets age (x increases), the fraction of sets still in service declines.

37.

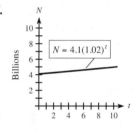

$N = 4.1(1.02)^t$

39.

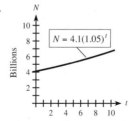

$N = 4.1(1.05)^t$

41. (a) Growth; $e > 1$ and the exponent is positive for $t > 0$.

(b)

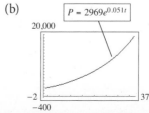

$P = 2969e^{0.051t}$

43. (a)

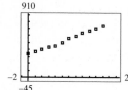

(b) $y = 342.8(1.037^x)$

(c)

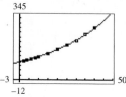

The model is an excellent fit to the data.

45. (a) $y = 492.4(1.070^x)$
(b) \$24,608 billion is an overestimate (c) 2021

47. (a) $y = 92.750(1.0275^x)$
(b)

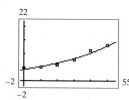

(c) \$198.23
(d) $x \approx 36.6$, during 2047

49. (a) $y = 4.10(1.02^x)$ (b) growth
(c)

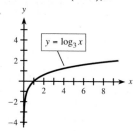

(d) 16.1 million

5.2 EXERCISES

1. $2^4 = 16$ **3.** $4^{1/2} = 2$ **5.** $x = 81$ **7.** $x = \frac{1}{4}$
9. $x = 16$ **11.** $x = 26.75$ **13.** $x \approx 2.013$
15. $\log_2 32 = 5$
17. $\log_4\left(\frac{1}{4}\right) = -1$
19. $3x + 5 = \ln(0.55); x \approx -1.866$
21.

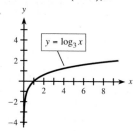

23.

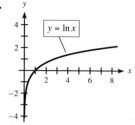

25.

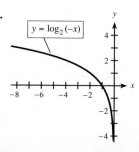

27. (a) 3 (b) -1 **29.** x **31.** -7 **33.** 3
35. (a) 4.9 (b) 0.4 (c) 12.4 (d) 0.9
37. $\log x - \log(x + 1)$ **39.** $\log_7 x + \frac{1}{3}\log_7(x + 4)$
41. $\ln(x/y)$ **43.** $\log_5[x^{1/2}(x + 1)]$
45. equivalent; Properties V and III
47. not equivalent; $\log(\sqrt[3]{8}/5)$
49. (a)

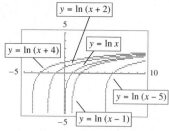

(b) For each c, the domain is $x > c$ and the vertical asymptote is at $x = c$.
(c) Each x-intercept is at $x = c + 1$.
(d) The graph of $y = f(x - c)$ is the graph of $y = f(x)$ shifted c units on the x-axis.

51. (a) 4.0875 (b) -0.1544
53.

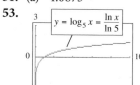

55.

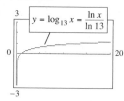

57. If $\log_a M = u$ and $\log_a N = v$, then $a^u = M$ and $a^v = N$. Therefore, $\log_a(M/N) = \log_a(a^u/a^v) = \log_a(a^{u-v}) = u - v = \log_a M - \log_a N$.
59. 63.1 times as severe **61.** 3.2 times as severe
63. 40
65. $L = 10\log(I/I_0)$

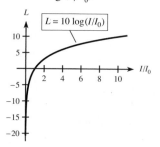

67. 0.1 and 1×10^{-14}
69. $\text{pH} = \log\dfrac{1}{[H^+]} = \log 1 - \log[H^+] = -\log[H^+]$
71. $\log_{1.02} 2 = 4t; t \approx 8.75$ years
73. (a)

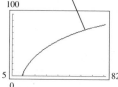

(b) 83.6 million (c) 2023

75. (a) $y = -13.0 + 11.9 \ln x$

(b)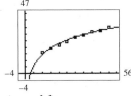

A good fit.

(c) 26.1%

77. $y = 20.0 + 20.8 \ln x$

(b) A good fit.   (c) 91%

5.3 EXERCISES

1. $\frac{5}{3}$ **3.** 2.943   **5.** 9.390   **7.** 18.971   **9.** 151.413

11. 6.679   **13.** $10^5 = 100,000$ **15.** 7

17. $5e^{10}$ **19.** $\dfrac{10^6}{2} = 5 \cdot 10^5$ **21.** 3

23. (a)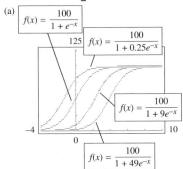

(b) Different c-values change the y-intercept and how the graph approaches the asymptote.

25. (a) 2038   (b) 4.9 months

27. (a) 13.86 years   (b) $105,850.00

29. 24.5 years   **31.** 128,402

33. $t \approx 13.3$, in 2019

35. (a) $4.98   (b) 8   **37.** $502

39. (a) $14,203.51   (b) $x \approx 164.6$

41. $x \approx 56.7$, in 2027

43. (a) $10,100.31   (b) 6.03 years

45. (a) $5469.03

(b) 7 years, 9 months (approximately)

47. (a) $142.5 million   (b) 35.3 years, in 2026

49. (a) $P(18) \approx 0.66$ means that in 2028, $1.00 is expected to purchase 66% of what it did in 2012.

(b) $x \approx 29.3$, in 2040

51. $x \approx 993.3$; about 993 units

53. (a) 600   (b) 2119   (c) 3000

(d)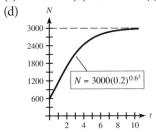

55. (a) 10   (b) 2.5 years

57. (a) 37   (b) 1.5 hours

59. (a) 52   (b) the 10th day

61. (a) 2%   (b) 20 months ($x = 19.6$)

63. (a) 0.23 km³   (b) 5.9 years

65. (a) 160.48 million

(b) $t \approx 84.7$; in 2025

67. (a) $y = \dfrac{120}{1 + 5.25e^{-0.0637x}}$

(b) 63.9 million

(c) 120 million

(d) $x \approx 36.8$, in 2037

CHAPTER 5 REVIEW EXERCISES

1. (a) $\log_2 y = x$ (b) $\log_3 2x = y$

2. (a) $7^{-2} = \frac{1}{49}$ (b) $4^{-1} = x$

3. 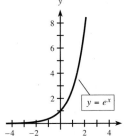   **4.**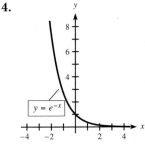

5. 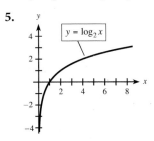   **6.**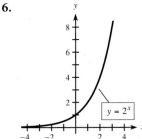

7. 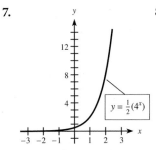   **8.**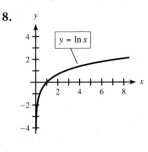

9. 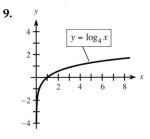   **10.**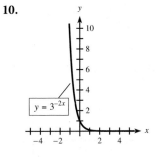

11.

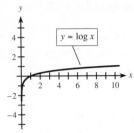

12.

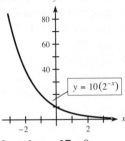

13. 0   **14.** 3   **15.** $\frac{1}{2}$ **16.** −1   **17.** 8

18. 1   **19.** 5   **20.** 3.15   **21.** −2.7

22. 0.6   **23.** 5.1   **24.** 15.6   **25.** $\log y + \log z$

26. $\frac{1}{2}\ln(x+1) - \frac{1}{2}\ln x$ **27.** no   **28.** −2

29. 5   **30.** 1   **31.** 0   **32.** 3.4939   **33.** −1.5845

34.

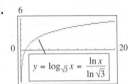

35.

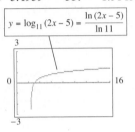

36. $x = 1.5$ **37.** $x \approx 51.224$

38. $x \approx 28.175$ **39.** $x \approx 40.236$

40. $x = 8$ **41.** $x = 14$ **42.** $x = 6$

43. Growth exponential, because the general outline has the same shape as a growth exponential.

44. (a) $C(15) \approx 139.19$ means that in 2025 goods that cost $100.00 in 2012 are expected to cost $139.19.

(b)

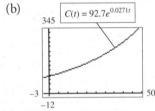

45. (a) $32,627.66

(b)

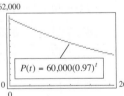

46. (a) $y = 42.1(1.04^x)$ (b)

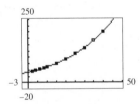

(c) $239.2 thousand

47. (a) $11,938   (b)

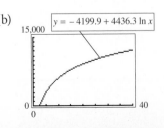

48. (a) $y = -363.3 + 166.9\ln x$

(b) 341.1 million

(c) $x \approx 96.9$, in 2047

49. (a) −3.9   (b) $0.14B_0$ (c) $0.004B_0$ (d) yes

50. (a) 27,441   (b) 12 weeks

51. 1366   **52.** 5.8 years

53. (a) $5532.77   (b) 5.13 years

54. logistic, because the graph begins like an exponential function but then grows at a slower rate

55. (a) 3000   (b) 8603   (c) 10,000

56. (a)

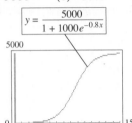

(b) 5   (c) 4969   (d) 10 days

CHAPTER 5 TEST

1.

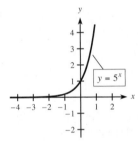

2.

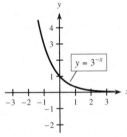

3.

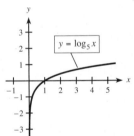

4.

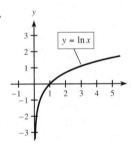

5.

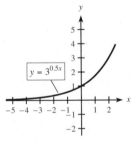

6.

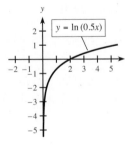

7.

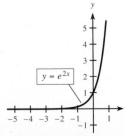

8.

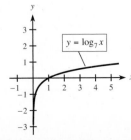

9. 54.598   **10.** 0.100   **11.** 1.386

12. 1.322   **13.** $x = 7^{3.1}, x \approx 416.681$

14. $\log_3(27) = 2x; x = 1.5$ **15.** $x \approx 0.203$

16. $x = 8$ **17.** 3   **18.** x^4 **19.** 3   **20.** x^2

21. $\ln M + \ln N$ **22.** $\ln(x^3 - 1) - \ln(x + 2)$

23. $\dfrac{\ln(x^3 + 1)}{\ln 4} \approx 0.721 \ln(x^3 + 1)$

24. $x \approx 38.679$

25. Decay exponential

26. (a) Decay exponential

(b) In 2030, the expected tons of CO_2 emissions per person is 14.8 tons.

27. (a) $11,943.51

(b) $t \approx 16.0$, in 2034

28. about 6.2 months

29. (a) about $16,716 billion

(b) $x \approx 39.8$; in 2025

30. (a) $y = 63.31(1.048^x)$

(b) 461,000

(c) $x \approx 59.0$, in 2069

6.1 EXERCISES

1. $r = 0.0625, I = 250, P = 1000, t = 4$

3. $P = 8000, S = 9600, I = 1600, r = 0.05, t = 4$

5. (a) $9600   (b) $19,600

7. (a) $30   (b) $1030

9. $864   **11.** $3850   **13.** 13%

15. (a) 5.13%   (b) 4.29%   **17.** $1631.07

19. $12,000   **21.** 10 years   **23.** pay on time

25. (a) $2120   (b) $2068.29 (nearest cent)

27. 3, 6, 9, 12, 15, 18, 21, 24, 27, 30

29. $-\frac{1}{3}, \frac{1}{5}, -\frac{1}{7}, \frac{1}{9}, -\frac{1}{11}, \frac{1}{13}$ **31.** $-1, -\frac{1}{4}, -\frac{1}{15}, 0; a_{10} = \frac{1}{20}$

33. (a) $d = 3, a_1 = 2$ (b) 11, 14, 17

35. (a) $d = \frac{3}{2}, a_1 = 3$ (b) $\frac{15}{2}, 9, \frac{21}{2}$

37. -35 **39.** 203   **41.** 2185   **43.** 1907.5

45. $-15,862.5$ **47.** 21, 34, 55   **49.** $4800

51. the job starting at $40,000 ($58,000 versus $57,600)

53. (a) $6000   (b) $9000   (c) plan II, by $3000

(d) $20,000   (e) $27,000   (f) plan II, by $7000

(g) plan II

6.2 EXERCISES

(Minor differences may occur because of rounding.)

1. $S = 3216.87$ = future value; principal = 2000; rate = 0.02; periods = 24

3. $P = 6049.97$ = principal; future value = 25,000; rate = 0.03; periods = 48

5. (a) 8%   (b) 7   (c) 2% = 0.02   (d) 28

7. (a) 9%   (b) 5   (c) $\left(\frac{9}{12}\right)\% = 0.0075$ (d) 60

9. $24,846.79   **11.** $S = \$4755.03; I = \$1555.03$

13. $13,322.91   **15.** $5583.95

17. $7309.98   **19.** $502.47

21. (a) $12,245.64   (b) $11,080.32

(c) A $\frac{1}{2}$% increase in the interest rate reduces the amount required by $1165.32.

23. $50.26 more at 8%   **25.** (a) 7.55%   (b) 6.18%

27. 8% compounded monthly, 8% compounded quarterly, 8% compounded annually

29. 8.2% continuously yields 8.55%. 8.4% compounded quarterly yields 8.67% and so is better.

31. The higher graph is for continuous compounding because its yield (its effective annual rate) is higher.

33. 37.02%   **35.** 3 years   **37.** 4%   **39.** $3996.02

41. (a) $2,124,876.38   (b) $480,087.44 more

43. $n \approx 20.5$; 21 quarters   **45.** $13,916.24

47.

	A	B	C
1		Future Value	(Yearly)
2	End of Year	Quarterly	Monthly
3	0	$5000.00	$5000.00
4	1	$5322.52	$5324.26
5	2	$5665.84	$5669.54
6	3	$6031.31	$6037.22
7	4	$6420.36	$6428.74
8	5	$6834.50	$6845.65
9	6	$7275.35	$7289.60
10	7	$7744.64	$7762.34
11	8	$8244.20	$8265.74
12	9	$8775.99	$8801.79
13	10	$9342.07	$9372.59

(a) from quarterly and monthly spreadsheets: after $6\frac{1}{2}$ years (26 quarters or 78 months)

(b) See the spreadsheet.

49. (a) 24, 48, 96   (b) 24, 16, $\frac{32}{3}$

51. $10(2^{12})$ **53.** $4 \cdot \left(\frac{3}{2}\right)^{15}$ **55.** $\dfrac{6(1 - 3^{17})}{-2}$

57. $\dfrac{3^{35} - 1}{2}$ **59.** $18\left[1-\left(\frac{2}{3}\right)^{18}\right]$

61. $350,580$ (approx.)   **63.** 24.4 million (approx.)

65. 35 years   **67.** 40.5 ft   **69.** $4096

71. 320,000   **73.** $7,231,366   **75.** $6; 5^6 = 15,625$

77. 305,175,780

6.3 EXERCISES

1. $S = \$285,129 =$ future value; $R = 2500$; $i = 0.02$; $n = 60$

3. $R = \$1426 =$ payment; $S = 80,000$; $i = 0.04$; $n = 30$

5. (a) The higher graph is $1120 per year.
  (b) $R = \$1000: S = \$73,105.94$;
    $R = \$1120: S = \$81,878.65$;
    Difference $= \$8772.71$

7. $7328.22   **9.** $1072.97   **11.** $1482.94

13. $n \approx 27.1$; 28 quarters   **15.** $4651.61

17. $1180.78   **19.** $4152.32   **21.** $3787.92

23. $n \approx 17.7$; 18 quarters

25. (a) ordinary annuity   (b) $4774.55

27. (a) annuity due   (b) $3974.73

29. (a) ordinary annuity   (b) $1083.40

31. (a) ordinary annuity   (b) $1130.51

33. (a) annuity due   (b) $290,976.81

35. (a) ordinary annuity   (b) $n \approx 108.5$; 109 months

37. (a) annuity due   (b) $235.16

39. (a) annuity due   (b) $26,517.13

41. (a) ordinary annuity   (b) $795.75

43. $53,677.40

45. (a) $n \approx 35$ quarters   (b) $1,062,412 (nearest dollar)

6.4 EXERCISES

1. $A_n = \$22,480 =$ present value; $R = 1300$;
  $i = 0.04$; $n = 30$

3. $R = \$809 =$ payment; $A_n = 135,000$; $i = 0.005$; $n = 360$

5. $69,913.77   **7.** $2,128,391   **9.** $4595.46

11. $n \approx 73.8$; 74 quarters   **13.** $1141.81; premium

15. (a) The higher graph corresponds to 8%.
  (b) $1500 (approximately)
  (c) With an interest rate of 10%, a present value of about $9000 is needed to purchase an annuity of $1000 for 25 years. If the interest rate is 8%, about $10,500 is needed.

17. Ordinary annuity—payments at the end of each period
  Annuity due—payments at the beginning of each period

19. $69,632.02   **21.** $445,962.23   **23.** $2145.59

25. (a) ordinary annuity   (b) $10,882.46

27. (a) annuity due   (b) $316,803.61

29. (a) ordinary annuity
  (b) Taking $500,000 and $140,000 payments for the next 10 years has a slightly higher present value: $1,506,436.24.

31. (a) annuity due   (b) $146,235.06

33. (a) ordinary annuity
  (b) $n \approx 1025.7$; 1026 months; about 85.5 yrs.

35. (a) annuity due   (b) $22,663.74

37. (a) ordinary annuity   (b) $11,810.24

39. (a) ordinary annuity   (b) $27,590.62

41. (a) $8629.16   (b) $9883.48

43. (a) $30,078.99   (b) $16,900   (c) $607.02
  (d) $36,421.20

45. (a) $4504.85   (b) $n \approx 21.9$; 22 withdrawals

47. $25,785.99   **49.** $74,993.20   **51.** $59,768.91

53. $1317.98   **55.** $1,290,673.16

57. (a) The spreadsheet below shows the payments for the first 12 months and the last 12 months. Full payments for $13\frac{1}{2}$ years.

	A	B	C	D
1	End of Month	Acct. Value	Payment	New Balance
2	0	$100000.00	$0.00	$100000.00
3	1	$100650.00	$1000.00	$99650.00
4	2	$100297.73	$1000.00	$99297.73
5	3	$99943.16	$1000.00	$98943.16
6	4	$99586.29	$1000.00	$98586.29
7	5	$99227.10	$1000.00	$98227.10
8	6	$98865.58	$1000.00	$97865.58
9	7	$98501.70	$1000.00	$97501.70
10	8	$98135.47	$1000.00	$97135.47
11	9	$97766.85	$1000.00	$96766.85
12	10	$97395.83	$1000.00	$96395.83
13	11	$97022.40	$1000.00	$96022.40
14	12	$96646.55	$1000.00	$95646.55
⋮	⋮	⋮	⋮	⋮
154	152	$10684.71	$1000.00	$9684.71
155	153	$9747.66	$1000.00	$8747.66
156	154	$8804.52	$1000.00	$7804.52
157	155	$7855.25	$1000.00	$6855.25
158	156	$6899.81	$1000.00	$5899.81
159	157	$5938.16	$1000.00	$4938.16
160	158	$4970.25	$1000.00	$3970.25
161	159	$3996.06	$1000.00	$2996.06
162	160	$3015.53	$1000.00	$2015.53
163	161	$2028.64	$1000.00	$1028.64
164	162	$1035.32	$1000.00	$35.32
165	163	$35.55	$35.55	$0.00

(b) The spreadsheet below shows the payments for the first 12 months and the last 12 months. Full payments for almost 4 years.

	A	B	C	D
1	End of Month	Acct. Value	Payment	New Balance
2	0	$100000.00	$0.00	$100000.00
3	1	$100650.00	$2500.00	$98150.00
4	2	$98787.98	$2500.00	$96287.98
5	3	$96913.85	$2500.00	$94413.85
6	4	$95027.54	$2500.00	$92527.54
7	5	$93128.97	$2500.00	$90628.97
8	6	$91218.05	$2500.00	$88718.05
9	7	$89294.72	$2500.00	$86794.72
10	8	$87358.89	$2500.00	$84858.89
11	9	$85410.47	$2500.00	$82910.47
12	10	$83449.39	$2500.00	$80949.39
13	11	$81475.56	$2500.00	$78975.56
14	12	$79488.90	$2500.00	$76988.90
⋮	⋮	⋮	⋮	⋮
38	36	$27734.95	$2500.00	$25234.95
39	37	$25398.98	$2500.00	$22898.98
40	38	$23047.83	$2500.00	$20547.83
41	39	$20681.39	$2500.00	$18181.39
42	40	$18299.57	$2500.00	$15799.57
43	41	$15902.26	$2500.00	$13402.26
44	42	$13489.38	$2500.00	$10989.38
45	43	$11060.81	$2500.00	$8560.81
46	44	$8616.45	$2500.00	$6116.45
47	45	$6156.21	$2500.00	$3656.21
48	46	$3679.98	$2500.00	$1179.98
49	47	$1187.65	$1187.65	$0.00

6.5 EXERCISES

1. (a) the 10-year loan, because the loan must be paid more quickly
   (b) the 25-year loan, because the loan is paid more slowly
3. $1288.29
5. $553.42
7. $31,035.34

9.
Period	Payment	Interest
1	$39,505.48	$9000.00
2	39,505.48	6254.51
3	39,505.47	3261.92
	118,516.43	18,516.43

Period	Balance Reduction	Unpaid Balance
		$100,000.00
1	$30,505.48	69,494.52
2	33,250.97	36,243.55
3	36,243.55	0.00
	100,000.00	

11.
Period	Payment	Interest
1	$5380.54	$600.00
2	5380.54	456.58
3	5380.54	308.87
4	5380.54	156.71
	21,522.16	1,522.16

Period	Balance Reduction	Unpaid Balance
		$20,000.00
1	$4780.54	15,219.46
2	4923.96	10,295.50
3	5071.67	5,223.83
4	5223.83	0.00
	20,000.00	

13. $8852.05
15. $11,571.67
17. (a) $17,436.92
    (b) $348,738.40 + $150,000 = $498,738.40
    (c) $148,738.40
19. (a) $11,714.19
    (b) $1,405,702.80 + $250,000 = $1,655,702.80
    (c) $405,702.80

21. $8903.25
23. (a) $276,991.32
    (b) $263,575.30
25. (a) $89,120.53
    (b) $6451.45
27. (a) $168,121.30
    (b) $1286.12
    (c) $56,282.40
29. (a) $368.43; $383.43
    (b) $n \approx 57$
    (c) $211.95
31. (a) $12,007,827.36
    (b) $246,017.20
    (c) $n \approx 86.9$; 87 payments (rather than 100)
33. (a) The line is the total amount paid ($644.30 per month × the number of months). The curve is the total amount paid toward the principal.
    (b)

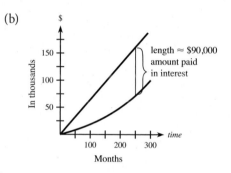

35.
	Rate	Payment	Total Interest
(a)	8%	$366.19	$2577.12
	8.5%	$369.72	$2746.56
(b)	6.75%	$518.88	$106,796.80
	7.25%	$545.74	$116,466.40

(c) The duration of the loan seems to have the greatest effect. It greatly influences payment size (for a $15,000 loan versus one for $80,000), and it also affects total interest paid.

37.
		Payment	Points	Total Paid
(a)	(i)	$738.99	–	$221,697
	(ii)	$722.81	$1000	$217,843
	(iii)	$706.78	$2000	$214,034

(b) The 7% loan with 2 points.
39. (a) $17,525.20
    (b) $508.76
    (c) $n \approx 33.2$; 34 quarters
    (d) $471.57

41. The spreadsheet shows the amortization schedule for the first 12 and the last 12 payments.

	A	B	C	D	E
1	Period	Payment	Interest	Bal. Reduction	Unpaid Bal.
2	0				$16700.00
3	1	$409.27	$114.12	$295.15	$16404.85
4	2	$409.27	$112.10	$297.17	$16107.68
5	3	$409.27	$110.07	$299.20	$15808.48
6	4	$409.27	$108.02	$301.25	$15507.23
7	5	$409.27	$105.97	$303.30	$15203.93
8	6	$409.27	$103.89	$305.38	$14898.55
9	7	$409.27	$101.81	$307.46	$14591.09
10	8	$409.27	$99.71	$309.56	$14281.52
11	9	$409.27	$97.59	$311.68	$13969.84
12	10	$409.27	$95.46	$313.81	$13656.03
13	11	$409.27	$93.32	$315.95	$13340.08
14	12	$409.27	$91.16	$318.11	$13021.97
	⋮	⋮	⋮	⋮	⋮
39	37	$409.27	$32.11	$377.16	$4322.49
40	38	$409.27	$29.54	$379.73	$3942.75
41	39	$409.27	$26.94	$382.33	$3560.43
42	40	$409.27	$24.33	$384.94	$3175.49
43	41	$409.27	$21.70	$387.57	$2787.91
44	42	$409.27	$19.05	$390.22	$2397.70
45	43	$409.27	$16.38	$392.89	$2004.81
46	44	$409.27	$13.70	$395.57	$1609.24
47	45	$409.27	$11.00	$398.27	$1210.97
48	46	$409.27	$8.27	$401.00	$809.97
49	47	$409.27	$5.53	$403.74	$406.24
50	48	$409.02	$2.78	$406.24	$0.00

CHAPTER 6 REVIEW EXERCISES

1. $1, \frac{1}{4}, \frac{1}{9}, \frac{1}{16}$

2. Arithmetic: (a) and (c)   (a) $d = -5$ (c) $d = \frac{1}{6}$

3. 235   **4.** 109   **5.** 315

6. Geometric: (a) and (b)   (a) $r = 8$ (b) $r = -\frac{3}{4}$

7. 8   **8.** 2, 391, $484\frac{4}{9}$

9. $10,880   **10.** $6\frac{2}{3}\%$ **11.** $2941.18   **12.** $4650

13. the $40,000 job ($490,000 versus $472,500)

14. (a) 40   (b) $2\% = 0.02$

15. (a) $S = P(1 + i)^n$ (b) $S = Pe^{rt}$

16. (b) monthly   **17.** $372.79   **18.** $14,510.26

19. $1616.07   **20.** $21,299.21

21. $n \approx 34.3$; 35 quarters

22. (a) 13.29%   (b) 14.21%

23. (a) 7.40%   (b) 7.47%

24. 2^{63} **25.** $2^{32} - 1$ **26.** $29,428.47

27. $6069.44   **28.** $31,194.18   **29.** $10,841.24

30. $n \approx 36$ quarters   **31.** $130,079.36   **32.** $12,007.09

33. (a) $11.828 million   (b) $161.5 million

34. $1726.85   **35.** $5390.77

36. $4053.54; discount   **37.** $n \approx 16$ half-years (8 years)

38. $88.85   **39.** $3443.61   **40.** $163,792.21

41.

Payment Number	Payment Amount	Interest	Balance Reduction	Unpaid Balance
57	$699.22	$594.01	$105.21	$94,936.99
58	$699.22	$593.36	$105.86	$94,831.13

42. (a) $401,134.67   (b) $45,596.63
(c) $22,376.50

43. $S = R\left[\dfrac{(1 + i)^n - 1}{i}\right]$ **44.** $I = Prt$

45. $A_n = R\left[\dfrac{1 - (1 + i)^{-n}}{i}\right]$ **46.** $S = P(1 + i)^n$

47. $A_n = R\left[\dfrac{1 - (1 + i)^{-n}}{i}\right]$, solved for R

48. $S = Pe^{rt}$ **49.** 16.32%   **50.** $213.81

51. (a) $1480   (b) $1601.03

52. $9319.64   **53.** 14.5 years   **54.** 79.4%

55. $381,195

56. (a) $4728.19   (b) $5398.07   (c) $1749.88
(d) 10.78%

57. $21,474.08   **58.** $12,162.06

59. Quarterly APY = 6.68%. This rate is better for the bank; it pays less interest. Continuous APY = 6.69%. This rate is better for the consumer, who earns more interest.

60. (a) $243,429.20
(b) $1825.11
(c) $n \approx 158.9$; 159 payments (rather than 180)
(d) $62,438.30

61. $32,834.69   **62.** $3,466.64

63. (a) $1185.51   (b) $355,653   (c) $171,653
(d) $\approx$ $156,366

64. (a) $95,164.21   (b) $1300.14

65. $994.08

66. Future value of IRA = $172,971.32
Present value needed = $2,321,520.10
Future value needed from deposits = $2,148,548.78
Deposits = $711.60

67. Regular payment = $64,337.43
Unpaid balance = $2,365,237.24
Number of $70,000 payments = $n \approx 46.1$
Savings $\approx$ $118,546

CHAPTER 6 TEST

1. 25.3 years (approximately)   **2.** $7999.41
3. 6.87%   **4.** $158,524.90   **5.** 33.53%
6. (a) $698.00   (b) $112,400
7. $2625   **8.** $815.47   **9.** 8.73%
10. $119,912.92   **11.** $40,552.00   **12.** $22,439.01
13. $6781.17   **14.** $n \approx 66.8$; 67 half-years
15. $116,909.12   **16.** $29,716.47
17. (a) $161,270.52   (b) $924.08   (c) $5788.80
18. $12,975.49; premium
19. (a) The difference between successive terms is always -5.5.
   (b) 23.8   (c) 8226.3
20. 1000 mg (approximately)
21. (a) $145,585.54 (with the $2000);
   $147,585.54 (without the $2000)
   (b) $n \approx 318.8$; 319 months
   Total interest = $243,738.13
   (c) $n \approx 317.2$; 318 months
   Total interest = $245,378.24
   (d) Paying the $2000 is slightly better; it saves about $1640 in interest.
22. $1688.02
23. (a) $279,841.35   (b) $13,124.75

7.1 EXERCISES

1. (a) $\frac{2}{5}$ (b) 0   (c) 1   **3.** $\frac{1}{4}$ **5.** 1
7. (a) $\frac{3}{10}$ (b) $\frac{1}{2}$ (c) $\frac{1}{5}$ (d) $\frac{3}{5}$ (e) $\frac{7}{10}$
9. (a) $\frac{1}{13}$ (b) $\frac{1}{2}$ (c) $\frac{1}{4}$
11. {HH, HT, TH, TT};   (a) $\frac{1}{4}$ (b) $\frac{1}{2}$ (c) $\frac{1}{4}$
13. (a) $\frac{1}{12}$ (b) $\frac{1}{12}$ (c) $\frac{1}{36}$ **15.** (a) $\frac{1}{2}$ (b) $\frac{5}{12}$
17. (a) 431/1200
   (b) If fair, $\Pr(6) = \frac{1}{6}$; 431/1200 not close to $\frac{1}{6}$, so not a fair die
19. (a) 2:3   (b) 3:2   **21.** (a) $\frac{1}{21}$ (b) $\frac{20}{21}$
23. 0.46   **25.** (a) 63/425   (b) 32/425
27. (a) R: 0.63   D: 0.41   I: 0.51
   (b) Republican
29. (a) 1/3601   (b) 100/3601   (c) 3500/3601
   (d) 30%
31. (a) 0.402   (b) 0.491   (c) 100%; yes
33. $S = \{A+, A-, B+, B-, AB+, AB-, O+, O-\}$; No.
   Type O+ is the most frequently occurring blood type.
35. (a) 0.04   (b) 0.96   **37.** (a) 0.13   (b) 0.87
39. 0.03   **41.** 0.75   **43.** $\frac{1}{3}$ **45.** $\frac{1}{3}$
47. 0.22; yes, 0.39 is much higher than 0.22   **49.** $\frac{3}{8}$
51. (a) no   (b) {BB, BG, GB, GG}   (c) $\frac{1}{2}$
53. $\frac{1}{8}$ **55.** $\frac{3}{125}$
57. $\Pr(A) = 0.000019554$, or about 2.0 accidents per 100,000
   $\Pr(B) = 0.000035919$, or about 3.6 accidents per 100,000
   $\Pr(C) = 0.000037679$, or about 3.8 accidents per 100,000
   Intersection C is the most dangerous.

59. (a) 557/1200   (b) 11/120
61. (a) boy: 1/5; girl: 4/5
   (b) boy: 0.4946; girl: 0.5054   (c) part (b)
63. 3/4   **65.** 3/3995 $\approx$ 0.00075   **67.** 3 to 1

7.2 EXERCISES

1. $\frac{1}{6}$ **3.** $\frac{2}{3}$ **5.** $\frac{2}{5}$ **7.** (a) $\frac{1}{7}$ (b) $\frac{5}{7}$
9. $\frac{3}{4}$ **11.** $\frac{2}{3}$ **13.** $\frac{10}{17}$ **15.** $\frac{2}{3}$
17. (a) $\frac{1}{2}$ (b) $\frac{1}{3}$ (c) $\frac{8}{9}$ (d) $\frac{1}{9}$
19. 0.54   **21.** (a) 362/425   (b) $\frac{66}{85}$
23. $\frac{17}{50}$ **25.** (a) $\frac{5}{6}$ (b) $\frac{1}{6}$
27. (a) 0.35   (b) 0.08   (c) 0.83
29. (a) 0.508   (b) 0.633   (c) 0.761
31. (a) 0.210   (b) 0.015   (c) 0.879
33. (a) $\frac{11}{12}$ (b) $\frac{5}{6}$ **35.** (a) $\frac{1}{2}$ (b) $\frac{7}{8}$ (c) $\frac{3}{4}$
37. 0.56   **39.** 0.965
41. (a) 0.72   (b) 0.84   (c) 0.61
43. $\frac{31}{40}$ **45.** 0.13

7.3 EXERCISES

1. (a) $\frac{1}{2}$ (b) $\frac{1}{13}$ **3.** (a) $\frac{1}{3}$ (b) $\frac{1}{3}$ **5.** $\frac{4}{7}$
7. (a) $\frac{2}{3}$ (b) $\frac{4}{9}$ (c) $\frac{3}{5}$ **9.** (a) $\frac{1}{4}$ (b) $\frac{1}{2}$
11. $\frac{1}{36}$ **13.** (a) $\frac{1}{8}$ (b) $\frac{7}{8}$
15. (a) $\frac{3}{50}$ (b) $\frac{1}{15}$
   (c) The events in part (a) are independent because the result of the first draw does not affect the probability for the second draw.
17. (a) $\frac{4}{25}$ (b) $\frac{9}{25}$ (c) $\frac{6}{25}$ (d) 0   **19.** $\frac{5}{68}$
21. (a) $\frac{1}{5}$ (b) $\frac{3}{5}$ (c) 0   **23.** (a) $\frac{1}{17}$ (b) 13/204
25. (a) $\frac{13}{17}$ (b) $\frac{4}{17}$ (c) $\frac{8}{51}$ **27.** $\frac{31}{52}$ **29.** $\frac{25}{96}$
31. $\frac{43}{50}$ **33.** $\frac{65}{87}$ **35.** 35/435 $= \frac{7}{87}$ **37.** $\frac{1}{10}$
39. 1/144,000,000   **41.** 0.004292   **43.** 0.06
45. 0.045   **47.** $(0.95)^5 = 0.774$ **49.** 0.06
51. (a) 0.366   (b) 0.634
53. (a) 0.4565   (b) 0.5435
55. (a) $\left(\frac{1}{3}\right)^3 \left(\frac{1}{5}\right)^4 = 1/16{,}875$
   (b) $\left(\frac{2}{3}\right)^3 \left(\frac{4}{5}\right)^4 = 2048/16{,}875$ (c) 14,827/16,875
57. 4/11; 4:7
59. (a) 364/365   (b) $\frac{1}{365}$
61. (a) 0.59   (b) 0.41

7.4 EXERCISES

1. $\frac{2}{5}$ **3.** (a) $\frac{2}{21}$ (b) $\frac{4}{21}$ (c) $\frac{23}{35}$
5. (a) $\frac{5}{42}$ (b) $\frac{10}{21}$ (c) $\frac{4}{9}$
7. (a) $\frac{1}{30}$ (b) $\frac{1}{2}$ (c) $\frac{5}{6}$ **9.** $\frac{3}{5}$
11. (a) $\frac{6}{25}$ (b) $\frac{9}{25}$ (c) $\frac{12}{25}$ (d) $\frac{19}{25}$
13. $\frac{2}{3}$ **15.** $\frac{2}{3}$ **17.** 0.3095
19. (a) 81/10,000   (b) 1323/5000
21. (a) $\frac{6}{35}$ (b) $\frac{6}{35}$ (c) $\frac{12}{35}$
23. (a) $\frac{4}{7}$ (b) $\frac{5}{14}$ (c) $\frac{7}{10}$ (d) $\frac{16}{25}$ **25.** $\frac{17}{45}$
27. 0.079   **29.** (a) 49/100   (b) $\frac{12}{49}$

7.5 EXERCISES

1. 360   **3.** 151,200   **5.** 1
7. (a) $6 \cdot 5 \cdot 4 \cdot 3 = 360$ (b) $6^4 = 1296$ **9.** $n!$
11. $n+1$ **13.** 16   **15.** 4950   **17.** 1   **19.** 1
21. 10   **23.** (a) 8   (b) 240   **25.** 604,800
27. 120   **29.** 24   **31.** 64   **33.** 720
35. $2^{10} = 1024$ **37.** $4(_{13}C_5) = 5148$
39. 10,816,000   **41.** 30,045,015   **43.** 792
45. 210   **47.** 2,891,999,880   **49.** 3,700,000

7.6 EXERCISES

1. $\frac{1}{120}$ **3.** (a) 120   (b) $\frac{1}{120}$
5. 0.639   **7.** (a) 1/10,000   (b) 1/5040
9. $1/10^6$ **11.** 0.000048   **13.** $1/10! = 1/3,628,800$
15. (a) $\frac{1}{22}$ (b) $\frac{6}{11}$ (c) $\frac{9}{22}$
17. 0.098   **19.** $\dfrac{_{90}C_{28} \cdot {}_{10}C_2}{_{100}C_{30}}$
21. (a) 0.119   (b) 0.0476   (c) 0.476
23. 0.0238   **25.** (a) 0.721   (b) 0.262   (c) 0.279
27. (a) $\frac{1}{3}$ (b) $\frac{1}{6}$ **29.** $\dfrac{_{20}C_{10}}{_{80}C_{10}} = 0.00000011$
31. (a) 0.033   (b) 0.633
33. (a) 0.0005   (b) 0.011   **35.** 0.00198
37. (a) $1/1000 = 0.001$
  (b) 444, 446, 464, 644, 466, 646, 664, 666
  (b) $1/8 = 0.125$

7.7 EXERCISES

1. can   **3.** cannot, sum $\neq$ 1
5. cannot, not square   **7.** can
9. [0.248   0.752]   **11.** [0.228   0.236   0.536]
13. [0.25   0.75]   **15.** [0.249   0.249   0.502]
17. $\left[\frac{1}{4} \quad \frac{3}{4}\right]$ **19.** $\left[\frac{1}{4} \quad \frac{1}{4} \quad \frac{1}{2}\right]$
21. [0.5   0.4   0.1]; [0.44   0.43   0.13];
  [0.431   0.43   0.139]; [0.4292   0.4291   0.1417]
23.
$$\begin{array}{c} \\ R \\ N \end{array} \begin{array}{cc} R & N \\ \begin{bmatrix} 0.8 & 0.2 \\ 0.3 & 0.7 \end{bmatrix} \end{array}$$
  25. 0.45
27.
$$\begin{array}{c} \\ A \\ F \\ V \end{array} \begin{array}{ccc} A & F & V \\ \begin{bmatrix} 0 & 0.7 & 0.3 \\ 0.6 & 0 & 0.4 \\ 0.8 & 0.2 & 0 \end{bmatrix} \end{array}$$
29. [0.3928   0.37   0.2372]
31. [46/113   38/113   29/113]
33.
$$\begin{array}{c} \\ r \\ u \end{array} \begin{array}{cc} r & u \\ \begin{bmatrix} 0.7 & 0.3 \\ 0.1 & 0.9 \end{bmatrix} \end{array}; [1/4 \quad 3/4]$$
35. $\left[\frac{1}{14} \quad \frac{3}{14} \quad \frac{5}{7}\right]$ **37.** $\left[\frac{4}{7} \quad \frac{2}{7} \quad \frac{1}{7}\right]$
39. [49/100   42/100   9/100]

CHAPTER 7 REVIEW EXERCISES

1. (a) $\frac{5}{9}$ (b) $\frac{1}{3}$ (c) $\frac{2}{9}$
2. (a) S_1 (b) $\frac{3}{4}$ (c) $\frac{1}{2}$ (d) $\frac{1}{6}$ (e) $\frac{2}{3}$
3. (a) 3:4   (b) 4:3   **4.** (a) $\frac{1}{4}$ (b) $\frac{1}{2}$ (c) $\frac{1}{4}$
5. (a) $\frac{3}{8}$ (b) $\frac{1}{8}$ (c) $\frac{3}{8}$ **6.** $\frac{2}{13}$ **7.** 16/169
8. $\frac{3}{4}$ **9.** $\frac{2}{13}$ **10.** $\frac{7}{13}$ **11.** (a) $\frac{2}{9}$ (b) $\frac{2}{3}$ (c) $\frac{7}{9}$
12. $\frac{2}{7}$ **13.** $\frac{1}{2}$ **14.** 7/320
15. 7/342   **16.** 3/14   **17.** $\frac{8}{15}$
18. (a) $\frac{3}{14}$ (b) $\frac{4}{7}$ (c) $\frac{3}{8}$ **19.** 49/89
20. 30   **21.** 35   **22.** 26^3 **23.** 56
24. (a) Not square   (b) The row sums are not 1.
25. [0.76   0.24], [0.496   0.504]
26. [0.2   0.8]
27. $\frac{5}{8}$ **28.** $\frac{1}{4}$ **29.** $\frac{29}{50}$
30. $\frac{5}{56}$ **31.** $\frac{33}{56}$ **32.** $\frac{15}{22}$ **33.** 0.72
34. (a) 63/2000   (b) $\frac{60}{63}$ **35.** 39/116   **36.** $4! = 24$
37. $_8P_4 = 1680$ **38.** $_{12}C_4 = 495$ **39.** $_8C_4 = 70$
40. (a) $_{12}C_2 = 66$ (b) $_{12}C_3 = 220$ **41.** 1,544,760
42. If her assumption about blood groups is accurate, there
  would be $4 \cdot 2 \cdot 4 \cdot 8 = 256$, not 288, unique groups.
43. $\frac{1}{24}$ **44.** $\frac{3}{500}$ **45.** $\frac{3}{1250}$
46. (a) 0.3398   (b) 0.1975   **47.** $\frac{1}{10}$
48. (a) $(_{10}C_5)(_2C_1)/_{12}C_6$
  (b) $\dfrac{(_{10}C_5)(_2C_1) + (_{10}C_4)(_2C_2)}{_{12}C_6}$
49. [0.135   0.51   0.355], [0.09675   0.3305   0.57275],
  [0.0640875   0.288275   0.6476375]
50. [12/265   68/265   37/53]

CHAPTER 7 TEST

1. (a) $\frac{4}{7}$ (b) $\frac{3}{7}$ **2.** (a) $\frac{2}{7}$ (b) $\frac{5}{7}$
3. (a) 0   (b) 1   **4.** $\frac{1}{7}$ **5.** $\frac{1}{7}$
6. (a) $\frac{2}{7}$ (b) $\frac{4}{7}$ **7.** $\frac{2}{7}$ **8.** $\frac{3}{7}$ **9.** $\frac{2}{3}$
10. 1/17,576   **11.** 0.2389   **12.** (a) $\frac{1}{5}$ (b) $\frac{1}{20}$
13. (a) $\frac{3}{95}$ (b) $\frac{6}{19}$ (c) $\frac{21}{38}$ (d) 0
14. (a) 5,245,786   (b) 1/5,245,786
15. (a) 2,118,760   (b) 1/2,118,760
16. 0.064   **17.** (a) 0.633   (b) 0.962   **18.** 0.229
19. (a) $\frac{1}{5}$ (b) $\frac{1}{14}$ (c) $\frac{13}{14}$ **20.** $\frac{3}{14}$
21. (a) 2^{10} (b) $\dfrac{1}{2^{10}}$ (c) $\frac{1}{3}$ (d) Change the code.
22. (a) $A = \begin{bmatrix} 0.80 & 0.20 \\ 0.07 & 0.93 \end{bmatrix}$ (b) [0.25566   0.74434];
  about 25.6%   (c) $\frac{7}{27}$; 25.9% of market

8.1 EXERCISES

1. (a) 1/6   (b) 5/6   (c) 18   (d) 0.045
3. 0.0595
5. (a) $\frac{1}{64}$ (b) $\frac{5}{16}$ (c) $\frac{15}{64}$ **7.** 0.0284
9. (a) 0.2304   (b) 0.0102   (c) 0.3174
11. 0.0585   **13.** 0.2759
15. (a) 0.375   (b) 0.0625

17. (a) 0.035 (b) 0.706
19. (a) $\frac{27}{64}$ (b) $\frac{27}{128}$ (c) $\frac{81}{256}$
21. (a) 0.0729 (b) 0.5905 (c) 0.9914
23. 0.2457 **25.** 0.0007
27. (a) 0.1323 (b) 0.0308
29. (a) 0.9044 (b) 0.0914 (c) 0.0043
31. (a) 0.8683 (b) 0.2099 **33.** 0.740

8.2 EXERCISES

1.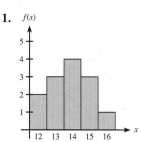
3.

5.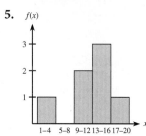
7.

9. 3 **11.** 13 **13.** 5 **15.** 10.5
17. mode = 2, median = 4.5, mean = 6
19. mode = 17, median = 18.5, mean = 23.5
21. mode = 5.3, median = 5.3, mean = 5.32
23. 12.21, 14.5, 14.5 **25.** 9 **27.** 14
29. 4, 8.5714, 2.9277 **31.** 14, 4.6667, 2.1602
33. 2.73, 1.35 **35.** 6.75, 2.96
37. (a)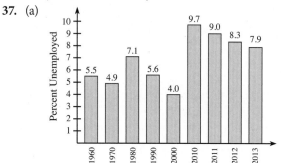

(b) $\bar{x} = 6.9$, $s = 1.98$
39.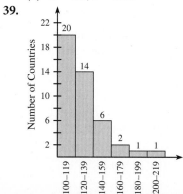

Mobile Phones per 100 Persons

41. The mean will give the highest measure.
43. The median will give the most representative average.
45. (a) $\bar{x} \approx 63.4\%$; $s \approx 19.3\%$ (b) Yes
47.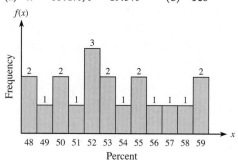

49. $\bar{x} = 3.32$ kg; $s = 0.677$ kg
51. (a) $60,000 (b) $36,000 (c) $32,000
53. (a) $23.33 (b) $5.14
55. (a) 35,434,000 (b) 13,312,000
57. (a)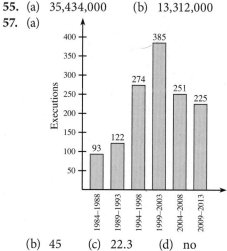

(b) 45 (c) 22.3 (d) no

8.3 EXERCISES

1. no; $\Pr(x) \neq 0$ **3.** yes; both conditions satisfied
5. yes; both conditions satisfied
7. no; $\Sigma \Pr(x) > 1$ **9.** $\frac{15}{8}$ **11.** 5
13. $\mu = \frac{13}{8}$, $\sigma^2 = 1.48$, $\sigma = 1.22$
15. $\mu = \frac{13}{3}$, $\sigma^2 = 2.22$, $\sigma = 1.49$
17. 3 **19.** 2
21. (a)

x	$\Pr(x)$
0	125/216
1	25/72
2	5/72
3	1/216

(b) $3\left(\frac{1}{6}\right) = \frac{1}{2}$ (c) $\sqrt{3\left(\frac{1}{6}\right)\left(\frac{5}{6}\right)} = \left(\frac{1}{6}\right)\sqrt{15}$
23. (a) 42 (b) 3.55
25. (a) 30 (b) 3.464
27. 125
29. $a^6 + 6a^5b + 15a^4b^2 + 20a^3b^3 + 15a^2b^4 + 6ab^5 + b^6$
31. $x^4 + 4x^3h + 6x^2h^2 + 4xh^3 + h^4$
33. 1.85
35. TV: 37,500; personal appearances: 35,300
37. −$67.33

39. Expect to lose $2 per game in the long run.
41. 100
43. E(cost with policy) = \$108,
E(cost without policy) = \$80;
save \$28 by "taking the chance"
45. no; some pipes may be more than 0.01 in. from 2 in.
even if average is 2 in.
47. (a) $100(0.10) = 10$ (b) $\sqrt{100(0.10)(0.90)} = 3$
49. (a) 60,000   (b) $\sqrt{24,000} = 154.919$
51. 59,690   **53.** (a) 4   (b) 1.79
55. 2, 1.41   **57.** 300

8.4 EXERCISES

1. 0.4641   **3.** 0.2257   **5.** 0.9153   **7.** 0.1070
9. 0.0166   **11.** 0.0228   **13.** 0.8849   **15.** 0.1915
17. 0.3944   **19.** 0.3830   **21.** 0.7745   **23.** 0.9772
25. 0.0668   **27.** -1.645 **29.** -0.3
31. (a) 0.3413   (b) 0.3944   **33.** 0.9876
35. (a) 0.4192   (b) 0.0228   (c) 0.0580   (d) 0.8965
37. (a) 0.0668   (b) 0.3085   (c) 0.3830
39. (a) 0.0475   (b) 0.2033   (c) 0.5934
41. (a) 0.0228   (b) 0.1587   (c) 0.8186
43. 86   **45.** 128.7 oz.   **47.** 1.544

8.5 EXERCISES

1. yes   **3.** no   **5.** 0.0668   **7.** 0.0001
9. 0.0521   **11.** 0.0110   **13.** 0.9890
15. 0.7324   **17.** 0.3520   **19.** 0.0443
21. 0.5398   **23.** 0.7852   **25.** 0.0129
27. 0.2514   **29.** 0.0011   **31.** 0.9990
33. 0.1272; 0.4364
35. (a) 0.0038   (b) yes; students were smarter or
questions were leaked   **37.** 0.0166

CHAPTER 8 REVIEW EXERCISES

1. 0.0774   **2.** (a) 0.3545   (b) 0.5534
3. 0.407
4. $f(x)$

5. 3
6. $\frac{77}{26} \approx 2.96$
7. 3

8. $f(x)$

9. 14
10. 14
11. 14.3

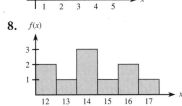

12. $\bar{x} = 3.86$; $s^2 = 6.81$; $s = 2.61$
13. $\bar{x} = 2$; $s^2 = 2.44$; $s = 1.56$
14. 2.4   **15.** yes   **16.** no; $\Sigma \Pr(x) \neq 1$
17. yes   **18.** no; $\Pr(x) \not\geq 0$ **19.** 2

20. (a) 4.125   (b) 2.7344   (c) 1.654
21. (a) $\frac{37}{12}$ (b) 0.9097   (c) 0.9538
22. $\mu = 4, \sigma = (2\sqrt{3})/3$ **23.** 3
24. $x^5 + 5x^4y + 10x^3y^2 + 10x^2y^3 + 5xy^4 + y^5$
25. 0.9165   **26.** 0.1498   **27.** 0.1039   **28.** 0.3413
29. 0.6826   **30.** 0.1359   **31.** 2.33   **32.** not good
33. good   **34.** 0.0151   **35.** 0.9625   **36.** 0.8475
37. 0.0119   **38.** 0.297   **39.** 0.16308   **40.** 0.2048
41. (a) $(99{,}999/100{,}000)^{99{,}999} \approx 0.37$
(b) $1 - (99{,}999/100{,}000)^{100{,}000} \approx 0.63$
42. $f(x)$

43. 30.3%
44. 8.35%
45. 455
46. 3
47. \$18.00
48. $-\$0.50$

49. (a) 1   (b) $\left(\frac{4}{5}\right)^4$
50. (a) 0.4772   (b) 0.1359   (c) 0.0228
51. 15%   **52.** 47 minutes   **53.** 0.3821
54. 0.4090   **55.** 0.0262   **56.** 0.1788

CHAPTER 8 TEST

1. (a) $\frac{40}{243}$ (b) $\frac{51}{243} = \frac{17}{81}$
2. (a) 4   (b) $\mu = 4, \sigma^2 = \frac{8}{3}, \sigma = \frac{2}{3}\sqrt{6}$
3. (i) For each $x, 0 \leq \Pr(x) \leq 1$ (ii) $\Sigma \Pr(x) = 1$
4. 5.1
5. $\mu = 16.7, \sigma^2 = 26.61, \sigma = 5.16$
6. $\bar{x} = 21.57$, median $= 21$, mode $= 21$
7. (a) 0.4706   (b) 0.8413   (c) 0.0669
8. (a) 0.3891   (b) 0.5418   (c) 0.1210
9. 38.4   **10.** 0.9554   **11.** 0.6331
12. $f(x)$

13. $\bar{x} = 48.3, s = 15.6$
14. (a) 38.5   (b) under 30; it would be lower
15. (a) 109.5   (b) Increase, unless a new technology
replaces mobile phones.
16. (a) 0.00003   (b) 30
17. 2 (1.8)   **18.** 5 (5.4)
19. 0 (0.054) with correct use
20. (a) 0.0158   (b) 0.0901   (c) 0.5383
21. 0.1814

9.1 EXERCISES

1. (a) -8 (b) -8
3. (a) 10 (b) does not exist
5. (a) 0 (b) -6
7. (a) does not exist (∞) (b) does not exist (∞)
  (c) does not exist (∞) (d) does not exist
9. (a) 3 (b) -6 (c) does not exist
  (d) -6

11.

x	$f(x)$
0.9	-2.9
0.99	-2.99
0.999	-2.999
1.001	-3.001
1.01	-3.01
1.1	-3.1

$\lim\limits_{x\to 1} f(x) = -3$

13.

x	$f(x)$
0.9	3.5
0.99	3.95
0.999	3.995
1.001	4.995999
1.01	4.9599
1.1	4.59

$\lim\limits_{x\to 1^-} f(x) = 4$ and $\lim\limits_{x\to 1^+} f(x) = 5$. These limits
differ so $\lim\limits_{x\to 1} f(x)$ does not exist

15. -1 **17.** -4 **19.** -2 **21.** 6
23. $3/4$ **25.** $3/2$ **27.** 0 **29.** does not exist
31. -3 **33.** does not exist **35.** does not exist
37. $3x^2$ **39.** $\frac{1}{30}$ **41.** does not exist
43. -4 **45.** 9
47.

a	$(1+a)^{1/a}$
0.1	2.5937
0.01	2.7048
0.001	2.7169
0.0001	2.7181
0.00001	2.71827
$\downarrow$	$\downarrow$
0	≈ 2.718

49. (a) 2 (b) 6 (c) -8 (d) $-\frac{1}{2}$
51. (a) -6 (b) -85 (c) $-33/17$
53. $150,000
55. (a) $32 (thousands) (b) $55.04 (thousands)
57. (a) $2800 (b) $700 (c) $560
59. (a) 1.52 units/hr (b) 0.85 units/hr (c) lunch
61. (a) $0; p \to 100^-$ means the water approaches not being
  treated (containing 100% or all of its impurities);
  the associated costs of nontreatment approach zero.
  (b) ∞ (c) no, because $C(0)$ is undefined
63. (a) $5081.25 (b) $5081.25 (c) $5081.25

65. $C(x) = \begin{cases} 12.76 + 15.96x & 0 \le x \le 10 \\ 172.36 + 13.56(x-10) & 10 < x \le 120 \\ 1675 & x > 120 \end{cases}$

$\lim\limits_{x\to 10} C(x) = 172.36$

67. 11,228.00. This corresponds to the Dow Jones opening
average.
69. (a) 33.7
  (b) This predicts the percent of obese Americans who
  are severely obese as the year approaches 2040.
  (c) Yes. The table shows increases of 3% to 6% per
  decade, and severe obesity is expected to get
  worse.

9.2 EXERCISES

1. (a) continuous
  (b) discontinuous; $f(1)$ does not exist
  (c) discontinuous; $\lim\limits_{x\to 3} f(x)$ does not exist
  (d) discontinuous; $f(0)$ does not exist and $\lim\limits_{x\to 0} f(x)$
  does not exist
3. continuous
5. discontinuous; $f(-3)$ does not exist
7. discontinuous; $\lim\limits_{x\to 2} f(x)$ does not exist
9. continuous
11. discontinuity at $x = -2$; $g(-2)$ and $\lim\limits_{x\to -2} g(x)$ do not
exist
13. continuous **15.** continuous
17. discontinuity at $x = -1$; $f(-1)$ does not exist
19. discontinuity at $x = 3$; $\lim\limits_{x\to 3} f(x)$ does not exist
21. vertical asymptote: $x = -2$;
  $\lim\limits_{x\to\infty} f(x) = 0$; $\lim\limits_{x\to -\infty} f(x) = 0$; $y = 0$
23. vertical asymptotes: $x = -2$, $x = 3$;
  $\lim\limits_{x\to\infty} f(x) = 2$; $\lim\limits_{x\to -\infty} f(x) = 2$; $y = 2$
25. (a) 0 (b) $y = 0$ is a horizontal asymptote.
27. (a) 1 (b) $y = 1$ is a horizontal asymptote.
29. (a) $5/3$ (b) $y = 5/3$ is a horizontal asymptote.
31. (a) does not exist (∞)
  (b) no horizontal asymptotes
33. (a)

$\lim\limits_{x\to +\infty} f(x) = 0.5$

  (b) The table indicates $\lim\limits_{x\to\infty} f(x) = 0.5$.
35. (a) $x = -1000$ (b) 1000
  (c) These values are so large that experimenting with
  windows may never locate them.

37. $\lim\limits_{x\to\infty} \dfrac{a_n + \dfrac{a_{n-1}}{x} + \cdots + \dfrac{a_1}{x^{n-1}} + \dfrac{a_0}{x^n}}{b_n + \dfrac{b_{n-1}}{x} + \cdots + \dfrac{b_1}{x^{n-1}} + \dfrac{b_0}{x^n}}$
$= \dfrac{a_n + 0 + \cdots + 0 + 0}{b_n + 0 + \cdots + 0 + 0} = \dfrac{a_n}{b_n}$

39. (a) no, not at $p = -8$ (b) yes (c) yes
  (d) $p > 0$
41. (a) yes, $q = -1$ (b) yes
43. (a) R/i (b) $10,000 **45.** yes, $0 \le p \le 100$

47. 100%; No, for p to approach 100% (as a limit) requires spending to increase without bound, which is impossible.

49. $R(x)$ is discontinuous at $x = 18,150; x = 73,800;$ $x = 148,850; x = 226,850; x = 405,100;$ and $x = 457,600$

51. (a) $158.80
   (b) $\lim_{x \to 100} C(x) = 38.80;\ \lim_{x \to 500} C(x) = 98.80$
   (c) yes

53. (a) $m(x) = 0.591x + 43.2;\ w(x) = 0.787x + 20.9$
   (b) $r(x) = \dfrac{0.591x + 43.2}{0.787x + 20.9}$
   (c) $\lim_{x \to 0} r(x) \approx 2.07$ means that for years approaching 1950 there were about 2.07 men per woman in the U.S. workforce.
   $\lim_{x \to 100} r(x) \approx 1.03$ means that for years approaching 2050 it is projected that there will be 1.03 men per woman in the U.S. workforce.
   (d) $\lim_{x \to \infty} r(x) \approx 0.751 \approx 3/4$ means that the long-term projection is for about 3 men per 4 women in the U.S. workforce.

9.3 EXERCISES

1. (a) 6   (b) 8   **3.** (a) $\frac{10}{3}$ (b) -5
5. (a) $-3.9, -3.99$ (b) $-4.1, -4.01$ (c) -4
7. (a) 32   (b) 32   (c) $(4, 64)$
9. (a) verification   (b) -8 (c) -8
   (d) $(-1, 5)$
11. (a) $P(1, 1), A(3, 0)$ (b) $-\frac{1}{2}$ (c) $-\frac{1}{2}$ (d) $-\frac{1}{2}$
13. (a) $P(1, 3), A(0, 3)$ (b) 0   (c) 0   (d) 0
15. (a) $f'(x) = 10x + 6$ (b) $10x + 6; -14$
   (c) -14
17. (a) $p'(q) = 4q + 1$ (b) $4q + 1; 41$ (c) 41
19. (a) 89.000024   (b) 89.0072   (c) ≈ 89
21. (a) 294.000008   (b) 294.0084   (c) ≈ 294
23. -31
25. (a) At A the slope is positive; at B it is negative.
   (b) $-1/3$
27. $f'(4) = 7/3; f(4) = -11$ **29.** $y = 5x - 14$
31. (a) a, b, d (b) c (c) A, C, E
33. (a) A, B, C, D (b) A, D
35. (a) $f'(x) = 2x + 1$ (b) $f'(2) = 5$
   (c) $y = 5x - 4$ (d)

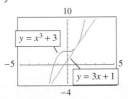

37. (a) $f'(x) = 3x^2$ (b) $f'(1) = 3$
   (c) $y = 3x + 1$
   (d)

39. (a) 43 dollars per unit   (b) 95.50 dollars per unit
   (c) The average cost per printer when 100 to 300 are produced (a) is $43 per printer, and the average cost when 300 to 600 are produced (b) is $95.50 per printer.
41. (a) $-100/3$ (b) $-4/3$
43. AB, AC, BC. Average rate is found from the slope of a segment; AB rises most slowly; BC is steepest.
45. (a) $R'(x) = \overline{MR} = 300 - 2x$
   (b) 200; the predicted change in revenue from selling the 51st unit is about $200.
   (c) -100; the predicted change in revenue from the 201st unit is about -100 dollars.
   (d) 0   (e) It changes from increasing to decreasing.
47. 200
49. (a) 100; the expected profit from the sale of the 201st car is $100.
   (b) -100; the expected profit from the sale of the 301st car is a loss of $100.
51. (a) 1.039
   (b) If humidity changes by 1%, the heat index will change by about $1.039°F$.
53. (a) Marginal revenue is given by the slope of the tangent line, which is steeper at 300 cell phones. Hence marginal revenue is greater for 300 cell phones.
   (b) Marginal revenue predicts the revenue from the next unit sold. Hence, the 301st item brings in more revenue because the marginal revenue for 300 cell phones is greater than for 700.

9.4 EXERCISES

1. $y' = 0$ **3.** $f'(t) = 1$
5. $y' = -8 + 4x = 4x - 8$ **7.** $f'(x) = 12x^3 - 6x^5$
9. $y' = 50x^4 - 9x^2 + 5$ **11.** $w'(z) = 7z^6 - 18z^5$
13. $g'(x) = 24x^{11} - 30x^5 + 36x^3 + 1$
15. (a) 30   (b) 30   **17.** (a) 6   (b) 6
19. $y' = -5x^{-6} - 8x^{-9} = -5/x^6 - 8/x^9$
21. $\dfrac{dz}{dt} = 11t^{8/3} - \frac{7}{2}t^{3/4} - \frac{1}{2}t^{-1/2}$
   $= 11\sqrt[3]{t^8} - \frac{7}{2}\sqrt[4]{t^3} - \dfrac{1}{2\sqrt{t}}$
23. $f'(x) = -4x^{-9/5} - \frac{8}{3}x^{-7/3}$
   $= \dfrac{-4}{\sqrt[5]{x^9}} - \dfrac{8}{3\sqrt[3]{x^7}}$
25. $g'(x) = \dfrac{-15}{x^6} - \dfrac{8}{x^5} + \dfrac{2}{\sqrt[3]{x^2}}$
27. $y = -7x + 10$ **29.** $y = 3$
31. $(1, -1), (5, 31)$ **33.** $(0, 9), (3, -18)$
35. (a) $-1/2$ (b) -0.5000 (to four decimal places)
37. (a) $f'(x) = 6x^2 + 5$
   (b)

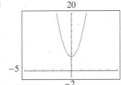

Graph of $f'(x)$ and numerical derivative of $f(x)$

39. (a) $h'(x) = -30x^{-4} + 4x^{-7/5} + 2x$

$$= \frac{-30}{x^4} + \frac{4}{\sqrt[5]{x^7}} + 2x$$

(b)

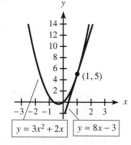

Graph of $h'(x)$ and numerical derivative of $h(x)$

41. (a) $y = 8x - 3$

(b)   (c) Yes

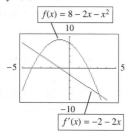

43. (a) $f'(x) = -2 - 2x$

(b)

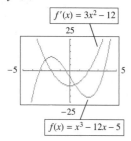

(c) $f'(x) = 0$ at $x = -1$; $f'(x) > 0$ for $x < -1$; $f'(x) < 0$ for $x > -1$

(d) $f(x)$ has a maximum when $x = -1$.
$f(x)$ rises for $x < -1$.
$f(x)$ falls for $x > -1$.

45. (a) $f'(x) = 3x^2 - 12$

(b)

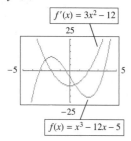

(c) $f'(x) = 0$ at $x = -2$ and $x = 2$
$f'(x) > 0$ for $x < -2$ and $x > 2$
$f'(x) < 0$ for $-2 < x < 2$

(d) $f(x)$ has a maximum when $x = -2$, a minimum when $x = 2$
$f(x)$ rises when $x < -2$ and when $x > 2$
$f(x)$ falls when $-2 < x < 2$

47. (a) 40; the expected change in revenue from the 301st unit is about $40

(b) -20; the expected change in revenue from the 601st unit is about -20 dollars

49. $\dfrac{dq}{dM} = \dfrac{3k}{4M^{1/4}}$

51. (a) -4; if the price changes to $26, the quantity demanded will change by approximately -4 units

(b) $-\frac{1}{2}$; if the price changes to $101, the quantity demanded will change by approximately $-\frac{1}{2}$ unit

53. (a) $\overline{C}'(x) = (-4000/x^2) + 0.1$ (b) 200

(c) $C'(200) = \overline{C}(200) = 95$

55. (a) -1200

(b) If the impurities change from 10% to 11%, then the cost decreases by about $1200.

57. (a) $WC = 45.0625 - 29.3375s^{0.16}$

(b) -0.31

(c) At 15° F, if the wind speed changes by $+1$ mph (to 26 mph), then the wind chill will change by approximately $-0.31°$F.

59. (a) $y = 0.0682x^2 + 1.76x + 96.0$

(b) $2.57 per year

(c) $\dfrac{dy}{dx} = 0.1364x + 1.76$

(d) $\dfrac{dy}{dx}\Big|_{x=10} = 3.124$; about $3.12 per year

(e) $126.81

61. (a) $P(t) = -0.0000738t^3 + 0.0102t^2 + 2.20t + 163$

(b) $P'(t) = -0.0002214t^2 + 0.0204t + 2.20$

(c) 2000: $P'(50) \approx 2.67$ means that for 2001, the U.S. population will rise by about 2.67 million people.
2025: $P'(75) \approx 2.48$ means that for 2026, the U.S. population is expected to rise by about 2.48 million people.

9.5 EXERCISES

1. $y' = 15x^2 - 14x - 6$

3. $f'(x) = (x^{12} + 3x^4 + 4)(6x^2) + (2x^3 - 1)(12x^{11} + 12x^3)$
$= 30x^{14} - 12x^{11} + 42x^6 - 12x^3 + 24x^2$

5. $y' = (7x^6 - 5x^4 + 2x^2 - 1)(36x^8 + 21x^6 - 10x + 3)$
$+ (4x^9 + 3x^7 - 5x^2 + 3x)(42x^5 - 20x^3 + 4x)$

7. $y' = (x^2 + x + 1)(\frac{1}{3}x^{-2/3} - x^{-1/2}) + (x^{1/3} - 2x^{1/2} + 5)(2x + 1)$

9. (a) 40   (b) 40

11. $\dfrac{dp}{dq} = \dfrac{2q^2 - 2q - 6}{(2q - 1)^2}$

13. $\dfrac{dy}{dx} = \dfrac{4x^5 - 4x^3 - 16x}{(x^4 - 2x^2 + 5)^2}$

15. $\dfrac{dz}{dx} = 2x + \dfrac{2x - x^2}{(1 - x - 2x^2)^2}$

17. $\dfrac{dp}{dq} = \dfrac{2q + 1}{\sqrt[3]{q^2}(1 - q)^2}$

19. $y' = \dfrac{2x^3 - 6x^2 - 8}{(x - 2)^2}$

21. (a) $\frac{3}{5}$ (b) $\frac{3}{5}$ **23.** $y = 44x - 32$

25. $y = \frac{10}{3}x - \frac{10}{3}$ **27.** 104

29. 1.3333 (to four decimal places)

31. (a) $f'(x) = 3x^2 - 6x - 24$

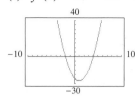

Graph of both $f'(x)$ and numerical derivative of $f(x)$

(b) Horizontal tangents where $f'(x) = 0$; at $x = -2$ and $x = 4$

(c)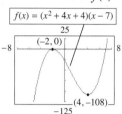

33. (a) $y' = \dfrac{x^2 - 4x}{(x - 2)^2}$

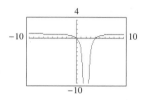

Graph of both y' and the numerical derivative of y

(b) Horizontal tangents where $y' = 0$; at $x = 0$ and $x = 4$

(c)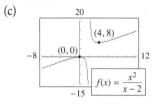

35. (a) $f'(x) = \dfrac{20x}{(x^2 + 1)^2}$ (b)

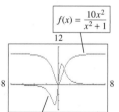

(c) $f' = 0$ at $x = 0$
  $f' > 0$ for $x > 0$
  $f' < 0$ for $x < 0$

(d) f has a minimum at $x = 0$.
  f is increasing for $x > 0$.
  f is decreasing for $x < 0$.

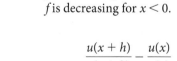

37. $f'(x) = \lim\limits_{h \to 0} \dfrac{\dfrac{u(x + h)}{v(x + h)} - \dfrac{u(x)}{v(x)}}{h}$

$= \lim\limits_{h \to 0} \dfrac{u(x + h)v(x) - u(x)v(x + h)}{h \cdot v(x)v(x + h)}$

$= \lim\limits_{h \to 0} \dfrac{u(x + h)v(x) - u(x)v(x) + u(x)v(x) - u(x)v(x + h)}{h \cdot v(x)v(x + h)}$

$= \lim\limits_{h \to 0} \dfrac{v(x)\left[\dfrac{u(x + h) - u(x)}{h}\right] - u(x)\left[\dfrac{v(x + h) - v(x)}{h}\right]}{v(x)v(x + h)}$

$= \dfrac{v(x)u'(x) - u(x)v'(x)}{[v(x)]^2}$

39. $C'(p) = 810{,}000/(100 - p)^2$

41. $R'(49) \approx 30.00$ The expected revenue from the sale of the next unit (the 50th) is about \$30.00.

43. $R'(5) = -50$ As the group changes by 1 person (to 31), the revenue will drop by about \$50.

45. $S = 1000x - x^2$

47. $\dfrac{dR}{dn} = \dfrac{r(1 - r)}{[1 + (n - 1)r]^2}$

49. (a) $P'(6) \approx 0.045$ During the next (7th) month of the campaign, the proportion of voters who recognize the candidate will change by about 0.045, or 4.5%.

(b) $P'(12) \approx -0.010$ During the next (13th) month of the campaign, the proportion of voters who recognize the candidate will drop by about 0.010, or 1%.

(c) It is better for $P'(t)$ to be positive—that is, to have increasing recognition.

51. (a) $f'(20) \approx -0.79$

(b) At 0°F, if the wind speed changes by 1 mph (to 21 mph), the wind chill will change by about $-0.79°$F.

53. (a) $B'(t) = (0.01t + 3)(0.0476t - 9.79)$
  $+ (0.01)(0.0238t^2 - 9.79t + 3100)$

(b) $B'(70) \approx 1.42$ means that in 2020 the number of beneficiaries will be changing at the rate of about 1.42 million per year.

(c) [2010, 2020]: 1.55
  [2020, 2030]: 1.39
  [2010, 2030]: 1.47
  The average rate over [2020, 2030] is best but is still off by almost 0.03 million per year.

55. (a) $p'(t) = \dfrac{2154.06}{(1.38t + 64.1)^2}$

(b) 2005: $p'(55) \approx 0.110$; 2020: $p'(70) \approx 0.083$

(c) $p'(55)$ means that in 2005, the percent of women in the workforce was changing about 0.110 percentage points per year. $p'(70)$ predicts the rate in 2020 will be about 0.083 percentage points per year.

9.6 EXERCISES

1. $\dfrac{dy}{du} = 3u^2, \dfrac{du}{dx} = 2x, \dfrac{dy}{dx} = 3u^2 \cdot 2x = 6x(x^2 + 1)^2$

3. $\dfrac{dy}{du} = 4u^3, \dfrac{du}{dx} = 8x - 1, \dfrac{dy}{dx} = 4u^3(8x - 1)$
  $= 4(8x - 1)(4x^2 - x + 8)^3$

5. $f'(x) = 20(3x^5 - 2)^{19}(15x^4) = 300x^4(3x^5 - 2)^{19}$

7. $h'(x) = 6(x^5 - 2x^3 + 5)^7(5x^4 - 6x^2)$
  $= 6x^2(5x^2 - 6)(x^5 - 2x^3 + 5)^7$

9. $s'(t) = 5 - 9(2t^4 + 7)^2(8t^3) = 5 - 72t^3(2t^4 + 7)^2$

11. $g'(x) = -2(x^4 - 5x)^{-3}(4x^3 - 5) = \dfrac{-2(4x^3 - 5)}{(x^4 - 5x)^3}$

13. $f'(s) = -12(2s^5 + 1)^{-5}(10s^4) = \dfrac{-120s^4}{(2s^5 + 1)^5}$

15. $g'(x) = -\dfrac{3}{4}(2x^3 + 3x + 5)^{-7/4}(6x^2 + 3)$
  $= \dfrac{-3(6x^2 + 3)}{4(2x^3 + 3x + 5)^{7/4}}$

17. $y' = \frac{1}{2}(3x^2 + 4x + 9)^{-1/2}(6x + 4)$

$= \frac{3x + 2}{\sqrt{3x^2 + 4x + 9}}$

19. $y' = \frac{66}{9}(x^3 - 7)^5(3x^2) = 22x^2(x^3 - 7)^5$

21. $\frac{dz}{dw} = \frac{15(3w + 1)^4 - 3}{7}$

23. (a) and (b) 96,768
25. (a) and (b) 2
27. $y = 3x - 5$ **29.** $9x - 5y = 2$
31. (a) $f'(x) = 6x(x^2 - 4)^2$
   (b)

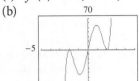

   (c) $x = 0, x = 2,$
       $x = -2$
   (d) $(0, -52),$
       $(2, 12),$
       $(-2, 12)$

Graph of both $f'(x)$ and numerical derivative of $f(x)$
   (e)

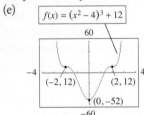

33. (a) $f'(x) = 8x(1 - x^2)^{1/3}$
   (b)

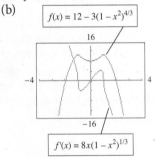

   (c) $f'(x) = 0$ at $x = -1, x = 0, x = 1$
       $f'(x) > 0$ for $x < -1$ and $0 < x < 1$
       $f'(x) < 0$ for $-1 < x < 0$ and $x > 1$
   (d) $f(x)$ has a maximum at $x = -1$ and $x = 1$, a
       minimum at $x = 0$.
       $f(x)$ is increasing for $x < -1$ and $0 < x < 1$.
       $f(x)$ is decreasing for $-1 < x < 0$ and $x > 1$.
35. (a) $y' = 2x^2$ (b) $y' = -2/x^4$

   (c) $y' = 2(2x)^2$ (d) $y' = \frac{-18}{(3x)^4}$

37. 120 in./sec
39. \$1499.85 (approximately); if a 101st unit is sold,
   revenue will change by about \$1499.85
41. (a) -0.114 (approximately)
   (b) If the price changes by \$1, to \$22, the weekly sales
       volume will change by approximately -0.114
       thousand units.

43. (a) $-\$3.20$ per unit
   (b) If the quantity demanded changes from 49 to 50
       units, the change in price will be about $-\$3.20$.

45. $\frac{dy}{dx} = \left(\frac{8k}{5}\right)(x - x_0)^{3/5}$

47. $\frac{dp}{dq} = -100(2q + 1)^{-3/2} = \frac{-100}{(2q + 1)^{3/2}}$

49. $\frac{dK_c}{dv} = 8(4v + 1)^{-1/2} = \frac{8}{\sqrt{4v + 1}}$

51. (a) \$658.75. If the interest changed from 6% to 7%,
       the amount of the investment would change by
       about \$658.75.
   (b) \$2156.94. If the interest rate changed from 12% to
       13%, the amount of the investment would change
       by about \$2156.94.
53. (a) 2008: $A'(8) \approx 126.3$; 2015: $A'(15) \approx 231.4$
       These mean that the total national expendi-
       tures for health are predicted to change by about
       \$126.3 billion from 2008 to 2009 and about
       \$231.4 billion from 2015 to 2016.
   (b) The average rate for 2014 to 2015 is best: \$228 bil-
       lion/year.
55. (a) 2005: $G'(5) \approx 1370.64$; 2015: $G'(15) \approx 934.56$
       These mean that the rate of change of the GDP
       was about \$1370.64 billion per year in 2005 and
       about \$934.5 billion per year in 2015.
   (b) 912.5 (billion per year)
   (c) 2010: $G'(10) \approx 1024.86$. The answer from part (b)
       is not a good approximation to $G'(10)$.

9.7 EXERCISES

1. 0   **3.** $4(-4x^{-5})$; $-16/x^5$
5. $15x^2 + 4(-x^{-2})$; $15x^2 - 4/x^2$
7. $(x^2 - 2)1 + (x + 4)(2x)$; $3x^2 + 8x - 2$
9. $\frac{u^2(3u^2) - (u^3 + 1)(2u)}{(u^2)^2}$; $(u^3 - 2)/u^3$
11. $(3x^2 - 4)(x^3 - 4x)^9$
13. $\frac{5}{3}x^3[3(4x^5 - 5)^2(20x^4)] + (4x^5 - 5)^3(5x^2)$;
   $5x^2(4x^5 - 5)^2(24x^5 - 5)$
15. $(x - 1)^2(2x) + (x^2 + 1)2(x - 1)$;
   $2(x - 1)(2x^2 - x + 1)$
17. $\frac{(x^2 + 1)3(x^2 - 4)^2(2x) - (x^2 - 4)^3(2x)}{(x^2 + 1)^2}$;
   $\frac{2x(x^2 - 4)^2(2x^2 + 7)}{(x^2 + 1)^2}$
19. $3[(q + 1)(q^3 - 3)]^2[(q + 1)3q^2 + (q^3 - 3)1]$;
   $3(4q^3 + 3q^2 - 3)[(q + 1)(q^3 - 3)]^2$
21. $4[x^2(x^2 + 3x)]^3[x^2(2x + 3) + (x^2 + 3x)(2x)]$;
   $4x^2(4x + 9)[x^2(x^2 + 3x)]^3 = 4x^{11}(4x + 9)(x + 3)^3$
23. $4\left(\frac{2x - 1}{x^2 + x}\right)^3\left[\frac{(x^2 + x)2 - (2x - 1)(2x + 1)}{(x^2 + x)^2}\right]$;
   $\frac{4(-2x^2 + 2x + 1)(2x - 1)^3}{(x^2 + x)^5}$

25. $(8x^4 + 3)^2 3(x^3 - 4x)^2(3x^2 - 4) +$
$(x^3 - 4x)^3 2(8x^4 + 3)(32x^3);$
$(8x^4 + 3)(x^3 - 4x)^2(136x^6 - 352x^4 + 27x^2 - 36)$

27. $\dfrac{(4 - x^2)\frac{1}{3}(x^2 + 5)^{-2/3}(2x) - (x^2 + 5)^{1/3}(-2x)}{(4 - x^2)^2};$

$\dfrac{2x(2x^2 + 19)}{3\sqrt[3]{(x^2 + 5)^2}(4 - x^2)^2}$

29. $(x^2)\frac{1}{4}(4x - 3)^{-3/4}(4) + (4x - 3)^{1/4}(2x);$
$(9x^2 - 6x)/\sqrt[4]{(4x - 3)^3}$

31. $(2x)\frac{1}{2}(x^3 + 1)^{-1/2}(3x^2) + (x^3 - 1)^{1/2}(2);$
$(5x^3 + 2)/\sqrt{x^3 + 1}$

33. (a) $F_1'(x) = 12x^3(x^4 + 1)^4$

(b) $F_2'(x) = \dfrac{-12x^3}{(x^4 + 1)^6}$

(c) $F_3'(x) = 12x^3(3x^4 + 1)^4$

(d) $F_4'(x) = \dfrac{-300x^3}{(5x^4 + 1)^6}$

35. $dP/dx = 90(3x + 1)^2$

37. (a) \$59,900 per camper
(b) An 11th camper sold would change revenue by about \$59,900.

39. $dC/dy = 1/\sqrt{y + 1} + 0.4$

41. $dV/dx = 144 - 96x + 12x^2$

43. -1.6; This means that from the 9th to the 10th week, sales are expected to decrease by about \$1600.

45. (a) 2005: \$350/year; 2015: \$549/year
(b) In 2015, the per capita expenditures for health care are predicted to be changing at a rate of about \$549 per year.
(c) Average rate = 380; This approximates the instantaneous rate in 2005 quite well.

9.8 EXERCISES

1. $f''(x) = 180x^8 - 360x^3 - 72x$ **3.** $g''(x) = 6x - 2x^{-3}$
5. $d^2y/dx^2 = 6x + \frac{1}{4}x^{-3/2}$ **7.** $d^3y/dx^3 = 60x^2 - 96$
9. $f'''(x) = 1008x^6 - 720x^3$ **11.** $d^3y/dx^3 = -6/x^4$
13. $d^2y/dx^2 = 20x^3 + \frac{1}{4}x^{-3/2}$ **15.** $f'''(x) = \frac{3}{8}(x + 1)^{-5/2}$
17. $d^4y/dx^4 = 0$ **19.** $f^{(4)}(x) = -15/(16x^{7/2})$
21. $y^{(4)} = 24(4x - 1)^{-5/2}$ **23.** $f^{(6)}(x) = -2(x + 1)^{-3}$
25. 26   **27.** 16.0000 (to four decimal places)
29. 0.0004261
31. (a) $f'(x) = 3x^2 - 6x$ $f''(x) = 6x - 6$

(b)

$f'(x) = 3x^2 - 6x$
$f''(x) = 6x - 6$
15
-3 5
-10
$f(x) = x^3 - 3x^2 + 5$

(c) $f''(x) = 0$ at $x = 1$
$f''(x) > 0$ for $x > 1$
$f''(x) < 0$ for $x < 1$

(d) $f'(x)$ has a minimum at $x = 1$.
$f'(x)$ is increasing for $x > 1$.
$f'(x)$ is decreasing for $x < 1$.
(e) $f''(x) < 0$ (f) $f''(x) > 0$.

33. $a = 0.12$ m/sec^2

35. -0.02 \$/unit per unit

37. (a) $\dfrac{dR}{dm} = mc - m^2$ (b) $\dfrac{d^2R}{dm^2} = c - 2m$

(c) second

39. (a) 0.0009 (approximately)
(b) When 1 more unit is sold (beyond 25), the marginal revenue will change by about 0.0009 thousand dollars per unit, or \$0.90 per unit.

41. (a) $S' = \dfrac{-3}{(t + 3)^2} + \dfrac{36}{(t + 3)^3}$ (b) $S''(15) = 0$

(c) After 15 weeks, the rate of change of the rate of sales is zero because the rate of sales reaches a minimum value.

43. (a) $W'(t) \approx 0.0447t^{1.11}$ (b) $W''(t) \approx 0.0497t^{0.11}$
(c) $W(50) \approx 81.5$, $W'(50) \approx 3.44$, and $W''(50) \approx 0.0764$. These mean that in 2025 the average annual wage is expected to be \$81,500 and changing at a rate of about \$3440 per year. Also that rate is expected to be changing about \$76.40 per year per year.

45. (a) $R(x) = -0.000190x^3 + 0.0519x^2 - 4.06x + 192$
(b) $R'(x) = -0.000570x^2 + 0.1038x - 4.06$
(c) $R''(x) = -0.00114x + 0.1038$
(d) $R'(90) \approx 0.665; R''(90) \approx 0.0012$
In 2040, the economic dependency ratio is expected to be changing at a rate of about 0.665 per year, and this rate is expected to be changing about 0.0012 per year per year.

47. (a) $f'(x) = 0.002592x^2 - 0.256x + 6.61$
(b) At age 25: about \$1830 per year and at age 55: about \$371 per year
(c) $f''(x) = 0.005184x - 0.256$
(d) At age 25: about $-\$126$ per year per year of age and at age 55: about \$29 per year per year of age
(e) The rate of change in the median income of 55-year-old workers is about \$371 per year of age, and it is increasing at a rate of about \$29 per year per year of age.

9.9 EXERCISES

1. (a) $\overline{MR} = 4$
(b) The sale of each additional item brings in \$4 revenue at all levels of production.

3. (a) \$3500; this is revenue from the sale of 100 units.
(b) $\overline{MR} = 36 - 0.02x$
(c) \$34; Revenue will increase by about \$34 if a 101st item is sold and by about \$102 if 3 additional units past 100 units are sold.
(d) Actual revenue from the sale of the 101st item is \$33.99.

5. (a) $R(x) = 80x - 0.4x^2$ (in hundreds of dollars)
   (b) 7500 subscribers ($x = 75$); $R = \$375{,}000$
   (c) Lower the price per month.
   (d) $\overline{MR} = R'(x) = 80 - 0.8x$; when $p = 50$, $x = 75$
       $\overline{MR}(75) = 20$ means that if the number of
       customers increased from 75 to 76 (hundred),
       the revenue would increase by about 20 (hun-
       dred dollars), or $2000. This means the company
       should try to increase subscribers by lowering its
       monthly charge.

7. (a)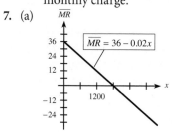
       (b) $x = 1800$
       (c) $\$32{,}400$

9. $\overline{MC} = 8$ **11.** $\overline{MC} = 13 + 2x$
13. $\overline{MC} = 3x^2 - 12x + 24$ **15.** $\overline{MC} = 27 + 3x^2$
17. (a) $10; the cost will increase by about $10.
   (b) $11
19. $46; the cost will increase by about $46. For 3
   additional units, the cost will increase by about $138.
21.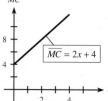

23. (a) The 101st item costs more. The tangent line slope
       is greater at $x = 100$ than at $x = 500$, and the slope
       of the tangent line gives the marginal cost and
       predicts the cost of the next item.
   (b) More efficient. As x increases, the slopes of the
       tangents decrease. This means that the costs of
       additional items decrease as x increases.
25. $\overline{MP} = 5$; This means that for each additional unit sold,
   profit changes by $5.
27. (a) $5600   (b) $\overline{MP} = 20 - 0.02x$
   (c) 10; profit will increase by about $10 if a 501st unit
       is sold.
   (d) 9.99; the sale of the 501st item results in a profit
       of $9.99.
29. (a) $P(x) = R(x) - C(x)$, so profit is the distance
       between $R(x)$ and $C(x)$ (when $R(x)$ is above $C(x)$).
       $P(100) < P(700) < P(400)$; $P(100) < 0$, so there is
       a loss when 100 units are sold.
   (b) This asks us to rank $\overline{MP}(100)$, $\overline{MP}(400)$,
       and $\overline{MP}(700)$. Because $\overline{MP} = \overline{MR} - \overline{MC}$,
       compare the slopes of the tangents to
       $R(x)$ and $C(x)$ at the three x-values. Thus
       $\overline{MP}(700) < \overline{MP}(400) < \overline{MP}(100)$. $\overline{MP}(700) < 0$
       because $C(x)$ is steeper than $R(x)$ at $x = 700$. At
       $x = 100$, $R(x)$ is much steeper than $C(x)$.

31. (a) $A < B < C$. Amount of profit is the height of the
       graph. There is a loss at A.
   (b) $C < B < A$. Marginal profit is the slope of the
       tangent to the graph. Marginals (slopes) are posi-
       tive at all three points.
33. (a)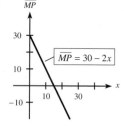
       (b) 15 hundred units
       (c) 15 hundred units
       (d) $25 thousand

35. 700   **37.** $9000

CHAPTER 9 REVIEW EXERCISES

1. (a) 2   (b) 2   **2.** (a) 0   (b) 0
3. (a) 2   (b) 1   **4.** (a) 2   (b) does not exist
5. (a) does not exist   (b) 2
6. (a) does not exist   (b) does not exist
7. 55   **8.** 0   **9.** -2 **10.** 4/5   **11.** $\frac{1}{2}$
12. $\frac{1}{5}$ **13.** no limit   **14.** 0   **15.** 4
16. no limit   **17.** 3   **18.** no limit   **19.** $6x$
20. $1 - 4x$ **21.** -14 **22.** 5
23. (a) yes   (b) no   **24.** (a) yes   (b) no
25. 2   **26.** no limit   **27.** 1   **28.** no   **29.** yes
30. yes   **31.** discontinuity at $x = 5$
32. discontinuity at $x = 2$ **33.** continuous
34. discontinuity at $x = 1$
35. (a) $x = 0, x = 1$ (b) 0   (c) 0
36. (a) $x = -1, x = 0$ (b) $\frac{1}{2}$ (c) $\frac{1}{2}$
37. $-2; y = -2$ is a horizontal asymptote.
38. $0; y = 0$ is a horizontal asymptote.
39. 7   **40.** true   **41.** false
42. $f'(x) = 6x + 2$ **43.** $f'(x) = 1 - 2x$
44. $[-1, 0]$; the segment over this interval is steeper.
45. (a) no   (b) no   **46.** (a) yes   (b) no
47. (a) -5.9171 (to four decimal places)   (b) -5.9
48. (a) 4.9/3   (b) 7   **49.** about $-1/4$
50. B, C, A: $B < 0$ and $C < 0$; the tangent line at $x = 6$ falls
   more steeply than the segment over $[2, 10]$.
51. $20x^4 - 18x^2$ **52.** $90x^8 - 30x^5 + 4$
53. 1   **54.** $1/(2\sqrt{x})$ **55.** 0   **56.** $-4/(3\sqrt[3]{x^4})$
57. $\dfrac{-1}{x^2} + \dfrac{1}{2\sqrt{x^3}}$ **58.** $\dfrac{-3}{x^3} - \dfrac{1}{3\sqrt[3]{x^2}}$
59. $y = 15x - 18$ **60.** $y = 34x - 48$
61. (a) $x = 0, x = 2$ (b) $(0, 1)$ $(2, -3)$
   (c)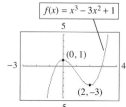
62. (a) $x = 0, x = 2, x = -2$
   (b) $(0, 8)$ $(2, -24)(-2, -24)$

(c)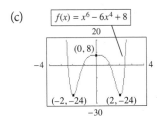

$f(x) = x^6 - 6x^4 + 8$

63. $9x^2 - 26x + 4$ **64.** $21x^6 + 4x^3 + 27x^2$

65. $\dfrac{15q^2}{(2q^3 + 1)^2}$ **66.** $\dfrac{1 - 3t}{[2\sqrt{t}(3t + 1)^2]}$ **67.** $\dfrac{9x + 2}{2\sqrt{x}}$

68. $\dfrac{5x^6 + 2x^4 + 20x^3 - 3x^2 - 4x}{(x^3 + 1)^2}$

69. $(9x^2 - 24x)(x^3 - 4x^2)^2$

70. $6(30x^5 + 24x^3)(5x^6 + 6x^4 + 5)^5$

71. $72x^3(2x^4 - 9)^8$ **72.** $\dfrac{-(3x^2 - 4)}{2\sqrt{(x^3 - 4x)^3}}$

73. $2x(2x^4 + 5)^7(34x^4 + 5)$ **74.** $\dfrac{-2(3x + 1)(x + 12)}{(x^2 - 4)^2}$

75. $36[(3x + 1)(2x^3 - 1)]^{11}(8x^3 + 2x^2 - 1)$

76. $\dfrac{3}{(1 - x)^4}$ **77.** $\dfrac{(2x^2 - 4)}{\sqrt{x^2 - 4}}$ **78.** $\dfrac{2x - 1}{(3x - 1)^{4/3}}$

79. $y'' = \dfrac{-1}{4}x^{-3/2} - 2$ **80.** $y'' = 12x^2 - 2/x^3$

81. $\dfrac{d^5y}{dx^5} = 0$ **82.** $\dfrac{d^5y}{dx^5} = -30(1 - x)$

83. $\dfrac{d^3y}{dx^3} = -4[(x^2 - 4)^{-3/2}]$ **84.** $\dfrac{d^4y}{dx^4} = \dfrac{2x(x^2 - 3)}{(x^2 + 1)^3}$

85. (a) $400,000 (b) $310,000 (c) $90,000

86. (a) $70,000; this is fixed costs.
   (b) $0; $x = 1000$ is break-even.

87. (a) $140 per unit
   (b) $\overline{C}(x) \to \infty$; the limit does not exist.

88. (a) $C(x) \to \infty$; the limit does not exist. As the number of units produced increases without bound, so does the total cost.
   (b) $60 per unit. As more units are produced, the average cost of each unit approaches $60.

89. The average annual percent change of (a) elderly men in the workforce is 0.0825 percentage points per year and of (b) elderly women in the workforce is 0.0775 percentage points per year.

90. (a) Annual average rate of change of percent of elderly men in the workforce:
   1970–1980: -0.78 percentage points per year
   2030–2040: -0.23 percentage points per year
   (b) Annual average rate of change of percent of elderly women in the workforce
   1970–1980: -0.16 percentage points per year
   2030–2040: -0.16 percentage points per year

91. (a) $x'(10) = -1$ means that if price changes from $10 to $11, the number of units demanded will change by about -1.
   (b) $x'(20) = -\frac{1}{4}$ means that if price changes from $20 to $21, the number of units demanded will change by about $-\frac{1}{4}$.

92. $h(100) \approx 4.15$; $h'(100) \approx 0.08$ means that when the updraft speed is 100 mph, the hail diameter is about 4.15 inches (softball-sized) and changing at the rate of 0.08 inch per mph of updraft.

93. The slope of the tangent at A gives $\overline{MR}(A)$. The tangent line at A is steeper (so has greater slope) than the tangent line at B. Hence, $\overline{MR}(A) > \overline{MR}(B)$, so the $(A + 1)$st unit will bring more revenue.

94. $R'(10) = 1570$. Raised. An 11th rent increase of $30 (and hence an 11th vacancy) would change revenue by about $1570.

95. (a) $P(20) = 23$ means productivity is 23 units per hour after 20 hours of training and experience.
   (b) $P'(20) \approx 1.4$ means that the 21st hour of training or experience will change productivity by about 1.4 units per hour.

96. $\dfrac{dq}{dp} = \dfrac{-p}{\sqrt{0.02p^2 + 500}}$

97. $x'(10) = \frac{1}{6}$ means if price changes from $10 to $11, the number of units supplied will change by about $\frac{1}{6}$.

98. $s''(t) = a = -2t^{-3/2}$; $s''(4) = -0.25$ ft/sec/sec

99. $P'(x) = 70 - 0.2x$; $P''(x) = -0.2$
   $P'(300) = 10$ means that the 301st unit brings in about $10 in profit.
   $P''(300) = -0.2$ means that marginal profit ($P'(x)$) is changing at the rate of -0.2 dollars per unit, per unit.

100. (a) $\overline{MC} = 6x + 6$ (b) 186
   (c) If a 31st unit is produced, costs will change by about $186.

101. $C'(4) = 53$ means that a 5th unit produced would change total costs by about $53.

102. (a) $\overline{MR} = 40 - 0.04x$ (b) $x = 1000$ units

103. $\overline{MP}(10) = 48$ means that if an 11th unit is sold, profit will change by about $48.

104. (a) $\overline{MR} = 80 - 0.08x$ (b) 72
   (c) If a 101st unit is sold, revenue will change by about $72.

105. $\dfrac{120x(x + 1)}{(2x + 1)^2}$ **106.** $\overline{MP} = 4500 - 3x^2$

107. $\overline{MP} = 16 - 0.2x$

108. (a) C: Tangent line to $R(x)$ has smallest slope at C, so $\overline{MR}(C)$ is smallest and the next item at C will earn the least revenue.
   (b) B: $R(x) > C(x)$ at both B and C. Distance between $R(x)$ and $C(x)$ gives the amount of profit and is greatest at B.
   (c) A: $\overline{MR}$ greatest at A and $\overline{MC}$ least at A, as seen from the slopes of the tangents. Hence $\overline{MP}(A)$ is greatest, so the next item at A will give the greatest profit.
   (d) C: $\overline{MC}(C) > \overline{MR}(C)$, as seen from the slopes of the tangents. Hence $\overline{MP}(C) < 0$, so the next unit sold reduces profit.

CHAPTER 9 TEST

1. (a) $\frac{3}{4}$ (b) $-8/5$ (c) $9/8$ (d) does not exist

2. (a) $f'(x) = \lim\limits_{h \to 0} \dfrac{f(x+h) - f(x)}{h}$

   (b) $f'(x) = 6x - 1$

3. $x = 0, x = 8$

4. (a) $\dfrac{dB}{dW} = 0.523$ (b) $p'(t) = 90t^9 - 42t^6 - 17$

   (c) $\dfrac{dy}{dx} = \dfrac{99x^2 - 24x^9}{(2x^7 + 11)^2}$

   (d) $f'(x) = (3x^5 - 2x + 3)(40x^9 + 40x^3) + (4x^{10} + 10x^4 - 17)(15x^4 - 2)$

   (e) $g'(x) = 9(10x^4 + 21x^2)(2x^5 + 7x^3 - 5)^{11}$

   (f) $y' = 2(8x^2 + 5x + 18)(2x + 5)^5$

   (g) $f'(x) = \dfrac{6}{\sqrt{x}} + \dfrac{20}{x^3}$

5. $\dfrac{d^3y}{dx^3} = 6 + 60x^{-6}$

6. (a) $y = -15x - 5$ (b) $(4, -90), (-2, 18)$

7. -15 8. (a) 2 (b) does not exist   (c) -4

9. $g(-2) = 8;\ \lim\limits_{x \to -2^-} g(x) = 8,\ \lim\limits_{x \to -2^+} g(x) = -8$

   $\therefore \lim\limits_{x \to -2} g(x)$ does not exist and $g(x)$ is not continuous at $x = -2$.

10. (a) $\overline{MR} = R'(x) = 250 - 0.02x$

    (b) $R(72) = 17{,}948.16$ means that when 72 units are sold, revenue is \$17,948.16.
    $R'(72) = 248.56$ means that the expected revenue from the 73rd unit is about \$248.56.

11. (a) $P(x) = 50x - 0.01x^2 - 10{,}000$

    (b) $\overline{MP} = 50 - 0.02x$

    (c) $\overline{MP}(1000) = 30$ means that the predicted profit from the sale of the 1001st unit is approximately \$30.

12. 104

13. (a) -5 (b) -1 (c) 4 (d) does not exist
    (e) 2 (f) $3/2$ (g) $-4, 1, 3, 6$
    (h) $-4, 3, 6$
    (i) $f'(-2) <$ average rate over $[-2, 2] < f'(2)$

14. (a) $2/3$ (b) -4 (c) $2/3$

15. (a) B: $R(x) > C(x)$ at B, so there is profit. Distance between $R(x)$ and $C(x)$ gives the amount of profit.
    (b) A: $C(x) > R(x)$
    (c) A and B: slope of $R(x)$ is greater than the slope of $C(x)$. Hence $\overline{MR} > \overline{MC}$ and $\overline{MP} > 0$.
    (d) C: Slope of $C(x)$ is greater than the slope of $R(x)$. Hence $\overline{MC} > \overline{MR}$ and $\overline{MP} < 0$.

10.1 EXERCISES

1. (a) $(1, 5)$ (b) $(4, 1)$ (c) $(-1, 2)$
3. (a) $(1, 5)$ (b) $(4, 1)$ (c) $(-1, 2)$
5. (a) $3, 7$ (b) $3 < x < 7$ (c) $x < 3, x > 7$
   (d) 7 (e) 3

7. (a) $x = 0, x = 4$
   (b)
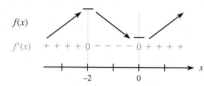

   min: $(4, -58)$; max: $(0, 6)$

9. (a) $x = -2, x = 0$

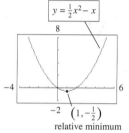

   (b) max: $(-2, 5)$; min: $(0, -11)$

11. (a) max: $(-1, 6)$; min: $(1, 2)$
    (b) $dy/dx = 3x^2 - 3;$ $x = 1, x = -1$
    (c) $(1, 2), (-1, 6)$
    (d) yes

13. (a) HPI: $(-1, -3)$
    (b) $dy/dx = 3x^2 + 6x + 3;$ $x = -1$
    (c) $(-1, -3)$
    (d) yes

15. (a) $\dfrac{dy}{dx} = x - 1$ (e)
    (b) $x = 1$
    (c) $\left(1, -\dfrac{1}{2}\right)$
    (d) decreasing: $x < 1$
        increasing: $x > 1$

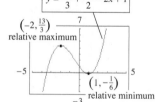

17. (a) $dy/dx = x^2 + x - 2$
    (b) $x = -2, x = 1$ (c) $\left(-2, \dfrac{13}{3}\right), \left(1, -\dfrac{1}{6}\right)$
    (d) increasing: $x < -2$ and $x > 1$
        decreasing: $-2 < x < 1$
    (e)

19. (a) $\dfrac{dy}{dx} = \dfrac{2}{3x^{1/3}}$ (b) $x = 0$ (c) $(0, 0)$
    (d) decreasing: $x < 0$
        increasing: $x > 0$
    (e)

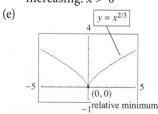

21. (a) $f'(x) = 0$ at $x = -\frac{1}{2}$
$f'(x) > 0$ for $x < -\frac{1}{2}$
$f'(x) < 0$ for $x > -\frac{1}{2}$
(b) $f'(x) = -1 - 2x$ verifies these conclusions.

23. (a) $f'(x) = 0$ at $x = 0, x = -3, x = 3$
$f'(x) > 0$ for $-3 < x < 3, x \neq 0$
$f'(x) < 0$ for $x < -3$ and $x > 3$
(b) $f'(x) = \frac{1}{3}x^2(9 - x^2)$ verifies these conclusions.

25. HPI $\left(1, \frac{4}{3}\right)$
no max or min

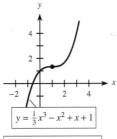

$y = \frac{1}{3}x^3 - x^2 + x + 1$

27. $(-6, 128)$ rel max;
$\left(4, -38\frac{2}{3}\right)$ rel min

$y = \frac{1}{3}x^3 + x^2 - 24x + 20$

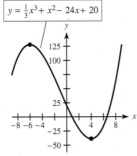

29. $(-1, 3)$ rel max;
$(1, -1)$ rel min;
HPI $(0, 1)$

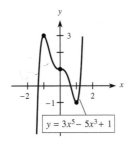

$y = 3x^5 - 5x^3 + 1$

31. $(1, 1)$ rel max;
$(0, 0), (2, 0)$ rel min

$y = (x^2 - 2x)^2$

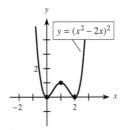

33. $(3, 4)$ rel max;
$(5, 0)$ rel min; HPI $(0, 0)$

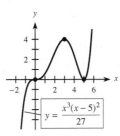

$y = \frac{x^3(x - 5)^2}{27}$

35. $(0, 0)$ rel max;
$(2, -4.8)$ rel min

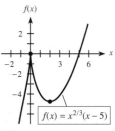

$f(x) = x^{2/3}(x - 5)$

37. $(50, 300,500), (100, 238,000)$
$0 \leq x \leq 150, 0 \leq y \leq 301,000$

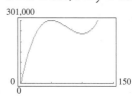

39. $(0, -40,000), (60, 4,280,000)$
$-20 \leq x \leq 90, \quad -500,000 \leq y \leq 5,000,000$

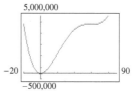

41. $(0, 2)$
$(0.1, 1.99975)$
$-0.1 \leq x \leq 0.2$
$1.9997 \leq y \leq 2.0007$

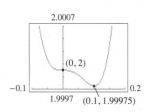

43. critical values: $x = -1, x = 2$
$f(x)$ increasing for $x < -1$ and $x > 2$
$f(x)$ decreasing for $-1 < x < 2$
rel max at $x = -1$; rel min at $x = 2$

possible graph of $f(x)$

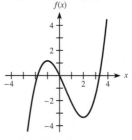

45. critical values: $x = 0, x = 3$
$f(x)$ increasing for $x > 3$
$f(x)$ decreasing for $x < 3, x \neq 0$
rel min at $x = 3$; HPI at $x = 0$

possible graph of $f(x)$

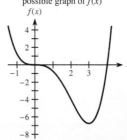

47. Graph on left is $f(x)$; on right is $f'(x)$ because $f(x)$ is increasing when $f'(x) > 0$ (i.e., above the x-axis) and $f(x)$ is decreasing when $f'(x) < 0$ (i.e., below the x-axis).

49. decreasing for $t \geq 0$

51. (a) $2 \pm \sqrt{13}$
(b) $2 + \sqrt{13} \approx 5.6$
(c) $0 \leq t < 2 + \sqrt{13}$
(d)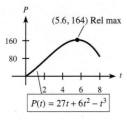

53. (a) $x = 5$ (b) $0 < x < 5$
(c) increasing for $x > 5$

55. (a) at $x = 150$, increasing; at $x = 250$, changing from increasing to decreasing; at $x = 350$, decreasing
(b) increasing for $x < 250$ (c) 250 units

57. (a) $t = 6$ (b) 6 weeks

59. (a) 10 (b) January 1

61. $x \approx 21.1$; in 2022 the model will achieve a maximum of 9.04 billion subscriberships

63. (a) $y = -0.000487x^3 - 0.162x^2 + 19.8x + 388$
(b) 912 million in 2020 (at $x \approx 49.9$)

65. (a) $M(t) = 0.0000889t^3 - 0.0111t^2 + 0.237t + 11.5$
(b) relative maximum at $(12.6, 12.9)$; relative minimum at $(70.7, 4.2)$
(c) No. The years corresponding to the critical points are 2023 and 2081. Despite the fact that the model is an excellent fit to the data given, it is unrealistic for the model to be accurate for more than 40 years beyond the data set.

10.2 EXERCISES

1. (a) concave down (b) concave up

3. (a, c) and (d, e) **5.** (c, d) and (e, f) **7.** c, d, e

9. concave up when $x > 2$; concave down when $x < 2$; POI at $x = 2$

11. concave up when $x < -2$ and $x > 1$
concave down when $-2 < x < 1$
points of inflection at $x = -2$ and $x = 1$

13. no points of inflection;
$(2, -2)$ min
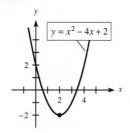

15. $\left(1, \frac{10}{3}\right)$ max; $(3, 2)$ min;
$\left(2, \frac{8}{3}\right)$ point of inflection
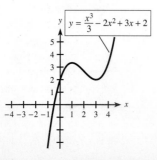

17. $(0, 0)$ rel max; $(2\sqrt{2}, -64)$, $(-2\sqrt{2}, -64)$ min; points of inflection: $(2\sqrt{6}/3, -320/9)$ and $(-2\sqrt{6}/3, -320/9)$
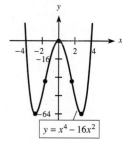

19. $(-2, 64)$ rel max; $(2, -64)$ rel min; points of inflection: $(-\sqrt{2}, 39.6)$, $(0, 0)$, and $(\sqrt{2}, -39.6)$
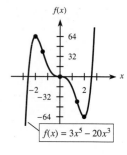

21. $(1, -3)$ min; points of inflection: $(-2, 7.6)$ and $(0, 0)$
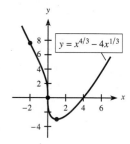

23. (a) $f''(x) = 0$ when $x = 1$
$f''(x) > 0$ when $x < 1$
$f''(x) < 0$ when $x > 1$
(b) rel max for $f'(x)$ at $x = 1$; no rel min
(c)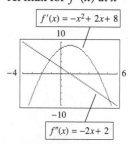

25. (a) concave up when $x < 2$; concave down when $x > 2$
(b) point of inflection at $x = 2$
(c)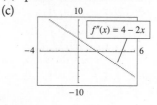

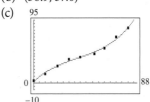

(d) possible graph of $f(x)$

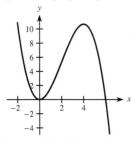

27. (a) G   (b) C   (c) F   (d) H   (e) I
29. (a) concave up when $x < 0$
         concave down when $x > 0$
         point of inflection at $x = 0$
    (b) concave up when $-1 < x < 1$
         concave down when $x < -1$ and $x > 1$
         POI at $x = -1$ and $x = 1$
    (c) concave up when $x > 0$
         concave down when $x < 0$
         point of inflection at $x = 0$
31. (a) $P'(t)$ (b) B   (c) C
33. (a) C   (b) right   (c) yes
35. (a) in an 8-hour shift, max when $t = 8$ (b) 4 hr
37. (a) 9 days   (b) 15 days
39. (a) (15.9, 100.3), (49.7, 85.3)
    (b) max $= 100.3$ at $t = 15.9$; min $= 85.3$ at $t = 49.7$.
         These mean that over the years 1985 to 2035,
         the energy use per capita reached a maximum
         of 100.3% of the 1995 use during 1996 and is
         expected to reach a minimum of 85.3% of the
         1995 use during 2030.
41. (a) $y = -0.000102x^3 + 0.00884x^2 + 1.43x + 57.9$
    (b) $x = 28.9$, during 1979
43. (a) $C(t) = 0.342t^3 - 6.83t^2 + 105t + 1330$
    (b) POI at $t = 6.66$, $C(t) = 1827$. These costs increase
         at a decreasing rate until $t = 6.66$ (in 2012) and
         increase at an increasing rate afterward.

10.3 EXERCISES

1. min -6 at $x = 2$, max $3.\overline{481}$ at $x = -2/3$
3. min -1 at $x = -2$, max 2 at $x = -1$
5. (a) $x = 1800$ units, $R = \$32,400$
    (b) $x = 1500$ units, $R = \$31,500$
7. $x = 20$ units, $R = \$24,000$ 9. 85 people
11. $p = \$95$, $R = \$451,250$
13. (a) max $= \$2100$ at $x = 10$
    (b) $\overline{R}(x) = \overline{MR}$ at $x = 10$
15. $x = 50$ units, $\overline{C} = \$43$
17. $x = 90$ units, $\overline{C} = \$18$
19. 10,000 units ($x = 100$), $\overline{C} = \$216$ per 100 units
21. $\overline{C}(x)$ has its minimum and $\overline{C}(x) = \overline{MC}$ at $x = 5$.
23. (a) A line from (0, 0) to $(x, C(x))$ has slope
         $C(x)/x = \overline{C}(x)$; this is minimized when the line
         has the least rise—that is, when the line is tangent
         to $C(x)$.
    (b) $x = 600$ units

25. $x = 80$ units, $P = \$280,000$
27. $x = 10\sqrt{15} \approx 39$ units, $P \approx \$71,181$ (using $x = 39$)
29. $x = 1000$ units, $P = \$39,700$
31. (a) B   (b) B   (c) B   (d) $\overline{MR} = \overline{MC}$
33. $\$860$ 35. $x = 600$ units, $P = \$495,000$
37. (a) 60   (b) $\$570$ (c) $\$9000$
39. (a) 1000 units   (b) $\$8066.67$ (approximately)
41. 2000 units priced at $90/unit; max profit is $90,000/wk
43. (a) $R(x) = 2x - 0.0004x^2$
         $P(x) = 1.8x - 0.0005x^2 - 800$
    (b) $p = \$1.28$, $x = 1800$, $P(1800) = \$820$
    (c) $p = \$1.25$, $x = 1875$, $P(1875) = \$817.19$
         Coastal would still provide sodas; profits almost
         the same.
45. (a) $y = 0.000252x^3 - 0.0279x^2 + 1.63x + 2.16$
    (b) (36.9, 37.0)
    (c)

         The *rate* of change of the number of beneficiaries was
         decreasing until 1987, after which the rate has been
         increasing. Hence, since 1987 the number of beneficia-
         ries has been increasing at an increasing rate.
47. (a) about mid-May
    (b) just after September 11, when the terrorists' planes
         crashed into the World Trade Center and the
         Pentagon
49. (a) 16.5
    (b) 1.9
    (c) Rise. As the number of workers per beneficiary
         drops, either the amount contributed by each
         worker must rise or support must diminish.

10.4 EXERCISES

1. (a) $x_1 = \$25$ million, $x_2 = \$13.846$ million
    (b) $\$38.846$ million
3. 100 trees   5. (a) 5   (b) 237.5   7. $\$50$
9. $m = c$ 11. 1 week   13. $t = 8, p = 45\%$
15. 240 ft   17. 300 ft $\times$ 150 ft
19. 20 ft long, $6\frac{2}{3}$ ft across (dividers run across)
21. 4 in. $\times$ 8 in. $\times$ 8 in. high   23. 30,000
25. 12,000   27. $x = 2$ 29. 3 weeks from now
31. 25 plates
33. (a) 2034 (at $t = 23.3$); max $R = 27.8$ billion barrels.
         This is the absolute maximum.
    (b) absolute min: (1, 19.5)
    (c) $t = 3.2$. The rate reaches its maximum at $t = 3.2$
         (in 2014); $R''(t)$ changes from positive to negative
         at $t = 3.2$. After this $R(t)$ increases at a decreasing
         rate until its maximum.

10.5 EXERCISES

1. (a) $x = 2$ (b) 1   (c) 1   (d) $y = 1$
3. (a) $x = 2, x = -2$ (b) 3   (c) 3   (d) $y = 3$
5. HA: $y = 2$; VA: $x = 3$
7. HA: $y = 0$; VA: $x = -2, x = 2$
9. HA: none; VA: none
11. HA: $y = 2$; VA: $x = 3$
no max, min, or points
of inflection

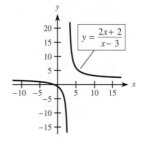

13. VA: $x = 0$;
$(-2, -4)$ rel max;
$(2, 4)$ rel min

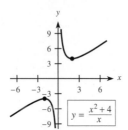

15. VA: $x = -1$; HA: $y = 0$;
$(0, 0)$ rel min; $(2, 4)$ rel max;

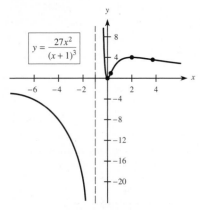

17. HA: $y = 0$; $(1, 8)$ rel max;
$(-1, -8)$ rel min

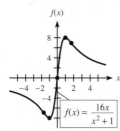

19. HA: $y = 0$; VA: $x = 1$;
$\left(-1, -\frac{1}{4}\right)$ rel min;
point of inflection: $(-2, -2/9)$

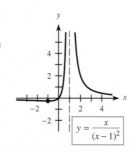

21. VA: $x = 3$;
$(2, -1)$ rel max;
$(4, 7)$ rel min

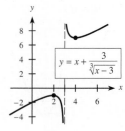

23. HA: $y = 0$;
VA: $x = 0$;
$(2, 0)$ rel min;
$(3, 1)$ rel max;
points of inflection:
$(1.87, 0.66)$,
$(4.13, 0.87)$

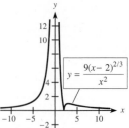

25. (a) HA: approx. $y = -2$; VA: approx. $x = 4$
   (b) HA: $y = -\frac{9}{4}$, VA: $x = \frac{17}{4}$
27. (a) HA: approx. $y = 2$;
     VA: approx. $x = 2.5, x = -2.5$
   (b) HA: $y = \frac{20}{9}$, VA: $x = \frac{7}{3}, x = -\frac{7}{3}$
29. $f(x) = \dfrac{x + 25}{x^2 + 1400}$
   (a)

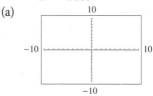

   (b) HA: $y = 0$; rel min $(-70, -0.0071)$;
        rel max $(20, 0.025)$
   (c) x: -500 to 400
        y: -0.01 to 0.03

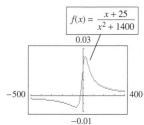

31. $f(x) = \dfrac{100(9 - x^2)}{x^2 + 100}$
   (a)

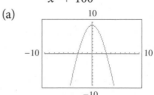

   (b) HA: $y = -100$;
        rel max $(0, 9)$
   (c) x: -75 to 75
        y: -120 to 20

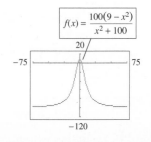

33. $f(x) = \dfrac{1000x - 4000}{x^2 - 10x - 2000}$

(a)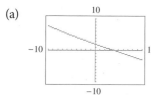

(b) HA: $y = 0$;
VA: $x = -40$,
$x = 50$; no max or min

(c) x: -200 to 200
y: -200 to 200

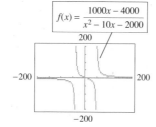

35. (a) none   (b) $C \geq 0$ (c) $p = 100$ (d) no

37. (a)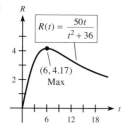

(b) 6 weeks
(c) 22 weeks after its release

39. (a) yes, $x = -1$
(b) no; domain is $x \geq 5$
(c) yes, $y = -58.5731$
(d) At $0°$F, as the wind speed increases, there is a limiting wind chill of about $-58.6°$F. This is meaningful because at high wind speeds, additional wind probably has little noticeable effect.

41. (a) $P = C$ (b) C (c) $P' = 0$ (d) 0

43. (a) 57.0
(b) The model predicts that in the long run, 57% of workers will be female.
(c) No. Vertical asymptote is only at $t \approx -46.4$.
(d) $p(t) > 0$ for $t > 0$ and $p(t)$ never exceeds 100, so the model is never inappropriate.

45. (a) No. Barometric pressure can drop off the scale (as shown), but it cannot decrease without bound. In fact, it must always be positive.
(b) See your library with regard to the "storm of the century" in March 1993.

CHAPTER 10 REVIEW EXERCISES

1. $(0, 0)$ max   **2.** $(2, -9)$ min   **3.** HPI $(1, 0)$
4. $\left(1, \frac{3}{2}\right)$ max, $\left(-1, -\frac{3}{2}\right)$ min

5. (a) $\frac{1}{3}, -1$
(b) $(-1, 0)$ rel max, $\left(\frac{1}{3}, -\frac{32}{27}\right)$ rel min
(c) none   (d)

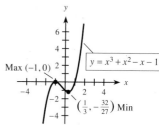

6. (a) $3, 0$ (b) $(3, 27)$ max   (c) $(0, 0)$
(d)

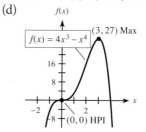

7. (a) $-1, 6$
(b) $(-1, 11)$ rel max, $(6, -160.5)$ rel min
(c) none
(d)

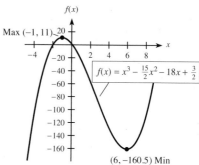

8. (a) $0, \pm 1$ (b) $(-1, 1)$ rel max, $(1, -3)$ rel min
(c) $(0, -1)$ (d)

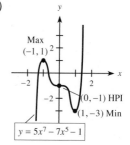

9. (a) 0   (b) $(0, -1)$ min   (c) none
(d)

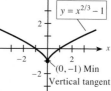

10. (a) 0, 1, 4
  (b) (0, 0) rel min, (1, 9) rel max, (4, 0) rel min
  (c) none (d)

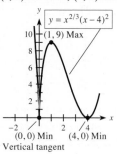

11. concave up
12. concave up when $x < -1$ and $x > 2$; concave down
  when $-1 < x < 2$; points of inflection at $(-1, -3)$ and
  $(2, -42)$
13. $(-1, 15)$ rel max; $(3, -17)$ rel min; point of inflection
  $(1, -1)$
14. $(-2, 16)$ rel max; $(2, -16)$ rel min;
  point of inflection $(0, 0)$

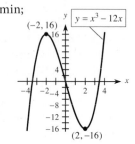

15. $(1, 4)$ rel max; $(-1, 0)$ rel min; points of
  inflection: $\left(\dfrac{1}{\sqrt{2}}, 2 + \dfrac{7}{4\sqrt{2}}\right)$, $(0, 2)$, and
  $\left(-\dfrac{1}{\sqrt{2}}, 2 - \dfrac{7}{4\sqrt{2}}\right)$

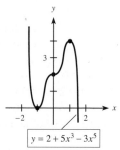

16. (a) (0, 0) absolute min; (140, 19,600) absolute max
  (b) (0, 0) absolute min; (100, 18,000) absolute max
17. (a) (50, 233,333) absolute max; (0, 0) absolute min
  (b) (64, 248,491) absolute max; (0, 0) absolute min
18. (a) $x = 1$ (b) $y = 0$ (c) 0 (d) 0
19. (a) $x = -1$ (b) $y = \frac{1}{2}$ (c) $\frac{1}{2}$ (d) $\frac{1}{2}$
20. HA: $y = \frac{3}{2}$, VA: $x = 2$
21. HA: $y = -1$; VA: $x = 1, x = -1$
22. (a) HA: $y = 3$; VA: $x = -2$
  (b) no max or min (c)

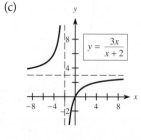

23. (a) HA: $y = 0$; VA: $x = 0$
  (b) (4, 1) max (c)

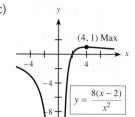

24. (a) HA: none; VA: $x = 1$
  (b) (0, 0) rel max; (2, 4) rel min
  (c)

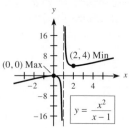

25. (a) $f'(x) > 0$ for $x < \frac{2}{3}$ (approximately) and $x > 2$
    $f'(x) < 0$ for about $\frac{2}{3} < x < 2$
    $f'(x) = 0$ at about $x = \frac{2}{3}$ and $x = 2$
  (b) $f''(x) > 0$ for $x > \frac{4}{3}$
    $f''(x) < 0$ for $x < \frac{4}{3}$
    $f''(x) = 0$ at $x = \frac{4}{3}$
  (c)
  (d)

26. (a) $f'(x) > 0$ for about $-13 < x < 0$ and $x > 7$
    $f'(x) < 0$ for about $x < -13$ and $0 < x < 7$
    $f'(x) = 0$ at about $x = 0, x = -13, x = 7$
  (b) $f''(x) > 0$ for about $x < -8$ and $x > 4$
    $f''(x) < 0$ for about $-8 < x < 4$
    $f''(x) = 0$ at about $x = -8$ and $x = 4$
  (c)

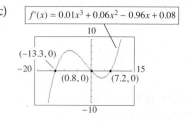

(d)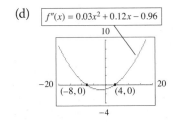
$$f''(x) = 0.03x^2 + 0.12x - 0.96$$

27. (a) $f(x)$ increasing for $x < -5$ and $x > 1$
    $f(x)$ decreasing for $-5 < x < 1$
    $f(x)$ has rel max at $x = -5$, rel min at $x = 1$
    (b) $f''(x) > 0$ for $x > -2$ (where $f'(x)$ increases)
    $f''(x) < 0$ for $x < -2$ (where $f'(x)$ decreases)
    $f''(x) = 0$ for $x = -2$
    (c)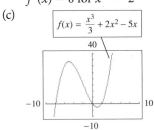
    $$f(x) = \frac{x^3}{3} + 2x^2 - 5x$$
    (d)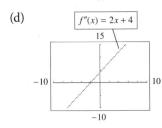
    $$f''(x) = 2x + 4$$

28. (a) $f(x)$ increasing for $x < 6$, $x \neq 0$
    $f(x)$ decreasing for $x > 6$
    $f(x)$ has rel max at $x = 6$, point of inflection at $x = 0$
    (b) $f''(x) > 0$ for $0 < x < 4$
    $f''(x) < 0$ for $x < 0$ and $x > 4$
    $f''(x) = 0$ at $x = 0$ and $x = 4$
    (c)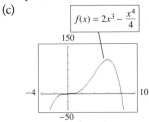
    $$f(x) = 2x^3 - \frac{x^4}{4}$$
    (d)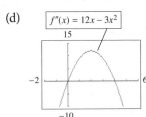
    $$f''(x) = 12x - 3x^2$$

29. (a) $f(x)$ is concave up for $x < 4$.
    $f(x)$ is concave down for $x > 4$.
    $f(x)$ has point of inflection at $x = 4$.

(b)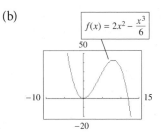
$$f(x) = 2x^2 - \frac{x^3}{6}$$

30. (a) $f(x)$ is concave up for $-3 < x < 2$.
    $f(x)$ is concave down for $x < -3$ and $x > 2$.
    $f(x)$ has points of inflection at $x = -3$ and $x = 2$.
    (b)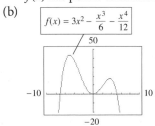
    $$f(x) = 3x^2 - \frac{x^3}{6} - \frac{x^4}{12}$$

31. $x = 5$ units, $\overline{C} = \$45$ per unit
32. (a) $x = 1600$ units, $R = \$25,600$
    (b) $x = 1500$ units, $R = \$25,500$
33. $P = \$54,000$ at $x = 100$ units   34. $x = 300$ units
35. $x = 150$ units   36. $x = 7$ units
37. $x = 500$ units, when $\overline{MP} = 0$ and changes from positive to negative.
38. 30 hours
39. (a) $I = 60$. The point of diminishing returns is located at the point of inflection (where bending changes).
    (b) $m = f(I)/I =$ the average output
    (c) The segment from $(0, 0)$ to $y = f(I)$ has maximum slope when it is tangent to $y = f(I)$, close to $I = 70$.
40. \$1040 per bike
41. \$1380 per bike
42. \$93,625 at 325 units
43. (a) 150   (b) \$650
44. \$208,490.67 at 64 units
45. $x = 1000$ mg   46. 10:00 A.M.
47. 325 in 2020   48. 20 mi from A, 10 mi from B
49. 4 ft $\times$ 4 ft   50. $8\frac{3}{4}$ in. $\times$ 10. in.
51. 500 mg
52. (a) $x \approx 7.09$; during 2008
    (b) point of inflection
53. 24,000
54. (a) vertical asymptote at $x = 0$
    (d)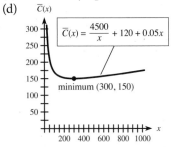
    $$\overline{C}(x) = \frac{4500}{x} + 120 + 0.05x$$
    minimum (300, 150)

55. (a) 3%
(b) $y = 38$. The long-term market share approaches 38%.

CHAPTER 10 TEST

1. max $(-3, 3)$; min $(-1, -1)$; POI $(-2, 1)$

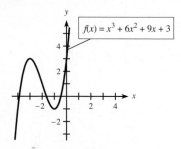

2. max $(3, 17)$; HPI $(0, -10)$; POI $(2, 6)$

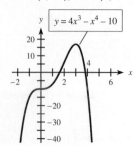

3. max $(0, -3)$;
min $(4, 5)$;
vertical asymptote $x = 2$

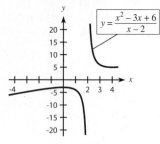

4. $\left(-\dfrac{1}{\sqrt{2}}, 0\right)$ and $\left(\dfrac{1}{\sqrt{2}}, \infty\right)$

5. $(0, 2)$, HPI; $\left(-\dfrac{1}{\sqrt{2}}, 3.237\right)$, $\left(\dfrac{1}{\sqrt{2}}, 0.763\right)$

6. max $(-1, 4)$; min $(1, 0)$
7. max 67 at $x = 8$; min -122 at $x = 5$
8. horizontal asymptote $y = 200$; vertical asymptote $x = -300$
9.

Point	f	f'	f''
A	$-$	$+$	$-$
B	$+$	$-$	0
C	$+$	0	$+$

10. (a) 2   (b) $x = -3$
(c) $y = 2$
11. local max at $(6, 10)$
12. (a) relative minimum: $(8.7, 19.6)$; relative maximum: $(51.5, 39.0)$

(b) The point $(8.7, 19.6)$ means that when $t = 8.7$ (during 1999), the aged dependency ratio reached a minimum of 19.6 aged individuals per 100 individuals ages 20–64. The point $(51.5, 39.0)$ means that when $t = 51.5$ (during 2042), the aged dependency ratio is expected to reach a maximum of 39.0 aged individuals per 100 individuals ages 20–64.

13. (a) $x = 7200$ (b) $518,100   **14.** 100 units
15. $250   **16.** $\frac{10}{3}$ centimeter   **17.** 28,000 units
18. (a) $y = -0.0000700x^3 + 0.00567x^2 + 0.863x + 16.0$
(b) $x \approx 27.0$; during 1977
(c) x-coordinate of the point of inflection

11.1 EXERCISES

1. $f'(x) = 4/x$

3. $y' = 1/x$

5. $y' = 4/x$

7. $f'(x) = \dfrac{4}{4x + 9}$

9. $y' = \dfrac{4x - 1}{2x^2 - x} + 3$

11. $dp/dq = 2q/(q^2 + 1)$

13. (a) $y' = \dfrac{1}{x} - \dfrac{1}{x - 1} = \dfrac{-1}{x(x - 1)}$

(b) $y' = \dfrac{-1}{x(x - 1)}$; $\ln\left(\dfrac{x}{x - 1}\right) = \ln(x) - \ln(x - 1)$

15. (a) $y' = \dfrac{2x}{3(x^2 - 1)}$

(b) $y' = \dfrac{2x}{3(x^2 - 1)}$; $\ln(x^2 - 1)^{1/3} = \frac{1}{3}\ln(x^2 - 1)$

17. (a) $y' = \dfrac{4}{4x - 1} - \dfrac{3}{x} = \dfrac{-8x + 3}{x(4x - 1)}$

(b) $y' = \dfrac{-8x + 3}{x(4x - 1)}$;

$\ln\left(\dfrac{4x - 1}{x^3}\right) = \ln(4x - 1) - 3\ln(x)$

19. $\dfrac{dp}{dq} = \dfrac{2q}{q^2 - 1} - \dfrac{1}{q} = \dfrac{q^2 + 1}{q(q^2 - 1)}$

21. $\dfrac{dy}{dt} = \dfrac{2t}{t^2 + 3} - \dfrac{1}{2}\left(\dfrac{-1}{1 - t}\right) = \dfrac{3 + 4t - 3t^2}{2(1 - t)(t^2 + 3)}$

23. $\dfrac{dy}{dx} = \dfrac{3}{x} + \dfrac{1}{2(x + 1)} = \dfrac{7x + 6}{2x(x + 1)}$

25. $y' = 1 - \dfrac{1}{x}$

27. $y' = (1 - \ln x)/x^2$

29. $y' = 8x^3/(x^4 + 3)$

31. $y' = \dfrac{4(\ln x)^3}{x}$

33. $y' = \dfrac{8x^3\ln(x^4 + 3)}{x^4 + 3}$

35. $y' = \dfrac{1}{x \ln 4}$

37. $y' = \dfrac{4x^3 - 12x^2}{(x^4 - 4x^3 + 1)\ln 6}$

39. rel min $(e^{-1}, -e^{-1})$

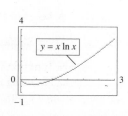

41. rel min $(2, 4 - 8 \ln 2)$

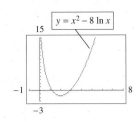

$y = x^2 - 8 \ln x$

43. (a) $\overline{MC} = \dfrac{400}{2x + 1}$

(b) $\overline{MC} = \dfrac{400}{401} \approx 1.0$; the approximate cost of the 201st unit is $1.00

(c) $\overline{MC} > 0$. Yes

45. (a) $\overline{MR} = \dfrac{2500[(x + 1) \ln (10x + 10) - x]}{(x + 1) \ln^2(10x + 10)}$

(b) 309.67; at 100 units, selling 1 additional unit yields about $309.67.

47. (a) -5.23 (b) -1.89 (c) increasing

49. A/B **51.** $dR/dI = 1/(I \ln 10)$

53. (a) $y = -31.7 + 18.7 \ln (x)$

(b) $y' = \dfrac{18.7}{x}$ (c) 0.47 percentage points per year

11.2 EXERCISES

1. $y' = 5e^x - 1$

3. $f'(x) = e^x - ex^{e-1}$

5. $g'(x) = 50e^{-0.1x}$

7. $y' = 3x^2e^{x^3}$

9. $y' = 36xe^{3x^2}$

11. $y' = 12x(x^2 + 1)^2e^{(x^2+1)^3}$

13. $y' = 3x^2$

15. $y' = e^{-1/x}/x^2$

17. $y' = \dfrac{2}{x^3}e^{-1/x^2} - 2xe^{-x^2}$

19. $ds/dt = te^t(t + 2)$

21. $y' = 4x^3e^{x^4} - 4e^{4x}$

23. $y' = \dfrac{4e^{4x}}{e^{4x} + 2}$

25. $y' = e^{-3x}/x - 3e^{-3x} \ln (2x)$

27. $y' = (2e^{5x} - 3)/e^{3x} = 2e^{2x} - 3e^{-3x}$

29. $y' = 30e^{3x}(e^{3x} + 4)^9$ **31.** $y' = 6^x \ln 6$

33. $y' = 4^{x^2}(2x \ln 4)$

35. (a) $y'(1) = 0$ (b) $y = e^{-1}$

37. (a) $z = 0$ (b)

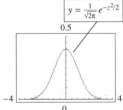

$y = \dfrac{1}{\sqrt{2\pi}} e^{-z^2/2}$

39. rel min at $x = 1, y = e$

41. rel max at $x = 0, y = -1$

43. (a) $(0.1) Pe^{0.1n}$ (b) $(0.1) Pe^{0.1}$

(c) Yes, because $e^{0.1n} > 1$ for any $n \geq 1$.

45. (a) $\dfrac{dS}{dt} = -50{,}000e^{-0.5t}$

(b) The function is a decay exponential. The derivative is always negative.

47. $40e \approx 108.73$ dollars per unit

49. (a) $\dfrac{dy}{dt} = 46.2e^{-0.462t}$

(b) 29.107 percent per hour

51. $\dfrac{dx}{dt} = -0.0684e^{-0.38t}$

53. 177.1 ($billion/year)   **55.** $\dfrac{dI}{dR} = 10^R \ln 10$

57. (a) $d'(t) = 0.1328e^{0.083t}$

(b) $d'(50) \approx 8.42$ billion dollars per year; $d'(125) \approx 4256$ billion dollars per year

59. $y' = \dfrac{98{,}990{,}100e^{-0.99t}}{(1 + 9999e^{-0.99t})^2}$

61. (a) 1.60; the rate of increase in this population was approximately 1.60 million per year in 1995.

(b) 0.808 million (808 thousand) per year

(c) The rate of increase in this population in 2040 is approximately half of the rate of increase in 1995.

63. (a) $y = 0.544(1.07^x)$

(b) $y' = 0.0368(1.07^x)$

(c) 0.55 percentage points per year

65. (a) $y' = 8.864(1.055^x)$

(b) $y'(35) \approx \$57.7$ billion

67. (a) $y = 42.1(1.04^x)$

(b) $y' = 1.65(1.04^x)$

(c) $y'(30) \approx 5.35$ thousand dollars per year

11.3 EXERCISES

1. $\frac{1}{2}$ **3.** $-\frac{1}{2}$ **5.** $-\frac{5}{3}$ **7.** $-x/(2y)$

9. $-(2x + 4)/(2y - 3)$ **11.** $y' = -x/y$

13. $y' = \dfrac{-y}{2x - 3y}$ **15.** $\dfrac{dp}{dq} = \dfrac{p^2}{4 - 2pq}$

17. $\dfrac{dy}{dx} = \dfrac{x(3x^3 - 2)}{3y^2(1 + y^2)}$ **19.** $\dfrac{dy}{dx} = \dfrac{4x^3 + 6x^2y^2 - 1}{-4x^3y - 3y^2}$

21. $\dfrac{dy}{dx} = \dfrac{(4x^3 + 9x^2y^2 - 8x - 12y)}{(18y + 12x - 6x^3y + 10y^4)}$ **23.** undefined

25. 1   **27.** $y = \frac{1}{2}x + 1$ **29.** $y = 4x + 5$

31. $\dfrac{dy}{dx} = \dfrac{1}{2xy}$ **33.** $\dfrac{dy}{dx} = \dfrac{-y}{2x \ln x}$ **35.** -15

37. $-1/x$ **39.** $\dfrac{-xy - 1}{x^2}$

41. $ye^x/(1 - e^x)$ **43.** $\frac{1}{3}$ **45.** $y = 3 - x$

47. (a) $(2, \sqrt{2}), (2, -\sqrt{2})$

(b) $(2 + 2\sqrt{2}, 0), (2 - 2\sqrt{2}, 0)$

49. (a) and (b) are verifications

(c) yes, because $x^2 + y^2 = 4$

51. $1/(2x\sqrt{x})$

53. max at $(0, 3)$; min at $(0, -3)$

55. $\frac{1}{2}$, so an additional 1 (thousand dollars) of advertising yields about $\frac{1}{2}$ (thousand) additional units

57. $-\frac{243}{128}$ hours of skilled labor per hour of unskilled labor

59. At $p = \$80$, $q = 49$ and $dq/dp = -\frac{5}{16}$, which means that if the price is increased to $81, quantity demanded will decrease by approximately $\frac{5}{16}$ unit.

61. $-0.000436y$ **63.** $\dfrac{dh}{dt} = -\dfrac{3}{44} - \dfrac{h}{12}$

11.4 EXERCISES

1. 36 **3.** $\frac{1}{8}$ **5.** $-\frac{24}{5}$ **7.** $\frac{7}{6}$

9. -5 if $z = 5$, -10 if $z = -5$

11. -80 units/sec **13.** 12π ft^2/min

15. $\frac{16}{27}$ in/sec **17.** 1798/day

19. $0.42/day

21. 430 units/month **23.** 36π mm^3/month

25. $\dfrac{\frac{dW}{dt}}{W} = 3\left(\dfrac{\frac{dL}{dt}}{L}\right)$ **27.** $\dfrac{\frac{dC}{dt}}{C} = 1.54\left(\dfrac{\frac{dW}{dt}}{W}\right)$

29. $\dfrac{1}{4\pi}$ micrometer/day **31.** $1/(20\pi)$ in/min

33. 0.75 ft/sec **35.** $-120\sqrt{6}$ mph ≈ -294 mph

37. approaching at 61.18 mph **39.** $\frac{1}{25}$ ft/hr

11.5 EXERCISES

1. (a) 1 (b) no change
3. (a) 84 (b) Revenue will decrease.
5. (a) $\frac{100}{99}$ (b) elastic (c) decrease
7. (a) 0.81 (b) inelastic (c) increase
9. (a) $\eta = 11.1$ (approximately) (b) elastic
11. (a) $\eta = \dfrac{375 - 3q}{q}$

(b) unitary: $q = 93.75$; inelastic: $q > 93.75$; elastic: $q < 93.75$

(c) As q increases over $0 < q < 93.75$, p decreases, so elastic demand means R increases. Similarly, R decreases for $q > 93.75$.

(d) maximum for R when $q = 93.75$; yes.

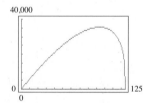

13. (a) $p = 250 - 0.125q$
(b) $\eta = \frac{2000}{q} - 1$
(c) $\eta \approx 2.33$; elastic. no
(d) $q = 1000$; $p = 125; max $R = $125,000$
15. $12/item **17.** $t = 350 **19.** $115/item
21. $483 per item; $40,100 **23.** $1100/item

CHAPTER 11 REVIEW EXERCISES

1. $dy/dx = 10(6x - 1)e^{3x^2 - x}$ **2.** $y' = 12/(4x + 11)$
3. $\dfrac{dp}{dq} = \dfrac{1}{q} - \dfrac{2q}{q^2 - 1}$ **4.** $dy/dx = e^{x^2}(2x^2 + 1)$
5. $f'(x) = 10e^{2x} + 4e^{-0.1x}$
6. $g'(x) = 18e^{3x+1}(2e^{3x+1} - 5)^2$
7. $\dfrac{ds}{dx} = \dfrac{9x^{11} - 6x^3}{x^{12} - 2x^4 + 5}$ **8.** $dw/dt = 2t\ln(t^2 + 1)$
9. $dy/dx = 3^{3x-3}\ln 3$ **10.** $dy/dx = \dfrac{1}{\ln 8}\left(\dfrac{10}{x}\right)$

11. $\dfrac{dy}{dx} = \dfrac{1 - \ln x}{x^2}$ **12.** $dy/dx = -2e^{-x}/(1 - e^{-x})^2$

13. $y = 12ex - 8e$, or $y \approx 32.62x - 21.75$

14. $y = 3x + 5$ **15.** $\dfrac{dy}{dx} = \dfrac{y}{x(10y - \ln x)}$

16. $dy/dx = ye^{xy}/(1 - xe^{xy})$ **17.** $dy/dx = 2/y$

18. $\dfrac{dy}{dx} = \dfrac{2(x + 1)}{3(1 - 2y)}$ **19.** $\dfrac{dy}{dx} = \dfrac{6x(1 + xy^2)}{y(5y^3 - 4x^3)}$

20. $d^2y/dx^2 = -(x^2 + y^2)/y^3 = -1/y^3$ **21.** $5/9$
22. $\left(-2, \pm\sqrt{\frac{2}{3}}\right)$ **23.** $3/4$ **24.** 11 square units/min

25. (a) $P'(t) = \dfrac{17.4}{t}$

(b) $P(58) \approx 166.8$; $P'(58) \approx 0.3$. These mean that in 2028 this population is projected to be about 166.8 million and growing about 0.3 million (300,000) per year.

26. (a) $L'(y) = \dfrac{7.10}{y}$

(b) $L(50) \approx 13.2$ and $L(80) \approx 16.5$. These mean that the expected number of additional years of life expectancy at age 65 is about 13.2 in 2000 and 16.5 in 2030.

(c) $L'(50) \approx 0.14$ and $L'(80) \approx 0.09$. These mean that the number of additional years of life expectancy at age 65 is expected to change at the rate of about 0.14 years of life per year in 2000 and 0.09 years of life per year in 2030.

27. (a) $S'(n) = 120e^{0.12n}$
(b) after 1 year: about $135.30 per year; after 10 years: about $398.41 per year

28. (a) $D'(t) \approx 229.7e^{0.02292t}$
(b) $D(20) \approx 15,850$ and $D'(20) \approx 363$. These mean that in 2030 total U.S. disposable income is expected to be about $15,850 billion and changing at the rate of about $363 billion per year.

29. (a) $-0.00001438A_0$ units/year
(b) $-0.00002876A_0$ units/year (c) less
30. $1200e \approx $3261.94 per unit
31. $-$1797.36 per year **32.** $-1/(25\pi)$ mm/min
33. $\frac{48}{25}$ ft/min **34.** $\dfrac{dS/dt}{S} = \dfrac{1}{3}\left(\dfrac{dA/dt}{A}\right)$ **35.** yes
36. $t = 1466.67, $T \approx $58,667$
37. $t = 880, $T = 3520
38. (a) 1 (b) no change
39. (a) $\frac{25}{12}$, elastic (b) revenue decreases
40. (a) 1 (b) no change
41. (a) [graph] (b) $q = 100$

$\eta(q) = \dfrac{2(100 - 0.5q)}{q}$

(c) max revenue at $q = 100$

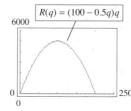

$R(q) = (100 - 0.5q)q$

(d) Revenue is maximized where elasticity is unitary.

CHAPTER 11 TEST

1. $y' = 15x^2 e^{x^3} + 2x$

2. $y' = \dfrac{12x^2}{x^3 + 1}$

3. $y' = \dfrac{12x^3}{x^4 + 1}$

4. $f'(x) = 20(3^{2x}) \ln 3$

5. $\dfrac{dS}{dt} = e^{t^4}(4t^4 + 1)$

6. $y' = \dfrac{e^{x^3 + 1}(3x^3 - 1)}{x^2}$

7. $y' = \dfrac{3 - 12 \ln x}{x^5}$

8. $g'(x) = \dfrac{8}{(4x + 7) \ln 5}$

9. $y' = \dfrac{-3x^3}{y}$

10. $-\dfrac{3}{2}$

11. $y' = \dfrac{-e^y}{xe^y - 10}$

12. \$1349.50 per week

13. $\eta = 3.71$; decreases

14. -0.05 unit per dollar

15. 586 units per day

16. (a) $y' = 81.778e^{0.062t}$

(b) 2005: $y'(5) \approx 111.5$ (billion dollars per year)

2020: $y'(20) \approx 282.6$ (billion dollars per year)

17. \$540

18. (a) $y = -12.97 + 11.85 \ln x$

(b) $y' = \dfrac{11.85}{x}$

(c) $y(25) \approx 25.2$ means the model estimates that about 25.2% of the U.S. population will have diabetes in 2025.

$y'(25) \approx 0.474$ predicts that in 2025 the percent of the U.S. population with diabetes will be changing by about 0.474 percentage points per year.

19. (a) $P'(t) = 1.078(0.9732^t) \cdot \ln (0.9732) \approx -0.02928(0.9732^t)$

(b) $P(18) \approx 0.661$ and $P'(18) \approx -0.018$. These mean that in 2028 the purchasing power of \$1 is about 66.1% of what it was in 2012 and is expected to be decreasing about 1.8% per year.

12.1 EXERCISES

1. $x^4 + C$

3. $\frac{1}{7}x^7 + C$

5. $\frac{1}{8}x^8 + C$

7. $2x^4 + C$

9. $27x + \frac{1}{14}x^{14} + C$

11. $3x - \frac{2}{5}x^{5/2} + C$

13. $\frac{1}{5}x^5 - 3x^3 + 3x + C$

15. $13x - 3x^2 + 3x^7 + C$

17. $2x + \frac{4}{3}x\sqrt{x} + C$

19. $\frac{24}{5}x \sqrt[4]{x} + C$

21. $-5/(3x^3) + C$

23. $\frac{3}{2} \sqrt[3]{x} + C$

25. $\dfrac{1}{4}x^4 - 4x - \dfrac{1}{x^5} + C$

27. $\dfrac{1}{10}x^{10} + \dfrac{1}{2x^2} + 3x^{2/3} + C$

29. $2x^8 - \frac{4}{3}x^6 + \frac{1}{4}x^4 + C$

31. $-1/x - 1/(2x^2) + C$

33.

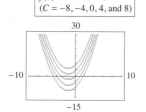

$f(x) = x^2 + 3x + C$
$(C = -8, -4, 0, 4, \text{and } 8)$

35. $f(x) = 18x^8 - 35x^4$

37. $\int \left(5 - \frac{1}{2}x\right) dx$

39. $\int (3x^2 - 6x)\, dx$

41. $R(x) = 30x - 0.2x^2$

43. $R(50) = \$22{,}125$

45. $P(t) = \frac{1}{4}t^4 + \frac{4}{3}t^3 + 6t$

47. (a) $x = t^{7/4}/1050$ (b) 0.96 ton

49. (a) $\overline{C}(x) = x/4 + 100/x + 30$ (b) \$56 per unit

51. (a) $E(t) = 20.61t^2 - 116.4t + 7398$

(b) \$13,314 per person

53. (a) The wind chill temperature decreases because

$\dfrac{dt}{dw} < 0$ for $w > 0$.

The rate increases because $\dfrac{d^2t}{dw^2} > 0$ for $w > 0$.

(b) $t = 48.12 - 27.2w^{0.16}$

55. (a) $t \approx 63.1$; in 2024

(b) $P(t) = -0.0000729t^3 + 0.0138t^2 + 1.98t + 181$

(c) 348 million

12.2 EXERCISES

1. $du = 10x^4\, dx$

3. (a) Power Rule cannot be used; need x^3 as a factor of du

(b) $\dfrac{1}{8}(5x^3 + 11)^8 + C$

5. $\frac{1}{4}(x^2 + 3)^4 + C$

7. $\frac{1}{5}(5x^3 + 11)^5 + C$

9. $\frac{1}{3}(3x - x^3)^3 + C$

11. $\frac{1}{28}(7x^4 + 12)^4 + C$

13. $\frac{1}{4}(4x - 1)^7 + C$

15. $-\frac{1}{6}(4x^6 + 15)^{-2} + C$

17. $\frac{1}{10}(x^2 - 2x + 5)^5 + C$

19. $-\frac{1}{8}(x^4 - 4x + 3)^{-4} + C$

21. $\frac{7}{6}(x^4 + 6)^{3/2} + C$

23. $\frac{3}{8}x^8 + \frac{6}{5}x^5 + \frac{3}{2}x^2 + C$

25. $10.8x^{10} - 12x^6 + 6x^2 + C$

27. $\frac{2}{9}(x^3 - 3x)^{3/2} + C$

29. $\dfrac{-1}{[10(2x^5 - 5)^3]} + C$

31. $\dfrac{-1}{[8(x^4 - 4x)^2]} + C$

33. $\frac{2}{3}\sqrt{x^3 - 6x^2 + 2} + C$

35. $f(x) = 70(7x - 13)^9$

37. (a) $f(x) = \frac{1}{8}(x^2 - 1)^4 + C$

(b)

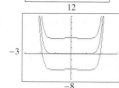

$f(x) = \frac{1}{8}(x^2 - 1)^4 + C$
$(C = -5, 0, 5)$

39. (a) $F(x) = \frac{15}{4}(2x - 1)^{2/5} + C$

(b)

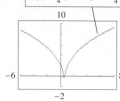

$F(x) = \frac{15}{4}(2x - 1)^{2/5} - \frac{7}{4}$

(c) $x = \frac{1}{2}$

(d) vertical

41. $\int \dfrac{8x(x^2 - 1)^{1/3}}{3} \, dx$

43. (b) $\dfrac{-7}{3(x^3 + 4)} + C$

(d) $\int (x^2 + 5)^{-4} \, dx$ (Many answers are possible.)

45. $R(x) = \dfrac{15}{2x + 1} + 30x - 15$

47. 3720 bricks

49. (a) $s = 10\sqrt{x + 1}$ (b) 50

51. (a) $A(t) = 100/(t + 10) - 1000/(t + 10)^2$

(b) 2.5 million

53. 7400

55. (a) $f(x) = \dfrac{-505}{0.743x + 6.97} + 36.2$

(b) 23.7%

57. (a) $p(t) = 56.19 - \dfrac{1561}{1.38t + 64.1}$

(b)

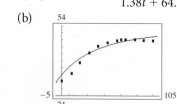

(c) The model is a good fit to the data.

12.3 EXERCISES

1. $e^{3x} + C$ **3.** $-e^{-x} + C$ **5.** $10,000e^{0.1x} + C$

7. $-1200e^{-0.7x} + C$ **9.** $\frac{1}{12}e^{3x^4} + C$

11. $-\frac{3}{2}e^{-2x} + C$ **13.** $\frac{1}{18}e^{3x^6 - 2} + C$

15. $\frac{1}{4}e^{4x} + 6/e^{x/2} + C$

17. $\ln |x^3 + 4| + C$ **19.** $\frac{1}{4}\ln |4z + 1| + C$

21. $\frac{3}{4}\ln |2x^4 + 1| + C$ **23.** $\frac{2}{5}\ln |5x^2 - 4| + C$

25. $\ln |x^3 - 2x| + C$ **27.** $\frac{1}{3}\ln |z^3 + 3z + 17| + C$

29. $\frac{1}{3}x^3 + \ln |x - 1| + C$ **31.** $x + \frac{1}{2}\ln |x^2 + 3| + C$

33. $f(x) = h(x), \int f(x) \, dx = g(x)$

35. $F(x) = -\ln |3 - x| + C$

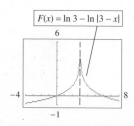

$F(x) = \ln 3 - \ln |3 - x|$

37. $f(x) = 1 + \dfrac{1}{x}; \displaystyle\int \left(1 + \dfrac{1}{x}\right) dx$

39. $f(x) = 5e^{-x} - 5xe^{-x}; \int (5e^{-x} - 5xe^{-x}) \, dx$

41. (c) $\frac{1}{3}\ln |x^3 + 3x^2 + 7| + C$;   (d) $\frac{5}{8}e^{2x^4} + C$

43. $1030.97   **45.** $n = n_0 e^{-Kt}$ **47.** 55

49. (a) $S = Pe^{0.1n}$ (b) ≈ 7 years

51. (a) $p = 95e^{-0.491t}$ (b) ≈ 90.45

53. (a) $l(t) = 11.028 + 14.304 \ln (t + 20)$

(b)

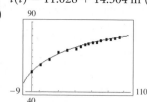

(c) The model is a very good fit to the data.

55. (a) Yes. The rate is an exponential that is always positive. Hence the function is always increasing.

(b) $C(t) = 80.39e^{0.0384t} + 0.6635$

(c) $C(35) \approx 308.91; C'(35) \approx 11.84$

For 2025, the model predicts that the CPI will be $308.91 and will be changing at a rate of about $11.84 per year.

12.4 EXERCISES

1. $C(x) = x^2 + 100x + 200$

3. $C(x) = 2x^2 + 2x + 80$ **5.** $3750

7. (a) $x = 3$ units is optimal level

(b) $P(x) = -4x^2 + 24x - 200$

(c) loss of $164

9. (a) profit of $3120   (b) 896 units

11. (a) $\overline{C}(x) = \frac{6}{x} + \frac{x}{6} + 8$ (b) $10.50

13. (a) and (b)

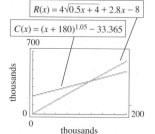

$R(x) = 4\sqrt{0.5x + 4} + 2.8x - 8$

$C(x) = (x + 180)^{1.05} - 33.365$

(c) Maximum profit is $114.743 thousand at $x = 200$ thousand units.

15. $C(y) = 0.80y + 7$

17. $C(y) = 0.3y + 0.4\sqrt{y} + 8$

19. $C(y) = 2\sqrt{y + 1} + 0.4y + 4$

21. $C(y) = 0.7y + 0.5e^{-2y} + 5.15$

23. $C(y) = 0.85y + 5.15$

25. $C(y) = 0.8y + \dfrac{2\sqrt{3y + 7}}{3} + 4.24$

12.5 EXERCISES

1. $4y - 2xy' = 4x^2 - 2x(2x) = 0$ ✔

3. $2y \, dx - x \, dy = 2(3x^2 + 1) \, dx - x(6x \, dx) = 2 \, dx$ ✔

5. $y = \frac{1}{2}e^{x^2+1} + C$

7. $y^2 = 2x^2 + C$

9. $y^3 = x^2 - x + C$

11. $y = e^{x-3} - e^{-3} + 2$

13. $y = \ln|x| - \frac{x^2}{2} + \frac{1}{2}$

15. $\frac{y^2}{2} = \frac{x^3}{3} + C$

17. $\frac{1}{2x^2} + \frac{y^2}{2} = C$

19. $\frac{1}{x} + y + \frac{y^3}{3} = C$

21. $\frac{1}{y} + \ln|x| = C$

23. $x^2 - y^2 = C$

25. $y = C(x + 1)$

27. $x^2 + 4\ln|x| + e^{-y^2} = C$

29. $3y^4 = 4x^3 - 1$

31. $2y = 3x + 4xy$ or $y = \dfrac{3x}{2 - 4x}$

33. $e^{2y} = x^2 - \dfrac{2}{x} + 2$

35. $y^2 + 1 = 5x$

37. $y = Cx^k$

39. (a) $x = 10{,}000e^{0.06t}$ (b) $\$10{,}618.37$; $\$13{,}498.59$

(c) 11.55 years

41. $P = 100{,}000e^{0.05t}$; 5%

43. ≈ 8.4 hours

45. $y = \dfrac{32}{(p + 8)^{2/5}}$

47. $\approx 23{,}100$ years

49. $x = 6(1 - e^{-0.05t})$

51. $x = 20 - 10e^{-0.025t}$

53. $V = 1.86e^{2 - 2e^{-0.1t}}$

55. $V = \dfrac{k^3t^3}{27}$

57. $t \approx 4.5$ hours

59. (a) $E(t) = 18.5e^{0.0164t}$

(b)

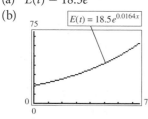

$E(t) = 18.5e^{0.0164x}$

The model looks similar, except at the right end, where it rises more sharply than the graph of the data.

61. (a) $P(t) = 80{,}000e^{-0.05t}$ (b) $\$37{,}789.32$

CHAPTER 12 REVIEW EXERCISES

1. $\frac{1}{7}x^7 + C$

2. $\frac{2}{3}x^{3/2} + C$

3. $3x^4 - x^3 + 2x^2 + 5x + C$ **4.** $\frac{7}{5}x^5 - \frac{14}{3}x^3 + 7x + C$

5. $\frac{7}{6}(x^2 - 1)^3 + C$

6. $\frac{1}{18}(x^3 - 3x^2)^6 + C$

7. $\frac{3}{8}x^8 + \frac{24}{5}x^5 + 24x^2 + C$ **8.** $\frac{5}{63}(3x^3 + 7)^7 + C$

9. $\frac{1}{3}\ln|x^3 + 1| + C$

10. $\dfrac{-1}{3(x^3 + 1)} + C$

11. $\frac{1}{2}(x^3 - 4)^{2/3} + C$

12. $\frac{1}{3}\ln|x^3 - 4| + C$

13. $\frac{1}{2}x^2 - \frac{1}{x} + C$

14. $\frac{1}{3}x^3 + \frac{1}{2}x^2 - 2x - \ln|x - 1| + C$

15. $\frac{1}{3}e^{y^3} + C$ **16.** $\frac{1}{39}(3x - 1)^{13} + C$

17. $\frac{1}{2}\ln|2x^3 - 7| + C$

18. $\dfrac{-5}{4e^{4x}} + C$

19. $x^4/4 - e^{3x}/3 + C$

20. $\frac{1}{2}e^{x^2+1} + C$

21. $\dfrac{-3}{40(5x^8 + 7)^2} + C$

22. $-\frac{7}{2}\sqrt{1 - x^4} + C$

23. $\frac{1}{4}e^{2x} - e^{-2x} + C$

24. $x^2/2 + 1/(x + 1) + C$

25. (a) $\frac{1}{10}(x^2 - 1)^5 + C$ (b) $\frac{1}{22}(x^2 - 1)^{11} + C$

(c) $\frac{3}{16}(x^2 - 1)^8 + C$ (d) $\frac{3}{2}(x^2 - 1)^{1/3} + C$

26. (a) $\ln|x^2 - 1| + C$ (b) $\dfrac{-1}{x^2 - 1} + C$

(c) $3\sqrt{x^2 - 1} + C$ (d) $\frac{3}{2}\ln|x^2 - 1| + C$

27. $y = C - 92e^{-0.05t}$

28. $y = 64x + 38x^2 - 12x^3 + C$

29. $(y - 3)^2 = 4x^2 + C$ **30.** $(y + 1)^2 = 2\ln|t| + C$

31. $e^y = \dfrac{x^2}{2} + C$ **32.** $y = Ct^4$

33. $3(y + 1)^2 = 2x^3 + 75$

34. $x^2 = y + y^2 + 4$ **35.** $\$28{,}800$ **36.** 472

37. $P(t) = 400[1 - 5/(t + 5) + 25/(t + 5)^2]$

38. $p = 1990.099 - 100{,}000/(t + 100)$

39. (a) $y = -60e^{-0.04t} + 60$ (b) 23%

40. $R(x) = 800\ln(x + 2) - 554.52$

41. (a) $\$1000$ (b) $C(x) = 3x^2 + 4x + 1000$

42. 80 units, $\$440$

43. $C(y) = \sqrt{2y + 16} + 0.6y + 4.5$

44. $C(y) = 0.8y - 0.05e^{-2y} + 7.85$ **45.** $W = CL^3$

46. (a) $\ln|P| = kt + C_1$ (b) $P = Ce^{kt}$

(c) $P = 50{,}000\,e^{0.1t}$ (d) The interest rate is $k = 0.10 = 10\%$.

47. ≈ 10.7 million years **48.** $x = 360(1 - e^{-t/30})$

49. $x = 600 - 500e^{-0.01t}$; ≈ 161 min

CHAPTER 12 TEST

1. $2x^3 + 4x^2 - 7x + C$

2. $11x - \dfrac{x^4}{2} + C$

3. $\dfrac{5x^3}{3} - 5x + C$

4. $4x + \frac{2}{3}x\sqrt{x} + \frac{1}{x} + C$

5. $\dfrac{(7 + 2x^3)^{10}}{10} + C$

6. $\dfrac{(4x^3 - 7)^{10}}{24} + C$

7. $-\frac{1}{6}(3x^2 - 6x + 1)^{-2} + C$ **8.** $e^x + 5\ln|x| - x + C$

9. $\dfrac{\ln|2s^4 - 5|}{8} + C$ **10.** $-10{,}000e^{-0.01x} + C$

11. $\frac{5}{8}e^{2y^4 - 1} + C$

12. $x^2 - x + \dfrac{1}{2}\ln|2x + 1| + C$

13. $6x^2 - 1 + 5e^x$ **14.** $\dfrac{1}{9}x^3 - \dfrac{5}{8}x + C$

15. $y = x^4 + x^3 + 4$ **16.** $y = \frac{1}{4}e^{4x} + \frac{7}{4}$

17. $y = \dfrac{4}{C - x^4}$ **18.** 157,498

19. $P(x) = 450x - 2x^2 - 300$

20. $C(y) = 0.78y + \sqrt{0.5y + 1} + 5.6$

21. about 332 days **22.** $x = 16 - 16e^{-t/40}$

.ₛ

.square units   **3.** 7.25 square units

.s square units   **7.** 11.25 square units

9. $S_L(10) = 4.08; S_R(10) = 5.28$

11. Both equal 14/3.

13. It would lie between $S_L(10)$ and $S_R(10)$. It would equal 14/3.

15. 3   **17.** 42   **19.** −5   **21.** 180   **23.** 11,315

25. $3 - \dfrac{3(n+1)}{n} + \dfrac{(n+1)(2n+1)}{2n^2} = \dfrac{2n^2 - 3n + 1}{2n^2}$

27. (a) $S = (n-1)/n$ (b) 9/10   (c) 99/100

    (d) 999/1000   (e) 1

29. (a) $S = \dfrac{(n+1)(2n+1)}{6n^2}$

    (b) 77/200 = 0.385   (c) 6767/20,000 ≈ 0.3384

    (d) 667,667/2,000,000 ≈ 0.3338   (e) $\frac{1}{3}$

31. $\frac{20}{3}$

33. (a) 8696 square units

    (b) This represents the total per capita out-of-pocket expenses for health care between 2013 and 2021.

35. There are approximately 90 squares under the curve, each representing 1 second by 10 mph, or

$$1 \sec \times \frac{10 \text{ mi}}{1 \text{ hr}} \times \frac{1 \text{ hr}}{3600 \text{ sec}} = \frac{1}{360} \text{ mile}.$$

The area under the curve is approximately $90\left(\frac{1}{360} \text{ mile}\right) = \frac{1}{4}$ mile.

37. 1550 square feet

39. 107.734 square units. This represents the total sulphur dioxide emissions (in millions of short tons) from electricity generation from 2010 to 2015.

13.2 EXERCISES

1. 18   **3.** 2   **5.** 60   **7.** $12\sqrt[3]{25}$ **9.** 0   **11.** 98

13. $-\frac{1}{10}$ **15.** 12,960   **17.** 0   **19.** 0   **21.** $\frac{49}{3}$

23. 2   **25.** $e^3/3 - 1/3$ **27.** 4   **29.** $\frac{8}{3}(1 - e^{-8})$

31. (a) $\frac{1}{6}\ln(112/31) \approx 0.2140853$ (b) 0.2140853

33. (a) $\frac{3}{2} + 3\ln 2 \approx 3.5794415$ (b) 3.5794415

35. (a) A, C (b) B

37. $\int_0^4 (2x - \frac{1}{2}x^2)dx$ (b) 16/3

39. (a) $\int_{-1}^0 (x^3 + 1)\,dx$ (b) 3/4

41. $\frac{1}{6}$ **43.** $\frac{1}{2}(e^9 - e)$

45. $\int_0^a g(x)\,dx > \int_0^a f(x)\,dx$; more area under $g(x)$

47. same absolute values, opposite signs

49. −6   **51.** 0   **53.** (a) $450,000   (b) $450,000

55. (a) $5390   (b) $2450

57. $20,405.39

59. 4146 represents the total million metric tons of CO_2 emissions from 2010 to 2020.

61. 0.04 cm³   **63.** 1222 (approximately)

65. 0.1808

67. (a) 0.5934   (b) 0.1733

69. (a) $G(t) = -0.157t^2 - 0.196t + 133$

    (b) 1267.9; the total amount of gasoline (1267.9 billion gallons) used by motor vehicles in the United States from 2014 to 2024.

13.3 EXERCISES

1. (a) $\int_0^2 (4 - x^2)\,dx$ (b) $\frac{16}{3}$

3. (a) $\int_1^8 [\sqrt[3]{x} - (2 - x)]\,dx$ (b) 28.75

5. (a) $\int_1^2 [(4 - x^2) - (\frac{1}{4}x^3 - 2)]\,dx$ (b) 131/48

7. (a) $(-1, 1), (2, 4)$ (b) $\int_{-1}^2 [(x + 2) - x^2]\,dx$

    (c) 9/2

9. (a) $(0, 0), \left(\frac{5}{2}, -\frac{15}{4}\right)$

    (b) $\int_0^{5/2} [(x - x^2) - (x^2 - 4x)]\,dx$ (c) $\frac{125}{24}$

11. (a) $(-2, -4), (0, 0), (2, 4)$

    (b) $\int_{-2}^0 [(x^3 - 2x) - 2x]\,dx + \int_0^2 [2x - (x^3 - 2x)]\,dx$

    (c) 8

13. $\frac{28}{3}$ **15.** $\frac{1}{4}$ **17.** $\frac{16}{3}$ **19.** $\frac{1}{3}$ **21.** $\frac{37}{12}$

23. $4 - 3\ln 3$ **25.** $\frac{8}{3}$ **27.** 6   **29.** 0   **31.** $-\frac{4}{9}$

33. $11.8\overline{3}$

35. average profit $= \dfrac{1}{x_1 - x_0}\displaystyle\int_{x_0}^{x_1} [R(x) - C(x)]\,dx$

37. (a) $1402 per unit   (b) $535,333.33

39. (a) 102.5 units   (b) 100 units

41. (a) 40.05 million/year   (b) 69.93 million/year

43. 147 mg

45. Black: 0.479; Hispanic: 0.459

The income distribution for Black households is more unequal than it is for Hispanic households.

47. 2012: 0.468; 2004: 0.455

2012 shows more income distribution inequality than 2004.

49. $G = \dfrac{p - 1}{p + 1}$

13.4 EXERCISES

1. $126,205.10   **3.** $346,664 (nearest dollar)

5. $506,000 (nearest thousand)

7. $18,660 (nearest dollar)

9. $82,155 (nearest dollar)

11. $PV = \$2,657,807$ (nearest dollar), $FV = \$3,771,608$ (nearest dollar)

13. $PV = \$190,519$ (nearest dollar), $FV = \$347,147$ (nearest dollar)

15. Gift Shoppe, $151,024; Wine Boutique, $141,093. The gift shop is a better buy.

17. $83.33   **19.** $161.89   **21.** (5, 56); $83.33

23. $11.50   **25.** $204.17   **27.** $2766.67

29. $17,839.58   **31.** $133.33   **33.** $2.50

35. $103.35

13.5 EXERCISES

1. formula 5: $\frac{1}{8} \ln |(4 + x)/(4 - x)| + C$
3. formula 11: $\frac{1}{3} \ln [(3 + \sqrt{10})/2]$
5. formula 14: $w(\ln w - 1) + C$
7. formula 12: $\frac{1}{3} + \frac{1}{4} \ln \left(\frac{3}{7}\right)$
9. formula 13: $\frac{1}{8} \ln \left| \dfrac{v}{3v + 8} \right| + C$
11. formula 7: $\frac{1}{2} [7\sqrt{24} - 25 \ln (7 + \sqrt{24}) + 25 \ln 5]$
13. formula 16: $\dfrac{(6w - 5)(4w + 5)^{3/2}}{60} + C$
15. formula 3: $\frac{1}{2}(5^{x^2})\log_5 e + C$
17. formula 1: $\frac{1}{3}(13^{3/2} - 8)$
19. formula 9: $-\frac{5}{2} \ln \left| \dfrac{2 + \sqrt{4 - 9x^2}}{3x} \right| + C$
21. formula 10: $\frac{1}{3} \ln |3x + \sqrt{9x^2 - 4}| + C$
23. formula 15: $\frac{3}{4} \left[\ln |2x - 5| - \dfrac{5}{2x - 5} \right] + C$
25. formula 8: $\frac{1}{3} \ln |3x + 1 + \sqrt{(3x + 1)^2 + 1}| + C$
27. formula 6: $\frac{1}{4} [10\sqrt{109} - \sqrt{10} + 9 \ln (10 + \sqrt{109})$
$- 9 \ln (1 + \sqrt{10})]$
29. formula 2: $-\frac{1}{6} \ln |7 - 3x^2| + C$
31. formula 8: $\frac{1}{2} \ln |2x + \sqrt{4x^2 + 7}| + C$
33. $2(e^{\sqrt{2}} - e) \approx 2.7899$
35. $\frac{1}{32} [\ln (9/5) - 4/9] \approx 0.004479$
37. $3391.10
39. (a) $C = \frac{1}{2}x\sqrt{x^2 + 9} + \frac{9}{2} \ln |x + \sqrt{x^2 + 9}| + 300$
$- \frac{9}{2} \ln 3$
(b) $314.94
41. $3882.9 thousand

13.6 EXERCISES

1. $\frac{1}{2}xe^{2x} - \frac{1}{4}e^{2x} + C$ 3. $\frac{1}{3}x^3 \ln x - \frac{1}{9}x^3 + C$
5. $\dfrac{104\sqrt{2}}{15}$ 7. $-(1 + \ln x)/x + C$ 9. 1
11. $\dfrac{x^2}{2} \ln (2x - 3) - \dfrac{1}{4}x^2 - \dfrac{3}{4}x - \dfrac{9}{8} \ln (2x - 3) + C$
13. $\frac{1}{5}(q^2 - 3)^{3/2}(q^2 + 2) + C$ 15. 282.4
17. $-e^{-x}(x^2 + 2x + 2) + C$ 19. $(9e^4 + 3)/2$
21. $\frac{1}{4}x^4 \ln^2 x - \frac{1}{8}x^4 \ln x + \frac{1}{32}x^4 + C$
23. $\frac{2}{15}(e^x + 1)^{3/2}(3e^x - 2) + C$ 25. II; $\frac{1}{2}e^{x^2} + C$
27. IV; $\frac{2}{3}(e^x + 1)^{3/2} + C$ 29. I; $-5e^{-4} + 1$
31. $2794.46   33. $34,836.73   35. 0.264
37. 166 million

13.7 EXERCISES

1. 1/5   3. 2   5. 1/e   7. diverges   9. diverges
11. 10   13. diverges   15. diverges   17. 0
19. 0   21. 0.5   23. 1/(2e)   25. $\frac{3}{2}$
27. $\int_{-\infty}^{\infty} f(x)\, dx = 1$ 29. $c = 1$ 31. $c = \frac{1}{4}$

33. 20
35. area $= \frac{8}{3}$
37. $\int_0^{\infty} Ae^{-rt}\, dt = A/r$
39. $2,400,000
41. $700,000
43. (a) 0.368   (b) 0.018
45. 0.147
47. (a) $500 \left[\dfrac{e^{-0.03b} + 0.03b - 1}{0.0009} \right]$
(b) The amount approaches ∞.

13.8 EXERCISES

1. $h = \frac{1}{2}; x_0 = 0, x_1 = \frac{1}{2}, x_2 = 1, x_3 = \frac{3}{2}, x_4 = 2$
3. $h = \frac{1}{2}; x_0 = 1, x_1 = \frac{3}{2}, x_2 = 2, x_3 = \frac{5}{2}, x_4 = 3, x_5 = \frac{7}{2},$
$x_6 = 4$
5. $h = 1; x_0 = -1, x_1 = 0, x_2 = 1, x_3 = 2, x_4 = 3, x_5 = 4$
7. (a) 9.13   (b) 9.00   (c) 9   (d) Simpson's
9. (a) 0.51   (b) 0.50   (c) $\frac{1}{2}$ (d) Simpson's
11. (a) 5.27   (b) 5.30   (c) 5.33   (d) Simpson's
13. (a) 3.283   (b) 3.240
15. (a) 0.743   (b) 0.747
17. (a) 7.132   (b) 7.197   19. 7.8   21. 10.3
23. 119.58 ($119,580)   25. $32,389.76
27. $14,133.33   29. 1222.35 (1222 units)
31. (a)

x	0	0.2	0.4	0.6	0.8	1
$L_a - L_b$	0	0.003	0.005	0.013	0.029	0

(b) 0.020
(c) positive; 1990
33. (a) Yes   (b) Simpson's   (c) 1586.67 ft^2

CHAPTER 13 REVIEW EXERCISES

1. 212   2. $\dfrac{3(n + 1)}{2n^2}$ 3. $\frac{91}{72}$ 4. 1   5. 1
6. 14   7. $\frac{248}{5}$ 8. $-\frac{205}{4}$ 9. $\frac{825}{4}$ 10. $\frac{4}{13}$
11. -2 12. $\frac{1}{6} \ln 47 - \frac{1}{6} \ln 9$ 13. $\frac{9}{2}$
14. $\ln 4 + \frac{14}{3}$ 15. 190/3   16. $\frac{1}{2} \ln 2$
17. $(1 - e^{-2})/2$ 18. $(e - 1)/2$ 19. 95/2
20. 36   21. $\frac{1}{4}$ 22. $\frac{1}{2}$
23. $\frac{1}{2} x\sqrt{x^2 - 4} - 2 \ln |x + \sqrt{x^2 - 4}| + C$
24. $2 \log_3 e$ 25. $\frac{1}{2}x^2(\ln x^2 - 1) + C$
26. $\frac{1}{2} \ln |x| - \frac{1}{2} \ln |3x + 2| + C$
27. $\frac{1}{6} x^6 \ln x - \frac{1}{36}x^6 + C$
28. $-e^{-2x}(x^2/2 + x/2 + 1/4) + C$
29. $2x\sqrt{x + 5} - \frac{4}{3}(x + 5)^{3/2} + C$
30. 1   31. diverges   32. -100
33. $\frac{5}{3}$ 34. $-\frac{1}{2}$
35. (a) $\frac{8}{9} \approx 0.889$ (b) 1.004   (c) 0.909
36. 3.135   37. 3.9
38. (a) $n = 5$ (b) $n = 6$ 39. $28,000
40. $1 - e^{-2.8} \approx 0.939$ 41. $1297.44   42. $76.60

43. 1969: 0.3737; 2000: 0.4264; more equally distributed in 1969

44. (a) $(7, 6)$ (b) \$7.33   **45.** \$24.50

46. \$1,621,803   **47.** (a) \$403,609   (b) \$602,114

48. \$217.42   **49.** \$10,066 (nearest dollar)

50. \$86,557.41

51. $C(x) = 3x + 30(x + 1)^2 \ln (x + 1)$
$\quad - 15(x + 1)^2 + 2015$

52. $e^{-1.4} \approx 0.247$

53. \$4000 thousand, or \$4 million

54. \$197,365   **55.** \$480,000 hundred, or \$48,000,000

CHAPTER 13 TEST

1. 3.496 (approximately)

2. (a) $5 - \dfrac{n + 1}{n}$ (b) 4

3. $\int_0^6 (12 + 4x - x^2)\, dx$; 72

4. (a) 4   (b) 3/4   (c) $\frac{5}{4} \ln 5$ (d) 7
$\quad$ (e) 0; limits of integration are the same
$\quad$ (f) $\frac{5}{6}(e^2 - 1)$

5. (a) $3xe^x - 3e^x + C$ (b) $\dfrac{x^2}{2} \ln (2x) - \dfrac{x^2}{4} + C$

6. -8

7. (a) $x[\ln (2x) - 1] + C$
$\quad$ (b) $\dfrac{2(9x + 14)(3x - 7)^{3/2}}{135} + C$

8. 16.089

9. (a) \$4000   (b) \$16,000/3

10. (a) \$961.18 thousand   (b) \$655.68 thousand
$\quad$ (c) \$1062.5 thousand

11. 125/6

12.
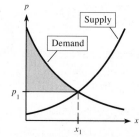

13. Before, 0.446; After, 0.19. The change decreases the difference in income.

14. (a) About 20.92 billion barrels
$\quad$ (b) About 2.067 billion barrels per year

15. About 2.9666   **16.** 6800 ft²

14.1 EXERCISES

1. $\{(x, y): x \text{ and } y \text{ are real numbers}\}$

3. $\{(x, y): x \text{ and } y \text{ are real numbers and } y \neq 0\}$

5. $\{(x, y): x \text{ and } y \text{ are real numbers and } 2x - y \neq 0\}$

7. $\{(p_1, p_2): p_1 \text{ and } p_2 \text{ are real numbers and } p_1 \geq 0\}$

9. -2 **11.** $\frac{5}{3}$ **13.** 2500   **15.** 36   **17.** 3

19. $\frac{1}{25} \ln (12)$ **21.** $\frac{13}{3}$

23. \$6640.23; the amount that results when \$2000 is invested for 20 years

25. 500; if the cost of placing an order is \$200, the number of items sold per week is 625, and the weekly holding cost per item is \$1, then the most economical order size is 500.

27. Max: $S \approx 112.5°F$; $A \approx 106.3°F$
$\quad$ Min: $S \approx 87.4°F$; $A \approx 77.6°F$

29. (a) \$752.80; when \$90,000 is borrowed for 20 years at 8%, the monthly payment is \$752.80.
$\quad$ (b) \$1622.82; when \$160,000 is borrowed for 15 years at 9%, the monthly payment is \$1622.82.

31. (a) $x = 4$ (b) $y = 2$
$\quad$ (c)
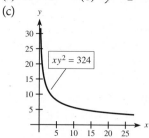

33. (a) 37,500 units
$\quad$ (b) $30(2K)^{1/4}(2L)^{3/4} = 30(2^{1/4})(2^{3/4})K^{1/4}L^{3/4} =$
$\quad\quad 2[30K^{1/4}L^{3/4}]$
$\quad$ (c)
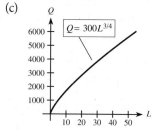

35. (a) 7200 units   (b) 5000 units   **37.** \$284,000

14.2 EXERCISES

1. $\dfrac{\partial z}{\partial x} = 4x^3 - 10x + 6$ $\dfrac{\partial z}{\partial y} = 9y^2 - 5$

3. $z_x = 3x^2 + 8xy$ $z_y = 4x^2 + 12y$

5. $\dfrac{\partial f}{\partial x} = 9x^2(x^3 + 2y^2)^2$ $\dfrac{\partial f}{\partial y} = 12y(x^3 + 2y^2)^2$

7. $f_x = 2x(2x^2 - 5y^2)^{-1/2}$ $f_y = -5y(2x^2 - 5y^2)^{-1/2}$

9. $\dfrac{\partial C}{\partial x} = -4y + 20xy$ $\dfrac{\partial C}{\partial y} = -4x + 10x^2$

11. $\dfrac{\partial Q}{\partial s} = \dfrac{2(t^2 + 3st - s^2)}{(s^2 + t^2)^2}$ $\dfrac{\partial Q}{\partial t} = \dfrac{3t^2 - 4st - 3s^2}{(s^2 + t^2)^2}$

13. $z_x = 2e^{2x} + \dfrac{y}{x}$ $z_y = \ln x$

15. $\dfrac{\partial f}{\partial x} = \dfrac{y}{xy + 1}$ $\dfrac{\partial f}{\partial y} = \dfrac{x}{xy + 1}$ **17.** 2

19. 7   **21.** -19

23. (a) 0   (b) $-2xz + 4$ (c) $2y$ (d) $-x^2$

25. (a) $8x_1 + 5x_2$ (b) $5x_1 + 12x_2$ (c) 1

27. (a) 2   (b) 0   (c) 0   (d) $-30y$

29. (a) $2y$ (b) $2x - 8y$ (c) $2x - 8y$ (d) $-8x$

31. (a) $2 + y^2e^{xy}$ (b) $xye^{xy} + e^{xy}$
     (c) $xye^{xy} + e^{xy}$ (d) x^2e^{xy}

33. (a) $1/x^2$ (b) 0 (c) 0 (d) $2 + 1/y^2$

35. -6 **37.** (a) $\frac{188}{4913}$ (b) $\frac{-188}{4913}$ **39.** $2 + 2e$

41. 0 **43.** (a) $24x$ (b) $24x$ (c) 0

45. (a) For a mortgage of $100,000 and an 8% interest rate, the monthly payment is $771.82.
     (b) The rate of change of the payment with respect to the interest rate is $66.25. That is, if the rate goes from 8% to 9% on a $100,000 mortgage, the approximate increase in the monthly payment is $66.25.

47. (a) If the number of items sold per week changes by 1, the most economical order quantity should also increase. $\dfrac{\partial Q}{\partial M} = \sqrt{\dfrac{K}{2Mh}} > 0$
     (b) If the weekly storage costs change by 1, the most economical order quantity should decrease.

$$\frac{\partial Q}{\partial h} = -\sqrt{\frac{KM}{2h^3}} < 0$$

49. (a) 23.912; If brand 2 is held constant and brand 1 is increased from 100 to 101 liters, approximately 24,000 additional insects will be killed.

51. (a) $2xy^2$ (b) $2x^2y$

53. $\dfrac{\partial Q}{\partial K} = 100$; If labor hours are held constant at 5832 and K changes by $1 (thousand) to $730,000, Q will change by about 100 units. $\dfrac{\partial Q}{\partial L} = 25$; If capital expenditures are held constant at $729,000 and L changes by 1 hour (to 5833), Q will change by about 25 units.

55. (a) $\dfrac{\partial WC}{\partial s} = 0.16s^{-0.84}(0.4275t - 35.75)$

     (b) At $t = 10$, $s = 25$, $\dfrac{\partial WC}{\partial s} \approx -0.34$
     This means that if wind speed changes by 1 mph (from 25 mph) while the temperature remains at 10°F, the wind chill temperature will change by about -0.34°F.

14.3 EXERCISES

1. (a) $105   (b) $C_x = 3$ means total costs would change by $3 if labor costs changed by $1 and raw material costs stayed the same.

3. (a) $2 + y/50$ (b) $4 + x/50$

5. (a) $25.78   (b) $74.80

7. (a) If y remains at 10, the expected change in cost for a 9th unit of X is about $36.
     (b) If x remains at 8, the expected change in cost for an 11th unit of Y is about $19.

9. (a) $\sqrt{y^2 + 1}$ dollars per unit
     (b) $xy/\sqrt{y^2 + 1}$ dollars per unit

11. (a) $1200y/(xy + 1)$ dollars per unit
     (b) $1200x/(xy + 1)$ dollars per unit

13. (a) $\sqrt{y/x}$ (b) $\sqrt{x/y}$

15. (a) $\ln(y + 1)/(2\sqrt{x})$ (b) $\sqrt{x}/(y + 1)$

17. $z = 1092$ crates (approximately)

19. $z_x = 3.6$; If 500 acres are planted, the approximate change in productivity from a 301st hour of labor is 3.6 crates.

21. (a) $z_x = \dfrac{240y^{2/5}}{x^{2/5}}$ (b)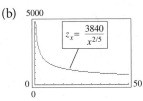

     (c) $z_y = \dfrac{160x^{3/5}}{y^{3/5}}$ (d)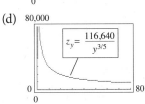

     (e) Both z_x and z_y are positive, so increases in both capital investment and work-hours result in increases in productivity. However, both are decreasing, so such increases have a diminishing effect on productivity.

23. $q_1 = 188$ units; $q_2 = 270$ units

25. any values for p_1 and p_2 that satisfy $6p_2 - 3p_1 = 100$ and that make q_1 and q_2 nonnegative, such as $p_1 = 10, $p_2 = $21\frac{2}{3}$

27. (a) -3 units per dollar   (b) -2 units per dollar
     (c) -6 units per dollar   (d) -5 units per dollar
     (e) complementary

29. (a) -50 units per dollar
     (b) $600/(p_B + 1)^2$ units per dollar
     (c) $-400/(p_B + 4)^2$ units per dollar
     (d) $400/(p_A + 4)^2$ units per dollar
     (e) competitive

31. (a) Competitive; as the price of one type of car declines, demand for the other declines
     (b) (i) $q_{NEW} = 2600 - p_{NEW}/30 + p_{USED}/15$

$$q_{USED} = 750 - 0.25\,p_{USED} + 0.0125\,p_{NEW}$$

        (ii) Since the mixed partials are both positive (1/15 and 0.0125), the products are competitive.

14.4 EXERCISES

1. max(0, 0, 9)   **3.** min(0, 0, 4)

5. saddle$(-2, -3, 16)$ **7.** min $(1, -2, 0)$

9. saddle$(1, -3, 8)$ **11.** max(12, 24, 456)

13. min$(-8, 6, -52)$

15. saddle(0, 0, 0); min(2, 2, -8)

17. $\hat{y} = 5.7x - 1.4$

19. $x = 5000$, $y = 128$; $P = $25,409.60

21. $x = \frac{20}{3}$, $y = \frac{10}{3}$; $W \approx 1926$ lb

23. $x = 28$, $y = 100$; $P = 5987.84$ tons

25. $x = 20$ thousand, $y = 30$ thousand;
$P = \$1900$ thousand
27. length $= 100$ in., width $= 100$ in., height $= 50$ in.
29. $x = 15$ thousand, $y = 24$ thousand;
$P = \$295$ thousand
31. (a) eat-in $= 2400$; take-out $= 3800$
  (b) eat-in @ \$3.60; take-out @ \$3.10; max profit $=$
    $\$12,480$
  (c) Change pricing; more profitable
33. (a) $\hat{y} = 0.81x - 2400$
  (b) $m = 0.81$; means that for every \$1 that males earn,
    females earn about \$0.81.
  (c) The slope would probably be smaller. Equal pay
    for women for equal work is not yet a reality, but
    much progress has been made since 1965.
35. (a) $\hat{y} = 0.06254x + 6.191$, x in years past 2000,
    $\hat{y}$ in billions
  (b) 7.317 billion
  (c) World population is changing about
    0.06254 billion persons per year past 2000.

14.5 EXERCISES

1. 18 at $(3, 3)$ **3.** 35 at $(3, 2)$ **5.** 32 at $(4, 2)$
7. -28 at $\left(3, \frac{5}{2}\right)$ **9.** 15 at $(5, 3)$
11. 3 at $(1, 1, 1)$ **13.** 1 at $(0, 1, 0)$
15. $x = 2, y = 2$
17. $x = 40, y = \frac{40}{3}$

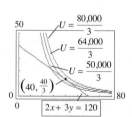

19. (a) $x = 400, y = 400$
  (b) $-\lambda = 1.6$; means that each additional dollar spent
    on production results in approximately 1.6 addi-
    tional units produced.
  (c)

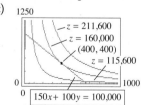

21. $x = 900, y = 300$; 900 units at plant X, 300 units at
  plant Y
23. $x = \$10,003.33, y = \$19,996.67$
25. length $= 100$ cm, width $= 100$ cm, height $= 50$ cm

CHAPTER 14 REVIEW EXERCISES

1. $\{(x, y): x \text{ and } y \text{ are real numbers and } y \neq 2x\}$
2. $\{(x, y): x \text{ and } y \text{ are real numbers with } y \geq 0 \text{ and } (x, y) \neq (0, 0)\}$
3. -5 **4.** 896,000
5. $15x^2 + 6y$ **6.** $24y^3 - 42x^3y^2$
7. $z_x = 8xy^3 + 1/y$; $z_y = 12x^2y^2 - x/y^2$

8. $z_x = x/\sqrt{x^2 + 2y^2}$; $z_y = 2y/\sqrt{x^2 + 2y^2}$
9. $z_x = -2y/(xy + 1)^3$; $z_y = -2x/(xy + 1)^3$
10. $z_x = 2xy^3e^{x^2y^3}$; $z_y = 3x^2y^2e^{x^2y^3}$
11. $z_x = ye^{xy} + y/x$; $z_y = xe^{xy} + \ln x$
12. $z_x = y$; $z_y = x$ **13.** -8 **14.** 8
15. (a) $2y$ (b) 0 (c) $2x - 3$ (d) $2x - 3$
16. (a) $18xy^4 - 2/y^2$ (b) $36x^3y^2 - 6x^2/y^4$
  (c) $36x^2y^3 + 4x/y^3$ (d) $36x^2y^3 + 4x/y^3$
17. (a) $2e^{y^2}$ (b) $4x^2y^2e^{y^2} + 2x^2e^{y^2}$
  (c) $4xye^{y^2}$ (d) $4xye^{y^2}$
18. (a) $-y^2/(xy + 1)^2$ (b) $-x^2/(xy + 1)^2$
  (c) $1/(xy + 1)^2$ (d) $1/(xy + 1)^2$
19. $\max(-8, 16, 208)$
20. saddles at $(2, -3, 38)$ and $(-2, 3, -38)$; min at
  $(2, 3, -70)$; max at $(-2, -3, 70)$
21. 80 at $(2, 8)$
22. 11,664 at $(6, 3)$
23. (a) $x^2y = 540$ (b) 3 units
24. (a) \$46,204
  (b) When the monthly contribution is \$250 and the
    interest rate is 7.8%, the accumulated value is
    about \$143,648.
  (c) When the contribution is \$250, if the interest
    rate changed from 7.8% to 8.8%, the approximate
    change in the account would be \$17,770.
  (d) $A_R \approx 574.59$ means that with an interest rate of
    7.8%, if the monthly contribution changed from
    \$250 to \$251, the approximate change in the accu-
    mulated value would be \$574.59.
25. (a) 8.996 thousand, or \$8996
  (b) 0.009; means that when the benefits are paid for
    20 years, if the account value changes from 1000
    to 1001 (thousand dollars), the monthly benefit
    increases by about 0.009 (thousand), or \$9.
  (c) -0.161; means that when the account value is
    \$1,000,000, if the duration of benefits changes
    from 20 to 21 years, the monthly benefit decreases
    by about \$161.
26. (a) If selling price is fixed, more dollars spent for
    advertising will increase sales.
  (b) If advertising dollars are fixed, an increase in the
    selling price will decrease sales.
27. (a) 280 dollars per unit of x
  (b) $2400/7$ dollars per unit of y
28. $\partial Q/\partial K = 81.92$ means that when capital expenditures increase by \$1000 (to \$626,000) and work-hours
  remain at 4096, output will change by about 8192 units;
  $\partial Q/\partial L = 37.5$ means that when labor hours change
  by 1 (to 4097) and capital expenditures remain at
  \$625,000, output will change by about 3750 units.
29. (a) -2 (b) -6 (c) complementary
30. competitive
31. $x = 20, y = 40$; $P = \$2000$
32. 200 units at plant I; 1500 units at plant II
33. $x = 10, y = 4$
34. (a) $x = 1000, y = 500$

(b) $-\lambda \approx 3.17$; means that each additional dollar spent on production results in approximately 3 additional units.

(c)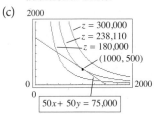

35. (a) $\hat{y} = 2375x + 39,630$ (b) $75,255

36. (a) $\hat{y} = 0.736x + 28.4$

(b) $m = 0.736$ means for every 1000 women entering the workforce, there are about 736 men entering it.

CHAPTER 14 TEST

1. (a) all pairs (x, y) with $y < x^2$ (b) 14

2. $z_x = 5 + 10y(xy + 1)^4$ $z_y = -18y + 10x(xy + 1)^4$
$z_{xx} = 40y^2(xy + 1)^3$ $z_{yy} = -18 + 40x^2(xy + 1)^3$
$z_{xy} = z_{yx} = 10(5xy + 1)(xy + 1)^3$

3. $(0, 2)$, a relative minimum; $(4, -6)$ and $(-4, -6)$, saddle points

4. (a) $1625 thousand

(b) 73.11; means that if capital investment increases from $10,000 to $11,000, the approximate change

in monthly production value will be $73.11 thousand, if labor hours remain at 1590.

(c) 0.56; means that if labor hours increase by 1 to 1591, the approximate change in monthly production value will be $0.56 thousand, if capital investment remains at $10,000.

5. (a) When $94,500 is borrowed for 25 years at 7%, the monthly payment is $667.91.

(b) If the percent goes from 7% to 8%, the approximate change in the monthly payment is $60.28, if the loan amount remains at $94,500 for 25 years.

(c) Negative. If the loan amount remains at $94,500 and the percent remains at 7%, increasing the time to pay off the loan will decrease the monthly payment, and vice versa.

6. $8xy\, e^{x^2y^2}(x^2y^2 + 1)$

7. Find $\dfrac{\partial q_1}{\partial p_2}$ and $\dfrac{\partial q_2}{\partial p_1}$ and compare their signs. Both positive means competitive. Both negative means complementary. These products are complementary.

8. $x = $7, y = $11; P = 5065

9. $x = 200, y = 100$

10. (a) $\hat{y} = 0.24x + 5.78$

(b) The fit is excellent.

(c) 16.6%

Index

0/0 indeterminate form, 540–541

a/0 form, 541
Abscissa of a graph, 65
Absolute maximum and minimum points, 656
Absolute value, 11–12
Absolute value function, 155
Acceleration, 165–166, 605, AP-31, AP-49
Addition. *See* Sums
Addition property
   equations, 53
   inequalities, 58
Additional improper integrals, 836–837
Additive identity, 10
Additive inverses, 10
Addresses, Excel, AP-27, AP-45
Algebraic concepts, 1–44
   algebraic fractions, 38–44
   factoring, 33–38
   integral exponents, 15–20
   operations with algebraic expressions, 27–33
   radicals and rational exponents, 21–27
   real numbers, 9–15
   sets, 2–9
Algebraic expressions, 27–33
Algebraic fractions, 38–44
Aligning the data, 54–55
Amortization formula, 404
Amortization of loans, 404–409, AP-12–AP-13, AP-37, AP-56
Amortization schedules, 406–407
Annual compounding, 370–371, 376
Annual percentage rate (APR), 404
Annual percentage yield, 375–376
Annuities, ordinary
   defined, 383
   future value of, 383–389, AP-12, AP-36, AP-55, AP-56
   present value of, 392–400, AP-12, AP-37, AP-56
Annuities due
   future value of, 388–389, AP-12, AP-36, AP-55, AP-56
   present value of, 397–398, AP-12, AP-37, AP-56
Antiderivatives. *See* Integrals
Antidifferentiation, 739. *See also* Integrals
APR (annual percentage rate), 404
Area between two curves, 807–813, AP-26
Area under a curve, 788–794
   defined, 788
   definite integrals relationship, 801–803
   estimating with technology, 793–794, AP-24–AP-25, AP-42, AP-61
   overview, 788–791
   standard normal curve, AP-65–AP-66
   summation notation, 791–793
Arithmetic sequences, 365–367, 790
Associative property, 10
Asymptotes
   of Gompertz curves, 346

graphing calculator viewing windows and, AP-21
   horizontal, 153, 154, 554, 675–681
   overview, 153–154
   vertical, 153, 154, 675, 676–681
Augmented matrices, 203
Average, 496. *See also* Mean; Median; Mode
Average cost
   defined, 148, 658
   graphs of, 154–155
   minimization of, 658–659
   using average value of a function, 812
Average rates of change
   linear functions, 138–139
   nonlinear functions, 560–561
Average value of a function, 811–813
Axis of symmetry, 132

Back substitution, 100
Bair, Frank, 864
Bar graphs, 492, AP-38, AP-57
Bases
   change-of-base formula, 336–337, 342, 698
   defined, 16
   exponential equations, 342
   logarithmic functions, 330–331, 694, AP-10, AP-35–AP-36, AP-54–AP-55
Basic feasible solutions, 274
Basic variables, 274, 276–277
Bayes' formula, 451–454
Bayes problems, 452
Bell-shaped curve, 323–324, AP-65–AP-66
Bernoulli (binomial) experiments, 487–489, AP-15
Best fit lines (linear regression), 161, AP-06, AP-31, AP-49
Binomial conjugates, 35–36
Binomial formula, 509, 573
Binomial probability, defined, 488
Binomial probability distributions
   with Excel, AP-38, AP-57
   with graphing calculators, AP-15
   normal curve approximation to, 522–525
   overview, 507–510
Binomial probability experiments, 487–489, AP-15
Binomial variables, 507
Binomials
   defined, 28
   products of, 29, 30
Bond pricing, 396–397
Bosch, W., 864
Boundary conditions, 772, 773
Break-even analysis, 57, 105–106, 142–143, 178
Budget constraints, 892
Business risk, 178

Calculators. *See* Graphing calculators
Calculus, fundamental theorem of, 798–800. *See also* Derivatives; Integrals
Capital value of an income stream, 834, 836

Cartesian coordinates, 65
Cells (spreadsheets), 88, AP-27, AP-45
Central Limit Theorem, 532
Central tendency, measures of. *See* Mean; Median; Mode
Chain Rule
    exponential functions, 703
    implicit differentiation, 710
    logarithmic functions, 695
    overview, 591–592
Change-of-base formula, 336–337, 342, 698, AP-11
Closed and bounded regions, 261, 262
Closed intervals, 10, 11
Closed Leontief model, 233–236
Cobb, L. G., 864
Cobb-Douglas production function, 713, 860, 895–897
Codes (encoding matrices), 197
Coefficient matrices, 203
Coefficient Rule, 575–576
Coefficients, 28
Column matrices, 183, 193
Columns, in matrices, 181
Combination formula, 460
Combinations
    with graphing calculators, AP-13, AP-14
    overview, 459–461
    probability and, 463–465, AP-14
Combining fractions, 41
Common differences, 365
Common factors, 33–34
Common logarithms, 330–331, 337–338
Common monomial factors, 33–34, 36
Common ratios, 378
Commutative law of addition for matrices, 184
Commutative property
    defined, 10
    matrix multiplication applicability, 195, 196, 221
Competitive markets
    marginal profit in, 612–613
    profit maximization in, 661
    taxation in, 727–729
Competitive products, 879
Complementary products, 879
Complements of events, 433, 434
Complements of sets, 4–5
Complete graphs, 86
Complex fractions, 41–42
Composite functions
    defined, 69
    derivatives of, 591–593
Compound interest
    annual compounding, 370–371, 376
    annual percentage yield, 375–376
    continuous compounding, 374–375, 376
    doubling time, 329, 331, 377
    exponential growth model of, 319
    overview, 369–370
    periodic compounding, 371–374, 376
    vs. simple interest, 372
Concave down curves, 644–645
Concave up curves, 644–645
Concavity
    graph observations, 650–651, AP-20, AP-41, AP-60
    overview, 644–645
    points of inflection and, 645–647, 649–650
    second-derivative test, 648–649
Conditional equations, 53
Conditional probability
    overview, 440–442
    product rule, 442–446
Conditional probability formula, 441
Conjugates of binomials, 35–36
Constant functions
    defined, 149
    derivatives of, 575–576
    slope of, 78, 149
Constant terms, 28
Constants of integration, 739, 742, 764–765
Constrained optimization
    with Excel, 896–897, AP-44, AP-63–AP-64
    overview, 892–897
    simplex method, 297–303
Constraints, defined, 260
Consumer's surplus, 818–821, 822
Consumption function, 768–769
Continuity
    differentiability and, 567–568
    functions of two or more variables, 862
    at a point, 549
Continuous compounding, 374–375, 376
Continuous functions
    average value of, 811–813
    limits at infinity and, 555
    overview, 549–552
Continuous income streams, 816–818
Continuous probability distributions, 803–804, 838–839
Converging integrals, 835
Conversion periods, 370–371, 377
Convex regions, 260
Coordinate axes, 64, 860–861
Coordinate planes, 861
Coordinate systems, 64, 860–861
Corners (graphical solutions)
    linear inequalities in two variables, 253, 254, 255–256
    linear programming, 261, 262
    simplex method, 273, 274, 275
Cost
    average, 148, 154–155, 658–659, 812
    fixed, 104, 764–765
    joint, 875–876
    marginal, 104–105, 610–612, 659, 764–765, 875–876
    minimization of, 292–293, 658–659, 668–669, 670–672
    total, 148, 154–155, 610–611, 764–767
    variable, 104
Cost functions
    average, 148, 658–659
    with inflation, 735
    joint, 875–876
    marginal, 104–105, 610–611, 764–766
    obtaining through integration, 764–766
    profit and, 103–104
    quadratic forms of, 131, 154–155
    tangent lines to, 626–627
Coupons (bonds), 396

Critical points
    with constraints, 892–893
    for functions of two variables, 882–885
    overview, 631–632
Critical values
    with constraints, 892–893
    graphing calculator capabilities, 635, 637–638, 699, 705–706,
        AP-19, AP-22, AP-23
    with logarithmic functions, 698–699
    overview, 631–632
Cube roots, 21
Cubic polynomial functions, 38, 151
Curve fitting, 161–166, AP-06, AP-30–AP-31, AP-48–AP-49
Curve sketching
    with Excel, 636, 639, AP-40, AP-59
    finding relative maxima and minima, 630–632, 633–635
    first-derivative test, 632–633
    with graphing calculators, 635, 637–638, AP-20
    horizontal points of inflection, 634–635
    sign diagrams, 632, 638–639
Curves. See specific types

Dantzig, George, 271
Data description. See Statistical data
Decay. See Exponential growth and decay
Decoding matrices, 216–220
Decreasing functions, 78
Decreasing on an interval, 630–631
Deferred annuities, 398–400
Definite integrals. See also Integrals
    area between two curves relationship, 807–813
    area under a curve relationship, 801–803
    business and economic applications, 816–822
    defined, 798
    with Excel, AP-42, AP-61
    fundamental theorem of calculus and, 798–800
    with graphing calculators, AP-25
    numerical integration, 841–846
    and probability, 804, 837–839
    properties of, 800–801
    Riemann sums and, 798
Degree of a polynomial, 28
Demand
    elasticity of, 723–727
    final, 230
    law of, 106
    market equilibrium and, 107–108, 144–146
    modeling with exponential and logarithmic functions, 345
    taxation effects, 108–109, 727–729
Demand functions
    as function of two or more variables, 878–879
    linear regression for, 161
Demand surface graphs, 878
Denominators, rationalizing, 24–25, 42–43
Dependent events, 445
Dependent systems of equations, 93
Dependent variables, 64, 860
Derivatives, 533–627
    approximation of, 567, 579–580, 605, AP-18–AP-19, AP-40, AP-59
    Chain Rule, 591–592, 695, 703, 710
    of composite functions, 591–593
    of constant functions, 575–576
        defined, 562–563
        of exponential functions, 702–706, AP-22
        formulas for, 597–600, 751, 755, 757
        graphs and, 678
        higher-order, 602–605
        implicit differentiation, 709–715, 717–721
        interpretations of, 566–567
        of logarithmic functions, 694–700, AP-22
        marginal cost and, 610–612
        marginal profit and, 612–614
        marginal revenue and, 566, 578–579, 608–610
        partial, 866–872
        Power Rule, 592–593, 597–600
        of powers of x, 573–575, 586
        Product Rule, 584, 597–600
        Quotient Rule, 584–586, 598, 868
        related rates, 717–721
        of sums and differences, 576–578
        undefined, 636
Determinants of matrices, 224–225
Difference of two squares factoring, 35–36
Difference quotients, 563
Differences
    derivatives of, 576, 577–578
    of functions, 68
    of matrices, 184, AP-07
Differentiability
    continuity and, 567–568
    defined, 562
Differential equations
    applications of, 775–777
    defined, 771
    with graphing calculators, AP-23
    separable, 773–775
    solutions to, 771–773
Differentials, 746
Differentiation. See Derivatives
Diminishing returns, 646–647
Discontinuous functions, 549–550
Discrete probability distributions, 503–506
Discrete random variables, 503
Disjoint sets, 2
Dispersion, measures of, 496–499, 506–507
Distributive law, 10, 28, 29–30
Diverging integrals, 835
Division. See Quotients
Domains
    of functions, 67–68, 860–861
    of relations, 63
Double inequalities, 10
Doubling time, 329, 331, 377
Dual maximization problems, 289–293
Duality, Principle of, 290

e
    as base for natural logarithm, 330–331, 694
    defined, 319
    derivation of, 375
    graphs of functions with base e, 320–322
Effective annual rate, 375–376
Elastic demand, 723, 725–726
Elasticity of demand, 723–727

Elementary row operations
   matrix inverses using, 217–219
   simplex matrices, 273–275, 276–277
   for solving systems of equations, 204–205
Elements
   of matrices, 182
   of sets, 2
Elimination solution of linear systems
   addition or subtraction of equations, 95–97
   Gauss-Jordan elimination, 203–212
   left-to-right elimination, 99–100
Empirical probability, 427
Empty sets, 2
Encoding matrices, 197, 216–220
Entries of a matrix, 182
Equal matrices, 183
Equal sets, 2
Equality, properties of, 53
Equally likely outcomes, 423
Equations. *See also* Systems of equations; *specific types*
   defined, 53
   of lines, 79–82
   modeling, AP-06
   solving with graphing calculators, 88–89, AP-02
   solving with spreadsheets, 89
Equilibrium prices, 106, 818–819. *See also* Market equilibrium
Equilibrium quantities, 106
Equiprobable sample spaces, 424
Equivalent equations, 53
Equivalent inequalities, 58
Equivalent matrices, 204
Equivalent systems of equations, 94
Error formulas
   Simpson's Rule, 844–845
   Trapezoidal Rule, 843
Even index of a radical, 21
Events
   empirical probabilities, 427
   independent, 444–446
   mutually exclusive, 436–437
   probability experiments, 423–425
Excel
   2003 guide, AP-27–AP-44
   2007 and 2010 guide, AP-45–AP-64
   amortization, 405, AP-37, AP-56
   area under a curve, 793–794
   bar graphs, 492, AP-38, AP-57
   basic use, AP-27, AP-28, AP-45, AP-46
   binomial experiments, 489
   binomial probabilities, AP-38, AP-57
   change-of-base formulas, 337
   constrained optimization problems, AP-44, AP-63–AP-64
   critical points and values, 636
   definite integral approximation, AP-42, AP-61
   derivative approximation, AP-40, AP-59
   exponential function graphing, 323, AP-34, AP-53
   exponential function modeling, 325–326, AP-34, AP-53
   finance package, 371, 372
   functions of two or more variables, 862, 896–897
   future value, 372, 374, 385, 386, 389, AP-36, AP-55
   graphing functions, AP-28, AP-43, AP-46, AP-62

graphing functions and derivatives, AP-41, AP-60
   indifference curves, 863
   inventory cost models, 671
   limits, AP-40, AP-59
   linear programming, 280–281, 282, AP-33, AP-51–AP-52
   logarithmic function graphing, AP-35, AP-36, AP-54, AP-55
   logarithmic function modeling, AP-35, AP-54
   lump sum future value, AP-36
   marginal profit, 614
   market equilibrium, 107
   matrix inverses, 219, 223–224, AP-32, AP-50
   matrix multiplication, 198–199, AP-31
   matrix operations, 188, AP-31
   maxima and minima with constraints, 896–897
   minimization problems, 293
   modeling, 165–166, AP-30–AP-31, AP-34, AP-35, AP-48–AP-49, AP-53, AP-54
   normal probabilities, 518, AP-39, AP-58
   numerical integrals, 843, 845
   optimal values, AP-41, AP-60
   payments, 405, AP-37, AP-56
   points of inflection, 650–651
   polynomial function graphing, 152, AP-30, AP-48
   present value, 393–394, AP-37, AP-56
   profit maximization, 662
   quadratic equation solving, 129, AP-29, AP-47
   quadratic function graphing, 136
   rational function graphing, 153, AP-30, AP-48
   relative maxima and minima, AP-40, AP-59
   sign diagrams, 638–639
   simplex method, 280–281, 282, 293, 303
   statistical capabilities, 498, AP-38–AP-39, AP-57–AP-58
   systems of equations, 97–98, AP-29, AP-32, AP-47
   zero of a function, 89, AP-28, AP-46
Expected value
   binomial distributions, 508
   continuous probability distributions, 838–839
   mean and, 504–506, 838–839
Exponential decay functions, 321–322
Exponential equations, solving, 341–342
Exponential form
   logarithmic functions, 330, 694, 759
   radicals, 21, 23
Exponential functions
   defined, 317
   derivatives of, 702–706, AP-22
   evaluation of, 319
   graphs of, 317–324, AP-10, AP-34, AP-53
   implicit differentiation, 714
   integrals involving, 755–757
   modeling with, 324–326, 357–358, AP-10, AP-34, AP-53
Exponential growth and decay, 318–321, 343–345
Exponents
   defined, 16
   integral, 15–20
   logarithms as, 330
   rational, 21–27

Face value, 363
Factorials, 458, AP-13
Factoring, 33–38, 123–125

Feasible regions, 260
Feasible solutions, 260–261, 273, 274
Fill-down and -across capabilities, 88
Final demand, 230
Firms, theory of, 103, 739
First-derivative test, 632–633, 651
First-order differential equations, 771. *See also* Differential
    equations
First-order partial derivatives, 866–871. *See also* Partial
    derivatives
Fitting integration formulas, 826–827
Fixed costs, 104, 764–765
Fixed-probability vectors. *See* Steady-state vectors
FOIL (First, Outside, Inside, Last) method, 29
Fourth-degree polynomial functions. *See* Quartic polynomial
    functions
Fractional equations, 55–56
Fractional exponents. *See* Rational exponents
Fractions, algebraic, 38–44
Frequency histograms, 492–493, 495, 496, AP-15
Frequency tables, 492, 495
Functions, 62–70. *See also specific types*
    business and economic applications, 103–109
    composite, 69–70
    defined, 63
    domains and ranges of, 67–68, 860–861
    graphing, 64–65, 149–150, 756–757, AP-02, AP-28,
        AP-46
    modeling using, 161–166, 324–326, 332–333, AP-10, AP-11,
        AP-35
    notation for, 66–67
    operations with, 68–69
    relations and, 63
Functions of two or more variables
    business and economic applications, 875–879
    with constraints, 892–897
    defined, 860
    domain of, 860–861
    evaluation of, 861–863
    graphing, 862, AP-43, AP-62
    partial derivatives of, 866–872
Fundamental Counting Principle, 457
Fundamental theorem of calculus, 798–800
Future value
    with annual compounding, 370–371
    of annuities due, 388–389, AP-12, AP-36, AP-55,
        AP-56
    with continuous compounding, 374–375
    of a continuous income stream, 818
    defined, 363
    interest rate effects, 373–374
    of a lump sum, 372, AP-11, AP-36, AP-55
    of ordinary annuities, 383–387, AP-12, AP-36, AP-55
    with periodic compounding, 371–374

Gauss-Jordan elimination
    nonsquare systems, 211–212
    systems with no solutions, 209–210
    systems with nonunique solutions, 208–211
    systems with unique solutions, 203–208, 210
General antiderivatives, 739

General form
    linear equations, 79, 82
    quadratic equations, 123, 126
    quadratic functions, 132
General solutions
    differential equations, 772, 775
    systems of equations, 209
Geometric sequences, 377–379
Gini coefficient of income, 808, 811, 815–816
Gompertz curves, 345–347, 779
Graphical solution methods
    linear equations in two variables, 88–89, 93–94
    linear inequalities in two variables, 252–254
    linear programming, 259–267
    minimization problems, 290
Graphing calculators
    amortization, 405, AP-12–AP-13
    annuities, 385, 389, AP-12
    area between two curves, AP-26
    area under a curve, 793–794, AP-24–AP-25
    basic operations, 13, AP-01
    binomial experiments, 489
    binomial probabilities, AP-15
    change-of-base formulas, 337
    combinations, AP-13, AP-14
    critical points and values, 635, 637–638, 699, 705–706, AP-19,
        AP-22, AP-23
    definite integrals, 804, 810, 828, AP-25
    differential equations, 773, AP-23
    discontinuity evaluation, 554–555
    exponential function derivatives, 705
    exponential function graphing, 318, 322–323, AP-10,
        AP-22
    exponential function modeling, 324
    factorials, 458, AP-13
    finance packages, 371–372
    frequency histograms, 493, AP-15
    function values, AP-02–AP-03
    future value, 372, 385, 387, 389, AP-11, AP-12
    graphing with, 86–89, AP-02
    graphs of a function and derivatives, AP-20
    graphs with asymptotes, 680–681
    implicit function graphing, 715
    indefinite integrals, AP-23
    intersection method, 146, AP-04
    inventory cost models, 671
    limits, 543–544, 554–555, AP-17–AP-18
    linear inequalities, AP-08–AP-09
    linear programming, 266–267, AP-09
    linear regression, 888
    logarithmic function graphing, AP-10, AP-11, AP-22
    market equilibrium, 146
    Markov chains, AP-14
    matrix data entry, AP-06
    matrix determinants, 224–225
    matrix inverses, 219, 221–223, 224–225, AP-08
    matrix operations, 185, 198, AP-07
    matrix solutions, AP-07, AP-08
    modeling, 162–165, AP-06
    normal distribution graphs, AP-16
    normal probabilities, 517, 518, AP-16

Graphing calculators (*Continued*)
numerical derivatives, 567, 579–580, 605, AP-18–AP-19
numerical integrals, 743, 804, 810, 828, 832, 843, 845
open Leontief model, 231–232
optimal values, AP-21, AP-22
payments, 405, AP-12–AP-13
permutations, 459, AP-13, AP-14
piecewise defined function graphing, 156, AP-05
points of inflection, 650
polynomial function graphing, 151–152, AP-05
present value, 393, AP-12
probability vectors, 471, 472
profit maximization, 662, 766
quadratic function graphing, 136, AP-04
rational function graphing, 154, AP-05
reducing augmented matrices, 206–207, 210–211
relative maxima and minima, AP-20
secant line and tangent line relationships, 568–569
sequence function applications, 366
single linear inequalities, 252
sinking funds, 387
solving exponential and logarithmic equations, 347
solving linear equations, 88–89, AP-03
solving quadratic equations, 128
solving systems of equations, 97, 98, AP-03
solving systems of linear inequalities, 256–257
statistical capabilities, 498, AP-15–AP-16
tangent lines, 712
third-degree polynomial graphing, 151
viewing windows, 637–638, 680–681, AP-20, AP-21
x-intercept method, 88–89, AP-03, AP-04
Graphing utilities, 85–89. *See also* Excel; Graphing
calculators
Graphs
with asymptotes, 676–678
defined, 65
demand surface graphs, 878
with Excel, AP-28, AP-43, AP-46, AP-62
of exponential functions, 317–324, AP-10, AP-34, AP-53
of functions, 64–65, 149–150, 756–757, AP-02, AP-28,
AP-46
of functions with two or more variables, 862, AP-43,
AP-62
of Gompertz curves, 345–347
of indifference curves, 862–863
of integrals, 756–757
with intercepts, 75–77
of linear equations in two variables, 85–89
of linear inequalities, 251–257, AP-08–AP-09
of lines, 75–77
of logarithmic functions, 330–332, AP-10, AP-11, AP-35,
AP-36, AP-54, AP-55
of logistic curves, 347, 348
of normal curves, 513, 519–520
of piecewise defined functions, 155–156, AP-05
of polynomial functions, AP-30, AP-48
of quadratic functions, 135–138, 146, AP-04
of rational functions, 152–155, 675–681, AP-05, AP-30,
AP-48
shifts in, 132, 150
of single linear inequalities, 252
of statistical data, 492–493

Gravy, producer's surplus as, 821
Gross production matrices
closed Leontief model, 234
open Leontief model, 230
parts-listing problems, 236
Grouped data, mean of, 495
Grouping, factoring by, 34
Growth and decay, exponential, 318–321, 343–345

Half-open intervals, 10, 11
Half-planes, 251
Hardy-Weinberg model, 476
Higher-order derivatives, 602–605
Higher-order partial derivatives, 871–872
Histograms
frequency, 492–493, 495, 496, AP-15
probability density, 504
Horizontal asymptotes
graphs with, 676–681
limits at infinity and, 554
overview, 153, 154
of rational functions, 675–676
Horizontal lines, 78, 80, 82
Horizontal points of inflection, 634–635
Horizontal tangents, 578, 713–714
Hypothesis testing, 532

Identities, defined, 53
Identity functions, 149
Identity matrices, 196, 204, 205–206
Implicit differentiation, 709–715, 717–721
Improper integrals, 834–839
Inclusion-exclusion principle, 435–436
Income inequality (Gini coefficient), 808, 811,
815–816
Income streams
capital value of, 834, 836
future value of, 818
present value of, 816, 817–818, 829, 832–833
total value of, 803, 816
Inconsistent systems of equations, 93
Increasing functions, 78
Increasing on an interval, 630–631
Indefinite integrals. *See also* Integrals
fundamental theorem of calculus and, 798–800
with graphing calculators, AP-23
overview, 739–743
Independent events, 444–446
Independent variables, 64, 860
Index of a radical, 21
Index of summation, 791
Indifference curves, 862–863
Indifference maps, 862–863
Inelastic demand, 723, 725–726
Inequalities
defined, 58
intervals and, 10–11
in one variable, 57–59
properties of, 58
in two variables, 251–257, AP-08–AP-09
Infinite sets, 2
Infinity, limits at, 552–554

Inflation
impact on cost and profit, 735
modeling with differential equations, 780
modeling with exponential functions, 322, 324–326, 755, 756
Inflection points, 634–635, 645–647, 649–651
Initial conditions, 772
Initial probability vectors, 469
Input-output problems
Leontief models, 229–235
parts-listing problems, 235–236
Instantaneous rates of change, 561–562, 566
Integral exponents, 15–20
Integral sign, 739
Integrals. *See also* Definite integrals
additional improper, 836–837
business and economic applications, 764–769
defined, 739
with exponential functions, 755–757
formulas for, 741–742, 751, 825–826, 829
fundamental theorem of calculus and, 798–800
improper, 834–839
indefinite, 739–743, 798–800, AP-23
integration by parts, 829–833
with logarithmic functions, 757–760
numerical, 743, 804, 810, 832, 841–846
with Power Rule, 747–752
of powers of x, 740–741
and probability, 804, 837–839
properties of, 741, 801–802
requiring division, 760
tables of, 825–828
Integrands, 739
Integration. *See also* Integrals
defined, 739
limits of, 801
by parts, 829–833
Intercepts, 75–77
Interest. *See* Compound interest; Simple interest
Intersection method, 146, AP-04
Intersections of events, 433–434
Intersections of sets, 3, 4
Intervals, 10–11
Inventory cost models, 670–671
Inverse functions, 334
Inverse matrices
determinants and, 224–225
with Excel, 219, 223, AP-32, AP-50
with graphing calculators, 219, 221–223, 224–225, AP-08
solving systems of equations with, 220–225, AP-08
of square matrices, 216–220
Irrational numbers, 318

Joint cost functions, 875–876

Karmarkar, N., 271
Keynesian analysis, 768

Lagrange multipliers, 892–894, AP-44, AP-63–AP-64
Law of demand, 106. *See also* Demand
Law of mass action, 352
Law of supply, 106. *See also* Supply
Lead variables, 100

Leading coefficients, 28
Least common denominators, 40–41
Least-squares method, 161, 887–889
Left-hand limits, 537, 541
Left-to-right elimination method, 99–100
Leontief, Wassily, 229
Leontief input-output models, 229–236
Like terms, 28
Limits, 535–545
algebraic evaluation of, 538–540, 543, 544
converging vs. diverging integrals and, 835–837
defined, 536
Excel capabilities, AP-40, AP-59
graphing calculator capabilities, 543–544, 554–555, AP-17–AP-18
at infinity, 552–554, AP-18
one-sided, 537–538
overview, 535–537
of piecewise defined functions, 543–544
of polynomial functions, 539–540
properties of, 538–540
of rational functions, 539, 540–542
Limits of integration, 801
Linear equations in one variable
calculator and spreadsheet solution methods, 88–89, AP-03
defined, 53
fractional equations, 55–56
properties of, 53
solution procedure, 53–56
stated problems, 57
Linear equations in three variables, 98–100
Linear equations in two variables
calculator and spreadsheet solution methods, 97–98, AP-29, AP-47
elimination solution method, 95–97
graphical solution method, 88–89, 93–94
graphing, 85–89
substitution solution method, 94–95
Linear functions, 75–82
defined, 75
equations of lines, 79–82
forms of, 82
intercepts, 75–77
as polynomials of degree 1, 151
rate of change, 77–79
slope of a line, 77–79
Linear inequalities in one variable, 57–59
Linear inequalities in two variables, 251–257, AP-08–AP-09
Linear programming. *See also* Simplex method
with Excel, 280–281, 282, AP-33, AP-51–AP-52
graphical methods, 259–267
with graphing calculators, 266–267, AP-09
impact of, 249
Linear regression, 160, 161, 886–889
Linear regression equation, 888
Lines
equations of, 79–82
graphs of, 75–77
horizontal, 78, 80, 82
parallel, 78–79
perpendicular, 79
slope of, 77–79
vertical, 78, 80, 82

Loans
  amortization of, 404–409, AP-12–AP-13, AP-37, AP-56
  refinancing of, 408–409
  simple interest, 362–364
  unpaid balance of, 407–408
Logarithmic equations, 330, 342–343
Logarithmic functions
  applications, 337–338
  defined, 329–330, 694
  derivatives of, 694–700, AP-22
  graphs of, 330–332, AP-10, AP-11, AP-35, AP-36, AP-54, AP-55
  implicit differentiation, 714
  integrals involving, 757–760
  modeling with, 332–333, AP-11, AP-35, AP-54
Logarithms. *See also* Logarithmic functions; Natural
      logarithms
  change-of-base formula, 336–337, 342, 698, AP-11
  common, 330–331, 337–338
  defined, 329
  as exponents, 330
  properties of, 333–336, 341, 696–697, 702, 703–704
Logistic curves, 347, 348
Logistic functions, 347–348, 358
Lorenz curve, 807–808, 811, 815–816
Lower limit of integration, 801
LPRE mnemonic for integration by parts, 830
Lump sums
  future value of, 372, AP-11, AP-36, AP-55
  present value of, 392–394, AP-12, AP-37, AP-56
  unpaid balance of a loan, 407–409
Luxury tax, 729

Mansfield, Edwin, 903
Marginal cost
  defined, 104–105
  as derivative, 610–612
  joint cost and, 875–876
  minimum average cost and, 659
  total cost from, 764–765
Marginal cost functions, 104–105, 610–611, 764–766
Marginal demand functions, 878–879
Marginal productivity, 876–878
Marginal productivity of money, 896
Marginal profit
  in competitive markets, 612–613
  defined, 104–105
  as derivative, 612–614
  profit maximization and, 659–662
Marginal propensity to consume, 764, 768–769
Marginal propensity to save, 764, 768–769
Marginal return to sales, 624
Marginal revenue
  defined, 104–105
  as derivative, 566, 578–579, 608–610
  nonlinear functions, 563–564
Marginal revenue functions, 608
Marginal utility, 874
Market equilibrium
  defined, 106
  determining with quadratic functions, 144–146
  overview, 106–108
  taxation effects, 108–109

Markov chains, 468–474
  defined, 468
  with graphing calculators, AP-14
  probability vectors, 469–472
  regular, 472
  steady-state vectors, 470–472, AP-14
  transition matrices, 468, 469, 472
Mass action, law of, 352
Mathematical models. *See* Modeling
Matrices, 179–236
  augmented, 203
  coefficient, 203
  column, 183, 193
  columns in, 181
  defined, 179
  determinants of, 224–225
  differences of, 184, AP-07
  elements of, 182
  equal, 183
  equations of, 220–225
  equivalent, 204
  with Excel, 188, 198–199, 219, 223, AP-31–AP-32, AP-49–AP-50
  with graphing calculators, 185, 198, 219, 221–223, 224–225,
      AP-06–AP-08
  gross production, 230, 234, 236
  identity, 196, 204, 205–206
  inverses of, 216–225, AP-08, AP-32, AP-50
  Leontief input-output models, 229–236
  multiplication of, 193–199, AP-07, AP-31, AP-49
  negative of, 184
  nonsquare systems, 211–212
  operations with, 181–188, 289, AP-07, AP-31, AP-49
  order of, 181–182
  reduced form of, 208
  row, 183, 193
  row operations for, 204
  rows of, 181
  simplex, 271–272
  square, 182, 216–220
  sums of, 183–184, 187–188, AP-07
  systems with nonunique solutions, 208–211
  systems with unique solutions, 203–208
  technology, 229, 231, 234
  transition, 468, 469, 472
  transposes of, 183, 289
  zero, 183
Matrix equations, 220–225
Maturity value, 396
Maxima and minima. *See also* Relative maxima and minima;
      Simplex method
  absolute, 656
  applications of, 626–627, 667–672
  constrained, 892–897, AP-44, AP-63–AP-64
  curve sketching, 630–639
  with Excel, AP-41, AP-60
  for functions of two or more variables, 882–886
  with graphing calculators, AP-21, AP-22
  with linear programming, 260–267
  points of inflection, 634–635, 645–647, 649–651
  relative, 630–639, 648–651, 883–886
  saddle points, 884–885
  tests for, 883

Maximizing profit. *See* Profit maximization

Maximizing revenue. *See* Revenue maximization

Maximum points, 132. *See also* Profit maximization; Revenue maximization

McMahon, T., 718

Mean
  of binomial distributions, 508
  of continuous probability distributions, 838–839
  defined, 494
  of discrete probability distributions, 504–505
  with Excel, AP-38, AP-39, AP-57, AP-58
  expected value and, 504–506, 838–839
  with graphing calculators, AP-15, AP-16
  of grouped data, 495–496, AP-16, AP-39, AP-58
  overview, 494–496
  of raw data, AP-15, AP-38, AP-57

Measures of central tendency. *See* Mean; Median; Mode

Measures of dispersion, 496–499, 506–507

Median, 494, 496, AP-39, AP-57

Members of sets, 2

Method of least squares, 161, 887–889

Minima. *See* Maxima and minima

Minimization problems, 288–293

Minimizing cost, 292–293, 658–659, 668–669, 670–672

Minimum points, 132

Mixed constraints, 297–303

Mode, 493–494, 496

Modeling
  choosing a function for, 161–166
  with Excel, AP-34, AP-35, AP-48–AP-49, AP-53, AP-54
  with exponential functions, 324–326, AP-10
  with graphing calculators, 162–165, AP-06, AP-10
  with logarithmic functions, 332–333, AP-11, AP-35, AP-54
  overview, 67, 161

Money on the table concept, 821

Monomial factors, 33–34, 36

Monomials, 28

Monopoly markets
  consumer's surplus in, 820
  profit maximization in, 660–661
  revenue in, 142–143

Multiplication
  derivatives of products, 584, 597–600
  of fractions, 39
  of functions, 68
  of matrices, 193–199, AP-07, AP-31, AP-49

Multiplication property
  equations, 53
  inequalities, 58

Multiplicative identity, 10

Multiplicative inverses, 10

Mutually exclusive events, 436–437

n factorial ($n!$), 458

Napier, John, 329

National consumption function, 768–769

Natural logarithms
  applications, 337
  change-of-base formula, 336–337, 342, 698, AP-11
  defined, 330–332, 694
  derivatives of, 695–696, 699–700
  implicit differentiation, 714
  properties of, 696–697

Natural numbers, set of, 2

Negative exponents, 17

Negative numbers, 10

Negative of a matrix, 184

Negative of a number, 10

Negative radicands, 21

Negative slopes, 78

Nominal annual rates, 370

Nonbasic variables, 274, 276–277

Nonreal solutions, 127

Nonsquare systems of equations, 211–212

Normal curve, 513

Normal curve approximation to the binomial distribution, 522–525

Normal probability distributions, 513–520
  with Excel, 518, AP-39, AP-58
  with graphing calculators, 517, 518, AP-16
  graphs of, 519–520
  properties of, 513
  standard normal probability curve, 323–324, AP-65–AP-66
  z-scores, 514–517

nth roots (principal roots), 21

nth term of a sequence
  arithmetic sequences, 365
  geometric sequences, 378

Null set, 2

Numerators, rationalizing, 25

Numerical derivatives
  with Excel, AP-40, AP-59
  with graphing calculators, 567, 579–580, 605, AP-18–AP-19

Numerical integration
  with graphing calculators, 743, 804, 810, 828, 832
  Simpson's Rule, 843–846
  Trapezoidal Rule, 841–843, 845–846

Objective functions, 260, 892

Octants, 861

Odds and probability, 428–429

One-container mixture problems, 776–777

One-sided limits, 537–538, 541

Open intervals, 10, 11

Open Leontief model, 230–233

Operating leverage, 178

Operations
  with algebraic expressions, 27–33
  with algebraic fractions, 39–43
  with functions, 68–69
  with matrices, 181–188, 289, AP-07, AP-31, AP-49
  with polynomials, 28–31
  with radicals, 23–25
  with rational exponents, 22–23
  with real numbers, 12
  with sets, 3–9
  with signed numbers, 12

Optimal values. *See* Maxima and minima

Optimization. *See also* Linear programming; Simplex method
  constrained, 297–303, 892–897, AP-44, AP-63–AP-64
  with functions of one variable, 655–662
  with functions of two or more variables, 882–889
  tangent lines and, 626–627

Order of a matrix, 181–182
Order of operations, 12–13
Ordered pairs, 63
Ordinary annuities
    future value of, 383–387, AP-12, AP-36, AP-55, AP-56
    present value of, 392–397, AP-12, AP-37, AP-56
    sinking funds as, 387
Ordinate of a graph, 65
Origin of a coordinate system, 64
Outcomes (probability experiments), 423
Outstanding principal of a loan, 407–408

Par value for bonds, 396
Parabolas, 131–139
    average rate of change, 138–139
    graphs of quadratic functions, 135–138
    revenue and profit maximization for, 144
    vertex of, 132–134
    zeros of quadratic functions and, 134
Paraboloids, 862
Parallel lines, 78–79
Partial derivatives, 866–872
Particular solutions to differential equations, 772. See also
        Differential equations
Parts-listing problems, 235–236
Payments
    amortization calculations, 404–405, AP-12–AP-13, AP-37, AP-56
    from annuities, 388–389, 395
Payoff amount, 407–408
Percent rates of change, 718–719
Perfect-square trinomials, 35–36
Periodic compounding, 371–374, 376
Periods, interest rate, 370–371, 377
Permutations
    with graphing calculators, 459, AP-13, AP-14
    overview, 458–459
    probability and, 463–465, AP-14
Perpendicular lines, 78–79
Perpetuities, 557–558, 836
Pie charts, 492
Piecewise defined functions
    continuous 555
    graphs of, 155–156, AP-05
    limits of, 543–544, AP-18
Pivot columns, 273
Pivot entries, 273–275, 276
Pivot rows, 273
Pivoting process, 276–277, 299, 300
Point of diminishing returns, 646–647
Point-plotting method, 65
Points of inflection
    concavity and, 645–647
    horizontal, 634–635
    second-derivative test and, 649–651
Point-slope form, 79, 82
Polynomial functions
    continuous, 555
    graphs of, 151–152, AP-05, AP-30, AP-48
    limits of, 539–540
    power functions as, 149
Polynomials
    defined, 28

degree of, 28
division of, 31–32
graphs of, 151–152
integrals of, 742
operations with, 28–31
polynomials in x, 28–31
products of, 29–30
properties of, 28
in quadratic form, 36
Positive integer exponents, 16
Positive radicands, 21
Positive slopes, 78
Power functions, 149, 150
Power Rule
    defined, 592–593
    for derivative formulas, 597–600
    for integration, 747–752
    for powers of a function of x, 746–752
Powers of a function of x, 746–752
Powers of x
    derivatives of, 573–575, 586
    integrals of, 740–741
Present value
    of annuities due, 397–398, AP-12, AP-37, AP-56
    of a continuous income stream, 816, 817–818, 829, 832–833
    of deferred annuities, 398–400
    defined, 363
    of lump sums, 392–394, AP-12, AP-37, AP-56
    of ordinary annuities, 392–397, AP-12, AP-37, AP-56
    with periodic compounding, 372–373
Price
    equilibrium, 106, 818–819
    as piecewise defined function, 155–156
Primal problems, 292
Principal (loans), 362, 407–408
Principal nth roots, 21
Principle of Duality, 290
Probability, 421–474
    Bayes' formula, 451–454
    complements of events, 433, 434
    conditional, 440–446
    counting techniques for, 463–465, AP-14
    empirical, 427
    of an event, 424, 463–465
    formulas summary, 454
    improper integrals and, 837–839
    inclusion-exclusion principle, 435–436
    integrals and, 804, 837–839
    Markov chains and, 467–474, AP-14
    mutually exclusive events, 436–437
    odds and, 428–429
    sample spaces, 423–426
    of single events, 423
    unions and intersections of events, 433–434
Probability density functions, 804, 837–838
Probability density histograms, 504
Probability distributions
    binomial, 487–489, 507–510, 522–525, AP-15, AP-38, AP-57
    discrete, 503–506
    normal, 513–520, AP-16, AP-39, AP-58

Probability experiments
  binomial, 487–489, AP-15
  equally likely outcomes, 423
  states, 468
Probability formulas, 454
Probability measures, 424
Probability trees
  Bayes' formula and, 451–454
  binomial experiments, 487
  overview, 443–444
  solving problems using, 450–451
Probability vectors, 469–472
Probability weights, 424
Producer's surplus, 821–822
Product Rule
  for derivatives of products, 584, 597–600
  for probability events, 442–446
Production functions
  Cobb-Douglas, 713, 860, 895–897
  overview, 876–878
Products. *See* Multiplication
Profit
  as application of indefinite integrals,
      764–767
  in competitive markets, 612–613, 661
  defined, 103
  function for, 103–104
  marginal, 104–105, 612–614, 659–662
  reinvestment of, 419
Profit maximization
  as application of indefinite integrals, 765–767
  with linear programming, 260–263, 272–281
  marginal profit and, 659–662
  parabolas and, 134
  with quadratic functions, 143–144
  vs. revenue maximization, 143–144

Quadratic discriminants, 127
Quadratic equations
  calculator and spreadsheet solution methods, 128–129,
      AP-04, AP-29, AP-47
  factoring methods, 33–38, 123–125
  quadratic formula, 125–128
  square root solution method, 125–126
Quadratic formula, 125–128
Quadratic functions
  business applications, 142–146
  general form of, 132
  graphs of, 135–138, 146, AP-04
  parabolas, 131–139, 144
  as second degree polynomials, 151
  zeros of, 128, 134
Quadratic polynomials, 36
Quartic polynomial functions, 151–152
Quotient Rule, 584–586, 598, 868
Quotients
  derivatives of, 584–586
  of fractions, 40
  of functions, 68
  of polynomials, 31
  of radicals, 24
  rewriting, 18

Radical sign, 21
Radicals, 21–27, 593
Radicands, 21, 127
Random events, 424
Random variables, 503, 803–804
Ranges
  of functions, 67–68
  of relations, 63
  of a set of numbers, 497
Rates of change
  average, 138–139, 560–561
  of derivatives, 604–605
  instantaneous, 561–562, 566
  of linear functions, 77–79
  percent rates of change, 718–719
  related rates, 717–721
  slope as, 77
  velocity as, 561–562, 566
Rational exponents, 21–27, 318
Rational functions
  asymptotes of, 675–676
  as continuous vs. discontinuous, 550, 555
  graphs of, 152–155, 675–681, AP-05, AP-30, AP-48
  limits of, 539, 540–542
Rationalizing the denominator, 24–25, 42–43
Rationalizing the numerator, 25
Real functions, 67–68
Real number line, 9
Real numbers, 9–15, 63
Real value, 398
Reciprocal functions, 152
Reciprocals, 10
Rectangular coordinates, 65
Reduced form of matrices, 208, AP-07
Reduction of augmented matrices. *See* Gauss-Jordan
      elimination
Refinancing loans, 408–409
Regression. *See* Linear regression
Regression line calculations, 886–889
Regular Markov chains, 472
Regular transition matrices, 472
Related rates, 717–721
Relations, functions and, 63
Relative maxima and minima. *See also* Maxima and
      minima
  curve sketching, 630–639
  defined, 630
  with Excel, AP-40, AP-59
  first-derivative test, 632–633, 651
  for functions of two or more variables, 883–886
  with graphing calculators, AP-20
  second-derivative test, 648–651
Repeated integration by parts, 831–832
Replacement (dependent events), 445
Revenue
  average rate of change of, 138–139
  elasticity and, 725–726
  function for, 103–104
  marginal, 104–105, 563–564, 566, 578–579,
      608–610
  in monopoly markets, 142–143
  total, 103–104

Revenue maximization
   marginal revenue and, 609–610, 657–658
   vs. profit maximization, 143–144
   quadratic function solutions, 137–138
   tax revenue maximization, 727–729
Richter scale, 337, 339
Riemann sums, 798
Right-hand limits, 537, 541
Root functions, 149
Roots of a number, 21. *See also* Radicals
Row matrices, 183, 193
Row operations. *See* Elementary row operations
Rows, of matrices, 181
Ruffner, James, 864

Saddle points, 884–885
Sample points, 424
Sample spaces, 423–426
Savings and consumption, 768–769
Scalar multiplication, 186–187
Scatter plots, 161–166, AP-06, AP-30–AP-31, AP-34–AP-35,
      AP-48–AP-49, AP-53–AP-54
Scott, W., 439
Secant lines, 564–565, 568–569
Second derivatives
   concavity and, 644–645
   with graphing calculators, AP-19
   overview, 603, 605
   partial, 871–872
   points of inflection and, 645–647
Second-degree polynomials, 36
Second-derivative test, 648–651
Selling bonds at a discount, 397
Selling bonds at a premium, 397
Separable differential equations, 773–775
Sequence functions, 364
Sequences
   arithmetic, 365–367, 790
   defined, 364
   geometric, 377–379
Sets, 2–9, 497
Shadow prices, 269, 282, 313–314
Shifts in graphs, 132, 150
Sigma notation, 791–793
Sign diagrams, 632, 638–639
Signed numbers, 12
Significant digits, 163
Simple interest, 362–364, 372
Simplex matrices, 271–272, 276–277
Simplex method
   dual problems, 289–293
   history, 271
   maximization application, 271–284, 290, 298–299,
      301–303
   minimization application, 288–293, 298, 299–301
   with mixed constraints, 297–303
   multiple solutions, 281–282
   no solutions, 282–283
   overview, 272
   procedure for, 272–278
   shadow prices, 283
   spreadsheet application, 280–281, 282

Simplex tableaux. *See* Simplex matrices
Simplifying fractions, 39
Simpson's Rule, 843–846
Simultaneous solutions, 93
Single event probabilities, 423
Sinking funds, 387
Siple, Paul, 874
Slack variables
   defined, 272
   in linear programming, 276, 279, 280, 290
   shadow prices and, 313–314
Slope of a line, 77–79
Slope of the tangent
   approximating, 567
   defined as a derivative, 565–566
   with implicit differentiation, 710–715
   partial derivatives and, 869–870
Slope-intercept form, 81–82
Solution of an equation, defined, 53
Solution regions, 252
Solution set of an equation, 53
Solving an equation. *See also* Matrices; Systems of equations
   defined, 53
   differential equations, 771–777, AP-23
   exponential equations, 341–342
   fractional equations, 55–56
   linear equations in one variable, 53–55
   linear equations in two variables, 88–89, 93–98
   matrix equations, 220–225
   quadratic equations, 33–38, 123–129, AP-04, AP-29,
      AP-47
Solving inequalities, 58–60, 251–257
Solving simultaneously, 93
Solving stated problems, 57
Special binomial products, 30
Special factorizations, 35–36
Special functions, 148–156
Spreadsheets. *See also* Excel
   amortization calculations, 405, 407
   annuity calculations, 394–395
   compound interest calculations, 371–372, 377
   overview, 88, AP-27, AP-45
   sequence function applications, 366
   solving systems of equations with, 97–98
Square matrices, 182, 216–220
Square root property, 125–126
Square roots, 21
Standard deviations
   of binomial distributions, 508
   with Excel, AP-38, AP-39, AP-57, AP-58
   with graphing calculators, AP-15, AP-16
   of probability distributions, 506–507
   of sample data, 497–498
   z-score relationship, 516–517
Standard maximization problems, 272
Standard minimization problems, 289
Standard normal distributions, 514
Standard normal probability curve, 323–324, AP-65–AP-66
Standard scores, 514–517
Standard viewing windows, 86–87
Stated problems, 57
States (probability experiments), 468

Statistical data, 491–499. *See also* Probability distributions
  Excel capabilities, 498, AP-38–AP-39, AP-57–AP-58
  graphing calculator capabilities, 498, AP-15, AP-16
  graphs of, 492–493
  hypothesis testing, 532
  measures of central tendency, 493–496
  measures of dispersion, 496–499
Steady-state vectors, 470–472, AP-14
Subscripts, matrix elements, 182
Subsets of sets, 2
Substitution property
  equations, 53
  inequalities, 58
Substitution solution method
  linear equations in two variables, 94–95
  quadratic equations, 145
Subtraction. *See* Differences
Sum formulas, 791–792
Summation notation, 791–793
Sums
  of arithmetic sequences, 365–366
  derivatives of, 576–578
  of functions, 68
  of geometric sequences, 378–379
  of matrices, 183–184, 187–188, AP-07
Supply
  law of, 106
  market equilibrium and, 107–108, 144–146
  taxation effects, 108–109, 727–729
Surpluses
  consumer's, 818–821, 822
  open Leontief model, 230
  producer's, 821–822
Symbols of grouping, 29
Systems of equations
  algebraic solution methods, 94–98
  defined, 93
  Gauss-Jordan elimination solution method, 203–212
  inverse matrix solution method, 220–225, AP-08
  left-to-right elimination method, 99–100
  linear equations in three variables, 98–100
  linear equations in two variables, 85–89, 93–98, AP-29, AP-47
  solving with Excel, AP-29, AP-32, AP-47, AP-50
  solving with graphing calculators, 97, 98, AP-03, AP-07, AP-08
  solving with matrices, 203–212
Systems of linear inequalities, 252–257

Tables of integrals, 825–828
Tangent lines
  horizontal and vertical, 578, 713–714
  optimization and, 626–627
  overview, 564–566
  for powers of x, 575
  secant line relationships, 568–569
  slope of, 565–566, 567, 710–715, 869–870
  to a surface, 869–870
Taxation effects, 108–109, 727–729
Technological equations
  closed Leontief model, 234
  open Leontief model, 230–231

Technology matrices
  closed Leontief model, 234
  defined, 229
  open Leontief model, 231
Term, of annuities due, 388
Terms
  constant, 28
  defined, 28
  of a sequence, 364
Test points, 251
Theoretical mean of a probability distribution. *See* Expected value
Theory of the firm, 103, 739
Third derivatives, 603
Third-degree polynomial functions, 38, 151
Third-order partial derivatives, 872
Thompson, d'Arcy, 582, 722
Thrall, R. M., 606
Three equations in three variables, 98–100
Time to reach a goal, 386–387
Time-dependent variables, 717–721
Total cost
  as application of indefinite integrals, 764–767
  defined, 148
  derivatives and, 610–611
  graphs of, 154–155
  as quadratic, 131, 154–155
Total revenue, 103–104
Transition matrices, 468, 469, 472
Transition probabilities, 468
Transitions (changes of state), 468
Transposes of matrices, 183, 289
Trapezoidal Rule, 841–843, 845–846
Trinomials, 28, 34–36
True annual percentage rate (APR), 404

Unbounded regions, 264–265
Undefined derivatives, 636
Unions of events, 433–434
Unions of sets, 4
Unitary elastic demand, 723, 725–726
Universal set, 3, 9
Unpaid balance of a loan, 407–408
Upper limit of integration, 801
Utility functions, 862–863, 874, 892, 895

Variable costs, 104
Variables
  basic vs. nonbasic, 274, 276–277
  binomial, 507
  defined, 27
  independent vs. dependent, 64, 860
  random, 503, 803–804
  slack, 272, 276, 279, 280, 290, 313–314
  time-dependent, 717–721
Variance
  of probability distributions, 506
  of sample data, 497
Vectors
  probability, 469–472
  row and column matrices as, 183
  steady-state, 470–472, AP-14
Velocity, 561–562, 566

Velocity function, 562
Venn diagrams, 3, 5–6
Vertex of a parabola, 132
Vertical asymptotes
    graphs with, 676–681
    overview, 153, 154
    of rational functions, 675, 676
Vertical lines, 78, 80, 82
Vertical tangents, 713–714
Vertical-line test, 65
Viewing windows (graphing calculators), 637–638,
        680–681, AP-20, AP-21
von Neumann, John, 290

Wilson's lot size formula, 864, 874
Worksheets, Excel, AP-27, AP-45

x-coordinates, 65
x-intercept method, 88–89, AP-03,
        AP-04
x-intercepts, 75, 88
x-min and max, 86

y-coordinates, 65
Yield rate of a bond, 396
y-intercepts, 75
y-min and max, 86

Zero exponents, 17
Zero matrices, 183
Zero of a function, 76, 88–89, 128, 134
Zero product property, 123
z-scores, 514–517